Separable Algebras

GRADUATE STUDIES IN MATHEMATICS 183

Separable Algebras

Timothy J. Ford

American Mathematical Society
Providence, Rhode Island

2010 *Mathematics Subject Classification.* Primary 16H05, 15B05, 13A15, 13C20, 14F20, 14B25, 16-01, 13-01.

For additional information and updates on this book, visit
www.ams.org/bookpages/gsm-183

Library of Congress Cataloging-in-Publication Data

Names: Ford, Timothy J., 1954-
Title: Separable algebras / Timothy J. Ford.
Description: Providence, Rhode Island : American Mathematical Society, [2017] | Series: Graduate studies in mathematics ; volume 183 | Includes bibliographical references and index.
Identifiers: LCCN 2017013677 | ISBN 9781470437701 (alk. paper)
Subjects: LCSH: Separable algebras–Textbooks. | Associative rings–Textbooks. | AMS: Associative rings and algebras – Algebras and orders – Separable algebras (e.g., quaternion algebras, Azumaya algebras, etc.). msc | Commutative algebra – Ring extensions and related topics – Étale and flat extensions; Henselization; Artin approximation. msc | Commutative algebra – Ring extensions and related topics – Galois theory. msc | Commutative algebra – General commutative ring theory – Ideals; multiplicative ideal theory. msc | Commutative algebra – Theory of modules and ideals – Class groups. msc | Algebraic geometry – (Co)homology theory – Étale and other Grothendieck topologies and (co)homologies. msc | Algebraic geometry – Local theory – Local structure of morphisms: Étale, flat, etc. msc | Associative rings and algebras – Instructional exposition (textbooks, tutorial papers, etc.). msc | Commutative algebra – Instructional exposition (textbooks, tutorial papers, etc.). msc
Classification: LCC QA251.5 .F67 2017 | DDC 512/.46–dc23 LC record available at https://lccn.loc.gov/2017013677

∞ The paper used in this book is acid-free and falls within the guidelines
established to ensure permanence and durability.
Visit the AMS home page at `http://www.ams.org/`

10 9 8 7 6 5 4 3 2 1 22 21 20 19 18 17

Dedicated to Frank DeMeyer

Contents

Preface

The purpose of this book is to provide a thorough introduction to the theory of separable algebras over commutative rings. After introducing the general theory of separable algebras and establishing their basic properties, many of the fundamental roles played by separable algebras are then studied. In particular, rigorous introductions and treatments of Azumaya algebras, the henselization of local rings, and the Galois theory of commutative rings are presented. Interwoven throughout these applications is the essential role played by étale algebras.

Recall that an extension of fields F/k is separable if every element of F is a root of a separable polynomial over k, hence a root of a polynomial with no repeated roots. For an algebra A over a commutative ring R, separability is not defined element-wise because in general there are not enough separable elements in A. This is the subject of Exercise 12.7.12. To generalize the definition of separability to algebras over commutative rings, it is necessary to raise the level of abstraction. The definition is not in terms of elements of A, but instead is based on a certain module structure of the ring A over the enveloping algebra $A \otimes_R A^o$ which is induced by the multiplication map $x \otimes y \mapsto xy$. Naturally, the definitions agree for a finite extension of fields. Therefore we could say that the theory of separable algebras has as its essence the study of separable polynomials. Keeping this in mind, the reader of this book can perhaps make some sense out of the many abstract structures that arise.

Almost from the start we see that this subject matter has strong ties to all areas of algebra, including Algebraic Number Theory, Ring Theory, Commutative Algebra and Algebraic Geometry. These fundamental connections are present even in the definition of separability itself. As alluded to above, A is defined to be a separable R-algebra if A is a projective module over $A \otimes_R A^o$. Before Auslander and Goldman applied it to algebras over commutative rings, this module theoretic definition had been applied by Hochschild and others to study separable algebras over fields. This definition is in fact the noncommutative analog of one of the equivalent conditions used by Grothendieck in SGA 1 (see [**Gro71**, Proposition I.3.1, p. 2]), namely, that the diagonal morphism is an open immersion. While not obvious at first, the connection between the two ideas is the following. When A is commutative, the multiplication map is a homomorphism of R-algebras $A \otimes_R A \to A$ which in the language of Algebraic Geometry defines the so-called diagonal morphism $\operatorname{Spec} A \to \operatorname{Spec} A \otimes_R A$. If A is commutative and finitely generated as an R-algebra, then one can show A is a projective module over $A \otimes_R A$ if and only if the diagonal morphism is an open immersion. This is the subject of Exercise 8.1.31.

In the chapters which follow, many important attributes of separable algebras are studied. It is shown that separability is preserved under a change of base ring. The tensor product of two separable algebras is a separable algebra. Separability is transitive: if A is separable over S, and S is separable over R, then A is separable over R. Any localization of R is separable over R and a homomorphic image of R is separable over R. In the language of Algebraic Geometry, this means an open immersion is separable, and a closed immersion is separable. Over a field k, the separable algebras turn out to be finite direct sums of matrix algebras over finite dimensional k-division algebras and the center of each division algebra appearing is a finite separable extension field of k. More generally, for any commutative ring R, A is separable over R if and only if A is separable over its center and its center is separable over R. This fact shows that the study of separability can be split into two parts: algebras which are commutative, and algebras which are central. Following the example set by Grothendieck in his papers on the Brauer group ([**Gro68a**], [**Gro68b**], [**Gro68c**]), central separable R-algebras are called Azumaya algebras. Over a field, an algebra is Azumaya if and only if it is central simple.

Many examples of separable algebras are included along the way. Among the first examples of a noncommutative separable R-algebra which is exhibited is the ring of n-by-n matrices over R, which is denoted $M_n(R)$. The importance of this example is illustrated by the fact that every Azumaya R-algebra A of constant rank is a twisted form of matrices for a faithfully flat extension. By this we mean that if A is an Azumaya R-algebra of constant rank n^2, then there is a faithfully flat R-algebra S such that upon extension of the ring of scalars to S, $A \otimes_R S$ is isomorphic to the matrix algebra $M_n(S)$. The trivial commutative separable extension of R of rank n is R^n, the direct sum of n copies of R. Every commutative separable R-algebra

S which is a finitely generated projective R-module of constant rank n is a twisted form of R^n. In other words, there is a faithfully flat R-algebra T such that $T \otimes_R S$ is isomorphic to T^n. An R-algebra is said to be étale if it is commutative, separable, flat, and finitely presented as an R-algebra. Chapter 9 is devoted to the study of étale algebras. The henselization and strict henselization of a local ring are constructed in Chapter 10. A central theme of SGA 4 ([**AGV72a**], [**AGV72b**], [**AGV73**]) is that faithfully flat étale R-algebras are the counterpart in Algebraic Geometry of covering spaces in the analytic topology and the strictly henselian local rings play the role of the stalks at the points. The examples mentioned above show that Azumaya R-algebras of constant rank are those R-algebras which become isomorphic to a matrix algebra upon restriction to a suitably refined étale covering and commutative separable finitely generated projective algebras of constant rank are those which locally (for the étale topology) are isomorphic to a trivial covering.

One of the goals of this book is to consolidate the most popularly accepted resources on this subject. It includes almost all of the results that are found in the following standard references: 1. The monograph of DeMeyer and Ingraham on separable algebras ([**DI71**]); 2. The monograph of Knus and Ojanguren on faithfully flat descent and Azumaya algebras ([**KO74b**]); 3. The monograph of Orzech and Small on the Brauer group ([**OS75**]); 4. Auslander and Goldman's original papers on the Brauer group and maximal orders ([**AG60a**]; [**AG60b**]); 5. Raynaud's monograph on henselian local rings ([**Ray70**]). Moreover, it includes almost all of those results contained in Saltman's "Lectures on Division Algebras" ([**Sal99**]) which are primarily about separable algebras.

A serious attempt has been made to include many examples and there is an emphasis on computations. Many nontrivial examples of rings are exhibited both in the text and in the exercises for which the computations of the Brauer group, the Picard group, and other important invariants are completely carried out. I apologize that there are not more. I owe Harley Flanders a debt of gratitude for impressing upon me the importance of computing examples. In the words of Hermann Weyl ([**Wey97**, preface to the first edition]), "the special problems in all their complexity constitute the stock and core of mathematics; and to master their difficulties requires on the whole the harder labor". To compute the Brauer group, the Picard group, or one of the other invariants of a ring emphasized in this book we quickly learn that the trivial examples are few and far between. Even for the most common rings the computations can be very difficult. Then we learn that for those interesting, nontrivial rings for which the computations can be completely carried out, one fact is certain — a lot of machinery is required. Even the computation of the Brauer group of the ring of rational integers $\mathbb{Z}$ requires a lot of work. The original proof does not appear until page 95 in Grothendieck's Brauer group papers and in this book, the proof does not come until the last chapter (Theorem 14.3.8).

In any book of this type, there is always a balance to be achieved between which theorems are stated and proved, which are stated without proof, and which are merely cited (or tacitly assumed). Throughout, our benchmark has been to place higher emphasis on those results which have direct applications to separable algebras. As standard references on general algebra, the books by Dummit and Foote ([**DF04**]), Hungerford ([**Hun80**]), Bourbaki ([**Bou89a**]), and Herstein ([**Her75**]) are recommended (in that order). Rotman's book [**Rot79**] is recommended as a standard reference for those results from homological algebra that are not proved here. For commutative algebra, the books by Atiyah and Macdonald ([**AM69**]), Matsumura ([**Mat80**]), Zariski and Samuel ([**ZS75a**] and [**ZS75b**]), and Bourbaki ([**Bou89b**]) are recommended (in that order).

Chapters 1, 2 and 3 consist of a review of much of the background results on rings, modules, and commutative algebra which are needed for the rest of the book. Proofs are given for many of these results. However, most are stated without proof for reference and ease of exposition in the rest of the text. Whenever a proof is omitted, a citation is provided for the reader. Chapter 3 includes a quick proof of the noncommutative version of Nakayama's Lemma, the Artin-Wedderburn Theorem, and some applications which will be necessary later on. Therefore, Chapters 4 through 14 contain the bulk of the subject matter. For instance, the definition of separable algebras together with proofs of their basic properties are presented in Chapter 4.

Chapter 5 covers the background material from homological algebra that we require. There is an entire section on group cohomology. The Amitsur complex is introduced and applied to prove the basic theorems on faithfully flat descent. The first properties of the Hochschild cohomology groups are derived. Amitsur cohomology and its first properties are presented. Included in this treatment are the pointed cohomology sets in degrees 0 and 1 for noncommutative coefficient groups. For example, it is shown that in degree one, Amitsur cohomology classifies the twisted forms of a module.

The main purpose of Chapter 6 is to introduce and prove the fundamental properties of lattices over integral domains. Included is a method for the construction of reflexive modules in terms of locally free modules which is based on a theorem of B. Auslander. The Weil divisor class group is defined and Nagata's Theorem is proved. For a noetherian normal integral domain R, the class group is identified with the group of reflexive fractional ideals of R in the quotient field, modulo principal ideals.

Chapter 7 begins with the definition and first properties of Azumaya algebras. After proving the commutator theorems, we define the Brauer group. Over a field, Azumaya algebras and central simple algebras are equivalent and the Brauer group parametrizes the finite dimensional central division algebras. The Picard group of invertible bimodules is defined. This leads to the proof of the Skolem-Noether Theorem. As an application of Hochschild cohomology we prove that Azumaya algebras can be lifted modulo an ideal that is contained in the nil radical.

The topics in Chapter 8 are mostly from Commutative Algebra. From our perspective, the motivation here is to derive more tests for separability. In particular, separability criteria are achieved by applications of Hochschild cohomology, derivations, and Kähler differentials. Consequently, a jacobian criterion, a local ring criterion, and a residue field criterion (all for separability) are then proved. Differential crossed product algebras, which are a type of cyclic crossed product for purely inseparable radical extensions, are introduced. When the ground field is infinite, a version of Emmy Noether's Normalization Lemma is proved which allows us to construct the underlying polynomial ring in such a way that it contains a separating transcendence base. Lastly, useful differential and jacobian criteria for regularity are derived.

Chapter 9 is a deeper investigation into the properties of smooth and étale algebras. Every algebra which is étale in a neighborhood is locally isomorphic to a standard étale algebra. In addition to proving the fundamental properties of étale algebras, we show that formally étale implies étale. There is an entire section devoted to the construction of radical extensions that ramify only along a reduced effective divisor.

The henselization and strict henselization of a local ring are constructed in Chapter 10. The existence of a faithfully flat étale splitting ring for an Azumaya algebra is demonstrated. We show that an Azumaya algebra of constant rank over a commutative ring is a form of matrices for a faithfully flat étale covering, an important characterization with many applications. Together with Artin's Refinement Theorem and the Skolem-Noether Theorem, this allows us to show that up to isomorphism the Azumaya R-algebras of constant rank n^2 are classified by the pointed set of Čech cohomology $\check{\mathrm{H}}^1_{\mathrm{et}}(R, \mathrm{PGL}_n)$. From here, the embedding of the Brauer group into the second étale Čech cohomology group is constructed.

Chapter 11 is a deeper study of Azumaya algebras. Viewing an Azumaya algebra as a form of matrices allows us to associate to an element of the algebra a number of invariants that are typically associated to a matrix. In particular, every element has a characteristic polynomial, a norm, and a trace. A proof due to D. Saltman is presented showing that the Brauer group is torsion. Sufficient conditions for a maximal order in a central simple algebra to be an Azumaya algebra are derived. The subgroup of the Brauer group containing algebras that are locally split is described. Following papers by Knus, Ojanguren and Saltman, Brauer groups in characteristic p are studied. Ojanguren's example of a nontrivial locally trivial Azumaya algebra is exhibited. A key step in the proof involves a construction of the rank three reflexive module that is not projective which is based on the local to global theorem of B. Auslander.

Chapter 12 is an introduction to Galois theory for commutative rings. Included are the theorems on Galois descent, the Fundamental Theorem of Galois Theory, the Embedding Theorem, the separable closure, and the Fundamental Theorem of Infinite Galois Theory. There is an entire section devoted to cyclic Galois extensions. In the Kummer context, the short exact sequence classifying the group of cyclic Galois extensions of degree n is derived.

Probably the most useful and practical method for constructing Azumaya algebras is the so-called crossed product. Chapter 13 includes treatments of the usual crossed product algebra, the usual cyclic crossed product algebra, the generalized crossed product algebra of Kanzaki, and the generalized cyclic crossed product algebra. The Brauer group is one of the most important arithmetic invariants of any commutative ring. For instance, if L/K is a finite Galois extension of fields with group G, then any central simple K-algebra split by L is Brauer equivalent to a crossed product algebra. The Crossed Product Theorem says there is an isomorphism between the Galois cohomology group $\mathrm{H}^2(G, L^*)$ with coefficients in the group of invertible elements of L, and the relative Brauer group $\mathrm{B}(L/K)$. For a Galois extension of commutative rings S/R, the crossed product map $\mathrm{H}^2(G, S^*) \to \mathrm{B}(S/R)$ is in general not one-to-one or onto. The precise description of the kernel and cokernel of this map is the so-called Chase, Harrison and Rosenberg seven term exact sequence of Galois cohomology. When S is a factorial noetherian integral domain, the crossed product map is an isomorphism. Therefore, it is no surprise that computations involving the groups in the seven term cohomology sequence rely on a good knowledge of the class group of the covering ring S. Nontrivial examples are exhibited for which all of the terms in the seven term exact sequence of Galois cohomology are computed (see, for example, Example 13.6.13, Example 13.6.15, and Exercise 13.6.16).

In recent years there has been a renewed interest in the computation of the Brauer group of algebraic varieties. Varieties of low dimension (algebraic curves and surfaces) play important roles because in many cases the computations can be completely carried out. For examples the reader is referred to [**Bri13**], [**CTW12**], [**CV15**], [**IS15**], [**vG05**], and their respective bibliographies. Not only is it important to compute the Brauer group, but there is also a desire to construct Azumaya algebras representing any nontrivial Brauer classes. A strong motivation is the role played by the Brauer group in the so-called Brauer-Manin obstruction to the Hasse Principle. This connection was originally drawn by Manin in [**Man71**]. Usually these examples involve computations of nontrivial Azumaya algebras that are split by a finite extension of the ground field. Such computations generally rely on a good knowledge of the class group of Weil divisors of the covering ring. It is this strong relationship between the class group of Weil divisors and the Brauer group that emphasizes the importance of the subject matter of Chapter 6. Exercise 13.6.18 contains an example that explores this phenomenon.

Chapter 14 contains some additional topics that did not seem to fit anywhere else. There is an entire section on the important corestriction map. A Mayer-Vietoris Sequence for the Brauer Group is derived. It is applied to compute the Brauer group of an affine algebraic curve and the Brauer group of a subring of a global field. An elementary example of a commutative ring whose Brauer group is cyclic of order n is exhibited.

I am especially grateful for the influence Frank DeMeyer and David Saltman have had on this project. It is a privilege to thank two of my students, Djordje Bulj and Drake Harmon, who read and corrected early versions of the manuscript. I take this opportunity to express my gratitude to my college teachers, J. Cisneros, V. Snook, and R. Sanders and my high school teachers, J. Wilson and H. Pickering. Finally, I sincerely thank Fook Loy, my wife and the jewel of my life, for her constant encouragement, help and support.

Chapter 1

Background Material on Rings and Modules

This chapter contains a review of some basic topics from general algebra. Proofs are frequently omitted. The reader should consider making a quick scan of this chapter upon first reading, and when necessary, refer back to the appropriate sections later. To avoid listing many special cases, a theorem will frequently be stated in a form that is more general than any given application may require.

1. Rings and Modules

We first establish some conventions and fundamental definitions. A *ring* is always associative and contains an identity element denoted by 1. In a ring which has at least two elements we require $0 \neq 1$. If $R \neq (0)$ and R has no zero divisors, then we say R is a *domain*. A commutative domain is called an *integral domain*. A domain in which every nonzero element is invertible is called a *division ring*. A commutative division ring is called a *field*. The set of all invertible elements in a ring R is a group which is denoted $\text{Units}(R)$ or R^* and is called *the group of units in* R. The ring of integers is denoted $\mathbb{Z}$, the field of rational numbers $\mathbb{Q}$, the field of real numbers $\mathbb{R}$, and the field of complex numbers $\mathbb{C}$. If R is any ring, then the ring of n-by-n matrices over R is denoted by $M_n(R)$. If R is a commutative ring and G a group, then the *group ring* or *group algebra* is denoted $R(G)$. If R is any ring, the *opposite ring of* R is denoted R^o. As an additive abelian group, R^o and R are equal. However, the multiplication of R^o is reversed from that of R.

Example 1.1.1. If k is a field, the ring of *quaternions* over k is the four-dimensional k-vector space with basis $1, i, j, k$, and with multiplication defined by extending the relations

$$\begin{aligned} i^2 = j^2 = k^2 &= -1 \\ ij = -ji &= k \end{aligned}$$

by associativity and distributivity. The reader should verify the following.

(1) If k is equal to either $\mathbb{Q}$ or $\mathbb{R}$, then the ring of quaternions is a division ring. Over $\mathbb{R}$, the ring of quaternions is denoted $\mathbb{H}$.

(2) If k is equal to $\mathbb{C}$, then the ring of quaternions is isomorphic to $M_2(k)$.

(3) If k is equal to $\mathbb{Z}/(2)$, the ring of quaternions is isomorphic to the group ring over the elementary 2-group of order 4, hence is commutative.

If A is a ring and B is a subset of A, then we say B is a *subring* of A if B contains both 0 and 1 and B is a ring under the addition and multiplication rules of A. The *center* of A is the set $Z(A) = \{x \in A \mid xy = yx \text{ for all } y \in A\}$. Then $Z(A)$ is a commutative ring which is a subring of A. If $x \in Z(R)$, then we say x is *central.* A homomorphism from the ring A to the ring B maps the identity of A to the identity of B. All modules are required to be unitary, and unless otherwise specified, a module will be a left module. An R-module M is said to be *finitely generated* in case there exists a finite subset $X = \{x_1, \dots, x_n\}$ of M and M is the the smallest submodule of M containing X.

Let R be a commutative ring. An *R-algebra* is a ring A together with a homomorphism of rings $\theta : R \to Z(A)$ mapping R into the center of A. We call θ the *structure homomorphism* of A. Then A is an R-module with action $ra = \theta(r)a$. We write $R \cdot 1$ for the image of θ. If B is a subring of A containing $R \cdot 1$, then we say B is an *R-subalgebra* of A. We say A is a *finitely generated* R-algebra in case there exists a finite subset $X = \{x_1, \dots, x_n\}$ of A and A is the the smallest subalgebra of A containing X and $R \cdot 1$.

1.1. Categories and Functors. A *category* consists of a collection of *objects* and a collection of *morphisms* between pairs of those objects. The composition of morphisms is defined and is again a morphism. For our purposes, a category will always be one of the following:

(1) The category whose objects are modules over a ring R and whose morphisms are homomorphisms of modules. By ${}_R\mathfrak{M}$ we denote the

category of all left R-modules together with R-module homomorphisms. By $\mathfrak{M}_R$ we denote the category of all right R-modules together with R-module homomorphisms. If A and B are R-modules, the set of all R-module homomorphisms from A to B is denoted $\operatorname{Hom}_R(A, B)$.

(2) The category whose objects are rings and whose morphisms are homomorphisms of rings. A subcategory would be the category whose objects are commutative rings.

(3) The category whose objects are finitely generated algebras over a fixed commutative ring R and whose morphisms are R-algebra homomorphisms.

(4) The category whose objects are sets and whose morphisms are functions.

(5) The category of pointed sets. A *pointed set* is a pair (X, x) where X is a nonempty set and x is a distinguished element of X called the *base point*. A morphism from a pointed set (X, x) to a pointed set (Y, y) is a function $f : X \to Y$ such that $f(x) = y$.

A *covariant functor* from a category $\mathfrak{C}$ to a category $\mathfrak{D}$ is a correspondence $\mathfrak{F} : \mathfrak{C} \to \mathfrak{D}$ which is a function on objects $A \mapsto \mathfrak{F}(A)$ and for any pair of objects $A, B \in \mathfrak{C}$, each f in $\operatorname{Hom}_{\mathfrak{C}}(A, B)$ is mapped to a function $\mathfrak{F}(f)$ in $\operatorname{Hom}_{\mathfrak{D}}(\mathfrak{F}(A), \mathfrak{F}(B))$ such that the following are satisfied

(1) If $1 : A \to A$ is the identity map, then $\mathfrak{F}(1) : \mathfrak{F}(A) \to \mathfrak{F}(A)$ is the identity map.

(2) Given a commutative triangle

$$\begin{array}{ccc} & B & \\ {}^{f}\nearrow & & \searrow^{g} \\ A & \xrightarrow{gf} & C \end{array}$$

in $\mathfrak{C}$, the triangle

$$\begin{array}{ccc} & \mathfrak{F}(B) & \\ {}^{\mathfrak{F}(f)}\nearrow & & \searrow^{\mathfrak{F}(g)} \\ \mathfrak{F}(A) & \xrightarrow{\mathfrak{F}(gf)} & \mathfrak{F}(C) \end{array}$$

commutes in $\mathfrak{D}$.

The definition of a *contravariant functor* is similar, except the arrows get reversed. That is, if $\mathfrak{F} : \mathfrak{C} \to \mathfrak{D}$ is a contravariant functor and f is an element of $\operatorname{Hom}_{\mathfrak{C}}(A, B)$, then $\mathfrak{F}(f)$ is in $\operatorname{Hom}_{\mathfrak{D}}(\mathfrak{F}(B), \mathfrak{F}(A))$.

If $\mathfrak{F} : \mathfrak{C} \to \mathfrak{D}$ is a covariant functor between categories of modules, then $\mathfrak{F}$ is *left exact* if for every short exact sequence

$$0 \to A \xrightarrow{\alpha} B \xrightarrow{\beta} C \to 0 \tag{1.1}$$

in $\mathfrak{C}$, the corresponding sequence

$$0 \to \mathfrak{F}(A) \xrightarrow{\mathfrak{F}(\alpha)} \mathfrak{F}(B) \xrightarrow{\mathfrak{F}(\beta)} \mathfrak{F}(C)$$

is exact in $\mathfrak{D}$. We say $\mathfrak{F}$ is *right exact* if for every short exact sequence (1.1) in $\mathfrak{C}$, the sequence

$$\mathfrak{F}(A) \xrightarrow{\mathfrak{F}(\alpha)} \mathfrak{F}(B) \xrightarrow{\mathfrak{F}(\beta)} \mathfrak{F}(C) \to 0$$

is exact in $\mathfrak{D}$.

If $\mathfrak{F} : \mathfrak{C} \to \mathfrak{D}$ is a contravariant functor between categories of modules, then $\mathfrak{F}$ is *left exact* if for every short exact sequence (1.1) in $\mathfrak{C}$, the sequence

$$0 \to \mathfrak{F}(C) \xrightarrow{\mathfrak{F}(\beta)} \mathfrak{F}(B) \xrightarrow{\mathfrak{F}(\alpha)} \mathfrak{F}(A)$$

is exact in $\mathfrak{D}$. We say the contravariant functor $\mathfrak{F}$ is *right exact* if for every short exact sequence (1.1) in $\mathfrak{C}$, the sequence

$$\mathfrak{F}(C) \xrightarrow{\mathfrak{F}(\beta)} \mathfrak{F}(B) \xrightarrow{\mathfrak{F}(\alpha)} \mathfrak{F}(A) \to 0$$

is exact in $\mathfrak{D}$.

If $\mathfrak{F} : \mathfrak{A} \to \mathfrak{C}$ is a covariant functor, we say that $\mathfrak{F}$ is *fully faithful* if

$$\operatorname{Hom}_{\mathfrak{A}}(A, B) \to \operatorname{Hom}_{\mathfrak{C}}(\mathfrak{F}(B), \mathfrak{F}(A))$$

is a one-to-one correspondence. We say that $\mathfrak{F}$ is *essentially surjective* if for every object C in $\mathfrak{C}$, there exists A in $\mathfrak{A}$ such that C is isomorphic to $\mathfrak{F}(A)$. Let $\mathfrak{C}$ and $\mathfrak{D}$ be categories of modules and suppose we have two covariant functors $\mathfrak{F}$ and $\mathfrak{F}'$ from $\mathfrak{C}$ to $\mathfrak{D}$. We say that $\mathfrak{F}$ and $\mathfrak{F}'$ are *naturally equivalent* if for every module M in $\mathfrak{C}$ there is an isomorphism φ_M in $\operatorname{Hom}_{\mathfrak{D}}\big(\mathfrak{F}(M), \mathfrak{F}'(M)\big)$ such that, for every pair of modules M and N in $\mathfrak{C}$ and any $f \in \operatorname{Hom}_{\mathfrak{C}}(M, N)$, the diagram

$$\begin{array}{ccc} \mathfrak{F}(M) & \xrightarrow{\mathfrak{F}(f)} & \mathfrak{F}(N) \\ {\scriptstyle\varphi_M}\downarrow & & \downarrow{\scriptstyle\varphi_N} \\ \mathfrak{F}'(M) & \xrightarrow{\mathfrak{F}'(f)} & \mathfrak{F}'(N) \end{array}$$

commutes. We denote by $I_{\mathfrak{C}}$ the identity functor on the category $\mathfrak{C}$ defined by $I_{\mathfrak{C}}(M) = M$ and $I_{\mathfrak{C}}(f) = f$, for modules M and maps f. Then we say two categories $\mathfrak{C}$ and $\mathfrak{D}$ are *equivalent* if there is a functor $\mathfrak{F} : \mathfrak{C} \to \mathfrak{D}$ and a functor $\mathfrak{G} : \mathfrak{D} \to \mathfrak{C}$ such that $\mathfrak{F} \circ \mathfrak{G}$ is naturally equivalent to $I_{\mathfrak{D}}$ and $\mathfrak{G} \circ \mathfrak{F}$ is naturally equivalent to $I_{\mathfrak{C}}$. The functors $\mathfrak{F}$ and $\mathfrak{G}$ are then referred to as *inverse equivalences*. For a proof of the next proposition, the reader is referred to a book on Category Theory. For example, see [**Bas68**, Proposition (1.1), p. 4].

Proposition 1.1.2. *A functor $\mathfrak{F} : \mathfrak{C} \to \mathfrak{D}$ establishes an equivalence of categories if and only if $\mathfrak{F}$ is fully faithful and essentially surjective.*

Example 1.1.3. Let R be a ring and R^o the opposite ring. Any $M \in {}_R\mathfrak{M}$ can be made into a right R^o-module by defining $m * r = rm$. The reader should verify that this defines a covariant functor ${}_R\mathfrak{M} \to \mathfrak{M}_{R^o}$ and that the category of left R-modules is equivalent to the category of right R^o-modules.

1.2. Progenerator Modules. As references for this section, we recommend [**DF04**] and [**Rot79**]. First we recall some of the basic properties of free modules. Let R be a ring and I any index set. Denote by R^I the R-module direct sum $\bigoplus_{i\in I} R$. If $I = \{1, 2, \dots, n\}$, then we write $R^{(n)}$ for R^I. If M is an R-module, then we say M is *free on the index set* I if M is isomorphic to R^I for some index set I. The Kronecker *delta function* is the function $\delta : I \times I \to \{0, 1\}$ defined by:

$$\delta_{ij} = \begin{cases} 1 & \text{if } i = j, \\ 0 & \text{otherwise.} \end{cases}$$

The *standard basis* for R^I is $\{e_i \in R^I \mid i \in I\}$ where $e_i(j) = \delta_{ij}$. An R-module M is free if and only if there exists a subset $\{b_i \mid i \in I\}$ of M satisfying the following.

(1) Each $x \in M$ can be written $x = \sum_{i\in I'} x_i b_i$, where $I' \subseteq I$ is finite and $x_i \in R$.
(2) If $I' \subseteq I$ is finite and $\sum_{i\in I'} r_i b_i = 0$, then each $r_i = 0$.

The set $\{b_i \mid i \in I\}$ is called a *basis* for M. If R is a commutative ring and M a free R-module with a finite basis $\{b_1, \dots, b_n\}$, then any other basis of M has n elements. In this case, we call n the *rank* of M and write $\operatorname{Rank}_R(M) = n$.

Lemma 1.1.4. *Let R be a ring and M an R-module.*

(1) *There exists a free R-module F and a surjective homomorphism $F \to M$.*
(2) *M is finitely generated if and only if M is the homomorphic image of a free R-module $R^{(n)}$ for some n.*

Proposition 1.1.5. *If R is a ring and M an R-module, then the following are equivalent.*

(1) *M is isomorphic as an R-module to a direct summand of a free R-module.*

(2) *Every short exact sequence of R-modules*

$$0 \to A \to B \xrightarrow{\beta} M \to 0$$

is split exact.

(3) *For any diagram of R-modules*

$$\begin{array}{ccccc} & & M & & \\ & \overset{\exists\psi}{\swarrow} & \downarrow \phi & & \\ A & \xrightarrow{\alpha} & B & \longrightarrow & 0 \end{array}$$

with the bottom row exact, there exists an R-module homomorphism $\psi : M \to A$ such that $\alpha\psi = \phi$.

(4) *M has a dual basis $\{(m_i, f_i) \mid i \in I\}$ for some index set I consisting of $m_i \in M$, $f_i \in \operatorname{Hom}_R(M, R)$ satisfying:*
   (a) *for each $m \in M$, $f_i(m) = 0$ for all but finitely many $i \in I$, and*
   (b) *for all $m \in M$, $m = \sum_{i \in I} f_i(m) m_i$.*

If M satisfies any of the equivalent conditions of Proposition 1.1.5, then we say M is a *projective* R-module. Proposition 1.1.5 (1) is trivially satisfied by a free R-module. Hence a free module is projective.

Example 1.1.6. Let D be a division ring and $R = M_2(D)$ the ring of two-by-two matrices over D. If we identify D with the scalar matrices, then R is a left vector space of dimension 4 over D. Let

$$e_1 = \begin{bmatrix} 1 & 0 \\ 0 & 0 \end{bmatrix}, \quad e_2 = \begin{bmatrix} 0 & 0 \\ 0 & 1 \end{bmatrix}.$$

The reader should verify the following facts.

(1) e_1 and e_2 are idempotents and $e_1 e_2 = e_2 e_1 = 0$. We say e_1 and e_2 are *orthogonal idempotents*.
(2) Re_1 is the set of all matrices with second column consisting of zeros.
(3) Re_2 is the set of all matrices with first column consisting of zeros.
(4) $\dim_D(Re_1) = \dim_D(Re_2) = 2$.
(5) $R = Re_1 \oplus Re_2$ as R-modules.

By (5), Re_1 and Re_2 are projective R-modules. It follows from Exercise 1.1.9 below that Re_1 and Re_2 are not free R-modules. For any $n \geq 2$, we will see in Example 3.2.4 below that $M_n(D)$ is a simple artinian ring. By Theorem 3.2.2, proved below, every module over $M_n(D)$ is projective. The method used above can be modified to show that $M_n(D)$ contains left ideals that are projective but not free.

Let R be a ring and M an R-module. We say that M is *of finite presentation* if there exists an exact sequence

$$R^{(m)} \to R^{(n)} \to M \to 0$$

for some m and n. The proof of Corollary 1.1.7 is left to the reader.

Corollary 1.1.7. *Let R be a ring and M a finitely generated projective R-module. Then M is of finite presentation over R. There exists a finitely generated projective R-module N such that $M \oplus N$ is a finitely generated free R-module.*

Let R be a ring and M an R-module. The reader should verify that $\operatorname{Hom}_R(M, R)$ is a right R-module by the action $(fr)(m) = f(m)r$ (see Lemma 1.3.17 below). Using this, the reader should verify that the set

$$\mathfrak{T}_R(M) = \left\{ \sum_{i=1}^{n} f_i(m_i) \mid n \geq 1, f_i \in \operatorname{Hom}_R(M, R), m_i \in M \right\} \tag{1.2}$$

is a two-sided ideal in R, which is called the *trace ideal of M in R*. We say that M is a *generator* over R in case $\mathfrak{T}_R(M) = R$. We say that an R-module M is a *progenerator over* R in case M is finitely generated, projective and a generator over R.

Let $\theta : R \to S$ be a homomorphism of rings and M a left S-module. Then R acts on M from the left by the rule $rx = \theta(r)x$. Using θ, an S-module M can be viewed as an R-module. Since S is a left S-module, θ makes S into a left R-module. In a similar way, θ can be used to make a right S-module into a right R-module.

Proposition 1.1.8. *Let $\theta : R \to S$ be a homomorphism of rings and M an S-module.*

(1) *(Projective over Projective is Projective) If S is projective as an R-module and M is a projective S-module, then M is projective as an R-module. Moreover, if M is finitely generated over S and S is finitely generated as an R-module, then M is finitely generated as an R-module.*

(2) *(A Generator over a Generator is a Generator) If S is a generator when viewed as an R-module and M is an S-generator, then M is an R-generator.*

(3) *(A Progenerator over a Progenerator is a Progenerator) If S is an R-progenerator and M is an S-progenerator, then M is an R-progenerator.*

Proof. (1): Let $\{(m_i, f_i) \mid i \in I\}$ be a dual basis for M over S and $\{(s_j, g_j) \mid j \in J\}$ a dual basis for S over R. The reader should verify that the functions

$g_j f_i$ in $\operatorname{Hom}_R(M,R)$ and the products $s_j m_i$ in M make up a dual basis for M over R.

(2): For some $m>0$ there exist $\{f_1,\dots,f_m\}$ in $\operatorname{Hom}_S(M,S)$ and $\{x_1,\dots,x_m\}$ in M such that we can write $1\in S$ as $1=\sum_{i=1}^m f_i(x_i)$. For some n there exist $\{g_1,\dots,g_n\}$ in $\operatorname{Hom}_R(S,R)$ and $\{s_1,\dots,s_n\}$ in S such that we can write $1\in R$ as $1=\sum_{j=1}^n g_j(s_j)$. For each (i,j), the composite function $g_j f_i$ is in $\operatorname{Hom}_R(M,R)$ and the product $s_j m_i$ is in M. Then

$$\sum_{i=1}^m\sum_{j=1}^n g_j f_i(s_j m_i)=\sum_{i=1}^m\sum_{j=1}^n g_j\big(s_j f_i(m_i)\big)$$
$$=\sum_{j=1}^n g_j\Big(s_j\sum_{i=1}^m f_i(m_i)\Big)=\sum_{j=1}^n g_j(s_j)=1$$

shows that the trace ideal is equal to R.

Part (3) follows immediately from Parts (1) and (2). □

1.3. Exercises.

Exercise 1.1.9. (Free over Free is Free) Let $\theta: R\to S$ be a homomorphism of rings such that S is free as an R-module. Let M be a free S-module. Prove:

(1) M is a free R module.

(2) If M has a finite basis over S, and S has a finite basis over R, then M has a finite basis over R. In this case, if R and S are both commutative, then $\operatorname{Rank}_R(M)=\operatorname{Rank}_S(M)\operatorname{Rank}_R(S)$.

(3) If R and S are fields, then $\dim_R(S)$ and $\dim_S(M)$ are both finite if and only if $\dim_R(M)$ is finite.

Exercise 1.1.10. Let R be a ring and $\{M_i \mid i\in I\}$ a family of R-modules. Prove that the direct sum $\bigoplus_{i\in I} M_i$ is projective over R if and only if each M_i is projective over R.

Exercise 1.1.11. Proving that a module is a generator can be a difficult task. This exercise contains conditions on a module that are equivalent to being a generator. For another characterization of generator modules, see Exercise 1.3.30. Let R be a ring and M a left R-module. Prove that the following are equivalent.

(1) M is an R-generator.

(2) The R-module R is the homomorphic image of a direct sum $M^{(n)}$ of finitely many copies of M.

(3) The R-module R is the homomorphic image of a direct sum M^I of copies of M over some index set I.

(4) Every left R-module A is the homomorphic image of a direct sum M^I of copies of M over some index set I.

Exercise 1.1.12. Let k be a field, x an indeterminate, and $n > 1$ an integer. Let $T = k[x]$, $S = k[x^n, x^{n+1}]$, and $R = k[x^n]$. For the tower of subrings $R \subseteq S \subseteq T$, prove:

(1) T is free over R of rank n.

(2) S is free over R of rank n.

(3) T is not free over S.

1.4. Nakayama's Lemma. In this section we prove Nakayama's Lemma for a finitely generated module over a commutative ring. Throughout this book we will encounter various forms of Nakayama's Lemma. The version in Lemma 1.1.13, as well as its proof are from [**DI71**]. Corollary 1.1.17 is a second version, and Theorem 3.1.2 below contains a form that is valid for noncommutative rings.

Let R be a ring, $A \subseteq R$ a left ideal of R, and M an R-module. Denote by AM the R-submodule of M generated by all elements of the form am, where $a \in A$ and $m \in M$. The *annihilator* of M in R is $\operatorname{annih}_R M = \{r \in R \mid rm = 0 \text{ for all } m \in M\}$. The reader should verify that $\operatorname{annih}_R M$ is a two-sided ideal in R. If $\operatorname{annih}_R M = (0)$, then we say M is *faithful.*

Lemma 1.1.13 (Nakayama's Lemma). *Let R be a commutative ring and M a finitely generated R-module. An ideal A of R has the property that $AM = M$ if and only if $A + \operatorname{annih}_R(M) = R$.*

Proof. Assume $A + \operatorname{annih}_R(M) = R$. Write $1 = \alpha + \beta$ for some $\alpha \in A$ and $\beta \in \operatorname{annih}_R(M)$. Given m in M, $m = 1m = (\alpha + \beta)m = \alpha m + \beta m = \alpha m$. Therefore $AM = M$.

Conversely, suppose $AM = M$. Choose a generating set $\{m_1, \dots, m_n\}$ for M over R. Define

$$\begin{aligned} M = M_1 &= Rm_1 + \cdots + Rm_n \\ M_2 &= Rm_2 + \cdots + Rm_n \\ &\vdots \\ M_n &= Rm_n \\ M_{n+1} &= 0. \end{aligned}$$

We prove that for every $i = 1, 2, \dots, n+1$, there exists α_i in A such that $(1 - \alpha_i)M \subseteq M_i$. Since $(1 - 0)M = M \subseteq M_1$, take $\alpha_1 = 0$. Proceed inductively. Suppose $i \geq 1$, $\alpha_i \in A$, and $(1-\alpha_i)M \subseteq M_i$. Then

$$\begin{aligned}(1-\alpha_i)M &= (1-\alpha_i)AM \\ &= A(1-\alpha_i)M \\ &\subseteq AM_i.\end{aligned}$$

In particular, $(1-\alpha_i)m_i \in AM_i = Am_i + Am_{i+1} + \cdots + Am_n$. So there exist elements $\alpha_{ii}, \dots, \alpha_{im}$ in A such that

$$(1-\alpha_i)m_i = \sum_{j=i}^{n} \alpha_{ij} m_j.$$

Subtracting shows that

$$(1-\alpha_i-\alpha_{ii})m_i = \sum_{j=i+1}^{n} \alpha_{ij} m_j$$

is in M_{i+1}. Consider

$$\begin{aligned}(1-\alpha_i)(1-\alpha_i-\alpha_{ii})M &= (1-\alpha_i-\alpha_{ii})\big((1-\alpha_i)M\big) \\ &\subseteq (1-\alpha_i-\alpha_{ii})M_i \\ &\subseteq M_{i+1}.\end{aligned}$$

Set $\alpha_{i+1} = -(-\alpha_i - \alpha_{ii} - \alpha_i + \alpha_i^2 + \alpha_i\alpha_{ii})$. Then $\alpha_{i+1} \in A$ and $(1-\alpha_{i+1})M \subseteq M_{i+1}$. By finite induction, $(1-\alpha_{n+1})M = 0$. Hence $1-\alpha_{n+1} \in \operatorname{annih}_R(M)$ and $1 \in A + \operatorname{annih}_R(M)$. □

Corollary 1.1.14. *Let R be a commutative ring and M a finitely generated R-module. If $\mathfrak{m}M = M$ for every maximal ideal $\mathfrak{m}$ of R, then $M = 0$.*

Proof. If $M \neq 0$, then $1 \notin \operatorname{annih}_R(M)$. Some maximal ideal $\mathfrak{m}$ contains $\operatorname{annih}_R(M)$. So $\mathfrak{m} + \operatorname{annih}_R(M) = \mathfrak{m} \neq R$. By Lemma 1.1.13, $\mathfrak{m}M \neq M$. □

Proposition 1.1.15. *Let R be a commutative ring and M a finitely generated and projective R-module. Then $\mathfrak{T}_R(M) \oplus \operatorname{annih}_R(M) = R$.*

Proof. There exists a dual basis $\{(m_i, f_i) \mid 1 \leq i \leq n\}$ for M. For each $m \in M$, we see that $m = f_1(m)m_1 + \cdots + f_n(m)m_n$ is in $\mathfrak{T}_R(M)M$. Then $\mathfrak{T}_R(M)M = M$. By Lemma 1.1.13, $\mathfrak{T}_R(M) + \operatorname{annih}_R(M) = R$. Therefore, $\mathfrak{T}_R(M)\operatorname{annih}_R(M) = \mathfrak{T}_R(M) \cap \operatorname{annih}_R(M)$. It suffices to show that $\mathfrak{T}_R(M)\operatorname{annih}_R(M) = 0$. A typical generator for $\mathfrak{T}_R(M)$ is $f(m)$ for some $m \in M$ and $f \in \operatorname{Hom}_R(M,R)$. Given $\alpha \in \operatorname{annih}_R(M)$, we see that $\alpha f(m) = f(\alpha m) = f(0) = 0$. □

For a commutative ring, Proposition 1.1.15 supplies us with a powerful criterion for showing that a finitely generated projective module is a generator. As a consequence, we have the following two useful descriptions for progenerator modules.

Corollary 1.1.16. *Let R be a commutative ring and M an R-module. Then the following are true.*

(1) *M is an R-progenerator if and only if M is finitely generated projective and faithful.*

(2) *Assume R has no idempotents except 0 and 1. Then M is an R-progenerator if and only if M is finitely generated, projective, and $M \neq (0)$.*

We end this section with the following useful variation of Nakayama's Lemma.

Corollary 1.1.17. *Let R be a commutative ring. Suppose I is an ideal in R, M is an R-module, and there exist submodules A and B of M such that $M = A + IB$. If*

(1) *I is nilpotent (that is, $I^n = 0$ for some $n > 0$), or*

(2) *I is contained in every maximal ideal of R and M is finitely generated,*

then $M = A$.

Proof. Notice that

$$M/A = \frac{A + IB}{A} \subseteq \frac{A + IM}{A} \subseteq I(M/A) \subseteq M/A.$$

Assuming (1) we get $M/A = I(M/A) = \cdots = I^n(M/A) = 0$. Now assume (2) and let $\mathfrak{m}$ be an arbitrary maximal ideal of R. Then $M/A = I(M/A) \subseteq \mathfrak{m}(M/A)$. By Corollary 1.1.14, $M/A = 0$. □

1.5. Exercise.

Exercise 1.1.18. A *local ring* is a commutative ring R such that R has exactly one maximal ideal. If R is a local ring with maximal ideal $\mathfrak{m}$, then $R/\mathfrak{m}$ is called the *residue field* of R. If $(R, \mathfrak{m})$ and $(S, \mathfrak{n})$ are local rings and $\phi : R \to S$ is a homomorphism of rings, then we say ϕ is a *local homomorphism of local rings* in case $\phi(\mathfrak{m}) \subseteq \mathfrak{n}$. Let $\phi : R \to S$ be a local homomorphism of local rings. Assume S is a finitely generated R-module and $\mathfrak{m}$ is the maximal ideal of R. Show that if the homomorphism $R/\mathfrak{m} \to S/\mathfrak{m}S$ induced by ϕ is an isomorphism, then ϕ is onto. (Hint: S is generated by $\phi(R)$ and $\mathfrak{m}S$.)

1.6. Module Direct Summands of Rings. If a ring R decomposes into a direct sum of finitely many two-sided ideals, then the direct summands are generated by central idempotents. The main result of this section, Lemma 1.1.20, considers the more general situation of when a left ideal of R is an R-module direct summand of R. This important utility lemma has many applications in the rest of the book. In fact, the connection between module direct summands of R and idempotents will play an essential role in the definition of separability given in Chapter 4.

An idempotent $e \in R$ is said to be *primitive* if e cannot be written as a sum of two nonzero orthogonal idempotents. Let I be a nonzero left ideal in R. Then I is a *minimal left ideal of* R if whenever J is a left ideal of R and $J \subseteq I$, then either $J = 0$, or $J = I$.

Example 1.1.19. Let F be a field and $R = M_2(F)$ the ring of two-by-two matrices over F. Let

$$e_1 = \begin{bmatrix} 1 & 0 \\ 0 & 0 \end{bmatrix}, \quad e_2 = \begin{bmatrix} 0 & 0 \\ 0 & 1 \end{bmatrix}.$$

We saw in Example 1.1.6 that e_1 and e_2 are orthogonal idempotents and $R = Re_1 \oplus Re_2$ as R-modules. The reader should verify that Re_1 is a minimal left ideal and e_1 is a primitive idempotent.

Lemma 1.1.20. *Let R be a ring and I a left ideal of R.*

(1) *I is an R-module direct summand of R if and only if $I = Re$ for some idempotent e.*

(2) *Let e be an idempotent in R. Then e is primitive if and only if Re cannot be written as an R-module direct sum of proper left ideals of R.*

(3) *If I is a minimal left ideal, then I is an R-module direct summand of R if and only if $I^2 \neq 0$.*

(4) *Suppose $R = I \oplus J$ where I and J are two-sided ideals. Then $I = Re$ for some central idempotent e, I is a ring, and e is the multiplicative identity for I.*

Proof. (1): Assume $R = I \oplus L$. Write $1 = e + f$ where $e \in I$ and $f \in L$. Then $e = e^2 + ef$. Now $ef = e - e^2 \in I \cap L = 0$. Likewise $fe = 0$. Also $e + f = 1 = 1^2 = (e+f)^2 = e^2 + f^2$. In the direct sum the representation of 1 is unique, so $e = e^2$ and $f = f^2$. Let $x \in I$. Then $x = x \cdot 1 = xe + xf$. But $xf = x - xe \in I \cap L = 0$. So $Re = I$. Conversely, assuming $e^2 = e$, we prove that Re is a direct summand of R. Then $0 = e - e^2 = e(1-e) = (1-e)e$. Also $(1-e)^2 = 1-e-e+e^2 = 1-e$. This shows $e, 1-e$ are orthogonal idempotents. Since $1 = e + (1-e)$ we have $R = Re + R(1-e)$. Let $x \in Re \cap R(1-e)$. Then $x = ae = b(1-e)$ for some $a, b \in R$. Then $xe = ae^2 = ae = x$ and again $xe = b(1-e)e = 0$. Therefore $R = Re \oplus R(1-e)$.

(2): Using the same ideas as in (1), we see that e is a sum of nonzero orthogonal idempotents if and only if Re decomposes into a direct sum of proper left ideals of R.

(3): Assume I is a minimal left ideal of R. Suppose $R = I \oplus L$ for some left ideal L of R. By (1), $I = Re$ for some idempotent e. Then $e = e^2 \in I^2$ so $I^2 \neq 0$. Conversely assume $I^2 \neq 0$. There is some $x \in I$ such that $Ix \neq 0$. But Ix is a left ideal of R and since I is minimal, we have $Ix = I$. For some $e \in I$, we have $ex = x$. Let $L = \operatorname{annih}_R(x) = \{r \in R \mid rx = 0\}$. Then L is a left ideal of R. Since $(1-e)x = x - ex = x - x = 0$ it follows that $1 - e \in L$. Therefore $1 = e + (1-e) \in I + L$ so $R = I + L$. Also, $e \in I$ and $ex = x \neq 0$ shows that $e \notin L$. Now $I \cap L$ is a left ideal in R and is contained in the minimal left ideal I. Since $I \cap L \neq I$, it follows that $I \cap L = 0$ which proves that $R = I \oplus L$ as R-modules.

(4): This part is left to the reader. □

Our first application of Lemma 1.1.20 is to prove the following unique decomposition theorem for commutative rings.

Theorem 1.1.21. *Let R be a commutative ring and assume R decomposes into an internal direct sum $R = Re_1 \oplus \cdots \oplus Re_n$, where each e_i is a primitive idempotent. Then this decomposition is unique in the sense that, if $R = Rf_1 \oplus \cdots \oplus Rf_p$ is another such decomposition of R, then $n = p$, and after rearranging, $e_1 = f_1, \ldots, e_n = f_n$.*

Proof. Any idempotent of $R = Re_1 \oplus \cdots \oplus Re_n$ is of the form $x_1 + \cdots + x_n$ where x_i is an idempotent in Re_i. By Lemma 1.1.20, the only idempotents of Re_i are 0 and e_i. Hence, R has exactly n primitive idempotents, namely $e_1, \ldots, e_n$. □

1.7. Exercises. All of the exercises below will have significant applications later in the text. The reader is encouraged to prove them before continuing to the next section.

Exercise 1.1.22. Let R be a ring and I a left ideal in R. Prove that the following are equivalent.

(1) R/I is a projective left R-module.

(2) The R-module sequence $0 \to I \to R \to R/I \to 0$ is split-exact.

(3) I is an R-module direct summand of R.

(4) There is an element $e \in R$ such that $1 - e \in I$ and $Ie = (0)$.

(5) There is an idempotent $e \in R$ such that $I = R(1-e)$.

For another condition that is equivalent to (1) – (5), see Exercise 1.3.15.

Exercise 1.1.23. Let A be an R-algebra and e an idempotent in A.

(1) Show that eAe is an R-algebra.

(2) Show that there is an R-module direct sum decomposition:

$$A = eAe \oplus eA(1-e) \oplus (1-e)Ae \oplus (1-e)A(1-e).$$

Exercise 1.1.24. Let R be a commutative ring and J an ideal in R. Prove:

(1) If J is a direct summand of R, then $J^2 = J$.

(2) If J is a finitely generated ideal, and $J^2 = J$, then J is a direct summand of R.

2. Polynomial Functions

This section contains some results of a technical nature. Polynomial functions arise when we consider the coefficients of the characteristic polynomial associated to an element in a finite dimensional algebra over a field. In particular, the norm map and trace map are examples of homogeneous polynomial functions. The resultant operator defines a polynomial function which will be used when we construct a maximal commutative separable subfield of a division algebra. The main theorem on C_1 fields will be applied to prove Tsen's Theorem (Corollary 11.1.9).

2.1. The Ring of Polynomial Functions on a Module. Let R be a commutative ring, M an R-module, and $M^* = \operatorname{Hom}_R(M, R)$ the dual of M. By $\operatorname{Map}(M, R)$ we denote the set of all functions $f : M \to R$. Then $\operatorname{Map}(M, R)$ can be turned into an R-algebra. The addition and multiplication operations are defined pointwise: $(f+g)(x) = f(x) + g(x)$, $(fg)(x) = f(x)g(x)$. An element a in R defines the constant function $a : M \to R$, where $a(x) = a$. We can view M^* as an R-submodule of $\operatorname{Map}(M, R)$. The R-subalgebra of $\operatorname{Map}(M, R)$ generated by the set M^* is denoted $R[M^*]$ and is called the *ring of polynomial functions* on M. If $d \geq 0$, then a polynomial function $f \in R[M^*]$ is said to be *homogeneous of degree* d, if $f(rx) = r^d f(x)$, for all $x \in M$ and $r \in R$. Proposition 1.2.1 shows that the ring $R[M^*]$ is in fact a coordinate-free way to generalize the usual ring of polynomial functions on a vector space.

Proposition 1.2.1. *Let k be an infinite field, and V a finite dimensional k-vector space. If $\dim_k(V) = n$, then $k[V^*] \cong k[x_1, \dots, x_n]$ as k-algebras.*

Proof. Let $\{(v_i, f_i) \mid 1 \leq i \leq n\}$ be a dual basis for V. Define $\theta : k[x_1, \dots, x_n] \to k[V^*]$ by $x_i \mapsto f_i$. The reader should verify that θ is one-to-one and onto. $\square$

Lemma 1.2.2. *Let R be a commutative ring and P a free R-module of finite rank n. Let $\phi \in \operatorname{Hom}_R(P,P)$. If the characteristic polynomial of ϕ is $p(x) = x^n + a_1x^{n-1} + \cdots + a_{n-1}x + a_n$, then for each $i = 1, \dots, n$, the assignment $\phi \mapsto (-1)^i a_i$ defines a polynomial function $N_i : \operatorname{Hom}_R(P,P) \to R$ which is homogeneous of degree i.*

Proof. Let B be a basis for P. Let $\phi \in \operatorname{Hom}_R(P,P)$, and $(\phi_{ij}) = M(\phi, B)$ the matrix of ϕ with respect to B. By Proposition 1.2.1, a polynomial function on $\operatorname{Hom}_R(P,P)$ corresponds to a polynomial in the n^2 indeterminates $\Phi = \{\phi_{ij} \mid 1 \le i \le n, 1 \le j \le n\}$. The characteristic polynomial of ϕ is given by the combinatorial formula for the determinant

$$\det(xI_n - (\phi_{ij})) = \sum_{\vec{\ell} \in S_n} \operatorname{sign}(\ell) b_{\ell_1,1} \cdots b_{\ell_n,n} \tag{1.3}$$

where $b_{ii} = x - \phi_{ii}$ and $b_{ij} = -\phi_{ij}$ if $i \ne j$. A typical summand in (1.3) can be written in the form

$$\operatorname{sign}(\ell) b_{\ell_1,1} \cdots b_{\ell_n,n} = (x - a_{i_1 i_1}) \cdots (x - a_{i_d i_d}) m$$

where m is a monomial in $k[\Phi]$ of degree $n - d$. Therefore, $b_{\ell_1,1} \cdots b_{\ell_n,n}$ is a polynomial in x of degree d and for $0 \le k < n$, the coefficient of x^k is a homogeneous polynomial of degree $n - k$ in $k[\Phi]$. □

Example 1.2.3. Let k be a field and A a k-algebra. Assume $\dim_k(A) = n$ is finite. Using the left regular representation, we can embed A as a k-subalgebra of $\operatorname{Hom}_k(A,A)$. As in Lemma 1.2.2, let $N_i : \operatorname{Hom}_k(A,A) \to k$ be the homogeneous polynomial function of degree i defined by the coefficient of x^{n-i} in the characteristic polynomial of ϕ. For each i, upon restriction to A, $N_i : A \to k$ defines a homogeneous polynomial function on A of degree i. In particular N_n is the norm $N_k^A : A \to k$ defined in Exercise 1.2.14, and N_1 the trace $T_k^A : A \to k$. Fix a k-basis $\alpha_1, \dots, \alpha_n$ for A. Then this basis can be extended to a basis for $\operatorname{Hom}_k(A,A)$ and N_i can be identified with a homogeneous polynomial in $k[x_1, \dots, x_n]$ of degree i.

2.2. Resultant of Two Polynomials. Assume $m \ge 0$, $n \ge 0$, and $m + n \ge 1$. Let $f = \sum_{i=0}^m f_i x^i$ and $g = \sum_{i=0}^n g_i x^i$ be two polynomials in $k[x]$, where k is a field. So the degree of f is at most m, and the degree of g is at most n. In the general case, m and n are both positive, and the *Sylvester* *matrix* of f and g is the $(m+n)$-by-$(m+n)$ matrix

$$\mathrm{Syl}(f,g) = \begin{bmatrix} f_m & f_{m-1} & f_{m-2} & \cdots & f_0 & & & & \cdots & \\ & f_m & f_{m-1} & \cdots & f_1 & f_0 & & & \cdots & \\ & & f_m & \cdots & f_2 & f_1 & f_0 & & \cdots & \\ & & & \vdots & & & & & & \\ & \cdots & & f_m & f_{m-1} & f_{m-2} & \cdots & f_0 & & \\ & \cdots & & & f_m & f_{m-1} & \cdots & f_1 & f_0 & \\ & \cdots & & & & f_m & \cdots & f_2 & f_1 & f_0 \\ g_n & g_{n-1} & g_{n-2} & \cdots & g_0 & & & & \cdots & \\ & g_n & g_{n-1} & \cdots & g_1 & g_0 & & & \cdots & \\ & & g_n & \cdots & g_2 & g_1 & g_0 & & \cdots & \\ & & & \vdots & & & & & & \\ & \cdots & & g_n & g_{n-1} & f_{m-2} & \cdots & g_0 & & \\ & \cdots & & & g_n & g_{n-1} & \cdots & g_1 & g_0 & \\ & \cdots & & & & g_n & \cdots & g_2 & g_1 & g_0 \end{bmatrix}$$

where blank spaces consist of zeros. The top n rows are constructed from the coefficients of f, shifted and padded with zeros. The bottom m rows are constructed from shifting the coefficients of g, and padding with zeros. In the degenerate case when $m = 0$, $\mathrm{Syl}(f,g)$ is defined to be the n-by-n diagonal matrix $f_0(E_{11} + \cdots + E_{nn})$. In the degenerate case when $n = 0$, $\mathrm{Syl}(f,g)$ is defined to be the m-by-m diagonal matrix $g_0(E_{11} + \cdots + E_{mm})$. The *resultant* of f and g, written $\mathrm{Res}(f,g)$, is the determinant of $\mathrm{Syl}(f,g)$.

Lemma 1.2.4. *In the above context, we view $f_0, \ldots, f_m, g_0, \ldots, g_n$ as variables. Then* $\mathrm{Res}(f,g)$ *satisfies the following:*

(1) $\mathrm{Res}(f,g)$ *is a polynomial in* $\mathbb{Z}[f_0, \ldots, f_m, g_0, \ldots, g_n]$ *which is homogeneous of degree $n+m$.*

(2) *For any constant $c \in \mathbb{Z}$,*

$$\mathrm{Res}(cf,g) = c^n \,\mathrm{Res}(f,g)$$
$$\mathrm{Res}(f,cg) = c^m \,\mathrm{Res}(f,g)$$

Thus, $\mathrm{Res}(f,g)$ *is homogeneous of degree n in $f_0, \ldots, f_m$ and homogeneous of degree m in $g_0, \ldots, g_n$.*

Proof. The proof is left to the reader. □

Lemma 1.2.5. *In the above context, the following are true.*

(1) *If* $\deg(f) < m$ *and* $\deg(g) < n$, *then* $\mathrm{Res}(f,g) = 0$.

(2) *If $m = 0$, then* $\mathrm{Res}(f,g) = f_0^n$.

(3) *If $n = 0$, then* $\mathrm{Res}(f,g) = g_0^m$.

(4) $\operatorname{Res}(f, g) = (-1)^{mn} \operatorname{Res}(g, f)$.

(5) *If* $\deg(f) = m$ *and* $d = \deg(g) < n$, *then* $\operatorname{Res}(f, g) = f_m^{n-d} \operatorname{Res}(f, h)$, *where* $h = g_d x^d + \cdots + g_1 x + g_0$.

Proof. Parts (1) – (4) are left to the reader.

(5): The Sylvester matrix has the form

$$\operatorname{Syl}(f, g) = \begin{bmatrix} T & * \\ 0 & \operatorname{Syl}(f, h) \end{bmatrix}$$

where T is an upper triangular matrix of size $(n-d)$-by-$(n-d)$ with diagonal $(f_m, \dots, f_m)$. □

Lemma 1.2.6. *In the context of Lemma* 1.2.5, *assume* $m \le n$ *and* $\deg(f) = m$. *Let* q *and* r *be the unique polynomials in* $k[x]$ *guaranteed by the Division Algorithm which satisfy:* $q = \sum_{i=0}^{n-m} q_i x^i$, $r = \sum_{i=0}^{m-1} r_i x^i$, *and* $g = qf + r$. *Then* $\operatorname{Res}(f, g) = f_m^{n-m+1} \operatorname{Res}(f, r)$.

Proof. Write $c = -q_{n-m} = -g_n / f_m$, and set $h = g + cx^{n-m} f = \sum_{i=0}^{n-1} h_i x^i$. Let $I_m = E_{11} + \cdots + E_{mm} \in M_m(k)$, $I_n = E_{11} + \cdots + E_{nn} \in M_n(k)$, and $I_{mn} = E_{11} + \cdots + E_{mm} \in M_{mn}(k)$. The product

$$\begin{bmatrix} I_n & 0 \\ cI_{mn} & I_m \end{bmatrix} \operatorname{Syl}(f, g) = \begin{bmatrix} f_m & * \\ & \operatorname{Syl}(f, h) \end{bmatrix}$$

corresponds to elementary row operations. Computing determinants, we find that $\operatorname{Res}(f, g) = f_m \operatorname{Res}(f, h)$. By induction on $n - m$, we are done. □

Theorem 1.2.7. *In the context of Lemma* 1.2.5, *assume* F/k *is an extension of fields such that in the unique factorization domain* $F[x]$ *both polynomials* f *and* g *have no irreducible factor of degree greater than one.*

(1) *If* $m = \deg(f) \ge 1$ *and* $f = f_m(x-\alpha_1) \cdots (x-\alpha_m)$ *is a factorization of* f *into a product of linear polynomials, then*

$$\operatorname{Res}(f, g) = f_m^n \prod_{i=1}^{m} g(\alpha_i).$$

(2) *If* $\deg(g) = n \ge 1$ *and* $g = g_n(x - \beta_1) \cdots (x - \beta_n)$ *is a factorization of* g *into a product of linear polynomials, then*

$$\operatorname{Res}(f, g) = (-1)^{mn} g_n^m \prod_{j=1}^{n} f(\beta_j).$$

(3) *Suppose* $\deg(f) = m \ge 1$ *and* $\deg(g) = n \ge 1$. *If* $f = f_m(x - \alpha_1) \cdots (x - \alpha_m)$ *and* $g = g_n(x - \beta_1) \cdots (x - \beta_n)$ *are factorizations of* f

and g into products of linear polynomials, then

$$\operatorname{Res}(f,g) = f_m^n g_n^m \prod_{i=1}^{m}\prod_{j=1}^{n}(\alpha_i - \beta_j).$$

Proof. We prove (1) and (2) simultaneously. The reader should verify that Part (3) follows from Parts (1) and (2).

The proof is by induction on $m+n$. The basis for the induction, which follows from Lemma 1.2.5, is when $n = 0$ or $m = 0$. Assume from now on that $1 \le m$ and $1 \le n$.

Case 1: $\deg(f) = m \ge 1$, and $\deg(g) = d < n$. If we set $h = \sum_{i=0}^{d} g_i x^i$, then by Lemma 1.2.5 (5), $\operatorname{Res}(f,g) = f_m^{n-d}\operatorname{Res}(f,g)$. By the induction hypothesis, $\operatorname{Res}(f,g) = f_m^{n-d}\operatorname{Res}(f,g) = f_m^{n-d} f_m^d \prod_{i=1}^{m} g(\alpha_i)$, which proves (1) in this case.

Case 2: $\deg(g) = n \ge 1$, and $\deg(f) = d < m$. In this case, Part (2) follows by Case 1 and Lemma 1.2.5 (4).

Case 3: Assume $\deg(f) = m \ge 1$ and $\deg(g) = n \ge 1$, and $m \le n$. As in Lemma 1.2.6, write $g = fq + r$, where $r = \sum_{i=0}^{m-1} r_i x^i$. By Lemma 1.2.6 and the induction hypothesis,

$$\begin{aligned}
\operatorname{Res}(f,g) &= f_m^{n-m+1}\operatorname{Res}(f,r)\\
&= f_m^{n-m+1} f_m^{m-1}\prod_{i=1}^{m} r(\alpha_i)\\
&= f_m^n \prod_{i=1}^{m} g(\alpha_i)
\end{aligned}$$

where the last equation follows since $r(\alpha_i) = g(\alpha_i) - f(\alpha_i)q(\alpha_i)$. In this case, we have proved Part (1). By

$$\begin{aligned}
\operatorname{Res}(f,g) &= f_m^n \prod_{i=1}^{m} g(\alpha_i)\\
&= f_m^n \prod_{i=1}^{m} g_n(\alpha_i - \beta_1)\cdots(\alpha_i - \beta_n)\\
&= g_n^m \prod_{j=1}^{n} f_m(\alpha_1 - \beta_j)\cdots(\alpha_m - \beta_j)\\
&= (-1)^{mn} g_n^m \prod_{j=1}^{n} f(\beta_j)
\end{aligned}$$

we see that Part (2) holds in Case 3.

Case 4: Assume $\deg(f) = m \geq 1$ and $\deg(g) = n \geq 1$, and $n \leq m$. Applying Lemma 1.2.5 (4), this reduces to Case 3. $\square$

Corollary 1.2.8. *In the above context,* $\operatorname{Res}(f, g) = 0$ *if and only if one of the following is satisfied:*

(1) $\deg(f) < m$ *and* $\deg(g) < n$.

(2) $(f, g) \neq k[x]$, *or equivalently, f and g have a common irreducible factor, or equivalently, f and g have a common root in some extension field F/k.*

Proof. If (1) is true, then the first column of $\operatorname{Syl}(f, g)$ is made up of zeros, so $\operatorname{Res}(f, g) = 0$. Otherwise, by Lemma 1.2.5, we can reduce to the case where $\deg(f) = m$. By Theorem 1.2.7 (1), $\operatorname{Res}(f, g) = 0$ if and only if f and g have a common root in some extension field F/k, which is equivalent to (2). $\square$

2.3. Polynomial Functions on an Algebraic Curve. A field K is said to be C_1 if every homogeneous polynomial of degree $d > 0$ in n variables, where $n > d$, has a nontrivial zero. The purpose of this section is to proof that an algebraic function field in one variable over an algebraically closed field is C_1. The proof we give is modeled after that of [**Gre69**, Chapter 3]. Fields which are C_1 are the subject of Corollaries 11.1.8 and 7.9.7.

Example 1.2.9. Let k be a field and A a k-algebra. Assume $\dim_k(A) = n$ is finite. Using the left regular representation, we can embed A as a k-subalgebra of $\operatorname{Hom}_k(A, A)$. As in Example 1.2.3, the norm $N_k^A : A \to k$ is a homogeneous polynomial function on A of degree n and the trace $T_k^A : A \to k$ is a homogeneous linear polynomial function on A. Fix a k-basis $\alpha_1, \ldots, \alpha_n$ for A. With respect to this basis, we identify A with affine n-space over k (see, for example, [**DF04**, Section 15.2]). That is, an element $a_1\alpha_1 + \cdots + a_n\alpha_n \in A$ corresponds to the point $(a_1, \ldots, a_n) \in \mathbb{A}_k^n$. With this identification, the norm $N_k^A : A \to k$ corresponds to a homogeneous polynomial in $k[x_1, \ldots, x_n]$ of degree n. Using Exercise 1.2.14 we see that an element α in A is invertible if and only if $N_k^A(\alpha) \neq 0$. If A is a division algebra over k, then the norm defines a homogeneous polynomial in $k[x_1, \ldots, x_n]$ of degree n with no nontrivial zeros. We should advise the reader that the norm used in this example is not the norm defined specifically for an Azumaya algebra (or a central simple algebra) in Section 11.1.1.

Proposition 1.2.10. *Let k be an algebraically closed field and $K = k(x)$ the field of rational functions over k in the single variable x. Then K is C_1.*

Proof. Let f be a homogeneous polynomial of degree d in $K[x_1, \ldots, x_n]$, where $n > d$. If necessary, we can multiply by a polynomial in $k[x]$, and assume f is in $k[x][x_1, \ldots, x_n]$. Expand $f = t_1 m_1 + \cdots + t_p m_p$ as a sum where each t_i is a polynomial in $k[x]$ and each m_i is a monomial of degree d in $k[x_1, \ldots, x_n]$ with coefficient 1. Let q be the maximum of $\{\deg(t_1), \ldots, \deg(t_p)\}$. Pick m such that $m(n-d) > q - n + 1$. This is possible, since $n > d$. Let $Y = \{y_{ij} \mid 1 \le i \le n, 0 \le j \le m\}$ be a set of indeterminates. For $1 \le i \le n$, consider the polynomials $p_i = \sum_{j=0}^{n} y_{ij} x^j$ in $k[Y][x]$. For any $e \ge 0$, $(p_i)^e$ is a polynomial in $k[Y][x]$ and in $(p_i)^e$ the coefficient of each power x^j is a homogeneous polynomial in $k[Y]$ of degree e. For each of the monomials m_r we have: $m_r(p_1, \ldots, p_n)$ is a polynomial in $k[Y][x]$, the coefficient of each power x^j is a homogeneous polynomial in $k[Y]$ of degree d, and the highest power of x that occurs has degree md. Therefore, $f(p_1, \ldots, p_n)$ is a polynomial in $k[Y][x]$ and the coefficient of each power x^j is a homogeneous polynomial in $k[Y]$ of degree d. Because q is the highest degree of all of the polynomials $t_1, \ldots, t_p$, the highest power of x appearing in $f(p_1, \ldots, p_n)$ has degree $q + md$ or less. Expand $f(p_1, \ldots, p_n)$ as a polynomial in x with coefficients in $k[Y]$:

$$f(p_1, \ldots, p_n) = \sum_{j=0}^{q+md} f_j(Y) x^j.$$

Consider the ideal $I = (f_0(Y), \ldots, f_{q+md}(Y))$ in $k[Y]$. Because the number of indeterminates, $(m+1)n$, is greater than the number of generators, $q + md + 1$, by Exercise 1.2.22, $Z(I)$ is an infinite subset of $\mathbb{A}_k^{(m+1)n}$. That is, there are elements a_{ij} in k, not all zero, defining $p_i = \sum_j a_{ij} x^j$ in $k[x]$ such that $f(p_1, \ldots, p_n) = 0$. $\square$

Lemma 1.2.11. *Let k be a field which is not algebraically closed. For any positive integer N, there exists $n \ge N$ and a homogeneous polynomial $f \in k[x_1, \ldots, x_n]$ of degree n such that $Z(f) = \{(0, \ldots, 0)\}$.*

Proof. Let K/k be an extension of fields and assume $\dim_k(K) = d > 1$. As in Example 1.2.9, the norm $N_k^K : K \to k$ can be used to define a homogeneous polynomial $\phi \in k[x_1, \ldots, x_d]$ of degree d which has only the trivial zero. Let $X = \{x_{ij} \mid 1 \le i \le d, 1 \le j \le d\}$ be a set of d^2 indeterminates. The notation $\phi^{(2)}$ denotes the polynomial

$$\phi^{(2)} = \phi(\phi \mid \phi \mid \cdots \mid \phi) = \phi(\phi(x_{11}, \ldots, x_{1d}), \ldots, \phi(x_{d1}, \ldots, x_{dd}))$$

in $k[X]$ obtained by substituting ϕ into itself. Then $\phi^{(2)}$ is homogeneous of degree d^2 in d^2 variables. Since ϕ has only the trivial zero, we see that $\phi^{(2)}$ also has only the trivial zero. Using the same notation,

$$\phi^{(3)} = \phi^{(2)}(\phi \mid \phi \mid \cdots \mid \phi)$$

is the polynomial obtained by substituting ϕ into $\phi^{(2)}$. Then $\phi^{(3)}$ is a homogeneous polynomial of degree d^3 in d^3 variables and has only the trivial zero. Recursively, for any $m > 1$ we construct a homogeneous polynomial $\phi^{(m)}$ of degree d^m in d^m variables which has only the trivial zero. Take f to be $\phi^{(m)}$, for any m such that $d^m \geq N$. □

Lemma 1.2.12. *Assume K is a C_1 field and d, m and n are positive integers. Let $f_1, \dots, f_m$ be homogeneous polynomials of degree d in $k[x_1, \dots, x_n]$. If $n > md$, then the zero set $Z(f_1, \dots, f_m)$ is not equal to the singleton set $\{(0, \dots, 0)\}$.*

Proof. By Lemma 1.2.11, there exist an integer $e > m$ and a homogeneous polynomial ϕ in $K[x_1, \dots, x_e]$ of degree e such that ϕ has only the trivial zero. Let $\lfloor e/m \rfloor$ be the greatest integer less than or equal to e/m. In ϕ substitute $\lfloor e/m \rfloor$ groups of $f_1, \dots, f_m$ for $x_1, \dots, x_{n\lfloor e/m \rfloor}$ according to the pattern

$$\psi = \phi(f_1, \dots, f_m \mid f_1, \dots, f_m \mid \cdots \mid f_1, \dots, f_m \mid 0, \dots, 0)$$

where the vertical bar signifies that a different set of n variables are used in the different groups $f_1, \dots, f_m$. So ψ is a polynomial in $n\lfloor e/m \rfloor$ variables with coefficients in K, and is homogeneous of degree de. Since $n - dm > 0$, by taking e larger if necessary, we can pick ϕ and e such that $e > m$ and $(n - dm)\lfloor e/m \rfloor > dm$. Therefore, $n\lfloor e/m \rfloor > dm(1 + \lfloor e/m \rfloor) \geq de$. Since K is C_1, ψ has a nontrivial zero. By the choice of ϕ, this implies there exists a nontrivial simultaneous zero for $f_1, \dots, f_m$. □

Theorem 1.2.13. *Let F be a field which is C_1. If K is an algebraic extension field of F, then K is C_1.*

Proof. Let f be a homogeneous polynomial of degree d in $K[x_1, \dots, x_n]$, with $d < n$. The coefficients of f generate a finite dimensional extension of F in K. For this reason, it suffices to prove the theorem under the assumption that $\dim_F(K) = m$ is finite. Let $\{\kappa_1, \dots, \kappa_m\}$ be an F-basis for K. Let $Y = \{y_{ij} \mid 1 \leq i \leq n, 1 \leq j \leq m\}$ be a set of nm new indeterminates. Let $p_i = y_{i1}\kappa_1 + \cdots + y_{im}\kappa_m$, for $1 \leq i \leq n$. Then $p_1, \dots, p_n$ are homogeneous linear polynomials in $K[Y]$. We can write $f(p_1, \dots, p_n) \in K[Y]$ as

$$f(p_1, \dots, p_n) = f_1\kappa_1 + \cdots + f_m\kappa_m$$

where each f_j is a homogeneous polynomial of degree d in $F[Y]$. If follows from Lemma 1.2.12 that $Z(f_1, \dots, f_m) \subseteq \mathbb{A}_F^{mn}$ contains a point (a_{ij}) such that not all of the a_{ij} are zero. Evaluating $p_1, \dots, p_n$ at (a_{ij}) shows that there exists a nontrivial zero for f in $\mathbb{A}_K^n$. □

2.4. Exercises.

Exercise 1.2.14. Let R be a commutative ring and suppose A is an R-algebra which is finitely generated and free of rank n as an R-module. Let

$\theta : A \to \operatorname{Hom}_R(A, A)$ be the left regular representation of A in $\operatorname{Hom}_R(A, A)$ which is defined by $\alpha \mapsto \ell_\alpha$. Define $T_R^A : A \to R$ by the assignment $\alpha \mapsto \operatorname{trace}(\ell_\alpha)$. We call T_R^A the *trace from A to R*. Define $N_R^A : A \to R$ by the assignment $\alpha \mapsto \det(\ell_\alpha)$. We call N_R^A the *norm from A to R*.

(1) Show that $T_R^A(r\alpha+s\beta) = rT_R^A(\alpha)+sT_R^A(\beta)$, if $r, s \in R$ and $\alpha, \beta \in A$.

(2) Show that $N_R^A(\alpha\beta) = N_R^A(\alpha)N_R^A(\beta)$ and $N_R^A(r\alpha) = r^n N_R^A(\alpha)$, if $r \in R$ and $\alpha, \beta \in A$.

Exercise 1.2.15. Let R be a commutative ring and M a finitely generated free R-module of rank n. Let $\phi \in \operatorname{Hom}_R(M, M)$. Show that if $\operatorname{char.poly}_R(\phi) = x^n + a_{n-1}x^{n-1} + \cdots + a_0$, then $\operatorname{trace}(\phi) = -a_{n-1}$ and $\det(\phi) = (-1)^n a_0$.

Exercise 1.2.16. Let k be a field, V a finitely generated vector space over k, and $\phi \in \operatorname{Hom}_k(V, V)$. Suppose $q = \operatorname{min.poly}_k(\phi) = x^m + a_{m-1}x^{m-1} + \cdots + a_0$ is irreducible in $k[x]$. Prove the following.

(1) $\operatorname{char.poly}_k(\phi) = q^r$ for some integer r.

(2) $\det(\phi) = (-1)^{mr} a_0^r$.

(3) $\operatorname{trace}(\phi) = -ra_{m-1}$.

Exercise 1.2.17. Let $\theta : R \to S$ be a homomorphism of commutative rings.

(1) Show that θ induces a homomorphism of rings $\theta : M_n(R) \to M_n(S)$.

(2) Show that $\theta(\det(M)) = \det(\theta(M))$, for every M in $M_n(R)$.

(3) Show that under the homomorphism of rings $R[x] \to S[x]$, the characteristic polynomial of M maps to the characteristic polynomial of $\theta(M)$. That is, $\theta(\operatorname{char.poly}_R(M)) = \operatorname{char.poly}_S(\theta(M))$.

Exercise 1.2.18. Let R be a commutative ring and $n \geq 1$. If $A \in M_n(R)$, show that the trace of A satisfies:

$$\sum_{i=1}^{n} \sum_{j=1}^{n} e_{ij} A e_{ji} = \operatorname{trace}(A) I_n$$

where e_{ij} denotes the elementary matrix and $I_n = e_{11} + \cdots + e_{nn}$ the identity matrix.

Exercise 1.2.19. Let R be a commutative ring and $A = M_n(R)$ the ring of n-by-n matrices over R. The so-called *trace pairing* $\tau : A \times A \to R$ is defined by $\tau(\alpha, \beta) = \operatorname{trace}(\alpha\beta)$. Show that τ satisfies these properties:

(1) $\tau(\alpha, \beta) = \tau(\beta, \alpha)$.

(2) $\tau(a_1\alpha_1 + a_2\alpha_2, \beta) = a_1\tau(\alpha_1, \beta) + a_2\tau(\alpha_2, \beta)$ for $a_1, a_2 \in R$.

(3) $\tau(\alpha, b_1\beta_1 + b_2\beta_2) = b_1\tau(\alpha, \beta_1) + b_2\tau(\alpha, \beta_2)$ for $b_1, b_2 \in R$.

(4) If $\alpha \neq 0$ is fixed, then $\tau(\alpha, \ ) : A \to R$ is nonzero. That is, there exists β such that $\tau(\alpha, \beta) \neq 0$.

We say that τ is a *symmetric nondegenerate bilinear form.*

Exercise 1.2.20. Let R be an integral domain.

(1) A polynomial f in $R[x]$ defines a function $f : R \to R$. If R is infinite, show that f is the zero function (that is, $f(a) = 0$ for all $a \in R$) if and only if f is the zero polynomial.

(2) A polynomial f in $R[x_1, \dots, x_r]$ defines a function $f : R^r \to R$. If R is infinite, use induction on r to show f is the zero function if and only if f is the zero polynomial.

Exercise 1.2.21. Let k be a field. Show that for any $n \geq 1$ there exists a polynomial $f \in k[x]$ of degree n such that f has no repeated roots.

Exercise 1.2.22. Let k be an algebraically closed field, $n \geq 1$, and $S = k[x_0, \dots, x_n]$. Then $S = S_0 \oplus S_1 \oplus S_2 \oplus \cdots$ is a graded ring. Let $S_+ = (x_0, x_1, \dots, x_n)$ be the maximal ideal generated by the monomials of degree one. Prove:

(1) Let $f \in S_d$ be a homogeneous polynomial of degree $d > 0$. If $(a_0, \dots, a_n)$ is in $Z(f)$, and $\alpha \in K$, then $(\alpha a_0, \dots, \alpha a_n) \in Z(f)$.

(2) Let $I = (f_1, \dots, f_m)$ be an ideal in S, where each f_i is homogeneous and has degree greater than or equal to one. If $\operatorname{Rad} I$, the nil radical of I, is not equal to S_+, then $Z(I)$ is infinite. (See Section 2.2 for the definition of $\operatorname{Rad} I$.)

Exercise 1.2.23. Let k be a field and F an extension field of k. Suppose α and β are elements of F that are algebraic over k. Using resultants, show that $\alpha + \beta$ and $\alpha\beta$ are algebraic over k. Show how to find the minimal polynomials for $\alpha + \beta$ and $\alpha\beta$.

3. Hom and Tensor

We assume the reader has some familiarity with the tensor product of modules as well as Hom groups. We give quick reviews of the definitions and list the fundamental properties. As basic references we recommend [**DF04**] and [**Hun80**].

3.1. Tensor Product. Let S and R be rings. If $M \in \mathfrak{M}_R$ and $M \in {}_S\mathfrak{M}$, then M is said to be a *left S right R bimodule* if $s(mr) = (sm)r$ for all possible $s \in S$, $m \in M$ and $r \in R$. Denote by ${}_S\mathfrak{M}_R$ the category of all left S right R bimodules. We say that M is a *left R left S bimodule* if M is both a left R-module and a left S-module and $r(sm) = s(rm)$ for all possible

$r \in R$, $m \in M$ and $s \in S$. For example, if R is a ring and I is an ideal in R, then the associative law for multiplication in R shows that I is a left R right R bimodule. If R is a commutative ring, any left R-module M can be made into a left R right R bimodule by defining mr to be rm. If M is in ${}_R\mathfrak{M}$, N is in $\mathfrak{M}_R$, and C is a $\mathbb{Z}$-module, then a function $f : M \times N \to C$ is said to be an *R-balanced map* if it satisfies

(1) $f(m_1 + m_2, n) = f(m_1, n) + f(m_2, n)$,
(2) $f(m, n_1 + n_2) = f(m, n_1) + f(m, n_2)$, and
(3) $f(mr, n) = f(m, rn)$,

for all possible $m_i \in M$, $n_i \in N$, $r \in R$. The tensor product of M and N over R consists of an abelian group denoted $M \otimes_R N$, and an R-balanced map $\tau : M \times N \to M \otimes_R N$ satisfying the following universal mapping property. If C is an abelian group and $f : M \times N \to C$ is R-balanced, then there exists a unique homomorphism $\phi : M \otimes_R N \to C$ such that $\phi\tau = f$. The reader should verify that the tensor product exists, is unique up to isomorphism, and satisfies the properties enumerated in the following lemma.

Lemma 1.3.1. *Let R and S be rings.*

(1) *If M and M' are in ${}_S\mathfrak{M}_R$, and N and N' are in ${}_R\mathfrak{M}$, then the following are true.*
   (a) *If $s \in S$, $m \in M$ and $n \in N$, the multiplication rule $s(m \otimes n) = sm \otimes n$ turns $M \otimes_R N$ into a left S-module.*
   (b) *Let $f : M \to M'$ be a homomorphism of left S right R bimodules and $g : N \to N'$ a homomorphism of left R-modules. Then $f \otimes g : M \otimes_R N \to M' \otimes_R N'$ is a homomorphism of left S-modules.*

(2) *Since R is a left R right R bimodule, given $M \in {}_R\mathfrak{M}$ we view $R \otimes_R M$ as a left R-module. In this case, $R \otimes_R M \cong M$ as left R-modules under the map $x \otimes y \mapsto xy$.*

(3) *Assume $L \in \mathfrak{M}_R$, $M \in {}_R\mathfrak{M}_S$ and $N \in {}_S\mathfrak{M}$. Then $(L \otimes_R M) \otimes_S N$ is isomorphic as an abelian group to $L \otimes_R (M \otimes_S N)$ under the map which sends $(x \otimes y) \otimes z$ to $x \otimes (y \otimes z)$.*

(4) *Let M and $\{M_i\}_{i \in I}$ be in $\mathfrak{M}_R$. Let N and $\{N_j\}_{j \in J}$ be in ${}_R\mathfrak{M}$. There are isomorphisms of abelian groups*

$$M \otimes_R \Big(\bigoplus_{j \in J} N_j\Big) \cong \bigoplus_{j \in J} (M \otimes_R N_j)$$

and

$$\Big(\bigoplus_{i \in I} M_i\Big) \otimes_R N \cong \bigoplus_{i \in I} (M_i \otimes_R N).$$

(5) *Let R be a ring, $M \in \mathfrak{M}_R$, and $N \in {}_R\mathfrak{M}$. Then $M \otimes_R N \cong N \otimes_{R^o} M$ under the map $x \otimes y \mapsto y \otimes x$.*

(6) *Let R be a ring and M a right R-module. Given a short exact sequence*

$$0 \to A \xrightarrow{\alpha} B \xrightarrow{\beta} C \to 0$$

in ${}_R\mathfrak{M}$, the sequence

$$M \otimes_R A \xrightarrow{1 \otimes \alpha} M \otimes_R B \xrightarrow{1 \otimes \beta} M \otimes_R C \to 0$$

is an exact sequence of $\mathbb{Z}$-modules.

If S is a ring and M is a left S right R bimodule, then by Lemma 1.3.1 it follows that $M \otimes_R (\cdot)$ defines a right exact covariant functor from ${}_R\mathfrak{M}$ to ${}_S\mathfrak{M}$. In case $M \otimes_R (\cdot)$ is also left exact, then we say M is a *flat* R-module. Let $\phi : R \to S$ be a homomorphism of rings. Then R acts on S from both the left and right by the rules $rx = \phi(r)x$ and $xr = x\phi(r)$. Using the ring structure of S, the reader should verify that S is a left S right R bimodule. Theorem 1.3.2 below describes a general technique, called *change of base*. The proof is left to the reader.

Theorem 1.3.2. *Let $\phi : R \to S$ be a homomorphism of rings. As mentioned above, ϕ makes S into a left S right R bimodule.*

(1) *The assignment $M \mapsto S \otimes_R M$ defines a right exact covariant functor ${}_R\mathfrak{M} \to {}_S\mathfrak{M}$ which satisfies:*
  (a) *R is mapped to S.*
  (b) *$\bigoplus_{i \in I} M_i \mapsto \bigoplus_{i \in I} (S \otimes_R M_i)$. That is, a direct sum is mapped to a direct sum.*
  (c) *The free module R^I is mapped to the free S-module S^I.*

(2) *If M is R-projective, then $S \otimes_R M$ is S-projective.*

(3) *If M is an R-generator, then $S \otimes_R M$ is a S-generator.*

(4) *If M is finitely generated over R, then $S \otimes_R M$ is finitely generated over S.*

(5) *If M is a flat R-module, then $S \otimes_R M$ is a flat S-module.*

For a commutative ring R, Proposition 1.3.3 states that tensor product defines a product on the category of R-progenerator modules. The proof is left to the reader.

Proposition 1.3.3. *Let R be a commutative ring and let M and N be two R-modules.*

(1) *If M and N are finitely generated over R, then so is $M \otimes_R N$.*

(2) *If M and N are projective over R, then so is $M \otimes_R N$.*

(3) *If* M *and* N *are generators over* R*, then so is* $M \otimes_R N$*.*

(4) *If* M *and* N *are progenerators over* R*, then so is* $M \otimes_R N$*.*

Proposition 1.3.4 provides partial converses to the four parts of Proposition 1.3.3.

Proposition 1.3.4. *Let* R *be a ring. Let* M *and* N *be left* R *right* R*-bimodules. Assume* $M \otimes_R N$ *is a left* R*-generator module. Then the following are true.*

(1) M *and* N *are both left* R*-generator modules.*

(2) *If* $M \otimes_R N$ *is projective as a left* R*-module, then* M *and* N *are both projective as left* R*-modules.*

(3) *If* $M \otimes_R N$ *is finitely generated as a left* R*-module, then* M *and* N *are both finitely generated as a left* R*-modules.*

(4) *If* $M \otimes_R N$ *is a left progenerator over* R*, then* M *and* N *are both left progenerators over* R*.*

If $M \otimes_R N$ *is a right* R*-generator module, then right hand versions of (1) – (4) hold for* M *and* N*.*

Proof. (1): By Exercise 1.1.11 there is a free R-module F_1 of finite rank and a homomorphism f_1 of left R-modules such that $f_1 : F_1 \otimes_R (M \otimes_R N) \to R$ is onto. By Lemma 1.1.4 there is a free R-module F_2 and a left R-module homomorphism f_2 such that $f_2 : F_2 \to M$ is onto. By Lemma 1.3.1,

$$F_2 \otimes_R N \xrightarrow{f_2 \otimes 1} M \otimes_R N \to 0$$

is exact. For the same reason,

$$F_1 \otimes_R (F_2 \otimes_R N) \xrightarrow{1 \otimes f_2 \otimes 1} F_1 \otimes_R (M \otimes_R N) \to 0$$

is exact. Since $F_1 \otimes_R F_2$ is a free R-module, Lemma 1.3.1 shows that $F_1 \otimes_R (F_2 \otimes_R N)$ is a direct sum of copies of N. Then $f_1 \circ (1 \otimes f_2 \otimes 1)$ maps a direct sum of copies of N onto R. By another application of Exercise 1.1.11, this shows N is a left R-module generator. The other case is left to the reader.

(2) and (3): By Part (1) and Exercise 1.1.11 there exist a free R-module F of finite rank and a left R-module epimorphism $f : N \otimes_R F \to R$. But f is split since R is projective over R. By Exercise 1.3.7,

$$M \otimes_R N \otimes_R F \xrightarrow{f \otimes 1} M \to 0$$

is split exact. If $M \otimes_R N$ is projective, then by Lemma 1.3.1 and Exercise 1.1.10, M is projective. If $M \otimes_R N$ is finitely generated, then so is M. The other cases are left to the reader. □

In proving Exercise 1.3.15, Corollary 1.3.5 will be helpful. In Corollary 2.6.6 it will be combined with a test for locally free modules over an integral domain to give us a useful flatness criterion. For a proof, see [**Rot79**, Corollary 3.58], for example.

Corollary 1.3.5. *Let R be any ring and M a finitely generated left R-module. Then M is projective if and only if M is of finite presentation and flat.*

3.2. Exercises.

Exercise 1.3.6. Let R be a ring and let R^I and R^J be free R-modules. Prove:

(1) $R^I \otimes_R R^J$ is a free R-module.

(2) Assume R is commutative. If A is a free R-module of rank m and B is a free R-module of rank n, then $A \otimes_R B$ is free of rank mn.

Exercise 1.3.7. Let

$$0 \to A \xrightarrow{\alpha} B \xrightarrow{\beta} C \to 0 \tag{1.4}$$

be a short exact sequence of left R-modules. Given a right R-module M, consider the sequence

$$0 \to M \otimes_R A \xrightarrow{1\otimes\alpha} M \otimes_R B \xrightarrow{1\otimes\beta} M \otimes_R C \to 0. \tag{1.5}$$

Prove:

(1) If (1.4) is split exact, then (1.5) is split exact.

(2) If M is a free right R-module, then (1.5) is exact, hence M is flat.

(3) If M is a projective right R-module, then (1.5) is exact, hence M is flat.

Exercise 1.3.8. Let R be a commutative ring. This exercise outlines a proof that the tensor product is a coproduct in the category of commutative R-algebras. Suppose A and B are R-algebras. Then A and B come with homomorphisms $\theta_1 : R \to A$ and $\theta_2 : R \to B$ satisfying $\operatorname{im}(\theta_1) \subseteq Z(A)$ and $\operatorname{im}(\theta_2) \subseteq Z(B)$. Begin by proving that $A \otimes_R B$ is an R-algebra with multiplication induced by $(x_1 \otimes y_1)(x_2 \otimes y_2) = x_1x_2 \otimes y_1y_2$.

(1) Show that there exist R-algebra homomorphisms $\rho_1 : A \to A \otimes_R B$ and $\rho_2 : B \to A \otimes_R B$ such that the diagram

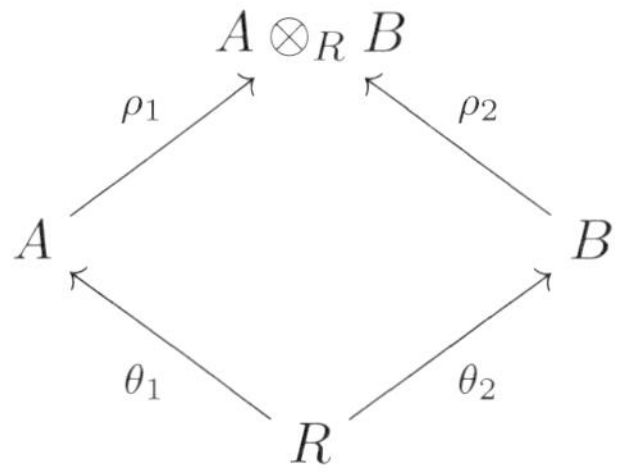

commutes. Show that elements in $\mathrm{im}(\rho_1)$ commute with elements in $\mathrm{im}(\rho_2)$. That is, $\rho_1(x)\rho_2(y) = \rho_2(y)\rho_1(x)$ for all $x \in A$, $y \in B$.

(2) Suppose there exist R-algebra homomorphisms $\alpha : A \to C$ and $\beta : B \to C$ such that $\mathrm{im}(\alpha)$ commutes with $\mathrm{im}(\beta)$. Show that there exists a unique R-algebra homomorphism $\gamma : A \otimes_R B \to C$ such that the diagram

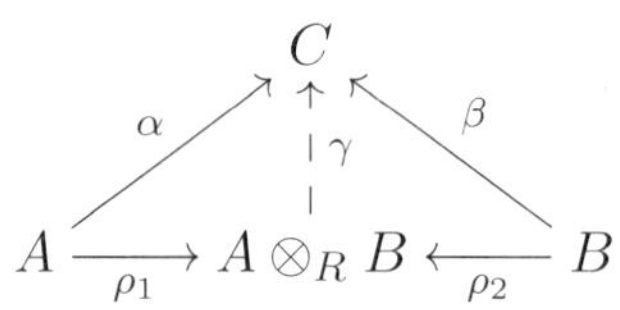

commutes.

(3) Show that if there exists an R-algebra homomorphism $\gamma : A \otimes_R B \to C$, then there exist R-algebra homomorphisms $\alpha : A \to C$ and $\beta : B \to C$ such that the image of α commutes with the image of β and the diagram of Part (2) commutes.

Exercise 1.3.9. Let S be a commutative R-algebra. Show that there is a well defined homomorphism of R-algebras $\mu : S \otimes_R S \to S$ which maps a typical element $\sum x_i \otimes y_i$ in the tensor algebra to $\sum x_i y_i$ in S.

Exercise 1.3.10. Let R be a commutative ring and let A and B be R-algebras. Prove that $A \otimes_R B \cong B \otimes_R A$ as R-algebras.

Exercise 1.3.11. Prove the following, if S and T are commutative R-algebras.

(1) If S and T are both finitely generated R-algebras, then $S \otimes_R T$ is a finitely generated R-algebra.

(2) If T is a finitely generated R-algebra, then $S \otimes_R T$ is a finitely generated S-algebra.

Exercise 1.3.12. Let $\theta : R \to S$ be a homomorphism of rings. Let $M \in \mathfrak{M}_S$ and $N \in {}_S\mathfrak{M}$. Via θ, M can be viewed as a right R-module and N as a left R-module. Show that θ induces a well defined $\mathbb{Z}$-module epimorphism

$M \otimes_R N \to M \otimes_S N$. (Note: The dual result, how a Hom group behaves when the ring in the middle is changed, is studied in Exercise 1.3.35.)

Exercise 1.3.13. Let $\theta : R \to S$ be a homomorphism of rings. Let $M \in \mathfrak{M}_R$ and $N \in {}_R\mathfrak{M}$, $M' \in \mathfrak{M}_S$ and $N' \in {}_S\mathfrak{M}$. Via θ, M' and N' are viewed as R-modules. In this context, let $f : M \to M'$ be a right R-module homomorphism and $g : N \to N'$ a left R-module homomorphism. Show that there is a well defined $\mathbb{Z}$-module homomorphism $M \otimes_R N \to M' \otimes_S N'$ which satisfies $x \otimes y \mapsto f(x) \otimes g(y)$.

Exercise 1.3.14. Let R be a commutative ring and S a commutative R-algebra. Let A be an S-algebra. Using Exercise 1.3.12, show that there is a well defined epimorphism of rings $A \otimes_R A \to A \otimes_S A$.

Exercise 1.3.15. Let R be a ring and I a left ideal in R. Prove that R/I is a projective left R-module if and only if I is finitely generated and R/I is a flat left R-module. This adds one more equivalent condition to the list in Exercise 1.1.22.

Exercise 1.3.16. Let k be a field and $n > 1$ an integer. Let $T = k[x, y]$, $S = k[x^n, xy, y^n]$, and $R = k[x^n, y^n]$. For the tower of subrings $R \subseteq S \subseteq T$, prove:

(1) T is free over R of rank n^2.

(2) S is free over R of rank n.

(3) T is not free over S. (Hint: Consider the quotient rings
$$S/(x^n, xy, y^n) \text{ and } T/(x^n, xy, y^n).)$$

For more properties of the ring $k[x^n, xy, y^n]$, see Exercise 6.2.9.

3.3. Hom Groups. If R is a ring and M and N are R-modules, then $\operatorname{Hom}_R(M, N)$ is the set of R-module homomorphisms from M to N. Then $\operatorname{Hom}_R(M, N)$ is an additive group under point-wise addition:
$$(f + g)(x) = f(x) + g(x).$$
If R is commutative, then $\operatorname{Hom}_R(M, N)$ can be turned into a left R-module by defining $(rf)(x) = rf(x)$. If R is noncommutative, then $\operatorname{Hom}_R(M, N)$ cannot be turned into an R-module per se. If S is another ring and M or N is a bimodule over R and S, then we can turn $\operatorname{Hom}_R(M, N)$ into an S-module. Lemma 1.3.17 lists four such possibilities. The proof is left to the reader.

Lemma 1.3.17. *Let R and S be rings.*

(1) *If M is a left R right S bimodule and N is a left R-module, then $\operatorname{Hom}_R(M, N)$ is a left S-module, with the action of S given by $(sf)(m) = f(ms)$.*

(2) *If M is a left R-module and N is a left R right S bimodule, then* $\operatorname{Hom}_R(M,N)$ *is a right S-module, with the action of S given by* $(fs)(m) = (f(m))s$.

(3) *If M is a left R left S bimodule and N is a left R-module, then* $\operatorname{Hom}_R(M,N)$ *is a right S-module, with the action of S given by* $(fs)(m) = f(sm)$.

(4) *If M is a left R-module and N is a left R left S bimodule, then* $\operatorname{Hom}_R(M,N)$ *is a left S-module, with the action of S given by* $(sf)(m) = s(f(m))$.

Let R be a ring and M a left R-module. Then $\operatorname{Hom}_R(M,M)$ is a ring where multiplication is composition of functions:

$$(fg)(x) = f(g(x)).$$

The ring $\operatorname{Hom}_R(M,M)$ acts from the left as a ring of functions on M. This multiplication makes M into a left $\operatorname{Hom}_R(M,M)$-module. If R is commutative, then $\operatorname{Hom}_R(M,M)$ is an R-algebra. The next two results are corollaries to Nakayama's Lemma (Lemma 1.1.13).

Corollary 1.3.18. *Let R be a commutative ring and M a finitely generated R-module. Let $f : M \to M$ be an R-module homomorphism such that f is onto. Then f is one-to-one.*

Proof. Let $R[x]$ be the polynomial ring in one variable over R. We turn M into an $R[x]$-module using f. Given $m \in M$ and $p(x) \in R[x]$, define

$$p(x) \cdot m = p(f)(m).$$

Since M is finitely generated over R, M is finitely generated over $R[x]$. Let I be the ideal in $R[x]$ generated by x. Then $IM = M$ because f is onto. By Lemma 1.1.13, $I + \operatorname{annih}_{R[x]} M = R[x]$. For some $p(x)x \in I$, $1 + p(x)x \in \operatorname{annih}_{R[x]} M$. Then $(1 - p(x)x)M = 0$ which says for each $m \in M$, $m = (p(f)f)(m)$. Then $p(f)f$ is the identity function, so f is one-to-one. □

Corollary 1.3.19. *Let R be a commutative ring, M an R-module, N a finitely generated R-module, and $f \in \operatorname{Hom}_R(M,N)$. Then f is onto if and only if for each maximal ideal $\mathfrak{m}$ in R, the induced map $\bar{f} : M/\mathfrak{m}M \to N/\mathfrak{m}N$ is onto.*

Proof. Let C denote the cokernel of f and let $\mathfrak{m}$ be an arbitrary maximal ideal of R. Since N is finitely generated, so is C. Tensor the exact sequence

$$M \xrightarrow{f} N \to C \to 0$$

with $(\cdot) \otimes_R R/\mathfrak{m}$ to get

$$M/\mathfrak{m}M \xrightarrow{\bar{f}} N/\mathfrak{m}N \to C/\mathfrak{m}C \to 0$$

which is exact since tensoring is right exact. If f is onto, then $C = 0$ so $\bar{f}$ is onto. Conversely if $\mathfrak{m}C = C$ for every $\mathfrak{m}$, then Corollary 1.1.14 implies $C = 0$. □

Given a fixed left R-module M, the assignment $N \mapsto \operatorname{Hom}_R(M, N)$ defines a covariant functor from R-modules to abelian groups. Proposition 1.3.20 summarizes some of the first properties of this functor. If M is in the second factor, then the assignment $N \mapsto \operatorname{Hom}_R(N, M)$ defines a contravariant functor, and in Proposition 1.3.21 we enumerate some of its basic properties. The proofs are left to the reader.

Proposition 1.3.20. *For a ring R and a left R-module M, the following are true.*

(1) $\operatorname{Hom}_R(M, \cdot)$ *is a left exact covariant functor from ${}_R\mathfrak{M}$ to ${}_\mathbb{Z}\mathfrak{M}$ which sends a left R module N to the abelian group* $\operatorname{Hom}_R(M, N)$*. Given any R-module homomorphism $f : A \to B$, the homomorphism of groups*

$$\operatorname{Hom}_R(M, A) \xrightarrow{\mathrm{H}_f} \operatorname{Hom}_R(M, B)$$

is defined by the assignment $g \mapsto fg$.

(2) *M is projective if and only if* $\operatorname{Hom}_R(M, \cdot)$ *is an exact functor.*

Proposition 1.3.21. *Given a ring R and a left R-module M,* $\operatorname{Hom}_R(\cdot, M)$ *defines a left exact contravariant functor from ${}_R\mathfrak{M}$ to ${}_\mathbb{Z}\mathfrak{M}$ which sends a left R module N to the abelian group* $\operatorname{Hom}_R(N, M)$*. Given any R-module homomorphism $f : A \to B$, the homomorphism of groups*

$$\operatorname{Hom}_R(B, M) \xrightarrow{\mathrm{H}_f} \operatorname{Hom}_R(A, M)$$

is defined by the assignment $g \mapsto gf$.

Lemma 1.3.22 is a partial converse to the left exactness property of the functor $\operatorname{Hom}_R(\cdot, M)$.

Lemma 1.3.22. *Let R be a ring. The sequence of R-modules*

$$A \xrightarrow{\alpha} B \xrightarrow{\beta} C$$

is exact, if for all R-modules M

$$\operatorname{Hom}_R(C, M) \xrightarrow{\mathrm{H}_\beta} \operatorname{Hom}_R(B, M) \xrightarrow{\mathrm{H}_\alpha} \operatorname{Hom}_R(A, M)$$

is an exact sequence of $\mathbb{Z}$-modules.

Proof. First we show $\operatorname{im}\alpha \subseteq \ker\beta$. Suppose there exists $a \in A$ such that $\beta\alpha a \neq 0$. We take M to be the nonzero module C. By assumption,

$$\operatorname{Hom}_R(C,C) \xrightarrow{\mathrm{H}_\beta} \operatorname{Hom}_R(B,C) \xrightarrow{\mathrm{H}_\alpha} \operatorname{Hom}_R(A,C)$$

is an exact sequence of $\mathbb{Z}$-modules. Let 1 denote the identity element in the ring $\operatorname{Hom}_R(C,C)$. By evaluating at the element a, we see that $\mathrm{H}_\alpha \mathrm{H}_\beta(1) \neq 0$, a contradiction. Now we show $\operatorname{im}\alpha \supseteq \ker\beta$. Suppose there exists $b \in B$ such that $\beta b = 0$ and $b \notin \operatorname{im}\alpha$. By Proposition 1.3.21, the exact sequence

$$A \xrightarrow{\alpha} B \xrightarrow{\pi} B/\operatorname{im}\alpha \to 0$$

gives rise to the exact sequence

$$0 \to \operatorname{Hom}_R(B/\operatorname{im}\alpha, B/\operatorname{im}\alpha) \xrightarrow{\mathrm{H}_\pi} \operatorname{Hom}_R(B, B/\operatorname{im}\alpha) \xrightarrow{\mathrm{H}_\alpha} \operatorname{Hom}_R(A, B/\operatorname{im}\alpha).$$

The identity map $1 \in \operatorname{Hom}_R(B/\operatorname{im}\alpha, B/\operatorname{im}\alpha)$ maps to the nonzero map $\pi = \mathrm{H}_\pi(1)$. Since $\mathrm{H}_\alpha(\pi) = \pi\alpha = 0$, we see that $\pi \in \ker \mathrm{H}_\alpha$. If we take M to be the nonzero module $B/\operatorname{im}\alpha$, then by assumption,

$$\operatorname{Hom}_R(C, B/\operatorname{im}\alpha) \xrightarrow{\mathrm{H}_\beta} \operatorname{Hom}_R(B, B/\operatorname{im}\alpha) \xrightarrow{\mathrm{H}_\alpha} \operatorname{Hom}_R(A, B/\operatorname{im}\alpha)$$

is an exact sequence of $\mathbb{Z}$-modules. So $\pi \in \operatorname{im}\mathrm{H}_\beta$. Therefore, there exists g in $\operatorname{Hom}_R(C, B/\operatorname{im}\alpha)$ such that $g\beta = \pi$. On the one hand we have $g\beta(b) = 0$. On the other hand we have $\pi(b) \neq 0$, a contradiction. □

Part (2) of Proposition 1.3.23 describes how the Hom functor commutes with a direct product in the second variable. Note however that a direct sum in the first variable is transformed into a direct product. As an important special case, it follows that Hom commutes with a finite direct sum (or finite direct product) in either variable. The proof is left to the reader.

Proposition 1.3.23. *Let R be a ring. Let M, N, $\{M_i \mid i \in I\}$ and $\{N_j \mid j \in J\}$ be R-modules.*

(1) *The map $f \mapsto f(1)$ defines an isomorphism $\operatorname{Hom}_R(R, M) \cong M$ of R-modules.*

(2) *There are isomorphisms*
 (a) $\operatorname{Hom}_R\left(\bigoplus_{i\in I} M_i, N\right) \cong \prod_{i\in I} \operatorname{Hom}_R(M_i, N)$ *and*
 (b) $\operatorname{Hom}_R\left(M, \prod_{j\in J} N_j\right) \cong \prod_{j\in J} \operatorname{Hom}_R(M, N_j)$
 of $\mathbb{Z}$-modules.

3.4. Hom Tensor Relations. In this section we prove several identities involving Hom groups and the tensor product. Following the convention of [**DI71**], we refer to these as "Hom Tensor Relations".

Theorem 1.3.24 (Adjoint Isomorphism). *Let R and S be rings.*

(1) *If $A \in {}_R\mathfrak{M}$, $B \in {}_S\mathfrak{M}_R$ and $C \in {}_S\mathfrak{M}$, then there is an isomorphism of $\mathbb{Z}$-modules*

$$\operatorname{Hom}_S(B \otimes_R A, C) \xrightarrow{\psi} \operatorname{Hom}_R(A, \operatorname{Hom}_S(B, C))$$

defined by $\psi(f)(a) = f(\cdot \otimes a)$.

(2) *If $A \in \mathfrak{M}_R$, $B \in {}_R\mathfrak{M}_S$ and $C \in \mathfrak{M}_S$, then there is an isomorphism of $\mathbb{Z}$-modules*

$$\operatorname{Hom}_S(A \otimes_R B, C) \xrightarrow{\phi} \operatorname{Hom}_R(A, \operatorname{Hom}_S(B, C))$$

defined by $\phi(f)(a) = f(a \otimes \cdot)$.

In both cases, the isomorphism is natural in both variables A and C. The "Tensor-Hom" pair, $(B \otimes_R (\cdot), \operatorname{Hom}_S(B, \cdot))$, is an adjoint pair.

Proof. See, for example, [**Rot79**, Theorem 2.11]. □

Lemma 1.3.25. *Let R and S be rings. Let $A \in {}_R\mathfrak{M}$ be finitely generated and projective. For any $B \in {}_R\mathfrak{M}_S$ and $C \in \mathfrak{M}_S$ there is a natural isomorphism*

$$\operatorname{Hom}_S(B, C) \otimes_R A \xrightarrow{\alpha} \operatorname{Hom}_S(\operatorname{Hom}_R(A, B), C)$$

of abelian groups. On generators, the map is defined by $\alpha(f \otimes a)(g) = f(g(a))$.

Proof. Note that $\operatorname{Hom}_S(B, C)$ is a right R-module by the action $(fr)(b) = f(rb)$ and $\operatorname{Hom}_R(A, B)$ is a right S-module by the action $(gs)(a) = g(a)s$. Given any (f, a) in $\operatorname{Hom}_S(B, C) \times A$, define $\phi(f, a) \in \operatorname{Hom}_S(\operatorname{Hom}_R(A, B), C)$ by $\phi(f, a)(g) = f(g(a))$. The reader should verify that ϕ is a well defined balanced map. Therefore α is a well defined group homomorphism. Also note that if $\psi : A \to A'$ is an R-module homomorphism, then the diagram

$$\begin{array}{ccc} \operatorname{Hom}_S(B, C) \otimes_R A & \xrightarrow{\alpha} & \operatorname{Hom}_S(\operatorname{Hom}_R(A, B), C) \\ {\scriptstyle 1 \otimes \psi} \downarrow & & \downarrow {\scriptstyle \mathrm{H}(\mathrm{H}(\psi))} \\ \operatorname{Hom}_S(B, C) \otimes_R A' & \xrightarrow{\alpha} & \operatorname{Hom}_S(\operatorname{Hom}_R(A', B), C) \end{array}$$

commutes. If $A = R$, then by Proposition 1.3.23 (1) we see that α is an isomorphism. If $A = R^n$ is finitely generated and free, then using Proposition 1.3.23 we see that α is an isomorphism. If A is a direct summand of a free R-module of finite rank, then the proof follows by combining the above results. □

Theorem 1.3.26. *Let R be a commutative ring and let A and B be R-algebras. Let M be a finitely generated projective A-module and N a finitely generated projective B-module. Then for any A-module M' and any B-module N', the mapping*

$$\operatorname{Hom}_A(M, M') \otimes_R \operatorname{Hom}_B(N, N') \xrightarrow{\psi} \operatorname{Hom}_{A\otimes_R B}(M \otimes_R N, M' \otimes_R N')$$

induced by $\psi(f \otimes g)(x \otimes y) = f(x) \otimes g(y)$ is an R-module isomorphism. If $M = M'$ and $N = N'$, then ψ is also a homomorphism of rings.

Proof. Define a function

$$\operatorname{Hom}_A(M, M') \times \operatorname{Hom}_B(N, N') \xrightarrow{\rho} \operatorname{Hom}_{A\otimes_R B}(M \otimes_R N, M' \otimes_R N')$$

by $\rho(f, g)(x \otimes y) = f(x) \otimes g(y)$. The equations

$$\begin{aligned} \rho(f_1 + f_2, g)(x \otimes y) &= (f_1 + f_2)(x) \otimes g(y) \\ &= (f_1(x) + f_2(x)) \otimes g(y) \\ &= f_1(x) \otimes g(y) + f_2(x) \otimes g(y) \\ &= \rho(f_1, g)(x \otimes y) + \rho(f_2, g)(x \otimes y) \\ &= \big(\rho(f_1, g) + \rho(f_2, g)\big)(x \otimes y) \end{aligned}$$

and

$$\begin{aligned} \rho(fr, g)(x \otimes y) &= (fr)(x) \otimes g(y) \\ &= f(x)r \otimes g(y) \\ &= f(x) \otimes rg(y) \\ &= f(x) \otimes (rg)(y) \\ &= \rho(f, rg)(x \otimes y) \end{aligned}$$

show that ρ is R-balanced. Therefore ψ is well defined. Now we show that ψ is an isomorphism. The method of proof is to reduce to the case where M and N are free modules.

Case 1: Show that ψ is an isomorphism if $M = A$ and $N = B$. By Proposition 1.3.23 (1), both sides are naturally isomorphic to $M' \otimes_R N'$.

Case 2: Show that ψ is an isomorphism if M is free of finite rank m over A and N is free of finite rank n over B. By Proposition 1.3.23, Exercise 1.3.6, and Case 1, both sides are naturally isomorphic to $(M' \otimes_R N')^{(mn)}$.

Case 3: The general case. By Proposition 1.1.5 (1), we can write $M \oplus L \cong F$ where F is a free A module of finite rank and $N \oplus K \cong G$ where G is a free B module of finite. Using Proposition 1.3.23 and Lemma 1.3.1

(1.6)
$$\operatorname{Hom}_A(F, M') \otimes_R \operatorname{Hom}_B(G, N') = \big(\operatorname{Hom}_A(M, M') \otimes_R \operatorname{Hom}_B(N, N')\big) \oplus H$$

is an internal direct sum of the left hand side for some submodule H. Likewise,

(1.7) $\mathrm{Hom}_{A\otimes_R B}(F\otimes_R G, M'\otimes_R N') = \mathrm{Hom}_{A\otimes_R B}(M\otimes_R N, M'\otimes_R N')\oplus H'$

is an internal direct sum of the right hand side, for some submodule H'. By Case 2, the natural map Ψ is an isomorphism between the left hand sides of (1.6) and (1.7). The restriction of Ψ gives the desired isomorphism ψ. □

Corollary 1.3.27 allows us to change the base ring for the Hom group from R to A, if the module N is finitely generated and projective. When A is a flat commutative R-algebra, see Proposition 2.4.10 for another important theorem on changing the base ring.

Corollary 1.3.27. *Let R be a commutative ring and N a finitely generated projective R-module. Let A be an R-algebra. Then*

$$A\otimes_R \mathrm{Hom}_R(N,N') \xrightarrow{\psi} \mathrm{Hom}_A(A\otimes_R N, A\otimes_R N')$$

is an R-module isomorphism for any R-module N'.

Proof. In Theorem 1.3.26, take $B = R$, $M = M' = A$. □

Corollary 1.3.28. *If R is commutative and M and N are finitely generated projective R-modules, then*

$$\mathrm{Hom}_R(M,M)\otimes_R \mathrm{Hom}_R(N,N) \xrightarrow{\psi} \mathrm{Hom}_R(M\otimes_R N, M\otimes_R N)$$

is an R-algebra isomorphism.

Proof. In Theorem 1.3.26, take $A = B = R$, $M = M'$ and $N = N'$. □

Theorem 1.3.29. *Let A and B be rings. Let L be a finitely generated and projective left A-module. Let M be a left A right B bimodule. Let N be a left B-module. Then*

$$\mathrm{Hom}_A(L,M)\otimes_B N \xrightarrow{\psi} \mathrm{Hom}_A(L, M\otimes_B N)$$

is a $\mathbb{Z}$-module isomorphism, where $\psi(f\otimes y)(x) = f(x)\otimes y$ for all $y\in N$ and $x\in L$.

Proof. By Lemma 1.3.17, $\mathrm{Hom}_A(L,M)$ is a right B-module by the action $(fb)(x) = f(x)b$. The reader should verify that ψ is balanced, hence well defined.

Case 1: Show that ψ is an isomorphism if $L = A$. This follows because both sides are naturally isomorphic to $M\otimes_B N$, by Proposition 1.3.23 (1).

Case 2: Show that ψ is an isomorphism if L is free of rank n over A. By Proposition 1.3.23 (2), Lemma 1.3.1 and Case 1, both sides are naturally isomorphic to $(M\otimes_R N)^{(n)}$.

Case 3: The general case. By Proposition 1.1.5 (1), we can write $L \oplus K \cong F$ where F is a free A module of rank n. Using Proposition 1.3.23 (2) and Lemma 1.3.1

$$\operatorname{Hom}_A(F, M) \otimes_B N = \operatorname{Hom}_A(L, M) \otimes_R N \oplus H \tag{1.8}$$

is an internal direct sum of the left hand side for some submodule H. Likewise,

$$\operatorname{Hom}_A(F, M \otimes_B N) = \operatorname{Hom}_A(L, M \otimes_R N) \oplus H' \tag{1.9}$$

is an internal direct sum of the right hand side, for some submodule H'. By Case 2, the natural map Ψ is an isomorphism between the left hand sides of (1.8) and (1.9). The restriction of Ψ gives the desired isomorphism ψ. □

3.5. Exercises.

Exercise 1.3.30. This exercise is based on [**Bas68**, Proposition 1.1(a), p. 52]. Let R be a ring and M a left R-module. The functor $\operatorname{Hom}_R(M, \cdot)$ from the category of left R-modules to the category of $\mathbb{Z}$-modules is said to be *faithful* in case for every R-module homomorphism $\beta : A \to B$, if $\beta \neq 0$, then there exists $h \in \operatorname{Hom}_R(M, A)$ such that $\beta h \neq 0$. This exercise outlines a proof that M is an R-generator if and only if the functor $\operatorname{Hom}_R(M, \cdot)$ is faithful.

(1) For any left R-module A, set $H = \operatorname{Hom}_R(M, A)$. Let M^H denote the direct sum of copies of M over the index set H. Show that there is an R-module homomorphism
$$\alpha : M^H \to A$$
defined by $\alpha(f) = \sum_{h \in H} h(f(h))$.

(2) Show that if $\operatorname{Hom}_R(M, \cdot)$ is faithful, then for any left R-module A, the map α defined in Part (1) is surjective. Conclude that M is an R-generator. (Hint: Let $\beta : A \to B$ be the cokernel of α. Show that the composition $M \xrightarrow{h} A \xrightarrow{\beta} B$ is the zero map for all $h \in H$.)

(3) Prove that if M is an R-generator, then $\operatorname{Hom}_R(M, \cdot)$ is faithful. (Hint: Use Exercise 1.1.11.)

Exercise 1.3.31. Let R be any ring and $\phi : A \to B$ a homomorphism of left R-modules. Prove that the following are equivalent.

(1) ϕ has a left inverse. That is, there exists an R-module homomorphism $\psi : B \to A$ such that $\psi\phi = 1_A$.

(2) The sequence $\operatorname{Hom}_R(B, M) \xrightarrow{\mathrm{H}_\phi} \operatorname{Hom}_R(A, M) \to 0$ is exact, for every left R-module M.

(3) The sequence $\operatorname{Hom}_R(B, A) \xrightarrow{\mathrm{H}_\phi} \operatorname{Hom}_R(A, A) \to 0$ is exact.

Exercise 1.3.32. Let R be a ring. View R as a left R-module. Show that there is an isomorphism of rings $\operatorname{Hom}_R(R, R) \cong R^o$, where R^o denotes the opposite ring.

Exercise 1.3.33. Let R be a ring, M a left R module, and N a right R-module. Prove the following.

(1) $M^* = \operatorname{Hom}_R(M, R)$ is a right R-module by the multiplication rule given in Lemma 1.3.17 (2).

(2) $N^* = \operatorname{Hom}_R(N, R)$ is a left R-module by the rule $(rf)(x) = rf(x)$.

(3) Let $M^{**} = \operatorname{Hom}_R(M^*, R)$ be the double dual of M. For $m \in M$, let $\varphi_m : M^* \to R$ be the "evaluation at m" map. That is, if $f \in M^*$, then $\varphi_m(f) = f(m)$. Prove that $\varphi_m \in M^{**}$, and that the assignment $m \mapsto \varphi_m$ defines a homomorphism of left R-modules $M \to M^{**}$.

Exercise 1.3.34. Let R be a ring. We say a left R-module M is *reflexive* in case the homomorphism $M \to M^{**}$ of Exercise 1.3.33 is an isomorphism. Prove the following.

(1) If $M_1, \dots, M_n$ are left R-modules, then the direct sum $\bigoplus_{i=1}^n M_i$ is reflexive if and only if each M_i is reflexive.

(2) A finitely generated free R-module is reflexive.

(3) A finitely generated projective R-module is reflexive.

(4) If P is a finitely generated projective R-module and M is a reflexive R-module, then $P \otimes_R M$ is reflexive.

(5) If R is a PID and M is a nonzero finitely generated torsion R-module, then M is not a reflexive R-module.

Exercise 1.3.35. Exercise 1.3.12 shows how the tensor group behaves when the ring in the middle is changed. The dual result for Hom groups is the object of this exercise. Let $\theta : R \to S$ be a homomorphism of rings. Let M and N be S-modules. Via θ, M and N can be viewed as R-modules. Show that θ induces a well defined $\mathbb{Z}$-module monomorphism $\operatorname{Hom}_S(M, N) \to \operatorname{Hom}_R(M, N)$.

4. Direct Limit and Inverse Limit

In this section we give the definitions and state (usually without proofs) the fundamental properties of the direct limit and the inverse limit. As references for this section, we recommend [**DF04**] and [**AM69**]. Perhaps the most fundamental tool in homological algebra is the so-called Snake Lemma which is stated below. The proof is left to the reader.

Theorem 1.4.1. *(The Snake Lemma) Let R be any ring and*

$$\begin{array}{ccccccc} & & A_1 & \xrightarrow{f_1} & A_2 & \xrightarrow{f_2} & A_3 & \longrightarrow & 0 \\ & & \downarrow{\scriptstyle \alpha} & & \downarrow{\scriptstyle \beta} & & \downarrow{\scriptstyle \gamma} & & \\ 0 & \longrightarrow & B_1 & \xrightarrow{g_1} & B_2 & \xrightarrow{g_2} & B_3 & & \end{array}$$

a commutative diagram of R-modules with exact rows. Then there is an exact sequence

$$\ker \alpha \xrightarrow{f_1^*} \ker \beta \xrightarrow{f_2^*} \ker \gamma \xrightarrow{\partial} \operatorname{coker} \alpha \xrightarrow{g_1^*} \operatorname{coker} \beta \xrightarrow{g_2^*} \operatorname{coker} \gamma.$$

If f_1 is one-to-one, then f_1^* is one-to-one. If g_2 is onto, then g_2^* is onto.

4.1. The Direct Limit. An index set I is called a *directed set* in case there is a reflexive transitive binary relation on I denoted $\leq$ such that for any two elements $i, j \in I$, there is an element $k \in I$ with $i \leq k$ and $j \leq k$. Let I be a directed set and $\mathfrak{C}$ a category. Usually $\mathfrak{C}$ will be a category of R-modules for some ring R. At other times $\mathfrak{C}$ will be a category of R-algebras for some commutative ring R. Suppose that for each $i \in I$ there is an object $A_i \in \mathfrak{C}$ and for each pair $i, j \in I$ such that $i \leq j$ there is a $\mathfrak{C}$-morphism $\phi_j^i : A_i \to A_j$ such that the following are satisfied.

(1) For each $i \in I$, $\phi_i^i : A_i \to A_i$ is the identity on A_i, and

(2) for all $i, j, k \in I$ with $i \leq j \leq k$, the diagram

$$\begin{array}{ccc} A_i & \xrightarrow{\phi_k^i} & A_k \\ {\scriptstyle \phi_j^i}\searrow & & \nearrow{\scriptstyle \phi_k^j} \\ & A_j & \end{array}$$

commutes.

Then the collection of objects and morphisms $\{A_i, \phi_j^i\}$ is called a *directed system* in $\mathfrak{C}$ with index set I.

Let $\{A_i, \phi_j^i\}$ be a directed system in $\mathfrak{C}$ for a directed index set I. The *direct limit* of this system, denoted $\varinjlim A_i$, is an object in $\mathfrak{C}$ together with a set of morphisms $\alpha_i : A_i \to \varinjlim A_i$ indexed by I such that the following are satisfied.

(1) For all $i \leq j$, $\alpha_i = \alpha_j \phi_j^i$, and

(2) $\varinjlim A_i$ satisfies the universal mapping property. Namely, if X is an object in $\mathfrak{C}$ and $f_i : A_i \to X$ is a set of morphisms indexed by I such that for all $i \leq j$, $f_i = f_j \phi_j^i$, then there exists a unique morphism

$\beta : \varinjlim A_i \to X$ such that the diagram

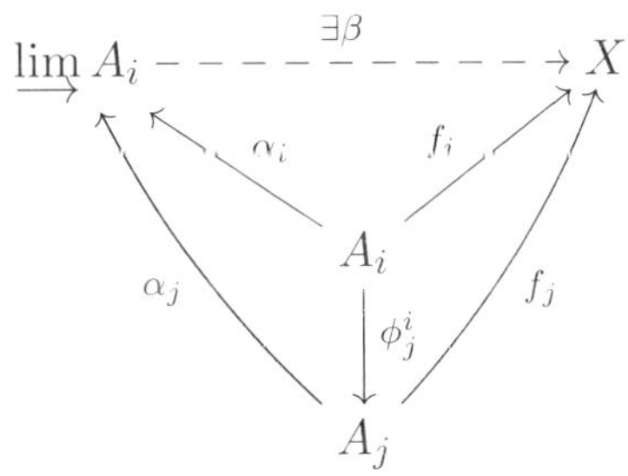

commutes for all $i \leq j$ in I.

Proposition 1.4.2 guarantees that for the categories in which we are interested, direct limits exist and are unique.

Proposition 1.4.2. *Let R be a ring.*

(1) *If $\{A_i, \phi^i_j\}$ is a directed system of R-modules for a directed index set I, then the direct limit $\varinjlim A_i$ exists. The direct limit is unique up to isomorphism.*

(2) *If R is a commutative ring and $\{A_i, \phi^i_j\}$ is a directed system of R-algebras for a directed index set I, then the direct limit $\varinjlim A_i$ exists and is unique up to isomorphism.*

Let R be a ring and I a directed index set. Suppose $\{A_i, \phi^i_j\}$ and $\{B_i, \psi^i_j\}$ are two directed systems of R-modules. A *morphism* from $\{A_i, \phi^i_j\}$ to $\{B_i, \psi^i_j\}$ is a set of R-module homomorphisms $\alpha = \{\alpha_i : A_i \to B_j\}_{i \in I}$ indexed by I such that the diagram

$$\begin{array}{ccc} A_i & \xrightarrow{\alpha_i} & B_i \\ {\scriptstyle \phi^i_j}\downarrow & & \downarrow{\scriptstyle \psi^i_j} \\ A_j & \xrightarrow{\alpha_j} & B_j \end{array}$$

commutes whenever $i \leq j$. Define $f_i : A_i \to \varinjlim B_i$ by composing α_i with the structure map $B_i \to \varinjlim B_i$. The universal mapping property guarantees a unique R-module homomorphism $\vec{\alpha} : \varinjlim A_i \to \varinjlim B_i$. According to the next theorem, the direct limit defines an exact functor on the category of directed systems of R-modules.

Theorem 1.4.3. *Let R be a ring, I a directed index set, and*

$$\{A_i, \phi^i_j\} \xrightarrow{\alpha} \{B_i, \psi^i_j\} \xrightarrow{\beta} \{C_i, \rho^i_j\}$$

a sequence of morphisms of directed systems of R-modules such that

$$0 \to A_i \xrightarrow{\alpha_i} B_i \xrightarrow{\beta_i} C_i \to 0$$

is exact for every $i \in I$. *Then*

$$0 \to \varinjlim A_i \xrightarrow{\vec{\alpha}} \varinjlim B_i \xrightarrow{\vec{\beta}} \varinjlim C_i \to 0$$

is an exact sequence of R*-modules.*

Let $\{R_i, \theta^i_j\}$ be a directed system of rings for a directed index set I. Each R_i can be viewed as a $\mathbb{Z}$-algebra, hence the direct limit $R = \varinjlim R_i$ exists, by Proposition 1.4.2. For the same index set I, let $\{M_i, \phi^i_j\}$ and $\{N_i, \psi^i_j\}$ be directed systems of $\mathbb{Z}$-modules such that each M_i is a right R_i-module and each N_i is a left R_i-module. For each $i \le j$, M_j and N_j are R_i-modules via $\theta^i_j : R_i \to R_j$. In this context, we also assume that the transition homomorphisms ϕ^i_j and ψ^i_j are R_i-linear:

$$\phi^i_j(ax) = \theta^i_j(a)\phi^i_j(x)$$
$$\psi^i_j(ax) = \theta^i_j(a)\phi^i_j(x)$$

for all $a \in R_i$, $x \in M_i$ and $y \in N_i$. By Exercise 1.3.13 there are $\mathbb{Z}$-module homomorphisms

$$\tau^i_j : M_i \otimes_{R_i} N_i \to M_j \otimes_{R_j} N_j$$

such that $\{M_i \otimes_{R_i} N_i, \tau^i_j\}$ is a directed system for I. Let $M = \varinjlim M_i$, $N = \varinjlim N_i$.

Proposition 1.4.4. *In the above context,* $\varinjlim M_i \otimes_{R_i} N_i = M \otimes_R N$.

Proof. By Exercise 1.3.13 there are $\mathbb{Z}$-module homomorphisms

$$\alpha_i : M_i \otimes_{R_i} N_i \to M \otimes_R N$$

such that $\alpha_i = \alpha_j \tau^i_j$. We show that $M \otimes_R N$ satisfies the universal mapping property. Suppose we are given $\mathbb{Z}$-module homomorphisms

$$f_i : M_i \otimes_{R_i} N_i \to X$$

such that $f_i = f_j \tau^i_j$. Suppose $(x, y) \in M \times N$. Then for some $i \in I$, (x, y) comes from $M_i \times N_i$. The reader should verify that $(x, y) \mapsto f_i(x \otimes y)$ defines an R-balanced map $M \times N \to X$. This induces $\beta : M \otimes_R N \to X$. By the universal mapping property for tensor products, β is unique and satisfies $\beta\alpha_i = f_i$. □

4.2. The Inverse Limit. Let $\mathfrak{C}$ be a category. Usually $\mathfrak{C}$ will be a category of modules or a category of algebras over a commutative ring. At other times $\mathfrak{C}$ will be a category of topological groups. Let I be an index set with a reflexive transitive binary relation denoted $\le$. (Do not assume I is a directed set.) Suppose that for each $i \in I$ there is an object $A_i \in \mathfrak{C}$ and for each pair $i, j \in I$ such that $i \le j$ there is a $\mathfrak{C}$-morphism $\phi^j_i : A_j \to A_i$ such that the following are satisfied.

(1) For each $i \in I$, $\phi_i^i : A_i \to A_i$ is the identity on A_i, and

(2) for all $i, j, k \in I$ with $i \le j \le k$, the diagram

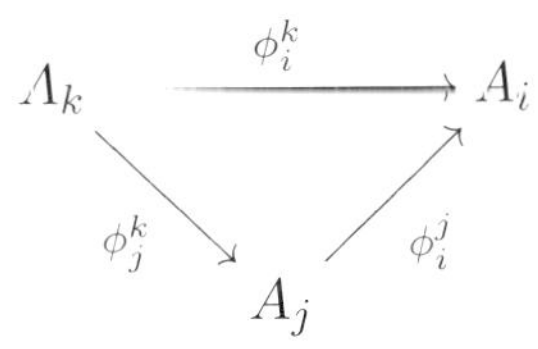

commutes.

Then the collection of objects and morphisms $\{A_i, \phi_i^j\}$ is called an *inverse system* in $\mathfrak{C}$ with index set I.

Let $\{A_i, \phi_i^j\}$ be an inverse system in $\mathfrak{C}$ for an index set I. The *inverse limit* of this system, denoted $\varprojlim A_i$, is an object in $\mathfrak{C}$ together with a set of morphisms $\alpha_i : \varprojlim A_i \to A_i$ indexed by I such that the following are satisfied.

(1) For all $i \le j$, $\alpha_i = \phi_i^j \alpha_j$, and

(2) $\varprojlim A_i$ satisfies the universal mapping property. Namely, if X is an object in $\mathfrak{C}$ and $f_i : X \to A_i$ is a set of morphisms indexed by I such that for all $i \le j$, $f_i = \phi_i^j f_j$, then there exists a unique morphism $\beta : X \to \varprojlim A_i$ such that the diagram

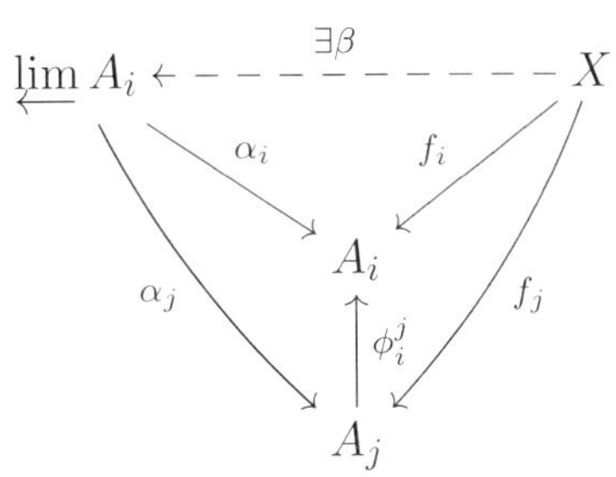

commutes for all $i \le j$ in I.

For the category of modules over a ring, or the category of algebras over a commutative ring, inverse limits exist and are unique.

Proposition 1.4.5. *Let R be a ring.*

(1) *If $\{A_i, \phi_i^j\}$ is an inverse system of R-modules for an index set I, then the inverse limit $\varprojlim A_i$ exists. The inverse limit is unique up to isomorphism.*

(2) *If R is a commutative ring and $\{A_i, \phi_i^j\}$ is an inverse system of R-algebras for an index set I, then the inverse limit $\varprojlim A_i$ exists and is unique up to isomorphism.*

Example 1.4.6. Let A be a ring. Suppose $f_1 : M_1 \to M_3$ and $f_2 : M_2 \to M_3$ are homomorphisms of left A-modules. Then the *pullback* (or *fiber product*) is defined to be $M = \{(x_1, x_2) \in M_1 \oplus M_2 \mid f_1(x_1) = f_2(x_2)\}$. Notice that M is the kernel of the A-module homomorphism $M_1 \oplus M_2 \to M_3$ defined by $(x_1, x_2) \mapsto f_1(x_1) - f_2(x_2)$, hence M is a left A-module. If h_1 and h_2 are induced by the coordinate projections, then

$$\begin{array}{ccc} M & \xrightarrow{h_2} & M_2 \\ {\scriptstyle h_1}\downarrow & & \downarrow{\scriptstyle f_2} \\ M_1 & \xrightarrow{f_1} & M_3 \end{array} \tag{1.10}$$

is a commutative diagram of A-modules. An important feature of the pullback is that it can be interpreted as an inverse limit. For the index set, take $I = \{1, 2, 3\}$ with the ordering $1 < 3$, $2 < 3$. The reader should verify that if f_1, f_2 are the transition homomorphisms, then $\{M_1, M_2, M_3\}$ is an inverse system and the inverse limit $\varprojlim M_i$ is isomorphic to the pullback M of (1.10). In particular, the pullback M satisfies the universal mapping property. That is, if N is an R-module and there exist h_1' and h_2' such that $f_1 h_1' = f_2 h_2'$, then there exists a unique morphism $\beta : N \to M$ such that the diagram

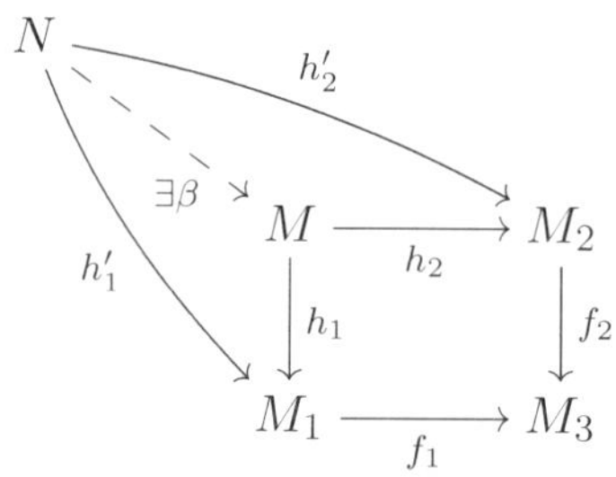

commutes. A commutative square of R-modules such as (1.10) is called a *cartesian square* (or *fiber product diagram*, or *pullback diagram*), if M is isomorphic to the pullback $\varprojlim M_i$. Let A_1, A_2, A_3 be rings. If $f_1 : A_1 \to A_3$ and $f_2 : A_2 \to A_3$ are homomorphisms, then the inverse limit $A = \varprojlim A_i$ with respect to the index set $I = \{1, 2, 3\}$ is a ring. As above, A can be identified with the pullback $A = \{(x_1, x_2) \in A_1 \oplus A_2 \mid f_1(x_1) = f_2(x_2)\}$.

4.3. Inverse Systems Indexed by Nonnegative Integers. For the index set $\mathbb{Z}_{\geq 0} = \{0, 1, 2, \dots\}$, the notation for an inverse system simplifies. Let R be any ring and $\{A_i, \phi_i^j\}$ an inverse system of R-modules for the index set $\{0, 1, 2, \dots\}$. Simply write ϕ_{i+1} for ϕ_i^{i+1}. Then for any $j > i$ we can multiply to get $\phi_i^j = \phi_{i+1}\phi_{i+2} \cdots \phi_j$. Using this notation, and Proposition 1.4.5, the inverse limit $\varprojlim A_i$ can be identified with the set of all sequences $(x_0, x_1, x_2, \dots)$ in $\prod_{n=0}^{\infty} A_n$ such that $x_n = \phi_{n+1} x_{n+1}$ for all $n \geq 0$.

Let R be a ring and suppose $\{A_i, \phi_{i+1}\}$ and $\{B_i, \psi_{i+1}\}$ are two inverse systems of R-modules indexed by $I = \{0, 1, 2, 3, \dots\}$. A *morphism* from $\{A_i, \phi_{i+1}\}$ to $\{B_i, \psi_{i+1}\}$ is a sequence of R-module homomorphisms $\alpha = \{\alpha_i : A_i \to B_j\}_{i \geq 0}$ such that the diagram

$$\begin{array}{ccc} A_{i+1} & \xrightarrow{\alpha_{i+1}} & B_{i+1} \\ {\scriptstyle \phi_{i+1}} \downarrow & & \downarrow {\scriptstyle \psi_{i+1}} \\ A_i & \xrightarrow{\alpha_i} & B_i \end{array}$$

commutes whenever $i \geq 0$. Define $f_i : \varprojlim A_i \to B_i$ by composing the structure map $\varprojlim A_i \to A_i$ with α_i. The universal mapping property guarantees a unique R-module homomorphism $\overleftarrow{\alpha} : \varprojlim A_i \to \varprojlim B_i$. For the category of inverse systems of R-modules, sufficient conditions for an inverse limit to be exact are described below in Proposition 1.4.8. To simplify its proof, we need another definition and a lemma. Define

$$d : \prod_{n=0}^{\infty} A_n \longrightarrow \prod_{n=0}^{\infty} A_n$$

by the formula $d(x_0, x_1, x_2, \dots) = (x_0 - \phi_1 x_1,\ x_1 - \phi_2 x_2,\ x_2 - \phi_3 x_3,\ \dots,\ x_n - \phi_{n+1} x_{n+1},\ \dots)$.

Lemma 1.4.7. *Let R be any ring and $\{A_i, \phi_{i+1}\}$ an inverse system of R-modules for the index set $\{0, 1, 2, \dots\}$. If $\phi_{n+1} : A_{n+1} \to A_n$ is onto for each $n \geq 0$, then there is an exact sequence*

$$0 \to \varprojlim A_n \to \prod_{n=0}^{\infty} A_n \xrightarrow{d} \prod_{n=0}^{\infty} A_n \to 0$$

where d is defined in the previous paragraph.

Proof. It follows at once that $\ker d = \varprojlim A_n$. Let $(y_0, y_1, y_2, \dots) \in \prod A_n$. To show that d is surjective, it is enough to solve the equations

$$\begin{aligned} x_0 - \phi_1 x_1 &= y_0 \\ x_1 - \phi_2 x_2 &= y_1 \\ &\vdots \\ x_n - \phi_{n+1} x_{n+1} &= y_n \end{aligned}$$

for $(x_0, x_1, x_2, \dots)$. This is possible because each ϕ_{n+1} is surjective. Simply take $x_0 = 0$, $x_1 = (\phi_1)^{-1}(-y_0)$, and recursively, $x_{n+1} = (\phi_{n+1})^{-1}(x_n - y_n)$. □

Proposition 1.4.8. *Let R be a ring, and*

$$\{A_i, \phi_{i+1}\} \xrightarrow{\alpha} \{B_i, \psi_{i+1}\} \xrightarrow{\beta} \{C_i, \rho_{i+1}\}$$

a sequence of morphisms of inverse systems of R-modules indexed by the nonnegative integers $\{0, 1, 2, 3, \dots\}$ such that

(1) $0 \to A_i \xrightarrow{\alpha_i} B_i \xrightarrow{\beta_i} C_i \to 0$ *is exact for every* $i \geq 0$, *and*

(2) $\phi_{i+1} : A_{i+1} \to A_i$ *is onto for every* $i \geq 0$.

Then

$$0 \to \varprojlim A_i \xrightarrow{\overleftarrow{\alpha}} \varprojlim B_i \xrightarrow{\overleftarrow{\beta}} \varprojlim C_i \to 0$$

is an exact sequence of R-modules.

Proof. The diagram

$$\begin{array}{ccccccccc} 0 & \longrightarrow & \prod A_n & \xrightarrow{\alpha} & \prod B_n & \xrightarrow{\beta} & \prod C_n & \longrightarrow & 0 \\ & & \downarrow d & & \downarrow d & & \downarrow d & & \\ 0 & \longrightarrow & \prod A_n & \xrightarrow{\alpha} & \prod B_n & \xrightarrow{\beta} & \prod C_n & \longrightarrow & 0 \end{array}$$

commutes and the rows are exact. By Lemma 1.4.7, the leftmost vertical map is onto. The rest of the proof follows from Theorem 1.4.1 and Lemma 1.4.7. □

Let R be a commutative ring, I an ideal in R and M an R-module. Then for all integers $n \geq 1$, I^n denotes the ideal generated by all products of the form $x_1 x_2 \cdots x_n$ where each x_i is in I. The chain of ideals $R \supseteq I^1 \supseteq I^2 \supseteq I^3 \supseteq \dots$ gives rise to the chain of submodules $M \supseteq I^1 M \supseteq I^2 M \supseteq I^3 M \supseteq \dots$. Then $I^{i+1}M \subseteq I^i M$ so there is a natural projection $\phi_{i+1} : M/I^{i+1}M \to M/I^i M$. The set of R-modules and homomorphisms $\{M/I^i M, \phi_{i+1}\}$ is an inverse system indexed by $\{1, 2, 3, 4, \dots\}$. The inverse limit of this system $\hat{M} = \varprojlim M/I^i M$, is called the *$I$-adic completion* of M. For each i, let $\eta_i : M \to M/I^i M$ be the natural projection. Clearly $\eta_i = \phi_{i+1}\eta_{i+1}$ so by the universal mapping property, there is a unique $\beta : M \to \hat{M}$ such that the diagram

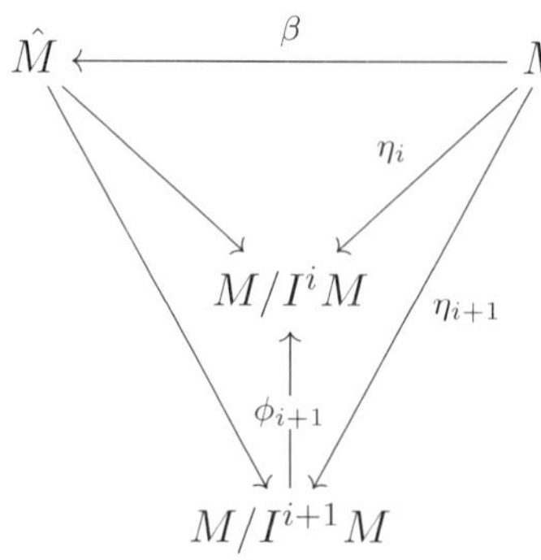

commutes. The proof of the next proposition is left to the reader.

Proposition 1.4.9. *Let I be an ideal in the commutative ring R and $\hat{R}$ the I-adic completion of R. Let M be an R-module and $\hat{M}$ the I-adic completion of M. Then the following are true.*

(1) *$\hat{M}$ is an $\hat{R}$-module.*

(2) *The natural map $\beta : M \to \hat{M}$ is one-to-one if and only if $\cap I^n M = 0$.*

4.4. Exercises.

Exercise 1.4.10. Let R be an arbitrary ring. Let I be an index set, $X = \{x_i\}_{i \in I}$ a set of indeterminates indexed by I. Let J be the set of all finite subsets of I, ordered by set inclusion. For each $\alpha \in J$, let $X_\alpha = \{x_j \mid j \in \alpha\}$. Show how to make the set of polynomial rings $\{R[X_\alpha]\}_{\alpha \in J}$ into a directed system of rings. Define $R[X] = \varinjlim R[X_\alpha]$ as the direct limit.

Exercise 1.4.11. Let A be an R-module.

(1) Suppose $A_0 \subseteq A_1 \subseteq A_2 \subseteq \dots$ is a chain of submodules of the R-module A. Show how to make $\{A_i\}$ into a directed system and prove that $\varinjlim A_i = \bigcup_i A_i$.

(2) Let $\{A_\alpha\}$ be the set of all finitely generated R-submodules of A. Show how to make $\{A_\alpha\}$ into a directed system and prove that $A = \varinjlim A_\alpha$.

Exercise 1.4.12. Let R be a commutative ring and A an R-algebra.

(1) Show that $A = \varinjlim A_\alpha$ where A_α runs over the set of all finitely generated R-subalgebras of A.

(2) Assume A is integral over R (see Section 3.3). Show that $A = \varinjlim A_\alpha$ where A_α runs over the set of all R-subalgebras of A such that A_α is finitely generated as an R-module.

(3) If $f \in A$, show that $A = \varinjlim A_\alpha$ where A_α runs over all finitely generated R-subalgebras of A such that $R[f] \subseteq A_\alpha \subseteq A$.

Exercise 1.4.13. Let A be a commutative ring and $R = A[x]$ the polynomial ring in one variable with coefficients in A. Let $I = Rx$ be the ideal in R generated by x. Show that the I-adic completion of R is isomorphic to the power series ring $A[[x]]$ in one variable over A.

Exercise 1.4.14. Let R be any ring and $\{A_i, \phi^i_j\}$ a directed system of flat R-modules for a directed index set I. Prove that the direct limit $\varinjlim A_i$ is a flat R-module.

Exercise 1.4.15. Let $\{R_i, \theta^i_j\}$ be a directed system of rings for a directed index set I. Let $R = \varinjlim R_i$ be the direct limit. As in Proposition 1.4.4, let $\{M_i, \phi^i_j\}$ a directed system of $\mathbb{Z}$-modules for the same index set I such that each M_i is a left R_i-module and the transition homomorphisms ϕ^i_j are R_i-module homomorphisms. If each M_i is a flat R_i-module, prove that $M = \varinjlim M_i$ is a flat R-module. (Hint: $\{R \otimes_{R_i} M_i, 1 \otimes \phi^i_j\}$ is a directed system of flat R-modules.)

Exercise 1.4.16. Let R be a ring and $\{A_i, \phi^i_j\}$ a directed system of R-modules for a directed index set I. Let P be a finitely generated projective R-module.

(1) Show that $\operatorname{Hom}_R(P, \varinjlim A_i) \cong \varinjlim \operatorname{Hom}_R(P, A_i)$. (Hint: Following the proof of Theorem 1.3.26, reduce to the case where P is free.)

(2) Show that $\operatorname{Hom}_R(P, \bigoplus_i A_i) \cong \bigoplus_i \operatorname{Hom}_R(P, A_i)$.

Exercise 1.4.17. Let R be a commutative ring and $\{A_i, \phi^i_j\}$ a directed system of R-algebras for a directed index set I. Show that an idempotent in $\varinjlim A_i$ comes from an idempotent in A_i, for some $i \in I$. In other words, if $e \in \varinjlim A_i$ and $e^2 = e$, then for some $i \in I$, there exists $e_i \in A_i$ such that $e_i^2 = e_i$ and if $\alpha_i : A_i \to \varinjlim A_i$ is the natural map, then $\alpha_i(e_i) = e$.

Exercise 1.4.18. In the context of the pullback diagram (1.10), prove the following:

(1) $\ker h_1 \cong \ker f_2$ and $\ker h_2 \cong \ker f_1$.

(2) If f_2 is onto, then h_1 is onto. If f_1 is onto, then h_2 is onto.

Exercise 1.4.19. Let A be a ring and let I and J be two-sided ideals in A. Show that

$$\begin{array}{ccc} \frac{A}{I\cap J} & \xrightarrow{h_2} & \frac{A}{J} \\ {\scriptstyle h_1}\downarrow & & \downarrow{\scriptstyle f_2} \\ \frac{A}{I} & \xrightarrow{f_1} & \frac{A}{I+J} \end{array}$$

is a cartesian square of rings, where all of the homomorphisms are the natural maps.

Exercise 1.4.20. Let B be a ring and I a two-sided ideal of B. Assume $A \subseteq B$ is a subring such that $I \subseteq A$. Show that

$$\begin{array}{ccc} A & \longrightarrow & B \\ {\scriptstyle h_1}\downarrow & & \downarrow{\scriptstyle f_2} \\ \frac{A}{I} & \xrightarrow{f_1} & \frac{B}{I} \end{array}$$

is a cartesian square of rings, where all of the homomorphisms are the natural maps.

5. The Morita Theorems

In the literature, the Morita Theorems are stated and proved by various authors in a number of ways and in different levels of generality. The approach we choose to follow is that used by [**DI71**]. Our functors are defined only on categories of modules. To define the functors, we will start with a ring R and a left R-module M. There are occasions when it is preferable to define the functor starting with a right R-module M. For example, when we apply the theorems in Section 7.8, it will be convenient to use a right handed Morita equivalence. Rather than state in this section more than one version of the theorems, we will leave it to the reader to make the necessary straightforward translations.

We begin by establishing some notation that will be in effect throughout this section. For any ring R and any left R-module M, set

$$M^* = \operatorname{Hom}_R(M, R)$$

and

$$S = \operatorname{Hom}_R(M, M).$$

Since R is a left R right R bimodule, by Lemma 1.3.17 (2), M^* is a right R-module under the operation $(fr)(m) = f(m)r$. Since S is a ring of R-module endomorphisms of M, M is a left S-module by $sm = s(m)$ and under this operation M is a left R left S bimodule. By Lemma 1.3.17 (3), we make M^* a right S-module by $(fs)(m) = f\big(s(m)\big)$, which is just composition of functions. The reader should verify that M^* is in fact a right R right S bimodule. It follows that we can form $M^* \otimes_R M$ and $M^* \otimes_S M$. Moreover $M^* \otimes_R M$ is a left S right S bimodule by virtue of M being a left R left S bimodule and M^* being a right R right S bimodule. Similarly $M^* \otimes_S M$ is a left R right R bimodule. Define

$$\theta_R : M^* \otimes_R M \to S = \operatorname{Hom}_R(M, M)$$

by the rule: $\theta_R(f \otimes m)(x) = f(x)m$. The reader should check that θ_R is both a left and a right S-module homomorphism. Define

$$\theta_S : M^* \otimes_S M \to R$$

by the rule: $\theta_S(f \otimes m) = f(m)$. The reader should verify that θ_S is a right and left R-module homomorphism whose image is the trace ideal $\mathfrak{T}_R(M)$ (see Eq.(1.2)).

Lemma 1.5.1. *In the above context,*

(1) *θ_R is onto if and only if M is finitely generated and projective. If θ_R is onto, it is one-to-one.*

(2) *θ_S is onto if and only if M is a generator. If θ_S is onto, it is one-to-one.*

Proof. (1): If θ_R is onto, then there exist $f_i \in M^*$ and $m_i \in M$ such that the identity map $1 : M \to M$ is equal to $\theta_R(\sum_{i=1}^n f_i \otimes m_i)$. That is, for every $x \in M$, $x = \sum_{i=1}^n f_i(x) m_i$. Then $\{(f_i, m_i)\}$ is a finite dual basis for M. By Proposition 1.1.5, M is finitely generated and projective. Conversely, if a finite dual basis exists, then the identity map $1 : M \to M$ is in the image of θ_R. Since θ_R is an S-module homomorphism, θ_R is onto.

To prove the second statement in (1), assume θ_R is onto. Let $f_1, \dots, f_n \in M^*$, $m_1, \dots, m_n \in M$ be a finite dual basis for M. Let $\alpha = \sum_j h_j \otimes n_j \in M^* \otimes_R M$ and $\theta_R(\alpha) = 0$. That is, $\sum_j h_j(x) n_j = 0$ for every x in M. In particular, for each f_i, we have

$$\Big[\sum_j h_j \cdot f_i(n_j)\Big](x) = \sum_j h_j(x) f_i(n_j) = \sum_j f_i\big(h_j(x) n_j\big) = f_i\big(\sum_j h_j(x) n_j\big) = f_i(0) = 0.$$

From this we get

$$\begin{aligned}\alpha = \sum_j h_j \otimes n_j &= \sum_j \big[h_j \otimes \big(\sum_i f_i(n_j) m_i\big)\big] = \sum_{i,j} h_j \otimes f_i(n_j) m_i \\ &= \sum_{i,j} \big(h_j \cdot f_i(n_j)\big) \otimes m_i = \sum_i \big[\big(\sum_j h_j \cdot f_i(n_j)\big) \otimes m_i\big] = \sum_i 0 \otimes m_i = 0.\end{aligned}$$

Hence, θ_R is one-to-one.

(2): Because the image of θ_S equals $\mathfrak{T}_R(M)$, the trace ideal of M, it is clear that θ_S is onto if and only if M is an R-generator.

To prove the second statement in (2), assume θ_S is onto. Let $\sum_j h_j \otimes n_j \in \ker \theta_S$. That is, $\sum_j h_j(n_j) = 0$. Since θ_S is onto, there exist $f_1, \dots, f_n$ in M^*, $m_1, \dots, m_n$ in M with $\sum_i f_i(m_i) = 1 \in R$. Notice that for each f_i and every $x \in M$,

$$\sum_j h_j \cdot \theta_R(f_i \otimes n_j)(x) = \sum_j h_j\big(f_i(x) n_j\big) = f_i(x) \sum_j h_j(n_j) = 0.$$

From this we get

$$\begin{aligned}\sum_j h_j \otimes n_j &= \sum_j h_j \otimes \big(\sum_i f_i(m_i)\big) n_j = \sum_j h_j \otimes \big(\sum_i f_i(m_i) n_j\big) \\ &= \sum_j h_j \otimes \big(\sum_i \theta_R(f_i \otimes n_j)(m_i)\big) = \sum_{i,j} h_j \otimes \theta_R(f_i \otimes n_j)(m_i) \\ &= \sum_i \big(\sum_j h_j \cdot \theta_R(f_i \otimes n_j)\big) \otimes (m_i) = \sum_i 0 \otimes m_i = 0.\end{aligned}$$

Therefore, θ_S is one-to-one. □

The main theorem of this section, Theorem 1.5.2, shows that if M is an R-progenerator, then the functors $(\,) \otimes_R M : \mathfrak{M}_R \to {}_S\mathfrak{M}$ and $M^* \otimes_S (\,) : {}_S\mathfrak{M} \to \mathfrak{M}_R$ are inverse equivalences. In this case, we say that the categories $\mathfrak{M}_R$ and ${}_S\mathfrak{M}$ are *Morita equivalent.*

Theorem 1.5.2. *Let R be any ring and let M be a left R-progenerator. Set $S = \operatorname{Hom}_R(M, M)$ and $M^* = \operatorname{Hom}_R(M, R)$. Then*

$$(\,) \otimes_R M : \mathfrak{M}_R \to {}_S\mathfrak{M}$$

and

$$M^* \otimes_S (\,) : {}_S\mathfrak{M} \to \mathfrak{M}_R$$

are inverse equivalences, establishing $\mathfrak{M}_R \sim {}_S\mathfrak{M}$.

Proof. Let L be any right R-module. Then, by the basic properties of the tensor product and Lemma 1.5.1 (2), we have

$$\begin{aligned} M^* \otimes_S (L \otimes_R M) &\cong M^* \otimes_S (M \otimes_{R^o} L) \\ &\cong (M^* \otimes_S M) \otimes_{R^o} L \cong R \otimes_{R^o} L \cong L \otimes_R R \cong L \end{aligned}$$

where the composite isomorphism is given by $f \otimes (l \otimes m) \mapsto l \cdot \theta_S(f \otimes m) = l \cdot f(m)$. This isomorphism allows one to verify that $(\,) \otimes_R M$ followed by $M^* \otimes_S (\,)$ is naturally equivalent to the identity functor on $\mathfrak{M}_R$. Likewise, given any left S-module N, the isomorphism of Lemma 1.5.1 (1) shows that

$$\begin{aligned} (M^* \otimes_S N) \otimes_R M &\cong (N \otimes_{S^o} M^*) \otimes_R M \\ &\cong N \otimes_{S^o} (M^* \otimes_R M) \cong N \otimes_{S^o} S \cong S \otimes_S N \cong N \end{aligned}$$

under the map $(f \otimes n) \otimes m \mapsto \theta_R(f \otimes m) \cdot n$. Again this gives us that $M^* \otimes_S (\,)$ followed by $(\,) \otimes_R M$ is naturally equivalent to the identity on ${}_S\mathfrak{M}$. □

Corollary 1.5.3. *In the context of Theorem 1.5.2, the following are true.*

(1) *$R \cong \operatorname{Hom}_S(M, M)$ (as rings) where r in R maps to "left multiplication by r".*

(2) *$M^* \cong \operatorname{Hom}_S(M, S)$ (as right S-modules) where f in M^* maps to the homomorphism $\theta_R\big(f \otimes (\,)\big)$.*

(3) *$M \cong \operatorname{Hom}_R(M^*, R) = M^{**}$ (as left R-modules) where m in M maps to the element in M^{**} which is "evaluation at m".*

(4) *$S^o \cong \operatorname{Hom}_R(M^*, M^*)$ (as rings) where s in S^o maps to "right multiplication by s".*

(5) *M is an S-progenerator.*

(6) *M^* is an R-progenerator.*

(7) *M^* is an S-progenerator.*

Proof. The fully faithful part of Proposition 1.1.2 applied to the functor $() \otimes_R M$ says that for any two right R-modules A and B, the assignment

$$\operatorname{Hom}_R(A, B) \to \operatorname{Hom}_S(A \otimes_R M, B \otimes_R M) \tag{1.11}$$

is a one-to-one correspondence. Under this equivalence, the right R-module R corresponds to the left S-module $R \otimes_R M \cong M$ and the right R-module M^* corresponds to the left S-module $M^* \otimes_R M \cong S$. For (1), use (1.11) with $A = B = R$. For (2), use (1.11) with $A = R$ and $B = M^*$. In each case, the reader should verify that the composite isomorphisms are the correct maps.

The fully faithful part of Proposition 1.1.2 applied to the functor $M^* \otimes_S () : {}_S\mathfrak{M} \to \mathfrak{M}_R$ says that for any two left S-modules C and D, the assignment

$$\operatorname{Hom}_S(C, D) \to \operatorname{Hom}_R(M^* \otimes_S C, M^* \otimes_S D) \tag{1.12}$$

is a one-to-one correspondence. By Proposition 1.3.23 (1), M is isomorphic to $\operatorname{Hom}_S(S, M)$. By (1.12) with $C = S$ and $D = M$, we get $\operatorname{Hom}_S(S, M) \cong \operatorname{Hom}_R(M^*, R) = M^{**}$, which is (3). For (4), use (1.12) with $C = D = S$. Since $M^* \otimes_S S \cong M^*$, we get the isomorphism of rings $\operatorname{Hom}_S(S, S) \cong \operatorname{Hom}_R(M^*, M^*)$. By Exercise 1.3.32, $S^o \cong \operatorname{Hom}_S(S, S)$ as rings. In each case, the reader should verify that the composite isomorphisms are the correct maps.

(5): Because M is an R-progenerator, we have $\theta_S : M^* \otimes_S M \cong R$ and $\theta_R : M^* \otimes_R M \cong S$. By (1) and (2) above, this gives rise to isomorphisms

$$\theta_S : \operatorname{Hom}_S(M, S) \otimes_S M \cong \operatorname{Hom}_S(M, M)$$

and

$$\theta_R : \operatorname{Hom}_S(M, S) \otimes_{\operatorname{Hom}_S(M,M)} M \cong S.$$

By Lemma 1.5.1 with R and S interchanged, it follows that M is an S-progenerator.

(6): Using $M^* \otimes_S M \cong R$ and substituting (3) and (4), we obtain

$$\begin{aligned} R &\cong M^* \otimes_S M \\ &\cong M^* \otimes_S \operatorname{Hom}_R(M^*, R) \\ &\cong \operatorname{Hom}_R(M^*, R) \otimes_{S^o} M^* \\ &\cong \operatorname{Hom}_R(M^*, R) \otimes_{\operatorname{Hom}_R(M^*,M^*)} M^*. \end{aligned} \tag{1.13}$$

We also have

$$\begin{aligned} \operatorname{Hom}_{R^o}(M^*, R^o) \otimes_{R^o} M^* &\cong M^* \otimes_R \operatorname{Hom}_R(M^*, R) \\ &\cong M^* \otimes_R M \\ &\cong S \\ &\cong \operatorname{Hom}_R(M^*, M^*) \\ &\cong \operatorname{Hom}_{R^o}(M^*, M^*) \end{aligned} \tag{1.14}$$

where the last isomorphism in (1.14) is set identity and M^* is considered as a left R^o-module since it is a right R-module. By Lemma 1.5.1 with M^* in place of M, we see that M^* is an R-generator by (1.13) and a finitely generated and projective left R^o-module by (1.14). This implies that M^* is a right R-progenerator.

(7): By (5), M is an S-progenerator. Applying (6) to the S-module M proves that $\operatorname{Hom}_S(M, S)$ is an S-progenerator. By Part (2), $\operatorname{Hom}_S(M, S) \cong M^*$. □

Corollary 1.5.4. *Let R, M and S be as in Theorem* 1.5.2.

(1) *For any two-sided ideal $\mathfrak{a}$ of R, $M^* \otimes_R (\mathfrak{a} \otimes_R M)$ is naturally isomorphic to the two-sided ideal of S consisting of all elements of the form $\sum_i \theta_R(f_i \otimes \alpha_i m_i)$, where $f_i \in M^*$, $\alpha_i \in \mathfrak{a}$, and $m_i \in M$.*

(2) *For any two-sided ideal $\mathfrak{b}$ of S, $M^* \otimes_S (\mathfrak{b} \otimes_S M)$ is naturally isomorphic to the two-sided ideal of R consisting of all elements of the form $\sum_i \theta_S(f_i \otimes \beta_i(n_i)) = \sum_i f_i(\beta_i(n_i))$, where $f_i \in M^*$, $\beta_i \in \mathfrak{b}$, and $n_i \in M$.*

The correspondences (1) and (2) are inverses of each other and establish a one-to-one, order preserving correspondence between the two-sided ideals of R and the two-sided ideals of S.

Proof. Since M and M^* are both R-projective, they are flat. The exact sequence $0 \to \mathfrak{a} \to R$ yields the exact sequence

$$0 \to M^* \otimes_R (\mathfrak{a} \otimes_R M) \to M^* \otimes_R (R \otimes_R M) \cong M^* \otimes_R M \cong S.$$

We consider $M^* \otimes_R (\mathfrak{a} \otimes_R M)$ as a subset of $M^* \otimes_R (R \otimes_R M)$. By θ_R, $M^* \otimes_R (R \otimes_R M)$ is isomorphic to S. This maps this submodule $M^* \otimes_R (\mathfrak{a} \otimes_R M)$ onto the ideal of S made up of elements of the form $\sum_i \theta_R(f_i \otimes \alpha_i m_i)$.

Likewise, M and M^* are S-projective. The exact sequence $0 \to \mathfrak{b} \to S$ yields the exact sequence

$$0 \to M^* \otimes_S (\mathfrak{b} \otimes_S M) \to M^* \otimes_S M \cong R.$$

We view $M^* \otimes_S (\mathfrak{b} \otimes_S M)$ as the ideal of R made up of elements looking like $\sum_i f_i(\beta_i(n_i))$. The reader should verify that the correspondences are inverses of each other. □

Corollary 1.5.5. *In the context of Theorem* 1.5.2, *let L be a right R-module and $L \otimes_R M$ its corresponding left S-module.*

(1) *L is finitely generated over R if and only if $L \otimes_R M$ is finitely generated over S.*

(2) *L is R-projective if and only if $L \otimes_R M$ is S-projective.*

(3) *L is an R-generator if and only if $L \otimes_R M$ is an S-generator.*

Proof. We apply Lemma 1.1.4 to write L as the homomorphic image of a free R-module. Suppose $R^I \to L \to 0$ is exact, for some index set, I. After tensoring with $(\cdot) \otimes_R M$, we get an exact sequence

$$M^I \to L \otimes_R M \to 0 \tag{1.15}$$

of S-modules. By Corollary 1.5.3 (5), M is finitely generated and projective as an S-module. For each biconditional, we prove only one direction. Each converse follows by categorical equivalence.

(1): If L is finitely generated over R, we may assume I is a finite set. In (1.15), $M^I = \bigoplus_{i \in I} M$ is a finite sum of finitely generated modules and is finitely generated. So $L \otimes_R M$ is finitely generated.

(2): If L is projective, then by Proposition 1.1.5 the exact sequence $R^I \to L \to 0$ splits. It follows that (1.15) also splits. By Exercise 1.1.10, the S-modules M^I and $L \otimes_R M$ are projective.

(3): Let L be an R-generator. Let $\delta : C \to D$ be a nonzero homomorphism of left S-modules. By Exercise 1.3.30 (3), to show that $L \otimes_R M$ is an S-generator it suffices to show that there exists an S-module homomorphism $f : L \otimes_R M \to C$ such that $\delta \circ f$ is nonzero. By the fully faithful part of Proposition 1.1.2, $1 \otimes \delta : M^* \otimes_S C \to M^* \otimes_S D$ is a nonzero homomorphism of right R-modules. Since L is an R-generator, by Exercise 1.3.30 (4), there exists an R-module homomorphism $\alpha : L \to M^* \otimes_S C$ such that $(1 \otimes \delta) \circ \alpha$ is nonzero. Again by Proposition 1.1.2, $\delta \circ (\alpha \otimes 1)$ is nonzero. □

5.1. Exercises.

Exercise 1.5.6. Let R be any ring and let M be a left R-progenerator. Set $S = \operatorname{Hom}_R(M, M)$. Prove that $(\ ) \otimes_R M : \mathfrak{M}_R \to {}_S\mathfrak{M}$ and $\operatorname{Hom}_S(M, \ ) : {}_S\mathfrak{M} \to \mathfrak{M}_R$ are inverse equivalences, establishing $\mathfrak{M}_R \sim {}_S\mathfrak{M}$. (Hint: Corollary 1.5.3 (2) and Theorem 1.3.29.)

Exercise 1.5.7. Let R be any ring. A left R-module M is said to be *faithfully flat* if M is flat and M has the property that $N \otimes_R M = 0$ implies $N = 0$. Show that a left R-progenerator is faithfully flat.

Chapter 2

Modules over Commutative Rings

This chapter, like Chapter 1, contains material of a preliminary nature that will be applied in subsequent chapters. Proofs that are readily available in the standard references are usually omitted. To avoid listing many special cases, a theorem will frequently be stated in a form that is more general than any given application may require. Some readers will probably prefer to make a quick scan through this chapter on the first pass, and return later as needed.

1. Localization of Modules and Rings

As a basic reference for this section, we recommend [**DF04**]. Throughout this section, R denotes a commutative ring. First we make a quick review of the localization construction and state some of the fundamental properties. Then in Section 2.1.1 we prove some local to global lemmas.

A *multiplicative subset* of R is a subset W that is closed under multiplication and such that $1 \in W$. Let M be an R-module and W a multiplicative subset of R. Define a relation on $M \times W$ by $(m_1, w_1) \sim (m_2, w_2)$ if and only if there exists $w \in W$ such that $w(w_2 m_1 - w_1 m_2) = 0$. The reader should verify that $\sim$ is an equivalence relation on $R \times W$. The set of equivalence classes is denoted $W^{-1}M$ and the equivalence class containing (m, w) is denoted by the fraction m/w. We call $W^{-1}M$ the *localization of* M *at* W. The assignment $m \mapsto m/1$ defines an R-module homomorphism $\sigma : M \to W^{-1}M$. The kernel of σ is equal to the the set of all $m \in M$ such that $wm = 0$ for some w in W. The multiplication rule $(r/w_1)(m/w_2) = (rm)/(w_1 w_2)$ makes

$W^{-1}M$ into a $W^{-1}R$-module. If M is an R-algebra, the multiplication rule $(m_1/w_1)(m_2/w_2) = (m_1m_2)/(w_1w_2)$ makes $W^{-1}M$ into an R-algebra. The assignment $\phi(m/w) = 1/w \otimes m$ defines a $W^{-1}R$-module isomorphism

$$W^{-1}M \xrightarrow{\phi} W^{-1}R \otimes_R M.$$

The inverse of ϕ is given by $a \otimes b \mapsto ab$. If $0 \in W$, then $W^{-1}M = (0)$.

Given a prime ideal P in R, $W = R - P = \{x \in R \mid x \notin P\}$ is a multiplicative set. The R-algebra $W^{-1}R$ is usually written R_P. The ideal generated by P in R_P is $PR_P = \{x/y \in R_P \mid x \in P, y \notin P\}$. If $x/y \notin PR_P$, then $x \notin P$ so $y/x \in R_P$ is the multiplicative inverse of x/y. Since the complement of PR_P consists of units, the ideal PR_P contains every noninvertible element. So PR_P is the unique maximal ideal of R_P. A local ring is a commutative ring that has a unique maximal ideal (Exercise 1.1.18). Hence R_P is a local ring with maximal ideal PR_P, which is called the *local ring of R at P*. The factor ring R_P/PR_P is a field and is also called the residue field of R_P. The factor ring R/P is an integral domain and by Exercise 2.1.9, R_P/PR_P is isomorphic to the quotient field of R/P.

Lemma 2.1.1 shows that a localization of a commutative ring R is a flat R-module. In general, a localization $W^{-1}R$ is not projective. For an example, see Exercise 2.1.14.

Lemma 2.1.1. *If R is a commutative ring and W a multiplicative set in R, then $W^{-1}R$ is a flat R-module.*

1.1. Local to Global Lemmas. Our first local to global result is Proposition 2.1.2. It simply says that if a module is locally trivial (for every maximal ideal), then it is trivial (globally). The proof can be found in [**DF04**].

Proposition 2.1.2. *Let R be a commutative ring and M an R-module. If $M_{\mathfrak{m}} = 0$ for every maximal ideal $\mathfrak{m}$ of R, then $M = 0$.*

Lemma 2.1.3 is a technical lemma whose proof is left to the reader.

Lemma 2.1.3. *Let R be a commutative ring, M a finitely generated R-module, and $W \subseteq R$ a multiplicative subset. Then $W^{-1}M = 0$ if and only if there exists $w \in W$ such that $wM = 0$.*

In the following, we write M_α instead of $M[\alpha^{-1}]$ for the localization of an R-module M with respect to the multiplicative set $\{1, \alpha, \alpha^2, \dots\}$. Lemmas 2.1.4 and 2.1.5 are called local to global lemmas because if $W \subseteq R$ is a multiplicative set and $\alpha \in W$, then $W^{-1}M$ is a localization of M_α. Lemma 2.1.4 provides sufficient conditions for which a local isomorphism of R-modules can be globalized.

Lemma 2.1.4. *Let R be a commutative ring and $\varphi : M \to N$ a homomorphism of R-modules. Let $W \subseteq R$ be a multiplicative subset and $\varphi_W : M \otimes_R W^{-1}R \to N \otimes_R W^{-1}R$.*

(1) *If φ_W is one-to-one and $\ker \varphi$ is a finitely generated R-module, then there exists $\alpha \in W$ such that $\varphi_\alpha : M_\alpha \to N_\alpha$ is one-to-one.*

(2) *If φ_W is onto and $\operatorname{coker} \varphi$ is a finitely generated R-module, then there exists $\beta \in W$ such that $\varphi_\beta : M_\beta \to N_\beta$ is onto.*

(3) *If φ_W is an isomorphism and both $\ker \varphi$ and $\operatorname{coker} \varphi$ are finitely generated R-modules, then there exists $w \in W$ such that $\varphi_w : M_w \to N_w$ is an isomorphism.*

Proof. Start with the exact sequence of R-modules

$$0 \to \ker(\varphi) \to M \xrightarrow{\varphi} N \to \operatorname{coker}(\varphi) \to 0. \tag{2.1}$$

Tensoring (2.1) with $(\cdot) \otimes_R R[W^{-1}]$ we get

$$0 \to W^{-1}\ker(\varphi) \to W^{-1}M \xrightarrow{\varphi_W} W^{-1}N \to W^{-1}\operatorname{coker}(\varphi) \to 0 \tag{2.2}$$

which is exact, by Lemma 2.1.1.

(1): If φ_W is one-to-one, then by Lemma 2.1.3 there is $\alpha \in W$ such that $\alpha(\ker(\varphi)) = 0$. Therefore, $\ker(\varphi) \otimes_R R[\alpha^{-1}] = 0$, and φ_α is one-to-one.

(2): If φ_W is onto, then by Lemma 2.1.3 there is an element β in W such that $\beta(\operatorname{coker}(\varphi)) = 0$. Therefore, $\operatorname{coker}(\varphi) \otimes_R R[\beta^{-1}] = 0$, and φ_β is onto.

(3): Let α be as in (1) and β as in (2). If we set $w = \alpha\beta$, then φ_w is an isomorphism of R_w-modules. □

Lemma 2.1.5 provides sufficient conditions for which a local isomorphism of R-algebras can be globalized.

Lemma 2.1.5. *Let R be a commutative ring. Let A and B be commutative R-algebras and $\varphi : A \to B$ an R-algebra homomorphism. Assume $\ker \varphi$ is a finitely generated ideal of A, and B is a finitely generated A-algebra. If $W \subseteq R$ is a multiplicative subset and $\varphi \otimes 1 : A \otimes_R W^{-1}R \to B \otimes_R W^{-1}R$ is an isomorphism of $W^{-1}R$-algebras, then there exists $w \in W$ such that $\varphi_w : A_w \to B_w$ is an isomorphism of R_w-algebras.*

Proof. Suppose $\ker \varphi = Ax_1 + \cdots + Ax_n$. By Lemma 2.1.3 there is $\alpha \in W$ such that $\alpha(Rx_1 + \cdots + Rx_n) = 0$. Therefore, $\alpha \ker \varphi = 0$. Suppose the A-algebra B is generated by $y_1, \dots, y_m$. By Lemma 2.1.3 there is $\beta \in W$ such that $\beta(Ry_1 + \cdots + Ry_m) \subseteq \varphi(A)$. If we set $w = \alpha\beta$, then $\varphi_w : A_w \to B_w$ is an isomorphism of R_w-algebras. □

Lemma 2.1.6 provides sufficient conditions for which an R-module that is free over the local ring at a prime ideal can be globalized to a free module.

Lemma 2.1.6. *Let R be a commutative ring and M an R-module of finite presentation. Let $\mathfrak{p} \in \operatorname{Spec} R$ and assume $M_{\mathfrak{p}} = M \otimes_R R_{\mathfrak{p}}$ is a free $R_{\mathfrak{p}}$-module. Then there exists $\alpha \in R - \mathfrak{p}$ such that M_α is a free R_α-module.*

Proof. Since M is finitely generated, we know that $M_{\mathfrak{p}}$ is free of finite rank. Pick a basis $\{m_1/\alpha_1, \ldots, m_n/\alpha_n\}$ for $M_{\mathfrak{p}}$ over $R_{\mathfrak{p}}$. Since $1/\alpha_1, \ldots, 1/\alpha_n$ are units in $R_{\mathfrak{p}}$, it follows that $\{m_1/1, \ldots, m_n/1\}$ is a basis for $M_{\mathfrak{p}}$ over $R_{\mathfrak{p}}$. Define $\varphi : R^n \to M$ by $(x_1, \ldots, x_n) \mapsto \sum_{i=1}^n x_i m_i$, and consider the exact sequence of R-modules

$$0 \to \ker \varphi \to R^n \xrightarrow{\varphi} M \to \operatorname{coker} \varphi \to 0. \tag{2.3}$$

Tensoring (2.3) with $(\cdot) \otimes_R R_{\mathfrak{p}}$, we get

$$0 \to (\ker \varphi)_{\mathfrak{p}} \to R^n_{\mathfrak{p}} \xrightarrow{\varphi_{\mathfrak{p}}} M_{\mathfrak{p}} \to (\operatorname{coker} \varphi)_{\mathfrak{p}} \to 0 \tag{2.4}$$

which is exact, by Lemma 2.1.1. As mentioned above, $M_{\mathfrak{p}}$ is free over $R_{\mathfrak{p}}$ with basis $\{m_1/1, \ldots, m_n/1\}$ and $\varphi_{\mathfrak{p}}$ maps the standard basis to this basis. That is, $\varphi_{\mathfrak{p}}$ is an isomorphism. So $0 = (\ker \varphi)_{\mathfrak{p}} = (\operatorname{coker} \varphi)_{\mathfrak{p}}$. Since M is finitely generated over R so is $\operatorname{coker} \varphi$. By Lemma 2.1.3 there exists $\beta \in R - \mathfrak{p}$ such that $\beta \cdot \operatorname{coker} \varphi = 0$. Then $(\operatorname{coker} \varphi)_\beta = 0$. Tensoring (2.3) with $(\ ) \otimes_R R_\beta$ we get the exact sequence

$$0 \to (\ker \varphi)_\beta \to R^n_\beta \xrightarrow{\varphi_\beta} M_\beta \to 0. \tag{2.5}$$

Since M is a finitely presented R-module, M_β is a finitely presented R_β-module. By Lemma 2.1.7 which is proved below, $(\ker \varphi)_\beta$ is a finitely generated R_β-module. Since $\beta \in R - \mathfrak{p}$, there exists a homomorphism of rings $R_\beta \to R_{\mathfrak{p}}$ so we can tensor (2.5) with $(\cdot) \otimes_{R_\beta} R_{\mathfrak{p}}$ to get (2.4) again. That is, $(\ker \varphi)_\beta \otimes_{R_\beta} R_{\mathfrak{p}} \cong (\ker \phi)_{\mathfrak{p}} = 0$. Lemma 2.1.3 says there exists μ/β^k in $R_\beta - \mathfrak{p} R_\beta$ such that $(\mu/\beta^k)(\ker \phi)_\beta = 0$. But β is a unit in R_β so this is equivalent to $\mu (\ker \phi)_\beta = 0$. It is easy to check that $R_{\mu\beta} = R[(\mu\beta)^{-1}] = (R_\beta)_\mu$. This means $0 = \left((\ker \phi)_\beta\right)_\mu = (\ker \phi)_{\beta\mu}$. We also have $(\operatorname{coker} \phi)_{\beta\mu} = 0$. Tensoring (2.3) with $R_{\mu\beta}$ results in $R^{(n)}_{\mu\beta} \cong M_{\mu\beta}$. □

Lemma 2.1.7, which was used in the proof of Lemma 2.1.6, provides useful criteria for proving that a submodule of a finitely generated module is finitely generated. Notice that the lemma is valid when the ring R is noncommutative.

Lemma 2.1.7. *Let R be any ring and*

$$0 \to A \xrightarrow{\alpha} B \xrightarrow{\beta} C \to 0$$

an exact sequence of R-modules.

(1) *If B is finitely generated, then C is finitely generated.*

(2) *If A and C are finitely generated, then B is finitely generated.*

(3) *If B is finitely generated and C is of finite presentation, then A is finitely generated.*

Proof. Parts (1) and (2) are left to the reader.

(3): Consider the commutative diagram

$$\begin{array}{ccccccccc} & & R^{(n)} & \xrightarrow{\phi} & R^{(n)} & \xrightarrow{\psi} & C & \longrightarrow & 0 \\ & & \downarrow \exists\rho & & \downarrow \exists\eta & & \downarrow = & & \\ 0 & \longrightarrow & A & \xrightarrow{\alpha} & B & \xrightarrow{\beta} & C & \longrightarrow & 0 \end{array} \tag{2.6}$$

where the top row exists because C is of finite presentation. The homomorphism η exists by Proposition 1.1.5 (3) because $R^{(n)}$ is projective. Now $\beta\eta\phi = \psi\phi = 0$ so $\operatorname{im}\eta\phi \subseteq \ker\beta = \operatorname{im}\alpha$. Again, since $R^{(n)}$ is projective there exists ρ making the diagram commute. Since B is finitely generated, so is $\operatorname{coker}\eta$ by Part (1). The Snake Lemma 1.4.1 applied to (2.6) says that $\operatorname{coker}\rho \cong \operatorname{coker}\eta$ so $\operatorname{coker}\rho$ is finitely generated. Because $\operatorname{im}\rho$ is finitely generated, the exact sequence

$$0 \to \operatorname{im}\rho \to A \to \operatorname{coker}\rho \to 0$$

and Part (2) show that A is finitely generated. □

1.2. Exercises.

Exercise 2.1.8. Suppose R is a commutative ring, $R = R_1 \oplus \cdots \oplus R_n$ is a direct sum, and $\pi_i : R \to R_i$ is the projection. Assume each R_i is a local ring with maximal ideal $\mathfrak{n}_i$. Let $\mathfrak{m}_i = \pi_i^{-1}(\mathfrak{n}_i)$. Prove the following.

(1) $\mathfrak{m}_1, \dots, \mathfrak{m}_n$ is the complete list of maximal ideals of R.

(2) π_i induces an isomorphism on local rings $R_{\mathfrak{m}_i} \cong R_i$.

(3) The natural homomorphism $R \to R_{\mathfrak{m}_1} \oplus \cdots \oplus R_{\mathfrak{m}_n}$ is an isomorphism.

Exercise 2.1.9. Let R be a commutative ring and P a prime ideal in R. Prove that $R_P/(PR_P)$ is isomorphic to the quotient field of R/P.

Exercise 2.1.10. Let $f : R \to S$ be a homomorphism of commutative rings and W a multiplicative subset of R. Prove the following.

(1) $f(W) \subseteq S$ is a multiplicative subset of S.

(2) If $Z = f(W)$ is the image of W, then $Z^{-1}S \cong W^{-1}S = S \otimes_R W^{-1}R$.

(3) If I is an ideal in R, then there are isomorphisms

$$W^{-1}(R/I) \cong (R/I) \otimes_R W^{-1}R \cong (W^{-1}R)/(I(W^{-1}R)).$$

Exercise 2.1.11. Let R be a commutative ring, P a prime ideal in R, and $\alpha \in R - P$. Prove that $R_P \cong (R_\alpha)_{PR_\alpha} \cong R_\alpha \otimes_R R_P$.

Exercise 2.1.12. Let $f : R \to S$ be a homomorphism of commutative rings. Let Q be a prime ideal in S and $P = f^{-1}(Q)$. Let $Q_P = Q \otimes_R R_P$ and $S_P = S \otimes_R R_P$. Prove the following.

(1) f induces a local homomorphism of local rings $g : R_P \to S_Q$. For the definition of local homomorphism of local rings, see Exercise 1.1.18.

(2) Q_P is a prime ideal of S_P.

(3) S_Q is isomorphic to the local ring of S_P at Q_P.

(4) The diagram

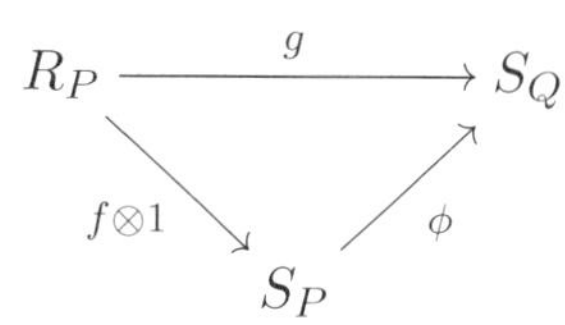

commutes, where ϕ is the localization map.

Exercise 2.1.13. (Local to Global Property for Idempotents) Let R be a commutative ring and $\mathfrak{p} \in \operatorname{Spec} R$. Let A be an R-algebra and e an idempotent in $A_\mathfrak{p}$. Show that there exists $\alpha \in R - \mathfrak{p}$ and an idempotent e_0 in $A_\alpha = A \otimes_R R[\alpha^{-1}]$ such that e is equal to the image of e_0 under the natural map $A_\alpha \to A_\mathfrak{p}$.

Exercise 2.1.14. Let R be a UFD which is not a field and α a nonzero element of R which is not invertible. Prove that the localization $W^{-1}R$ is not projective. (Hint: Show that $\operatorname{Hom}_R(R[\alpha^{-1}], R) = (0)$.)

2. The Prime Spectrum of a Commutative Ring

This section contains a brief review of the fundamental properties of the prime ideal spectrum of a commutative ring which is furnished with the Zariski topology. The notation and results established here play a central role throughout the rest of the book. We assume the reader has some familiarity with the prime spectrum, as well as a basic working knowledge of point set topology. As a basic reference, we suggest [**DF04**].

Let R be a commutative ring. The *prime ideal spectrum* of R, denoted $\operatorname{Spec} R$, is the set of all prime ideals in R. The *maximal ideal spectrum* of R, denoted $\operatorname{Max} R$, is the set of all maximal ideals in R. Given a subset $L \subseteq R$, let

$$V(L) = \{P \in \operatorname{Spec} R \mid P \supseteq L\}.$$

Given any collection $\{L_i\}$ of subsets of R, the reader should verify that

(1) $V(\{1\}) = \emptyset$, $V(\{0\}) = \operatorname{Spec} R$,
(2) $\bigcap_i V(L_i) = V(\bigcup L_i)$, and
(3) $V(L_1) \cup V(L_2) = V\big(\{x_1 x_2 \mid x_1 \in L_1, x_2 \in L_2\}\big)$.

The collection of sets $\{V(L) \mid L \subseteq R\}$ make up the closed sets for a topology on $\operatorname{Spec} R$, called the *Zariski topology*. Given a nonempty subset $Y \subseteq \operatorname{Spec} R$, let

$$I(Y) = \bigcap_{P \in Y} P.$$

Being an intersection of ideals, $I(Y)$ is an ideal. By definition, we take $I(\emptyset)$ to be the unit ideal R.

Lemma 2.2.1. *Let R be a commutative ring and W a multiplicative subset of $R - \{0\}$. Then there exists a prime ideal $P \in \operatorname{Spec} R$ such that $P \cap W = \emptyset$.*

Proof. Let $\mathcal{S} = \{I \subseteq R \mid I \text{ is an ideal and } I \cap W = \emptyset\}$. Then $(0) \in \mathcal{S}$. By Zorn's Lemma, $\mathcal{S}$ has a maximal element, say P. To see that P is a prime ideal, assume $x \notin P$ and $y \notin P$ and show that $xy \notin P$. By maximality of P we know $(Rx + P) \cap W \neq \emptyset$. Hence, there exists $a \in R$ and $u \in W$ such that $ax - u \in P$. Likewise, $(Ry + P) \cap W \neq \emptyset$, so there exists $b \in R$ and $v \in W$ such that $by - v \in P$. Multiplying, we find that $abxy - uv \in P$. Since $uv \in W$ and $P \cap W = \emptyset$, this implies $xy \notin P$. □

If R is a commutative ring and $\mathfrak{a}$ is an ideal in R, then the *nil radical of* $\mathfrak{a}$ is defined to be the set

$$\operatorname{Rad}(\mathfrak{a}) = \{x \in R \mid x^n \in \mathfrak{a} \text{ for some } n > 0\}.$$

Notice that $\mathfrak{a}$ is always a subset of $\operatorname{Rad}(\mathfrak{a})$ and $\mathfrak{a}$ is called a *radical ideal* if $\mathfrak{a} = \operatorname{Rad}(\mathfrak{a})$. When the ring R is ambiguous, we write $\operatorname{Rad}_R(\mathfrak{a})$ for $\operatorname{Rad}(\mathfrak{a})$. The nil radical of the zero ideal, $\operatorname{Rad}_R(0)$, is the set of all nilpotent elements in R. The next lemma draws the connection between the prime ideals containing $\mathfrak{a}$ and the nil radical of $\mathfrak{a}$.

Lemma 2.2.2. *Let R be a commutative ring and $\mathfrak{a}$ an ideal in R. Then $\operatorname{Rad}(\mathfrak{a})$ is an ideal in R which contains $\mathfrak{a}$ and*

$$\operatorname{Rad}(\mathfrak{a}) = \operatorname{Rad}_R(\mathfrak{a}) = I(V(\mathfrak{a})) = \bigcap_{P \in V(\mathfrak{a})} P.$$

In particular, $\operatorname{Rad}(0) = \operatorname{Rad}_R(0) = \bigcap_{P \in \operatorname{Spec} R} P$.

Proof. We begin by proving the last statement. Pick $x \in \operatorname{Rad}(0)$. Fix $P \in \operatorname{Spec} R$. If $x^n = 0$, then either $x = 0$ or $n \geq 2$. If $n \geq 2$, then $x \cdot x^{n-1} \in P$ so $x \in P$ or $x^{n-1} \in P$. Inductively, $x \in P$ so $\operatorname{Rad}(0) \subseteq P$. If

$x \notin \operatorname{Rad}(0)$, let $W = \{1, x, x^2, \dots\}$. Lemma 2.2.1 shows that there exists $P \in \operatorname{Spec} R$ such that $x \notin P$.

Now we prove the first statement. Under the natural map $\eta : R \to R/\mathfrak{a}$ there is a one-to-one correspondence between ideals of R containing $\mathfrak{a}$ and ideals of $R/\mathfrak{a}$. Under this correspondence, prime ideals correspond to prime ideals. To get the nil radical $\operatorname{Rad}_R \mathfrak{a}$, we lift the nil radical $\operatorname{Rad}_{R/\mathfrak{a}}(0)$. □

Let X be a topological space and Y a nonempty subset. We say Y is *irreducible* if whenever $Y \subseteq Y_1 \cup Y_2$ and Y_1, Y_2 are closed subsets of X, then $Y \subseteq Y_1$, or $Y \subseteq Y_2$. We say Y is *connected* if whenever $Y \subseteq Y_1 \cup Y_2$ and Y_1, Y_2 are disjoint closed subsets of X, then $Y \subseteq Y_1$, or $Y \subseteq Y_2$. The empty set is not considered to be irreducible or connected. Notice that an irreducible set is connected. If Z is a subset of the topological space X, then the *closure* of Z, denoted $\bar{Z}$, is the smallest closed subset of X that contains Z. Equivalently, $\bar{Z}$ is equal to the intersection of all closed sets that contain Z.

Lemma 2.2.3. *Let X be a topological space.*

(1) *If X is irreducible and $U \subseteq X$ is a nonempty open of X, then U is irreducible and dense.*

(2) *Let Z be a subset of X and denote by $\bar{Z}$ the closure of Z in X. Then Z is irreducible if and only if $\bar{Z}$ is irreducible.*

(3) *If X is irreducible, then X is connected.*

Proof. The proof is left to the reader. □

Lemma 2.2.4. *Let $\mathfrak{a}$ be an ideal in R and Y a subset of* $\operatorname{Spec} R$. *Then*

(1) $V(\mathfrak{a}) = V(\operatorname{Rad}\mathfrak{a})$, *and*

(2) $V(I(Y)) = \bar{Y}$, *the closure of Y in the Zariski topology.*

Proof. (1): Since $\mathfrak{a} \subseteq \operatorname{Rad}\mathfrak{a}$, it follows that $V(\mathfrak{a}) \supseteq V(\operatorname{Rad}\mathfrak{a})$. Conversely, if $P \in \operatorname{Spec} R$ and $P \supseteq \mathfrak{a}$, then by Lemma 2.2.2, $P \supseteq \operatorname{Rad}\mathfrak{a}$. Then $P \in V(\operatorname{Rad}\mathfrak{a})$.

(2): Since $V(I(Y))$ is closed we have $V(I(Y)) \supseteq \bar{Y}$. Since $\bar{Y}$ is closed, $\bar{Y} = V(\mathfrak{a})$ for some ideal $\mathfrak{a}$. Since $Y \subseteq \bar{Y}$, $I(Y) \supseteq I(\bar{Y}) = I(V(\mathfrak{a})) = \operatorname{Rad}\mathfrak{a} \supseteq \mathfrak{a}$. Thus, $V(I(Y)) \subseteq V(\mathfrak{a}) = \bar{Y}$. □

Corollary 2.2.5. *There is a one-to-one order-reversing correspondence between closed subsets of* $\operatorname{Spec} R$ *and radical ideals in R given by $Y \mapsto I(Y)$ and $\mathfrak{a} \mapsto V(\mathfrak{a})$. Under this correspondence, irreducible closed subsets correspond to prime ideals, and closed points correspond to maximal ideals.*

Proof. For the first part, see [**DF04**, Proposition 15.5.54], for example. The last part is proved in Lemma 2.2.6. □

Lemma 2.2.6. *Let R be a commutative ring and Y a subset of* Spec R. *Then Y is irreducible if and only if $P = I(Y)$ is a prime ideal in R. If Z is an irreducible closed subset of* Spec R, *then $P = I(Z)$ is the unique minimal element of Z, and is called the* generic point *of Z.*

Proof. Suppose Y is irreducible. Assume $x, y \in R$ and $xy \in I(Y)$. Notice that $Y \subseteq \bar{Y} = V(I(Y)) \subseteq V(xy) = V(x) \cup V(y)$. Since Y is irreducible, $Y \subseteq V(x)$ or $Y \subseteq V(y)$. Therefore, $x \in I(Y)$, or $y \in I(Y)$. This shows $I(Y)$ is a prime ideal. Conversely, assume $P = I(Y)$ is a prime ideal of R. The singleton set $\{P\}$ is irreducible, and by Lemma 2.2.3 the closure of $\{P\}$ is irreducible. By Lemma 2.2.4, the closure of $\{P\}$ is equal to $V(P)$, which is equal to $\bar{Y}$. By Lemma 2.2.3, Y is irreducible. The rest is left to the reader. □

Let R be a commutative ring. If $\alpha \in R$, the *basic open subset* of Spec R associated to α is

$$U(\alpha) = \operatorname{Spec} R - V(\alpha) = \{Q \in \operatorname{Spec} R \mid \alpha \notin Q\}.$$

By Lemma 2.2.7 (2) we see that the collection of all basic open sets $\{U(\alpha) \mid \alpha \in R\}$ is a basis for the Zariski topology on Spec R.

Lemma 2.2.7. *Let R be a commutative ring.*

(1) *Let $\alpha, \beta \in R$. The following are equivalent:*
  (a) $V(\alpha) = V(\beta)$.
  (b) $U(\alpha) = U(\beta)$.
  (c) *There exist $a \geq 1$, $b \geq 1$ such that $\alpha^a \in R\beta$ and $\beta^b \in R\alpha$.*

(2) *If I is an ideal in R, then*
$$\operatorname{Spec} R - V(I) = \bigcup_{\alpha \in I} U(\alpha)$$
Every open set can be written as a union of basic open sets.

Proof. (1): By Lemma 2.2.2, Rad $(R\alpha) = I(V(\alpha))$, and by Lemma 2.2.4, $V(\operatorname{Rad}(R\alpha)) = V(\alpha)$. So $V(\alpha) = V(\beta)$ if and only if $\operatorname{Rad}(R\alpha) = \operatorname{Rad}(R\beta)$ which is true if and only if there exist $a \geq 1$, $b \geq 1$ such that $\alpha^a \in R\beta$ and $\beta^b \in R\alpha$. The rest is left to the reader. □

We end this section with a series of results that describe the fundamental correspondence that exists between idempotents in R, ideals that are direct summands of R, and subsets of Spec R that are both open and closed. This correlation is the subject of Theorem 2.2.9 and its two corollaries. Playing a central role in this development is Lemma 1.1.20, which we proved in Chapter 1. First we show that the basic open set associated to an idempotent is also a closed set.

Lemma 2.2.8. *Let R be a commutative ring and* $\operatorname{idemp}(R) = \{e \in R \mid e = e^2\}$ *the set of all idempotents of R.*

(1) *If $e \in \operatorname{idemp}(R)$, then the closed set $V(1-e)$ is equal to the open set $U(e)$.*
(2) *Let $e, f \in \operatorname{idemp}(R)$. Then $V(e) = V(f)$ if and only if $e = f$.*
(3) *Let $e, f \in \operatorname{idemp}(R)$. Then $Re = Rf$ if and only if $e = f$.*

Proof. (1): Let $P \in \operatorname{Spec} R$. Since $e(1-e) = 0$, either $e \in P$, or $1 - e \in P$. Since $1 = e + (1-e)$, P does not contain both e and $1-e$.

(2): Assume $V(e) = V(f)$. By Lemma 2.2.7, there exist $a \geq 1$, $b \geq 1$ such that $e = e^a \in Rf$ and $f = f^b \in Re$. Write $e = xf$ and $f = ye$ for some $x, y \in R$. Then $e = xf = xf^2 = (xf)f = ef = eye = ye^2 = ye = f$.

(3): $Re = Rf$ implies $V(e) = V(f)$, which by Part (2) implies $e = f$. □

Theorem 2.2.9. *Let R be a commutative ring. Define $\mathcal{C}$ to be the set of all subsets $Y \subseteq \operatorname{Spec} R$ such that Y is both open and closed. Define $\mathcal{D}$ to be the set of all ideals A in R such that A is an R-module direct summand of R. Then there are one-to-one correspondences:*

$$\gamma : \operatorname{idemp}(R) \to \mathcal{C},$$

defined by $e \mapsto V(1-e) = U(e)$, and

$$\delta : \operatorname{idemp}(R) \to \mathcal{D},$$

defined by $e \mapsto Re$.

Proof. Parts (1) and (2) of Lemma 2.2.8 show that γ is well defined and one-to-one. By Lemma 1.1.20 (1), δ is well defined and onto. By Lemma 2.2.8 (3), δ is one-to-one. It remains to be proven that γ is onto. Assume A_1, A_2 are ideals in R, $X_1 = V(A_1)$, $X_2 = V(A_2)$, $X_1 \cup X_2 = \operatorname{Spec} R$, $X_1 \cap X_2 = \emptyset$. We prove that $X_i = V(e_i)$ for some $e_i \in \operatorname{idemp}(R)$. Since $\emptyset = X_1 \cap X_2 = V(A_1 + A_2)$, we know A_1 and A_2 are comaximal. Therefore, $A_1A_2 = A_1 \cap A_2$. Since $\operatorname{Spec} R = X_1 \cup X_2 = V(A_1A_2) = V(A_1 \cap A_2)$, Lemma 2.2.2 implies

$$A_1 \cap A_2 \subseteq \bigcap_{P \in \operatorname{Spec} R} P = \operatorname{Rad}_R(0).$$

That is, $A_1 \cap A_2$ consists of nilpotent elements. Write $1 = \alpha_1 + \alpha_2$, where $\alpha_i \in A_i$. Then $R = R\alpha_1 + R\alpha_2$ so $R\alpha_1$ and $R\alpha_2$ are comaximal. Also $R\alpha_1 \cap R\alpha_2 = R\alpha_1\alpha_2 \subseteq A_1 \cap A_2 \subseteq \operatorname{Rad}_R(0)$. So there exists $m > 0$ such that $(\alpha_1\alpha_2)^m = 0$. Then $R\alpha_1^m$ and $R\alpha_2^m$ are comaximal (Exercise 2.2.18) and

$R\alpha_1^m \cap R\alpha_2^m = (0)$. Therefore R is isomorphic to the internal direct sum $R \cong R\alpha_1^m \oplus R\alpha_2^m$. By Lemma 1.1.20 there are orthogonal idempotents $e_1, e_2 \in R$ such that $1 = e_1 + e_2$ and $Re_i = R\alpha_i^m$. Then $\operatorname{Spec} R = V(e_1) \cup V(e_2)$ and $V(e_1) \cap V(e_2) = \emptyset$. Moreover, $V(e_i) \supseteq V(R\alpha_i^m) \supseteq V(A_i) = X_i$. From this it follows that $X_i = V(e_i)$, hence γ is onto. □

Corollary 2.2.10. *Suppose R is a commutative ring and* $\operatorname{Spec} R = X_1 \cup \cdots \cup X_r$, *where each X_i is a nonempty closed subset and $X_i \cap X_j = \emptyset$ whenever $i \neq j$. Then there are idempotents $e_1, \dots, e_r$ in R such that $X_i = U(e_i) = V(1 - e_i)$ is homeomorphic to* $\operatorname{Spec} Re_i$, *and $R = Re_1 \oplus \cdots \oplus Re_r$. In particular, the topological space* $\operatorname{Spec} R$ *is connected if and only if* 0 *and* 1 *are the only idempotents of R.*

Proof. By Theorem 2.2.9 there are unique idempotents $e_1, \dots, e_r$ in R such that $X_i = U(e_i) = V(1-e_i)$. Since $R = Re_i \oplus R(1-e_i)$, the map $\pi_i : R \to Re_i$ defined by $x \mapsto xe_i$ is a homomorphism of rings with kernel $R(1 - e_i)$. By Exercise 2.2.14, π_i induces a homeomorphism $\operatorname{Spec} Re_i \to X_i$. If $i \neq j$, then $V(1 - e_i) \cap V(1 - e_j) = X_i \cap X_j = \emptyset$. It follows that the ideals $R(1 - e_i)$ are pairwise relatively prime. By the Chinese Remainder Theorem, the direct sum map

$$R \xrightarrow{\phi} Re_1 \oplus \cdots \oplus Re_r$$

is onto. By Exercise 2.2.17, the kernel of ϕ is the principal ideal generated by the product $(1 - e_1) \cdots (1 - e_r)$. But $X = X_1 \cup \cdots \cup X_r = V((1 - e_1) \cdots (1 - e_r))$. Therefore, $(1 - e_1) \cdots (1 - e_r) \in \operatorname{Rad}_R(0)$. Since the only nilpotent idempotent is 0, ϕ is an isomorphism. □

Corollary 2.2.11. *Let e be an idempotent of R. The following are equivalent:*

(1) *e is a primitive idempotent.*

(2) *$V(1 - e) = U(e)$ is a connected component of* $\operatorname{Spec} R$.

(3) 0 *and* 1 *are the only idempotents of the ring Re.*

Proof. (1) is equivalent to (3): This follows from Lemma 1.1.20 (2).

(2) is equivalent to (3): Since $R = Re \oplus R(1 - e)$, it follows from Exercise 2.2.14 that $V(1 - e)$ is homeomorphic to $\operatorname{Spec} Re$. The rest follows from Corollary 2.2.10. □

We say a commutative ring R is *connected* if $\operatorname{Spec} R$ is connected. By the above, R is connected if and only if R has only two idempotents.

2.1. Exercises. In the following, R always denotes a commutative ring.

Exercise 2.2.12. If $\theta : R \to S$ is a homomorphism of commutative rings, show that $P \mapsto \theta^{-1}(P)$ induces a function $\theta^\sharp : \operatorname{Spec} S \to \operatorname{Spec} R$ which is continuous for the Zariski topology. Show that if $\sigma : S \to T$ is another homomorphism, then $(\sigma\theta)^\sharp = \theta^\sharp \sigma^\sharp$.

Exercise 2.2.13. For the following, let I and J be ideals in R. Prove that the nil radical satisfies the following properties.

(1) $I \subseteq \operatorname{Rad} I$

(2) $\operatorname{Rad}(\operatorname{Rad} I) = \operatorname{Rad} I$

(3) $\operatorname{Rad}(IJ) = \operatorname{Rad}(I \cap J) = \operatorname{Rad} I \cap \operatorname{Rad} J$

(4) $\operatorname{Rad} I = R$ if and only if $I = R$

(5) $\operatorname{Rad}(I + J) = \operatorname{Rad}(\operatorname{Rad} I + \operatorname{Rad} J)$

(6) If $P \in \operatorname{Spec} R$, then for all $n > 0$, $P = \operatorname{Rad}(P^n)$.

(7) $I + J = R$ if and only if $\operatorname{Rad} I + \operatorname{Rad} J = R$.

Exercise 2.2.14. Let $I \subsetneq R$ be an ideal in R, $\eta : R \to R/I$ the natural map, and $\eta^\sharp : \operatorname{Spec}(R/I) \to \operatorname{Spec} R$ the continuous map of Exercise 2.2.12. Prove the following.

(1) $\eta^\sharp$ is a one-to-one order-preserving correspondence between the prime ideals of R/I and $V(I)$.

(2) There is a one-to-one correspondence between radical ideals in R/I and radical ideals in R containing I.

(3) Under $\eta^\sharp$ the image of a closed set is a closed set.

(4) $\eta^\sharp : \operatorname{Spec}(R/I) \to V(I)$ is a homeomorphism.

(5) If $I \subseteq \operatorname{Rad}_R(0)$, then $\eta^\sharp : \operatorname{Spec}(R/I) \to \operatorname{Spec}(R)$ is a homeomorphism.

Exercise 2.2.15. Let W be a multiplicative set in R and $\theta : R \to W^{-1}R$ the localization. For any subset $S \subseteq W^{-1}R$, use the intersection notation $S \cap R = \theta^{-1}(S)$ for the preimage. Prove the following.

(1) If J is an ideal in $W^{-1}R$, then $J = W^{-1}(J \cap R)$.

(2) The continuous map $\theta^\sharp : \operatorname{Spec}(W^{-1}R) \to \operatorname{Spec}(R)$ is one-to-one.

(3) If $P \in \operatorname{Spec} R$ and $P \cap W = \emptyset$, then $W^{-1}P$ is a prime ideal in $W^{-1}R$.

(4) The image of $\theta^\sharp : \operatorname{Spec}(W^{-1}R) \to \operatorname{Spec}(R)$ consists of those prime ideals in R that are disjoint from W.

(5) If $P \in \operatorname{Spec} R$, there is a one-to-one correspondence between prime ideals in R_P and prime ideals of R contained in P.

Exercise 2.2.16. Show that $\operatorname{Spec} R$ is quasi-compact. That is, every open cover of $\operatorname{Spec} R$ has a finite subcover.

Exercise 2.2.17. Let $I_1, I_2, \dots, I_n$ be pairwise comaximal ideals in R. Show that $I_1 I_2 \cdots I_n = I_1 \cap I_2 \cap \cdots \cap I_n$.

Exercise 2.2.18. Show that if I and J are comaximal ideals in R, then for every $m \geq 1$ and $n \geq 1$, I^m and J^n are comaximal. In this case $I^m J^n = I^m \cap J^n$.

3. Finitely Generated Projective Modules

Throughout this section, R is a commutative ring. Our goal in this section is to generalize the notion of rank of a free module to projective modules. We show that if R is connected, then there is a well defined notion of rank associated with any finitely generated projective R-module. For a free module, this rank agrees with the rank we already defined in Section 1.1.2.

Our first main result in this section is a proof that every finitely generated projective module over a local ring is free. Hence for local rings, there is a well defined notion of rank for finitely generated projective modules. Notice that the next two proofs are both applications of Nakayama's Lemma.

Lemma 2.3.1. *Let R be a commutative ring and I an ideal in R. Let M be an R-module. If*

(1) *I is nilpotent, or*
(2) *I is contained in every maximal ideal of R and M is finitely generated,*

then a subset $X \subseteq M$ generates M as an R-module if and only if the image of X generates M/IM as an R/I-module.

Proof. Let $\eta : M \to M/IM$. Suppose $X \subseteq M$ and let T be the R-submodule of M spanned by X. Then $\eta(T) = (T + IM)/IM$ is spanned by $\eta(X)$. If $T = M$, then $\eta(T) = M/IM$. Conversely, $\eta(T) = M/IM$ implies $M = T + IM$. By Corollary 1.1.17, this implies $M = T$. $\square$

Proposition 2.3.2. *Let R be a commutative local ring. If P is a finitely generated projective R-module, then P is free of finite rank. If $\mathfrak{m}$ is the maximal ideal of R and $\{x_i + \mathfrak{m}P \mid 1 \leq i \leq n\}$ is a basis for the vector space $P/\mathfrak{m}P$ over the residue field $R/\mathfrak{m}$, then $\{x_1, \dots, x_n\}$ is a basis for P over R. It follows that* $\operatorname{Rank}_R(P) = \dim_{R/\mathfrak{m}}(P/\mathfrak{m}P)$.

Proof. Define $\phi : R^{(n)} \to P$ by $\phi(\alpha_1, \dots, \alpha_n) = \sum_{i=1}^n \alpha_i x_i$. The goal is to show that ϕ is onto and one-to-one, in that order. Denote by T the image of ϕ. Then $T = Rx_1 + \cdots + Rx_n$ which is the submodule of P generated by $\{x_1, \dots, x_n\}$. It follows from Lemma 2.3.1 that ϕ is onto. To show that ϕ is one-to-one we prove that $\ker \phi = 0$. Since P is R-projective, the sequence

$$0 \to \ker \phi \to R^{(n)} \xrightarrow{\phi} P \to 0$$

is split exact. Therefore, $\ker \phi$ is a finitely generated projective R-module. Upon tensoring with $(\ ) \otimes_R R/\mathfrak{m}$, ϕ becomes the isomorphism $(R/\mathfrak{m})^{(n)} \cong P/\mathfrak{m}P$. By Exercise 1.3.7,

$$0 \to \ker \phi \otimes_R R/\mathfrak{m} \to (R/\mathfrak{m})^{(n)} \xrightarrow{\phi} P/\mathfrak{m}P \to 0$$

is split exact. Therefore, $\ker \phi \otimes_R R/\mathfrak{m} = 0$, or in other words $\mathfrak{m}(\ker \phi) = \ker \phi$. By Corollary 1.1.14, $\ker \phi = (0)$. □

Before continuing our discussion of ranks, we state and prove a corollary of Lemma 2.3.1 and Proposition 2.3.2.

Corollary 2.3.3. *Let R be a commutative local ring with residue field k. Let $\psi : M \to N$ be a homomorphism of R-modules, where M is finitely generated and N is finitely generated and free. Then*

$$0 \to M \xrightarrow{\psi} N$$

is split exact if and only if $\psi \otimes 1 : M \otimes_R k \to N \otimes_R k$ is one-to-one.

Proof. Assume $\psi \otimes 1$ is one-to-one. By Proposition 2.3.2 we can pick a generating set $\{x_1, \dots, x_n\}$ for the R-module M such that $\{x_1 \otimes 1, \dots, x_n \otimes 1\}$ is a basis for the k-vector space $M \otimes_R k$. Define $\pi : R^{(n)} \to M$ by mapping the ith standard basis vector to x_i. Then $\pi \otimes 1 : k^{(n)} \to M \otimes_R k$ is an isomorphism. The diagram

$$\begin{array}{ccccc} R^{(n)} & \xrightarrow{\pi} & M & \xrightarrow{\psi} & N \\ \downarrow & & \downarrow & & \downarrow \\ k^{(n)} & \xrightarrow{\pi \otimes 1} & M \otimes_R k & \xrightarrow{\psi \otimes 1} & N \otimes_R k \end{array}$$

commutes. The composite map $\psi\pi \otimes 1$ is one-to-one. By Exercise 2.3.11, there is an R-module homomorphism $\tau : N \to R^{(n)}$ which is a left inverse for $\psi\pi$. Since π is onto, it follows that $\pi\tau$ is a left inverse for ψ.

Conversely, if ψ has a left inverse, then clearly $\psi \otimes 1$ is one-to-one. □

Let M be a finitely generated projective module over a commutative ring R. For any prime ideal P of R, the localization M_P is a finitely generated projective R_P-module (Theorem 1.3.2). Therefore M_P is a finitely generated free R_P-module (Proposition 2.3.2), and M_P has a well defined rank. If there is an integer $n \geq 0$ such that $n = \operatorname{Rank}_{R_P}(M_P)$ for all $P \in \operatorname{Spec} R$, then we say M has *constant rank* and write $\operatorname{Rank}_R(M) = n$. We now prove that the constant rank property is preserved by change of base.

Proposition 2.3.4. *Let R be a commutative ring and S a commutative R-algebra. If M is a finitely generated projective R-module of constant rank $\operatorname{Rank}_R(M) = n$, then $M \otimes_R S$ is a finitely generated projective S-module of constant rank and $\operatorname{Rank}_S(M) = n$.*

Proof. By Theorem 1.3.2, $M \otimes_R S$ is a finitely generated projective S-module. Let $\theta : R \to S$ be the structure map. Let $Q \in \operatorname{Spec} S$ and $P = \theta^{-1}(Q) \in \operatorname{Spec} R$. Then by Exercise 2.1.12, θ extends to a local homomorphism of local rings $\theta : R_P \to S_Q$. The proof follows from the string of S_Q-module isomorphisms

$$(M \otimes_R S) \otimes_S S_Q \cong M \otimes_R (S \otimes_S S_Q) \cong M \otimes_R S_Q \cong M \otimes_R (R_P \otimes_{R_P} S_Q)$$
$$\cong (M \otimes_R R_P) \otimes_{R_P} S_Q \cong (R_P)^{(n)} \otimes_{R_P} S_Q \cong (S_Q)^{(n)} .$$

□

In the following we write R_α for the localization of R at the multiplicative set $\{1, \alpha, \alpha^2, \dots\}$. If M is a finitely generated projective R-module and $P \in \operatorname{Spec} R$, then as an application of the local to global lemmas we show that there is a basic open neighborhood $U(\alpha)$ of P such that M_α has constant rank on $U(\alpha)$.

Theorem 2.3.5. *Let R be a commutative ring and M a finitely generated projective R-module.*

(1) *Given $P \in \operatorname{Spec} R$ there exists $\alpha \in R - P$ such that M_α is a free R_α-module.*
(2) *If α is as in (1), then the values $\operatorname{Rank}_{R_Q}(M_Q)$ are constant for all $Q \in U(\alpha)$.*
(3) *The map*
$$\operatorname{Spec} R \xrightarrow{\phi} \{0, 1, 2, \dots\}$$
defined by $P \mapsto \operatorname{Rank}_{R_P} M_P$ is continuous if $\{0, 1, 2, \dots\}$ is given the discrete topology (that is, the topology where every subset is closed, or equivalently, "points are open").

Proof. (1): By Proposition 2.3.2 we know that M_P is a free module over R_P. By Corollary 1.1.7, M is an R-module of finite presentation. An application of Lemma 2.1.6 completes the proof.

(2): If $Q \in U(\alpha)$, then $\alpha \in R - Q$. By Exercise 2.1.11, $R_Q = (R_\alpha)_{QR_\alpha}$. Since M_α is R_α-free of rank n, it follows from Proposition 2.3.4 that M_Q is R_Q-free of rank n.

(3): We need to prove that for every $n \geq 0$, the preimage $\phi^{-1}(n)$ is open in $\operatorname{Spec} R$. Let $P \in \operatorname{Spec} R$ such that $\operatorname{Rank}_{R_P} M_P = n$. It is enough to find an open neighborhood of P in the preimage of n. By Part (1), there exists $\alpha \in R - P$ such that M_α is free of rank n over R_α. Since $U(\alpha)$ is an open neighborhood of P in $\operatorname{Spec} R$, it is enough to show that $\operatorname{Rank}_{R_Q}(M_Q) = n$ for all $Q \in U(\alpha)$. This shows that (3) follows from (2). □

Let M be a finitely generated projective R-module. Combining Theorem 2.3.5 with the decomposition theorem for $\operatorname{Spec} R$ (Corollary 2.2.10) yields the following unique decomposition theorem for M. In the terminology of [**McD84**, Theorem IV.27], the elements $e_1, \dots, e_t$ that appear in Corollary 2.3.6 are called the *structure idempotents* of M.

Corollary 2.3.6. *Let R be a commutative ring and M a finitely generated projective R-module. Then there are idempotents $e_1, \dots, e_t$ in R satisfying the following.*

(1) $R = Re_1 \oplus \cdots \oplus Re_t$.

(2) $M = Me_1 \oplus \cdots \oplus Me_t$.

(3) *If $R_i = Re_i$ and $M_i = M \otimes_R R_i$, then M_i is a finitely generated projective R_i-module of constant rank.*

(4) *If $\operatorname{Rank}_{R_i}(M_i) = n_i$, then $n_1, \dots, n_t$ are distinct.*

(5) *The integer t and the idempotents $e_1, \dots, e_t$ are uniquely determined by M.*

Proof. The rank function $\phi : \operatorname{Spec} R \to \{0, 1, 2, \dots\}$ defined by $\phi(P) = \operatorname{Rank}_P(M)$ is continuous and locally constant (Theorem 2.3.5). Let $U_n = \phi^{-1}(\{n\})$ for each $n \geq 0$. Then $\{U_n \mid n \geq 0\}$ is a collection of subsets of $\operatorname{Spec} R$ each of which is open and closed. Moreover, the sets U_n are pairwise disjoint. Since $\operatorname{Spec} R$ is quasi-compact (Exercise 2.2.16) the image of ϕ is a finite set, say $\{n_1, \dots, n_t\}$. Let $e_1, \dots, e_t$ be the idempotents in R corresponding to the disjoint union $\operatorname{Spec} R = U_{n_1} \cup \cdots \cup U_{n_t}$ (Corollary 2.2.10). The rest is left to the reader. □

3.1. Exercises.

For the following, R always denotes a commutative ring.

Exercise 2.3.7. Let L and M be finitely generated projective R-modules. Assume L and M both have constant rank. Prove:

(1) $L \oplus M$ has constant rank and $\operatorname{Rank}_R(L \oplus M)$ is equal to the sum $\operatorname{Rank}_R(L) + \operatorname{Rank}_R(M)$.

(2) $L \otimes_R M$ has constant rank and $\operatorname{Rank}_R(L \otimes_R M)$ is equal to the product $\operatorname{Rank}_R(L) \operatorname{Rank}_R(M)$.

(3) $\operatorname{Hom}_R(L, M)$ has constant rank and $\operatorname{Rank}_R(\operatorname{Hom}_R(L, M))$ is equal to the product $\operatorname{Rank}_R(L)\operatorname{Rank}_R(M)$.

Exercise 2.3.8. Let $f : R \to S$ be a homomorphism of commutative rings and $P \subset \operatorname{Spec} R$. Let $k_P = R_P/PR_P$ be the residue field and $S_P = S \otimes_R R_P$. Let $Q \in \operatorname{Spec} S$ such that $P = f^{-1}(Q)$. Prove:

(1) $S \otimes_R k_P \cong S_P/PS_P$.

(2) $Q \otimes_R k_P$ is a prime ideal of $S \otimes_R k_P$ and QS_P/PS_P is the corresponding prime ideal of S_P/PS_P.

(3) The localization of S_P/PS_P at QS_P/PS_P is S_Q/PS_Q.

(4) The localization of $S \otimes_R k_P$ at the prime ideal $Q \otimes_R k_P$ is $S_Q \otimes_R k_P$.

Exercise 2.3.9. Let $f : R \to S$ be a homomorphism of commutative rings and let $f^\sharp : \operatorname{Spec} S \to \operatorname{Spec} R$ the continuous map of Exercise 2.2.12. Assume

(a) $f^\sharp$ is one-to-one,

(b) the image of $f^\sharp$ is an open subset of $\operatorname{Spec} R$, and

(c) for every $\mathfrak{q} \in \operatorname{Spec} S$, if $\mathfrak{p} = \mathfrak{q} \cap R$, then the natural map $R_\mathfrak{p} \to S_\mathfrak{q}$ is an isomorphism.

If (a), (b) and (c) are satisfied, then we say $f^\sharp$ is an *open immersion*. Under these hypotheses, prove the following.

(1) For every $\mathfrak{q} \in \operatorname{Spec} S$, if $\mathfrak{p} = \mathfrak{q} \cap R$, then $S \otimes_R R_\mathfrak{p}$ is isomorphic to $S_\mathfrak{q}$.

(2) If $\alpha \in R$ and $U(\alpha)$ is a nonempty basic open subset of the image of $f^\sharp$, then $R[\alpha^{-1}]$ is isomorphic to $S \otimes_R R[\alpha^{-1}]$.

Exercise 2.3.10. Let M and N be finitely generated projective R-modules and $\varphi : M \to N$ an R-module homomorphism. Let $\mathfrak{p} \in \operatorname{Spec} R$ and assume $\varphi \otimes 1 : M \otimes_R R_\mathfrak{p} \to N \otimes_R R_\mathfrak{p}$ is an isomorphism. Show that there exists $\alpha \in R - \mathfrak{p}$ such that $\varphi \otimes 1 : M \otimes_R R_\alpha \to N \otimes_R R_\alpha$ is an isomorphism.

Exercise 2.3.11. Let R be a commutative local ring with residue field k. Let M and N be finitely generated free R-modules and $\psi : M \to N$ a homomorphism of R-modules. Show that if $\psi \otimes 1 : M \otimes_R k \to N \otimes_R k$ is one-to-one, then ψ has a left inverse. That is, there exists an R-module homomorphism $\sigma : N \to M$ such that $\sigma\psi = 1$ is the identity mapping on M.

Exercise 2.3.12. Let S be a commutative R-algebra that as an R-module is a progenerator. Show that if $\operatorname{Spec} R$ is connected, then the number of connected components of $\operatorname{Spec} S$ is bounded by $\operatorname{Rank}_R(S)$, hence is finite.

Exercise 2.3.13. Let R_1 and R_2 be rings (not necessarily commutative) and $S = R_1 \oplus R_2$ the direct sum. Let M be a left S-module. Using the projection maps $\pi_i : S \to R_i$, show that the R_i-modules $M_i = R_i \otimes_S M$ are S-modules. Show that M is isomorphic as an S-module to the direct sum $M_1 \oplus M_2$.

4. Faithfully Flat Modules and Algebras

As general references for this section, we suggest [**Mat80**] and [**Rot79**]. Recall that in Section 1.3.1 we defined a left R-module N to be flat if the functor $(\,) \otimes_R N$ is both left and right exact. In Exercise 1.5.7 we defined N to be faithfully flat if N is flat, and N has the property that for any right R-module M, $M \otimes_R N = 0$ implies $M = 0$. If R is a commutative ring, then Lemma 2.4.1 below adds two more necessary and sufficient conditions for N to be faithfully flat. In the remainder of the book, we will have many important applications which are based on properties of faithfully flat R-algebras. For example, the theory of faithfully flat descent (see Section 5.3) forms the basis for the formulation of Amitsur cohomology (see Section 5.5).

Lemma 2.4.1. *Let R be a commutative ring and N an R-module. The following are equivalent.*

(1) *A sequence of R-modules $0 \to A \to B \to C \to 0$ is exact if and only if $0 \to A \otimes_R N \to B \otimes_R N \to C \otimes_R N \to 0$ is exact.*
(2) *N is flat and for any R-module M, if $M \otimes_R N = 0$, then $M = 0$.*
(3) *N is flat and for every maximal ideal $\mathfrak{m}$ of R, $N \neq \mathfrak{m}N$.*

Example 2.4.2. Let R be a commutative ring and N an R-module. By the Morita Theorems, (see Exercise 1.5.7), if N is a progenerator, then N is faithfully flat. If N is projective, then N is flat (Exercise 1.3.7) but not necessarily faithfully flat. For example, suppose the ring $R = I \oplus J$ is an internal direct sum of two nonzero ideals I and J. Then $IJ = 0$, $I^2 = I$, $J^2 = J$ and $I + J = R$. The sequence $0 \to I \to 0$ is not exact. Tensor with $(\cdot) \otimes_R J$ and get the exact sequence $0 \to 0 \to 0$. So J is not faithfully flat.

Proposition 2.4.3. *Let R be a commutative ring. The R-module*

$$E = \bigoplus_{\mathfrak{m} \in \operatorname{Max} R} R_{\mathfrak{m}}$$

is faithfully flat.

Proof. Each $R_{\mathfrak{m}}$ is flat by Lemma 2.1.1, so E is flat by Exercise 2.4.15. For every maximal ideal $\mathfrak{m}$ of R, $\mathfrak{m}R_{\mathfrak{m}} \neq R_{\mathfrak{m}}$ so $\mathfrak{m}E \neq E$. By Lemma 2.4.1, we are done. □

Lemma 2.4.4. *Let $\theta : R \to S$ be a homomorphism of commutative rings such that S is a faithfully flat R-module. Then the following are true.*

(1) *For any R-module M, the map $M \to M \otimes_R S$ defined by $x \mapsto x \otimes 1$ is one-to-one. In particular, θ is one-to-one, so we can view $R = \theta(R)$ as a subring of S.*

(2) *For any ideal $I \subseteq R$, $IS \cap R = I$.*

(3) *The continuous map $\theta^\sharp : \operatorname{Spec} S \to \operatorname{Spec} R$ of Exercise 2.2.12 is onto.*

Lemma 2.4.5. *If $\theta : R \to S$ is a homomorphism of commutative rings, then the following are equivalent.*

(1) *S is faithfully flat as an R-module.*

(2) *S is a flat R-module and the continuous map $\theta^\sharp : \operatorname{Spec} S \to \operatorname{Spec} R$ is onto.*

(3) *S is a flat R-module and for each maximal ideal $\mathfrak{m}$ of R, there is a maximal ideal $\mathfrak{n}$ of S such that $\mathfrak{n} \cap R = \mathfrak{m}$.*

Proposition 2.4.6. *Let R be a commutative ring and $\epsilon : R \to A$ a homomorphism of rings such that ϵ makes A into a progenerator R-module.*

(1) *Under ϵ, R is mapped isomorphically onto an R-module direct summand of A.*

(2) *If B is a subring of A containing the image of ϵ, then the image of ϵ is an R-module direct summand of B.*

(3) *A is faithfully flat as an R-module.*

Proof. (1): By Corollary 1.1.16, A is R-faithful. Therefore the sequence

$$0 \to R \xrightarrow{\epsilon} A$$

is exact. By Exercise 1.3.31, there is a left inverse for ϵ if and only if

$$\operatorname{Hom}_R(A, R) \xrightarrow{\mathrm{H}_\epsilon} \operatorname{Hom}_R(R, R) \to 0 \tag{2.7}$$

is exact. Let $\mathfrak{m}$ be a maximal ideal in R. By Theorem 1.3.2, $A \otimes_R R/\mathfrak{m} = A/\mathfrak{m}A$ is a progenerator over the field $R/\mathfrak{m}$. In other words, $A/\mathfrak{m}A$ is a nonzero finite dimensional vector space over $R/\mathfrak{m}$. The diagram

$$\begin{array}{ccccc}
R/\mathfrak{m} \otimes_R \operatorname{Hom}_R(A, R) & \xrightarrow{1\otimes \mathrm{H}_\epsilon} & R/\mathfrak{m} \otimes_R \operatorname{Hom}_R(R, R) & \longrightarrow & 0 \\
\downarrow \cong & & \downarrow \cong & & \\
\operatorname{Hom}_{R/\mathfrak{m}}(A/\mathfrak{m}A, R/\mathfrak{m}) & \xrightarrow{\mathrm{H}_\epsilon} & \operatorname{Hom}_{R/\mathfrak{m}}(R/\mathfrak{m}, R/\mathfrak{m}) & \longrightarrow & 0
\end{array}$$

commutes. The bottom row is exact since $0 \to R/\mathfrak{m} \to A/\mathfrak{m}A$ is split exact over $R/\mathfrak{m}$. The vertical maps are isomorphisms by Corollary 1.3.27. Therefore the top row is exact. Corollary 1.3.19 says that (2.7) is exact. This proves (1).

(2): Assume $R \subseteq B \subseteq A$ is a tower of subrings. If $\sigma : A \to R$ is a left inverse for $\epsilon : R \to A$, then σ restricts to a left inverse for $R \to B$.

(3): This follows from Exercise 1.5.7. □

Proposition 2.4.7 and its corollary show that flatness is a local property.

Proposition 2.4.7. *Let R be a commutative ring and A an R-module. The following are equivalent.*

(1) *A is a flat R-module.*

(2) *A_p is a flat R_p-module, for every $p \in \operatorname{Spec} R$.*

(3) *$A_{\mathfrak{m}}$ is a flat $R_{\mathfrak{m}}$-module, for every $\mathfrak{m} \in \operatorname{Max} R$.*

Proof. It follows from Theorem 1.3.2 that (1) implies (2). It is immediate that (2) implies (3).

(3) implies (1): Denote by S the exact sequence

$$0 \to M \xrightarrow{\alpha} N \xrightarrow{\beta} P \to 0$$

of R-modules. Let $\mathfrak{m} \in \operatorname{Max} R$. Because $R_{\mathfrak{m}}$ is flat over R and $A_{\mathfrak{m}}$ is flat over $R_{\mathfrak{m}}$,

$$(S) \otimes_R R_{\mathfrak{m}} \otimes_{R_{\mathfrak{m}}} A_{\mathfrak{m}} = (S) \otimes_R A_{\mathfrak{m}}$$

is an exact sequence. Take the direct sum over all $\mathfrak{m}$. The reader should verify that

$$(S) \otimes_R \Big(\bigoplus_{\mathfrak{m} \in \operatorname{Max} R} A_{\mathfrak{m}} \Big) = (S) \otimes_R A \otimes_R \Big(\bigoplus_{\mathfrak{m} \in \operatorname{Max} R} R_{\mathfrak{m}} \Big)$$

is exact. By Proposition 2.4.3,

$$E = \bigoplus_{\mathfrak{m} \in \operatorname{Max} R} R_{\mathfrak{m}}$$

is a faithfully flat R-module, so $(S) \otimes_R A$ is exact. □

Proposition 2.4.8. *Let $f : R \to S$ be a homomorphism of commutative rings. The following are equivalent.*

(1) *S is a flat R-algebra.*

(2) *$S_{\mathfrak{q}}$ is a flat $R_{\mathfrak{p}}$-algebra, for every $\mathfrak{q} \in \operatorname{Spec} S$, if $f^{-1}(\mathfrak{q}) = \mathfrak{p}$.*

(3) *$S_{\mathfrak{m}}$ is a flat $R_{\mathfrak{p}}$-algebra, for every $\mathfrak{m} \in \operatorname{Max} S$, if $f^{-1}(\mathfrak{m}) = \mathfrak{p}$.*

Proof. The proof is left to the reader. □

If R is any ring, M is any left R-module, and I is a right ideal in R, the multiplication map $\mu : I \otimes_R M \to M$ is defined by $r \otimes x \mapsto rx$. The image of μ, which is denoted IM, is a $\mathbb{Z}$-submodule of M. If I is a two-sided ideal, then IM is an R-submodule of M. The following finiteness criterion for flatness will be applied in Chapter 9.

Corollary 2.4.9. *Let R be any ring and M a left R-module. The following are equivalent.*

(1) *M is a flat R-module.*

(2) *For every right ideal I of R, the sequence*

$$0 \to I \otimes_R M \xrightarrow{\mu} M \to M/IM \to 0$$

is an exact sequence of $\mathbb{Z}$-modules.

(3) *For every finitely generated right ideal I of R, the sequence*

$$0 \to I \otimes_R M \xrightarrow{\mu} M \to M/IM \to 0$$

is an exact sequence of $\mathbb{Z}$-modules.

(4) *If there exist elements $a_1, \dots, a_r$ in R and $x_1, \dots, x_r$ in M such that $\sum_i a_i x_i = 0$, then there exist an integer s, elements $\{b_{ij} \in R \mid 1 \leq i \leq r, 1 \leq j \leq s\}$ in R, and $y_1, \dots, y_s$ in M satisfying $\sum_i a_i b_{ij} = 0$ for all j and $x_i = \sum_j b_{ij} y_j$ for all i.*

The following proposition specifies sufficient conditions for changing the base ring of the Hom group. For a version of Proposition 2.4.10 that applies when the module M is finitely generated projective and S is noncommutative, see Corollary 1.3.27.

Proposition 2.4.10. *Let S be a flat commutative R-algebra. Let M be a finitely presented R-module and N any R-module. The natural map*

$$S \otimes_R \operatorname{Hom}_R(M, N) \xrightarrow{\alpha} \operatorname{Hom}_S(S \otimes_R M, S \otimes_R N)$$

is an isomorphism of S-modules.

Proof. If M is free, then this follows from Corollary 1.3.27. There is some n and an exact sequence of R-modules

$$R^{(n)} \to R^{(n)} \to M \to 0. \tag{2.8}$$

The functor $S \otimes_R (\cdot)$ is right exact, so

$$S^{(n)} \to S^{(n)} \to S \otimes_R M \to 0 \tag{2.9}$$

is an exact sequence of S-modules. By Proposition 1.3.21, the contravariant functor $\operatorname{Hom}_R(\cdot, N)$ is left exact. Applying it to (2.8), we get the exact sequence

$$0 \to \operatorname{Hom}_R(M, N) \to \operatorname{Hom}_R(R^{(n)}, N) \to \operatorname{Hom}_R(R^{(n)}, N). \tag{2.10}$$

By Proposition 1.3.23 (1), and Proposition 1.3.23, (2.10) can be written as

$$0 \to \operatorname{Hom}_R(M, N) \to N^{(n)} \to N^{(n)}. \tag{2.11}$$

Tensoring (2.11) with the flat module S gives the exact sequence

$$0 \to S \otimes_R \operatorname{Hom}_R(M, N) \to (S \otimes_R N)^{(n)} \to (S \otimes_R N)^{(n)}. \tag{2.12}$$

Now apply the left exact functor $\operatorname{Hom}_S(\cdot, S \otimes_R N)$ to (2.9) and simplify using Proposition 1.3.23. This gives the exact sequence

$$0 \to \operatorname{Hom}_S(S \otimes_R M, S \otimes_R N) \to (S \otimes_R N)^{(n)} \to (S \otimes_R N)^{(n)}. \tag{2.13}$$

Combining (2.12) and (2.13) with the natural maps yields a commutative diagram

$$\begin{array}{ccccccc} 0 & \longrightarrow & S \otimes_R \operatorname{Hom}_R(M, N) & \xrightarrow{f_1} & (S \otimes_R N)^{(n)} & \xrightarrow{f_2} & (S \otimes_R N)^{(n)} \\ & & \downarrow{\scriptstyle \alpha} & & \downarrow{\scriptstyle \beta}\,= & & \downarrow{\scriptstyle \gamma}\,= \\ 0 & \longrightarrow & \operatorname{Hom}_S(S \otimes_R M, S \otimes_R N) & \xrightarrow{g_1} & (S \otimes_R N)^{(n)} & \xrightarrow{g_2} & (S \otimes_R N)^{(n)} \end{array}$$

with exact rows. Since f_1 and β are one-to-one, α is one-to-one. To see that α is onto, let x be an element of the lower left corner. Set $y = \beta^{-1}(g_1(x))$. Then $\gamma(f_2(y)) = g_2(\beta(y)) = g_2(g_1(x)) = 0$. So $y = f_1(z)$ for some z in the upper left corner. Then $x = \alpha(z)$. Note that this also follows from a slight variation of the Snake Lemma 1.4.1. □

Proposition 2.4.11. *Let S be a flat commutative R-algebra and A an R-algebra. Let M be a finitely presented A-module and N any A-module. The natural map*

$$S \otimes_R \operatorname{Hom}_A(M, N) \xrightarrow{\alpha} \operatorname{Hom}_{S \otimes_R A}(S \otimes_R M, S \otimes_R N)$$

is an isomorphism of S-modules.

Proof. The proof is left to the reader. □

4.1. Exercises.

Exercise 2.4.12. Let R be a commutative ring, let M and N be R-modules, and $f \in \operatorname{Hom}_R(M, N)$. For any prime ideal $P \in \operatorname{Spec} R$ there is the R_P-module homomorphism $f_P : M_P \to N_P$ obtained by "localizing at P". Prove:

(1) f is one-to-one if and only if f_P is one-to-one for all $P \in \operatorname{Max} R$.
(2) f is onto if and only if f_P is onto for all $P \in \operatorname{Max} R$.

Exercise 2.4.13. Let R be a commutative ring. Let M and N be finitely generated and projective R-modules of constant rank such that $\operatorname{Rank}_R(M) = \operatorname{Rank}_R(N)$. Let $f \in \operatorname{Hom}_R(M, N)$. Show that if f is onto, then f is one-to-one.

Exercise 2.4.14. (Faithfully Flat Is Preserved by a Change of Base) If A is a commutative R-algebra and M is a faithfully flat R-module, show that $A \otimes_R M$ is a faithfully flat A-module.

Exercise 2.4.15. Let R be a ring and $\{M_i \mid i \in I\}$ a set of right R-modules. Show that the direct sum $\bigoplus_{i\in I} M_i$ is a flat R-module if and only if each M_i is a flat R-module.

Exercise 2.4.16. (Flat over Flat is Flat) Let $\theta : R \to A$ be a homomorphism of rings and M a left A-module. Using θ, view A as a left R-right A-bimodule and M as a left R-module. Show that if A is a flat R-module, and M is a flat A-module, then M is a flat R-module.

Exercise 2.4.17. (Faithfully Flat over Faithfully Flat is Faithfully Flat) Prove that if A is a commutative faithfully flat R-algebra and M is a faithfully flat A-module, then M is a faithfully flat R-module.

Exercise 2.4.18. Let R be a ring, $M \in {}_R\mathfrak{M}_R$ and $N \in {}_R\mathfrak{M}$.

(1) If M and N are flat left R-modules, show that $M \otimes_R N$ is a flat left R-module.

(2) Assume R is commutative. Show that if M and N are faithfully flat R-modules, then $M \otimes_R N$ is a faithfully flat R-module.

Exercise 2.4.19. Let $\theta : R \to S$ be a local homomorphism of local rings. Show that if S is a flat R-algebra, then S is faithfully flat.

Exercise 2.4.20. Let R be a commutative ring and $\{\alpha_i \mid i \in I\}$ a subset of $R - (0)$. Let $S = \prod_{i\in I} R[\alpha_i^{-1}]$. Then S is an R-algebra, where the structure homomorphism is the unique map $R \to S$ which commutes with each natural map $R \to R[\alpha_i^{-1}]$. Show that the following are equivalent.

(1) S is a faithfully flat R-module.

(2) There exists a finite subset $\{i_1, \dots, i_n\} \subseteq I$ such that the finite direct sum $R[\alpha_{i_1}^{-1}] \oplus \cdots \oplus R[\alpha_{i_n}^{-1}]$ is faithfully flat over R.

(3) There exists a finite subset $\{i_1, \dots, i_n\} \subseteq I$ such that $R = R\alpha_{i_1} + \cdots + R\alpha_{i_n}$.

Exercise 2.4.21. Let R be a commutative ring and I an ideal of R which is contained in the nil radical of R. Show that R/I is a flat R-algebra if and only if $I = (0)$.

5. Chain Conditions

We assemble in this section some background results on rings, modules, and topological spaces satisfying either an ascending chain condition or a descending chain condition. References are provided for those proofs that are omitted.

Let S be a set which is partially ordered by $\leq$. We say that S satisfies the *minimum condition* if every nonempty subset of S contains a minimal element. We say that S satisfies the *maximum condition* if every nonempty subset of S contains a maximal element. We say that S satisfies the *descending chain condition* (DCC) if every chain in S of the form $\{\dots, x_3 \leq x_2 \leq x_1 \leq x_0\}$ is eventually constant. That is, there is a subscript n such that $x_n = x_i$ for all $i \geq n$. We say that S satisfies the *ascending chain condition* (ACC) if every chain in S of the form $\{x_0 \leq x_1 \leq x_2 \leq x_3, \dots\}$ is eventually constant. The reader should verify that S satisfies the descending chain condition (DCC) if and only if S satisfies the minimum condition. Likewise, S satisfies the ascending chain condition (ACC) if and only if S satisfies the maximum condition.

Let R be any ring and M an R-module. We say M is *noetherian* if M satisfies the ACC on submodules (equivalently, M satisfies the maximum condition on submodules). The reader should verify that an R-module M is noetherian if and only if every submodule of M is finitely generated. We say M is *artinian* if M satisfies the DCC on submodules (equivalently, M satisfies the minimum condition on submodules). The ring R is said to be (left) *noetherian* if R is noetherian when viewed as a left R-module. In this case we say R satisfies the ACC on left ideals. The ring R is said to be (left) *artinian* if R is artinian when viewed as a left R-module. In this case we say R satisfies the DCC on left ideals.

Lemma 2.5.1. *Let R be any ring and*

$$0 \to A \xrightarrow{\alpha} B \xrightarrow{\beta} C \to 0$$

a short exact sequence of R-modules.

(1) *The following are equivalent.*
    (a) *B satisfies the ACC on submodules.*
    (b) *A and C satisfy the ACC on submodules.*
(2) *The following are equivalent.*
    (a) *B satisfies the DCC on submodules.*
    (b) *A and C satisfy the DCC on submodules.*

Proof. See, for example, [**DF04**] or [**Hun80**]. □

Corollary 2.5.2. *If R is a noetherian ring and M is a finitely generated R-module, then*

(1) *M satisfies the ACC on submodules,*
(2) *M is finitely presented,*

(3) *M satisfies the maximum condition on submodules, and*

(4) *every submodule of M is finitely generated.*

Proof. See, for example, [**DF04**] or [**Hun80**]. □

A topological space X is said to be *noetherian* if X satisfies the ascending chain condition on open sets. Notice that X is noetherian if and only if X satisfies the descending chain condition on closed sets.

Proposition 2.5.3. *Let X be a noetherian topological space and Z a nonempty closed subset of X.*

(1) *There are unique irreducible closed subsets $Z_1, \dots, Z_r$ such that $Z = Z_1 \cup \dots \cup Z_r$ and $Z_i \not\subseteq Z_j$ for all $i \neq j$. The sets Z_i are called the* irreducible components *of Z.*

(2) *There are unique connected closed subsets $Y_1, \dots, Y_c$ such that $Z = Y_1 \cup \dots \cup Y_c$ and $Y_i \cap Y_j = \emptyset$ for all $i \neq j$. The sets Y_i are called the* connected components *of Z.*

(3) *The number of connected components is less than or equal to the number of irreducible components.*

Proof. (1): See [**Har77**, Proposition I.1.5], or [**DF04**, Proposition 15.2.17]. The rest is left to the reader. □

Proposition 2.5.4. *Let R be a commutative noetherian ring.*

(1) Spec R *is a noetherian topological space.*

(2) Spec R *has a finite number of irreducible components.*

(3) Spec R *has a finite number of connected components.*

Proof. This follows from Corollary 2.2.5 and Proposition 2.5.3. □

Corollary 2.5.5. *Let R be a commutative noetherian ring and I an ideal of R which is not the unit ideal. There is a one-to-one correspondence between the irreducible components of $V(I)$ and the minimal prime over-ideals of I given by $Z \mapsto I(Z)$.*

Proof. Let $V(I) = Z_1 \cup \dots \cup Z_r$ be the decomposition into irreducible components, which exists by Propositions 2.5.4 and 2.5.3. For each i, let $P_i = I(Z_i)$. By Lemma 2.2.6, each of the ideals $P_1, \dots, P_r$ is prime. First we show that each P_i is minimal. Assume $I \subseteq Q \subseteq P_i$, for some prime Q. Then $V(I) \supseteq V(Q) \supseteq Z_i$. By Lemma 2.2.6, $V(Q)$ is irreducible. By the uniqueness part of Proposition 2.5.3, $V(Q) = Z_i$. Therefore, $Q = I(V(Q)) = P_i$. Now let P be a minimal prime over-ideal of I. We show that P is equal to one of $P_1, \dots, P_r$. By Lemma 2.2.6, $V(P)$ is an irreducible subset of $V(I)$.

Since $V(P) \subseteq Z_1 \cup \cdots \cup Z_r$, $V(P) \subseteq Z_i$, for some i. Therefore, $I \subseteq P_i \subseteq P$. Since P is minimal, $P = P_i$. □

Theorem 2.5.6. *Let R be a commutative noetherian ring. Then there exist primitive idempotents $e_1, \dots, e_n$ in R such that R is the internal direct sum $R = Re_1 \oplus \cdots \oplus Re_n$. This decomposition is unique in the sense that, if $R = Rf_1 \oplus \cdots \oplus Rf_p$ is another such decomposition of R, then $n = p$, and after rearranging, $e_1 = f_1$, ..., $e_n = f_n$.*

Proof. Let $\operatorname{Spec} R = X_1 \cup \cdots \cup X_n$ be the decomposition into connected components, which exists by Propositions 2.5.4 and 2.5.3. By Corollary 2.2.10 there are idempotents $e_1, \dots, e_n$ in R such that $X_i = U(e_i) = V(1 - e_i)$ is homeomorphic to $\operatorname{Spec} Re_i$, and $R = Re_1 \oplus \cdots \oplus Re_n$. Corollary 2.2.11 implies each e_i is a primitive idempotent. The uniqueness claim comes from Theorem 1.1.21. □

Before ending this section, we include for reference some definitions that will be needed for later chapters, as well as in some of the exercises below. Let R be any ring and M an R-module. We say M is *simple* if $M \neq (0)$ and (0) is a maximal submodule of M. So if M is a simple module, then (0) and M are the only submodules. Let R be any ring and M an R-module. Suppose there is a strictly descending finite chain of submodules

$$M = M_0 \supsetneq M_1 \supsetneq M_2 \supsetneq \cdots \supsetneq M_n = 0$$

starting with $M = M_0$ and ending with $M_n = 0$. The *length* of the chain is n. A *composition series* for M is a chain such that M_i / M_{i+1} is simple. If M has no composition series, then we say the length of M is infinite and write $\ell(M) = \infty$. Otherwise, the length of M, denoted $\ell(M)$, is defined to be the minimum of the lengths of all composition series of M. The following proposition will be required in Chapter 3. Its proof can be found, for example, in [**AM69**].

Proposition 2.5.7. *Let R be any ring and M an R-module. The following are equivalent.*

(1) *M has a composition series.*

(2) *M satisfies both the ACC and the DCC on submodules.*

5.1. Exercises.

Exercise 2.5.8. Let $R_1, \dots, R_n$ be rings. Prove that the direct sum $R_1 \oplus \cdots \oplus R_n$ is an artinian ring if and only if each R_i is an artinian ring.

Exercise 2.5.9. Prove the following, if R is an artinian ring.

(1) If M is a finitely generated R-module, then M satisfies the DCC on submodules.

(2) If I is a two-sided ideal in R, then R/I is artinian.
(3) If R is commutative and W is a multiplicative set in R, then $W^{-1}R$ is artinian.
(4) If R is an artinian domain, then R is a division ring.

Exercise 2.5.10. Let $\theta : R \to S$ be a homomorphism of commutative rings such that S is a faithfully flat R algebra. Show that if S is artinian, then R is artinian. Show that if S is noetherian, then R is noetherian.

Exercise 2.5.11. Let R be a commutative ring and M and N two R-modules. Prove:

(1) If R is noetherian, and M and N are finitely generated, then $\operatorname{Hom}_R(M, N)$ is a finitely generated R-module.
(2) If R is an integral domain, and N is torsion free, then $\operatorname{Hom}_R(M, N)$ is torsion free.

Exercise 2.5.12. Let D be a division ring and V a finite dimensional vector space over D. Prove:

(1) V is a simple module if and only if $\dim_D(V) = 1$.
(2) $\dim_D(V) = \ell(V)$.

Exercise 2.5.13. Let D be a division ring and V a vector space over D. Prove that the following are equivalent.

(1) V is finite dimensional over D.
(2) V is a D-module of finite length.
(3) V satisfies the ACC on submodules.
(4) V satisfies the DCC on submodules.

Exercise 2.5.14. Let D be a division ring. Prove:

(1) The ring $M_n(D)$ of all n-by-n matrices over D is both artinian and noetherian.
(2) If M is a finite dimensional D-vector space, then the ring of endomorphisms $\operatorname{Hom}_D(M, M)$ is both artinian and noetherian.

Exercise 2.5.15. Let k be a field and R a k-algebra which is finite dimensional as a k-vector space. Prove that the ring R is both artinian and noetherian.

Exercise 2.5.16. By the Hilbert Basis Theorem, a finitely generated $\mathbb{Z}$-algebra is a noetherian ring (see, for example, [**Hun80**, Theorem VIII.4.9]). Let R be a commutative ring. Viewing R as a $\mathbb{Z}$-algebra, show that $R = \varinjlim R_\alpha$, where $\{R_\alpha\}$ is a directed system of noetherian subrings of R.

Exercise 2.5.17. Let R be a commutative local ring with maximal ideal $\mathfrak{m}$. Show that there is a directed system $\{R_\alpha\}$ of noetherian local subrings of R satisfying the following:

(1) The maximal ideal of R_α is $\mathfrak{m}_\alpha = \mathfrak{m} \cap R_\alpha$.
(2) $R = \varinjlim R_\alpha$.
(3) $\mathfrak{m} = \varinjlim \mathfrak{m}_\alpha$.
(4) $R/\mathfrak{m} = \varinjlim (R_\alpha/\mathfrak{m}_\alpha)$.

6. Faithfully Flat Base Change

This section contains background results on modules over commutative rings. We first prove the fundamental theorem on faithfully flat base change. These results are then applied to show that a module is finitely generated projective if and only if it is locally free of finite rank. Lastly, we show that if R is a commutative ring, then the tensor product operation defines an abelian group on the set of all isomorphism classes of R-progenerator modules with rank one. This group is called the Picard group of R.

6.1. Fundamental Theorem on Faithfully Flat Base Change. Let S be a commutative faithfully flat R-algebra. Suppose A is either an R-module, or R-algebra. The faithfully flat base change results proved in this section are all of the type: "A has property P over R if $S \otimes_R A$ has property P over S". These powerful results allow us to prove that A has certain properties by restricting to a larger base ring S, where presumably the structure of $S \otimes_R A$ is easier to understand.

Theorem 2.6.1. *Let S be a commutative faithfully flat R-algebra and M an R-module.*

(1) *M is finitely generated over R if and only if $S \otimes_R M$ is finitely generated over S.*
(2) *M is of finite presentation over R if and only if $S \otimes_R M$ is of finite presentation over S.*
(3) *M is finitely generated projective over R if and only if $S \otimes_R M$ is finitely generated projective over S.*
(4) *M is flat over R if and only if $S \otimes_R M$ is flat over S.*
(5) *M is faithfully flat over R if and only if $S \otimes_R M$ is faithfully flat over S.*
(6) *M is a generator module over R if and only if $S \otimes_R M$ is a generator over S.*
(7) *M is faithful over R if and only if $S \otimes_R M$ is faithful over S.*

Proof. (1): If M is finitely generated, then Theorem 1.3.2 shows $S \otimes_R M$ is finitely generated. Conversely, choose generators $\{t_1, \dots, t_n\}$ for $S \otimes_R M$. After breaking up summations and factoring out elements of S, we can assume each t_i looks like $1 \otimes x_i$ where $x_i \in M$. Consider the sequence

$$R^{(n)} \to M \to 0 \tag{2.14}$$

which is defined by $(r_1, \dots, r_n) \mapsto \sum r_i x_i$. Tensoring (2.14) with S gives the sequence

$$S^{(n)} \to S \otimes_R M \to 0$$

which is exact. Since S is faithfully flat, (2.14) is exact.

(2): Assume M is finitely presented. Suppose $R^{(n)} \to R^{(n)} \to M \to 0$ is exact. Tensoring is right exact, so $S^{(n)} \to S^{(n)} \to S \otimes_R M \to 0$ is exact. Therefore $S \otimes_R M$ is finitely presented. Conversely assume $S \otimes_R M$ is finitely presented. By Part (1), M is finitely generated over R. Suppose $\phi : R^{(n)} \to M$ is onto. Let $N = \ker \phi$. It is enough to show that N is finitely generated. Since

$$0 \to N \to R^{(n)} \xrightarrow{\phi} M \to 0$$

is exact and S is faithfully flat,

$$0 \to S \otimes_R N \to S^{(n)} \xrightarrow{1 \otimes \phi} S \otimes_R M \to 0$$

is exact. By Lemma 2.1.7 (3), $S \otimes_R N$ is finitely generated over S. Part (1) says that N is finitely generated over R.

(3): If M is finitely generated and projective over R, then Theorem 1.3.2 says the same holds for $S \otimes_R M$ over S. Conversely, suppose $S \otimes_R M$ is finitely generated and projective over S. By Corollary 1.1.7, $S \otimes_R M$ is of finite presentation over S. By Part (2), M is of finite presentation over R. To show that M is R-projective, by Proposition 1.3.20 (2) it is enough to show $\operatorname{Hom}_R(M, \cdot)$ is a right exact functor. Start with an exact sequence

$$A \xrightarrow{\alpha} B \to 0 \tag{2.15}$$

of R-modules. It is enough to show

$$\operatorname{Hom}_R(M, A) \xrightarrow{\mathrm{H}_\alpha} \operatorname{Hom}_R(M, B) \to 0 \tag{2.16}$$

is exact. Since S is faithfully flat over R, it is enough to show

$$S \otimes_R \operatorname{Hom}_R(M, A) \xrightarrow{1 \otimes \mathrm{H}_\alpha} S \otimes_R \operatorname{Hom}_R(M, B) \to 0 \tag{2.17}$$

is exact. Tensoring is right exact, so tensoring (2.15) with $S \otimes_R (\cdot)$ gives the exact sequence

$$S \otimes_R A \xrightarrow{1 \otimes \alpha} S \otimes_R B \to 0. \tag{2.18}$$

Since we are assuming $S \otimes_R M$ is S-projective, by Proposition 1.3.20 (2) we can apply the functor $\operatorname{Hom}_S(S \otimes_R M, \cdot)$ to (2.18) yielding

$$\operatorname{Hom}_S(S \otimes_R M, S \otimes_R A) \xrightarrow{\mathrm{H}_{1\otimes\alpha}} \operatorname{Hom}_S(S \otimes_R M, S \otimes_R B) \to 0 \tag{2.19}$$

which is exact. Combine (2.17) and (2.19) to get the commutative diagram

$$\begin{array}{ccc} S \otimes_R \operatorname{Hom}_R(M, A) & \xrightarrow{1\otimes \mathrm{H}_\alpha} & S \otimes_R \operatorname{Hom}_R(M, B) & \\ \downarrow \cong & & \downarrow \cong & \\ \operatorname{Hom}_S(S \otimes_R M, S \otimes_R A) & \xrightarrow{\mathrm{H}_{1\otimes\alpha}} & \operatorname{Hom}_S(S \otimes_R M, S \otimes_R B) & \longrightarrow 0 \end{array}$$

where the vertical maps are the natural maps from Proposition 2.4.10. Since the bottom row is exact and the vertical maps are isomorphisms, it follows that $1 \otimes \mathrm{H}_\alpha$ is onto.

(4): Assume $M \otimes_R S$ is a flat S-module. By Exercise 2.4.16, $M \otimes_R S$ is flat over R. Let

$$0 \to A \to B \to C \to 0$$

be an exact sequence of R-modules. Then

$$0 \to A \otimes_R M \otimes_R S \to B \otimes_R M \otimes_R S \to C \otimes_R M \otimes_R S \to 0$$

is an exact sequence of R-modules. Since S is faithfully flat over R,

$$0 \to A \otimes_R M \to B \otimes_R M \to C \otimes_R M \to 0$$

is an exact sequence of R-modules.

Parts (5), (6) and (7) are left to the reader. □

Definition 2.6.2. Let R be a commutative ring and A an R-algebra. If the structure homomorphism $R \to Z(A)$ from R to the center of A is an isomorphism, then we say A is a *central* R-algebra.

Proposition 2.6.3. *Let R be a commutative ring and A an R-algebra.*

(1) *If S is a commutative faithfully flat R-algebra and $A \otimes_R S$ is a central S-algebra, then A is a central R-algebra.*

(2) *If $A_{\mathfrak{m}} = A \otimes_R R_{\mathfrak{m}}$ is a central $R_{\mathfrak{m}}$-algebra for every maximal ideal $\mathfrak{m}$ of R, then A is a central R-algebra.*

Proof. (1): Assume $A \otimes_R S$ is a central S-algebra. Since S is flat over R, $Z(A) \otimes_R S \to A \otimes_R S$ is one-to-one. By hypothesis, the composite map

$$R \otimes_R S \to Z(A) \otimes_R S \to Z(A \otimes_R S)$$

is an isomorphism. Since S is faithfully flat over R, $R \to Z(A)$ is an isomorphism.

(2): Let $\mathfrak{m}$ be a maximal ideal of R. Since $R_{\mathfrak{m}}$ is a flat R-module, $Z(A) \otimes_R R_{\mathfrak{m}} \to A_{\mathfrak{m}}$ is one-to-one. Clearly, $Z(A) \otimes_R R_{\mathfrak{m}} \subseteq Z(A_{\mathfrak{m}})$. We are given that the composite map

$$R_{\mathfrak{m}} \to Z(A) \otimes_R R_{\mathfrak{m}} \subseteq Z(A_{\mathfrak{m}})$$

is an isomorphism. Therefore, $R_{\mathfrak{m}} \to Z(A) \otimes_R R_{\mathfrak{m}}$ is an isomorphism. By Exercise 2.4.12, $R \to Z(A)$ is an isomorphism. □

6.2. Locally Free Finite Rank is Finitely Generated Projective. Let R be a commutative ring and M an R-module. Then M is *locally free of finite rank* if there exist elements $f_1, \dots, f_n$ in R such that $R = Rf_1 + \cdots + Rf_n$ and for each i, $M_{f_i} = M \otimes_R R_{f_i}$ is free of finite rank over R_{f_i}.

Proposition 2.6.4. *Let R be a commutative ring and M an R-module. The following are equivalent.*

(1) *M is finitely generated projective.*
(2) *M is locally free of finite rank.*
(3) *M is an R-module of finite presentation and for each $\mathfrak{p} \in \operatorname{Spec} R$, $M_{\mathfrak{p}}$ is a free $R_{\mathfrak{p}}$-module.*
(4) *M is an R-module of finite presentation and for each $\mathfrak{m} \in \operatorname{Max} R$, $M_{\mathfrak{m}}$ is a free $R_{\mathfrak{m}}$-module.*

Proof. It follows from Corollary 1.1.7 and Proposition 2.3.2 that (1) implies (3). It is trivial that (3) implies (4).

(4) implies (2): Using Lemma 2.1.6, for each $\mathfrak{m} \in \operatorname{Max} R$, choose $\alpha_{\mathfrak{m}} \in R - \mathfrak{m}$ such that $M_{\alpha_{\mathfrak{m}}} = M \otimes_R R_{\alpha_{\mathfrak{m}}}$ is free of finite rank over $R_{\alpha_{\mathfrak{m}}}$. Let $U(\alpha_{\mathfrak{m}}) = \operatorname{Spec} R - V(\alpha_{\mathfrak{m}})$ be the basic open set associated to $\alpha_{\mathfrak{m}}$. Since $U(\alpha_{\mathfrak{m}})$ is an open neighborhood of $\mathfrak{m}$, we have an open cover $\{U(\alpha_{\mathfrak{m}}) \mid \mathfrak{m} \in \operatorname{Max} R\}$ of $\operatorname{Spec} R$ (Corollary 2.2.5). By Exercise 2.2.16, there is a finite subset of $\{\alpha_{\mathfrak{m}} \mid \mathfrak{m} \in \operatorname{Max} R\}$, say $\{\alpha_1, \dots, \alpha_n\}$, such that $\{U(\alpha_1), \dots, U(\alpha_n)\}$ is an open cover of $\operatorname{Spec} R$. For each i, M_{α_i} is free of finite rank over R_{α_i} which proves M is locally free of finite rank.

(2) implies (1): Assume $\{U(f_1), \dots, U(f_n)\}$ is an open cover of $\operatorname{Spec} R$ and that for each i, M_{f_i} is free of rank N_i over R_{f_i}. Define N to be the maximum of $\{N_1, \dots, N_n\}$. Then

$$F_i = M_{f_i} \oplus R_{f_i}^{(N-N_i)}$$

is free of rank N over R_{f_i}. Set $S = \bigoplus_i R_{f_i}$. Then $R \to S$ is faithfully flat (Exercise 2.6.15). Set $F = \bigoplus_i F_i$. Then F is free over S of rank N and $M \otimes_R S = \bigoplus_i M_{f_i}$ is a direct summand of F (Exercise 2.6.13). Now apply Theorem 2.6.1 (3). □

Let R be a commutative ring. For any prime ideal $\mathfrak{p} \in \operatorname{Spec}(R)$, write $k_{\mathfrak{p}}$ for the residue field $R_{\mathfrak{p}}/\mathfrak{p}R_{\mathfrak{p}}$. If M is a finitely generated R-module, then M can be used to define a rank function $\operatorname{Spec} R \to \{0, 1, 2, \dots\}$, where $\mathfrak{p} \mapsto \dim_{k_{\mathfrak{p}}}(M \otimes_R k_{\mathfrak{p}})$. The next two corollaries to Proposition 2.6.4 utilize this rank function to give us a powerful test for locally free modules and for flatness over an integral domain.

Corollary 2.6.5. *Let R be an integral domain with quotient field K. For each maximal ideal $\mathfrak{m} \in \operatorname{Max}(R)$, write $k_{\mathfrak{m}}$ for $R/\mathfrak{m}$. The following are equivalent for any finitely generated R-module M.*

(1) *M is a locally free R-module of constant rank n.*

(2) $\dim_K(M \otimes_R K) = n$ *and for every* $\mathfrak{m} \in \operatorname{Max}(R)$, $\dim_{k_{\mathfrak{m}}}(M/\mathfrak{m}M) = n$.

Proof. (1) implies (2): If $M \cong R^{(n)}$, then $M \otimes_R k_{\mathfrak{m}} \cong k_{\mathfrak{m}}^{(n)}$ and $M \otimes_R K \cong K^{(n)}$.

(2) implies (1): Let $\mathfrak{m}$ be a maximal ideal of R and write $M_{\mathfrak{m}}$ for $M \otimes_R R_{\mathfrak{m}}$. Since $M/\mathfrak{m}M$ is free of dimension n over $k_{\mathfrak{m}}$, there exist $x_1, \dots, x_n$ in $M_{\mathfrak{m}}$ which restrict to a $k_{\mathfrak{m}}$-basis under the natural map $M_{\mathfrak{m}} \to M/\mathfrak{m}M$. For some $\alpha \in R - \mathfrak{m}$, the finite set $x_1, \dots, x_n$ is in the image of the natural map $M_\alpha \to M_{\mathfrak{m}}$. Define $\theta : R_\alpha^{(n)} \to M_\alpha$ by mapping the standard basis vector e_i to x_i. By Lemma 2.3.1, $M_{\mathfrak{m}}$ is generated by $x_1, \dots, x_n$ as an $R_{\mathfrak{m}}$-module. Therefore, upon localizing θ at the maximal ideal $\mathfrak{m}R_\alpha$, it becomes onto. Because the cokernel of θ is a finitely generated R_α-module, by Lemma 2.1.4, there exists $\beta \in R_\alpha - \mathfrak{m}R_\alpha$ such that if we replace α with $\alpha\beta$, then θ is onto. The diagram

$$\begin{array}{ccccccccc}
0 & \longrightarrow & \ker\theta & \longrightarrow & R_\alpha^{(n)} & \xrightarrow{\theta} & M_\alpha & \longrightarrow & 0 \\
 & & \downarrow{\scriptstyle\alpha} & & \downarrow{\scriptstyle\beta} & & \downarrow & & \\
0 & \longrightarrow & \ker\theta \otimes_R K & \longrightarrow & K^{(n)} & \xrightarrow{\theta\otimes 1} & M \otimes_R K & \longrightarrow & 0
\end{array}$$

commutes, where the second row is obtained by tensoring the top row with $(\ ) \otimes_R K$. Since the top row is exact, by Lemma 2.1.1 so is the second row. Since R is an integral domain, $R \to K$ is one-to-one. Therefore β is one-to-one. Since $M \otimes K$ has dimension n and $\theta \otimes 1$ is onto, it follows that $\ker\theta \otimes_R K = 0$. The Snake Lemma implies that $\ker\theta = 0$. We have shown that every maximal ideal $\mathfrak{m} \in \operatorname{Max}(R)$ has a basic open neighborhood $U(\alpha)$ such that M_α is a free R_α-module of rank n. The argument that was used to show (4) implies (2) in Proposition 2.6.4 can now be applied to finish the proof. $\square$

Corollary 2.6.6. *Let R be an integral domain with quotient field K and M a finitely generated R-module. Then the following are equivalent.*

(1) *M is of finite presentation and flat.*

(2) *M is an R-progenerator.*

(3) *There exists $n > 0$ such that $\dim_K(M \otimes_R K) = n$ and for every maximal ideal $\mathfrak{m}$ in $\operatorname{Max}(R)$, $\dim_{k_{\mathfrak{m}}}(M/\mathfrak{m}M) = n$.*

Proof. By Corollaries 1.3.5 and 1.1.16, (1) and (2) are equivalent. Proposition 2.6.4, Corollary 2.6.5, and Corollary 1.1.16 imply that (2) and (3) are equivalent. □

6.3. Invertible Modules and the Picard Group. In this section we show that to any commutative ring R is associated an abelian group $\operatorname{Pic}(R)$, which is called the Picard group of R. The Picard group of R parametrizes up to isomorphism the R-progenerator modules of constant rank one. The assignment $R \mapsto \operatorname{Pic}(R)$ defines a covariant functor from the category of commutative rings to the category of abelian groups. In Section 6.4 we will see that the Picard group of a noetherian integrally closed integral domain can be identified with a subgroup of the divisor class group.

Lemma 2.6.7. *Let M be a finitely generated projective faithful module over the commutative ring R. Then the following are equivalent:*

(1) $\operatorname{Rank}_R(M) = 1$.

(2) $\operatorname{Rank}_R(M^*) = 1$.

(3) $\operatorname{Hom}_R(M, M) \cong R$.

(4) $M^* \otimes_R M \cong R$.

(5) *For some R-module N, there is an isomorphism $M \otimes_R N \cong R$.*

Proof. The hypotheses on M imply that M is an R-progenerator module.

(1) is equivalent to (2): By Corollary 1.3.27, for each prime ideal $P \in \operatorname{Spec} R$, $M^* \otimes_R R_P = \operatorname{Hom}_R(M, R) \otimes_R R_P \cong \operatorname{Hom}_{R_P}(M_P, R_P)$. Since M is finitely generated projective, M_P is free of finite rank by Proposition 2.3.2. Assume $M_P \cong R_P^{(n)}$. Then $(M_P)^* = \operatorname{Hom}_{R_P}(M_P, R_P) \cong \operatorname{Hom}_{R_P}(R_P^{(n)}, R_P) \cong \operatorname{Hom}_{R_P}(R_P, R_P)^{(n)} \cong R_P^{(n)}$. Therefore $\operatorname{Rank}_R(M) = 1$ if and only if $\operatorname{Rank}_R(M^*) = 1$.

(1) implies (3): Let $\phi : R \to \operatorname{Hom}_R(M, M)$ be the left regular representation of R in $\operatorname{Hom}_R(M, M)$. Then $\phi(r)(x) = rx$. For each prime ideal $P \in \operatorname{Spec} R$, $M_P \cong R_P$. Therefore localizing ϕ at P yields

$$R_P \xrightarrow{\phi_P} \operatorname{Hom}_{R_P}(M_P, M_P) \cong \operatorname{Hom}_{R_P}(R_P, R_P) \cong R_P$$

which is an isomorphism. By Exercise 2.4.12, ϕ is an isomorphism.

(3) is equivalent to (4): It follows from Lemma 1.5.1 (1) that $M^* \otimes_R M$ is isomorphic to $\operatorname{Hom}_R(M, M)$.

(4) implies (5): If we take N to be M^*, then this is immediate.

(5) implies (1): Fix a prime $P \in \operatorname{Spec} R$. Then $M_P \cong R_P^{(n)}$ for some n. Now $R_P \cong R_P \otimes_R (M \otimes_R N) \cong M_P \otimes_R N \cong R_P^{(n)} \otimes_R N \cong N_P^{(n)}$. Therefore, N_P is finitely generated projective over the local ring R_P, hence is free of finite rank over R_P. Then $n = 1$. Since P was arbitrary, we are done. □

Definition 2.6.8. If M is an R-module that satisfies any of the equivalent properties of Lemma 2.6.7, then we say M is *invertible*. Given a commutative ring R let $\operatorname{Pic}(R)$ be the set of isomorphism classes of invertible R-modules. The isomorphism class containing a module M is denoted by $|M|$. As stated in Proposition 2.6.9 below, $\operatorname{Pic}(R)$ is an abelian group, which is called the *Picard group* of R. A generalization of the Picard group for a noncommutative R-algebra A is defined in Section 7.8.1. When R is an integral domain with field of fractions K, a description of $\operatorname{Pic}(R)$ in terms of invertible fractional ideals of R in K is made in Exercise 6.4.15.

Proposition 2.6.9. *Under the binary operation* $|P| \cdot |Q| = |P \otimes_R Q|$, $\operatorname{Pic}(R)$ *is an abelian group. The identity element is the class* $|R|$. *The inverse of* $|M| \in \operatorname{Pic}(R)$ *is* $|M^*|$. *The assignment* $R \mapsto \operatorname{Pic}(R)$ *defines a covariant functor from the category of commutative rings to the category of abelian groups.*

Proof. The proof is left to the reader. □

Example 2.6.10. Let k be any field. Let x and y be indeterminates. Let f be the polynomial $f = y^2 - x(x^2 - 1)$. Let R be the factor ring

$$R = \frac{k[x, y]}{(y^2 - x(x^2 - 1))}.$$

Then R is an integral domain. Let M be the maximal ideal of R generated by x and y. If we invert x^2-1, then $x = y^2(x^2-1)^{-1}$, so M becomes principal. If we invert x, then M becomes the unit ideal, and is principal. Since $R(x^2-1)$ and $R(x)$ are comaximal, there is an open cover $U(x^2-1) \cup U(x) = \operatorname{Spec} R$ on which M is locally free of rank 1. Proposition 2.6.4 shows that $|M| \in \operatorname{Pic} R$. Note that M^2 is generated by x^2, xy, y^2. But an ideal that contains x^2 and y^2 also contains x. We see that M^2 is generated by x, hence is free of rank one. The map

$$M \otimes_R M \to M^2$$
$$a \otimes b \mapsto ab$$

is R-linear. Since this map is onto and both sides are projective of rank one, it is an isomorphism. This proves that $M^* \cong M$ and $|M|^{-1} = |M|$.

Example 2.6.11. If R is a commutative ring with the property that every progenerator module is free, then $\operatorname{Pic}(R)$ contains just one element, namely $|R|$. Using the notation of abelian groups, we usually write $\operatorname{Pic}(R) = (0)$ in this case, despite the obvious inconsistency. For example, $\operatorname{Pic}(R) = (0)$ in each of the following cases.

(1) R is a field.

(2) R is a local ring (Proposition 2.3.2).

(3) R is a principal ideal domain.

We saw in Example 2.6.11 that the Picard group of a principal ideal domain is trivial. More generally, we will see in Chapter 6 that if R is a Dedekind domain with field of fractions K (see Section 6.4), then the Picard group $\operatorname{Pic}(R)$ is isomorphic to the quotient $\operatorname{Frac}(R)/\operatorname{Prin}(R)$, where $\operatorname{Frac}(R)$ denotes the group of all fractional ideals of R in K, and $\operatorname{Prin}(R)$ is the group of all principal R-submodules of K. In [**Cla66**], L. Claborn proved that given any abelian group G, there is a Dedekind domain R such that $\operatorname{Pic}(R) \cong G$ (see also [**Fos73**, Theorem 14.10]). Included throughout the following chapters are more examples of rings for which the Picard group can be completely determined. See Exercise 2.6.18 for one such example.

Before ending this section, we include for reference the definition of an operator that will be needed for later chapters, as well as in Exercise 2.6.17 below. Let R be a ring and M a left R-module. If I and J are submodules of M, then the *module quotient* is

$$I : J = \{r \in R \mid rJ \subseteq I\}.$$

The reader should verify that

$$I : J = \operatorname{annih}_R((I + J)/I) = \operatorname{annih}_R(J/(I \cap J))$$

hence $I : J$ is a two-sided ideal in R. The reader should be advised that the colon notation does not specify the ring R or the module M, hence is ambiguous. For example, if A is a commutative ring and R a subring of A, then the *conductor* from A to R is

$$R : A = \{\alpha \in A \mid \alpha A \subseteq R\}.$$

The reader should verify that $R : A$ is an A-submodule of R, hence it is an ideal of both R and A. The following technical lemma will be required in Chapter 6. Its proof can be found, for example, in [**Mat80**].

Lemma 2.6.12. *Let S be a commutative flat R-algebra. If I and J are ideals in R, then*

(1) $(I \cap J)S = IS \cap JS$.

(2) *If J is finitely generated, then* $(I : J)S = (IS : JS)$.

6.4. Exercises.

Exercise 2.6.13. Let R_1 and R_2 be rings and let $S = R_1 \oplus R_2$ be the direct sum. Let M_1 be an R_1-module and M_2 an R_2-module and let $M = M_1 \oplus M_2$. Prove:

(1) M is an S-module.

(2) If M_i is free of rank N over R_i for each i, then M is free of rank N over S.

(3) If M_i is finitely generated and projective over R_i for each i, then M is finitely generated and projective over S.

Exercise 2.6.14. Let R_1 and R_2 be commutative rings. Show that $\operatorname{Pic}(R_1 \oplus R_2)$ is isomorphic to $\operatorname{Pic}(R_1) \oplus \operatorname{Pic}(R_2)$.

Exercise 2.6.15. Let R be a commutative ring. Assume $f_1, \dots, f_n$ are elements of R that generate the unit ideal. That is, $R = Rf_1 + \cdots + Rf_n$. Let $S = R_{f_1} \oplus \cdots \oplus R_{f_n}$ be the direct sum. Let $\theta : R \to S$ be defined by $\theta(x) = (x/1, \dots, x/1)$. Show that S is a faithfully flat R-module.

Exercise 2.6.16. Let R be a commutative ring and M a finitely generated projective R-module of constant rank n. Show that there exist elements $f_1, \dots, f_m$ of R satisfying the following:

(1) $R = Rf_1 + \cdots + Rf_m$.

(2) If $S = R_{f_1} \oplus \cdots \oplus R_{f_m}$, then $M \otimes_R S$ is a free S-module of rank n.

Exercise 2.6.17. Let k be a field and $A = k[x]$ the polynomial ring over k in one variable. Let $R = k[x^2, x^3]$ be the k-subalgebra of A generated by x^2 and x^3. Show:

(1) R and A have the same quotient field, namely $K = k(x)$.

(2) A is a finitely generated R-module.

(3) The conductor ideal from A to R is $\mathfrak{m} = (x^2, x^3)$ which is a maximal ideal of R.

(4) Use Corollary 2.6.6 to show that A is not flat over R. (Hint: Consider $R/\mathfrak{m}$ and $A/\mathfrak{m}A$.)

(5) The rings $R[x^{-2}]$ and $A[x^{-2}]$ are equal, hence the extension $R \to A$ is flat upon localization to the nonempty basic open set $U(x^2)$.

For more properties of the rings R and A, see Exercises 2.6.18 and 8.1.35.

Exercise 2.6.18. This exercise is a continuation of Exercise 2.6.17. Let k be a field, $A = k[x]$ and $R = k[x^2, x^3]$. Show:

(1) For each $\alpha \in k$, set $P_\alpha = R(1 - \alpha x) + \mathfrak{m}$. Then P_α is an R-submodule of A, and P_α is isomorphic to R if and only if $\alpha = 0$.

(2) The R-module homomorphisms $P_\alpha \otimes_R P_\beta \to P_\alpha P_\beta \to P_{\alpha+\beta}$ are isomorphisms. (Hints: $x^4 \in \mathfrak{m}^2$, $x^3 \in P_\alpha \mathfrak{m}$, $x^2 \in P_\alpha \mathfrak{m}$, $1-(\alpha+\beta)x \in P_\alpha P_\beta$.)

(3) $\operatorname{Pic} R$ contains a subgroup isomorphic to the additive group k.

(4) $\operatorname{Pic} R \cong k$. (This is a challenge and may involve tools not yet covered in this text. See Exercise 14.2.19 for a method involving the Mayer-Vietoris sequence.)

(5) A is equal to the integral closure of R in K. (As in (4), this is a challenge. You can attempt to do it now, or come back to it later. See Section 3.3.)

Exercise 2.6.19. Let k be a field, $n > 1$ an integer, $T = k[x, y]$, $S = k[x^n, xy, y^n]$, and $S \to T$ the set containment map. Using Corollary 2.6.6 and Exercise 1.3.16, show that T is not flat over S. See Exercise 4.4.19 for more properties of the extension T/S.

Chapter 3

The Wedderburn-Artin Theorem

This chapter, like Chapters 1 and 2, contains material of a preliminary nature that will be applied in subsequent chapters. Proofs that are readily available in the standard references are usually omitted. For brevity, a general form of a theorem may be stated in lieu of several special cases. Some readers will probably prefer to make a quick scan through this chapter on the first pass, and return later as needed. The main subject of this chapter is the Wedderburn-Artin Theorem which classifies simple rings as matrix rings over division rings. This important class of rings will reappear in Chapter 4 where it is shown that a simple ring is a separable algebra over its center. Also included in this chapter are sections on integrality and completion.

1. The Jacobson Radical and Nakayama's Lemma

Let R be any ring and M a left R-module. If N is a submodule of M, then N is called *maximal* in case $N \neq M$ and whenever there is a submodule P such that $N \subseteq P \subseteq M$, then $N = P$ or $P = M$. If $N \subseteq M$ is a maximal submodule of M, then N/M is simple. The *Jacobson radical* of M is

$$\begin{aligned} \mathrm{J}(M) &= \bigcap\{N \mid N \text{ is a maximal submodule of } M\} \\ &= \bigcap\{\ker f \mid f \in \operatorname{Hom}_R(M, S) \text{ and } S \text{ is simple}\}. \end{aligned}$$

By $\mathrm{J}(R)$ we denote the Jacobson radical of R viewed as a left R-module. Then $\mathrm{J}(R)$ is equal to the intersection of all maximal left ideals of R.

Lemma 3.1.1. *$\mathrm{J}(R)$ is a two-sided ideal of R.*

Proof. For any R-module M, let $g \in \operatorname{Hom}_R(M, M)$, let S be any simple R-module and let $f \in \operatorname{Hom}_R(M, S)$. Then $f \circ g \in \operatorname{Hom}_R(M, S)$ so $\mathrm{J}(M) \subseteq \ker(f \circ g)$. Then $f(g(\mathrm{J}(M))) = 0$ for all f. That is, $g(\mathrm{J}(M)) \subseteq \mathrm{J}(M)$. Given $r \in R$, let ρ_r be the homomorphism in $\operatorname{Hom}_R(R, R)$ defined by "right multiplication by r". Then $\rho_r(\mathrm{J}(R)) = \mathrm{J}(R) \cdot r \subseteq \mathrm{J}(R)$. □

We now prove the counterpart of Lemma 1.1.13 and its corollaries for rings that are not necessarily commutative.

Theorem 3.1.2 (Nakayama's Lemma). *Let R be any ring and I a left ideal of R. The following are equivalent.*

(1) $I \subseteq \mathrm{J}(R)$.

(2) $1 + I = \{1 + x \mid x \in I\} \subseteq \operatorname{Units}(R)$.

(3) *If M is a finitely generated left R-module and $IM = M$, then $M = 0$.*

(4) *If M is a finitely generated left R-module and N is a submodule of M and $IM + N = M$, then $N = M$.*

Proof. (1) implies (2): Let $x \in I$. Assume $1 + x$ has no left inverse. Then $R(1 + x) \neq R$. By Zorn's Lemma, $R(1 + x)$ is contained in some maximal left ideal L of R. Then $1 + x = y \in L$. But $I \subseteq \mathrm{J}(R) \subseteq L$. So $x \in L$. Therefore $1 = y - x \in L$. This contradiction means there exists $u \in R$ such that $u(1 + x) = 1$. We show u has a left inverse. Since $1 = u + ux$, $u = 1 - ux = 1 + (-u)x \in 1 + I$ and by the previous argument, u has a left inverse. Then $u \in \operatorname{Units}(R)$ and $1 + x = u^{-1}$.

(2) implies (1): If L is a maximal left ideal and L does not contain I, then $I + L = R$, so $1 = x + y$ for some $x \in I$ and $y \in L$. Hence $y = 1 - x = 1 + (-x) \in 1 + I \subseteq \operatorname{Units}(R)$, a contradiction.

(1) plus (2) implies (3): Assume $IM = M$ and prove that $M = 0$. Now $I \subseteq \mathrm{J}(R)$ and $IM = M$ implies $\mathrm{J}(R)M \subseteq M = IM \subseteq \mathrm{J}(R)M$. Therefore $\mathrm{J}(R)M = M$. Assume $M \neq 0$. Pick a generating set $\{x_1, \ldots, x_n\}$ for M with $n \geq 1$ minimal. A typical element of M looks like $\sum_{i=1}^n r_i x_i$, $r_i \in R$. A typical element of $\mathrm{J}(R)M$ looks like $\sum_{i=1}^n a_i r_i x_i$, $a_i \in \mathrm{J}(R)$. By Lemma 3.1.1, $b_i = a_i r_i \in \mathrm{J}(R)$, so each element of $\mathrm{J}(R)M$ can be written in the form $\sum_{i=1}^n b_i x_i$, $b_i \in \mathrm{J}(R)$. In particular, $x_1 = \sum_{i=1}^n b_i x_i$, some $b_i \in \mathrm{J}(R)$. Then $x_1(1 - b_1) = \sum_{i=2}^n b_i x_i$. Now $1 - b_1 \in 1 + I$, so $1 - b_1$ is a unit. This shows that M is generated by $x_2, \ldots, x_n$. This contradiction implies $M = 0$.

(3) implies (4): Since M is finitely generated so is M/N. Then

$$I(M/N) = \frac{IM + N}{N} = M/N$$

and by (3) we conclude that $M/N = 0$, or $N = M$.

(4) implies (1): If L is a maximal left ideal of R and L does not contain I, then $I + L = R$. We now apply (4) with $L = N$, $R = M$. Since $IR \supseteq I$ we have $IR + L = R$. Thus $L = R$, which is a contradiction. $\square$

If R is a commutative ring, recall that the nil radical of R, $\operatorname{Rad}(R)$, is equal to the intersection of all prime ideals $\mathfrak{p}$ in $\operatorname{Spec} R$ (Lemma 2.2.2). Since $\operatorname{Max}(R) \subseteq \operatorname{Spec} R$, we have the set containment

$$\bigcap_{\mathfrak{p}\in\operatorname{Spec} R} \mathfrak{p} \subseteq \bigcap_{\mathfrak{m}\in\operatorname{Max} R} \mathfrak{m}.$$

Hence $\operatorname{Rad}(R) \subseteq \operatorname{J}(R)$. In general, we have the following.

Corollary 3.1.3. *If I is a left ideal of R which consists of nilpotent elements, then $I \subseteq \operatorname{J}(R)$.*

Proof. Let $a \in I$ and assume $a^n = 0$ for some $n \geq 1$. Then

$$(1 - a)(1 + a + a^2 + \cdots + a^{n-1}) = 1.$$

So $1 + I \subseteq \operatorname{Units}(R)$. $\square$

Corollary 3.1.4. *If R is artinian, then $\operatorname{J}(R)$ is nilpotent.*

Proof. Consider the chain of left ideals

$$\operatorname{J}(R) \supseteq \operatorname{J}(R)^2 \supseteq \operatorname{J}(R)^3 \supseteq \dots.$$

Since R is artinian, the minimum condition is satisfied on left ideals. There is some $n \geq 1$ such that $\operatorname{J}(R)^n = \operatorname{J}(R)^{n+1}$. Assume $\operatorname{J}(R)^n \neq 0$. Consider the set $\mathcal{L}$ of all finitely generated left ideals L such that $\operatorname{J}(R)^n L \neq 0$. Since $\operatorname{J}(R)^n = \operatorname{J}(R)^n \operatorname{J}(R) \neq 0$, there exists $a \in \operatorname{J}(R)$ such that $\operatorname{J}(R)^n Ra \neq 0$. Since $Ra \in \mathcal{L}$, the set is nonempty. Pick a minimal element L of $\mathcal{L}$. Now $\operatorname{J}(R)^n L \subseteq L$. Since $L \neq 0$, Theorem 3.1.2 (3) says $\operatorname{J}(R)^n L$ is a proper subset of L. But $\operatorname{J}(R)^n (\operatorname{J}(R)^n L) = \operatorname{J}(R)^{2n} L = \operatorname{J}(R)^n L \neq 0$. There exists $a \in \operatorname{J}(R)^n L$ such that $\operatorname{J}(R)^n Ra \neq 0$. So $Ra \in \mathcal{L}$. But $Ra \subseteq \operatorname{J}(R)^n L \subsetneq L$. This is a contradiction, because L is minimal. We conclude $\operatorname{J}(R)^n = 0$. $\square$

Corollary 3.1.5. *Let R be a ring.*

(1) $\operatorname{J}(R) = \bigcap\{I \mid I$ *is a maximal right ideal of* $R\}$.
(2) *If M is a maximal two-sided ideal of R, then $\operatorname{J}(R) \subseteq M$.*
(3) *If $f :\cdot R \to S$ is an epimorphism of rings, then $f(\operatorname{J}(R)) \subseteq \operatorname{J}(S)$.*
(4) *If R is commutative and A is an R-algebra which is finitely generated as an R-module, then $\operatorname{J}(R)A \subseteq \operatorname{J}(A)$.*

Proof. (1): Let $\operatorname{J}_r(R)$ denote the intersection of all maximal right ideals in R. By Lemma 3.1.1, both $\operatorname{J}_r(R)$ and $\operatorname{J}(R)$ are two-sided ideals of R. It follows from Theorem 3.1.2 (2) that $1+\operatorname{J}(R)$ consists of units of R. Applying a right hand version of Theorem 3.1.2 to the right ideal $\mathrm{J}(R)$, we conclude that $\mathrm{J}(R) \subseteq \mathrm{J}_r(R)$. The converse follows by symmetry.

(2): Assume the contrary. The ideal $\mathrm{J}(R) + M$ is a two-sided ideal of R. Since M is maximal, $\mathrm{J}(R) + M = R$. By Theorem 3.1.2 (4), $M = R$, a contradiction.

(3): Let $x \in \mathrm{J}(R)$ and $a \in R$. By Theorem 3.1.2, $1 + ax \in \mathrm{Units}(R)$, so $f(1 + ax) = 1 + f(a)f(x) \in \mathrm{Units}(S)$. Therefore the left ideal $Sf(x)$ is contained in $\mathrm{J}(S)$.

(4): Let M be a finitely generated left A-module. Then M is finitely generated as an R-module. If $(\mathrm{J}(R)A)M = M$, then $\mathrm{J}(R)(AM) = \mathrm{J}(R)M = M$. By (1) implies (3) of Theorem 3.1.2, $M = 0$. By (3) implies (1) of Theorem 3.1.2, $\mathrm{J}(R)A \subseteq \mathrm{J}(A)$. □

1.1. Exercises.

Exercise 3.1.6. Let R be a ring, I an ideal contained in $\mathrm{J}(R)$, and $\eta : R \to R/I$ the natural map. Prove:

(1) If $\eta(r)$ is a unit in R/I, then r is a unit in R.
(2) The natural map $\eta : \mathrm{Units}(R) \to \mathrm{Units}(R/I)$ is onto and the kernel is $1 + I$.

Exercise 3.1.7. Let $R = R_1 \oplus \cdots \oplus R_n$ be a direct sum, where each R_i is a commutative local ring. Show that if M is a finitely generated projective R-module of constant rank r, then M is a free R-module of rank r.

Exercise 3.1.8. A *semilocal ring* is a commutative ring R that has only a finite number of maximal ideals. Let R be a semilocal ring. Prove:

(1) $R/J(R)$ is isomorphic to a finite direct sum of fields.
(2) If M is a finitely generated projective R-module of constant rank r, then M is a free R-module of rank r. (Hint: Mimic the proof of Proposition 2.3.2.)

Exercise 3.1.9. Let R be a ring. Prove that $\mathrm{J}(M_n(R)) = M_n(\mathrm{J}(R))$. (Hint: First show that if S is a simple left R-module, then S^n is a simple left $M_n(R)$-module.)

2. Semisimple Modules and Semisimple Rings

Let R be a ring and M an R-module. If M satisfies any of the equivalent properties of Theorem 3.2.1, then M is said to be a *semisimple* R-module. The ring R is said to be *semisimple* if R is a semisimple left R-module. More necessary and sufficient conditions for a ring to be semisimple are provided by Theorem 3.2.2.

Theorem 3.2.1. *Let R be a ring and M a nonzero R-module. The following are equivalent.*

(1) $M = \bigoplus_{i\in I} M_i$ *is the internal direct sum of a family of simple submodules* $\{M_i \mid i \in I\}$.

(2) $M = \sum_{i\in I} M_i$ *is the sum of a family of simple submodules* $\{M_i \mid i \in I\}$.

(3) *Every submodule of M is a direct summand of M.*

Proof. It is immediate that (1) implies (2).

(2) implies (1): Assume $M = \sum_{i\in I} M_i$ and each M_i is a simple submodule of M. By Zorn's Lemma, choose a maximal subset $J \subseteq I$ such that the sum $\sum_{i\in J} M_i$ is a direct sum. Assume $\sum_{i\in J} M_i \neq M$. Then there is some $k \in I$ such that M_k is not contained in $\sum_{i\in J} M_i$. Since M_k is simple,

$$\Big(\sum_{i\in J} M_i\Big) \bigcap M_k = 0.$$

In this case, the sum $\left(\sum_{i\in J} M_i\right) + M_k$ is a direct sum which contradicts the choice of J.

(1) plus (2) implies (3): Then M is an internal direct sum of simple submodules $\{M_i \mid i \in I\}$. Let N be a submodule of M. If $N = M$, then we are done. Assume $N \neq M$. For each $i \in I$, $M_i \cap N$ is a submodule of M_i hence $M_i \cap N = 0$ or $M_i \cap N = M_i$. Then for some $k \in I$ we have $M_k \cap N = 0$. Choose a maximal subset $J \subseteq I$ such that

$$\Big(\sum_{i\in J} M_i\Big) \bigcap N = 0. \tag{3.1}$$

Let

$$N' = \Big(\sum_{i\in J} M_i\Big) + N.$$

If $N' = M$, then $M = \left(\sum_{i\in J} M_i\right) \oplus N$ and we are done. Otherwise for some index $k \in I$, $M_k \cap N' = 0$. Consider

$$x \in \Big(\sum_{i\in J} M_i + M_k\Big) \cap N.$$

Write $x = x_0 + x_k$ where $x_0 \in \sum_{i\in J} M_i$ and $x_k \in M_k$. So $x_k = x - x_0 \in N' \cap M_k = 0$. By (3.1) we see that $x = 0$. Then $J \cup \{k\}$ satisfies (3.1) which contradicts the choice of J.

(3) implies (2): Let $\{M_i \mid i \in I\}$ be the family of all simple submodules of M. Set $N = \sum_i M_i$. Assume $N \neq M$. By (3), $M = N \oplus N'$ for some nonzero submodule N'. To finish the proof, it is enough to show the existence of a simple submodule of N'. Let $x \in N' - (0)$. Being a direct summand of M, N' satisfies (3) (the reader should verify this). Therefore

$N' = Rx \oplus N''$. Let L be a maximal left ideal of R such that L contains $\operatorname{annih}_R(x)$. Then R/L is simple. The diagram

$$\begin{array}{ccccccccc} 0 & \longrightarrow & L & \longrightarrow & R & \longrightarrow & R/L & \longrightarrow & 0 \\ & & \downarrow{\scriptstyle \alpha} & & \downarrow{\scriptstyle \beta} & & \downarrow{\scriptstyle \eta} & & \\ 0 & \longrightarrow & Lx & \longrightarrow & Rx & \longrightarrow & Rx/Lx & \longrightarrow & 0 \end{array}$$

commutes. The rows are exact. The vertical maps α and β are onto, therefore η is onto. Since $x \notin Lx$, we know Rx/Lx is not zero. Then η is not the zero map. Since R/L is simple, η is an isomorphism. Applying (3) to Rx gives $Rx = Lx \oplus S$ where $S \cong Rx/Lx$ is a simple R-submodule of Rx. But then N' contains S, so we are done. □

Theorem 3.2.2. *Let R be a ring. The following conditions are equivalent.*

(1) *Every left R-module is projective.*
(2) *Every short exact sequence of left R-modules splits.*
(3) *Every left R-module is semisimple.*
(4) *R is a semisimple ring. That is, R is semisimple when viewed as a left R-module.*
(5) *R is artinian and $\mathrm{J}(R) = 0$.*

Proof. The reader should verify that (3) implies (4) and that the first three statements are pairwise equivalent.

(4) implies (1): Let M be a left R-module. Let $I = M$ and $F = R^I$. As in the proof of Proposition 1.1.5, there is an R-module homomorphism $\pi : F \to M$ which is surjective. Because R is semisimple, R is the internal direct sum of simple R-submodules. So F is an internal direct sum of simple R-modules. So F is semisimple and $\ker \pi$ is a direct summand of F. Then $F \cong \ker \pi \oplus M$, hence M is projective.

(4) implies (5): Since $\mathrm{J}(R)$ is a submodule of R, it is an internal direct summand of R. For some left ideal L we have $R = \mathrm{J}(R) \oplus L$. By Lemma 1.1.20, $\mathrm{J}(R) = Re_1$ and $L = Re_2$ and $e_1e_2 = 0$ and $1 = e_1 + e_2$. By Theorem 3.1.2, e_2 is a unit in R. Therefore $e_1 = 0$ and $\mathrm{J}(R) = 0$. To show that R is artinian, assume $I_1 \supseteq I_2 \supseteq I_3 \dots$ is a descending chain of ideals. Since R is semisimple as an R-module, I_1 is a direct summand of R, and we can write $R = L_0 \oplus I_1$. Also, I_2 is a direct summand of I_1, so $R = L_0 \oplus L_1 \oplus I_2$. For each index i, I_{i+1} is a direct summand of I_i and we can write $I_i = L_i \oplus I_{i+1}$. Each $L_i = Re_i$ for some idempotent e_i and $\bigoplus_{i=1}^{\infty} L_i$ is a direct summand of R. That is,

$$R = \left(\bigoplus_{i=1}^{\infty} L_i \right) \oplus L$$

for some L. The representation of 1 in the direct sum involves only a finite number of the e_i, and the rest are 0.

(5) implies (4): We show that R is the direct sum of a finite collection of minimal left ideals and apply Theorem 3.2.1 (1). Let L_1 be a minimal left ideal of R. This exists since R is artinian. Since $\mathrm{J}(R) = 0$ it follows from Corollary 3.1.3 that $L_1^2 \neq 0$. By Lemma 1.1.20 (3), there is a left ideal I_1 and $R = L_1 \oplus I_1$. If $I_1 = 0$, then we are done. Otherwise, by the minimum condition, there is a minimal left ideal L_2 of R contained in I_1. Again from Lemma 1.1.20 we have $R = L_2 \oplus J$ for some J. There exists an R-module homomorphism $\pi : R \to L_2$ which splits $L_2 \subseteq R$. The restriction of π to I_1 is therefore a left inverse of $L_2 \subseteq I_1$. Therefore, $I_1 = L_2 \oplus I_2$, where $I_2 = \{x \in I_1 \mid \pi(x) = 0\} = I_1 \cap \ker \pi$. Hence $R = L_1 \oplus L_2 \oplus I_2$ where L_1, L_2 are minimal ideals in R. If $I_2 = 0$, then we are done. Otherwise we continue inductively to get $R = L_1 \oplus \cdots \oplus L_n \oplus I_n$ where each L_i is a minimal left ideal. After a finite number of iterations, the process terminates with $I_n = 0$ because R is artinian and $I_1 \supseteq I_2 \supseteq \cdots \supseteq I_n$ is a descending chain of ideals. □

Example 3.2.3. Let R be an artinian ring. Then $R/\mathrm{J}(R)$ satisfies Theorem 3.2.2 (5), hence is semisimple.

2.1. Simple Rings and the Wedderburn-Artin Theorem. A ring R is called *simple* if R is artinian and the only two-sided ideals of R are 0 and R. Since $\mathrm{J}(R)$ is a two-sided ideal, a simple ring satisfies Theorem 3.2.2 (5) hence is semisimple.

Example 3.2.4. Let D be a division ring and M a finite dimensional D-vector space. Let $S = \mathrm{Hom}_D(M, M)$. By Exercise 2.5.14, S is artinian. By Corollary 1.5.4 it follows that there is a one-to-one correspondence between two-sided ideals of D and two-sided ideals of S. Since D is a simple ring, it follows that S is a simple ring. We prove the converse of this fact in Theorem 3.2.6.

Theorem 3.2.5. *Let A be an artinian ring and let R be a semisimple ring.*

(1) *Every simple left R-module is isomorphic to a minimal left ideal of R.*
(2) *R is a finite direct sum of simple rings.*
(3) *R is simple if and only if all simple left R-modules are isomorphic.*
(4) *If A is simple, then every nonzero A-module is faithful.*
(5) *If there exists a simple faithful A-module, then A is simple.*

Proof. (1): Let R be a semisimple ring. By the proof of Theorem 3.2.2 there are idempotents $e_1, \dots, e_n$ such that each Re_i is a minimal left ideal of R and $R = Re_1 \oplus \cdots \oplus Re_n$. Let S be any simple left R-module. Let x be a nonzero element of S. Then for some e_i we have $e_i x \neq 0$. The R-module homomorphism $Re_i \to S$ defined by $re_i \mapsto re_i x$ is an isomorphism because both modules are simple. This proves (1).

(2): Let $S_1, \dots, S_m$ denote representatives for the distinct isomorphism classes of simple left R-modules. By (1) there are only finitely many such isomorphism classes. For each i, define

$$R_i = \sum_j \{L_{ij} \mid L_{ij} \text{ is a left ideal of } R \text{ and } L_{ij} \cong S_i\}.$$

We proceed in four steps to show that $R = R_1 \oplus \cdots \oplus R_m$ and each R_i is a simple ring.

Step 1: R_i is a two-sided ideal. By definition, R_i is a left ideal of R. Pick any L_{ij}. Let $r \in R$ and consider the R-module homomorphism $\rho_r : L_{ij} \to R$ which is "right multiplication by r". Since L_{ij} is simple, either $\ker \rho_r = L_{ij}$ and $L_{ij} r \subseteq L_{ij}$, or $\ker \rho_r = 0$ and $L_{ij} \cong L_{ij} r$. In the latter case, the left ideal L_{ij} is isomorphic to some L_{ik}. In both cases, $L_{ij} r \subseteq R_i$ which shows $R_i r \subseteq R_i$ and R_i is a two-sided ideal of R.

Step 2: Let L be a minimal left ideal of R contained in R_i. We show that $L \cong S_i$. Since L is idempotent generated, there is some $e \in L$ such that $e^2 = e \neq 0$. Since $e \in L \subseteq R_i$, the R-module homomorphism $\rho_e : R_i \to L$ is nonzero. Since R_i is generated by the ideals L_{ij}, there is some j such that $L_{ij} e \neq 0$. The map $\rho_e : L_{ij} \to L$ is an isomorphism. Therefore $L \cong S_i$.

Step 3: $R = R_1 \oplus \cdots \oplus R_m$. Clearly $R = R_1 + \cdots + R_m$. For contradiction's sake, assume $R_1 \cap (R_2 + \cdots + R_m) \neq 0$. Let L be a minimal left ideal of R contained in $R_1 \cap (R_2 + \cdots + R_m)$. By Step 2, $L \cong S_1$. There is an idempotent e such that $L = Re$. As in Step 2, the map $\rho_e : R_2 + \cdots + R_m \to L$ is nonzero. Hence there exists L_{ik} such that $2 \leq i \leq m$ and $\rho_e : L_{ik} \to L$ is an isomorphism. This is a contradiction, since S_1 and S_i are not isomorphic. Therefore $R_1 \cap (R_2 + \cdots + R_m) = 0$. By induction on m, this step is done.

Step 4: Fix i and show that R_i is simple. By Theorem 3.2.2, R is artinian. Let I be a nonzero two-sided ideal in R_i. To show $I = R_i$, the plan is to show I contains each of the ideals L_{ij}. By Step 3, ideals of R_i are also ideals in R. In particular, I is a two-sided ideal in R. Let L be any minimal left ideal of R contained in I. By Step 2, $L = L_{ik}$ for some k. There exists an idempotent e such that $L_{ik} = Re$. Let L_{ij} be another minimal left ideal in R_i. There is an R-module isomorphism $\phi : I_{ik} \cong I_{ij}$. We have $L_{ij} = \operatorname{im} \phi = \{\phi(re) \mid r \in R\} = \{\phi(ree) \mid r \in R\} = \{re\phi(e) \mid r \in R\}$. Since e belongs to the two-sided ideal I, $L_{ij} \subseteq I$. Thus $I = R_i$.

(4): Assume A is simple. Let M be any nonzero left A-module. Let $I = \mathrm{annih}_A(M)$, a two-sided ideal of A. Since $1 \notin I$, it follows that $I \neq A$. Therefore $I = 0$ and M is faithful.

(3): By (2) we can write $R = R_1 \oplus \cdots \oplus R_m$ as a direct sum of simple rings. If all simple left R-modules are isomorphic, then $m = 1$ and R is simple. Now suppose R is simple and L is a simple left R-module. We know that $m = 1$, otherwise R_1 is a proper two-sided ideal. Then $L \cong L_{1j}$ for some j and all simple left R-modules are isomorphic.

(5): Assume A is artinian and S is a simple faithful left A-module. Since S is simple, $\mathrm{J}(A)S$ is either 0 or S. Since S is simple and faithful, S is nonzero and generated by one element. By Theorem 3.1.2 (3) we know $\mathrm{J}(A)S \neq S$. So $\mathrm{J}(A)S = 0$. Since S is faithful, $\mathrm{J}(A) = 0$. This proves A is semisimple. By (2) $A = A_1 \oplus \cdots \oplus A_n$ where each A_i is a two-sided ideal of A. Assume $n \geq 2$. By (1), we assume without loss of generality that $S \cong S_1$. Then $A_1 S = S$. Since the ideals are two-sided, $A_2 A_1 \subseteq A_1 \cap A_2 = 0$. Therefore $0 = (A_2 A_1)S = A_2(A_1 S) = A_2 S$. So $A_2 \subseteq \mathrm{annih}_A(S)$. This contradiction implies $n = 1$, and A is simple. □

Theorem 3.2.6 (Wedderburn-Artin). *Let R be a simple ring. Then R is isomorphic to the endomorphism ring $\mathrm{Hom}_D(M, M)$ for a finite dimensional vector space M over a division ring D. The division ring D and the dimension $\dim_D(M)$ are uniquely determined by R.*

Proof. Since R is semisimple we can write R as an internal direct sum $R = L_1 \oplus \cdots \oplus L_n$ where each L_i is a minimal left ideal of R. But R is simple, so $L_1 \cong \ldots \cong L_n$ by Theorem 3.2.5. Set $M = L_1$ and $D = \mathrm{Hom}_R(M, M)$. By Exercise 3.2.16, D is a division ring. Since $L_1 = Re_1$ for some idempotent e_1, M is finitely generated. By Theorem 3.2.2, M is projective. The trace ideal $\mathfrak{T}_R(M)$ of Eq. (1.2) is a two-sided ideal of R. Since R is simple, M is a generator over R. By Morita Theory, Corollary 1.5.3 (1), $R \cong \mathrm{Hom}_D(M, M)$. By Corollary 1.5.3 (5), M is a finitely generated D-vector space.

To prove the uniqueness claims, assume D' is another division ring and M' is a finite dimensional D'-vector space such that $\mathrm{Hom}_D(M, M)$ and $\mathrm{Hom}_{D'}(M', M')$ are isomorphic. By Morita Theory, $D' \cong \mathrm{Hom}_R(M', M')$ and M' is an R-progenerator. We know M' is a simple R-module, otherwise M' would have a nontrivial direct summand and $\mathrm{Hom}_R(M', M')$ would contain noninvertible elements. Since R is simple, by Theorem 3.2.5, $M \cong M'$ as R-modules. □

If R and D are as in Theorem 3.2.6, then we call D the *division ring component of R*.

2.2. Commutative Artinian Rings. We assemble in this section proofs of some important properties of artinian rings. Notice that in Theorem 3.2.7 we do not assume the ring is commutative.

Theorem 3.2.7. *Let R be an artinian ring and M an R-module. If M is artinian, then M is noetherian. In particular, R is a noetherian ring.*

Proof. Let $J = \mathrm{J}(R)$ denote the Jacobson radical of R. Then R/J is a semisimple ring, by Example 3.2.3. By Lemma 2.5.1, since M is artinian, so are the submodules J^nM and the quotient modules $J^nM/J^{n+1}M$, for all $n \geq 0$. Consequently, the quotient module $J^nM/J^{n+1}M$ is artinian over R/J. By Exercise 3.2.14, $J^nM/J^{n+1}M$ is noetherian as a R/J-module. Therefore, $J^nM/J^{n+1}M$ is noetherian as an R-module. For each $n \geq 0$, the sequence

$$0 \to J^{n+1}M \to J^nM \to \frac{J^nM}{J^{n+1}M} \to 0$$

is exact. By Corollary 3.1.4, for some r, we have $J^{r+1} = (0)$. Taking $n = r$ in the exact sequence, Lemma 2.5.1 implies J^rM is noetherian. A finite induction argument using Lemma 2.5.1 and the exact sequence proves J^nM is noetherian for $n = r, \dots, 1, 0$. □

Lemma 3.2.8. *Let R be a commutative noetherian local ring with maximal ideal $\mathfrak{m}$. If $\mathfrak{m}$ is the only prime ideal of R, then R is artinian.*

Proof. By Lemma 2.2.2, $I(V(0)) = \mathrm{Rad}_R(0) = \mathfrak{m}$. Therefore, $\mathfrak{m}^n = (0)$, for some $n \geq 1$. Consider the filtration

$$R \supseteq \mathfrak{m} \supseteq \mathfrak{m}^2 \supseteq \cdots \supseteq \mathfrak{m}^{n-1} \supseteq (0).$$

Each factor $\mathfrak{m}^i/\mathfrak{m}^{i+1}$ is finitely generated as an R-module, hence is finitely generated as a vector space over the field $R/\mathfrak{m}$. Since $\mathfrak{m}$ annihilates $\mathfrak{m}^i/\mathfrak{m}^{i+1}$, the R-submodules of $\mathfrak{m}^i/\mathfrak{m}^{i+1}$ correspond to $R/\mathfrak{m}$-subspaces. By Exercise 2.5.13, $\mathfrak{m}^i/\mathfrak{m}^{i+1}$ satisfies DCC as an $R/\mathfrak{m}$-vector space, hence as an R-module. In particular, $\mathfrak{m}^{n-1}$ satisfies DCC as an R-module. A finite induction argument using Lemma 2.5.1 and the exact sequences

$$0 \to \mathfrak{m}^{i+1} \to \mathfrak{m}^i \to \mathfrak{m}^i/\mathfrak{m}^{i+1} \to 0$$

shows that each R-module $\mathfrak{m}^i$ has the DCC on submodules. In particular, R is artinian. □

Proposition 3.2.9. *Let R be a commutative artinian ring. The following are true.*

(1) *Every prime ideal of R is maximal.*
(2) *The nil radical $\mathrm{Rad}_R(0)$ is equal to the Jacobson radical $\mathrm{J}(R)$.*

(3) *There are only finitely many maximal ideals in R, hence R is semilocal (Exercise 3.1.8).*

(4) *The nil radical* $\operatorname{Rad}_R(0)$ *is nilpotent.*

Proof. (1): Let P be a prime ideal in R. Then R/P is an artinian integral domain. By Exercise 2.5.9, R/P is a field.

(2): This part is left to the reader.

(3): Theorem 3.2.7 implies R is noetherian, and Proposition 2.5.4 implies $\operatorname{Spec} R$ has only a finite number of irreducible components. By Corollary 2.5.5, the irreducible components of $\operatorname{Spec} R$ correspond to the minimal primes of R. It follows from Part (1) that every prime ideal in R is minimal. Therefore, $\operatorname{Spec} R$ is finite.

(4): By Corollary 3.1.4, the Jacobson radical of an artinian ring is nilpotent. □

Proposition 3.2.10. *Let R be a commutative ring. The following are equivalent.*

(1) *R is artinian.*

(2) *R is noetherian and every prime ideal is maximal. In other words, R has Krull dimension zero.*

(3) *R is an R-module of finite length.*

Proof. By Proposition 2.5.7, it is enough to show (1) and (2) are equivalent.

(1) implies (2): By Theorem 3.2.7, R is noetherian. By Proposition 3.2.9, every prime ideal of R is maximal.

(2) implies (1): By Theorem 2.5.6, R has a decomposition $R = R_1 \oplus \cdots \oplus R_n$ where each R_i has only two idempotents. By Exercise 2.5.8 it suffices to show each R_i is artinian. Therefore, assume $\operatorname{Spec} R$ is connected. By Proposition 2.5.3, $\operatorname{Spec} R$ decomposes into a union of a finite number of irreducible closed subsets. Each prime ideal of R is maximal, so the irreducible components of $\operatorname{Spec} R$ are closed points. Since we are assuming $\operatorname{Spec} R$ is connected, this proves R is a local ring. By Lemma 3.2.8, R is artinian. □

Theorem 3.2.11. *Let R be a commutative artinian ring.*

(1) *$R = R_1 \oplus R_2 \oplus \cdots \oplus R_n$ where each R_i is a local artinian ring.*

(2) *The rings R_i in Part (1) are uniquely determined up to isomorphism.*

(3) *If $\mathfrak{m}_1, \dots \mathfrak{m}_n$ is the complete list of prime ideals in* $\operatorname{Spec} R$, *then the natural homomorphism $R \to R_{\mathfrak{m}_1} \oplus \cdots \oplus R_{\mathfrak{m}_n}$ is an isomorphism.*

Proof. (1): By Proposition 3.2.9, $\operatorname{Max} R = \operatorname{Spec} R = \{\mathfrak{m}_1, \dots, \mathfrak{m}_n\}$ is a finite set. So the topological space $\operatorname{Spec} R$ has the discrete topology. By Theorem 2.5.6, R can be written as a direct sum $R = R_1 \oplus \cdots \oplus R_r$ where $\operatorname{Spec} R_i$ is connected. Since the topology is discrete, this implies $\operatorname{Spec} R_i$ is a singleton set, hence R_i is a local ring. This also proves $n = r$.

(2): A local ring has only two idempotents, so this follows from Theorem 1.1.21.

(3): This follows from the decomposition $R \cong R_1 \oplus \cdots \oplus R_n$ of Part (1), by an application of Exercise 2.1.8. □

Corollary 3.2.12. *Let R be a commutative artinian ring. If M is a finitely generated projective R module of constant rank r, then M is a free R-module of rank r.*

Proof. By Theorem 3.2.11, R is the finite direct sum of local rings. By Exercise 3.1.7, M is a free module of rank r. □

Corollary 3.2.13. *Let R be a commutative ring and S a commutative R-algebra which is finitely generated and projective as an R-module. Let M be a finitely generated projective S-module. Let $\mathfrak{p}$ be a prime ideal in* $\operatorname{Spec} R$ *such that* $\operatorname{Rank}_{S_\mathfrak{p}}(M_\mathfrak{p}) = s$ *is defined. Then*

$$\operatorname{Rank}_{R_\mathfrak{p}}(M_\mathfrak{p}) = \operatorname{Rank}_{R_\mathfrak{p}}(S_\mathfrak{p}) \operatorname{Rank}_{S_\mathfrak{p}}(M_\mathfrak{p}).$$

Proof. Let $k = R_\mathfrak{p}/\mathfrak{p}R_\mathfrak{p}$ be the residue field of $R_\mathfrak{p}$. Then $S \otimes_R k$ is a finite dimensional k-algebra, hence is artinian. By Corollary 3.2.12, $M \otimes_R k = M \otimes_S (S \otimes_R k)$ is a free $S \otimes_R k$-module of constant rank s. Exercise 1.1.9 applies to the trio k, $S \otimes_R k$, $M \otimes_R k$. Applying Proposition 2.3.2 we get the rank formula over the local ring $R_\mathfrak{p}$. □

2.3. Exercises.

Exercise 3.2.14. Let R be a semisimple ring and M an R-module. Prove that M is artinian if and only if M is noetherian.

Exercise 3.2.15. Prove the converse of Theorem 3.2.5 (2). That is, a finite direct sum of simple rings is a semisimple ring.

Exercise 3.2.16. Prove the following form of Schur's Lemma: If R is a ring and M is a simple left R-module, then $S = \operatorname{Hom}_R(M, M)$ is a division ring.

Exercise 3.2.17. Let k be a field and R a commutative artinian finitely generated k-algebra. Prove that R is finite dimensional as a k-vector space. (Hint: Use Theorem 3.2.11 to reduce to the case where R is local artinian. Consider the chain $R \supseteq \mathfrak{m} \supseteq \mathfrak{m}^2 \supseteq \cdots \supseteq \mathfrak{m}^k \supseteq 0$. Show that each factor $\mathfrak{m}^i/\mathfrak{m}^{i+1}$ is a finitely generated vector space over k.)

Exercise 3.2.18. Let R be a commutative noetherian ring, $\mathfrak{m}$ a maximal ideal in R, and $n \geq 1$.

(1) Prove that $R/\mathfrak{m}^n$ is a local artinian ring.

(2) Prove that the natural map $R/\mathfrak{m}^n \to R_{\mathfrak{m}}/\mathfrak{m}^n R_{\mathfrak{m}}$ is an isomorphism.

3. Integral Extensions

General references for this section are [**AM69**] and [**Mat80**]. Let R be a commutative ring and A an R-algebra. An element $a \in A$ is said to be *integral* over R in case there exists a monic polynomial $p \in R[x]$ such that $p(a) = 0$. If every element of A is integral over R, then we say A is *integral* over R. The reader should verify that any homomorphic image of R is integral over R. The R-algebra A comes with a structure homomorphism $\theta : R \to Z(A)$. Assume θ is one-to-one, or equivalently, A is a faithful R-module. Then we identify R with $\theta(R)$ which is a subring of A. In this case, if every element of A is integral over R, we say A/R is an *integral extension.* If every element of A that is integral over R is an element of R, then we say R is *integrally closed* in A. An integral domain R is said to be integrally closed if R is integrally closed in its quotient field.

If A is an R-algebra which is R-faithful, and $a \in A$, then the R-subalgebra of A generated by a is denoted $R[a]$. Since $R \subseteq Z(A)$, $R[a]$ is commutative, and the substitution homomorphism $R[x] \to A$ defined by $x \mapsto a$ is an R-algebra homomorphism with image $R[a]$.

Proposition 3.3.1. *Let A be a faithful R-algebra, and $a \in A$. The following are equivalent.*

(1) *a is integral over R.*

(2) *$R[a]$ is a finitely generated R-module.*

(3) *There is an R-subalgebra B of A such that $R[a] \subseteq B \subseteq A$ and B is a finitely generated R-module.*

(4) *There exists a faithful $R[a]$-module which is finitely generated as an R-module.*

Lemma 3.3.2. *Let A be a faithful integral R-algebra and assume A has no zero divisors.*

(1) *If A is a division ring and C is a commutative R-subalgebra of A, then C is a field.*

(2) *If R is a field, then A is a division ring.*

Proof. Assume A is a division ring and $x \in C - (0)$. Then $x^{-1} \in A$ is integral over R. There exist $n \geq 1$ and $r_i \in R$ such that

$$x^{-n} + r_{n-1}x^{1-n} + \cdots + r_1 x^{-1} + r_0 = 0.$$

Multiply by x^{n-1} and get

$$x^{-1} + r_{n-1} + r_{n-2}x + \cdots + r_1 x^{n-2} + r_0 x^{n-1} = 0$$

which shows $x^{-1} \in C$. Thus C is a field.

Assume R is a field and $y \in A - (0)$. Then y is integral over R. There exist $n \geq 1$ and $r_i \in R$ such that

$$y^n + r_{n-1}y^{n-1} + \cdots + r_1 y + r_0 = 0.$$

If we choose the degree n to be minimal among all such dependence relations for y, then we can assume $r_0 \neq 0$. Since R is a field, divide this

$$r_0 = -y(y^{n-1} + r_{n-1}y^{n-2} + \cdots + r_1)$$

by r_0 to see that

$$y^{-1} = -r_0^{-1}(y^{n-1} + r_{n-1}y^{n-2} + \cdots + r_1)$$

is an element of A, so A is a division ring. □

Lemma 3.3.3 (Gauss' Lemma). *Let R be an integrally closed integral domain with quotient field K. Let $f \in R[x]$ be a monic polynomial, and suppose there is a factorization $f = gh$, where g, h are monic polynomials in $K[x]$. Then both g and h are in $R[x]$.*

Proof. Let L/K be an extension of fields such that L is a splitting field for f over K. Let S be the integral closure of R in L. Since f splits in $L[x]$, so does g. Write $g = \prod(x - \alpha_i)$. Each α_i is a root of f, hence is integral over R, hence lies in S. This shows that every coefficient of g is in S. So each coefficient of g is in $S \cap K$ which is equal to R since R is integrally closed in K. So $g \in R[x]$. The same argument applies to h. □

Theorem 3.3.4. *Let R be an integrally closed integral domain with quotient field K. Let A be a finite dimensional K-algebra. An element $\alpha \in A$ is integral over R if and only if* $\operatorname{min.poly}_K(\alpha) \in R[x]$.

Proof. Let $f = \operatorname{min.poly}_K(\alpha) \in K[x]$. Assume α is integral over R. Then there exists a monic polynomial $g \in R[x]$ such that $g(\alpha) = 0$. In this case, f divides g in $K[x]$. There is a factorization $g = fh$ for some monic polynomial $h \in K[x]$. By Gauss' Lemma 3.3.3, both f and h lie in $R[x]$. □

The following version of the Cohen-Seidenberg Theorem will be very useful. A proof can be found in [**Mat80**].

Theorem 3.3.5. *Assume B is a commutative faithful integral A-algebra.*

(1) *The natural map* $\theta^\sharp : \operatorname{Spec} B \to \operatorname{Spec} A$ *is onto.*

(2) *If* $p \in \operatorname{Spec} A$ *and* $q_1, q_2 \in \operatorname{Spec} B$ *are two primes in* B *lying over* p, *then* q_1 *is not a subset of* q_2.

(3) *(Going Up Holds) For any* p_1, p_2 *in* $\operatorname{Spec} A$ *such that* $p_1 \subsetneq p_2$, *and for any* $q_1 \in \operatorname{Spec} B$ *lying over* p_1, *there exists* $q_2 \in \operatorname{Spec} B$ *lying over* p_2 *such that* $q_1 \subsetneq q_2$.

(4) *If* A *is a local ring with maximal ideal* p, *then the prime ideals of* B *lying over* p *are precisely the maximal ideals of* B.

For (5) and (6) assume A *and* B *are integral domains, that* K *is the quotient field of* A *and that* A *is integrally closed in* K.

(5) *(Going Down Holds) For any* p_1, p_2 *in* $\operatorname{Spec} A$ *such that* $p_1 \subsetneq p_2$, *and for any* $q_2 \in \operatorname{Spec} B$ *lying over* p_2, *there exists* $q_1 \in \operatorname{Spec} B$ *lying over* p_1 *such that* $q_1 \subsetneq q_2$.

(6) *If* L *is a normal extension field of* K, *and* B *is equal to the integral closure of* A *in* L, *then any two prime ideals of* B *lying over the same prime* $p \in \operatorname{Spec} A$ *are conjugate to each other by some automorphism* $\sigma \in \operatorname{Aut}_K(L)$.

Corollary 3.3.6. *Let* R *be a local ring and* S *a commutative* R*-algebra which is faithful and finitely generated as an* R*-module. Then* S *is semilocal (Exercise* 3.1.8*).*

Proof. Let $\mathfrak{m}$ be the maximal ideal of R. By Theorem 3.3.5 (4), the maximal ideals of S correspond to the maximal ideals of $S/\mathfrak{m}S$. Because $S/\mathfrak{m}S$ is finite dimensional over $R/\mathfrak{m}$, it is artinian (Exercise 2.5.15). By Proposition 3.2.9, $S/\mathfrak{m}S$ is semilocal. □

Proposition 3.3.7. *Let* R *be an integral domain with quotient field* K. *The following are equivalent.*

(1) R *is integrally closed in* K.

(2) *For each* $P \in \operatorname{Spec} R$, R_P *is integrally closed in* K.

(3) *For each* $P \in \operatorname{Max} R$, R_P *is integrally closed in* K.

Definition 3.3.8. Let R be an integral domain with quotient field K. If R is integrally closed in K, then we say R is *normal*. Let $u \in K$. We say u is *almost integral over* R in case there exists $r \in R - (0)$ such that $ru^n \in R$ for all $n > 0$. We say R is *completely normal* in case the set of all elements in K that are almost integral over R is equal to R itself. By Proposition 3.3.7, an integral domain is integrally closed if and only if for each prime ideal the local ring is integrally closed. This motives the next definition. We say a commutative ring R is a *normal ring* in case R_P is a normal local integral domain for each $P \in \operatorname{Spec} R$.

Lemma 3.3.9. *Let R be an integral domain with quotient field K.*

(1) *If $u \in K$ and u is integral over R, then u is almost integral over R.*

(2) *If $u, v \in K$ are both almost integral over R, then $u + v$ and uv are almost integral over R.*

(3) *If R is noetherian and $u \in K$, then u is almost integral over R if and only if u is integral over R.*

Lemma 3.3.10. *Let R be a commutative noetherian ring with the property that $R_\mathfrak{m}$ is an integral domain, for each maximal ideal $\mathfrak{m} \in \operatorname{Max} R$. Let $P_1, \dots, P_n$ be the distinct minimal primes of R.*

(1) *The natural map*

$$R \xrightarrow{\phi} R/P_1 \oplus \cdots \oplus R/P_n$$

is an isomorphism.

(2) *The nil radical of R, $\operatorname{Rad}_R(0)$, is equal to (0).*

(3) *R is a normal ring if and only if each ring R/P_i is a normal integral domain.*

Lemma 3.3.11. *Let S be a commutative faithfully flat R-algebra.*

(1) *If S is an integrally closed integral domain, then R is an integrally closed integral domain.*

(2) *If S is a normal ring, then R is a normal ring.*

Proof. The proof is left to the reader. □

4. Completion of a Linear Topological Module

In this section we review some theorems on the completion of rings and modules with respect to a linear topology. General references for this section are [**AM69**], [**Mat80**] and [**Bou89b**]. We assume the reader is familiar with the basic results from [**AM69**, Chapter 11] and [**Mat80**, Chapter 11].

Let R be a ring and M an R-module. A *filtration* of M is any nonincreasing chain of submodules

$$M = M_0 \supseteq M_1 \supseteq M_2 \supseteq M_3 \dots.$$

Using the set of submodules $\{M_n\}_{n \geq 0}$ in a filtration, we define a topology on M. Given any $x \in M$, a base for the neighborhoods of x is the set $\{x + M_n \mid n \geq 0\}$. The *linear topology* on M defined by the filtration $\{M_n\}_{n \geq 0}$ is the smallest topology on M containing all of the open sets $\{x + M_n \mid x \in M, n \geq 0\}$. If L is a submodule of M and $\eta : M \to M/L$ is the natural map, then the chain $\{\eta(M_n)\}_{n \geq 0} = \{(M_n + L)/L\}_{n \geq 0}$ is a filtration of M/L that induces a linear topology on M/L. The chain of submodules

$\{M_n \cap L\}_{n\geq 0}$ is a filtration of L which induces a linear topology on L. We say that M is *separated* (that is, *Hausdorff*) if for any two distinct points $x, y \in M$, there are neighborhoods $x \in U$ and $y \in V$ such that $U \cap V = \emptyset$.

Lemma 3.4.1. *Let R be a ring, M an R-module with a filtration $\{M_n\}_{n\geq 0}$, and L a submodule. With respect to the linear topology defined by this filtration, the following are true.*

(1) *Each set M_n is open and closed.*
(2) *Addition on M is continuous.*
(3) *The natural maps $0 \to L \xrightarrow{\subseteq} M \xrightarrow{\eta} M/L \to 0$ are continuous.*
(4) *For each n, M/M_n has the discrete topology, which is to say "points are open".*

Lemma 3.4.2. *Let $\{M_n\}_{n\geq 0}$ be a filtration of the R-module M. Let $N = \bigcap_{n\geq 0} M_n$. Then*

(1) *N is the closure of $\{0\}$.*
(2) *M is separated if and only if $N = 0$.*
(3) *If L is a submodule of M, then M/L is separated if and only if L is closed.*

Let $\{M_n\}_{n\geq 0}$ be a filtration of the R-module M. A sequence (x_ν) of elements of M is a *Cauchy sequence* if for every open submodule U there exists $n_0 \geq 0$ such that $x_\mu - x_\nu \in U$ for all $\mu \geq n_0$ and all $\nu \geq n_0$. Since U is a submodule, this is equivalent to $x_{\nu+1} - x_\nu \in U$ for all $\nu \geq n_0$. A point x is a *limit* of a sequence (x_ν) if for every open submodule U there exists $n_0 \geq 0$ such that $x - x_\nu \in U$ for all $\nu \geq n_0$. We say M is *complete* if every Cauchy sequence has a limit. We say that two Cauchy sequences (x_ν) and (y_ν) are *equivalent* and write $(x_\nu) \sim (y_\nu)$ if 0 is a limit of $(x_\nu - y_\nu)$.

Lemma 3.4.3. *In the above context, let C denote the set of all Cauchy sequences in M.*

(1) *The relation $\sim$ is an equivalence relation on C.*
(2) *If $(x_\nu) \in C$ and $(y_\nu) \in C$, then $(x_\nu + y_\nu) \in C$.*
(3) *If $(x_\nu) \sim (x'_\nu) \in C$ and $(y_\nu) \sim y'_\nu) \in C$, then $(x_\nu + y_\nu) \sim (x'_\nu + y'_\nu) \in C$.*
(4) *If $(x_\nu) \in C$ and $r \in R$, then $(rx_\nu) \in C$.*
(5) *If $(x_\nu) \sim (x'_\nu) \in C$ and $r \in R$, then $(rx_\nu) \sim (rx'_\nu) \in C$.*

Let $\{M_n\}_{n\geq 0}$ be a filtration of the R-module M. Let M^* denote the set of all equivalence classes of Cauchy sequences in M. We call M^* the *topological completion* of M. Then Lemma 3.4.3 says that M^* is an R-module.

For any $x \in M$, the constant sequence (x) is a Cauchy sequence, so $x \mapsto (x)$ defines an R-module homomorphism $\eta : M \to M^*$. The reader should verify that the kernel of η is the subgroup N of Lemma 3.4.2. Therefore η is one-to-one if and only if M is separated. A Cauchy sequence is in the image of η if it has a limit in M, hence M is complete if the natural map $\eta : M \to M^*$ is onto. For M to be separated and complete it is necessary and sufficient that η be an isomorphism, which is true if and only if every Cauchy sequence has a unique limit in M.

Proposition 3.4.4. *Let $\{M_n\}_{n\geq 0}$ be a filtration of the R-module M and M^* the topological completion. Then M^* is isomorphic to $\varprojlim M/M_n$ as R-modules.*

Proof. For any n the natural map $\eta_n : M \to M/M_n$ is continuous and maps a Cauchy sequence (x_ν) in M to a Cauchy sequence $(\eta_n(x_\nu))$ in M/M_n. As M/M_n has the discrete topology, $(\eta_n(x_\nu))$ is eventually constant, hence has a limit. Two equivalent Cauchy sequences will have the same limit in M/M_n, so there is a well defined continuous R-module homomorphism $f_n : M^* \to M/M_n$ defined by $(x_\nu) \mapsto \varinjlim(\eta_n(x_\nu))$. According to the universal mapping property, there is a unique R-module homomorphism $\beta : M^* \to \varprojlim M/M_n$. A Cauchy sequence is in the kernel of β if and only if it is equivalent to 0. Therefore, β is one-to-one. By Proposition 1.4.5, we can view the inverse limit as a submodule of the direct product. If the inverse limit is given the topology it inherits from the direct product of the discrete spaces $\prod M/M_n$, then β is continuous. An element of the inverse limit can be viewed as $(x_n) \in \prod M/M_n$ such that $x_n = \phi_{n+1}(x_{n+1})$ for all n, where $\phi_{n+1} : M/M_{n+1} \to M/M_n$ is the natural map. In this case, $x_{n+1} - x_n \in M_n$ so (x_n) is the image under η of a Cauchy sequence in M. This shows β is onto, and therefore β is an isomorphism. □

Suppose that $\{A_n\}$ is a filtration for the R-module A, and that $\{B_n\}$ is a filtration for B. A *morphism* from $\{A_n\}$ to $\{B_n\}$ is an R-module homomorphism $\alpha : A \to B$ such that for each $n \geq 0$, $\alpha(A_n) \subseteq B_n$. In this case α induces a commutative square

$$\begin{array}{ccc} A/A_{n+1} & \xrightarrow{\alpha} & B/B_{n+1} \\ \downarrow{\scriptstyle \phi_{n+1}} & & \downarrow{\scriptstyle \psi_{n+1}} \\ A/A_n & \xrightarrow{\alpha} & B/B_n \end{array}$$

for each $n \geq 0$. Hence there is a morphism of inverse systems $\alpha : \{A/A_n\} \to \{B/B_n\}$ which induces a homomorphism $\varprojlim A/A_n \to \varprojlim B/B_n$.

Proposition 3.4.5. *If*

$$\{A_n\} \xrightarrow{\alpha} \{B_n\} \xrightarrow{\beta} \{C_n\}$$

is a sequence of morphisms of R-modules equipped with filtrations, such that for every $n \geq 0$ the sequence

$$0 \to A_n \xrightarrow{\alpha} B_n \xrightarrow{\beta} C_n \to 0$$

is an exact sequence of R-modules. Then

$$0 \to \varprojlim A/A_n \xrightarrow{\overleftarrow{\alpha}} \varprojlim B/B_n \xrightarrow{\overleftarrow{\beta}} \varprojlim C/C_n \to 0$$

is an exact sequence of R-modules.

Proof. The proof follows by applying Proposition 1.4.8 to the exact sequence of morphisms of inverse systems $\{A/A_n\} \xrightarrow{\alpha} \{A/B_n\} \xrightarrow{\beta} \{C/C_n\}$. □

Corollary 3.4.6. *Let $\{B_n\}$ be a filtration for the R-module B. Suppose*

$$0 \to A \xrightarrow{\alpha} B \xrightarrow{\beta} C \to 0$$

is an exact sequence of R-modules. Give A the filtration $\{A_n\} = \{\alpha^{-1}(B_n)\}$ and C the filtration $\{C_n\} = \{\beta(B_n)\}$. Then the sequence of completions

$$0 \to A^* \xrightarrow{\alpha^*} B^* \xrightarrow{\beta^*} C^* \to 0$$

is an exact sequence of R-modules.

Proof. The proof follows from Propositions 3.4.5 and 3.4.4. □

Corollary 3.4.7. *Let $\{M_n\}$ be a filtration for the R-module M and M^* the topological completion.*

(1) *For each $n \geq 0$ we have $M^*/M_n^* \cong M/M_n$.*
(2) *With respect to the filtration $\{M_n^*\}$, the R-module M^* is complete and separated. That is, $M^* \cong (M^*)^*$.*

Proof. (1): This follows from Corollary 3.4.6 applied to the exact sequence $0 \to M_n \to M \to M/M_n \to 0$. Since M/M_n has the discrete topology, $M/M_n \cong (M/M_n)^*$.

(2): This follows from Part (1) by taking inverse limits. □

Proposition 3.4.8. *Let R be a ring and I a two-sided ideal in R such that R is separated and complete with respect to the I-adic topology. Then I is contained in the Jacobson radical of R. Equivalently, $1 + x$ is a unit of R for every $x \in I$.*

Proof. By Theorem 3.1.2, it is enough to prove that $1-x$ is invertible for every $x \in I$. Since the I-adic topology on R is separated, $\cap I^n = 0$. The sequence $s = (1, 1+x, 1+x+x^2, 1+x+x^2+x^3, \dots)$ is a Cauchy sequence in R, hence converges in R. Now $(1-x)s = 1-(x, x^2, x^3, \dots)$ is equal to 1 since the Cauchy sequence $(x, x^2, x^3, \dots)$ converges to 0. □

4.1. Graded Rings and Graded Modules. A *graded ring* is a commutative ring R which under addition is the internal direct sum $R = \bigoplus_{n=0}^{\infty} R_n$ of a set of additive subgroups $\{R_n\}_{n\geq 0}$ satisfying the property that $R_i R_j \subseteq R_{i+j}$ for all $i, j \geq 0$. The reader should verify that R_0 is a subring of R and each R_n is an R_0-module. An element of R_n is said to be *homogeneous of degree* n. The set $R_+ = \bigoplus_{n=1}^{\infty} R_n$ is an ideal of R, and is called the *exceptional ideal* of R.

Let R be a graded ring. A *graded R-module* is an R-module M which under addition is the internal direct sum $M = \bigoplus_{n\in\mathbb{Z}} M_n$ of a set of additive subgroups $\{M_n\}_{n\in\mathbb{Z}}$ and such that $R_i M_j \subseteq M_{i+j}$ for all pairs i, j. The reader should verify that each M_n is an R_0-module. Any $x \in M_n$ is said to be *homogeneous* of degree n. Every $y \in M$ can be written uniquely as a finite sum $y = \sum_{n=-d}^{d} y_n$ where $y_n \in M_n$. We call the elements $y_{-d}, \dots, y_0, \dots, y_d$ the homogeneous components of y. Let M and N be graded R-modules and $\theta : M \to N$ an R-module homomorphism. We say θ is a *homomorphism of graded R-modules* if for every $n \in \mathbb{Z}$ we have $\theta(M_n) \subseteq N_n$.

Proposition 3.4.9. *Let R be a graded ring. The following are equivalent.*

(1) *R is a noetherian ring.*

(2) *R_0 is a noetherian ring and R is a finitely generated R_0-algebra.*

Example 3.4.10. Let R be a commutative ring. Suppose we have a filtration $J = \{J_n\}_{n\geq 0}$ of R by ideals

$$R = J_0 \supseteq J_1 \supseteq J_2 \supseteq \dots$$

such that for all $m, n \geq 0$ we have $J_m J_n \subseteq J_{m+n}$. The graded ring associated to this filtration is

$$\operatorname{gr}_J(R) = \bigoplus_{n=0}^{\infty} \frac{J_n}{J_{n+1}} = \frac{R}{J_1} \oplus \frac{J_1}{J_2} \oplus \dots \oplus \frac{J_n}{J_{n+1}} \oplus \dots.$$

The reader should verify that the product $J_m J_n \subseteq J_{m+n}$ turns $\operatorname{gr}_J(R)$ into a graded ring. When I is an ideal of R, the I-adic filtration

$$R = I^0 \supseteq I^1 \supseteq I^2 \supseteq \dots$$

has the associated graded ring $\operatorname{gr}_I(R) = \bigoplus_{n\geq 0} I^n/I^{n+1}$. The reader should verify that $\operatorname{gr}_I(R)$ is an R/I-algebra which is generated by the set of homogeneous elements of degree one, $\operatorname{gr}_I(R)_1 = I/I^2$.

Let R be a commutative ring and $J = \{J_n\}_{n\geq 0}$ a filtration of R by ideals. Let M be an R-module which also has a filtration $\{M_n\}_{n\geq 0}$. We say that M is a *filtered* R-module, or that the filtrations of R and M are *compatible*, if $J_i M_j \subseteq M_{i+j}$, for all $i \geq 0$ and $j \geq 0$. If the filtration of R is defined by an ideal I, then M is a filtered R-module if $IM_n \subseteq M_{n+1}$ for all $n \geq 0$. In this case, we also say the filtration $\{M_n\}_{n\geq 0}$ is an *I-filtration*. If $IM_n = M_{n+1}$ for all sufficiently large n, then we say the filtration is a *stable I-filtration*. For example, let R be a commutative ring, and I an ideal in R. Then the I-adic filtration of R and the I-adic filtration $\{I^n M\}$ of M are compatible. Moreover, $\{I^n M\}$ is a stable I-filtration of M.

Example 3.4.11. Let R be a commutative ring and I an ideal of R. Let M be an R module and $F = \{M_n\}_{n\geq 0}$ an I-filtration of M. Set $\operatorname{gr}_F(M) = \bigoplus_{n=0}^{\infty} M_n/M_{n+1}$. The reader should verify that $\operatorname{gr}_F(M)$ is a graded $\operatorname{gr}_I(R)$-module. We call this the *associated graded module* for the I-filtration F of M. The graded $\operatorname{gr}_I(R)$-module associated to the I-adic filtration $\{I^n M\}_{n\geq 0}$ is denoted $\operatorname{gr}_I(M)$.

Proposition 3.4.12. *Let R be a commutative ring, I an ideal in R, and M a finitely generated R-module. Let $\hat{R}$ be the I-adic completion of R and $\hat{M}$ the I-adic completion of M. Then the following are true.*

(1) *$\hat{R} \otimes_R M \to \hat{M}$ is onto.*

(2) *If M is finitely presented, then $\hat{R} \otimes_R M \cong \hat{M}$.*

(3) *If R is noetherian, then $\hat{R}$ is a flat R-module and $\hat{R} \otimes_R M \cong \hat{M}$.*

Proposition 3.4.13. *Let R be a commutative ring with a filtration $J = \{J_n\}_{n\geq 0}$ by ideals under which R is complete. Let M be a filtered R-module with filtration $\{M_n\}_{n\geq 0}$ under which M is separated.*

(1) *If the graded $\operatorname{gr}_J(R)$-module $\operatorname{gr}(M)$ is finitely generated, then M is a finitely generated R-module.*

(2) *If every graded $\operatorname{gr}_J(R)$-submodule of $\operatorname{gr}(M)$ is finitely generated, then the R-module M satisfies the ACC on submodules (in other words, M is noetherian).*

Corollary 3.4.14. *Let R be a commutative noetherian ring.*

(1) *If I is an ideal of R, then the I-adic completion of R is noetherian.*

(2) *If $S = R[[x_1, \dots, x_m]]$ is the power series ring over R in m variables, then S is noetherian.*

Corollary 3.4.15. *Let R be a commutative ring with a filtration by ideals $\{J_n\}_{n\geq 0}$. Let M be a filtered R-module with filtration $\{M_n\}_{n\geq 0}$. Assume that R is complete and that M is separated. Let F be a finitely generated submodule of M. If $M_k = M_{k+1} + J_k F$ for all $k \geq 0$, then $F = M$.*

Proof. Let $\{x_1, \ldots, x_m\}$ be a generating set for the R-module F, which we view as a subset of $M = M_0$. Let ξ_i be the image of x_i in M/M_1. Let F_1 be the kernel of $F \to M/M_1$. For all $k \geq 0$, $J_k F \subseteq M_k$. By hypothesis, the natural map $\eta_k : J_k F \to M_k/M_{k+1}$ is onto. Since $J_k F_1 + J_{k+1}F \subseteq M_{k+1}$, $(J_k/J_{k+1})(F/F_1) \to M_k/M_{k+1}$ is onto. Therefore, the graded $\operatorname{gr}_J(R)$-module $\operatorname{gr}(M)$ is generated by the finite set $\{\xi_1, \ldots, \xi_m\}$. By Proposition 3.4.13, M is generated by $\{x_1, \ldots, x_m\}$. □

4.2. Lifting of Idempotents. For any ring R, the set of idempotents of R is denoted $\operatorname{idemp}(R)$. The homomorphic image of an idempotent is an idempotent, so given a homomorphism of rings $A \to B$, there is a function $\operatorname{idemp}(A) \to \operatorname{idemp}(B)$. In Chapter 10, the definition of a henselian local ring is based on this functor. In the following corollaries to Nakayama's Lemma we derive sufficient conditions on A and B for the function on idempotents to be onto. As a consequence, we are able to deduce sufficient conditions on A and B such that any finitely generated projective B-module lifts to a finitely generated projective A-module.

Corollary 3.4.16. *Let R be a ring and I a two-sided ideal of R such that $I \subseteq \mathrm{J}(R)$.*

(1) *If R is a commutative ring, then the natural map* $\operatorname{idemp}(R) \to \operatorname{idemp}(R/I)$ *is one-to-one.*
(2) *If I consists of nilpotent elements, then* $\operatorname{idemp}(R) \to \operatorname{idemp}(R/I)$ *is onto.*
(3) *If R is separated and complete with respect to the I-adic topology (that is, $R \to \varprojlim R/I^n$ is an isomorphism), then* $\operatorname{idemp}(R) \to \operatorname{idemp}(R/I)$ *is onto.*

Proof. (1): Let $e_0, e_1 \in \operatorname{idemp}(R)$ and assume $x = e_0 - e_1 \in I$. We show that $x = 0$. Consider

$$
\begin{aligned}
x^3 &= e_0^3 - 3e_0^2 e_1 + 3e_0 e_1^2 - e_1^3 \\
&= e_0 - 3e_0 e_1 + 3e_0 e_1 - e_1 \\
&= e_0 - e_1 \\
&= x.
\end{aligned}
$$

Then $x(x^2 - 1) = 0$. By Theorem 3.1.2, $x^2 - 1$ is a unit, which implies that $x = 0$.

(2): Assume I consists of nilpotent elements. If $x \in R$, denote by $\bar{x}$ the image of x in R/I. Assume $\bar{x}^2 = \bar{x}$. It follows that $(1 - \bar{x})^2 = 1 - \bar{x}$. Since $x - x^2 \in I$, for some $n > 0$ we have $(x - x^2)^n = x^n(1 - x)^n = 0$. Set $e_0 = x^n$ and $e_1 = (1 - x)^n$. Then $e_0 e_1 = e_1 e_0 = 0$, $\bar{e}_0 = \bar{x}^n = \bar{x}$, and

$\bar{e}_1 = (1-\bar{x})^n = 1-\bar{x}$. This says that $e_0 + e_1 - 1 \in I$, so by Theorem 3.1.2, $u = e_0 + e_1$ is a unit in R. We have $1 = e_0u^{-1} + e_1u^{-1} = u^{-1}e_0 + u^{-1}e_1$, hence $e_0 = e_0^2u^{-1} = u^{-1}e_0^2$, and $e_0u = e_0^2 = ue_0$. We have shown that e_0 commutes with u. From this it follows that e_0u^{-1} is an idempotent of R. Since $\bar{u} = 1$, $\bar{e}_0\bar{u}^{-1} = \bar{x}$.

(3): Let $\bar{x} \in R/I$ be an idempotent. For $n \geq 1$, I/I^n is nilpotent. By (2), $\operatorname{idemp}(R/I^n) \to \operatorname{idemp}(R/I)$ is onto for $n > 1$. Set $e_1 = x$. By induction, there is a sequence $(\bar{e}_i)$ in $\prod_i R/I^i$ such that $e_n^2 - e_n \in I^n$ and $e_{n+1} - e_n \in I^n$. So $(\bar{e}_i)$ is an idempotent in $R = \varprojlim R/I^n$ which maps to $\bar{x}$ in R/I. □

Corollary 3.4.17. *Let R be a commutative ring and I an ideal in R such that R is separated and complete with respect to the I-adic topology. Let A be an R-algebra which is integral over R.*

(1) *If A is an R-module of finite presentation, then the natural map* $\operatorname{idemp}(A) \to \operatorname{idemp}(A \otimes_R (R/I))$ *is onto. That is, an idempotent $\bar{e}$ in A/IA lifts to an idempotent e in A.*

(2) *If A is commutative, then* $\operatorname{idemp}(A) \to \operatorname{idemp}(A \otimes_R (R/I))$ *is onto.*

Proof. (1): Assume that A is an R-module of finite presentation. We are given that $R \to \varprojlim R/I^n$ is an isomorphism. By Proposition 3.4.12, $A \to \varprojlim A/(I^nA)$ is an isomorphism, so A is separated and complete in the IA-adic topology. By Proposition 3.4.8, IA is contained in the Jacobson radical of A. By Corollary 3.4.16 (3), there is an idempotent e in A which maps onto $\bar{e}$.

(2): First we reduce to the case where A is generated as an R-algebra by a single element. Let $a \in A$ be a preimage of $\bar{e}$. Let C be the R-subalgebra of A generated by a. Then A is a faithful C-algebra which is integral over C. By Theorem 3.3.5, $\operatorname{Spec} A \to \operatorname{Spec} C$ is onto. The reader should verify that $\operatorname{Spec} \bar{A} \to \operatorname{Spec} \bar{C}$ is onto as well, where $\bar{C} = C/IC$. Write $\bar{a}$ for the image of a in $\bar{C}$. Under the natural map $\bar{C} \to \bar{A}$, we have $\bar{a} \mapsto \bar{e}$. The reader should verify that $\operatorname{Spec} \bar{C} = V(\bar{a}) \cup V(1-\bar{a})$, so by Corollary 2.2.10 there is a unique idempotent $\bar{f}$ in $\bar{C}$ such that $V(\bar{a}) = V(\bar{f})$. From this it follows that $\bar{f} \mapsto \bar{e}$. If there exists an idempotent f in C that lifts $\bar{f}$, then using $C \to A$, we get a lifting of $\bar{e}$.

Now assume A is generated as an R-algebra by a single element a. Then a is integral over R. Let $p \in R[x]$ be a monic polynomial such that $p(a) = 0$. Let $C = R[x]/(p)$. Then C is a finitely generated free R-module. Let J be the kernel of the natural projection $C \to A$. Let $\{J_\alpha\}$ be the directed system of all finitely generated ideals in C such that $J_\alpha \subseteq J$. Then $C_\alpha = C/J_\alpha$ is an R-module of finite presentation, for each α, and $A = \varinjlim C_\alpha$. Therefore, $\bar{A} = A/IA = \varinjlim C_\alpha/IC_\alpha = \varinjlim \bar{C}_\alpha$. By Exercise 1.4.17, an idempotent $\bar{e}$

in $\bar{A}$ comes from an idempotent $\bar{e}_\alpha$ in $\bar{C}_\alpha$, for some α. By (1) we can lift $\bar{e}_\alpha$ to an idempotent $e_\alpha \in C_\alpha$. Using $C_\alpha \to A$, we get a lifting of $\bar{e}$ to an idempotent in A. □

Using Corollary 3.4.16, we give sufficient conditions for lifting a finitely generated projective module. A projective module is a direct summand of a free module. Hence, to show that projectives lift under a homomorphism, it is enough to show that free modules and idempotent matrices lift.

Proposition 3.4.18. *Let R be a ring and I a two-sided ideal of R such that $I \subseteq \mathrm{J}(R)$ and R is separated and complete with respect to the I-adic topology (that is, $R \to \varprojlim R/I^n$ is an isomorphism).*

(1) *If Q is a finitely generated projective R/I-module, then there is a finitely generated projective R-module P such that $Q \cong P \otimes_R (R/I)$.*

(2) *If $g : Q_1 \to Q_2$ is a homomorphism of finitely generated projective R/I-modules, then g lifts to a homomorphism $f : P_1 \to P_2$ of finitely generated projective R-modules.*

(3) *If Q is an R/I-progenerator module, then there is an R-progenerator module P such that $Q \cong P \otimes_R (R/I)$.*

Proof. (1): For some $m > 0$, there is an isomorphism $(R/I)^m \cong Q \oplus Q_0$. Let $\bar{e}$ be the idempotent matrix in $M_m(R/I)$ such that $Q \cong \mathrm{im}(\bar{e})$ and $Q_0 \cong \ker(\bar{e})$. Since $\varprojlim M_n(R/I^n) = M_n(\varprojlim R/I^n) = M_n(R)$, by Corollary 3.4.16, we can lift $\bar{e}$ to an idempotent $e \in M_n(R)$. If we set $P = \mathrm{im}(e)$, then $Q \cong P \otimes_R (R/I)$.

(2): Using (1), there are projective R-modules P_i such that $Q_i \cong P_i \otimes_R (R/I)$. Combined with g, there is a diagram

$$\begin{array}{ccccc} P_1 & \longrightarrow & Q_1 & \longrightarrow & 0 \\ \downarrow{\scriptstyle \exists f} & & \downarrow{\scriptstyle g} & & \\ P_2 & \longrightarrow & Q_2 & \longrightarrow & 0 \end{array}$$

where the rows are exact. Since P_1 is a projective R-module, there exists a map f which makes the diagram commutative (Proposition 1.1.5).

(3): This part is left to the reader. □

Chapter 4

Separable Algebras, Definition and First Properties

This chapter begins with the definition of a separable algebra over a commutative ring. Some of the first properties satisfied by separable algebras are derived. We show that separability is preserved under a change of base ring. We show that separability is transitive. Any localization of R is separable over R and a homomorphic image of R is separable over R. Over a field k, the separable algebras turn out to be finite direct sums of matrix algebras over finite dimensional k-division algebras and the center of each division algebra appearing is a finite separable extension field of k. More generally, for any commutative ring R, A is separable over R if and only if A is separable over its center and its center is separable over R. This means that the study of separability can be split into the study of algebras which are commutative, and algebras which are central. There are many important examples presented in this section. For instance the ring of n-by-n matrices over R is separable over R. If G is a finite group and the order of G is invertible in R, then the group ring $R(G)$ is separable over R. The last two sections emphasize commutative separable R-algebras.

1. Separable Algebra, the Definition

In this section the notion of a separable algebra over a commutative ring is defined and a few basic properties of separable algebras are proved. Most of the results in this section are due to Auslander and Goldman [**AG60a**], including the definition of separability. Out treatment closely follows that of [**DI71**].

Recall that an extension of fields F/k is separable if every element of F is a root of a separable polynomial over k. To generalize the definition of separability to an algebra A over an arbitrary commutative ring R, it is necessary to raise the level of abstraction. The definition of separability given below is not in terms of elements of A, but instead is based on a certain module structure of the ring A over the ring $A \otimes_R A^o$. We will see in Section 4.5.2 that for an extension of fields, the definitions agree. For later reference, much of the new terminology and notation have been assembled into the following long definition.

Definition 4.1.1. Let A be an algebra over the commutative ring R. A *two-sided A/R-module* is a left A right A bimodule M such that the two induced R-actions are equal. That is, for all $a, b \in A$, $r \in R$, $x \in M$ we have

$$\begin{aligned} (ax)b &= a(xb), \text{ and} \\ rx = (r \cdot 1)x &= x(r \cdot 1) = xr. \end{aligned} \tag{4.1}$$

The *enveloping algebra* of A is $A^e = A \otimes_R A^o$. Then A^e is an R-algebra. If M is a left A^e-module, then we can make M into a two-sided A/R-module by defining

$$\begin{aligned} ax &= a \otimes 1 \cdot x, \text{ and} \\ xa &= 1 \otimes a \cdot x. \end{aligned} \tag{4.2}$$

Conversely, any two-sided A/R-module can be turned into a left A^e-module in the same way. The reader should verify that these module actions are well defined. The R-algebra A is a two-sided A/R-module, hence by (4.2) we view A as a left A^e-module. The left regular representation of A^e as a ring of R-module endomorphisms of A induces an R-algebra homomorphism

$$\varphi : A^e \to \operatorname{Hom}_R(A, A) \tag{4.3}$$

where an element α of A^e is mapped to the element $\varphi(\alpha)$ of $\operatorname{Hom}_R(A, A)$ which is "left multiplication by α". Specifically, if $\alpha = \sum a_i \otimes b_i$, then for any $x \in A$, $\varphi(\alpha)(x) = \alpha \cdot x = \sum_i a_i x b_i$. The map φ will be called the *enveloping homomorphism* of A. Viewing A^e as a left A^e-module, the assignment $x \mapsto x \cdot 1$ defines an A^e-module homomorphism $\mu : A^e \to A$. Notice that $\mu(a \otimes b) = ab$ and μ is onto. When A is commutative, μ is a homomorphism of R-algebras (see Exercise 1.3.9). Denote by $J_{A/R}$ the kernel of μ. Then

$$0 \to J_{A/R} \to A^e \xrightarrow{\mu} A \to 0 \tag{4.4}$$

is an exact sequence of A^e-modules. Since $J_{A/R}$ is an A^e-submodule of A^e, it is a left ideal of A^e. Notice that $\mu(a \otimes 1 - 1 \otimes a) = 0$, so $a \otimes 1 - 1 \otimes a \in J_{A/R}$. In Exercise 4.1.10 the reader is asked to prove that the left ideal $J_{A/R}$ is generated by elements of the form $a \otimes 1 - 1 \otimes a$.

Proposition 4.1.2. *Let R be a commutative ring and A an R-algebra. In the above context, the following are equivalent.*

(1) *A is projective as a left A^e-module.*
(2) *Eq. (4.4) is a split exact sequence of left A^e-modules.*
(3) *There is an element $e \in A^e$ such that $\mu(e) = 1$ and $J_{A/R}e = 0$.*
(4) *There is an idempotent $e \in A^e$ such that $J_{A/R}$ is equal to the principal left ideal in A^e generated by $1 - e$.*

Proof. The proof follows directly from Exercise 1.1.22. □

It was Auslander and Goldman who first recognized that the equivalent conditions enumerated in Proposition 4.1.2 provide the appropriate framework for defining when an algebra over a commutative ring is separable. For an extension of fields, the above conditions were already recognized as being equivalent to the element-wise definition.

Definition 4.1.3. Let R be a commutative ring and A an R-algebra. If A satisfies any of the equivalent properties of Proposition 4.1.2, then we say A is a *separable* R-algebra. Notice that the same element e works for both (3) and (4). The element $e \in A^e$ is called a *separability idempotent* for A. If A is commutative, then a separability idempotent is unique, if it exists (Exercise 4.1.8).

If M is any two-sided A/R-module, the *centralizer of A in M* is

$$M^A = \{x \in M \mid ax = xa, \text{ for all } a \in A\}. \tag{4.5}$$

The reader should verify that M^A is an R-submodule of M and the assignment $M \mapsto M^A$ defines a covariant functor from the category of left A^e-modules to the category of left R-modules. The rest of this section is devoted to deriving a test for separability based on this functor.

Lemma 4.1.4. *Let R be a commutative ring, A an R-algebra, and M a left A^e-module. Then the following are true.*

(1) *There is an isomorphism of R-modules $\operatorname{Hom}_{A^e}(A, M) \cong M^A$ which is defined by $f \mapsto f(1)$.*
(2) *If $g : M \to N$ is an A^e-module homomorphism, then the diagram*

$$\begin{array}{ccc} \operatorname{Hom}_{A^e}(A, M) & \xrightarrow{g\circ(\cdot)} & \operatorname{Hom}_{A^e}(A, N) \\ \downarrow & & \downarrow \\ M^A & \xrightarrow{\quad g \quad} & N^A \end{array}$$

commutes.

(3) *The functors* $\operatorname{Hom}_{A^e}(A, \cdot)$ *and* $(\cdot)^A$ *are naturally isomorphic and both are left exact.*

Proof. Let $f \in \operatorname{Hom}_{A^e}(A, M)$. For any $a \in A$ we have

$$a \cdot f(1) = a \otimes 1 \cdot f(1) = f(a \otimes 1 \cdot 1) = f(a) = f(1 \otimes a \cdot 1) = 1 \otimes a \cdot f(1) = f(1) \cdot a$$

which shows $f(1) \in M^A$. Conversely, suppose $x \in M^A$. Define $\rho_x : A \to M$ to be "right multiplication by x". In other words, $\rho_x(a) = ax$. It follows from

$$\begin{aligned}\rho_x(b \otimes c \cdot a) = \rho_x(bac) = (bac)x = (b \otimes c \cdot a)x \\ = b \otimes c \cdot (a \otimes 1 \cdot x) = b \otimes c \cdot (ax) = b \otimes c \cdot \rho_x(a)\end{aligned}$$

that ρ_x is A^e-linear. Since $\rho_x(1) = x$ and $\rho_{f(1)}(x) = xf(1) = f(x)$, these are inverses of each other. The rest is left to the reader. □

Corollary 4.1.5. *In the context of Lemma* 4.1.4, *the correspondence* $f \mapsto f(1)$ *defines an isomorphism of rings* $\operatorname{Hom}_{A^e}(A, A) \cong Z(A)$.

Proof. Take $M = A$ in Lemma 4.1.4 and note that $A^A = Z(A)$. □

Corollary 4.1.6. *In the context of Lemma* 4.1.4, *let*

$$0 : J_{A/R} = \{x \in A^e \mid yx = 0, \textit{ for all } y \in J_{A/R}\}$$

be the right annihilator of $J_{A/R}$ *in* A^e. *Then the following are true.*

(1) $\operatorname{Hom}_{A^e}(A, A^e) \cong 0 : J_{A/R}$.

(2) *If* A *is* R*-separable, then* $\mu(0 : J_{A/R}) = Z(A)$.

Proof. By Exercise 4.1.10, $J_{A/R}$ is generated as a left ideal by elements of the form $a \otimes 1 - 1 \otimes a$. Take $M = A^e$ in Lemma 4.1.4. Then

$$\begin{aligned}\operatorname{Hom}_{A^e}(A, A^e) &\cong \left(A^e\right)^A \\ &= \{x \in A^e \mid (a \otimes 1 - 1 \otimes a)x = 0, \text{ for all } a \in A\} \\ &= 0 : J_{A/R}.\end{aligned}$$

If A is R-separable, then A is A^e-projective and by Proposition 1.3.20, the functor $\operatorname{Hom}_{A^e}(A, \cdot)$ is exact. The exact sequence

$$A^e \xrightarrow{\mu} A \to 0$$

gives rise to the exact sequence

$$\operatorname{Hom}_{A^e}(A, A^e) \xrightarrow{\mu \circ ()} \operatorname{Hom}_{A^e}(A, A) \to 0.$$

By Lemma 4.1.4 and Part (1), we get $\mu(0 : J_{A/R}) = Z(A)$. □

We end this section with another corollary to Lemma 4.1.4 which provides an important necessary and sufficient condition for an R-algebra A to be separable.

Corollary 4.1.7. *An R-algebra A is separable if and only if $(\cdot)^A$ is a right exact functor.*

Proof. By Proposition 1.3.20, the functor $\operatorname{Hom}_{A^e}(A, \cdot)$ is right exact if and only if A is a projective A^e-module. □

1.1. Exercises.

Exercise 4.1.8. Prove that if S is a commutative separable R-algebra, then the separability idempotent is unique. (Hint: Lemma 2.2.8.)

Exercise 4.1.9. Let R be a commutative ring. Prove:

(1) R is a separable R-algebra. The zero ring, (0), is a separable R-algebra.
(2) If $W \subseteq R$ is a multiplicative set, then the localization R_W is a separable R-algebra.
(3) If I is an ideal in R, then R/I is a separable R-algebra.

Exercise 4.1.10. In the context of Eq. (4.4), prove that $J_{A/R}$ is the left ideal in $A \otimes_R A^o$ generated by the set $\{a \otimes 1 - 1 \otimes a \mid a \in A\}$.

Exercise 4.1.11. Let R be a commutative ring.

(1) Let $R \oplus R$ be the ring direct sum of two copies of R. Let $e_1 = (1, 0)$ and $e_2 = (0, 1)$ be the orthogonal idempotents in $R \oplus R$. Use Exercise 4.1.10 to show that $e = e_1 \otimes e_1 + e_2 \otimes e_2$ is a separability idempotent. Hence, $R \oplus R$ is separable over R.
(2) Let $R^n = R \oplus \cdots \oplus R$ be the ring direct sum of n copies of R. Show that R^n is separable over R. We call R^n the *trivial* commutative separable extension of R of rank n. The trivial extension R^n is revisited in Section 4.6.3.

Exercise 4.1.12. Show that $\mathbb{C}$ is separable over $\mathbb{R}$. (Hint: Use Exercise 4.1.10 to show that $\frac{1}{2}(1 \otimes 1 - i \otimes i)$ is a separability idempotent.)

Exercise 4.1.13. Let R be any commutative ring in which 2 is invertible. The ring of quaternions over R is the free R-module $\Lambda = R1 + Ri + Rj + Rij$ with basis $1, i, j, ij$. Multiplication is defined by extending the relations

$$i^2 = j^2 = (ij)^2 = -1, \quad ij = -ji$$

by associativity and distributivity (see Example 1.1.1).

(1) Show that the center of Λ is equal to R.

(2) Show that Λ is isomorphic to Λ^o. (Hint: $a + bi + cj + dij \mapsto a - bi - cj - dij$ is an R-algebra isomorphism.)

(3) Show that Λ is a separable R-algebra. (Hint: $e = \frac{1}{4}(1 \otimes 1 - i \otimes i - j \otimes j - ij \otimes ij)$ is a separability idempotent.)

Exercise 4.1.14. Show that if A is a separable R-algebra and e is a separability idempotent, then $(A \otimes_R A^o)e = (A \otimes_R 1)e = (1 \otimes_R A^o)e$.

Exercise 4.1.15. Prove the following generalization of Lemma 4.1.4. Let R be a commutative ring, A an R-algebra, and S a commutative R-subalgebra of A. If M is a left $S \otimes_R A^o$-module, then the assignment $f \mapsto f(1)$ induces an isomorphism of R-modules $\operatorname{Hom}_{S\otimes_R A^o}(A, M) \cong M^S$. If $g : M \to N$ is a homomorphism of left $S \otimes_R A^o$-modules, then the diagram

$$\begin{array}{ccc} \operatorname{Hom}_{S\otimes_R A^o}(A, M) & \xrightarrow{g\circ(\cdot)} & \operatorname{Hom}_{S\otimes_R A^o}(A, N) \\ \downarrow & & \downarrow \\ M^S & \xrightarrow{g} & N^S \end{array}$$

commutes. The functors $\operatorname{Hom}_{S\otimes_R A^o}(A, \cdot)$ and $(\cdot)^S$ are naturally isomorphic and both are left exact.

Exercise 4.1.16. Let A be an R-algebra with structure homomorphism $\theta : R \to Z(A)$. Show that if I is an ideal in R and $I \subseteq \ker\theta$, then A is R-separable if and only if A is R/I-separable.

Exercise 4.1.17 ([**OS75**, Example 2.5 (c)]). In $\mathbb{C} = \mathbb{R}[i]$, let ζ be a primitive cube root of 1. Let $R = \mathbb{Z}[\sqrt{3}i]$ and $A = \mathbb{Z}[\zeta]$, both viewed as subrings of $\mathbb{C}$.

(1) Show that R is a subring of A. Show that $2\zeta \in R$.

(2) Show that A is separable over R. (Hint: $e = 1 \otimes 1 + 1 \otimes \zeta - \zeta \otimes 1$ is a separability idempotent.)

(3) Show that A is not flat over R.

2. Examples of Separable Algebras

Three standard examples of separable algebras are presented below. The first is a matrix algebra, the second a group algebra, and the third a quadratic Kummer extension.

Example 4.2.1. Let R be a commutative ring. In this example we show that $M_n(R)$, the ring of n-by-n matrices over R, is R-separable. Let e_{ij} be the elementary matrix having a single 1 in position (i, j) and 0 elsewhere.

Notice that

$$e_{k\ell}e_{ij} = \begin{cases} e_{kj} & \text{if } \ell = i, \\ 0 & \text{otherwise.} \end{cases}$$

Fix j and define

$$e = \sum_{i=1}^{n} e_{ij} \otimes e_{ji}$$

which is an element of the enveloping algebra of $M_n(R)$. Then

$$\mu(e) = \sum_i e_{ij}e_{ji} = \sum_i e_{ii} = 1.$$

For any k and l we see that

$$(e_{kl} \otimes 1 - 1 \otimes e_{kl})e = \sum_i (e_{kl}e_{ij} \otimes e_{ji} - e_{ij} \otimes e_{ji}e_{kl}) = e_{kj} \otimes e_{jl} - e_{kj} \otimes e_{jl} = 0.$$

Since the matrices e_{kl} generate $M_n(R)$ as an R-module, Exercise 4.1.10 shows that $J_{A/R}e = 0$. By Proposition 4.1.2 we see that $M_n(R)$ is a separable R-algebra and e is a separability idempotent.

As we will see later, matrix algebras over commutative rings play a fundamental role in the study of noncommutative separable algebras. For example, if A is a central separable R-algebra, then in Theorem 7.1.4 we show that A is an R-progenerator. If A is of constant rank, then in Corollary 10.3.10 we show that there is a faithfully flat R-algebra S such that upon extension of the ring of scalars to S, $A \otimes_R S$ is isomorphic to a matrix algebra $M_n(S)$. In the terminology of Remark 5.5.11, we say that the central separable R-algebra A is a *twisted form of matrices.*

Example 4.2.2. Let G be a finite group of order n and R a commutative ring such that $n = n \cdot 1$ is a unit in R. Starting with the identity element, let $G = \{1 = \sigma_1, \sigma_2, \dots, \sigma_n\}$ be an enumeration of the elements of G. Let $R(G) = R \cdot 1 \oplus R \cdot \sigma_2 \oplus \cdots \oplus R \cdot \sigma_n$ be the group algebra. In this example we show that $R(G)$ is R-separable. Let

$$e = \frac{1}{n} \sum_{\sigma \in G} \sigma \otimes \sigma^{-1}$$

which is an element of the enveloping algebra of $R(G)$. Then

$$\mu(e) = \frac{1}{n} \sum_{\sigma \in G} \sigma\sigma^{-1} = \frac{1}{n} \sum_{\sigma \in G} 1 = 1.$$

If we fix any $\tau \in G$, then as sets we have $G = \{\sigma\tau \mid \sigma \in G\}$, hence

$$(\tau \otimes 1)e = \frac{1}{n}\sum_{\sigma \in G} \tau\sigma \otimes \sigma^{-1} = \frac{1}{n}\sum_{\rho \in G} \rho \otimes \rho^{-1}\tau$$
$$= \frac{1}{n}\sum_{\rho \in G} \rho \otimes \tau * \rho^{-1} = (1 \otimes \tau)e$$

where we write $x * y = yx$ as the product in the opposite algebra. The group algebra $R(G)$ is generated over R by the basis elements $\tau \in G$. This together with Exercise 4.1.10 and Proposition 4.1.2 shows that e is a separability idempotent for $R(G)$. Hence, the group algebra $R(G)$ is a separable R-algebra. For the converse of this result see Exercise 4.5.15.

Example 4.2.3. Let R be an integral domain with quotient field K and assume $2 = 2 \cdot 1$ is a unit in R. In this example, we see that an element of order two in $\operatorname{Pic} R$, the Picard group of R, gives rise to a separable quadratic extension of R. Let M be an invertible R-module (M is projective and has rank one). Since M is flat, $M \to M \otimes_R K$ is one-to-one. Hence, we identify M with a finitely generated R-submodule I of $K \cong M \otimes_R K$. Assume that the class of I in $\operatorname{Pic}(R)$ has order 1 or 2. By Proposition 2.6.9, $I \otimes_R I \cong R$ is free. The multiplication map $K \otimes_R K \to K$ of Exercise 1.3.9 induces an R-module homomorphism $\phi : I \otimes_R I \to I^2$. Therefore, the image of ϕ is principal. Since K is a faithful R-module, for some $\alpha \in K - (0)$, $I^2 = R\alpha$ is isomorphic to R. Since the domain and range of ϕ are both free of rank one, ϕ is an isomorphism of R-modules. Let S be the R-module direct sum $R \oplus I$. We turn S into a commutative R-algebra using α to define a multiplication operation:

$$(a \oplus b)(c \oplus d) = (ac + \alpha^{-1}(bd)) \oplus (ad + cb).$$

The reader should verify that this multiplication rule is associative, commutative, distributes over addition, and that $1 \oplus 0$ is the identity element.

We show S is separable by constructing the separability idempotent for S. By assumption, there exist elements $a_1, \dots, a_n, b_1, \dots, b_n$ in I and $\sum_i a_i b_i = \alpha$. In S define two sequences

$$x_1 = 0 \oplus a_1, \dots, x_n = 0 \oplus a_n, x_{n+1} = 1 \oplus 0$$

and

$$y_1 = 0 \oplus b_1, \dots, y_n = 0 \oplus b_n, y_{n+1} = 1 \oplus 0.$$

Notice that

$$\begin{aligned}\sum_{i=1}^{n+1} x_i y_i &= x_1 y_1 + \cdots + x_n y_n + x_{n+1} y_{n+1} \\ &= \left(\alpha^{-1}(a_1 b_1 + \cdots + a_n b_n) + 1\right) \oplus 0 \\ &= 2 \oplus 0.\end{aligned}$$

In the enveloping algebra S^e, define

$$e = \frac{1}{2} \sum_{i=1}^{n+1} x_i \otimes y_i.$$

By the above,

$$\mu(e) = \frac{1}{2} \sum_i x_i y_i = 1 \oplus 0 = 1.$$

By Exercise 4.1.10, $J_{S/R}$ is generated by elements of the form $x \otimes 1 - 1 \otimes x$, where $x \in S = R \oplus I$. Since $a \otimes 1 - 1 \otimes a = 0$, if $a \in R$, it follows that $J_{S/R}$ is generated by elements of the form $x \otimes 1 - 1 \otimes x$, where $x \in 0 \oplus I$. Notice that $(0 \oplus I)^2 \subseteq R \oplus 0$. Therefore, if $x \in 0 \oplus I$, then

$$\begin{aligned}x \otimes 1 \cdot e &= \frac{1}{2}\left(\sum_{j=1}^{n} x x_j \otimes y_j + x \otimes 1\right) \\ &= \frac{1}{2}\left(\sum_{j=1}^{n} 1 \otimes x x_j y_j + x \otimes 1\right) \\ &= \frac{1}{2}\left(1 \otimes \left(\sum_{j=1}^{n} x_j y_j\right) \cdot 1 \otimes x + x \otimes 1\right) \\ &= \frac{1}{2}(1 \otimes x + x \otimes 1)\end{aligned}$$

which by a similar argument is equal to $1 \otimes x \cdot e$. Then $J_{S/R} e = (0)$. By Proposition 4.1.2, e is a separability idempotent for S and S is separable over R. This example is a small part of Kummer Theory. In Example 12.9.6 we generalize this example to elements of order n in the Picard group. We also show that S is a Galois extension of R with a cyclic group of order two.

3. Separable Algebras Under a Change of Base Ring

We begin this section with a general result on the tensor product of separable algebras which has two important consequences. First it shows that for a commutative ring R, tensor product defines a product on the category of separable R-algebras. At the same time, it shows that the property of an algebra being separable is preserved under a change of base ring. The first results on descent of separable algebras (Proposition 4.3.3 and its corollaries) are also proved. The results in this section have their origins in [**AG60a**]. Our approach very closely follows that of [**DI71**].

Theorem 4.3.1. *Let R be a commutative ring and S_1 and S_2 commutative R-algebras. Let A_1 be a separable S_1-algebra and A_2 a separable S_2-algebra. Then $A_1 \otimes_R A_2$ is separable over $S_1 \otimes_R S_2$.*

Proof. We show that $(\cdot)^{A_1 \otimes_R A_2}$ is an exact functor on two-sided $A_1 \otimes_R A_2/S_1 \otimes_R S_2$-modules and then apply Corollary 4.1.7. Start with an exact sequence

$$M \xrightarrow{f} N \to 0$$

of two-sided $A_1 \otimes_R A_2/S_1 \otimes_R S_2$-modules. The diagram of ring homomorphisms

$$\begin{array}{ccc} A_1 & \longrightarrow & A_1 \otimes_R A_2 \\ \uparrow & & \uparrow \\ S_1 & \longrightarrow & S_1 \otimes_R S_2 \end{array}$$

commutes so M and N can be turned into two-sided A_1/S_1-modules. Since A_1 is separable over S_1, the sequence

$$(M)^{A_1} \xrightarrow{f} (N)^{A_1} \to 0$$

is exact. From Exercise 1.3.8, the diagram

$$\begin{array}{ccccc} & & A_1 \otimes_R A_2 & & \\ & \overset{\rho_1}{\nearrow} & & \overset{\rho_2}{\nwarrow} & \\ A_1 & & & & A_2 \\ & \nwarrow & & \nearrow & \\ & & R & & \end{array}$$

commutes and every element in $\operatorname{im}(\rho_1)$ commutes with every element in $\operatorname{im}(\rho_2)$. So we turn M^{A_1} and N^{A_1} into two-sided A_2/S_2-modules. Since A_2 is separable over S_2, the sequence

$$\left(M^{A_1}\right)^{A_2} \xrightarrow{f} \left(N^{A_1}\right)^{A_2} \to 0$$

is exact. As a ring $A_1 \otimes_R A_2$ is generated by the images of ρ_1 and ρ_2. So $\left(M^{A_1}\right)^{A_2} \subseteq M^{A_1 \otimes_R A_2}$. Conversely, $M^{A_1 \otimes_R A_2} \subseteq M^{A_1 \otimes_R 1} \cap M^{1 \otimes_R A_2} = \left(M^{A_1}\right)^{A_2}$. □

Corollary 4.3.2. *If A and B are separable R-algebras and S is a commutative R-algebra, then*

(1) *$A \otimes_R B$ is a separable R-algebra, and*

(2) *$A \otimes_R S$ is a separable S-algebra.*

Proposition 4.3.3 and its corollaries are examples of "Descent Theorems". They allow us to conclude that an algebra is separable, if it becomes separable under a "sufficiently nice" change of base. See Section 8.1.5 for more such theorems.

Proposition 4.3.3. *Let R be a commutative ring and S_1 and S_2 commutative R-algebras. Let A_1 be any S_1-algebra and A_2 any S_2-algebra such that $A_1 \otimes_R A_2$ is separable over $S_1 \otimes_R S_2$. If A_2 is faithful as an R-module and $R \cdot 1$ is an R-module direct summand of A_2, then A_1 is separable over S_1.*

Proof. We show that $(\cdot)^{A_1}$ is right exact and apply Corollary 4.1.7. Let M be a two-sided A_1/S_1-module. The reader should verify that $M \otimes_R A_2$ is therefore a two-sided $A_1 \otimes_R A_2/S_1 \otimes_R S_2$-module. By our hypothesis, the sequence of natural maps $0 \to R \to A_2$ splits. That is, $A_2 = L \oplus R \cdot 1$ as R-modules and there is an isomorphism

$$M \otimes_R A_2 = M \otimes_R (L \oplus R \cdot 1) \cong (M \otimes_R L) \oplus (M \otimes_R R \cdot 1).$$

The reader should verify that in fact $M \otimes_R R \cdot 1$ is a two-sided A_1/S_1-module direct summand of $M \otimes_R A_2$, hence there is a projection

$$M \otimes_R A_2 \xrightarrow{\pi} M \otimes_R R \cdot 1 \tag{4.6}$$

of two-sided A_1/S_1-modules. Applying the functor $(\cdot)^{A_1}$ to (4.6), we get the R-module homomorphism

$$(M \otimes_R A_2)^{A_1} \xrightarrow{\pi} (M \otimes_R R \cdot 1)^{A_1}.$$

Since $\left(M \otimes_R A_2\right)^{A_1 \otimes_R A_2} \subseteq (M \otimes_R A_2)^{A_1}$, the map π restricted to $\left(M \otimes_R A_2\right)^{A_1 \otimes_R A_2}$ takes values in $(M \otimes_R R \cdot 1)^{A_1}$. Using the fact that A_2 is R-faithful, the reader should verify that $M^{A_1} \otimes_R R \cdot 1 = (M \otimes_R R \cdot 1)^{A_1}$ and the sequence

$$\left(M \otimes_R A_2\right)^{A_1 \otimes_R A_2} \xrightarrow{\pi} M^{A_1} \otimes_R R \cdot 1 \to 0 \tag{4.7}$$

is exact. Consider an arbitrary exact sequence

$$M \xrightarrow{f} N \to 0 \tag{4.8}$$

of two-sided A_1/S_1-modules. Combine (4.7) with (4.8) to get the diagram

$$\begin{array}{ccccc} \left(M \otimes_R A_2\right)^{A_1 \otimes_R A_2} & \xrightarrow{f \otimes 1} & \left(N \otimes_R A_2\right)^{A_1 \otimes_R A_2} & \longrightarrow & 0 \\ \downarrow \pi & & \downarrow \pi & & \\ M^{A_1} \otimes_R R \cdot 1 & \xrightarrow{f \otimes 1} & N^{A_1} \otimes_R R \cdot 1 & \longrightarrow & 0 \end{array} \tag{4.9}$$

which commutes. The functor $(\cdot) \otimes_R A_2$ is always right exact, and by assumption the functor $(\cdot)^{A_1 \otimes_R A_2}$ is right exact. Therefore the top row of

(4.9) is exact. By (4.7), π is onto, which implies the bottom row of (4.9) is exact. Since $R \to R \cdot 1$ is an isomorphism, $f : M^{A_1} \to N^{A_1}$ is onto. □

Corollary 4.3.4. *Let A_1 and A_2 be R-algebras such that A_2 is faithful over R, and $R \cdot 1$ is an R-module direct summand of A_2. If $A_1 \otimes_R A_2$ is separable over R, then A_1 is separable over R.*

Proof. Take $S_1 = S_2 = R$ in Proposition 4.3.3. □

Corollary 4.3.5. *Let S be a commutative faithful R-algebra such that $R \cdot 1$ is an R-module direct summand of S. Let A be an R-algebra such that $A \otimes_R S$ is S-separable. Then the following are true.*

(1) *A is R-separable.*
(2) *If the image of $R \otimes_R S \to A \otimes_R S$ is equal to the center of $A \otimes_R S$, then $R \cdot 1$ is equal to the center of A.*

Proof. For the first part, take $A_1 = A$, $A_2 = S_2 = S$ and $S_1 = R$ in Proposition 4.3.3. For the second part, notice that

$$1 \otimes_R S = Z(A \otimes_R S) = \left(A \otimes_R S\right)^{A \otimes_R S}$$

maps onto $A^A = Z(A)$ by the proof of Proposition 4.3.3. But the projection map π is the left inverse to $R \to S$ which has image $R \cdot 1$. Hence $1 \otimes S$ projects onto $1 \otimes R \cong R \cdot 1$. □

Proposition 4.3.6. *If A is a separable R-algebra and I is a two-sided ideal of A, then A/I is a separable R-algebra. Moreover,*

$$Z(A/I) = \frac{Z(A) + I}{I}.$$

Proof. Let M be a two-sided $(A/I)/R$-module. Then M can be viewed as a two-sided A/R-module using the natural homomorphism $\eta : A \to A/I$. Then $M^A = M^{A/I}$. Then A/I is R-separable by Corollary 4.1.7. Now

$$A \to A/I \to 0$$

is an exact sequence of two-sided A/R-modules. Since A is R-separable,

$$A^A \to (A/I)^A \to 0$$

is exact. So $Z(A/I)$ is the image under η of $Z(A)$. □

Corollary 4.3.7. *Let A_1 be an R_1-algebra and A_2 an R_2-algebra, where R_1 and R_2 are commutative rings. Then $A_1 \oplus A_2$ is a separable $R_1 \oplus R_2$-algebra if and only if both A_1 and A_2 are separable over R_1 and R_2 respectively.*

Proof. The proof follows from Corollary 4.1.7 and Proposition 4.3.6. □

4. Homomorphisms of Separable Algebras

The first three results proved below will all play fundamental roles in the study of separable algebras. Theorem 4.4.3 is quite general and more technical than the other two. It applies to any homomorphism of R-algebras $\theta : A \to B$, provided A is separable. Theorem 4.4.1 is a theorem of permanence which allows us to show that a module is A-projective, if it is R-projective. Theorem 4.4.2 shows transitivity of separability, plus another descent criterion for separability. As general references for this section, we recommend [**DI71**], [**Mil80**, § I.3], [**AG60a**], and [**KO74b**]. For the most part, our presentation follows the first two.

Theorem 4.4.1. *Let R be a commutative ring and A a separable R-algebra. By the structure homomorphism $\theta : R \to A$, any left A-module M inherits the structure of a left R-module.*

(1) *Let $0 \to L \to N \xrightarrow{\eta} M \to 0$ be an exact sequence of left A-modules. If the sequence is split exact in ${}_R\mathfrak{M}$, then it is split-exact in ${}_A\mathfrak{M}$.*
(2) *Let M be a left A-module. If M is R-projective, then M is A-projective.*

Proof. By Proposition 1.1.5, (2) follows from (1). Suppose there exists an R-module homomorphism $\psi : M \to N$ with $\eta\psi = 1_M$. Since both N and M are left A-modules, Lemma 1.3.17 shows that $\operatorname{Hom}_R(M, N)$ can be given the structure of a left A^e-module under the operation induced by

$$\big[(x \otimes y) \cdot f\big](m) = x \cdot f(y \cdot m),$$

where $x \otimes y \in A \otimes_R A^o$, $f \in \operatorname{Hom}_R(M, N)$, and $m \in M$. Since A is R-separable, let $e \in A^e$ be a separability idempotent for A. Define $\psi' = e\psi$. That is, if $e = \sum_i x_i \otimes y_i$ and $m \in M$, then

$$\psi'(m) = \sum_i x_i\psi(y_i m).$$

Since η is an A-module homomorphism and $\mu(e) = 1$, we have

$$\eta\psi'(m) = \eta\big(\sum_i x_i\psi(y_i m)\big) = \sum_i x_i\eta \cdot \psi(y_i m) = \sum_i x_i y_i m = m,$$

for all $m \in M$. Since $J_{A/R}e = 0$, we have $(a\otimes 1 - 1\otimes a)\psi' = (a\otimes 1 - 1\otimes a)e\cdot\psi = 0$, for all $a \in A$. It follows that $a\psi'(m) = a \otimes 1 \cdot \psi'(m) = 1 \otimes a \cdot \psi'(m) = \psi'(am)$, for all $a \in A$, $m \in M$. □

Theorem 4.4.2. *Let S be a commutative R-algebra and let A be an S-algebra. Then A is also an R-algebra.*

(1) *(Separable over Separable is Separable) If S is separable over R and A is separable over S, then A is separable over R.*

(2) *If A is separable over R, then A is separable over S.*

(3) *If A is separable over R and A is an S-progenerator, then S is separable over R.*

Proof. (1): Any two-sided A/R-module M is also a two-sided S/R-module. Given any $x \in M^S$, $a \in A$ and $s \in S$, the equations

$$\begin{aligned} s \cdot (a \cdot x) &= a \cdot (s \cdot x) \\ &= a \cdot (x \cdot s) \\ &= (a \cdot x) \cdot s \end{aligned}$$

show that $ax \in M^S$. It follows that M^S is a two-sided A/S-module, with $\left(M^S\right)^A = M^A$. For any two-sided A/R-modules M and N, if

$$M \xrightarrow{f} N \to 0$$

is exact, then by Corollary 4.1.7 applied to the separable R-algebra S, it follows that

$$M^S \xrightarrow{f} N^S \to 0$$

is exact. But $\left(M^S\right)^A = M^A$ and $\left(N^S\right)^A = N^A$. By Corollary 4.1.7 applied to the separable S-algebra A, it follows that

$$M^A \xrightarrow{f} N^A \to 0$$

is exact. Hence A is R-separable, which proves (1).

(2): In the commutative diagram

$$\begin{array}{ccccccccc} 0 & \longrightarrow & J_{A/R} & \longrightarrow & A \otimes_R A^o & \xrightarrow{\mu} & A & \longrightarrow & 0 \\ & & \downarrow & & \downarrow & & \downarrow{=} & & \\ 0 & \longrightarrow & J_{A/S} & \longrightarrow & A \otimes_S A^o & \xrightarrow{\mu} & A & \longrightarrow & 0 \end{array}$$

all of the vertical maps are onto (Exercise 1.3.12). A separability idempotent for A/R maps to a separability idempotent for A/S.

(3): By Part (2), A is separable over S. Since A is S-projective, so is A^o. The reader should verify (for example, by an argument involving dual bases) that $A \otimes_R A^o$ is projective over $S \otimes_R S$. Because A is separable over R, A is projective as a left $A \otimes_R A^o$-module under the μ-action. By Proposition 1.1.8, it follows that A is projective as a left $S \otimes_R S$-module. By Proposition 2.4.6, $S \cdot 1$ is an S-module direct summand of A, so we can write $A = S \oplus L$ for some L. It follows that S is also an $S \otimes_R S$-module direct summand of A under the μ-action. Hence S is $S \otimes_R S$-projective and S is R-separable. $\square$

Before stating the next theorem, it is necessary to establish some notation. Let R be a commutative ring and $\theta : A \to B$ an R-algebra homomorphism. Assuming A is separable over R, we consider properties of θ under various hypotheses on B. Consider the commutative diagram

$$\text{(4.10)} \qquad \begin{array}{ccc} A \otimes_R B^0 & \xrightarrow{\quad\gamma\quad} & B \\ & {\scriptstyle \theta\otimes 1}\searrow \quad \nearrow{\scriptstyle \mu} & \\ & B \otimes_R B^o & \end{array}$$

where γ is defined to be the R-algebra homomorphism $\theta \otimes 1$, followed by the left $B \otimes_R B^o$-module homomorphism μ. Therefore, all of the terms in (4.10) can be viewed as left $A \otimes_R B^o$-modules. Notice that $\gamma(x \otimes y) = \theta(x)y$, hence the left $A \otimes_R B^o$-module action on B is given by $(a \otimes b) \cdot x = \theta(a)xb$. We emphasize that γ is not a homomorphism of rings unless the image of θ is a subring of the center of B.

Theorem 4.4.3. *Let R be a commutative ring and $\theta : A \to B$ an R-algebra homomorphism. If A is separable over R, then the following are true.*

(1) *The sequence of left $A \otimes_R B^o$-modules*

$$A \otimes_R B^o \xrightarrow{\gamma} B \to 0$$

is split-exact. The kernel of γ is idempotent generated, and B is projective as a left $A \otimes_R B^o$-module.

(2) *If B is a flat left R-module, then B is a flat left A-module.*

(3) *If B is a projective left R-module, then B is a projective left A-module.*

(4) *If A is commutative,* $\mathrm{im}(\theta) \subseteq Z(B)$, *and B is separable over R, then B is separable over A.*

Proof. (1): Since A/R is separable, there is a split-exact sequence

$$\text{(4.11)} \qquad 0 \to J_{A/R} \to A^e \xrightarrow{\mu} A \to 0$$

of left A^e-modules. The R-algebra homomorphism $1 \otimes \theta : A^e \to A \otimes_R B^o$ allows us to view $A \otimes_R B^o$ as a left $A \otimes_R B^o$ right A^e-bimodule. Applying the functor $(A \otimes_R B^o) \otimes_{A^e} (\,)$ to sequence (4.11) yields the split-exact sequence

$$\text{(4.12)} \qquad 0 \to (A \otimes_R B^o) \otimes_{A^e} J_{A/R} \to$$
$$(A \otimes_R B^o) \otimes_{A^e} A^e \xrightarrow{1\otimes\mu} (A \otimes_R B^o) \otimes_{A^e} A \to 0$$

of left $A \otimes_R B^o$-modules. By Lemma 1.3.1, the middle term in (4.12) is isomorphic to $A \otimes_R B^o$. Define $\phi : B \to (A \otimes_R B^o) \otimes_{A^e} A$ by $x \mapsto 1 \otimes x \otimes 1$. The reader should verify that ϕ is onto. Notice $a \otimes b \cdot \phi(x) = a \otimes b \cdot 1 \otimes x \otimes 1 = a \otimes xb \otimes 1 = 1 \otimes xb \otimes a = 1 \otimes \theta(a)xb \otimes 1 = \phi(a \otimes b \cdot x)$, so ϕ is a well defined

$A \otimes_R B^o$-module epimorphism. To see that ϕ is one-to-one, consider the $\mathbb{Z}$-module homomorphisms

(4.13)

$$B \xrightarrow{\phi} (A \otimes_R B^o) \otimes_{A^e} A \xrightarrow{\theta \otimes 1 \otimes \theta} (B \otimes_R B^o) \otimes_{A^e} B \xrightarrow{\xi} (B \otimes_R B^o) \otimes_{B^e} B \xrightarrow{\cong} B$$

where ξ is from Exercise 1.3.12, and the last isomorphism is Lemma 1.3.1. In (4.13), the composite map is the identity on B. This shows ϕ is an isomorphism, hence the last term in (4.12) is isomorphic to B. The reader should verify that γ is the map induced by $1 \otimes \mu$, and that

$$0 \to \ker(\gamma) \to A \otimes_R B^o \xrightarrow{\gamma} B \to 0$$

is a split-exact sequence of left $A \otimes_R B^o$-modules. The kernel of γ is idempotent generated, by Lemma 1.1.20. This proves (1).

(2): Since B is a flat left R-module, $A \otimes_R B^o$ is a flat left A-module, by Theorem 1.3.2. By Exercise 1.3.7, a projective module is flat. Part (1) and Exercise 2.4.16 imply that B is a flat left A-module.

(3): This can be proved using the method of Part (2). Alternatively, this follows from Theorem 4.4.1.

(4): This is Theorem 4.4.2 (2). □

Corollary 4.4.4. *Let A be a ring and $Z(A)$ the center of A. Let R and S be subrings of the center of A forming the tower of subrings*

$$R \subseteq S \subseteq Z(A) \subseteq A.$$

Then any two of the following statements imply the third.

(1) *S is a separable R-algebra and a finitely generated projective R-module.*

(2) *A is a separable S-algebra and a finitely generated projective S-module.*

(3) *A is a separable R-algebra and a finitely generated projective R-module.*

Proof. (1) plus (2) implies (3): This follows from Proposition 1.1.8 and Theorem 4.4.2 (1).

(1) plus (3) implies (2): Since A is a finitely generated R-module, A is a finitely generated S-module. Since A is projective over R and S is separable over R, by Theorem 4.4.1, A is projective over S. Since A is separable over R, by Theorem 4.4.2 (2), A is separable over S.

(2) plus (3) implies (1): By Theorem 4.4.2 (3), S is separable over R. By Proposition 2.4.6, $S \cdot 1$ is a S-module direct summand of A. Therefore,

the R-module S is isomorphic to a direct summand of the R-progenerator A. This shows that S is a finitely generated projective R-module. □

The next finiteness criterion for a separable algebra over a commutative ring is due to O. Villamayor and D. Zelinsky ([**VZ66**]).

Proposition 4.4.5. *Let R be a commutative ring and A a separable R-algebra which is projective as an R-module. Then the following are true.*

(1) *A is finitely generated as an R-module.*

(2) *If A is a faithful R-module, then A is an R-module progenerator.*

Proof. Part (2) follows from Part (1) and Corollary 1.1.16. We now prove (1). Since A and A^o are identical as R-modules, it is enough to show A^o is finitely generated. Let $\{f_i, a_i\}$ be a dual basis for A^o over R with $a_i \in A^o$ and $f_i \in \operatorname{Hom}_R(A^o, R)$. For every $a \in A^o$, $f_i(a) = 0$ for almost all i and

$$a = \sum_i f_i(a)a_i.$$

Identify $A \otimes_R R$ with A, and consider $1_A \otimes f_i$ as an element of $\operatorname{Hom}_A(A^e, A)$. The set $\{1_A \otimes f_i, 1 \otimes a_i\}$ forms a dual basis for A^e as a projective left A-module. That is,

$$u = \sum_i (1_A \otimes f_i)(u) \cdot (1 \otimes a_i)$$

for all $u \in A^e$. Applying the multiplication map μ and setting $u = (1 \otimes a)e$ where e is a separability idempotent for A over R, we obtain

$$a = \mu\big((1 \otimes a)e\big) = \sum_i \Big[(1_A \otimes f_i)\big((1 \otimes a)e\big)\Big] \cdot a_i \tag{4.14}$$

for each $a \in A^o$. Since

$$(1_A \otimes f_i)\big((1 \otimes a)e\big) = (1_A \otimes f_i)\big((a \otimes 1)e\big) = (a \otimes 1)\big((1_A \otimes f_i)(e)\big)$$

the set of subscripts i for which $(1_A \otimes f_i)\big((1 \otimes a)e\big)$ is not equal to zero is contained in the finite set of subscripts for which $(1_A \otimes f_i)(e)$ is not equal to zero. This latter set is independent of a. Therefore the summation (4.14) may be taken over a fixed finite set. Writing

$$e = \sum_j x_j \otimes y_j$$

we have from (4.14) that $a = \sum_{i,j} x_j f_i(y_j a) a_i = \sum_{i,j} f_i(y_j a) x_j a_i$ for each $a \in A^o$. This shows that the finite set $\{x_j a_i\}$ generates A^o over R. □

The following result is attributed to F. DeMeyer.

Corollary 4.4.6. *Assume A is a separable R-algebra which as an R-module is faithful and projective. Then the following are true.*

(1) *A is an R-module progenerator.*

(2) *If S is a commutative separable R-subalgebra of A, then S is an R-module progenerator.*

Proof. Part (1) follows from Proposition 4.4.5. We prove (2). Identify R with the subring $R \cdot 1$ of A. Since A is finitely generated over R, A is finitely generated over S. By Theorem 4.4.1, A is a projective S-module. Since A is faithful over S, by Corollary 1.1.16, A is an S-progenerator. By Proposition 2.4.6, $S \cdot 1$ is an S-module direct summand of A. Therefore, $S \cdot 1$ is an R-module direct summand of A. Then S is finitely generated and projective over R, and by Corollary 1.1.16, S is an R-progenerator. □

Corollary 4.4.7. *Let S and A be separable R-algebras and $f : S \to A$ an R-algebra homomorphism. Assume the image of f is a commutative R-subalgebra of A, and A is an R-module progenerator. The following are true:*

(1) *The diagram of R-algebra homomorphisms*

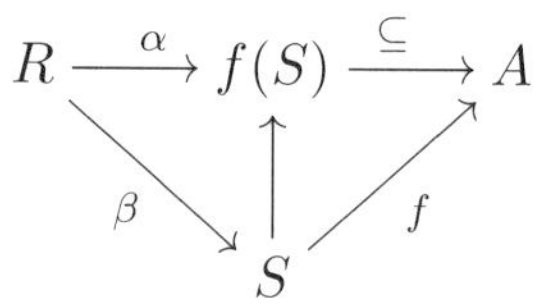

commutes.

(2) *α and β are one-to-one.*

(3) *The kernel of f is idempotent generated.*

(4) *If S is commutative and connected, then f is a monomorphism.*

Proof. By Proposition 2.4.6, α is one-to-one. By Proposition 4.3.6, the image of f is a commutative separable R-subalgebra of A. By Corollary 4.4.6, $f(S)$ is an R-progenerator. By Theorem 4.4.1, $f(S)$ is projective over S. By Exercise 1.1.22, the kernel of f is idempotent generated. The rest is left to the reader. □

Let A be an R-algebra with structure homomorphism $\theta : R \to A$. Let $\sigma : A \to R$ be an R-algebra homomorphism such that $\sigma\theta = 1$. We say σ is a section to θ. If A is separable over R, then Corollary 4.4.8 shows that σ is determined by a central idempotent of A. It is easy to construct examples for which A is not separable and the kernel of a section is not idempotent generated. The reader is encourage to complete Exercise 4.4.12 before continuing.

Corollary 4.4.8. *Let R be a commutative ring and A a separable R-algebra. Then the following are true.*

(1) *There is a one-to-one correspondence between the set of all R-algebra homomorphisms $\sigma : A \to R$, and the set of all central idempotents e in A such that the composite mapping*

$$R \to Re \to Ae$$

is one-to-one and onto. In this case $\sigma(e) = 1$ and $\sigma(x)e = xe$ for all $x \in A$.

(2) *Suppose R is connected, $\sigma_1, \dots, \sigma_n$ are distinct R-algebra homomorphisms from A to R, and $e_1, \dots, e_n$ are the corresponding idempotents. Then*
  (a) *$\sigma_j(e_i) = 0$ if $i \neq j$, and*
  (b) *$e_i e_j = 0$, if $i \neq j$.*

Proof. (1): Let $\theta : R \to A$ be the structure homomorphism. Let e be a central idempotent in A and $\pi : A \to Ae$ the canonical projection map. The diagram

$$\begin{array}{ccc} R & \xrightarrow{\alpha} & Re \\ \theta\downarrow & & \downarrow\subseteq \\ A & \xrightarrow{\pi} & Ae \end{array}$$

of R-algebra homomorphisms commutes, where $\alpha(x) = xe$. If $Re = Ae$ and α is one-to-one, then $\alpha^{-1}\pi$ is an R-algebra homomorphism.

Conversely, assume $\sigma : A \to R$ is an R-algebra homomorphism. By Theorem 4.4.1, σ makes R into a projective A-module. By Exercise 1.1.22 $\ker\sigma$ is an A-module direct summand of A, hence $\ker\sigma = Ae_0$ for some idempotent $e_0 \in A$. Since $\ker\sigma$ is a two-sided ideal of A, $e = 1 - e_0$ is a central idempotent by Lemma 1.1.20. The rest is left to the reader.

(2): Since R is connected, $\sigma_j(e_i)$ is equal to either 0 or 1. Suppose $\sigma_j(e_i) = 1$. Then for every $x \in A$, $\sigma_j(x) = \sigma_j(x)\sigma_j(e_i) = \sigma_j(xe_i) = \sigma_j(\sigma_i(x)e_i) = \sigma_i(x)\sigma_j(e_i) = \sigma_i(x)$ which implies $i = j$. This proves (a). Lastly, $\sigma_j(x)e_j = xe_j$ for all $x \in A$ implies $\sigma_j(e_i)e_j = e_ie_j$. This proves (b). □

4.1. Exercises.

Exercise 4.4.9. Let $f : R \to S$ be a homomorphism of commutative rings. Let $q \in \operatorname{Spec} S$ and $p = f^{-1}(q)$. Show that if S is a separable R-algebra, then S_q is a separable R_p-algebra.

Exercise 4.4.10. Let R be a commutative ring. Let A_1 and A_2 be R-algebras. Prove that $A_1 \oplus A_2$ is separable over R if and only if A_1 and A_2 are separable over R.

Exercise 4.4.11. Let k be a field and x an indeterminate.

(1) Show that $A = k[x]/(x^2)$ is not separable over k. (Hint: Show that A^e is a local ring. What are the candidates for e?)
(2) Show that $k[x]/(x^n)$ is separable over k if and only if $n = 1$.
(3) Suppose $f \in k[x]$ is a nonconstant polynomial such that each irreducible factor of f has degree one. Show that $k[x]/(f)$ is separable over k if and only if f has no repeated roots.
(4) Suppose $f \in k[x]$ is a nonconstant polynomial and F is a splitting field for f over k. Show that $k[x]/(f)$ is separable over k if and only if f has no repeated roots in F. Equivalently, show that $k[x]/(f)$ is separable over k if and only if $\gcd(f, f') = 1$.

Exercise 4.4.12. Let k be a field, R a commutative ring, and x an indeterminate.

(1) Show that $A = k[x]$ is not separable over k. (Hint: A^e is an integral domain.)
(2) Show that $R[x]$ is not separable over R.
(3) Show that Corollary 4.4.8 (1) is false if A is not separable. (Hint: Let $a \in R$ and define $\sigma_a : R[x] \to R$ to be the substitution map defined by $x \mapsto a$.)

Exercise 4.4.13. Let $A = \mathbb{Z}[i]$ be the ring of gaussian integers. Then up to isomorphism A is equal to $\mathbb{Z}[x]/(x^2+1)$. Show that A is not separable over $\mathbb{Z}$. (Hints: Use Corollary 4.3.2. Take $S = \mathbb{Z}/2$ and apply the argument of Exercise 4.4.11 to show $A \otimes \mathbb{Z}/2$ is not separable over the field $\mathbb{Z}/2$. We say that A *ramifies* at the prime 2.)

Exercise 4.4.14. Let R be an integral domain in which 2 is invertible. Let a be a unit of R and define $S = R[\sqrt{a}]$ to be R with the square root of a adjoined. That is, $S = R[x]/(x^2 - a)$.

(1) Show that S is a faithfully flat R-algebra.
(2) Show that the $\sqrt{a} \mapsto -\sqrt{a}$ induces an R-algebra automorphism $\sigma : S \to S$. (Hint: first define $\sigma : R[x] \to R[x]$ by $\sigma(x) = -x$ and show that this induces the desired automorphism on S.)
(3) The trace map $T : S \to R$ is defined by $T(z) = z + \sigma(z)$. Show that T is an R-module homomorphism and the image of T is R. Show that the kernel of T is the R-submodule generated by $\sqrt{a}$. Conclude that $S \cong R \cdot 1 \oplus R\sqrt{a}$ as R-modules.
(4) Let $\mathfrak{m}$ be any maximal ideal in S. Show that $\mathfrak{m}$ does not contain the R-submodule $R\sqrt{a}$.
(5) Show that S is a separable R-algebra. (Hint: $e = \frac{1}{2}(1 \otimes 1 + \sqrt{a} \otimes \frac{1}{\sqrt{a}})$ is a separability idempotent.)

For a generalization of this exercise and Exercise 4.4.15, see Exercise 4.6.20.

Exercise 4.4.15. Let R be an integral domain in which 2 is a unit. Let $a \in R$ and $S = R[\sqrt{a}] = R[x]/(x^2 - a)$. Show:

(1) If $a = b^2$ and b is a unit in R, then $S \cong R \oplus R$ as R-algebras.

(2) If a is not a unit in R, then S is not separable over R.

Exercise 4.4.16. Let the Cartesian plane $\mathbb{R}^2$ have the usual metric space topology. Let X be the x-axis and $\pi : \mathbb{R}^2 \to X$ the standard projection map defined by $\pi(x, y) = x$.

(1) Let $S = \mathbb{R}[x, y]/(x^2 - y^2)$ and $R = \mathbb{R}[x]$. Show that S is faithfully flat over R, but is not separable. Geometrically, S corresponds to two intersecting lines and R corresponds to the x-axis. In $\mathbb{R}^2$ let Y denote the two lines $x = \pm y$. The projection $\pi : Y \to X$ of Y onto the x-axis is two-to-one everywhere except at the origin, hence is not a local homeomorphism.

(2) Let $S = \mathbb{R}[x, y]/(x^2 + y^2 - 1)$. Show that S is faithfully flat over $R = \mathbb{R}[x]$, but is not separable. Geometrically, S corresponds to a circle of radius 1 and R corresponds to the x-axis. In $\mathbb{R}^2$ let Y denote the circle $x^2 + y^2 = 1$. The projection $\pi : Y \to X$ of Y onto the x-axis is two-to-one everywhere except at the points where $x = \pm 1$, hence is not a local homeomorphism.

Exercise 4.4.17. Let R be a commutative ring in which 2 is invertible. Let α and β be units of R. Define a generalized quaternion algebra over R by the following rules. Start with the free R-module $A = R{\cdot}1 + R{\cdot}u + R{\cdot}v + R{\cdot}uv$ with basis $1, u, v, uv$. Multiplication in A is defined by extending the relations

$$u^2 = \alpha,\ v^2 = \beta,\ uv = -vu$$

by associativity and distributivity (see Exercise 4.1.13).

(1) Show that the center of A is equal to R.

(2) Show that A is isomorphic to A^o.

(3) Show that A is a separable R-algebra. (Hint: $e = \frac{1}{4}(1 \otimes 1 + u \otimes u^{-1} + v \otimes v^{-1} + uv \otimes (uv)^{-1})$ is a separability idempotent.)

(4) Assume moreover that R is an integral domain which contains an element i such that $i^2 = -1$. Assume $\alpha = a^2$ and $\beta = b^2$ for some a, b in R. Show that A is isomorphic to the ring $M_2(R)$ of two-by-two matrices over R. (Hint: Define the map $A \to M_2(R)$ on generators by

$$u \mapsto \begin{bmatrix} 0 & -ia \\ ia & 0 \end{bmatrix}, \quad v \mapsto \begin{bmatrix} 0 & b \\ b & 0 \end{bmatrix}.$$

Show that this definition extends to A.)

(5) Show that if $R = \mathbb{C}$, then $A \cong M_2(\mathbb{C})$ for every choice of α and β.

The algebra A is an example of a symbol algebra. See Section 13.4.1 for the general construction.

Exercise 4.4.18. Let S be a commutative separable R-algebra. For $n \geq 1$, let $T^n_R(S) = S \otimes_R S \otimes_R \cdots \otimes_R S$ be the tensor product of n copies of S. View $T^n_R(S)$ as an S-algebra by the homomorphism $\rho : S \to T^n_R(S)$, where $\rho(s) = s \otimes 1 \otimes \cdots \otimes 1$. Let $\mu : T^n_R(S) \to S$ be the product map, where $\mu(x_1 \otimes \cdots \otimes x_n) = x_1 \cdots x_n$.

(1) Show that μ is an S-algebra homomorphism and the kernel of μ is idempotent generated.

(2) Show that there is an idempotent $e \in T^n_R(S)$ such that $Se = (S \otimes_R 1 \otimes_R \cdots \otimes_R 1)e = T^n_R(S)e$.

Exercise 4.4.19. Let k be a field, $n > 1$ an integer, $T = k[x, y]$, and $S = k[x^n, xy, y^n]$ (see Exercise 2.6.19).

(1) Let $\mathfrak{p} = (y)$ be the principal ideal in T generated by y. If $\mathfrak{q} = \mathfrak{p} \cap S$, show that $\mathfrak{q} = (xy, y^n)$.

(2) Show that $S/\mathfrak{q} \to T/\mathfrak{p}$ is not separable.

(3) Show that $S/\mathfrak{q} \to T/\mathfrak{q}T$ is not separable.

(4) Conclude that T is not separable over S.

See Exercise 6.2.9 for more properties of the ring S.

Exercise 4.4.20. Let A and B be commutative rings. Let I be an ideal in A, J an ideal in B and assume $\sigma : A/I \to B/J$ is an isomorphism. If f_1 and f_2 are the natural maps, let R be the pullback which makes the diagram

$$\begin{array}{ccccc} R & & \xrightarrow{h_1} & & B \\ {\scriptstyle h_2}\downarrow & & & & \downarrow{\scriptstyle f_2} \\ A & \xrightarrow{f_1} & A/I & \xrightarrow{\sigma} & B/J \end{array}$$

into a cartesian square. Prove that $R \to A \oplus B$ is separable. In Algebraic Geometry, R corresponds to the scheme obtained by gluing two schemes along a closed subscheme.

Exercise 4.4.21. Let k be a field. Let $f_1 : k[x] \to k$ be defined by $x \mapsto 0$. Let $f_2 : k[y] \to k$ be defined by $y \mapsto 0$. As in Exercise 4.4.20, if R is the pullback that makes the diagram

$$\begin{array}{ccccc} R & & \xrightarrow{h_1} & & k[y] \\ {\scriptstyle h_2}\downarrow & & & & \downarrow{\scriptstyle f_2} \\ k[x] & \xrightarrow{f_1} & k & \xrightarrow{=} & k \end{array}$$

a cartesian square, then $k[x] \oplus k[y]$ is separable over R.

(1) Show that R can be identified with $\{(\alpha, \beta) \in k[x] \oplus k[y] \mid \alpha(0) = \beta(0)\}$ and that R is isomorphic to $k[x, y]/(xy)$.

(2) Show that $k[x] \oplus k[y]$ is not flat over R.

For more properties of the ring R, see Exercise 14.2.17.

5. Separable Algebras over a Field

In this section, we study separable algebras over a field k. Before proceeding, we mention several highlights that appear below. First, is the proof that a k-algebra is central separable if and only if it is central simple (Corollary 4.5.4). Second, we prove two decomposition theorems for separable k-algebras. Every separable algebra over k is shown to have a unique decomposition as a finite direct sum of matrix algebras over division algebras whose centers are separable field extensions of k (Theorem 4.5.7). Any commutative separable k-algebra has a unique decomposition into a finite direct sum of separable field extensions of k (Corollary 4.5.8). In addition, we prove a Primitive Element Theorem for commutative separable k-algebras if k is infinite. General references for the material in this section are [**Sal99**], [**DI71**] and [**KO74b**].

5.1. Central Simple Equals Central Separable. Let k be a field and A a k-algebra. We say A is a *central simple k-algebra* if these three conditions are met:

(1) A is a simple ring.

(2) A is a central k-algebra.

(3) $\dim_k(A) < \infty$.

It follows from Example 3.2.4 that the ring of matrices $M_n(k)$ over a field k is a central simple k-algebra. If A is a central simple k-algebra, then by Theorem 3.2.6 we know $A \cong \operatorname{Hom}_D(E, E)$ where D is a division ring and E is a finite dimensional D-vector space. Since A is k-central, so is D. Since $\dim_k(A)$ is finite, so is $\dim_k(D)$.

Proposition 4.5.1. *Let k be an algebraically closed field and A a central simple k-algebra. Then $A \cong M_n(k)$ for some n.*

Proof. Let D be the division algebra component of A. Let $\alpha \in D$. Because D is a finite dimensional division algebra over k, $k[\alpha]$ is an algebraic field extension of k. Because k is algebraically closed, $\alpha \in k$. Therefore, $k = D$. □

Theorem 4.5.2. *Let k be a field and let A and B be simple k-algebras. If A is a central simple k-algebra, then*

(1) *$A \otimes_k B$ is a simple ring.*
(2) *$Z(A \otimes_k B) = Z(B)$.*

Proof. (1): Let I be a nonzero two-sided ideal in $A \otimes_k B$. Let x be a nonzero element of I. Then there are $a_1, \dots, a_n$ in A and there are k-linearly independent $b_1, \dots, b_n$ in B such that $x = \sum_{i=1}^n a_i \otimes b_i$. Choose x such that n is minimal. Since A is simple, the principal ideal Aa_1A is the unit ideal. Pick $r_1, \dots, r_m, s_1, \dots, s_m$ in A such that $\sum_j r_j a_1 s_j = 1$. Since $(r_j \otimes 1)x(s_j \otimes 1) \in I$ for each j,

$$
\begin{aligned}
y &= \sum_j (r_j \otimes 1)\, x\, (s_j \otimes 1) \\
&= \sum_j \Big((r_j \otimes 1) \Big(\sum_i a_i \otimes b_i \Big) (s_j \otimes 1) \Big) \\
&= \sum_j \sum_i (r_j a_i s_j \otimes b_i) \\
&= \sum_i \Big(\Big(\sum_j r_j a_i s_j \Big) \otimes b_i \Big) \\
&= 1 \otimes b_1 + a_2' \otimes b_2 + \cdots + a_n' \otimes b_n
\end{aligned}
$$

is an element of I for some $a_2', \dots, a_n'$ in A. For all $a \in A$ we have

$$
\begin{aligned}
&(a \otimes 1)y - y(a \otimes 1) = \\
&(a \otimes b_1 + aa_2' \otimes b_2 + \cdots + aa_n' \otimes b_n) - (a \otimes b_1 + a_2'a \otimes b_2 + \cdots + a_n'a \otimes b_n) \\
&\qquad = (aa_2' - a_2'a) \otimes b_2 + \cdots + (aa_n' - a_n'a) \otimes b_n
\end{aligned}
$$

is in I. Because the length n of x was minimal, $(a \otimes 1)y - y(a \otimes 1) = 0$. Because $b_1, \dots, b_n$ are k-linearly independent in B, it follows that $1 \otimes b_1, \dots, 1 \otimes b_n$ are A-linearly independent in $A \otimes_k B$. It follows that $aa_i' = a_i'a$ for all $a \in A$ and all $2 \le i \le n$. That is to say, each a_i' is in $Z(A) = k$. In that case we can write

$$
y = 1 \otimes b_1 + 1 \otimes a_2'b_2 + \cdots + 1 \otimes a_n'b_n = 1 \otimes (b_1 + a_2'b_2 + \cdots + a_n'b_n) = 1 \otimes b
$$

where b is nonzero because $b_1 \neq 0$ and the set $\{b_i\}$ is k-linearly independent. Since B is simple, there exist $u_1, \dots, u_p, v_1, \dots, v_p \in B$ such that $\sum_j u_j b v_j =$

1. Now $y = 1 \otimes b$ is in the ideal I, so

$$\sum_j \big((1 \otimes u_j)(1 \otimes b)(1 \otimes v_j)\big) = 1 \otimes \sum_j u_j b v_j = 1 \otimes 1$$

is in I. This shows $I = A \otimes_k B$.

(2): It is easy to see that $1 \otimes_k Z(B) \subseteq Z(A \otimes_k B)$. Let $x \in Z(A \otimes_k B)$. Assume $x \neq 0$ and write $x = \sum_{i=1}^n a_i \otimes b_i$ where we assume $b_1, \dots, b_n$ are linearly independent over k. For each $a \in A$ we have

$$\begin{aligned} 0 &= (a \otimes 1)x - x(a \otimes 1) \\ &= aa_1 \otimes b_1 + \cdots + aa_n \otimes b_n - (a_1 a \otimes b_1 + \cdots + a_n a \otimes b_n) \\ &= (aa_1 - a_1 a) \otimes b_1 + \cdots + (aa_n - a_n a) \otimes b_n. \end{aligned}$$

Since $1 \otimes b_i$ are A-linearly independent in $A \otimes_k B$, we conclude that $aa_i = a_i a$ for each i. That is, each a_i is in $Z(A) = k$. Therefore, $x = 1 \otimes b$. It is now easy to verify that $b \in Z(B)$. □

Proposition 4.5.3. *Let k be a field and A a finite dimensional k-algebra. Then A is a central simple k-algebra if and only if the enveloping homomorphism $\varphi : A^e \to \operatorname{Hom}_k(A, A)$ of Definition 4.1.1 is an isomorphism.*

Proof. If A is a central simple k-algebra, then so is A^o. By Theorem 4.5.2, it follows that A^e is a central simple k-algebra. Therefore φ is one-to-one and counting dimensions over k proves that φ is onto. Conversely, suppose that φ is an isomorphism. Since $\operatorname{Hom}_k(A, A)$ is isomorphic to a ring of matrices $M_n(k)$, it is a central simple k-algebra by Example 3.2.4. If I is a two-sided ideal of A, then $I \otimes_k A^o$ is an ideal in A^e. So I is either (0) or A. If $\alpha \in Z(A)$, then $\alpha \otimes 1 \in Z(A^e)$ so $\varphi(\alpha \otimes 1) \in k$. Since φ is a k-algebra isomorphism, $\alpha \otimes 1 \in k \cdot 1 \otimes 1$. It follows that $\alpha \in k$. □

Corollary 4.5.4. *Let k be a field and A a k-algebra.*

(1) *If A is a separable k-algebra, then A is a finite dimensional k-vector space.*

(2) *A is a central simple k-algebra if and only if A is a central separable k-algebra.*

Proof. Part (1) follows from Proposition 4.4.5.

(2): Assume A is a central simple k-algebra. Let K be an algebraic closure of k. Then by Theorem 4.5.2, $A \otimes_k K$ is a central simple K-algebra. By Proposition 4.5.1, $A \otimes_k K \cong M_n(K)$ for some n. By Example 4.2.1, $A \otimes_k K$ is a central separable K-algebra. By Corollary 4.3.5, A is a separable k-algebra. Conversely assume A is a central separable k-algebra. Then $Z(A) = k$ and by Part (1), A is finite dimensional over k. Any left A-module is a k-vector space, hence is projective as a k-module. By Theorem 4.4.1, every left A-module is projective. By Theorem 3.2.2, A is semisimple. By Theorem 3.2.5, A is a finite direct sum of simple rings. Since the center of A is the field k, it follows that A is simple. □

5.2. Unique Decomposition Theorems. Before proving the unique decomposition theorems for separable algebras over a field, we prove that an extension of fields is separable in the usual sense (for example, in the sense of [**DF04**]) if and only if it is separable in the sense of Definition 4.1.3. First we need a lemma.

Lemma 4.5.5. *Let k be a field and A a k-algebra. The following are equivalent.*

(1) *A is a separable k-algebra.*

(2) *A is finite dimensional over k and if K/k is any field extension of k, then $A \otimes_k K$ is semisimple.*

Proof. (1) implies (2): By Proposition 4.4.5, A is finite dimensional over k. By Corollary 4.3.2, $A \otimes_k K$ is a separable K-algebra. Every $A \otimes_k K$-module is free over K. By Theorem 4.4.1, every $A \otimes_k K$-module is projective. By Theorem 3.2.2, $A \otimes_k K$ is semisimple.

(2) implies (1): Let $\bar{k}$ be the algebraic closure of k. By Theorem 3.2.5 (2), $A \otimes_k \bar{k} = R_1 \oplus \cdots \oplus R_n$ is a direct sum of a finite number of simple rings R_i. Each R_i is finite dimensional over $\bar{k}$. Since $\bar{k}$ is algebraically closed, the center of R_i is $\bar{k}$. By Corollary 4.5.4, each R_i is central separable over $\bar{k}$. Therefore $A \otimes_k \bar{k}$ is separable over $\bar{k} \oplus \cdots \oplus \bar{k}$. By Exercise 4.1.11, $\bar{k} \oplus \cdots \oplus \bar{k}$ is separable over $\bar{k}$. By Theorem 4.4.2 (1), $A \otimes_k \bar{k}$ is separable over $\bar{k}$. By Corollary 4.3.5, A is separable over k. □

Theorem 4.5.6. *Let k be a field and F a finite dimensional extension field of k. Then F is a separable k-algebra if and only if F/k is a separable field extension.*

Proof. Assume F is a separable field extension of k. Then there is a finite set of separable elements $u_1, \ldots, u_m$ in F such that $F = k(u_1, \ldots, u_m)$. By Theorem 4.4.2 (1), it is enough to assume $F = k[x]/(f(x))$ is a simple extension and prove that F is a separable k-algebra. Let K/k be a splitting field for $f(x)$. In $K[x]$ we have the factorization $f(x) = (x - \alpha_1) \ldots (x - \alpha_n)$ where the roots α_i are distinct. The Chinese Remainder Theorem shows that $F \otimes_k K \cong K[x]/(f(x))$ is isomorphic to a direct sum of n copies of K. By Exercise 4.1.11, $F \otimes_k K$ is separable over K. By Corollary 4.3.5, F is a separable k-algebra.

Conversely assume F/k is not a separable extension of fields and let S be the separable closure of k in F. Let p be the characteristic of k. Since

F/S is purely inseparable, there exists $u \in F$, $n \geq 1$, and $\alpha \in S$ such that the irreducible polynomial of u over S is $\operatorname{Irr.poly}_S(u) = x^{p^n} - \alpha$. Consider the element $t = u \otimes 1 - 1 \otimes u$ in $F \otimes_S F$. It is easy to see that t is nonzero and that $t^{p^n} = 0$. Therefore the ring $F \otimes_S F$ is not semisimple. By Lemma 4.5.5, F is not a separable S-algebra. By Theorem 4.4.2 (2), F is not a separable k-algebra. □

Theorem 4.5.7. *Let k be a field and A a k-algebra. Then A is a separable k-algebra if and only if A is isomorphic to a finite direct sum of matrix rings $M_{n_i}(D_i)$ where each D_i is a finite dimensional k-division algebra such that the center $Z(D_i)$ is a finite separable extension field of k.*

Proof. If A is separable over k, then by Lemma 4.5.5, A is semisimple. It follows from Theorem 3.2.5 (2) that $A = A_1 \oplus \cdots \oplus A_m$ is a direct sum of a finite number of simple rings A_i. By Exercise 4.4.10, A_i is separable over k, for each i. By Theorem 3.2.6, $A_i \cong M_{n_i}(D_i)$ where D_i is a finite dimensional k-division algebra. The center of A_i is $Z(D_i)$ and by Theorem 4.4.2 (3), $Z(D_i)$ is separable over k. By Theorem 4.5.6, $Z(D_i)/k$ is a finite separable field extension.

For the converse, suppose K/k is a finite separable field extension and D is a finite dimensional K-central division algebra. Then by Example 3.2.4 and Theorem 4.5.2, $M_n(D) \cong \operatorname{Hom}_K(K^{(n)}, K^{(n)}) \otimes_K D$ is K-central simple. By Corollary 4.5.4, $M_n(D)$ is K-central separable. By Theorem 4.5.6 and Theorem 4.4.2 (1), $M_n(D)$ is separable over k. The part of the claim concerning a finite direct sum now follows from Exercise 4.4.10. □

Corollary 4.5.8. *Let k be a field and A a commutative k-algebra. Then the following are true.*

(1) *A is separable over k if and only if A is isomorphic to a finite direct sum of fields $K_1 \oplus \cdots \oplus K_n$ where each K_i is a finite separable extension field of k.*

(2) *If k is infinite and A is separable over k, then there is a monic polynomial $f(x) \in k[x]$ such that $\gcd(f, f') = 1$ and A is isomorphic to $k[x]/(f(x))$ as a k-algebra. There is a primitive element $\alpha \in A$ such that A is generated as a k-algebra by α.*

Proof. Part (1) follows from Theorem 4.5.7.

(2): By Part (1), there is a k-algebra isomorphism $A \cong K_1 \oplus \cdots \oplus K_n$, where each K_i is a separable extension field of k. By the Primitive Element Theorem (for example, [**DF04**, Theorem 14.4.25]), $K_i \cong k[x]/(p_i(x))$, for some irreducible monic separable polynomial $p_i(x) \in k[x]$. By induction, assume $n \geq 2$ and there is a monic polynomial $f(x)$ such that $\gcd(f, f') = 1$ and $K_2 \oplus \cdots \oplus K_n$ is isomorphic to $k[x]/(f(x))$ as a k-algebra. Let F be a splitting field for $f(x)p_1(x)$. Let $\{u_1, \dots, u_r\}$ be all the roots of $f(x)p_1(x)$ in F. Assume $p_1(u_1) = 0$. Since k is infinite, pick $a \in k$ such that a is not in the set $\{0, u_2 - u_1, \dots, u_r - u_1\}$. So $p_1(x-1)$ is a monic irreducible separable polynomial in $k[x]$ and $a + u_1$ is a root of $p_1(x-a)$ but not a root of $f(x)$. Therefore, $p_1(x-a)$ does not divide $f(x)$. Hence $p_1(x-a)f(x)$ is a separable polynomial. By the Chinese Remainder Theorem,

$$\frac{k[x]}{(p_1(x-a)f(x))} \to \frac{k[x]}{(p_1(x-a))} \bigoplus \frac{k[x]}{(f(x))}$$

is an isomorphism. But the k-algebra on the right is isomorphic to A. □

Before we finish this section, we consider algebras over a finite field. The goal is to prove Theorem 4.5.11, a kind of "Primitive Element Theorem" for commutative separable algebras over a finite field which is due to T. McKenzie ([**McK98**]). In order to prove this, we need some combinatorial results that show there are enough irreducible polynomials. For the rest of this section, p will be a fixed prime number and $\mathbb{F}_p = \mathbb{Z}/p$ the prime field of order p.

Theorem 4.5.9. *The factorization of the polynomial $x^{p^n} - x$ in $\mathbb{F}_p[x]$ into irreducible factors is equal to the product of all the monic irreducible polynomials of degree d where d runs through all divisors of n.*

Proof. See, for example, [**DF04**, Proposition 14.3.18]. □

Theorem 4.5.10. *Let $\psi(n)$ denote the number of distinct monic irreducible polynomials of degree n in $\mathbb{F}_p$. Then the following are true.*

(1) $\psi(n) = \frac{1}{n}\sum_{d|n} \mu(d) p^{n/d} = \frac{1}{n}\sum_{d|n} \mu\left(\frac{n}{d}\right) p^d$, *where μ is the Möbius function.*

(2) $\psi(n) > \frac{p^n}{2n}$.

Proof. (1): By Theorem 4.5.9, $p^n = \sum_{d|n} d\psi(d)$. Now apply the Möbius Inversion Formula.

(2): The reader should verify the identities:

$$\begin{aligned}
n\psi(n) &= p^n + \sum_{d|n, d<n} \mu\left(\frac{n}{d}\right) p^d \\
&\geq p^n - \sum_{d|n, d<n} p^d \\
&\geq p^n - \sum_{1 \leq d \leq n/2} p^d \\
&\geq p^n - p^{\lfloor n/2 \rfloor + 1}
\end{aligned}$$

where $\lfloor n/2 \rfloor$ is the greatest integer less than $n/2$. If $n > 2$, then $\lfloor n/2 \rfloor + 1 \le n - 1$, so

$$\psi(n) > \frac{1}{n}\left(p^n - p^{n-1}\right) = \frac{p^n}{n}\left(1 - \frac{1}{p}\right) \ge \frac{p^n}{2n}.$$

If $n = 2$, the formula can be derived from $\psi(2) = (1/2)(p^2 - p)$. □

Theorem 4.5.11. *If k is a finite field and A is a commutative separable k-algebra, then there is a monic polynomial $f(x) \in k[x]$ such that* $\gcd(f, f') = 1$ *and A is isomorphic to a k-subalgebra of $k[x]/(f(x))$.*

Proof. By Corollary 4.5.8 (1), there is a k-algebra isomorphism $A \cong K_1 \oplus \cdots \oplus K_n$, where each K_i is a separable extension field of k. Let $d_i = \dim_k(K_i)$, and $d = \operatorname{LCM}(d_1, \ldots, d_n)$. By Exercise 4.5.18 there exists a polynomial $f(x) \in k[x]$ such that $\gcd(f, f') = 1$ and $k[x]/(f(x))$ is isomorphic to the direct sum $F \oplus \cdots \oplus F$ of n copies of the field F, where $\dim_k(F) = dm$, for some $m \ge 1$. Then F contains a subfield isomorphic to K_i, and we can embed A into $k[x]/(f(x))$. □

5.3. The Skolem-Noether Theorem. If R is a commutative ring with trivial Picard group, then in Section 7.8.2 we prove the counterpart of Theorem 4.5.12 and its corollary for a central separable R-algebra A. The proof given below is based on [**Rei75**, Theorem (7.21)].

Theorem 4.5.12 (Skolem-Noether). *Let A be a central simple k-algebra. Let B and $\tilde{B}$ be two simple k-subalgebras of A and $\varphi : B \to \tilde{B}$ a k-algebra isomorphism. Then φ extends to an inner automorphism of A. That is, there exists an invertible $u \in A$ such that $\varphi(x) = uxu^{-1}$, for all $x \in B$.*

Proof. By Theorem 3.2.6, if M is a minimal left ideal of A, then $D = \operatorname{Hom}_A(M, M)$ is a division ring, and $A \cong \operatorname{Hom}_D(M, M)$. For $a \in A$, let $\lambda_a : M \to M$ be "left multiplication by a". For all $x \in M$, $d \in D$, $b \in B$, we have $\lambda_d \lambda_b x = \lambda_b \lambda_b x$. Therefore, we can make M into a left $D \otimes_k B$-module by $d \otimes b \cdot x = dbx$. Using φ, define a second left $D \otimes_k B$-module structure on M by $d \otimes b \cdot x = d\varphi(b)x$. Denote this module by ${}_\varphi M$. By Theorem 4.5.2, $D \otimes_k B$ is a simple ring. It follows from Theorem 3.2.1 and Theorem 3.2.5 that V and ${}_\varphi M$ are isomorphic $D \otimes_k B$-modules. Therefore, there exists an isomorphism $\theta \in \operatorname{Hom}_k(M, M)$ satisfying:

$$\theta(d \otimes b \cdot x) = d \otimes b \cdot \theta(x) = d\varphi(b)\theta(x).$$

For $b = 1$, this implies $\theta(dx) = d\theta(x)$, so $\theta \in \operatorname{Hom}_D(M, M) = A$. That is, $\theta = \lambda_u$, for some invertible $u \in A$. The equation above becomes

$$u(db)x = d\varphi(b)ux.$$

If $d = 1$, this becomes: $ubx = \varphi(b)ux$. Since M is a faithful module (Theorem 3.2.5), this proves $\varphi(b) = ubu^{-1}$. □

Corollary 4.5.13. *Let k be a field and A a central simple k-algebra. If $\theta : A \to A$ is a k-algebra homomorphism, then θ is an inner automorphism of A.*

Proof. Since A is simple, the kernel of θ is the zero ideal, hence θ is one-to-one. The image of θ has dimension $\dim_k(A)$, hence θ is onto. □

5.4. Exercises.

Exercise 4.5.14. Let k be a field and G a finite group of order $[G : 1]$.

(1) Prove Maschke's Theorem: If $[G : 1]$ is invertible in k, then the group algebra $k(G)$ is semisimple.

(2) This exercise contains an outline of a proof of the converse to Maschke's Theorem. In the group algebra $k(G)$, let $t = \sum_{\sigma \in G} \sigma$. Let $I = k(G)t$ be the left ideal generated by t.
   (a) Show that I is equal to kt.
   (b) Show that if the characteristic of k divides $[G : 1]$, then $I^2 = 0$. Conclude that I is not a $k(G)$-module direct summand of $k(G)$.
   (c) Show that if the group algebra $k(G)$ is semisimple, then $[G : 1]$ is invertible in k.

Exercise 4.5.15. The purpose of this exercise is to prove the converse of Example 4.2.2. Let R be a commutative ring and G a finite group of order $[G : 1]$. Show that if the group algebra $R(G)$ is separable over R, then $[G : 1]$ is invertible in R. (Hint: If $\mathfrak{m}$ is a maximal ideal in R which contains $[G : 1]$, then by Exercise 4.5.14, the group algebra $(R/\mathfrak{m})(G)$ is not semisimple.)

Exercise 4.5.16. Let $\theta : R \to S$ be a local homomorphism of local rings and assume θ makes S into a separable R-algebra. Let $\mathfrak{m}$ be the maximal ideal of R, $\mathfrak{n}$ the maximal ideal of S, and $R/\mathfrak{m} \to S/\mathfrak{n}$ the corresponding extension of residue fields. Prove that $\mathfrak{m}S = \mathfrak{n}$, $S \otimes_R R/\mathfrak{m} = S/\mathfrak{n}$, and $R/\mathfrak{m} \to S/\mathfrak{n}$ is a finite separable extension of fields.

Exercise 4.5.17. Let K be a finite field of order p^d. As in Theorem 4.5.10, let $\psi(n)$ be the number of irreducible monic polynomials of degree n in $\mathbb{F}_p[x]$. If $d \mid n$, show that there are at least $\psi(n)$ irreducible monic polynomials of degree n/d in $K[x]$.

Exercise 4.5.18. Let k be a finite field and K/k a finite dimensional extension of fields, with $\dim_k K = d$. Let n be an arbitrary positive integer and $A = K \oplus \cdots \oplus K$ the direct sum of n copies of K.

(1) Show that if there exists a surjective k-algebra homomorphism $f : k[x] \to A$, then there exist at least n distinct irreducible monic polynomials in $k[x]$ of degree d.

(2) Find an example of k and A such that the k-algebra A is not the homomorphic image of $k[x]$.

(3) Show that for some integer $m \geq 1$ there exist n distinct irreducible monic polynomials $h_1, \dots, h_n$ in $k[x]$ such that each h_i has degree md.

(4) Show that for some integer $m \geq 1$, if F/k is a finite extension field with $\dim_k F = md$, then the direct sum $F \oplus \cdots \oplus F$ of n copies of F is the homomorphic image of $k[x]$. Show that m can be chosen to be relatively prime to d.

(5) Show that there is a separable polynomial $g \in k[x]$ such that A is isomorphic to a subalgebra of $k[x]/(g)$.

Exercise 4.5.19. Let R be a commutative ring and Λ a separable R-algebra. Prove:

(1) If $\mathrm{J}(\Lambda)$ denotes the Jacobson radical of the ring Λ, then

$$\mathrm{J}(\Lambda) \subseteq \bigcap_{\mathfrak{m} \in \operatorname{Max} R} \mathfrak{m}\Lambda.$$

(2) If Λ is commutative and $\mathfrak{m} \in \operatorname{Max}(R)$, then

$$\mathfrak{m}\Lambda = \bigcap_{\substack{\mathfrak{p} \in \operatorname{Max} \Lambda \\ \mathfrak{m}\Lambda \subseteq p}} \mathfrak{p}$$

and the set $\{\mathfrak{p} \in \operatorname{Max} \Lambda \mid \mathfrak{m}\Lambda \subseteq p\}$ is finite.

(3) If Λ is commutative and integral over R, then

$$\mathrm{J}(\Lambda) = \bigcap_{\mathfrak{m} \in \operatorname{Max} R} \mathfrak{m}\Lambda.$$

(Hint: Reduce to the case where Λ is faithful and apply Theorem 3.3.5.) See Exercise 7.4.11 for the case when Λ is noncommutative.

Exercise 4.5.20. Let k be a field and A the subring of $M_2(k)$ consisting of all 2-by-2 matrices of the form $\begin{pmatrix} a & 0 \\ b & c \end{pmatrix}$.

(1) Show that A is k-central.

(2) Compute the Jacobson radical of A.

(3) Show that A is not separable over k.

(4) Show that there exist two distinct k-algebra homomorphisms $A \to k$. Compare this with Corollary 4.4.8 (1).

6. Commutative Separable Algebras

Throughout this section we study separable algebras which are themselves commutative. General references for the material in this section are [**Sal99**] and [**DI71**].

6.1. Separable Extensions of Commutative Rings. If S is a commutative ring and R is a subring of S, then we say S/R is an *extension of commutative rings*. A monic polynomial $f(x)$ in $R[x]$ is said to be *separable* in case $R[x]/(f(x))$ is separable over R. The next proposition is a criterion for a polynomial to be separable. Since it is based on formal derivatives, it is a valuable test for computing examples of separable extensions. A generalization of Proposition 4.6.1 (2) will be proved in Proposition 8.3.8. The proof given below is based on [**Sal99**, Proposition 2.16].

Proposition 4.6.1. *Let R be a commutative ring and $f(x)$ a monic polynomial in $R[x]$. Let $I = (f(x), f'(x))$ be the ideal of $R[x]$ generated by $f(x)$ and the formal derivative, $f'(x)$. Let $S = R[x]/(f(x))$. Then the following are true.*

(1) *S is a free R-module and* $\operatorname{Rank}_R(S) = \deg(f(x))$.
(2) *S is separable over R if and only if the ideal I is the unit ideal.*

Proof. Part (1) is left to the reader.

(2): Assume I is not the unit ideal of $R[x]$. By (1), $R[x]/I$ is a finitely generated R-module. By Nakayama's Lemma (Corollary 1.1.14), there is a maximal ideal $\mathfrak{m}$ in R such that

$$(R[x]/I) \otimes_R (R/\mathfrak{m}) = \frac{(R/\mathfrak{m})[x]}{(f, f')}$$

is nonzero. Let $k = R/\mathfrak{m}$. Then in $k[x]$, (f, f') is not the unit ideal. By Exercise 4.4.11, $S \otimes_R k$ is not separable over k. By Corollary 4.3.2, S is not separable over R.

Now we prove the converse of (2). In $R[x, y]$, $y - x$ is monic in y and linear, so the Division Algorithm applies. Upon dividing $f(y) - f(x)$ by $y - x$ one finds the remainder is 0. We can write $f(y) = f(x) + (y - x)q(x, y)$. Compute the derivative with respect to y: $f'(y) = q(x, y) + (y - x)q_y(x, y)$.

By assumption, there are $u(y), v(y) \in R[y]$ such that

$$
\begin{aligned}
1 &= f(y)u(y) + f'(y)v(y) \\
&= \big(f(x) + (y-x)q(x,y)\big)u(y) + \big(q(x,y) + (y-x)q_y(x,y)\big)v(y) \qquad (4.15)\\
&= (y-x)\big(q(x,y)u(y) + q_y(x,y)v(y)\big) + f(x)u(y) + q(x,y)v(y).
\end{aligned}
$$

Under the mapping $R[x,y] \to S[y]$, all of the polynomials above represent elements in $S[y]$. Consider the principal ideals $A = (y-x)$, $B = (q(x,y))$ in $S[y]$. By (4.15), A and B are comaximal in $S[y]$. By Exercise 2.2.18 $A \cap B = AB$. But in $S[y]$ the equation $f(y) = (y-x)q(x,y)$ holds. The Chinese Remainder Theorem implies

$$
\frac{S[y]}{(f(y))} \xrightarrow{\phi_1 \oplus \phi_2} \frac{S[y]}{(y-x)} \bigoplus \frac{S[y]}{(q(x,y))} \qquad (4.16)
$$

is an isomorphism. To interpret the map $\mu : S \otimes_R S \to S$ of Definition 4.1.1, it is convenient to write the generators of the three copies of S as x, y, and z. Then $\mu(x \otimes 1) = \mu(1 \otimes y) = z$. The diagram

$$
\begin{array}{ccccc}
S \otimes_R S & \longrightarrow & \frac{R[x]}{(f(x))} \otimes_R \frac{R[y]}{(f(y))} & \longrightarrow & S \otimes_R \frac{R[y]}{(f(y))} \\
\downarrow \mu & & \downarrow \mu & & \downarrow \phi_1 \\
S & \longrightarrow & \frac{R[z]}{(f(z))} & \longrightarrow & \frac{S[y]}{(y-x)}
\end{array}
$$

commutes, the horizontal maps are isomorphisms. As we have already seen in (4.16), the kernel of ϕ_1 is idempotent generated. □

Let R be a local ring with maximal ideal $\mathfrak{m}$ and residue field k. An important question is whether a separable k-algebra comes from a separable R-algebra by reduction modulo $\mathfrak{m}$. Corollary 4.6.2 is the first main result on this subject. Partial converses to Corollary 4.6.2 appear in Theorem 9.3.9, Corollary 4.6.3, and Corollary 12.7.3.

Corollary 4.6.2. *Let R be a local ring with maximal ideal $\mathfrak{m}$ and residue field k. Let F be a finite dimensional commutative k-algebra such that $\dim_k(F) = n$. Assume F is generated as a k-algebra by a primitive element u. Then there is a commutative faithful R-algebra S satisfying the following.*

(1) *S is a free R-module of rank n.*
(2) *S is generated as an R-algebra by a primitive element a.*
(3) *$S \otimes_R k$ is isomorphic to F.*
(4) *If F is a field, then S is a local ring and $\mathfrak{m}S$ is the maximal ideal of S.*
(5) *If F/k is a separable extension, then S/R is separable.*

Proof. Let $\theta : k[x] \to F$ be defined by $x \mapsto u$. Let $f \in k[x]$ be the monic polynomial that generates the kernel of θ. Since θ is onto, f has degree n. Lift f to a monic polynomial $g \in R[x]$. Set $S = R[x]/(g)$.

(1): The proof of this part is left to the reader.

(2): Take a to be the image of x.

(3) and (4): $S \otimes_R k = k[x]/(f) = F$.

(5): Under the map $R[x] \to k[x]$, the ideal (g, g') in $R[x]$ restricts to the ideal (f, f') in $k[x]$. Since F/k is separable, Proposition 4.6.1 implies $(f, f') = k[x]$. Since $R[x]/(g, g')$ is a finitely generated R-module, Nakayama's Lemma (Corollary 1.1.14) implies $(g, g') = R[x]$. It follows from Proposition 4.6.1 that S/R is separable. □

Corollary 4.6.3. *Let $\theta : R \to S$ be a local homomorphism of local rings such that S is a separable R-algebra and finitely generated as an R-module. Then S is a homomorphic image of $R[x]$. That is, S is generated as an R-algebra by a primitive element a.*

Proof. Let $\mathfrak{m}$ be the maximal ideal of R, and k the residue field. By Exercise 4.5.16, $\mathfrak{m}S$ is equal to the maximal ideal of S, and $S/\mathfrak{m}S$ is a finite separable extension field of k. By the Primitive Element Theorem (for example, [**DF04**, Theorem 14.4.25]), $S/\mathfrak{m}S = k(u)$ is a simple extension. Define $\phi : R[x] \to S$ by $x \mapsto a$, where $a \in S$ is a preimage of u. Then $R[x] \otimes_R k \to S \otimes_R k$ is onto, S is generated as an R-module by $\operatorname{im}(\phi)$ and $\mathfrak{m}S$, and Nakayama's Lemma (Corollary 1.1.17) implies ϕ is onto. □

If the residue field of R is infinite, then Corollary 4.6.4, which is due to G. Janusz [**Jan66**], shows that in Corollary 4.6.3 it is not necessary to assume S is local.

Corollary 4.6.4. *Let R be a local ring with infinite residue field k. If S is a separable R-algebra which is finitely generated as an R-module, then S is a homomorphic image of $R[x]$. That is, S is generated as an R-algebra by a primitive element a.*

Proof. By Corollary 4.5.8, there is a monic separable polynomial $f \in k[x]$ such that $\gcd(f, f') = 1$ and $k[x]/(f) \cong S \otimes_R k$. The rest of the proof is as in Corollary 4.6.3. □

6.2. Separability and the Trace. Given an extension of commutative rings, our goal is to derive a separability criterion based on the trace function. The fundamental theorem on this subject is Theorem 4.6.7. The theorem and its corollaries are based on [**DI71**, Section 3.2]. As an application of the trace, a finiteness theorem for the integral closure of an integral domain is proved.

Definition 4.6.5. Let A be any R-algebra. Let M be a left A-module which as an R-module is finitely generated and projective. Let $x_1, \dots, x_m \in M$ and $f_1, \dots, f_m \in \operatorname{Hom}_R(M, R)$ be a dual basis for the R-module M. Define $T_R^{A,M} : A \to R$ by the rule

$$T_R^{A,M}(x) = \sum_{i=1}^{m} f_i(xx_i).$$

The reader should verify that $T_R^{A,M} \in \operatorname{Hom}_R(A, R)$. We call $T_R^{A,M}$ the *trace from A to R afforded by M*. The reader is asked to prove in Exercise 4.6.12 that $T_R^{A,M}$ is independent of the choice of a dual basis for M. When $M = A$, we simplify the notation and write T_R^A. The reader should verify that $T_R^R(x) = x$ for all $x \in R$.

Example 4.6.6. Let F/k be a Galois extension of fields with finite group G. The reader should verify that the trace map is given by $T_k^F(x) = \sum_{\sigma \in G} \sigma(x)$ for all $x \in S$.

Theorem 4.6.7. *Let S/R be an extension of commutative rings. Then S is finitely generated as an R-module, projective, and separable over R if and only if there exists an element $T \in \operatorname{Hom}_R(S, R)$ and elements $x_1, \dots, x_n$, $y_1, \dots, y_n$ in S satisfying*

(1) $\sum_{j=1}^{n} x_j y_j = 1$, *and*

(2) $\sum_{j=1}^{n} x_j T(y_j x) = x$ *for all* $x \in S$.

Moreover, the map T is always equal to T_R^S, the trace map from S to R.

Proof. Assume S is a finitely generated R-module, projective, and separable over R. Pick a dual basis $\{a_1, \dots, a_m\}$, $\{f_1, \dots, f_m\}$ for the R-module S. The trace map from S to R is given by

$$T_R^S(x) = \sum_{j=1}^{m} f_j(xa_j)$$

for all $x \in S$. Since S is a finitely generated, projective extension of R, by Theorem 1.3.2, $S \otimes S$ is a finitely generated projective extension of $S \otimes 1$. A dual basis for $S \otimes S$ over $S \otimes 1$ is $\{1 \otimes a_1, \dots, 1 \otimes a_m\}$, $\{1 \otimes f_1, \dots, 1 \otimes f_m\}$ and the trace map from $S \otimes S$ to $S \otimes 1$ is equal to $T_{S\otimes 1}^{S\otimes S} = 1 \otimes T_R^S$. Since S is a separable extension of R, $S \otimes S$ is a separable extension of $S \otimes 1$, by Corollary 4.3.2. Let e be a separability idempotent for S over R. Under the homomorphism μ of Proposition 4.1.2, $(S \otimes 1)e \cong S \otimes 1$. By Proposition 4.1.2, as $S \otimes S$-modules, we have $S \otimes S \cong J_{A/R} \oplus (S \otimes 1)e \cong$

$J_{A/R} \oplus (S \otimes 1)$. Exercise 4.6.13 allows us to write the trace from $S \otimes S$ to $S \otimes 1$ as the sum

$$T^{S\otimes S}_{S\otimes 1} = T^{J_{A/R}}_{S\otimes 1} + T^{S\otimes 1}_{S\otimes 1},$$

where $T^{J_{A/R}}_{S\otimes 1}$ is the restriction of the trace map to $J_{A/R}$ and $T^{S\otimes 1}_{S\otimes 1}$ is the restriction to $(S \otimes 1)e$. To compute $T^{S\otimes 1}_{S\otimes 1}$, use the dual basis $\{e, \sigma\}$ where $\sigma : (S \otimes 1)e \to S \otimes 1$ is the isomorphism defined by $\sigma(e) = 1$. For any $x \in S$,

$$\begin{aligned} T^{S\otimes S}_{S\otimes 1}((x \otimes 1)e) &= T^{J_{A/R}}_{S\otimes 1}((x \otimes 1)e) + T^{S\otimes 1}_{S\otimes 1}((x \otimes 1)e) \\ &= T^{S\otimes 1}_{S\otimes 1}((x \otimes 1)e) = x \otimes 1. \end{aligned}$$

Now let $x \in S$ and let $e = \sum_{j=1}^n x_j \otimes y_j$. Then (1) follows from $\mu(e) = \sum_{j=1}^n x_j y_j = 1$ and (2) follows from applying μ to both sides of

$$\begin{aligned} x \otimes 1 = T^{S\otimes S}_{S\otimes 1}((x \otimes 1) \cdot e) &= T^{S\otimes S}_{S\otimes 1}((1 \otimes x) \cdot e) \\ &= (1 \otimes T^S_R)\left(\sum_{j=1}^n x_j \otimes y_j x\right) = \sum_{j=1}^n x_j \otimes T^S_R(y_j x). \end{aligned}$$

Conversely, suppose we are given $T \in \operatorname{Hom}_R(S, R)$ and a set of elements $x_1, \dots, x_n, y_1, \dots, y_n$ in S satisfying (1) and (2). The reader should verify that the assignment $s \mapsto T(y_j s)$ defines an element $T(y_j \cdot)$ in $\operatorname{Hom}_R(S, R)$. The set $\{x_1, \dots, x_n\}$, $\{T(y_1 \cdot), \dots, T(y_n \cdot)\}$ forms a dual basis for S over R. Therefore S is a finitely generated, projective R-module. Define an element in $S \otimes_R S$ by $e = \sum_{j=1}^n x_j \otimes y_j$. If μ is as in Proposition 4.1.2, $\mu(e) = \sum_{j=1}^n x_j y_j = 1$. For any $x \in S$,

$$\begin{aligned} (1 \otimes x)e = \sum_{j=1}^n x_j \otimes y_j x &= \sum_{j=1}^n \sum_{i=1}^n x_j \otimes x_i T(y_i y_j x) \\ &= \sum_{i=1}^n \sum_{j=1}^n x_j T(y_j y_i x) \otimes x_i = \sum_{i=1}^n x y_i \otimes x_i. \end{aligned}$$

If $x = 1$, then we see $e = \sum_{j=1}^n x_j \otimes y_j = \sum_{i=1}^n y_i \otimes x_i$. It follows that $(1 \otimes x)e = (x \otimes 1)e$ and by Proposition 4.1.2, S is separable over R.

Lastly, the set $\{x_1, \dots, x_n\}$, $\{T(y_1 \cdot), \dots, T(y_n \cdot)\}$ is a dual basis for S over R, so by Exercise 4.6.12, $T^S_R(x) = \sum_{j=1}^n T(y_j x x_j) = T(x \sum_{j=1}^n x_j y_j) = T(x)$. □

Assume S/R is an extension of commutative rings. There is an R-algebra embedding $\theta : S \to \operatorname{Hom}_R(S, S)$ given by $\alpha \mapsto \ell_\alpha$ where ℓ_α is "left multiplication by α". Using Lemma 1.3.17, we turn $\operatorname{Hom}_R(S, R)$ into a right S-module. In fact, for every $f \in \operatorname{Hom}_R(S, R)$ and $\alpha \in S$, $f\alpha$ is defined to be $f \circ \ell_\alpha$.

Corollary 4.6.8. *Let S/R be an extension of commutative rings such that S is a finitely generated projective R-module. Then S is separable over R if and only if the trace map T_R^S from S to R is a free right S-module generator of* $\operatorname{Hom}_R(S, R)$.

Proof. Assume S/R is a separable extension of R which is a finitely generated projective R-module. Let $x_1, \dots, x_n, y_1, \dots, y_n$ be elements in R guaranteed by Theorem 4.6.7. For any $f \in \operatorname{Hom}_R(S, R)$ and any $x \in S$

$$f(x) = \sum_{j=1}^{n} f\big(x_j T_R^S(y_j x)\big) = \sum_{j=1}^{n} T_R^S(y_j x) f(x_j) = T_R^S\big(\sum_{j=1}^{n} y_j x f(x_j)\big).$$

Let $\alpha = \sum_{j=1}^{n} f(x_j) y_j$. Then $f(x) = T_R^S(\alpha x)$, for all $x \in S$, which shows that $f = T_R^S \circ \ell_\alpha$. If $T_R^S \alpha = 0$ in $\operatorname{Hom}_R(S, R)$, then by Theorem 4.6.7 (2) we have $0 = \sum_{j=1}^{n} x_j T_R^S(y_j \alpha) = \alpha$. This shows that the assignment $\alpha \mapsto T_R^S \alpha$ defines an S-module isomorphism $S \cong \operatorname{Hom}_R(S, R)$.

Conversely suppose $x_1, \dots, x_m, f_1, \dots, f_m$ is a dual basis for S over R. Assuming T_R^S generates $\operatorname{Hom}_R(S, R)$ as an S-module, there exist $y_1, \dots, y_m$ in S such that $f_j = T_R^S \circ \ell_{y_j}$. We prove that (1) and (2) of Theorem 4.6.7 are satisfied. For any $x \in S$,

$$x = \sum_{j=1}^{m} f_j(x) x_j = \sum_{j=1}^{m} x_j T_R^S(y_j x)$$

which is (2). For any $z \in S$

$$\begin{aligned} T_R^S\Big(\big(1 - \sum_{j=1}^{m} x_j y_j\big) z\Big) &= T_R^S(z) - T_R^S\Big(\sum_{j=1}^{m} x_j y_j z\Big) \\ &= \sum_{j=1}^{m} f_j(z x_j) - T_R^S\Big(\sum_{j=1}^{m} x_j y_j z\Big) = \sum_{j=1}^{m} T_R^S(y_j z x_j) - T_R^S\Big(\sum_{j=1}^{m} x_j y_j z\Big) \\ &= T_R^S\Big(\sum_{j=1}^{m} y_j x_j z\Big) - T_R^S\Big(\sum_{j=1}^{m} x_j y_j z\Big) = 0. \end{aligned}$$

Since T_R^S is a free generator of $\operatorname{Hom}_R(S, R)$, we conclude that $\sum_{j=1}^{m} x_j y_j = 1$. This proves (1). □

Corollary 4.6.9. *If S is a separable extension of R which is a finitely generated projective R-module, and T_R^S is the trace map from S to R, then there is an element $c \in S$ with $T_R^S(c) = 1$. Moreover $R \cdot c$ is an R-module direct summand of S.*

Proof. By hypothesis, S is a finitely generated projective faithful R-module. By Corollary 1.1.16, S is an R-progenerator module. There exist elements

$f_1, \dots, f_n$ in $\operatorname{Hom}_R(S, R)$ and $x_1, \dots, x_n$ in S with $1 = \sum_{j=1}^n f_j(x_j)$. By Corollary 4.6.8, for each j there is a unique element $a_j \in S$ such that $f_j(x) = T_R^S(a_j x)$ for all $x \in S$. Set $c = \sum_{j=1}^n a_j x_j$. Then $T_R^S(c) = 1$. The R-module homomorphism $R \to S$ which is defined by $1 \mapsto c$ is split by the trace map $T_R^S : S \to R$. □

As an application of the trace, we prove the following result on the integral closure of an integral domain in a separable extension of its quotient field. In the terminology of Section 6.3.1, Theorem 4.6.10 says that S is an R-lattice in L. When R is a finitely generated algebra over a field, see Theorem 8.4.7 for a stronger version of Theorem 4.6.10. The proof given below is based on [**AM69**, Theorem 5.17].

Theorem 4.6.10. *Let R be an integral domain which is integrally closed in its quotient field K. Let L/K be a finite separable field extension and let S be the integral closure of R in L. Then L is equal to the quotient field of S and there exist bases $\{\lambda_1, \dots, \lambda_n\}$ and $\{\mu_1, \dots, \mu_n\}$ for L/K such that $R\lambda_1 + \dots + R\lambda_n \subseteq S \subseteq R\mu_1 + \dots + R\mu_n$. If R is noetherian, then S is a finitely generated R-module.*

Proof. Every $\lambda \in L$ is algebraic over K. There is an equation $r_m\lambda^m + \dots + r_1\lambda + r_0 = 0$, where each r_i is in R. Multiply by r_m^{m-1} to get $(r_m\lambda)^m + \dots + r_1 r_m^{m-2}(r_m\lambda) + r_0 r_m^{m-1} = 0$. This shows that $r_m\lambda$ is integral over R, hence is in S. There exists a basis $\lambda_1, \dots, \lambda_n$ for L/K such that each λ_i is in S. By Exercise 4.6.16 there is a K-basis $\mu_1, \dots, \mu_n$ for L such that $T_K^L(\mu_i\lambda_j) = \delta_{ij}$ (the Kronecker delta function). Let s be an arbitrary element of S. View s as an element of L and write $s = \alpha_1\mu_1 + \dots + \alpha_n\mu_n$, where each $\alpha_j \in K$. Since $\lambda_i \in S$, we have $s\lambda_i \in S$. By Theorem 3.3.4, the minimal polynomial $\operatorname{min.poly}_K(s\lambda_i)$ is in $R[x]$. By Exercise 4.6.15, $T_K^L(s\lambda_i)$ is, up to multiplication by an integer, a coefficient of $\operatorname{min.poly}_K(s\lambda_i)$. This shows $T_K^L(s\lambda_i) \in R$. But

$$T_K^L(s\lambda_i) = T_K^L\Big(\sum_{j=1}^n \alpha_j\lambda_i\mu_j\Big) = \sum_{j=1}^n T_K^L(\alpha_j\lambda_i\mu_j) = \sum_{j=1}^n \alpha_j T_K^L(\lambda_i\mu_j) = \alpha_i$$

which shows that each α_i is in R. It follows that $S \subseteq R\mu_1 + \dots + R\mu_n$. If R is noetherian, then by Corollary 2.5.2, S is a finitely generated R-module. □

6.3. Twisted Form of the Trivial Extension. Let R be a commutative ring and $n \geq 1$. We write R^n for the direct sum $R \oplus \dots \oplus R$. By Exercise 4.1.11, R^n is separable and is called the trivial commutative extension of R of rank n. Let S be an R-algebra and T a faithfully flat R-algebra. Then we say S is a twisted form of R^n for the extension T/R if $S \otimes_R T \cong R^n \otimes_R T \cong T^n$ (see Definition 5.5.6). In terms of twisted forms of

R^n, Proposition 4.6.11 is another criterion for separability. The proof given below is based on [**Sal99**, Proposition 2.18].

Proposition 4.6.11. *Let S be a commutative R-algebra. The following are cquivalcnt:*

(1) *S is a separable R-algebra and an R-module progenerator of constant rank n.*

(2) *There is a commutative separable R-algebra T which is an R-module progenerator of constant rank $n!$ and $S \otimes_R T \cong T^n$ as T-algebras.*

(3) *There is a faithfully flat R-algebra T such that $S \otimes_R T \cong T^n$ as T-algebras.*

Proof. (1) implies (2): Let $e \in S \otimes_R S$ be a separability idempotent. Then $S \otimes_R S = (S \otimes_R S)e \oplus (S \otimes_R S)(1-e)$ and $(S \otimes_R S)e \cong S$. By Proposition 2.3.4, $S \otimes_R S$ is an S-module progenerator of rank n. Using Exercise 4.4.10 one can check that $(S \otimes_R S)(1-e)$ is separable over S and is an S-module progenerator of constant rank $n-1$. If $n = 1$, then we take $T = S$. Otherwise, inductively, there is a commutative separable S-algebra T which is an S-module progenerator of rank $(n-1)!$ such that $(S \otimes_R S)(1-e) \otimes_S T \cong T^{n-1}$. The reader should verify that T is a separable R-algebra, an R-module progenerator of rank $n!$, and $S \otimes_R T \cong T^n$.

(2) implies (3): By Proposition 2.4.6, T is faithfully flat.

(3) implies (1): We are given that T is faithfully flat over R and $S \otimes_R T \cong T^n$. Using this and Theorem 2.6.1, the reader should verify that S is an R-module which is a progenerator of constant rank n. A projective dual basis for S over R gives rise to a dual basis for $S \otimes_R T$, so the trace $T_T^{S \otimes_R T}$ is $T_R^S \otimes 1$. By Proposition 2.4.11, we see that $T_R^S \otimes 1$ is a free right $S \otimes_R T$-module generator of $\operatorname{Hom}_R(S, R) \otimes_R T$. Using the fact that T is faithfully flat over R, the reader should verify that T_R^S is a free right S-module generator for $\operatorname{Hom}_R(S, R)$. Corollary 4.6.8 implies S is separable over R. □

6.4. Exercises.

Exercise 4.6.12. In Definition 4.6.5 the trace map from A to R afforded by M, $T_R^{A,M}$, was defined using a dual basis for M. Prove that the function $T_R^{A,M}$ is independent of the choice of dual basis for M.

Exercise 4.6.13. Let A be an R-algebra and M a left A-module which is a finitely generated projective R-module. If $M = M_1 \oplus M_2$ as A-modules, prove that

$$T_R^{A,M} = T_R^{A,M_1} + T_R^{A,M_2}.$$

Exercise 4.6.14. Let A be an R-algebra which is finitely generated and free as an R-module. Show that the trace mapping T_R^S defined in Exercise 1.2.14 is equal to the trace mapping defined in Definition 4.6.5.

Exercise 4.6.15. Let k be a field and A a finite dimensional k-algebra. Suppose $\alpha \in A$ and $\operatorname{min.poly}_k(\alpha) = x^m + a_{m-1}x^{m-1} + \cdots + a_1 + a_0$ is irreducible in $k[x]$. Prove that $T_k^A(\alpha) = ra_{m-1}$ for some integer r.

Exercise 4.6.16. Let S be a commutative faithful R-algebra which is a finitely generated free R-module of rank n. Let $\lambda_1, \dots, \lambda_n$ be a free basis for S over R. For each i, let $\pi_i \in \operatorname{Hom}_R(S, R)$ be the projection onto the coefficient of λ_i. Prove:

(1) The trace map is given by $T_R^S(z) = \sum_{i=1}^n \pi_i(z\lambda_i)$.

(2) The following are equivalent:
   (a) S is separable over R.
   (b) There exist $\mu_1, \dots, \mu_n$ in S such that $T_R^S \cdot \mu_i = \pi_i$.

(3) If S/R is separable, then the elements $\mu_1, \dots, \mu_n$ appearing in (2) make up a free R-basis for S and $T_R^S(\mu_i\lambda_j) = \delta_{ij}$ (the Kronecker delta function).

Exercise 4.6.17. Let R be a commutative ring and P a finitely generated projective R-module. By Lemma 1.5.1, $\theta_R : P^* \otimes_R P \to \operatorname{Hom}_R(P, P)$ is an isomorphism of R-modules, where $\theta_R(f \otimes p)(x) = f(x)p$.

(1) Define $T : P^* \otimes_R P \to R$ by $T(f \otimes p) = f(p)$. Show that T is an R-module homomorphism.

(2) Assume P is free and finitely generated. Show that the map T induces a map $T : \operatorname{Hom}_R(P, P) \to R$ which is equal to the trace map of Exercise 1.2.14 and the trace map of Definition 4.6.5.

Exercise 4.6.18. Let $R \subseteq T \subseteq S$ be a tower of extensions of commutative rings such that T/R is finitely generated and projective and S/T is finitely generated and projective. Let T_T^S and T_R^T be the trace maps from S to T and T to R respectively. Prove that $T_R^T T_T^S = T_R^S$.

Exercise 4.6.19. Let R be a connected commutative ring and S a commutative separable R-algebra that as an R-module is a progenerator of rank n. Prove that there exists a commutative R-algebra T that satisfies:

(1) T is connected.

(2) T is separable over R.

(3) T is an R-module progenerator.

(4) $S \otimes_R T \cong T^n$.

(Hints: Start with the algebra T constructed in Proposition 4.6.11. By Exercise 2.3.12, $\operatorname{Spec} T$ has only finitely many connected components. Show that T can be replaced with one of its connected components.)

Exercise 4.6.20. This exercise generalizes Exercises 4.4.14 and 4.4.15. Let $n \geq 2$ be an integer and R a commutative ring. Prove the following for $S = R[x]/(x^n - a)$.

(1) S is free of rank n as an R-module with basis $1, x, \dots, x^{n-1}$.

(2) If na is a unit of R, then x is a unit of S and S is a separable R-algebra. (Hint: $e = \frac{1}{n}\sum_{i=0}^{n-1} x^i \otimes x^{-i}$ is a separability idempotent.)

(3) If n is not a unit of R, then S is not separable over R.

(4) If a is not a unit of R, then S is not separable over R.

For the application of this exercise to Kummer Theory, see Example 12.9.2.

7. Formally Unramified, Smooth and Étale Algebras

The notions defined below in Definition 4.7.1 (formally unramified, formally smooth, and formally étale) were first introduced in EGA IV ([**Gro64**], [**Gro65**], [**Gro66**], and [**Gro67**]). In this section we derive sufficient conditions for a separable algebra to have each of these properties. We will return to these important topics in Sections 8.3.2, 9.2.2 and 9.2.3. A general reference for the material in this section is [**Mil80**, Chapter I].

Definition 4.7.1. Suppose S and A are two commutative R-algebras, and I is an ideal in A such that $I^2 = (0)$. There is a map on sets

$$\beta : \operatorname{Hom}_{R\text{-alg}}(S, A) \to \operatorname{Hom}_{R\text{-alg}}(S, A/I)$$

which is induced by the natural map $A \to A/I$ on R-algebras. We say S is *formally unramified* if β is one-to-one for all choices of A and I. We say S is *formally smooth* if β is onto for all choices of A and I. We say S is *formally étale* if β is a one-to-one correspondence for all choices of A and I.

The next two results show that a commutative separable algebra is formally unramified and a commutative separable flat algebra is formally étale. When S is finitely generated, a converse to Corollary 4.7.2 is proved in Theorem 8.3.6.

Corollary 4.7.2. *(Separable and Commutative is Formally Unramified) Let R be a commutative ring, and S a commutative separable R-algebra. Suppose A is a commutative R-algebra, I is a nilpotent ideal in A, and $g : S \to A/I$ is an R-algebra homomorphism. If there exists an R-algebra homomorphism*

$\gamma : S \to A$ such that the diagram

$$\begin{array}{ccc} S & \xrightarrow{g} & A/I \\ \uparrow & \overset{\gamma}{\searrow} & \uparrow \\ R & \longrightarrow & A \end{array}$$

commutes, then γ is unique.

Proof. By Corollary 4.3.2, $S \otimes_R A$ is separable over A and $S \otimes_R (A/I)$ is separable over A/I. The diagram

$$\begin{array}{ccccc} S \otimes_R A & \xrightarrow{\gamma \otimes 1} & A \otimes_R A & \xrightarrow{\mu_A} & A \\ \downarrow{\scriptstyle \delta} & & \downarrow & & \downarrow \\ S \otimes_R (A/I) & \xrightarrow{g \otimes 1} & (A/I) \otimes_R (A/I) & \xrightarrow{\mu_{A/I}} & A/I \end{array}$$

commutes. The composite mapping $\mu_A \circ \gamma \otimes 1$ is an A-algebra homomorphism, and by Corollary 4.4.8 is completely determined by an idempotent in $S \otimes_R A$. By the same argument, $\mu_{A/I} \circ g \otimes 1$ is an A/I-algebra homomorphism, and is completely determined by an idempotent in $S \otimes_R (A/I)$. The kernel of δ is generated by elements of the form $x \otimes y$, where $x \in S$, $y \in I$. Therefore, the kernel of δ is contained in the nil radical of $A \otimes_R S$. By Corollary 3.4.16, $\operatorname{idemp}(S \otimes_R A) \to \operatorname{idemp}(S \otimes_R (A/I))$ is a one-to-one correspondence. □

When R is noetherian and S is finitely generated, a converse to Corollary 4.7.3 is proved in Theorem 9.2.19.

Corollary 4.7.3 (Separable, Flat, and Commutative is Formally Étale). *Let R be a commutative ring and S a commutative R-algebra which is separable and flat. Then the following universal mapping property is satisfied. Suppose A is a commutative R-algebra, I is a nilpotent ideal in A, and $g : S \to A/I$ is an R-algebra homomorphism. Then there exists a unique R-algebra homomorphism $\gamma : S \to A$ such that the diagram*

$$\begin{array}{ccc} S & \xrightarrow{g} & A/I \\ \uparrow & \overset{\exists \gamma}{\searrow} & \uparrow \\ R & \longrightarrow & A \end{array}$$

commutes.

Proof. By Corollary 4.7.2, γ is unique, if it exists. By Theorem 1.3.2 and Corollary 4.3.2, $S \otimes_R A$ is separable and flat over A and $S \otimes_R (A/I)$ is separable and flat over A/I. Since S is flat over R, the diagram

$$\begin{array}{ccccccccc} 0 & \longrightarrow & I & \longrightarrow & A & \longrightarrow & A/I & \longrightarrow & 0 \\ & & \downarrow & & \downarrow \alpha & & \downarrow \beta & & \\ 0 & \longrightarrow & I \otimes_R S & \longrightarrow & A \otimes_R S & \xrightarrow{\delta} & A/I \otimes_R S & \longrightarrow & 0 \end{array}$$

commutes and the rows are exact sequences of $\mathbb{Z}$-modules. The kernel of δ is generated by elements of the form $x \otimes y$, where $x \in I$, $y \in S$. Therefore, the ideal $I \otimes_R S$ is contained in the nil radical of $A \otimes_R S$. The map g followed by the multiplication map induces $\sigma : A/I \otimes_R S \to A/I$, which is a section to β. By Corollary 4.4.8 the kernel of σ is idempotent generated. Let $\bar{e}$ be the idempotent in $A/I \otimes_R S$ such that $\ker \sigma$ is generated by $1 - e$. By Corollary 3.4.16, $\operatorname{idemp}(A \otimes_R S) \to \operatorname{idemp}(A/I \otimes_R S)$ is a one-to-one correspondence. Let e be the idempotent in $A \otimes_R S$ which maps onto $\bar{e}$. Consider the commutative diagram:

$$\begin{array}{ccccccccc} 0 & \longrightarrow & I & \longrightarrow & A & \longrightarrow & A/I & \longrightarrow & 0 \\ & & \downarrow & & \downarrow \alpha & & \downarrow \beta & & \\ 0 & \longrightarrow & I \otimes_R S & \longrightarrow & A \otimes_R S & \xrightarrow{\delta} & A/I \otimes_R S & \longrightarrow & 0 \\ & & \downarrow & & \downarrow \epsilon & & \downarrow \zeta & & \\ 0 & \longrightarrow & (I \otimes_R S)e & \longrightarrow & (A \otimes_R S)e & \xrightarrow{\lambda} & (A/I \otimes_R S)\bar{e} & \longrightarrow & 0. \end{array}$$

We are given that $\zeta\beta$ is an isomorphism. Therefore, the A-module $(A \otimes_R S)e$ is generated by the image of $\epsilon\alpha$ and $I(1 \otimes_R S)e$. By Corollary 1.1.17, $\epsilon\alpha$ is onto. The map ϵ is a localization, hence is flat. Therefore, $\epsilon\alpha$ is flat. But $\ker(\epsilon\alpha)$ is contained in I, which consists of nilpotent elements. By Exercise 2.4.21, $\ker(\epsilon\alpha) = (0)$. Hence $\epsilon\alpha$ is an isomorphism and the desired map γ is the composition

$$1 \otimes_R S \to A \otimes_R S \xrightarrow{\epsilon} (A \otimes_R S)e \xrightarrow{(\epsilon\alpha)^{-1}} A.$$

The diagram is commutative, so γ agrees with g. □

In Remark 10.1.11 we show how to apply Proposition 4.7.4 to prove Hensel's Lemma.

Proposition 4.7.4. *Let R be a commutative ring and I an ideal in R such that R is separated and complete with respect to the I-adic topology. Let S be a commutative finitely generated separable flat R-algebra. For any R-algebra homomorphism $\sigma_0 : S \to R/I$, there exists an R-algebra homomorphism*

$\sigma : S \to R$ such that the diagram

$$\begin{array}{ccc} S & \xrightarrow{\sigma_0} & R/I \\ \uparrow & \searrow^{\exists \sigma} & \uparrow \\ R & \xrightarrow{=} & R \end{array}$$

commutes.

Proof. By the universal mapping property, it suffices to construct R-algebra maps $\sigma_i : S \to R/I^{i+1}$ such that for all $i \geq 1$ the diagram

$$\begin{array}{ccc} S & \xrightarrow{\sigma_i} & R/I^{i+1} \\ & {\scriptstyle \sigma_{i-1}} \searrow & \downarrow \\ & & R/I^i \end{array}$$

commutes. The map σ_0 is given. Inductively, assume σ_{i-1} exists. Since I^i/I^{i+1} is a nilpotent ideal in R/I^{i+1}, σ_i exists, by Corollary 4.7.3. $\square$

Chapter 5

Background Material on Homological Algebra

This chapter contains those tools which we will require from group cohomology, Hochschild cohomology, and Amitsur cohomology. Our treatment of the subject is limited to a utilitarian point of view. The reader is assumed to have a basic working knowledge of homological algebra, including familiarity with the definitions and techniques of derived functors, Ext groups, and Tor groups. There is a section on faithfully flat descent, as well as a short section on the properties of the tensor algebra associated to a module. As general references on Homological Algebra, we recommend [**Rot79**], [**CE56**], or another introductory book.

1. Group Cohomology

Let G be a group, written multiplicatively, with identity element denoted 1. Let $\mathbb{Z}G$ denote the group ring. A left $\mathbb{Z}G$-module is also called a *G-module*. The augmentation map $\epsilon : \mathbb{Z}G \to \mathbb{Z}$ is the homomorphism of rings induced by $G \to \langle 1 \rangle$. Via ϵ, any $\mathbb{Z}$-module A can be made into a *trivial G-module*. In this case, for every $x \in A$ and $\sigma \in G$ we have $\sigma x = x$. That is, every $\sigma \in G$ acts as the trivial automorphism of A. In particular, ϵ induces the trivial left $\mathbb{Z}G$-module structure on $\mathbb{Z}$.

Let G be a group and A a left G-module. For $n \geq 0$, the *nth cohomology group of G with coefficients in A* is defined to be

$$\mathrm{H}^n(G, A) = \mathrm{Ext}^n_{\mathbb{Z}G}(\mathbb{Z}, A)$$

where $\mathbb{Z}$ has the trivial left $\mathbb{Z}G$-module structure. The groups $\mathrm{H}^n(G, A)$ are isomorphic to the right derived groups of the left exact contravariant functor $\mathrm{Hom}_{\mathbb{Z}G}(\cdot, A)$, as well as the right derived groups of the left exact covariant functor $\mathrm{Hom}_{\mathbb{Z}G}(\mathbb{Z}, \cdot)$. Exercise 5.1.21 shows that the groups $\mathrm{H}^n(G, A)$ are also isomorphic to the right derived groups of the left exact covariant functor $A \mapsto A^G$.

Example 5.1.1. Let $G = \langle 1 \rangle$ be the trivial group and A a $\mathbb{Z}$-module. Since $\mathbb{Z}$ is a projective $\mathbb{Z}$-module, the cohomology of G with coefficients in A is

$$\mathrm{H}^n(\langle 1 \rangle, A) = \mathrm{Ext}^n_{\mathbb{Z}}(\mathbb{Z}, A) = \begin{cases} \mathrm{Hom}_{\mathbb{Z}}(\mathbb{Z}, A) = A & \text{if } n = 0, \\ 0 & \text{if } n > 0. \end{cases}$$

Now we describe the constructions of the two standard resolutions of $\mathbb{Z}$ by free G-modules. In Section 5.1.1 we apply these two resolutions to derive the usual cocycle and coboundary relations in low degrees. The easiest way to distinguish the two resolutions is to remember that their names are based on the cocycles they define. The unnormalized standard resolution gives rise to unnormalized cocycles, the normalized standard resolution gives rise to normalized cocycles. When we define crossed product algebras in Section 12.1, it will be convenient to use normalized 2-cocycles.

For the proofs which are omitted, the reader is referred to the references mentioned in the introduction to this chapter. When G is a finite cyclic group, there is another useful free resolution of $\mathbb{Z}$ given in Lemma 5.1.16.

For any $r \geq 1$, let $G^r = \prod_{i=1}^r G$ be the product of r copies of G. To denote the elements of G^r, it is customary to use brackets instead of parentheses. Elements of G^r are written as $[\sigma_1, \dots, \sigma_r]$, or sometimes as $[\sigma_0, \dots, \sigma_{r-1}]$.

Definition 5.1.2. For $n \geq 1$, we define Q_n to be the free $\mathbb{Z}G$-module on the index set G^n. The basis for Q_n is the set $\{[\sigma_1, \dots, \sigma_n] \mid \sigma_i \in G\}$. For consistency, define Q_0 to be the free $\mathbb{Z}G$-module on the singleton set $\{[\,]\}$. For $n \geq 1$ define a boundary map $d_n : Q_n \to Q_{n-1}$ by specifying its value on a typical basis element:

$$\begin{aligned} d_n[\sigma_1, \dots, \sigma_n] = \sigma_1[\sigma_2, \dots, \sigma_n] + \\ \sum_{i=1}^{n-1} (-1)^i [\sigma_1, \dots, \sigma_{i-1}, \sigma_i\sigma_{i+1}, \sigma_{i+2}, \dots, \sigma_n] \\ + (-1)^n [\sigma_1, \dots, \sigma_{n-1}]. \end{aligned}$$

Theorem 5.1.3 says that when augmented by ϵ, we have a resolution of $\mathbb{Z}$ by free $\mathbb{Z}G$-modules. This complex will be denoted $Q_\bullet$ and is called the *unnormalized, or homogeneous standard resolution.*

Theorem 5.1.3. *The sequence*

$$\cdots \to Q_n \xrightarrow{d_n} Q_{n-1} \xrightarrow{d_{n-1}} \cdots \to Q_1 \xrightarrow{d_1} Q_0 \xrightarrow{\epsilon} \mathbb{Z} \to 0$$

of Definition 5.1.2 *is a free resolution of the* $\mathbb{Z}G$ *module* $\mathbb{Z}$.

Definition 5.1.4. Let $G_1 = G - \langle 1 \rangle = \{\sigma \in G \mid \sigma \neq 1\}$. For $n \geq 1$ define B_n to be the $\mathbb{Z}G$-submodule of Q_n (see Definition 5.1.2) generated by those basis elements $[\sigma_1, \dots, \sigma_n]$ which belong to G_1^n. We take $B_0 = Q_0$, the free module on $[\,]$. The set inclusion map $G_1^n \subseteq G^n$ induces an idempotent $\eta_n \in \operatorname{Hom}_{\mathbb{Z}G}(Q_n, Q_n)$ which projects Q_n onto B_n. The boundary map $d_n : B_n \to B_{n-1}$ is defined to be the inclusion map $B_n \subseteq Q_n$ followed by the boundary map $d_n : Q_n \to Q_{n-1}$ of Definition 5.1.2 followed by η_{n-1}. By construction, the diagram

$$\begin{array}{ccccc} B_n & & \xrightarrow{\quad d_n \quad} & & B_{n-1} \\ \downarrow{\scriptstyle\subseteq} & & & & \downarrow{\scriptstyle\subseteq} \\ Q_n & \xrightarrow{d_n} & Q_{n-1} & \xrightarrow{\eta_{n-1}} & Q_{n-1} \end{array}$$

commutes. Theorem 5.1.5 states that when augmented with $\epsilon : B_0 \to \mathbb{Z}$, the complex $B_\bullet$ is a free $\mathbb{Z}G$-module resolution of $\mathbb{Z}$. This complex is called the *bar resolution*, or *normalized standard resolution*.

Theorem 5.1.5. *The sequence*

$$\cdots \to B_n \xrightarrow{d_n} B_{n-1} \xrightarrow{d_{n-1}} \cdots \to B_1 \xrightarrow{d_1} B_0 \xrightarrow{\epsilon} \mathbb{Z} \to 0$$

of Definition 5.1.4 *is a free resolution of the* $\mathbb{Z}G$*-module* $\mathbb{Z}$.

1.1. Cocycle and Coboundary Groups in Low Degree. The purpose of this section is to derive generators and relations for the groups $\mathrm{H}^n(G, A)$ for $n = 0, 1, 2, 3$. We do this twice. First we derive the so-called homogeneous cochain complex in Eq.(5.1). This is based on the unnormalized resolution of $\mathbb{Z}$ and is the classical approach, because it was inspired by the homology of a simplicial complex. The cocycles and coboundaries in low degrees are described in Proposition 5.1.6. Secondly, starting with the bar resolution (or normalized resolution) of $\mathbb{Z}$, we describe the cocycles and coboundaries. These are listed in Corollary 5.1.7 and are the familiar normalized factor sets that are useful for the crossed product construction.

Let A be a $\mathbb{Z}G$-module. So A is an abelian group with binary operation written additively, and G acts as a group on A. The cohomology groups $\mathrm{H}^n(G, A)$ are defined to be $\operatorname{Ext}^n_{\mathbb{Z}G}(\mathbb{Z}, A)$. If $Q_\bullet \to \mathbb{Z}$ is the standard (homogeneous) resolution from Definition 5.1.2, and $B_\bullet \to \mathbb{Z}$ is the bar resolution from Definition 5.1.4, then

$$\begin{aligned}\mathrm{H}^n(G, A) &= \mathrm{Ext}^n_{\mathbb{Z}G}(\mathbb{Z}, A)\\ &= \mathrm{H}^n(\mathrm{Hom}_{\mathbb{Z}G}(Q_\bullet, A))\\ &= \mathrm{H}^n(\mathrm{Hom}_{\mathbb{Z}G}(B_\bullet, A)).\end{aligned}$$

Notice that $\mathrm{H}^n(\mathrm{Hom}_{\mathbb{Z}G}(Q_\bullet, A))$ is an abelian group, where functions are added point-wise: $(f + g)(x) = f(x) + g(x)$. Since $Q_0 = \mathbb{Z}G$, we have $\mathrm{Hom}_{\mathbb{Z}G}(Q_0, A) = A$ (Proposition 1.3.23 (1)). For $n \geq 1$, because Q_n is the free $\mathbb{Z}G$-module on G^n, we can identify $\mathrm{Hom}_{\mathbb{Z}G}(Q_n, A)$ with $\mathrm{Map}(G^n, A)$, the set of all functions mapping G^n to A. The cochain map

$$\mathrm{Hom}_{\mathbb{Z}G}(Q_{n-1}, A) \xrightarrow{d^{n-1}} \mathrm{Hom}_{\mathbb{Z}G}(Q_n, A)$$

is defined by $d^{n-1}(f) = fd_n$. Using the formula for the boundary d_n in Definition 5.1.2, on a typical basis element of Q_n we have

$$\begin{aligned}(5.1)\quad d^{n-1}(f)[\sigma_1, \dots, \sigma_n] &= fd_n[\sigma_1, \dots, \sigma_n]\\ &= \sigma_1 f[\sigma_2, \dots, \sigma_n] + \sum_{i=1}^{n-1}(-1)^i f[\sigma_1, \dots, \sigma_i\sigma_{i+1}, \dots, \sigma_n]\\ &\qquad + (-1)^n f[\sigma_1, \dots, \sigma_{n-1}].\end{aligned}$$

In the first summand, we have used the fact that f is $\mathbb{Z}G$-linear. For all $n \geq 0$,

$$\mathrm{H}^n(G, A) = \mathrm{Z}^n(G, A)/\,\mathrm{B}^n(G, A)$$

where $\mathrm{Z}^n(G, A) = \ker d^n$, and $\mathrm{B}^n(G, A) = \mathrm{im}\, d^{n-1}$. By convention, $d^{-1} = 0$ and $\mathrm{B}^0(G, A) = 0$.

Proposition 5.1.6. *In the above context,*

(1) $\mathrm{H}^0(G, A) = \mathrm{Z}^0(G, A) = A^G$ *is the subset of* A *fixed by* G.

(2) $\mathrm{Z}^1(G, A)$ *is the set of all functions* $f : G \to A$ *such that*

$$f(\sigma\tau) = f(\sigma) + \sigma f(\tau),$$

for all $(\sigma, \tau) \in G^2$.

(3) $\mathrm{B}^1(G, A)$ *is the set of all functions* $f : G \to A$ *such that there exists* $x \in A$ *and* $f(\sigma) = \sigma(x) - x$, *for all* $\sigma \in G$.

(4) $\mathrm{Z}^2(G, A)$ *is the set of all functions* $f : G \times G \to A$ *such that*

$$f(\rho, \sigma) + f(\rho\sigma, \tau) = \rho f(\sigma, \tau) + f(\rho, \sigma\tau),$$

for all $(\rho, \sigma, \tau) \in G^3$.

(5) $\mathrm{B}^2(G, A)$ *is the set of all functions* $f : G \times G \to A$ *such that there exists* $g : G \to A$ *and* $f(\sigma, \tau) = \sigma g(\tau) - g(\sigma\tau) + g(\sigma)$, *for all* $(\sigma, \tau) \in G^2$.

Proof. The proof follows straight from (5.1) and the definitions. □

Corollary 5.1.7. *In the above context, the normalized cocycles and coboundaries in degrees* 1 *and* 2 *are:*

(1) $Z^1(G, A)$ *is the set of all functions* $f : G \to A$ *such that* $f(1) = 0$, *and*
$$f(\sigma\tau) = f(\sigma) + \sigma f(\tau),$$
for all $(\sigma, \tau) \in G^2$.

(2) $Z^2(G, A)$ *is the set of all functions* $f : G \times G \to A$ *such that*
$$f(\rho, \sigma) + f(\rho\sigma, \tau) = \rho f(\sigma, \tau) + f(\rho, \sigma\tau),$$
and $f(1, \tau) = f(\sigma, 1) = 0$, *for all* $(\rho, \sigma, \tau) \in G^3$.

(3) $B^2(G, A)$ *is the set of all functions* $f : G \times G \to A$ *such that there exists* $g : G \to A$ *where* $g(1) = 0$ *and* $f(\sigma, \tau) = \sigma g(\tau) - g(\sigma\tau) + g(\sigma)$, *for all* $(\sigma, \tau) \in G^2$.

Proof. Let $B_\bullet \to \mathbb{Z}$ be the bar resolution. In (5.1), d_n is zero whenever 1 appears in the n-tuple. Notice that elements of $B^1(G, A)$ are always normalized. The details are left to the reader. □

Remark 5.1.8. For the record, we mention that the group $Z^3(G, A)$ is the set of all $f : G^3 \to A$ such that the 3-cocycle identity
$$f(\sigma_1\sigma_2, \sigma_3, \sigma_4) + f(\sigma_1, \sigma_2, \sigma_3\sigma_4) = \\ f(\sigma_1, \sigma_2, \sigma_3) + \sigma_1 f(\sigma_2, \sigma_3, \sigma_4) + f(\sigma_1, \sigma_2\sigma_3, \sigma_4)$$
is satisfied for all $(\sigma_1, \sigma_2, \sigma_3, \sigma_4) \in G^4$. Moreover, to compute $H^3(G, A)$, normalized cocycles can be used. That is, $f(\sigma_1, \sigma_2, 1) = f(\sigma_1, 1, \sigma_3) = f(1, \sigma_2, \sigma_3) = 0$. The set of 3-coboundaries, $B^3(G, A)$, consists of all $f : G^3 \to A$ for which there exists $g : G \times G \to A$ and
$$f(\rho, \sigma, \tau) = \rho g(\sigma, \tau) - g(\rho\sigma, \tau) + g(\rho, \sigma\tau) - g(\sigma, \tau)$$
for all $(\rho, \sigma, \tau) \in G^3$.

1.2. Applications and Computations. We begin by proving the basic functorial properties of group cohomology. When the group G is a finite cyclic group, we derive the common formulas for $H^n(G, A)$ in terms of the norm and difference maps on A. As an application, we prove Hilbert's Theorem 90.

Definition 5.1.9. Let G be a group.

(1) If $\theta : G \to K$ is a homomorphism of groups and A is a $\mathbb{Z}K$-module, then the ring homomorphism $\theta : \mathbb{Z}G \to \mathbb{Z}K$ makes A into a $\mathbb{Z}G$-module.

(2) If H is a subgroup of G and A is a $\mathbb{Z}H$-module, then $\operatorname{Hom}_{\mathbb{Z}H}(\mathbb{Z}G, A)$ is a left $\mathbb{Z}G$-module (see Lemma 1.3.17 (1)) which is called the *induced G-module.*

Theorem 5.1.10 (Shapiro's Lemma). *Let G be a group, H a subgroup of G, and A a $\mathbb{Z}H$-module. There are isomorphisms*

$$\mathrm{H}^n(H, A) \cong \mathrm{H}^n(G, \operatorname{Hom}_{\mathbb{Z}H}(\mathbb{Z}G, A))$$

which are induced by the adjoint isomorphism of Theorem 1.3.24.

Proof. Since $\mathbb{Z}G$ is a free left $\mathbb{Z}H$-module, this follows directly from the isomorphism

$$\operatorname{Ext}^n_{\mathbb{Z}H}(\mathbb{Z}G \otimes_{\mathbb{Z}G} \mathbb{Z}, A) \cong \operatorname{Ext}^n_{\mathbb{Z}G}(\mathbb{Z}, \operatorname{Hom}_{\mathbb{Z}H}(\mathbb{Z}G, A))$$

which is induced by the adjoint isomorphism of Theorem 1.3.24. It is also of interest to know how this map is defined on cochains. Let $Q_\bullet \to Z$ be the standard resolution of $\mathbb{Z}$ as a $\mathbb{Z}G$-module. By Exercise 1.1.9, $Q_\bullet \to Z$ is also a free resolution of $\mathbb{Z}$ as a $\mathbb{Z}H$-module. The adjoint isomorphism

$$\operatorname{Hom}_{\mathbb{Z}H}(Q_n, A) \xrightarrow{\phi} \operatorname{Hom}_{\mathbb{Z}G}(Q_n, \operatorname{Hom}_{\mathbb{Z}H}(\mathbb{Z}G, A))$$

maps an n-cochain f to ϕf. If y comes from Q_n, then $(\phi f)(y)$ is the element of $\operatorname{Hom}_{\mathbb{Z}H}(\mathbb{Z}G, A)$ defined by $(\phi f)(y)(x) = f(xy)$. The details are left to the reader. □

Now we show that group cohomology is a covariant functor in the second variable, and a contravariant functor in the first variable.

Lemma 5.1.11. *Let G be a group and A a $\mathbb{Z}G$-module.*

(1) *If $\psi : A \to B$ is a homomorphism of $\mathbb{Z}G$-modules, then ψ induces a homomorphism*

$$\mathrm{H}^n(G, A) \to \mathrm{H}^n(G, B)$$

of abelian groups, for each $n \ge 0$.

(2) *If $\theta : H \to G$ is a homomorphism of groups, then θ induces a homomorphism*

$$\mathrm{H}^n(G, A) \to \mathrm{H}^n(H, A)$$

of abelian groups, for each $n \ge 0$.

Proof. Part (1) follows from the fact that $\operatorname{Ext}^n_{\mathbb{Z}G}(\mathbb{Z}, \cdot)$ is a covariant functor.

(2): Let $(Q_G)_\bullet \to \mathbb{Z}$ be the standard resolution for the $\mathbb{Z}G$-module $\mathbb{Z}$, and $(Q_H)_\bullet \to \mathbb{Z}$ the counterpart for the $\mathbb{Z}H$-module. The homomorphism $\theta : H \to G$ induces a homomorphism $H^n \to G^n$, for each n. For each n,

$(Q_H)_n$ is free on H^n and $(Q_H)_n$ is free on G^n. Hence there is an induced morphism

$$\theta : (Q_G)_\bullet \to (Q_H)_\bullet \tag{5.2}$$

of complexes. Now suppose A is a $\mathbb{Z}G$-module, which is made into a $\mathbb{Z}H$-module by virtue of $\theta : \mathbb{Z}H \to \mathbb{Z}G$. There are morphisms of complexes

$$\operatorname{Hom}_{\mathbb{Z}G}((Q_G)_\bullet, A) \to \operatorname{Hom}_{\mathbb{Z}G}((Q_H)_\bullet, A) \to \operatorname{Hom}_{\mathbb{Z}H}((Q_H)_\bullet, A)$$

where the first morphism is induced by the functor $\operatorname{Hom}_{\mathbb{Z}G}(\cdot, A)$ applied to the morphism (5.2) and the second is induced by the map defined in Exercise 1.3.35. □

The next (long) definition describes the two standard homomorphisms on group cohomology associated to a subgroup $H \subseteq G$. They are called the restriction and corestriction homomorphisms. When H is normal, there is a third homomorphism, called inflation.

Definition 5.1.12. Let G be a group and A a $\mathbb{Z}G$-module.

(1) If H is a subgroup of G, then the homomorphism of abelian groups

$$\operatorname{Res} : \mathrm{H}^n(G, A) \to \mathrm{H}^n(H, A)$$

defined in Lemma 5.1.11 (2) is called the *restriction homomorphism.* Suppose $f : G^n \to A$ is an n-cocycle in $\mathrm{Z}^n(G, A)$. Viewing H^n as a subset of G^n, the restriction of f defines $g : H^n \to A$ which is an n-cocycle in $\mathrm{Z}^n(H, A)$. The restriction homomorphism maps the cohomology class $\bar{f}$ to $\bar{g}$.

(2) If N is a normal subgroup of G, then A^N can be made into a $\mathbb{Z}(G/N)$-module. The multiplication rule is induced by $(gN)x = gx$. The natural map $\eta : G \to G/N$ and the set inclusion $\iota : A^N \to A$ induce homomorphisms

$$\begin{array}{ccc} \mathrm{H}^n(G/N, A^N) & \xrightarrow{\quad\operatorname{Inf}\quad} & \mathrm{H}^n(G, A) \\ & {\scriptstyle\eta}\searrow \quad \nearrow{\scriptstyle\iota} & \\ & \mathrm{H}^n(G, A^N) & \end{array}$$

and the composite map, Inf , is called the *inflation homomorphism.* Suppose $f : (G/H)^n \to A^H$ is an n-cocycle in $\mathrm{Z}^n(G/H, A^H)$. Define $g : G^n \to A$ by the rule $g(\sigma_1, \dots, \sigma_n) = f(\bar{\sigma_1}, \dots, \bar{\sigma_n})$, where $\bar{\sigma_i}$ is the coset represented by σ_i in G/H. Then g is an n-cocycle in $\mathrm{Z}^n(G, A)$, and the inflation homomorphism maps the cohomology class $\bar{f}$ to $\bar{g}$.

(3) Suppose H is a subgroup of G of finite index $[G : H] = m$ and $x_1, \ldots, x_m$ is a full set of left coset representatives for H. Let A be a left $\mathbb{Z}G$-module. The reader should verify that the map

$$\operatorname{Hom}_{\mathbb{Z}H}(\mathbb{Z}G, A) \xrightarrow{\psi} A$$

defined by $\psi(f) = \sum_{i=1}^{m} x_i f(x_i^{-1})$ is a homomorphism of $\mathbb{Z}G$-modules and does not depend on the choices of $x_1, \ldots, x_m$. This defines a homomorphism on cohomology groups

$$\mathrm{H}^n(G, \operatorname{Hom}_{\mathbb{Z}H}(\mathbb{Z}G, A)) \xrightarrow{\psi} \mathrm{H}^n(G, A).$$

The *corestriction homomorphism*, denoted Cor, is defined by composing ψ with the isomorphism from Theorem 5.1.10. By definition, the diagram

$$\begin{array}{ccc} \mathrm{H}^n(H, A) & \xrightarrow{\quad\mathrm{Cor}\quad} & \mathrm{H}^n(G, A) \\ & {\scriptstyle\cong}\searrow \quad \nearrow{\scriptstyle\psi} & \\ & \mathrm{H}^n(G, \operatorname{Hom}_{\mathbb{Z}H}(\mathbb{Z}G, A)) & \end{array}$$

commutes. Using the description of the isomorphism in the proof of Theorem 5.1.10, we can describe the corestriction map on n-cocycles. Let f be a cocycle in $\operatorname{Hom}_{\mathbb{Z}H}(Q_n, A)$ defining a cohomology class c in $\mathrm{H}^n(H, A)$. Then $\mathrm{Cor}(f)$ is a cocycle in $\operatorname{Hom}_{\mathbb{Z}G}(Q_n, A)$ which represents a cohomology class $\mathrm{Cor}(c)$ in $\mathrm{H}^n(G, A)$. If $y \in Q_n$, then

$$\mathrm{Cor}(f)(y) = \sum_{i=1}^{m} x_i \phi(f)(y)(x_i^{-1}) = \sum_{i=1}^{m} x_i f(x_i^{-1} y).$$

For example, consider the $n = 0$ case. From Proposition 5.1.6, $\mathrm{Z}^0(H, A) = A^H$. Then f is a constant valued function, say $f(x) = a$. In this case, $\mathrm{Cor}(f)$ is the constant valued function $\sum_{i=1}^{m} x_i a$.

Theorem 5.1.13. *Let H be a subgroup of G of finite index $[G : H] = m$. If A is a left $\mathbb{Z}G$-module, then*

$$\mathrm{Cor}\,\mathrm{Res}\,\mathrm{H}^n(G, A) = m\,\mathrm{H}^n(G, A).$$

Proof. Let f be a cocycle in $\operatorname{Hom}_{\mathbb{Z}H}(Q_n, A)$ defining a cohomology class c in $\mathrm{H}^n(H, A)$. If f is in the image of $\mathrm{Res} : \mathrm{H}^n(G, A)$, then f is $\mathbb{Z}G$-linear. We now employ the description of the corestriction given above in Definition 5.1.12 (3). For any $y \in Q_n$,

$$\mathrm{Cor}(f)(y) = \sum_{i=1}^{m} x_i f(x_i^{-1} y) = \sum_{i=1}^{m} x_i x_i^{-1} f(y) = m f(y)$$

which proves the claim. □

Corollary 5.1.14. *If G is a finite group of order m and A is any $\mathbb{Z}G$-module, then $m\,\mathrm{H}^n(G,A) = 0$ for all $n \geq 1$.*

Proof. If $H = \langle 1 \rangle$, then $[G : H] = m$. By Theorem 5.1.13, the diagram

$$\begin{array}{ccc} \mathrm{H}^n(G,A) & \xrightarrow{\quad m \quad} & \mathrm{H}^n(G,A) \\ & {\scriptstyle\mathrm{Res}}\searrow \quad \nearrow{\scriptstyle\mathrm{Cor}} & \\ & \mathrm{H}^n(H,A) & \end{array}$$

commutes, where the horizontal map is "multiplication by m". As we saw in Example 5.1.1, the group $\mathrm{Ext}^n_{\mathbb{Z}}(\mathbb{Z}, A) = \mathrm{H}^n(\langle 1 \rangle, A)$ is trivial for $n \geq 1$. $\square$

The next lemma provides a convenient description of the induced G-module as a tensor product.

Lemma 5.1.15. *Let H be a subgroup of G of finite index $[G : H] = m$ and $x_1, \ldots, x_m$ a full set of left coset representatives for H. If A is a left $\mathbb{Z}H$-module, then there is an isomorphism of $\mathbb{Z}G$-modules*

$$\mathrm{Hom}_{\mathbb{Z}H}(\mathbb{Z}G, A) \xrightarrow{\psi} \mathbb{Z}G \otimes_{\mathbb{Z}H} A$$

defined by $\psi(f) = \sum_{i=1}^m x_i \otimes f(x_i^{-1})$.

Proof. The reader should verify that the map ψ does not depend on the choices for $x_1, \ldots, x_m$. Notice that $\mathbb{Z}G \cong \bigoplus_{i=1}^m x_i \mathbb{Z}H$ as right $\mathbb{Z}H$-modules. By Lemma 1.3.1,

$$\mathbb{Z}G \otimes_{\mathbb{Z}H} A \cong \bigoplus_{i=1}^m x_i \otimes_{\mathbb{Z}H} A$$

as left $\mathbb{Z}$-modules. Also, $\mathbb{Z}G \cong \bigoplus_{i=1}^m \mathbb{Z}H x_i^{-1}$ as left $\mathbb{Z}H$-modules. By Proposition 1.3.23,

$$\mathrm{Hom}_{\mathbb{Z}H}(\mathbb{Z}G, A) \cong \bigoplus_{i=1}^m \mathrm{Hom}_{\mathbb{Z}H}(\mathbb{Z}H x_i^{-1}, A)$$

as left $\mathbb{Z}$-modules. The reader should verify that f in $\mathrm{Hom}_{\mathbb{Z}H}(\mathbb{Z}Hx_i^{-1}, A)$ is mapped by ψ to $x_i \otimes f(x_i^{-1})$ and hence ψ is bijective. We check that ψ is $\mathbb{Z}G$-linear. Let $g \in G$. Right multiplication by g is a permutation of the right cosets of H. For each i, there is a unique i' and $h_i \in H$ such that $x_i^{-1} g = h_i x_{i'}^{-1}$, or equivalently $x_i h_i = g x_{i'}$. Let $f \in \mathrm{Hom}_{\mathbb{Z}H}(\mathbb{Z}G, A)$. For $x \in \mathbb{Z}G$, $(gf)(x) = f(xg)$. Therefore, $\psi(gf) = \sum x_i \otimes f(x_i^{-1} g) = \sum x_i \otimes f(h_i x_{i'}^{-1}) = \sum x_i h_i \otimes f(x_{i'}^{-1}) = \sum g x_{i'} \otimes f(x_{i'}^{-1}) = g\psi(f)$. $\square$

When G is a finite cyclic group, the next lemma shows that there is a free resolution of $\mathbb{Z}$ by $\mathbb{Z}G$-modules all of rank one.

Lemma 5.1.16. *Let $G = \langle\sigma\rangle$ be a finite cyclic group of order m. In $\mathbb{Z}G$, let $D = \sigma - 1$, and $N = 1 + \sigma + \cdots + \sigma^{m-1}$. Then multiplication by D and N, together with the augmentation map ϵ define an exact sequence*

$$\cdots \xrightarrow{N} \mathbb{Z}G \xrightarrow{D} \mathbb{Z}G \xrightarrow{N} \mathbb{Z}G \xrightarrow{D} \mathbb{Z}G \xrightarrow{\rho} \mathbb{Z} \to 0$$

which is a free resolution of the trivial $\mathbb{Z}G$-module $\mathbb{Z}$.

Proof. The maps are $\mathbb{Z}G$-module homomorphisms because G is abelian. The reader should verify that the kernel of ϵ is equal to the image of D. The sequence is a complex, since $DN = ND = 0$. Let $x = \sum a_i\sigma^i$ be a typical element of $\mathbb{Z}G$. Then

$$\begin{aligned} x &= a_0 + a_1\sigma + a_2\sigma^2 + \cdots + a_{m-1}\sigma^{m-1} \\ \sigma x &= a_{m-1} + a_0\sigma + a_1\sigma^2 + \cdots + a_{m-2}\sigma^{m-1} \\ \sigma^2 x &= a_{m-2} + a_{m-1}\sigma + a_0\sigma^2 + \cdots + a_{m-3}\sigma^{m-1} \\ &\vdots \\ \sigma^{m-1}x &= a_1 + a_2\sigma + a_3\sigma^2 + \cdots + a_0\sigma^{m-1}. \end{aligned} \tag{5.3}$$

If $x = \sigma x$, then (5.3) shows that $a_0 = a_1 = \cdots = a_{m-1}$, hence $x = Na_0$. Thus $\ker D = \operatorname{im} N$. It follows from (5.3) that $Nx = (\sum_i a_i)N$. If $Nx = 0$, then $\sum_{a_i} = 0$. Hence, the kernel of N is equal to the kernel of ϵ. Thus $\ker N = \operatorname{im} D$. □

Let $G = \langle\sigma\rangle$ be a finite cyclic group of order m. Keeping the notation of Lemma 5.1.16, let $D = \sigma - 1$, and $N = 1 + \sigma + \cdots + \sigma^{m-1}$. For any $\mathbb{Z}G$-module A, left multiplication by D and N define $\mathbb{Z}G$-module endomorphisms $D : A \to A$, $N : A \to A$. The images are denoted DA and NA, respectively. The kernel of D is A^G, and the kernel of N is denoted ${}_NA = \{x \in A \mid Nx = 0\}$. The reader should verify that the groups DA, NA and ${}_NA$ do not depend on the choice of σ.

Theorem 5.1.17. *Let G be a finite cyclic group. For any $\mathbb{Z}G$-module A,*

$$\mathrm{H}^n(G, A) = \begin{cases} A^G & \text{if } n = 0, \\ {}_NA/DA & \text{if } n \text{ is odd}, \\ A^G/NA & \text{if } n > 0 \text{ is even}. \end{cases}$$

Proof. The proof follows from an application of the functor $\operatorname{Hom}_{\mathbb{Z}G}(\cdot, A)$ to the resolution of $\mathbb{Z}$ in Lemma 5.1.16. □

Corollary 5.1.18. *If G is a finite cyclic group of order m and A is a trivial $\mathbb{Z}G$-module, then*

$$\mathrm{H}^n(G, A) = \begin{cases} A & \text{if } n = 0, \\ {}_mA & \text{if } n \text{ is odd}, \\ A/mA & \text{if } n > 0 \text{ is even}, \end{cases}$$

where ${}_mA = \{x \in A \mid mx = 0\}$, *and* $mA = \{mx \mid x \in A\}$.

Proof. The map D is the zero operator on A, and N is the "multiplication by m" operator. □

Corollary 5.1.19. *Let $G = \langle\sigma\rangle$ be a finite cyclic group of order n and A a $\mathbb{Z}G$-module (written multiplicatively). If $m \mid n$, $\tau = \sigma^{n/m}$, and $H = \langle\tau\rangle$ is the subgroup of order m, then the image of the inflation homomorphism (Definition* 5.1.12 *(2))*

$$\mathrm{Inf} : \mathrm{H}^2(G/H, A^H) \to \mathrm{H}^2(G, A)$$

is divisible by m. That is, for any $z \in \mathrm{H}^2(G/H, A^H)$, there exists $y \in \mathrm{H}^2(G, A)$ such that $\mathrm{Inf}(z) = y^m$.

Proof. Let $\bar{z} \in \mathrm{H}^2(G/H, A^H)$. Write $\bar{\sigma}$ for the coset represented by σ in G/H. By Exercise 5.1.28, there is $a \in A^G$ such that $\bar{z}$ is represented by a 2-cocycle $z : (G/H) \times (G/H) \to A^H$ of the form

$$z(\bar{\sigma}^i, \bar{\sigma}^j) = \begin{cases} 1 & \text{if } i + j < n/m, \\ a & \text{if } i + j \geq n/m, \end{cases}$$

for $0 \leq i, j < n/m$. The image of $\bar{z}$ under the inflation homomorphism is represented by the 2-cocycle $\xi : G \times G \to A$ defined by $\xi(\sigma^i, \sigma^j) = z(\bar{\sigma}^i, \bar{\sigma}^j)$. By Exercise 5.1.28, there is an isomorphism $\mathrm{H}^2(G, A) \to A^G/NA$ which is induced by $\xi \mapsto a_\xi$, where

$$\begin{aligned} a_\xi &= \prod_{j=0}^{n-1} \xi(\sigma^j, \sigma) \\ &= \prod_{k=0}^{m-1} \prod_{i=0}^{n/m-1} \xi((\sigma^{n/m})^k \sigma^i, \sigma) \\ &= a^m. \end{aligned}$$

By Exercise 5.1.28, ξ is cohomologous to χ_a^m, where

$$\chi_a(\sigma^i, \sigma^j) = \begin{cases} 1 & \text{if } i + j < n, \\ a & \text{if } i + j \geq n, \end{cases}$$

for $0 \leq i, j < n$. □

We end this section with an application to the Galois cohomology of fields. Let F be a field and G a finite group of automorphisms of F. Write F^+ for the additive group of F, and F^* for the group of units. A generalization of Theorem 5.1.20 (1) is proved in Corollary 13.3.2. A generalization of Theorem 5.1.20 (2) is proved in Theorem 12.2.10 (4).

Theorem 5.1.20. *Let F be a field and G a finite group of automorphisms of F.*

(1) *(Hilbert's Theorem 90)* $\mathrm{H}^1(G, F^*) = \langle 1 \rangle$.
(2) *For all $n \geq 1$,* $\mathrm{H}^n(G, F^+) = (0)$.

Proof. (1): Let $f \in \mathrm{Z}^1(G, F^*)$ be a 1-cocycle. By Proposition 5.1.6, we can assume $f : G \to F^*$ and $f(\sigma\tau) = f(\sigma)\sigma f(\tau)$. The elements of G are linearly independent over F (see, for example, [**Her75**, Theorem 5.6.1]). There exists $x \in F$ such that

$$\alpha = \sum_{\tau \in G} f(\tau)\tau(x) \neq 0.$$

In other words, α is a unit in F. For any $\sigma \in G$, we see that $\sigma(\alpha)$ is equal to $\sum_{\tau \in G} \sigma f(\tau)\sigma\tau(x)$. By the 1-cocycle identity,

$$\sigma(\alpha) = \Big(\sum_{\tau \in G} \sigma f(\sigma\tau)\sigma\tau(x)\Big) f(\sigma)^{-1} = \alpha f(\sigma)^{-1}.$$

Therefore, $f(\sigma) = \alpha/\sigma(\alpha)$, for all $\sigma \in G$, which proves f is the 1-coboundary defined by α.

(2): By Exercise 5.1.22, $F^+ \cong \mathbb{Z}G \otimes_{\mathbb{Z}} k^+$. Now apply Exercise 5.1.24 (1). □

1.3. Exercises.

Exercise 5.1.21. Let G be a group. Prove the following.

(1) If $\mathfrak{F}^G(A)$ is defined to be A^G, then $\mathfrak{F}^G : {}_{\mathbb{Z}G}\mathfrak{M} \to {}_{\mathbb{Z}}\mathfrak{M}$ is a left exact covariant functor.
(2) For every $A \in {}_{\mathbb{Z}G}\mathfrak{M}$, the assignment $f \mapsto f(1)$ induces an isomorphism of abelian groups $\operatorname{Hom}_{\mathbb{Z}G}(\mathbb{Z}, A) \cong A^G$.
(3) The functors $\mathfrak{F}^G$ and $\operatorname{Hom}_{\mathbb{Z}G}(\mathbb{Z}, \cdot)$ are naturally equivalent.
(4) The cohomology groups $\mathrm{H}^n(G, A)$ are isomorphic to the right derived groups $R^n\mathfrak{F}^G(A)$.

Exercise 5.1.22. Let F/k be a Galois extension of fields with finite group G.

(1) Show that the additive group F^+ is a $\mathbb{Z}G$-module.

(2) Show that there is an isomorphism of $\mathbb{Z}G$-modules $\phi : \mathbb{Z}G \otimes_{\mathbb{Z}} k^{+} \to F^{+}$. (Hint: By the Primitive Element Theorem, $F = k(\alpha)$ for some element α. Define $\phi(\sigma \otimes a) = \sigma(\alpha)a$.)

Exercise 5.1.23. Let $G = \operatorname{Aut}_{\mathbb{R}}(\mathbb{C})$ be the Galois group of $\mathbb{C}/\mathbb{R}$. Prove that

$$\mathrm{H}^n(G, \mathbb{C}^*) = \begin{cases} \mathbb{R}^* & \text{if } n = 0, \\ \langle 1 \rangle & \text{if } n \text{ is odd}, \\ \langle -1 \rangle & \text{if } n \text{ is even}. \end{cases}$$

Exercise 5.1.24. Let G be a finite group.

(1) Prove that the induced $\mathbb{Z}G$-module $\mathbb{Z}G \otimes_{\mathbb{Z}} A$ has trivial cohomology, for any abelian group A. That is, $\mathrm{H}^n(G, \mathbb{Z}G \otimes_{\mathbb{Z}} A) = (0)$, for all $n > 0$. (Hint: Use Lemma 5.1.15, Theorem 5.1.10, and Example 5.1.1.)

(2) In [**Ser79**], a $\mathbb{Z}G$-module M is said to be *relatively projective* if M is a $\mathbb{Z}G$-module direct summand of an induced G-module $\mathbb{Z}G \otimes_{\mathbb{Z}} A$ for some $\mathbb{Z}$-module A. Prove that $\mathrm{H}^n(G, M) = (0)$, for all $n > 0$, if M is relatively projective. The reader is also referred to [**CE56**] where such modules are called *weakly projective*.

Exercise 5.1.25. Let G be a finite group and $\{A_i \mid i \in I\}$ a collection of $\mathbb{Z}G$-modules. Show that if $\mathrm{H}^1(G, A_i) = 0$ for each $i \in I$, then $\mathrm{H}^1(G, \bigoplus_i A_i) = 0$.

Exercise 5.1.26. Let G be a finite group and $\{A_i \mid i \in I\}$ a collection of $\mathbb{Z}G$-modules. Show that for all $r \geq 0$, $\mathrm{H}^r(G, \bigoplus_i A_i) = \bigoplus_i \mathrm{H}^r(G, A_i)$. (Hint: Apply Exercise 1.4.16 to the bar resolution of $\mathbb{Z}$.)

Exercise 5.1.27. Let $G = \langle \sigma \rangle$ be a finite cyclic group of order n. Let A be a left $\mathbb{Z}G$-module (written multiplicatively). In this exercise we outline a proof that $\mathrm{H}^1(G, A) \cong {}_N A/DA$ (Theorem 5.1.17) by exhibiting the isomorphism on normalized 1-cocycles. Let $\mathrm{Z}^1(G, A)$ be the normalized 1-cocycles and $\mathrm{B}^1(G, A)$ the normalized 1-coboundaries, as defined in Corollary 5.1.7. Define a function $\theta : \mathrm{Z}^1(G, A) \to A$ by the rule $\theta(\xi) = \xi(\sigma)$. Define a second function $\chi : {}_N A \to \operatorname{Map}(G, A)$ by the rule $b \mapsto \chi_b$, where $\chi_b(\sigma^i) = b\sigma(b) \cdots \sigma^{i-1}(b)$, for all $0 < i$. Prove the following.

(1) θ induces a homomorphism of groups $\mathrm{H}^1(G, A) \to {}_N A/DA$.

(2) χ induces a homomorphism of groups ${}_N A/DA \to \mathrm{H}^1(G, A)$.

(3) The homomorphisms of (1) and (2) are inverses of each other.

Exercise 5.1.28. Let $G = \langle \sigma \rangle$ be a finite cyclic group of order n. Let A be a left $\mathbb{Z}G$-module (written multiplicatively). In this exercise we outline a proof that $\mathrm{H}^2(G, A) \cong A^G/NA$ (Theorem 5.1.17) by exhibiting the isomorphism on normalized 2-cocycles. Let $\mathrm{Z}^2(G, A)$ be the normalized 2-cocycles and

$\mathrm{B}^2(G, A)$ the 2-coboundaries, as defined in Corollary 5.1.7. Define a function $\theta : \mathrm{Z}^2(G, A) \to A$ by the rule

$$\theta(\xi) = a_\xi = \prod_{j=0}^{n-1} \xi(\sigma^j, \sigma).$$

Define another function $\phi : A^G \to \mathrm{Map}(G \times G, A)$ by the rule $\phi(a) = \phi_a$, where

$$\phi_a(\sigma^i, \sigma^j) = \begin{cases} 1 & \text{if } i + j < n, \\ a & \text{if } i + j \geq n, \end{cases}$$

for all $0 \leq i, j \leq n - 1$. Prove the following.

(1) θ and ϕ are homomorphisms of groups.
(2) The image of θ is contained in A^G.
(3) If $\xi \in \mathrm{B}^2(G, A)$, then a_ξ is in NA.
(4) θ induces a homomorphism of groups $\mathrm{H}^2(G, A) \to A^G/NA$.
(5) If $a \in A^G$, then $\phi_a(\sigma^i, \sigma^j)\phi_a(\sigma^{i+j}, \sigma^k) = \phi_a(\sigma^j, \sigma^k)\phi_a(\sigma^i, \sigma^{j+k})$. Therefore, the image of ϕ is contained in $\mathrm{Z}^2(G, A)$.
(6) Let $b \in A$, and assume $a = b\sigma(b) \cdots \sigma^{n-1}(b) = N(b)$. Define $\chi : G \to A$ by

$$\chi(\sigma^i) = \begin{cases} 1 & i = 0, \\ b\sigma(b) \cdots \sigma^{i-1}(b) & 0 < i < n. \end{cases}$$

Then ϕ_a is the 2-coboundary defined by χ.
(7) ϕ induces a homomorphism of groups $A^G/NA \to \mathrm{H}^2(G, A)$.
(8) The homomorphisms of (4) and (7) are inverses of each other, hence $\mathrm{H}^2(G, A) \cong A^G/NA$.

Exercise 5.1.29. Let G be a group, H a normal subgroup of G, and A a left $\mathbb{Z}G$-module.

(1) If $n \geq 1$, show that $\mathrm{Res} \circ \mathrm{Inf} : \mathrm{H}^n(G/H, A^H) \to \mathrm{H}^n(H, A)$ is the zero map. (Hint: Use normalized cocycles and Definition 5.1.12.)
(2) Show that the sequence

$$0 \to \mathrm{H}^1(G/H, A^H) \xrightarrow{\mathrm{Inf}} \mathrm{H}^1(G, A) \xrightarrow{\mathrm{Res}} \mathrm{H}^1(H, A)$$

is exact.

Exercise 5.1.30. In this exercise we construct an example of a $\mathbb{Z}/n$-module which is not free. Let $G = \langle \sigma \rangle$ be a finite cyclic group of order n and $M = \mathbb{Z}^{(n-1)}$ the free $\mathbb{Z}$-module of rank $n-1$ with standard basis $e_1, \dots, e_{n-1}$. Let C be the $(n-1)$-by-$(n-1)$ companion matrix of the cyclotomic polynomial $x^{n-1} + x^{n-2} + \cdots + x + 1$. Let $\sigma : M \to M$ be the homomorphism defined by C with respect to the standard basis. Show that this makes M into a left $\mathbb{Z}G$-module and

$$\mathrm{H}^r(G, M) = \begin{cases} 0 & \text{if } r \text{ is even,} \\ \mathbb{Z}/n & \text{if } r \text{ is odd.} \end{cases}$$

Exercise 5.1.31. Let F be a finite field, G a group of automorphisms of F, and $k = F^G$. Prove:

$$\mathrm{H}^n(G, F^*) = \begin{cases} k^* & \text{if } n = 0, \\ \langle 1 \rangle & \text{if } n > 0. \end{cases}$$

2. The Tensor Algebra of a Module

In this short section we define the so-called tensor algebra of a module and derive the fundamental properties which it satisfies. Start with a commutative ring R, an R-algebra A, and a two-sided A/R-module M (Definition 4.1.1). For $n \geq 0$ we define two-sided A/R-modules $T^n(M)$ as follows. Define $T^0(M)$ to be A, which is a two-sided A/R-module (see Definition 4.1.1). If $n > 0$, define $T^n(M)$ to be $M^{\otimes n}$ by which we mean $M \otimes_A \cdots \otimes_A M$, the tensor product of n copies of M. The multiplication rule

$$(a \otimes b)(x_1 \otimes x_2 \otimes \cdots \otimes x_n) = ax_1 \otimes x_2 \otimes \cdots \otimes x_n b$$

makes $T^n(M)$ into a two-sided A/R-module. The *tensor algebra* of M, denoted $T(M)$, is the graded R-algebra defined by the following rules.

(1) As a graded R-module, $T(M)$ is equal to $\bigoplus_{n \geq 0} T^n(M)$.

(2) The product rule on $T(M)$ is induced on homogeneous components by

$$T^i(M) \otimes_A T^j(M) \xrightarrow{\eta_{i,j}} T^{i+j}(M)$$

which is a two-sided A/R-module isomorphism.

The reader should verify that $T(M)$ is a graded R-algebra, the identity mapping of A onto $T^0(M)$ is a natural R-algebra homomorphism $\tau^0 : A \to T(M)$, and the identity mapping of M onto $T^1(M)$ is a two-sided A/R-module homomorphism $\tau^1 : M \to T(M)$. In case the rings A and R are ambiguous, we write $T^n_{A/R}(M)$ instead of $T^n(M)$ and $T_{A/R}(M)$ instead of $T(M)$. If $A = R$, we sometimes write $T^n_R(M)$ instead of $T^n_{A/R}(M)$ and $T_R(M)$ instead of $T_{A/R}(M)$.

Proposition 5.2.1. *Let A be an R-algebra and M a two-sided A/R-module. The tensor algebra satisfies the following.*

(1) *The R-algebra $T(M)$ is generated by the set $T^0(M) + T^1(M)$.*

(2) *(Universal Mapping Property) For any R-algebra homomorphism $\theta : A \to B$ and two-sided A/R-module homomorphism $f : M \to B$, there exists a unique homomorphism ϕ of both R-algebras and two-sided A/R-modules such that the diagram of R-algebras*

$$\begin{array}{ccc} A & \xrightarrow{\tau^0} & T(M) \\ & {\scriptstyle\theta}\searrow & \swarrow{\scriptstyle\exists\phi} \\ & B & \end{array}$$

commutes and the diagram of two-sided A/R-modules

$$\begin{array}{ccc} M & \xrightarrow{\tau^1} & T(M) \\ & {\scriptstyle f}\searrow & \swarrow{\scriptstyle\exists\phi} \\ & B & \end{array}$$

commutes. Up to an isomorphism of both R-algebras and two-sided A/R-modules, $T(M)$ is uniquely determined by this mapping property.

(3) *If $\theta : M \to N$ is a homomorphism of two-sided A/R-modules, then there exists a unique homomorphism $T(\theta)$ of both graded R-algebras and two-sided A/R-modules such that the diagram*

$$\begin{array}{ccc} M & \xrightarrow{\tau_M} & T(M) \\ {\scriptstyle\theta}\downarrow & & \downarrow{\scriptstyle T(\theta)} \\ N & \xrightarrow{\tau_N} & T(N) \end{array}$$

commutes.

(4) *The assignment $M \mapsto T(M)$ defines a covariant functor from the category of two-sided A/R-modules to the category of graded R-algebras which are also two-sided A/R-modules. The assignment $M \mapsto T^n(M)$ defines a covariant functor from the category of two-sided A/R-modules to the category of two-sided A/R-modules.*

(5) *Given an exact sequence of two-sided A/R-modules*

$$0 \to K \to M \xrightarrow{\theta} N \to 0$$

the graded R-algebra homomorphism $T(\theta) : T(M) \to T(N)$ is onto, and the kernel of $T(\theta)$ is the ideal in $T(M)$ generated by the image of K in $T^1(M)$.

(6) *If $R \to S$ is a homomorphism of commutative rings, then for all $n \geq 0$ there is an isomorphism of two-sided $(S \otimes_R A)/S$-modules*

$$S \otimes_R T^n_{A/R}(M) \cong T^n_{S \otimes_R A/S}(S \otimes_R M)$$

and an isomorphism

$$S \otimes_R T_{A/R}(M) \cong T_{S \otimes_R A/S}(S \otimes_R M)$$

of both graded S-algebras and two-sided $(S \otimes_R A)/S$-modules.

Proof. Parts (1), (4) and (6) are left to the reader.

(2): Notice that

$$\phi(x) = \begin{cases} \theta(x) & \text{for all } x \in T^0(M), \\ f(x) & \text{for all } x \in T^1(M), \end{cases}$$

and $T^0(M) + T^1(M)$ contains a generating set for the R-algebra $T(M)$. The rest is left to the reader.

(3): This follows from Part (2) applied to the composite map $M \to N \to T(N)$.

(5): Applying Lemma 5.2.2 and induction on n shows that $T^n(\theta) : T^n(M) \to T^n(N)$ is onto. Since $T(\theta)(K) = 0$, it is clear that the ideal generated by K is in the kernel of $T(\theta)$. Applying Lemma 5.2.2 and induction on n shows that the kernel of $T^n(\theta) : T^n(M) \to T^n(N)$ is generated by elements of the form $x_1 \otimes x_2 \otimes \cdots \otimes x_n$ where at least one of the x_i is in K. Elements of this form are in the two-sided ideal of $T(M)$ generated by K. □

Lemma 5.2.2. *Let R be any ring. Let*

$$0 \to A \xrightarrow{\alpha} B \xrightarrow{\beta} C \to 0$$

be an exact sequence in $\mathfrak{M}_R$ and

$$0 \to D \xrightarrow{\delta} E \xrightarrow{\epsilon} F \to 0$$

an exact sequence in ${}_R\mathfrak{M}$. Then

$$(A \otimes_R E) \oplus (B \otimes_R D) \xrightarrow{\alpha \otimes 1 + 1 \otimes \delta} B \otimes_R E \xrightarrow{\beta \otimes \epsilon} C \otimes_R F \to 0$$

is an exact sequence of abelian groups.

Proof. By Lemma 1.3.1, tensoring is right exact. The diagram

$$\begin{array}{ccccccc}
A \otimes_R D & \xrightarrow{\alpha\otimes 1} & B \otimes_R D & \xrightarrow{\beta\otimes 1} & C \otimes_R D & \longrightarrow & 0 \\
\downarrow{\scriptstyle 1\otimes\delta} & & \downarrow{\scriptstyle 1\otimes\delta} & & \downarrow{\scriptstyle 1\otimes\delta} & & \\
A \otimes_R E & \xrightarrow{\alpha\otimes 1} & B \otimes_R E & \xrightarrow{\beta\otimes 1} & C \otimes_R E & \longrightarrow & 0 \\
\downarrow{\scriptstyle 1\otimes\epsilon} & & \downarrow{\scriptstyle 1\otimes\epsilon} & & \downarrow{\scriptstyle 1\otimes\epsilon} & & \\
A \otimes_R F & \xrightarrow{\alpha\otimes 1} & B \otimes_R F & \xrightarrow{\beta\otimes 1} & C \otimes_R F & \longrightarrow & 0 \\
\downarrow & & \downarrow & & \downarrow & & \\
0 & & 0 & & 0 & &
\end{array}$$

commutes and the rows and columns are exact. It is clear that $\beta \otimes \epsilon$ is onto and the image of $\alpha \otimes 1 + 1 \otimes \delta$ is contained in the kernel of $\beta \otimes \epsilon$. Given $z \in \ker(\beta \otimes \epsilon)$, we need to show that $z \in \operatorname{im}(\alpha \otimes 1 + 1 \otimes \delta)$. The reader should verify that there exists $w \in B \otimes_R D$ such that $(1 \otimes \delta)(\beta \otimes 1)(w) = (\beta \otimes 1)(z)$. The proof follows from the fact that $(1 \otimes \delta)(w) - z \in \ker(\beta \otimes 1) = \operatorname{im}(\alpha \otimes 1)$. □

2.1. Exercises.

Exercise 5.2.3. Let R be a commutative ring. Let $M = Ra$ be a free R-module of rank 1 with generator a. Show that there is an isomorphism of R-algebras $T(M) \to R[x]$ defined by the assignment $a \mapsto x$.

Exercise 5.2.4. Let R be a commutative ring. Let M be a rank one R-progenerator. Use Proposition 2.6.4, Exercise 2.6.15, and Exercise 5.2.3 to prove that the tensor algebra $T(M)$ is commutative.

Exercise 5.2.5. Let R be a commutative ring, S a commutative R-algebra, and M an S-module. Show that $T^n_R(M)$ is a left $T^n_R(S)$-module where the multiplication rule is $(s_1 \otimes \cdots \otimes s_n)(x_1 \otimes \cdots \otimes x_n) = (s_1x_1 \otimes \cdots \otimes s_nx_n)$. Prove the following.

(1) If M is a finitely generated S-module, then $T^n_R(M)$ is a finitely generated $T^n_R(S)$-module.

(2) If M is a projective S-module, then $T^n_R(M)$ is a projective $T^n_R(S)$-module.

(3) If M is an S-module generator, then $T^n_R(M)$ is a $T^n_R(S)$-module generator.

(4) If A is an S-algebra, then $T^n_R(A)$ is a $T^n_R(S)$-algebra.

3. Theory of Faithfully Flat Descent

Let S be a commutative faithfully flat R-algebra. The topics in this section are motivated by the following descent problem for S-modules. If M is an S-module, then what are necessary and sufficient conditions such that M is isomorphic to $N \otimes_R S$, for some R-module N? The theory of faithfully flat descent arose as an attempt to answer this and related questions.

The technique of faithfully flat descent was developed by A. Grothendieck in a series of Bourbaki seminars over the period 1959–1962. The bulk of the book [**Gro62**] consists of copies of typewritten lecture notes that are based on these lectures. Subsequently, these lectures were rewritten and published as [**Gro71**, Exposé VIII] (see also [**Gro03**]). The presentation of this topic appearing in [**OS75**, Chapter 10] is attributed to mimeographed copies of lecture notes made by M. Artin at M.I.T. in 1966. The approach used in this section closely follows that of Knus and Ojanguren [**KO74b**].

3.1. The Amitsur Complex. Let $\theta : R \to S$ be a homomorphism of commutative rings. Let $\{M_i \mid i \in I\}$ be a family of R-modules. For any $n+1$-tuple $(i_0, \ldots, i_n)$ in $I^{(n+1)}$, and for any j such that $0 \le j \le n+1$, there is an R-module homomorphism

$$M_{i_0} \otimes_R \cdots \otimes_R M_{i_n} \xrightarrow{e_j} M_{i_0} \otimes_R \cdots \otimes_R M_{i_{j-1}} \otimes_R S \otimes_R M_{i_j} \otimes_R \cdots \otimes_R M_{i_n}$$
$$(x_0 \otimes \cdots \otimes x_n) \mapsto x_0 \otimes \cdots \otimes x_{j-1} \otimes 1 \otimes x_j \otimes \cdots \otimes x_n.$$

By $S^{\otimes r}$ we denote $S \otimes_R \cdots \otimes_R S$, the tensor product of r copies of S. The *Amitsur complex* for S/R is

$$0 \to R \xrightarrow{\theta} S \xrightarrow{d^0} S^{\otimes 2} \xrightarrow{d^1} S^{\otimes 3} \xrightarrow{d^2} \cdots$$

where the coboundary map $d^r : S^{\otimes(r+1)} \to S^{\otimes(r+2)}$ is defined to be

$$d^r = \sum_{i=0}^{r+1} (-1)^i e_i.$$

Denote this complex by $C^\bullet(S/R)$. The reader should verify that $e_j e_i = e_{i+1} e_j$ for all $j \le i$, and that this is a complex of R-modules.

Proposition 5.3.1. *Let S be a commutative faithfully flat R algebra.*

(1) *The Amitsur complex $C^\bullet(S/R)$ is an exact sequence.*

(2) *If M is any R-module, then the complex $M \otimes_R C^\bullet(S/R)$*

$$0 \to M \xrightarrow{1\otimes\theta} M \otimes_R S \xrightarrow{1\otimes d^0} M \otimes S^{\otimes 2} \xrightarrow{1\otimes d^1} M \otimes S^{\otimes 3} \xrightarrow{1\otimes d^2} \cdots$$

is an exact sequence.

Proof. (1): Step 1: Show that $C^\bullet(S/R)$ is exact if there exists an R-module homomorphism $\sigma : S \to R$ which is a left inverse for the structure homomorphism $\theta : R \to S$. This is true for example, if S is faithful and $R \cdot 1$ is an R-direct summand of S. Define a homotopy operator $k^r : S^{\otimes(r+2)} \to S^{\otimes(r+1)}$ by $k^r(x_0 \otimes \cdots \otimes x_{r+1}) = \sigma(x_0)x_1 \otimes \cdots \otimes x_{r+1}$. It follows from

$$k^r d^r(x_0 \otimes \cdots \otimes x_r) = k^r \sum_{i=0}^{r} (-1)^r e_i(x_0 \otimes \cdots \otimes x_r)$$
$$= x_0 \otimes \cdots \otimes x_r - \sigma(x_0) \otimes x_1 \otimes \cdots \otimes x_r + \sigma(x_0)x_1 \otimes 1 \otimes \cdots \otimes x_r + \cdots$$

and

$$d^{r-1}k^{r-1}(x_0 \otimes \cdots \otimes x_r) = d^{r-1}(\sigma(x_0)x_1 \otimes \cdots \otimes x_r)$$
$$= 1 \otimes \sigma(x_0)x_1 \otimes \cdots \otimes x_r - \sigma(x_0)x_1 \otimes 1 \otimes \cdots \otimes x_r + \cdots$$

that $k^r d^r + d^{r-1}k^{r-1}$ is the identity map on $S^{\otimes(r+1)}$. Since k is a contracting homotopy, the complex is an exact sequence.

Step 2: If T is another commutative R-algebra, then $C^\bullet(S \otimes_R T/T)$, the Amitsur complex for $S \otimes_R T$ over T, is obtained by applying the functor $(\cdot) \otimes_R T$ to the complex $C^\bullet(S/R)$. This is because $S^{\otimes r} \otimes_R T \cong (S \otimes_R T)^{\otimes r}$.

Step 3: Let $\rho : S \to S \otimes_R S$ by $a \mapsto a \otimes 1$. Define $\mu : S \otimes_R S \to S$ by $\mu(a \otimes b) = ab$. Then μ is a left inverse for ρ and by Step 1 the Amitsur complex $C^\bullet(S \otimes_R S/S)$ for $\rho : S \to S \otimes_R S$ is exact. Since $C^\bullet(S \otimes_R S/S)$ is exact and S is faithfully flat, by Step 2 applied to S, it follows that $C^\bullet(S/R)$ is exact.

(2): As in (1), assume there is a section and construct a contracting homotopy. The rest is left to the reader. □

3.2. The Descent of Elements.

Example 5.3.2. Let R be a commutative ring and $\alpha_1, \dots, \alpha_n$ a set of n elements of R such that $R = R\alpha_1 + \cdots + R\alpha_n$. For the localization of R with respect to the multiplicative set $\{\alpha^n \mid n \geq 0\}$, write R_α instead of $R[\alpha^{-1}]$. Because $R = R\alpha_1 + \cdots + R\alpha_n$, $U(\alpha_1), \dots, U(\alpha_n)$ is an open cover for the Zariski topology on $\operatorname{Spec} R$. By Exercise 2.4.20, $S = \bigoplus_{i=1}^n R_{\alpha_i}$ is faithfully flat over R. We identify $R_{\alpha_i} \otimes_R R_{\alpha_j}$ with $R_{\alpha_i\alpha_j}$. Then the Amitsur complex $C^\bullet(S/R)$ looks like

$$0 \to R \xrightarrow{\theta} \bigoplus_{i \in I_n} R_{\alpha_i} \xrightarrow{d^0} \bigoplus_{(i,j) \in I_n^2} R_{\alpha_i\alpha_j} \xrightarrow{d^1} \bigoplus_{(i,j,k) \in I_n^3} R_{\alpha_i\alpha_j\alpha_k} \xrightarrow{d^2} \cdots$$

where $I_n = \{1, \dots, n\}$. By Proposition 5.3.1, this sequence is exact, so we know that an element $y \in R$ is completely determined by a set of local data $x = (x_1, \dots, x_n) \in S$ such that $x_i = x_j$ in $R_{\alpha_i\alpha_j}$.

The element y can be constructed from the local data x and the elements α_i. For some $p \geq 0$ there exist $y_1, \dots, y_n$ in R such that $x_i = y_i \alpha_i^{-p}$. Assuming $d^0(x) = 0$, there exists $q \geq 0$ such that for all i, j pairs

$$(\alpha_i \alpha_j)^q (\alpha_j^p y_i - \alpha_i^p y_j) = 0.$$

Since $R = R\alpha_1^{q+p} + \cdots + R\alpha_n^{q+p}$, there exist $g_i \in R$ such that $1 = g_1 \alpha_1^{q+p} + \cdots + g_n \alpha_n^{q+p}$. Set $y = g_1 \alpha_1^q y_1 + \cdots + g_n \alpha_n^q y_n$. The reader should verify that $y = y_j \alpha_j^{-p} = x_j$ in R_{α_j}, hence $\theta(y) = x$.

Example 5.3.3. Let R be a commutative ring, P a finitely generated projective R-module, and $\phi \in \operatorname{Hom}_R(P, P)$. In this example, we show how to construct the *characteristic polynomial* of ϕ. Let $\alpha_1, \dots, \alpha_n$ be a set of n elements of R such that $R = R\alpha_1 + \cdots + R\alpha_n$ and $P_{\alpha_i} = P \otimes_R R_{\alpha_i}$ is free of finite rank over R_{α_i}. Let $S = \bigoplus_{i=1}^n R_{\alpha_i}$ and as in Example 5.3.2 identify $S^{\otimes 2} = \bigoplus_{(i,j)} R_{\alpha_i \alpha_j}$. Then $S[x] = S \otimes_R R[x] = \bigoplus_{i=1}^n R_{\alpha_i}[x]$ and $S^{\otimes 2}[x] = \bigoplus_{(i,j)} R_{\alpha_i \alpha_j}[x]$. The Amitsur complex $C^\bullet(S[x]/R[x])$ becomes

$$0 \to R[x] \xrightarrow{\theta} \bigoplus_{i \in I_n} R_{\alpha_i}[x] \xrightarrow{d^0} \bigoplus_{(i,j) \in I_n^2} R_{\alpha_i \alpha_j}[x] \xrightarrow{d^1} \cdots$$

which is an exact sequence, because $S[x]$ is faithfully flat over $R[x]$.

For each i, let $\phi_i = \phi \otimes 1 \in \operatorname{Hom}_{R_{\alpha_i}}(P_{\alpha_i}, P_{\alpha_i})$. The characteristic polynomial $p_i(x) = \operatorname{char.poly}_{R_{\alpha_i}}(\phi_i)$ can be defined as a determinant of $x - \phi_i$ and does not depend on the choice of a basis of P_{α_i}. The polynomial $p_i(x)$ is an element of $R_{\alpha_i}[x]$. We remark that the determinant operator commutes with change of base ring. In other words, if $\theta : A \to B$ is a homomorphism of commutative rings and M is a matrix in $M_n(A)$, then $\det(\theta(M)) = \theta(\det(M))$. Therefore, if $\phi_{ij} = \phi \otimes 1 \in \operatorname{Hom}_{R_{\alpha_i \alpha_j}}(P_{\alpha_i \alpha_j}, P_{\alpha_i \alpha_j})$, then in $R_{\alpha_i \alpha_j}[x]$ the identities $\operatorname{char.poly}_{R_{\alpha_i}}(\phi_i) = \operatorname{char.poly}_{R_{\alpha_i \alpha_j}}(\phi_{ij}) = \operatorname{char.poly}_{R_{\alpha_j}}(\phi_j)$ are satisfied. Therefore $(p_1(x), \dots, p_n(x))$ is in the kernel of d^0. So the local data $(p_1(x), \dots, p_n(x))$ descend to a polynomial $p(x)$ in $R[x]$. The polynomial $p(x)$ is usually denoted by $\operatorname{char.poly}_R(\phi)$ and is called the characteristic polynomial of ϕ.

Now we show that the polynomial $\operatorname{char.poly}_R(\phi)$ just constructed does not depend on the open cover of $\operatorname{Spec} R$. Let $\beta_1, \dots, \beta_m$ be another set of elements in R that generated the unit ideal and such that P_{β_j} is free over R_{β_j} for each j. Let $T = \bigoplus_{j=1}^m R_{\beta_j}$ and by the above method, let $q(x)$ be the characteristic polynomial of ϕ constructed using the faithfully flat R-algebra T. We show that $q(x)$ is equal to the polynomial $p(x)$ which was constructing initially. Notice that $S \otimes_R T$ is a faithfully flat R-algebra and we can identify $S \otimes_R T = \bigoplus_{(i,j)} R_{\alpha_i \beta_j}$. The image of $p(x)$ in $R_{\alpha_i \beta_j}[x]$ is equal to the image of $q(x)$ in $R_{\alpha_i\beta_j}[x]$. Since the Amitsur complex $C^\bullet(S \otimes_R T[x]/R[x])$ is exact, this proves $p(x) = q(x)$.

Now we prove that the Cayley-Hamilton Theorem applies to $p(x) = \operatorname{char.poly}_R(\phi)$. Since S is faithfully flat over R, by Proposition 5.3.1, the sequence

$$0 \to \operatorname{Hom}_R(P, P) \xrightarrow{\theta} \operatorname{Hom}_R(P, P) \otimes_R S$$

is exact. We identify $\operatorname{Hom}_R(P, P) \otimes_R S$ with $\bigoplus_{i=1}^n \operatorname{Hom}_{R_{\alpha_i}}(P_{\alpha_i}, P_{\alpha_i})$. The image of $p(\phi)$ under θ is $(p_1(\phi_1), \dots, p_n(\phi_n))$. By the Cayley-Hamilton Theorem (see, for example, [**DF04**, Proposition 12.2.20]), this image is $(0, \dots, 0)$, which means $p(\phi) = 0$.

If $\operatorname{Rank}_R(P) = n$ is defined, then the characteristic polynomial will have constant degree n. Let $\operatorname{char.poly}_R(\phi) = x^n + a_{n-1}x^{n-1} + \cdots + a_1 x + a_0$. Define the *determinant* of ϕ to be $\det(\phi) = (-1)^n a_0$ and the *trace* of ϕ to be $\operatorname{trace}(\phi) = -a_{n-1}$. The reader should verify that $\det(\phi\psi) = \det(\phi)\det(\psi)$.

3.3. Descent of Homomorphisms. Let S be a commutative R-algebra and M and N a pair of R-modules. The goal is to find sufficient conditions on a homomorphism $g \in \operatorname{Hom}_S(M \otimes_R S, N \otimes_R S)$ such that $g = f \otimes 1$ for some $f \in \operatorname{Hom}_R(M, N)$. The maps $e_i : S \to S \otimes_R S$ defined by $e_0(s) = 1 \otimes s$ and $e_1(s) = s \otimes 1$ are both R-algebra homomorphisms. Therefore, $S \otimes_R S$ is an S-algebra in two different ways. Tensoring e_i with $(M \otimes_R S) \otimes_S ()$ we get the maps of Paragraph 5.3.1

$$e_i : M \otimes_R S \to (M \otimes_R S) \otimes_S (S \otimes_R S) \cong M \otimes_R S \otimes_R S$$

where $e_0(x \otimes s) = x \otimes 1 \otimes s$ and $e_1(x \otimes s) = x \otimes s \otimes 1$. Assign the appellation $\mathfrak{F}_i$ to the functor "tensoring with the S-algebra $e_i : S \to S \otimes_R S$". There is a commutative square

$$\begin{array}{ccc} M \otimes_R S & \xrightarrow{e_i} & M \otimes_R S \otimes_R S \\ {\scriptstyle g}\downarrow & & \downarrow{\scriptstyle \mathfrak{F}_i(g)} \\ N \otimes_R S & \xrightarrow{e_i} & N \otimes_R S \otimes_R S \end{array}$$

for $i = 0, 1$ and $\mathfrak{F}_i(g)$ is an $S \otimes_R S$-module homomorphism.

Proposition 5.3.4. *Let R be a commutative ring, S a faithfully flat commutative R-algebra, and M and N a pair of R-modules. The sequence*

$$0 \to \operatorname{Hom}_R(M, N) \xrightarrow{\mathfrak{F}} \operatorname{Hom}_S(M \otimes_R S, N \otimes_R S) \xrightarrow{\mathfrak{F}_0 - \mathfrak{F}_1} \operatorname{Hom}_{S \otimes_R S}(M \otimes_R S \otimes_R S, N \otimes_R S \otimes_R S)$$

is exact, where $\mathfrak{F}(f) = f \otimes 1$ and $\mathfrak{F}_0$, $\mathfrak{F}_1$ are defined in the previous paragraph.

Proof. Since each $\mathfrak{F}_i$ is an additive functor, $\mathfrak{F}_0 - \mathfrak{F}_1$ is a $\mathbb{Z}$-module homomorphism. If $f \in \operatorname{Hom}_R(M, N)$, then the diagram

$$\begin{array}{ccccccccc} 0 & \longrightarrow & M & \longrightarrow & M \otimes_R S & \xrightarrow{d^0} & M \otimes_R S \otimes_R S \\ & & \downarrow f & & \downarrow f\otimes 1 & & \downarrow f\otimes 1 \otimes 1 \\ 0 & \longrightarrow & N & \longrightarrow & N \otimes_R S & \xrightarrow{d^0} & N \otimes_R S \otimes_R S \end{array}$$

commutes and the rows are exact. Therefore, $\mathfrak{F}$ is one-to-one and $f \otimes 1 \otimes 1 = 0$. To complete the proof, we show $\ker(\mathfrak{F}_0 - \mathfrak{F}_1) \subseteq \operatorname{im}(\mathfrak{F})$. Let $g \in \operatorname{Hom}_S(M \otimes_R S, N \otimes_R S)$ and assume $\mathfrak{F}_0(g) = \mathfrak{F}_1(g)$. Given $m \in M$ we have $e_0(m \otimes 1) = e_1(m \otimes 1)$, so

$$e_0 g(m\otimes 1) = \mathfrak{F}_0(g)e_0(m\otimes 1) = \mathfrak{F}_0(g)e_1(m\otimes 1) = \mathfrak{F}_1(g)e_1(m\otimes 1) = e_1 g(m\otimes 1).$$

By Proposition 5.3.1, this proves that $g(m\otimes 1) \in N\otimes_R 1$. Define $f : M \to N$ by $f(m) = g(m \otimes 1)$. Then $g = \mathfrak{F}(f)$. $\square$

Example 5.3.5. Let R be a commutative ring and P a finitely generated projective R-module. By Lemma 1.5.1, $\theta_R : P^* \otimes_R P \to \operatorname{Hom}_R(P, P)$ is an isomorphism of R-modules, where $\theta_R(f \otimes p)(x) = f(x)p$. Define $T : P^* \otimes_R P \to R$ by $T(f \otimes p) = f(p)$. By Exercise 4.6.17, this induces an R-module homomorphism $T : \operatorname{Hom}_R(P, P) \to R$ which is equal to the trace map of Exercise 1.2.14 and the trace map of Definition 4.6.5, when P is free. As in Example 5.3.2, let $R \to S$ be a faithfully flat R-algebra such that $P \otimes_R S$ is free. Upon change of base, $T \otimes 1 : \operatorname{Hom}_S(P \otimes_R S, P \otimes_R S) \to S$ is the trace map of Exercise 1.2.14. By Proposition 5.3.4, the map T is equal to the trace map of Definition 4.6.5. Assuming P has constant rank, we also see that T is equal to the trace defined in Example 5.3.3 using the characteristic polynomial.

3.4. Descent of Modules. Let $\theta : R \to S$ be a homomorphism of commutative rings. Given S-modules A, B, C and D and an $S \otimes_R S$-module homomorphism $f : A \otimes_R B \to C \otimes_R D$, there are three $S \otimes_R S \otimes_R S$-module homomorphisms

$$\begin{aligned} f_1 &: S \otimes_R A \otimes_R B \to S \otimes_R C \otimes_R D \\ f_2 &: A \otimes_R S \otimes_R B \to C \otimes_R S \otimes_R D \\ f_3 &: A \otimes_R B \otimes_R S \to C \otimes_R D \otimes_R S \end{aligned}$$

where f_i is obtained by tensoring f with the identity map on S in position i. We employ this construction in the following setting. Start with any S-module M. Then $S \otimes_R M$ and $M \otimes_R S$ are two $S \otimes_R S$-modules. Then an $S \otimes_R S$-module homomorphism $g : S \otimes_R M \to M \otimes_R S$ gives rise to three

$S \otimes_R S \otimes_R S$-module homomorphisms

$$\begin{aligned} g_1 &: S \otimes_R S \otimes_R M \to S \otimes_R M \otimes_R S \\ g_2 &: S \otimes_R S \otimes_R M \to M \otimes_R S \otimes_R S \\ g_3 &: S \otimes_R M \otimes_R S \to M \otimes_R S \otimes_R S. \end{aligned}$$

The ring homomorphism θ induces $\theta : M \to S \otimes_R M$, where $x \mapsto 1 \otimes x$. Let $\mu : M \otimes_R S \to M$ be the multiplication map, where $x \otimes s \mapsto sx$. The composition

$$S \otimes_R M \xrightarrow{g} M \otimes_R S \xrightarrow{\mu} M \xrightarrow{\theta} S \otimes_R M$$

upon restriction to $\operatorname{im}\theta$ induces an S-module homomorphism which will be denoted by $\bar{g} : 1 \otimes_R M \to 1 \otimes_R M$. Then $\bar{g}(1 \otimes m) = 1 \otimes \mu g(1 \otimes m)$.

Proposition 5.3.6. *Let $\theta : R \to S$ be a homomorphism of commutative rings, M an S-module and $g : S \otimes_R M \to M \otimes_R S$ an $S \otimes_R S$-module homomorphism. The following are equivalent.*

(1) *$\bar{g}$ is the identity map on $1 \otimes_R M$ and $g_2 = g_3 g_1$.*

(2) *g is an isomorphism of $S \otimes_R S$-modules and $g_2 = g_3 g_1$.*

Proof. (1) implies (2): Let $\tau : M \otimes_R S \to S \otimes_R M$ be the twist map defined by $x \otimes s \mapsto s \otimes x$. The reader should verify that $\tilde{g} = t^{-1} g \tau$ is an $S \otimes_R S$-module homomorphism. We show that $\tilde{g}$ is the inverse of g. Let $m \in M$. Then $1 \otimes m$ is a typical generator for the $S \otimes_R S$-module $S \otimes_R M$. If we write $g(1 \otimes m) = \sum_i m_i \otimes s_i$, then since $\bar{g}$ is the identity map,

$$1 \otimes m = \bar{g}(1 \otimes m) = 1 \otimes \sum_i m_i s_i.$$

Next write $g(1 \otimes m_i) = \sum_j m_{ij} \otimes t_{ij}$. We have

$$\begin{aligned} \tilde{g}\big(g(1 \otimes m)\big) &= \tilde{g}\Big(\sum_i m_i \otimes s_i\Big) \\ &= \sum_i \tilde{g}(m_i \otimes s_i) \\ &= \sum_i (1 \otimes s_i)\tilde{g}(m_i \otimes 1) \\ &= \sum_i (1 \otimes s_i) \sum_j t_{ij} \otimes m_{ij} \\ &= \sum_i \sum_j t_{ij} \otimes s_i m_{ij}. \end{aligned}$$

Let $\omega : M \otimes_R S \otimes_R S \to S \otimes_R M$ be the function $x \otimes a \otimes b \mapsto a \otimes xb$ which multiplies the two extreme factors. Since $g_2 = g_3 g_1$,

$$\omega\left(g_2(1 \otimes 1 \otimes m)\right) = \omega\Big(\sum_i m_i \otimes 1 \otimes s_i\Big) = 1 \otimes \sum_i m_i s_i = 1 \otimes m$$

is equal to

$$\begin{aligned}\omega g_3 g_1(1 \otimes 1 \otimes m) &= \sum_i \omega g_3(1 \otimes m_i \otimes s_i) \\ &= \sum_i \sum_j \omega(m_{ij} \otimes t_{ij} \otimes s_i) = \sum_i \sum_j t_{ij} \otimes m_{ij} s_i\end{aligned}$$

which is equal to $\tilde{g}g(1 \otimes m)$. This proves that $\tilde{g}g$ is the identity map on $S \otimes_R M$. The reader should verify that $g\tilde{g}$ is the identity map on $M \otimes_R S$.

(2) implies (1): We are given an isomorphism $g : S \otimes_R M \to M \otimes_R S$. Let $m \in M$ and write $g(1 \otimes m) = \sum_i m_i \otimes s_i$. Then

$$\bar{g}(1 \otimes m) = 1 \otimes \mu g(1 \otimes m) = 1 \otimes \sum_i m_i s_i.$$

Since g is one-to-one, it is enough to show $g(1 \otimes m) = g\big(1 \otimes \sum_i m_i s_i\big)$. Write $g(1 \otimes m_i) = \sum_j m_{ij} \otimes t_{ij}$. We have

$$g\big(1 \otimes \sum_i m_i s_i\big) = \sum_i g(1 \otimes m_i)(1 \otimes s_i) = \sum_i \sum_j m_{ij} \otimes t_{ij} s_i.$$

Let $\omega : M \otimes_R S \otimes_R S \to M \otimes_R S$ be the function $x \otimes a \otimes b \mapsto x \otimes ab$ which multiplies the last two factors. Since $g_2 = g_3 g_1$,

$$\omega g_2(1 \otimes 1 \otimes m) = \omega\Big(\sum_i m_i \otimes 1 \otimes s_i\Big) = \sum_i m_i \otimes s_i$$

is equal to

$$\begin{aligned}\omega g_3 g_1(1 \otimes 1 \otimes m) &= \sum_i \omega g_3(1 \otimes m_i \otimes s_i) \\ &= \sum_i \sum_j \omega(m_{ij} \otimes t_{ij} \otimes s_i) = \sum_i \sum_j m_{ij} \otimes t_{ij} s_i.\end{aligned}$$

It follows from these computations that $g(1 \otimes m) = g\big(1 \otimes \sum_i m_i s_i\big)$. □

If one of the equivalent properties of Proposition 5.3.6 is satisfied, then we say g is a *descent datum* for M over S.

Theorem 5.3.7 (The Theorem of Faithfully Flat Descent). *Let S be a commutative faithfully flat R-algebra. Let M be an S-module and let $g : S \otimes_R M \to M \otimes_R S$ be a descent datum for M over S. Then there exists* *an R-module N and an isomorphism $\nu : N \otimes_R S \to M$ of S-modules such that the diagram of $S \otimes_R S$-modules*

$$\text{(5.4)}\qquad \begin{array}{ccc} S \otimes_R N \otimes_R S & \xrightarrow{1\otimes\nu} & S \otimes_R M \\ \tau\downarrow & & \downarrow g \\ N \otimes_R S \otimes_R S & \xrightarrow{\nu\otimes 1} & M \otimes_R S \end{array}$$

commutes, where $\tau(a\otimes b\otimes c) = b\otimes a\otimes c$. Up to isomorphism, these properties uniquely determine the module N and the isomorphism ν.

Proof. (Existence.) Set $N = \{x \in M \mid x \otimes 1 = g(1 \otimes x)\}$ and let $\nu : N \otimes_R S \to M$ be the multiplication map $\nu(x \otimes s) = xs$. We show that N and ν have the desired properties. Notice that N is the kernel of the R-module homomorphism $ge_0 - e_1 : M \to M \otimes_R S$, hence the sequence

$$\text{(5.5)}\qquad 0 \to N \to M \xrightarrow{ge_0 - e_1} M \otimes_R S$$

is exact and N is an R-module. Over $S \otimes_R S$, the module $S \otimes_R N \otimes_R S$ is generated by elements of the form $1 \otimes x \otimes 1$, for $x \in N$. Diagram (5.4) commutes since

$$g\big((1 \otimes \nu)(1 \otimes x \otimes 1)\big) = g(1 \otimes x) = x \otimes 1 = \\ (\nu \otimes 1)(x \otimes 1 \otimes 1) = (\nu \otimes 1)\big(\tau(1 \otimes x \otimes 1)\big).$$

The diagram of S-module homomorphisms

$$\text{(5.6)}\qquad \begin{array}{ccc} S \otimes_R M & \xrightarrow{1\otimes e_1} & S \otimes_R M \otimes_R S \\ g\downarrow & & \downarrow g_3 = g\otimes 1 \\ M \otimes_R S & \xrightarrow{1\otimes e_1 = e_2} & M \otimes_R S \otimes_R S \end{array}$$

commutes, since

$$g_3\big((1 \otimes e_1)(a \otimes x)\big) = g_3(a \otimes x \otimes 1) = g(a \otimes x) \otimes 1 = e_2\big(g(a \otimes x)\big).$$

Since $g_2 = g_3 g_1$, it follows that

$$g_3\big((1 \otimes ge_0)(a \otimes x)\big) = g_3\big(a \otimes g(1 \otimes x)\big) = \\ g_3 g_1(a \otimes 1 \otimes x) = g_2(a \otimes 1 \otimes x) = e_1 g(a \otimes x).$$

Therefore, the diagram of S-module homomorphisms

$$\text{(5.7)}\qquad \begin{array}{ccc} S \otimes_R M & \xrightarrow{1\otimes ge_0} & S \otimes_R M \otimes_R S \\ g\downarrow & & \downarrow g_3 = g\otimes 1 \\ M \otimes_R S & \xrightarrow{1\otimes e_0 = e_1} & M \otimes_R S \otimes_R S \end{array}$$

commutes. Consider the diagram of S-module homomorphisms

$$
\begin{array}{ccccccc}
0 \longrightarrow & S \otimes_R N & \longrightarrow & S \otimes_R M & \xrightarrow{1\otimes(ge_0 - e_1)} & S \otimes_R M \otimes_R S \\
& \downarrow \phi & & \downarrow g & & \downarrow g_3 \\
0 \longrightarrow & M & \xrightarrow{1\otimes\theta} & M \otimes_R S & \xrightarrow{1\otimes(e_0 - e_1)} & M \otimes_R S \otimes_R S.
\end{array}
\tag{5.8}
$$

The top row of (5.8) is exact, because it is obtained by applying the exact functor $S \otimes_R ()$ to the exact sequence (5.5). The bottom row of (5.8) is exact by Proposition 5.3.1. The diagram (5.8) commutes because it is constructed from the commutative diagrams (5.6) and (5.7). Since g and g_3 are isomorphisms, the S-module homomorphism ϕ exists and is an isomorphism, by Theorem 1.4.1. For $x \in N$, $\phi(1 \otimes x) = x$, hence ϕ agrees with ν. This proves ν is an isomorphism.

(Uniqueness.) Suppose K is another R-module and $\kappa : K \otimes_R S \to M$ the corresponding S-module isomorphism. Consider the commutative diagram

$$
\begin{array}{ccccc}
S \otimes_R K \otimes_R S & \xrightarrow{1\otimes\kappa} & S \otimes_R M & \xleftarrow{1\otimes\nu} & S \otimes_R N \otimes_R S \\
\downarrow \tau & & \downarrow g & & \downarrow \tau \\
K \otimes_R S \otimes_R S & \xrightarrow{\kappa\otimes 1} & M \otimes_R S & \xleftarrow{\nu\otimes 1} & N \otimes_R S \otimes_R S.
\end{array}
$$

In the notation of Proposition 5.3.4, this says

$$
\mathfrak{F}_0(\nu^{-1}\kappa) = \tau\Big((1 \otimes \nu^{-1}\kappa)\big(\tau^{-1}(x \otimes a \otimes b)\big)\Big)
$$

is equal to

$$
\mathfrak{F}_1(\nu^{-1}\kappa) = \big((\nu^{-1}\kappa)(x \otimes a)\big) \otimes b.
$$

By Proposition 5.3.4, there exists $\lambda \in \operatorname{Hom}_R(K, N)$ such that $\nu^{-1}\kappa = \lambda \otimes 1$. Since S is faithfully flat over R and $\nu^{-1}\kappa$ is an isomorphism, $\lambda : K \to N$ is an R-module isomorphism. Lastly, $\kappa = \nu(\lambda \otimes 1)$. □

Remark 5.3.8. Theorem 5.3.7 is sometimes stated from the opposite point of view. That is, the role of the descent datum is played by the function $h = g^{-1}$. Then $h : M \otimes_R S \to S \otimes_R M$ is an $S \otimes_R S$-module isomorphism which satisfies the 1-cocycle identity $h_1 h_3 = h_2$. Then $N = \{x \in M \mid h(x \otimes 1) = 1 \otimes x\}$, $\nu : N \otimes_R S \to M$ is the multiplication map $\nu(x \otimes s) = xs$, and $h = (1 \otimes \nu)(\nu \otimes 1)^{-1}$.

Example 5.3.9. Let R be a commutative ring and $\alpha_1, \dots, \alpha_n$ a set of n elements of R such that $R = R\alpha_1 + \cdots + R\alpha_n$. For the localization of R with respect to the multiplicative set $\{\alpha^n \mid n \geq 0\}$, write R_α instead of $R[\alpha^{-1}]$. By Exercise 2.4.20, $S = \bigoplus_{i=1}^n R_{\alpha_i}$ is faithfully flat over R. We identify $R_{\alpha_i} \otimes_R R_{\alpha_j}$ with $R_{\alpha_i\alpha_j}$. Then $S \otimes_R S = \bigoplus_{(i,j)\in I_n^2} R_{\alpha_i\alpha_j}$, where

$I_n = \{1, \ldots, n\}$. Suppose for each i that M_i is an R_{α_i}-module. Then $M = \bigoplus_{i=1}^n M_i$ is an S-module. We have

$$S \otimes_R M = \bigoplus_{(i,j)\in I_n^2} R_{\alpha_i} \otimes_R M_j$$

and

$$M \otimes_R S = \bigoplus_{(i,j)\in I_n^2} M_i \otimes_R R_{\alpha_j}.$$

A descent datum $g : S \otimes_R M \to M \otimes_R S$ consists of a collection of $R_{\alpha_i\alpha_j}$-module isomorphisms

$$R_{\alpha_i} \otimes_R M_j \xrightarrow{g_{ij}} M_i \otimes_R R_{\alpha_j}$$

where $(i,j) \in I_n^2$. The identity $g_2 = g_3 g_1$ is equivalent to the statement that the diagram of $R_{\alpha_i\alpha_j\alpha_k}$-module homomorphisms

$$\begin{array}{ccc} R_{\alpha_i} \otimes_R R_{\alpha_j} \otimes_R M_k & \xrightarrow{g_{ik}\otimes 1} & M_i \otimes_R R_{\alpha_j} \otimes_R R_{\alpha_k} \\ & {\scriptstyle g_{jk}\otimes 1} \searrow \quad \nearrow {\scriptstyle g_{ij}\otimes 1} & \\ & R_{\alpha_i} \otimes_R M_j \otimes_R R_{\alpha_k} & \end{array}$$

commutes for all triples $(i,j,k) \in I_n^3$. If a descent datum exists, then by Theorem 5.3.7, there is an R-module N and for each i an isomorphism $M_i \cong N \otimes_R R_{\alpha_i}$ of R_{α_i}-modules.

3.5. Descent of Algebras. Let R be a commutative ring and S a faithfully flat commutative R-algebra. Let N be an R-module such that the S-module $N_S = N \otimes_R S$ has a multiplication operation which is defined by an S-module homomorphism $\mu : N_S \otimes_S N_S \to N_S$. If we identify $N_S \otimes_S N_S$ with $N \otimes_R N \otimes_R S$, then μ belongs to $\operatorname{Hom}_S(N \otimes_R N \otimes_R S, N \otimes_R S)$. By Proposition 5.3.4, the homomorphism μ descends to a unique R-module homomorphism $N \otimes_R N \to N$ if and only if $\mathfrak{F}_0(\mu)$ and $\mathfrak{F}_1(\mu)$ induce equal multiplication operations on $N \otimes_R S \otimes_R S$.

Theorem 5.3.10. *Let S be a commutative faithfully flat R-algebra. Let B be an S-algebra and $g : S \otimes_R B \to B \otimes_R S$ a descent datum for B over S such that g is an isomorphism of $S \otimes_R S$-algebras. Then there exists an R-algebra A and an isomorphism $\nu : A \otimes_R S \to B$ of S-algebras.*

Proof. The existence and uniqueness of the R-module A and the S-module isomorphism $\nu : A \otimes_R S \to B$ are guaranteed by Theorem 5.3.7. The diagram

$$\begin{array}{ccc} S \otimes_R A \otimes_R S & \xrightarrow{1\otimes\nu} & S \otimes_R B \\ {\scriptstyle\tau}\downarrow & & \downarrow{\scriptstyle g} \\ A \otimes_R S \otimes_R S & \xrightarrow{\nu\otimes 1} & B \otimes_R S \end{array} \tag{5.9}$$

commutes, where $\tau(a \otimes b \otimes c) = b \otimes a \otimes c$. The commutative square

$$\begin{array}{ccc} S \otimes_R (A \otimes_R A) \otimes_R S & \xrightarrow{1\otimes(\nu\otimes_S\nu)} & S \otimes_R (B \otimes_R B) \\ {\scriptstyle\tau}\downarrow & & \downarrow{\scriptstyle g\otimes_S g} \\ (A \otimes_R A) \otimes_R S \otimes_R S & \xrightarrow{(\nu\otimes_S\nu)\otimes 1} & (B \otimes_R B) \otimes_R S \end{array} \tag{5.10}$$

is the counterpart of (5.9) for $A \otimes_R A \otimes_R S \cong B \otimes_S B$. Since g is an $S \otimes_R S$-algebra isomorphism, the diagram

$$\begin{array}{ccc} S \otimes_R B \otimes_R B = (S \otimes_R B) \otimes_S (S \otimes_R B) & \longrightarrow & S \otimes_R B \\ {\scriptstyle g\otimes_S g}\downarrow & & \downarrow{\scriptstyle g} \\ B \otimes_R B \otimes_R S = (B \otimes_R S) \otimes_S (B \otimes_R S) & \longrightarrow & B \otimes_R S \end{array} \tag{5.11}$$

commutes, where the horizontal arrows are the multiplication maps. The multiplication μ on $A_S = A \otimes_R S$ is defined by the multiplication operation on B and the S-algebra isomorphism ν. By definition of μ, the diagram

$$\begin{array}{ccc} A \otimes_R A \otimes_R S = A_S \otimes_S A_S & \xrightarrow{\mu} & A_S \\ {\scriptstyle\nu\otimes_S\nu}\downarrow & & \downarrow{\scriptstyle\nu} \\ B \otimes_S B & \longrightarrow & B \end{array} \tag{5.12}$$

commutes, where the bottom arrow is multiplication in B. As was mentioned in the paragraph preceding the theorem, it suffices to show that $\mathfrak{F}_0(\mu)$ and $\mathfrak{F}_1(\mu)$ induce equal multiplication operations on $A \otimes_R S \otimes_R S$. Apply either functor $\mathfrak{F}_i$ to the commutative square (5.12) to get the commutative square

$$\begin{array}{ccc} (A \otimes_R A) \otimes_R S \otimes_R S & \xrightarrow{\mathfrak{F}_i(\mu)} & A \otimes_R S \otimes_R S \\ {\scriptstyle(\nu\otimes_S\nu)\otimes 1}\downarrow & & \downarrow{\scriptstyle\nu\otimes 1} \\ B \otimes_S B \otimes_R S & \longrightarrow & B \otimes_R S \end{array} \tag{5.13}$$

Combine diagrams (5.9), (5.10), (5.11), and (5.13) to get the commutative diagram

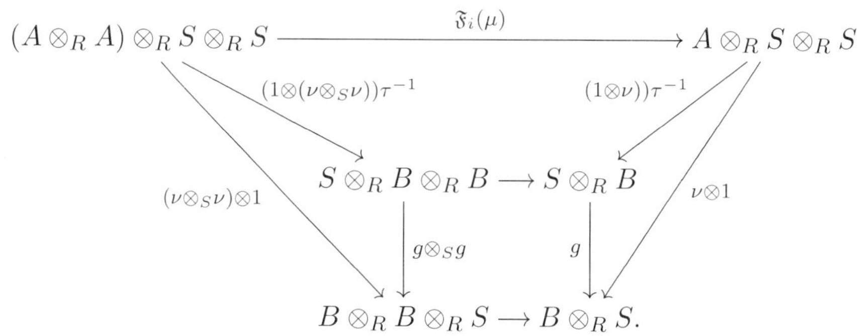

This diagram commutes with either $\mathfrak{F}_0(\mu)$ or $\mathfrak{F}_1(\mu)$ in the top row. Therefore the multiplication on A_S descends to a multiplication on A. The associative, commutative and distributive laws hold in A because they hold in A_S. $\square$

4. Hochschild Cohomology

The cohomology groups defined in this section were introduced by Hochschild in [**Hoc45**] and were applied to study separable algebras over a field. A criterion for separability based on Hochschild cohomology will be proved in Theorem 8.1.19 below. Results in this section will be applied in the proof of Theorem 7.9.1, where we consider the problem of lifting an algebra modulo an ideal. A general reference for this section is [**CE56**, Chapter IX].

Definition 5.4.1. Let R be a commutative ring, A an R-algebra, and $A^e = A \otimes_R A^o$ the enveloping algebra (Definition 4.1.1). If M is a two-sided A/R-module (Definition 4.1.1), then the *nth Hochschild cohomology group of A with coefficients in M* is defined to be

$$\mathrm{H}^n(A, M) = \mathrm{Ext}^n_{A^e}(A, M)$$

where we make M into a left A^e-module by $a \otimes b \cdot x = axb$.

Let R be a commutative ring and A an R-algebra. We construct a chain complex $S_\bullet(A) \to A$ of A^e-modules. When A is a projective R-module, $S_\bullet(A)$ is a projective resolution of A as a left A^e-module, and is called the standard resolution. The standard resolution is applied to compute the Hochschild cohomology groups $\mathrm{H}^n(A, M)$.

Denote by $A^{\otimes n}$ the tensor product $A \otimes_R \cdots \otimes_R A$ of n copies of A. For $n \geq 0$, define left A^e-modules by

$$S_n(A) = \begin{cases} A^e = A \otimes_R A^o & \text{if } n = 0, \\ A \otimes_R (A^{\otimes n}) \otimes_R A & \text{if } n > 0. \end{cases} \tag{5.14}$$

Note that $S_n(A)$ is a left A^e-module by $a \otimes b \cdot x = axb$. For notational convenience, we define $S_{-1}(A)$ to be A. For $n \geq 0$ and for $0 \leq i \leq n$, let $\mu_{n,i} : S_n \to S_{n-1}$ be define by

$$\mu_{n,i}(x_0 \otimes \cdots \otimes x_i \otimes \cdots \otimes x_{n+1}) = x_0 \otimes \cdots \otimes x_i x_{i+1} \otimes \cdots \otimes x_{n+1}.$$

Then $\mu_{n,i}$ is defined by tensoring the multiplication map $\mu : A^e \to A$ in the ith factor with the identity map elsewhere. Define boundary maps $d_n : S_n \to S_{n-1}$ by

$$d_n = \sum_{i=0}^{n} (-1)^n \mu_{n,i}. \tag{5.15}$$

Since μ is an A^e-module homomorphism, it follows that $\mu_{n,i}$ and d_n are A^e-module homomorphisms.

Lemma 5.4.2. *In the above context,*

$$\cdots \to S_n(A) \xrightarrow{d_n} S_{n-1}(A) \to \cdots \to S_1 \xrightarrow{d_1} S_0 \xrightarrow{\mu} A \to 0$$

is an exact sequence. If A is projective as an R-module, then $S_\bullet(A) \to A$ is a projective resolution of A as a left A^e-module.

Proof. By a slight variation of Theorem 1.3.2, we see that if A is a projective R-module, then $S_n(A)$ is a projective A^e-module. We must show that $d_{n-1}d_n = 0$, and that the homology of the complex is (0). For $n \geq -1$ define $k_n : S_n(A) \to S_{n+1}(A)$ by $k_n(x) = 1 \otimes x$. For all $n \geq 0$ and $x \in S_n(A)$, we see that

$$\begin{aligned} d_{n+1}k_n(x) &= d_{n+1}(1 \otimes x) \\ &= x + \sum_{i=1}^{n+1} (-1)^i \mu_{n+1,i}(1 \otimes x) \\ &= x - \sum_{i=0}^{n} (-1)^i 1 \otimes \mu_{n,i}(x) \end{aligned}$$

and

$$\begin{aligned} k_{n-1}d_n(x) &= k_{n+1} \sum_{i=0}^{n} (-1)^i \mu_{n,i}(x) \\ &= \sum_{i=0}^{n} (-1)^i 1 \otimes \mu_{n,i}(x). \end{aligned}$$

Therefore, the contracting homotopy relations

$$d_{n+1}k_n + k_{n-1}d_n = 1$$

are satisfied. Now we show that $d_{n-1}d_n = 0$. The proof is by induction on n. For $n = 1$,

$$\mu d_1(x \otimes y \otimes z) = \mu(xy \otimes z - x \otimes yz) = (xy)z - x(yz) = 0$$

by the associative property for multiplication in A. Inductively, assume $n \geq 2$ and that $d_{n-2}d_{n-1} = 0$. Notice that the image of k_{n-1} contains a generating set for S_n as an A^e-module. Using the identity $d_n k_{n-1} + k_{n-2}d_{n-1} = 1_{S_{n-1}}$, we get

$$\begin{aligned}
d_{n-1}d_n k_{n-1} &= d_{n-1}(1_{S_{n-1}} - k_{n-2}d_{n-1}) \\
&= d_{n-1}1_{S_{n-1}} - d_{n-1}k_{n-2}d_{n-1}) \\
&= d_{n-1} - (1_{S_{n-1}} - k_{n-3}d_{n-2})d_{n-1} \\
&= d_{n-1} - d_{n-1} + k_{n-3}d_{n-2}d_{n-1} \\
&= 0,
\end{aligned}$$

which completes the proof. □

Let A be an R-algebra which is projective as an R-module. Let M be a left A^e-module. Recall that in Definition 5.4.1 the Hochschild cohomology group $\mathrm{H}^n(A, M)$ is defined to be $\mathrm{Ext}^n_{A^e}(A, M)$. The projective resolution $S_\bullet(A) \to A$ of Lemma 5.4.2 is called the *standard complex of* A. From (5.14) we have

$$S_n(A) = A \otimes_R T^n(A) \otimes_R A = A^e \otimes_R T^n(A)$$

where $T^n(A) = A^{\otimes n}$ is the nth tensor module of A over R. The adjoint isomorphism of Theorem 1.3.24 (1) implies

$$\begin{aligned}
\mathrm{Hom}_{A^e}(S_n(A), M) &\cong \mathrm{Hom}_{A^e}(A^e \otimes_R T^n(A), M) \\
&\cong \mathrm{Hom}_R(T^n(A), \mathrm{Hom}_{A^e}(A^e, M)) \\
&\cong \mathrm{Hom}_R(T^n(A), M).
\end{aligned}$$

The Hochschild cohomology groups are the homology groups of the truncated complex

$$\begin{aligned}
\mathrm{Hom}_{A^e}(S_\bullet(A), M) &= \mathrm{Hom}_{A^e}(A^e \otimes_R T^\bullet(A), M) \\
&= \mathrm{Hom}_R(T^\bullet(A), M).
\end{aligned}$$

The terms of low degree are

$$\begin{aligned}
(5.16)\quad 0 \to M &\xrightarrow{\delta^0} \mathrm{Hom}_R(A, M) \xrightarrow{\delta^1} \mathrm{Hom}_R(A \otimes_R A, M) \\
&\xrightarrow{\delta^2} \mathrm{Hom}_R(A \otimes_R A \otimes_R A, M) \xrightarrow{\delta^3} \mathrm{Hom}_R(A^{\otimes 4}, M) \to \cdots.
\end{aligned}$$

A tedious computation involving (5.16), the boundary maps d_n from (5.15), the adjoint isomorphism, and the Hom functor results in a formula for the coboundary maps. Let $f \in \operatorname{Hom}_R(A^{\otimes n}, M)$ be an n-cochain. Then

$$(5.17)\quad (\delta^n f)(x_1 \otimes \dots \otimes x_{n+1}) = x_1 f(x_2 \otimes \dots x_{n+1}) + \sum_{i=1}^{n} (-1)^i f(x_1 \otimes \dots \otimes x_i x_{i+1} \otimes \dots \otimes x_{n+1}) + (-1)^{n+1} f(x_1 \otimes \dots \otimes x_n) x_{n+1}.$$

5. Amitsur Cohomology

The cohomology defined in this section was first used by S. Amitsur in [**Ami59**]. It is the basis of the Čech cohomology which was introduced by Grothendieck and Cartier for schemes (see Section 10.4). In Theorem 13.3.18 we show that for a finite Galois extension, Amitsur cohomology agrees with Galois cohomology. General references for the material presented here are [**Hoo66**], [**KO74b**] and [**OS75**].

5.1. The Definition and First Properties. Let S be a commutative R-algebra. By $S^{\otimes r}$ we denote $S \otimes_R \cdots \otimes_R S$, the tensor product of r copies of S. As in Section 5.3.1, for $0 \le j \le n+1$, there is an R-algebra homomorphism

$$S^{\otimes(n+1)} \xrightarrow{e_j} S^{\otimes(n+2)}$$
$$(x_0 \otimes \cdots \otimes x_n) \mapsto x_0 \otimes \cdots \otimes x_{j-1} \otimes 1 \otimes x_j \otimes \cdots \otimes x_n.$$

Let $\mathfrak{F}$ be a covariant functor from the category of commutative R-algebras to the category of abelian groups. The *Amitsur complex* for S/R with coefficients in $\mathfrak{F}$ is

$$(5.18)\qquad 1 \to \mathfrak{F}(S) \xrightarrow{d^0} \mathfrak{F}(S^{\otimes 2}) \xrightarrow{d^1} \mathfrak{F}(S^{\otimes 3}) \xrightarrow{d^2} \cdots$$

where the coboundary map $d^r : \mathfrak{F}(S^{\otimes(r+1)}) \to \mathfrak{F}(S^{\otimes(r+2)})$ is defined to be

$$d^r = \prod_{i=0}^{r+1} \mathfrak{F}(e_i)^{(-1)^i}.$$

Denote this complex by $C^\bullet(S/R, \mathfrak{F})$. Since $e_j e_i = e_{i+1} e_j$ for all $j \le i$, the reader should verify that (5.18) is a complex of abelian groups.

Definition 5.5.1. In the cochain complex (5.18), the kernel of d^n is the group of n-cocycles, $\mathrm{Z}^n(S/R, \mathfrak{F}) = \ker d^n$. The image of d^{n-1} is the group of n-coboundaries, $\mathrm{B}^n(S/R, \mathfrak{F}) = \operatorname{im} d^{n-1}$. The group of cocycles modulo the coboundaries is

$$\mathrm{H}^n(S/R, \mathfrak{F}) = \mathrm{Z}^n(S/R, \mathfrak{F}) / \mathrm{B}^n(S/R, \mathfrak{F})$$

which is called the *nth Amitsur cohomology group of S/R with coefficients in $\mathfrak{F}$*.

Example 5.5.2. In degrees 0 and 1, we have

$$\begin{aligned} \mathrm{Z}^0(S/R,\mathfrak{F}) &= \mathrm{H}^0(S/R,\mathfrak{F}) \\ &= \{\alpha \in \mathfrak{F}(S) \mid \mathfrak{F}(e_0)(\alpha) = \mathfrak{F}(e_1)(\alpha)\} \\ \mathrm{B}^1(S/R,\mathfrak{F}) &= \{\mathfrak{F}(e_0)(\alpha)\mathfrak{F}(e_1)(\alpha^{-1}) \mid \alpha \in \mathfrak{F}(S)\} \\ \mathrm{Z}^1(S/R,\mathfrak{F}) &= \{\alpha \in \mathfrak{F}(S \otimes_R S) \mid \mathfrak{F}(e_2)(\alpha)\mathfrak{F}(e_0)(\alpha) = \mathfrak{F}(e_1)(\alpha)\}. \end{aligned} \tag{5.19}$$

Example 5.5.3. For any commutative R-algebra S, let $\mathbb{G}_a(S)$ be the additive abelian group of S. If S is faithfully flat, then by Proposition 5.3.1,

$$\mathrm{H}^n(S/R,\mathbb{G}_a) = \begin{cases} \mathbb{G}_a(R) & \text{if } n = 0, \\ 0 & \text{if } n \geq 1. \end{cases}$$

Definition 5.5.4. When $\mathfrak{F}$ is nonabelian, the cohomology is defined using the relations of (5.19). In this case, the result is not a group, but a pointed set. Let $\mathfrak{F}$ be a functor from the category of commutative R-algebras to the category of groups. We define

$$\mathrm{H}^0(S/R,\mathfrak{F}) = \{\alpha \in \mathfrak{F}(S) \mid \mathfrak{F}(e_0)(\alpha) = \mathfrak{F}(e_1)(\alpha)\}$$

with base point being the group identity of $\mathfrak{F}(S)$. We define

$$\mathrm{Z}^1(S/R,\mathfrak{F}) = \{\alpha \in \mathfrak{F}(S \otimes_R S) \mid \mathfrak{F}(e_2)(\alpha)\mathfrak{F}(e_0)(\alpha) = \mathfrak{F}(e_1)(\alpha)\}$$

with base point being the group identity of $\mathfrak{F}(S \otimes_R S)$. Now define a relation on $\mathrm{Z}^1(S/R,\mathfrak{F})$ by $\alpha \sim \beta$ if there exists $\gamma \in \mathfrak{F}(S)$ such that

$$\alpha = \mathfrak{F}(e_1)(\gamma)\beta\mathfrak{F}(e_0)(\gamma^{-1}).$$

The reader should verify that $\sim$ is an equivalence relation. We define $\mathrm{H}^1(S/R,\mathfrak{F})$ to be the set of equivalence classes $\mathrm{Z}^1(S/R,\mathfrak{F})/\sim$, with base point being the equivalence class containing the group identity of $\mathfrak{F}(S \otimes_R S)$. When the functor $\mathfrak{F}$ takes its values in the category of abelian groups, it is clear that this definition agrees with Definition 5.5.1 for $n = 0, 1$.

Theorem 5.5.5. *Suppose*

$$\begin{array}{ccc} S & \xrightarrow{f} & S' \\ {\scriptstyle\theta}\uparrow & & \uparrow{\scriptstyle\theta'} \\ R & \xrightarrow{\phi} & R' \end{array}$$

is a commutative diagram of homomorphisms of commutative R-algebras. Let $\mathfrak{F}$ be a functor from the category of commutative R-algebras to the category of abelian groups. Then f induces homomorphisms

$$f^* : \mathrm{H}^n(S/R,\mathfrak{F}) \to \mathrm{H}^n(S'/R',\mathfrak{F})$$

for $n \geq 0$. Moreover, f^* is independent of f. That is, if $g : S \to S'$ is another such homomorphism, then $f^* = g^*$. If $\mathfrak{F}$ is a functor that takes its values in the category of nonabelian groups, then the above is true for $n = 0, 1$, where f^* is a morphism of pointed sets.

Proof. Since $\mathfrak{F}$ is a functor, and the diagram of algebra homomorphisms commutes, f induces a morphism of cochain complexes $f : \mathfrak{F}(S^{\otimes n}) \to \mathfrak{F}((S')^{\otimes n})$. Consequently, there are homomorphisms $f^* : \mathrm{H}^n(S/R, \mathfrak{F}) \to \mathrm{H}^n(S'/R', \mathfrak{F})$.

Case 1: Assume $\mathfrak{F}$ is abelian and use additive notation in the groups $\mathfrak{F}(\cdot)$. It is enough to show that the two morphisms f and g between $\mathfrak{F}(S^{\otimes n})$ and $\mathfrak{F}((S')^{\otimes n})$ are homotopic (see, for example, [**Rot79**, Theorem 6.8]). We define $k^n : \mathfrak{F}(S^{\otimes(n+1)}) \to \mathfrak{F}((S')^{\otimes n})$ and show that

$$(f^*)^n - (g^*)^n = d^{n-1}k^n + k^{n+1}d^n \tag{5.20}$$

for $n \geq 1$. For $0 \leq i < n$ define $k_i^n : S^{\otimes(n+1)} \to (S')^{\otimes n}$ by

$$k_i^n(s_0 \otimes \cdots \otimes s_n) = f(s_0) \otimes \cdots \otimes f(s_i)g(s_{i+1}) \otimes \cdots \otimes g(s_n). \tag{5.21}$$

Then each k_i^n is an R-algebra homomorphism (Exercises 1.3.8 and 1.3.14). The homotopy operator is defined by $k^n = \sum_{i=0}^{n-1}(-1)^i \mathfrak{F}(k_i^n)$. We define auxiliary R-algebra homomorphisms $h_i^n : S^{\otimes(n+1)} \to (S')^{\otimes(n+1)}$ by

$$\begin{aligned} h_i^n&(s_0 \otimes \cdots \otimes s_n) \\ &= \begin{cases} g(s_0) \otimes \cdots \otimes g(s_n) & \text{if } i = 0, \\ f(s_0) \otimes \cdots \otimes f(s_{i-1}) \otimes g(s_i) \otimes \cdots \otimes g(s_n) & \text{if } 1 \leq i \leq n, \\ f(s_0) \otimes \cdots \otimes f(s_n) & \text{if } i = n+1. \end{cases} \end{aligned} \tag{5.22}$$

The reader should verify the relations

$$k_j^{n+1} e_i = \begin{cases} e_{i-1}k_j^n & \text{if } j < i-1, \\ h_i^n & \text{if } i-1 \leq j \leq i, \\ e_i k_{j-1}^n & \text{if } i < j. \end{cases} \tag{5.23}$$

Starting with the rightmost term in (5.20),

$$\begin{aligned} k^{n+1}d^n &= \sum_{j=0}^{n}\sum_{i=0}^{n+1}(-1)^j(-1)^i \mathfrak{F}(k_j^{n+1})\mathfrak{F}(e_i) \\ &= \sum_{j=0}^{n}\sum_{i=0}^{n+1}(-1)^{j+i}\mathfrak{F}(k_j^{n+1}e_i) \end{aligned}$$

Using (5.23), we get

$$
\begin{aligned}
k^{n+1}d^n &= \sum_{i=2}^{n+1}\sum_{j=0}^{i-2}(-1)^{j+i}\mathfrak{F}(e_{i-1}k_j^n) \\
&\quad + \sum_{i=1}^{n+1}(-1)^{i-1+i}\mathfrak{F}(h_i) + \sum_{i=0}^{n}(-1)^{i+i}\mathfrak{F}(h_i) \\
&\quad + \sum_{i=0}^{n-1}\sum_{j=i+1}^{n}(-1)^{j+i}\mathfrak{F}(e_ik_{j-1}^n)) \\
&= \sum_{i=2}^{n+1}\sum_{j=0}^{i-2}(-1)^{j+i}\mathfrak{F}(e_{i-1}k_j^n) \\
&\quad + \mathfrak{F}(h_0) - \mathfrak{F}(h_{n+1}) \\
&\quad + \sum_{i=0}^{n-1}\sum_{j=i+1}^{n}(-1)^{j+i}\mathfrak{F}(e_ik_{j-1}^n)) \\
&= \sum_{j=0}^{n-1}(-1)^{j+n+1}\mathfrak{F}(e_nk_j^n) \\
&\quad + \sum_{i=1}^{n-1}\left(\sum_{j=0}^{i-1}(-1)^{j+i+1}\mathfrak{F}(e_ik_j^n)) + \sum_{j=i}^{n-1}(-1)^{j+i+1}\mathfrak{F}(e_ik_j^n))\right) \\
&\quad + \sum_{j=0}^{n-1}(-1)^{j}\mathfrak{F}(e_0k_j^n)) \\
&\quad + \mathfrak{F}(h_0) - \mathfrak{F}(h_{n+1}) \\
&= (g^*)^n - (f^*)^n - \sum_{i=0}^{n}\sum_{j=0}^{n-1}(-1)^{j+i}\mathfrak{F}(e_ik_j^n)) \\
&= (g^*)^n - (f^*)^n - d^{n-1}k^n,
\end{aligned}
$$

which proves the theorem when $\mathfrak{F}$ is abelian.

Case 2: Assume $\mathfrak{F}$ is nonabelian (written multiplicatively) and $n = 0$. Let $k_0^1 : S \otimes_R S \to S'$ be as in (5.21). Note that $k_0^1 e_0 = g$ and $k_0^1 e_1 = f$. If $\alpha \in \mathrm{Z}^0(S/R, \mathfrak{F})$, then $\mathfrak{F}(g)\alpha = \mathfrak{F}(k_0^1 e_0)\alpha = \mathfrak{F}(k_0^1)\mathfrak{F}(e_0)\alpha = \mathfrak{F}(k_0^1)\mathfrak{F}(e_1)\alpha = \mathfrak{F}(k_0^1 e_1)\alpha = \mathfrak{F}(f)\alpha$.

Case 3: Assume $\mathfrak{F}$ is nonabelian (written multiplicatively) and $n = 1$. Let k_0^2 and k_1^2 be the R-algebra homomorphisms defined in (5.21). If $\alpha \in$

$\mathrm{Z}^1(S/R, \mathfrak{F})$, then on the one hand,

$$
\begin{aligned}
\mathfrak{F}(f \otimes g)(\alpha) &= \mathfrak{F}(h_1^1)(\alpha) && \text{(by (5.22))} \\
&= \mathfrak{F}(k_0^2 e_1)(\alpha) && \text{(by (5.23))} \\
&= \mathfrak{F}(k_0^2 e_2)(\alpha)\mathfrak{F}(k_0^2 e_0)(\alpha) && \text{(since } \alpha \in \mathrm{Z}^1(S/R, \mathfrak{F})) \\
&= \mathfrak{F}(e_1 k_0^1)(\alpha)\mathfrak{F}(h_0^1)(\alpha) && \text{(by (5.23))} \\
&= \mathfrak{F}(e_1 k_0^1)(\alpha)\mathfrak{F}(g \otimes g)(\alpha) && \text{(by (5.22))}.
\end{aligned}
\tag{5.24}
$$

On the other hand,

$$
\begin{aligned}
\mathfrak{F}(f \otimes g)(\alpha) &= \mathfrak{F}(h_1^1)(\alpha) && \text{(by (5.22))} \\
&= \mathfrak{F}(k_1^2 e_1)(\alpha) && \text{(by (5.23))} \\
&= \mathfrak{F}(k_1^2 e_2)(\alpha)\mathfrak{F}(k_1^2 e_0)(\alpha) && \text{(since } \alpha \in \mathrm{Z}^1(S/R, \mathfrak{F})) \\
&= \mathfrak{F}(h_2^1)(\alpha)\mathfrak{F}(e_0 k_0^1)(\alpha) && \text{(by (5.23))} \\
&= \mathfrak{F}(f \otimes f)(\alpha)\mathfrak{F}(e_0 k_0^1)(\alpha) && \text{(by (5.22))}.
\end{aligned}
\tag{5.25}
$$

Set $\gamma = \mathfrak{F}(k_0^1)(\alpha)$. Combining (5.24) and (5.25),

$$\mathfrak{F}(f \otimes f)(\alpha) = \mathfrak{F}(e_1)(\gamma)\mathfrak{F}(g \otimes g)(\alpha)\mathfrak{F}(e_0)(\gamma^{-1})$$

which shows $\mathfrak{F}(f \otimes f) \sim \mathfrak{F}(g \otimes g)$. □

5.2. Twisted Forms. Let R be a commutative ring and $\mathfrak{C}_{\mathrm{fl}}(R)$ the category of isomorphism classes of faithfully flat R-algebras. If A is an R-module (or R-algebra), let $\mathrm{Aut}(A)$ denote the functor from $\mathfrak{C}_{\mathrm{fl}}(R)$ to the category of groups defined by $S \mapsto \mathrm{Aut}_S(A \otimes_R S)$.

Definition 5.5.6. Let R be a commutative ring and A a fixed R-module (or R-algebra). Given an R-module (or R-algebra) B and a faithfully flat R-algebra S, we say B is a *twisted form of A for the extension S/R* if there exists an isomorphism of S-algebras $B \otimes_R S \cong A \otimes_R S$.

Proposition 5.5.7. *In the above context, the pointed set $\mathrm{H}^1(S/R, \mathrm{Aut}(A))$ classifies up to R-module (or R-algebra) isomorphism the twisted forms of A for the extension S/R.*

Proof. Suppose B is a twisted form of A for the extension S/R, and $\beta : B \otimes_R S \to A \otimes_R S$ is an S-module isomorphism. In a switch from the notation of Proposition 5.3.4, we write β_i instead of $\mathfrak{F}_i(\beta)$. Define $\theta \in \mathrm{Aut}_{S \otimes_R S}(A \otimes_R S \otimes_R S)$ by $\theta = \beta_1 \beta_0^{-1}$. Therefore θ is the map that makes

the diagram

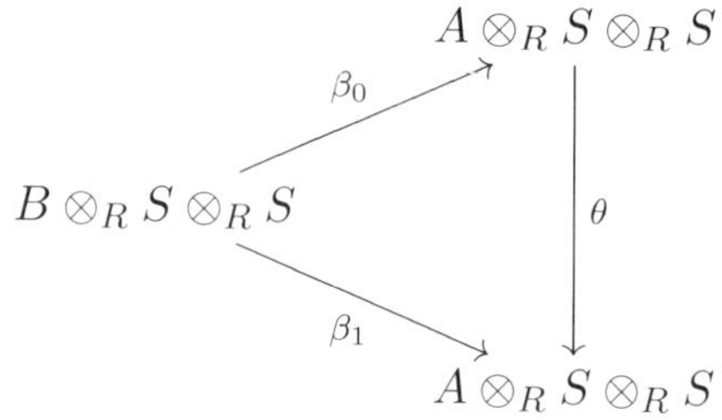

commutes. The reader should verify the identities: $(\beta_0)_0 = (\beta_0)_1$, $(\beta_0)_2 = (\beta_1)_0$, $(\beta_1)_1 = (\beta_1)_2$. Therefore,

$$\begin{aligned}\theta_2\theta_0 &= (\beta_1\beta_0^{-1})_2(\beta_1\beta_0^{-1})_0\\ &= (\beta_1)_2(\beta_0^{-1})_2(\beta_1)_0(\beta_0^{-1})_0 = (\beta_1)_1(\beta_0^{-1})_1 = (\beta_1\beta_0^{-1})_1 = \theta_1.\end{aligned}$$

So θ is a 1-cocycle. To show that the cohomology class of θ depends only on B, suppose $\alpha : B \otimes_R S \to A \otimes_R S$ is another S-module isomorphism, and $\phi = \alpha_1\alpha_0^{-1}$. Set $\gamma = \alpha\beta^{-1}$. Then γ is an S-module automorphism of $A \otimes_R S$. We have $\gamma_1\theta\gamma_0^{-1} = \gamma_1(\beta_1\beta_0^{-1})\gamma_0^{-1} = \alpha_1\beta_1^{-1}(\beta_1\beta_0^{-1})\beta_0\alpha_0^{-1} = \alpha_1\alpha_0^{-1} = \phi$. Therefore, ϕ is cohomologous to θ.

Let $\theta \in \operatorname{Aut}_{S\otimes_R S}(A \otimes_R S \otimes S)$ and assume θ defines a 1-cocycle in $\mathrm{Z}^1(S/R, \operatorname{Aut}(A))$. In a switch from the notation of Section 5.5.1, write θ_i instead of $\mathfrak{F}(e_i)(\theta)$. Then $\theta_2\theta_0 = \theta_1$. As in Section 5.3.3, for $i = 0, 1$ there are R-module homomorphisms $e_i : A \otimes_R S \to A \otimes_R S \otimes_R S$. Define

$$\begin{aligned}B &= \left\{\sum a_i \otimes s_i \in A \otimes_R S \mid \theta\left(\sum a_i \otimes 1 \otimes s_i\right) = \sum a_i \otimes s_i \otimes 1\right\}\\ &= \ker\left\{\theta e_0 - e_1 : A \otimes_R S \to A \otimes_R S \otimes_R S\right\}.\end{aligned}$$

Then B is an R-module. Define $\beta : B \otimes_R S \to A \otimes_R S$ to be the multiplication map, $\beta((\sum a_i \otimes s_i) \otimes s) = \sum a_i \otimes s_i s$. As in the proof of Theorem 5.3.7, the reader should verify that β is an isomorphism of S-modules and $\theta = \beta_1\beta_0^{-1}$. Therefore B is a twisted form of A for the extension S/R.

To see that B depends only on the cohomology class of θ, suppose ϕ is a 1-cocycle that is cohomologous to θ. Then there is $\gamma \in \operatorname{Aut}(A \otimes_R S)$ such that $\gamma_1\theta\gamma_0^{-1} = \phi$. Since ϕ is a descent datum, there is an R-module C, and an isomorphism $\alpha : C \otimes_R S \to A \otimes_R S$ such that $\phi = \alpha_1\alpha_0^{-1}$. It follows from

$$\begin{aligned}\phi &= \gamma_1\theta\gamma_0^{-1}\\ \alpha_1\alpha_0^{-1} &= \gamma_1\beta_1\beta_0^{-1}\gamma_0^{-1}\\ \alpha_0^{-1}\gamma_0\beta_0 &= \alpha_1^{-1}\gamma_1\beta_1\end{aligned}$$

that $(\alpha^{-1}\gamma\beta)_0 = (\alpha^{-1}\gamma\beta)_1$. In the notation of Proposition 5.3.4, this implies $\mathfrak{F}_0(\alpha^{-1}\gamma\beta) = \mathfrak{F}_0(\alpha^{-1}\gamma\beta)$. This implies the isomorphism $\alpha^{-1}\gamma\beta : B \otimes_R S \to C \otimes_R S$ of S-modules comes from an isomorphism $B \cong C$ of R-modules. $\square$

As an application of Proposition 5.5.7, we classify the twisted forms of a finitely generated free module. Let R be a commutative ring and denote by R^n the direct sum of n copies of R. Let S be a commutative faithfully flat R-algebra. A free module of rank n is a projective module of rank n. It follows from Theorem 2.6.1 that a twisted form of R^n for S/R is a projective module of rank n. The group $\operatorname{Aut}_S(R^n \otimes_R S) = \operatorname{Aut}_S(S^n)$ is isomorphic to the group of invertible matrices in $M_n(S)$. The group of invertible n-by-n matrices over S is also denoted $\mathrm{GL}_n(S)$ and is called the *general linear group*. We also denote by GL_n the functor from $\mathfrak{C}_{\mathrm{fl}}$ to the category of groups defined by $S \mapsto \mathrm{GL}_n(S)$.

Corollary 5.5.8. *Let S be a commutative faithfully flat R-algebra.*

(1) *The twisted forms of the free R-module of rank n for S/R are classified up to isomorphism by the pointed set $\mathrm{H}^1(S/R, \mathrm{GL}_n)$.*
(2) *If R is a ring for which finitely generated projective modules are free, then $\mathrm{H}^1(S/R, \mathrm{GL}_n) = \{1\}$. This is true, for instance, if R is a local ring or a principal ideal domain.*

For $n = 1$, the general linear group $\mathrm{GL}_1(S)$ is equal to $S^* = \mathbb{G}_m(S)$, the group of units of the ring S. Since S is commutative, $\mathbb{G}_m(S)$ is an abelian group and the pointed set $\mathrm{H}^1(S/R, \mathrm{GL}_n)$ is a group. By Corollary 5.5.8, $\mathrm{H}^1(S/R, \mathbb{G}_m)$ classifies the group of rank one projective R-modules P such that $P \otimes_R S \cong S$. This and Proposition 2.6.9 proves Corollary 5.5.9.

Corollary 5.5.9. *In the above context, the group $\mathrm{H}^1(S/R, \mathbb{G}_m)$ is isomorphic to the kernel of the natural homomorphism $\operatorname{Pic} R \to \operatorname{Pic} S$.*

Remark 5.5.10. Consider a twisted form of a finitely generated free R-algebra. Let $R^n = R \oplus \cdots \oplus R$ be the trivial commutative separable extension of R of rank n. Let S be a commutative faithfully flat R-algebra. It follows from Proposition 4.6.11 that an R-algebra B is a separable R-algebra which is an R-module progenerator of constant rank n, if and only if there exists such an R-algebra S and B is a twisted form of R^n for S/R.

Remark 5.5.11. Consider a twisted form of a matrix algebra. If S is a commutative R-algebra, then $M_n(R) \otimes_R S = M_n(S)$. Let $\operatorname{Aut}(M_n)$ denote the functor from $\mathfrak{C}_{\mathrm{fl}}$, the category of faithfully flat R-algebras, to the category of groups, defined by $S \mapsto \operatorname{Aut}_S(M_n(S))$. Now let S be a commutative faithfully flat R-algebra. By Proposition 5.5.7, $\mathrm{H}^1(S/R, \operatorname{Aut}(M_n))$ classifies the twisted forms of $M_n(R)$ for S/R. We will see in Corollary 10.3.10 below that an R-algebra is a central separable R-algebra of constant rank n^2, if and only if there exists such an R-algebra S and B is a twisted form of $M_n(R)$ for S/R.

Chapter 6

The Divisor Class Group

This chapter contains an introduction to the theory of divisor classes of a noetherian normal integral domain R with field of fractions K. There are two approaches to defining the so-called class group of R. First, in Section 6.2, the class group is defined in terms of Weil divisor classes. Later, in Section 6.4, it is interpreted as the group of classes of reflexive fractional ideals of R in K. We see that the Picard group of R can be interpreted as the subgroup of the divisor class group generated by projective fractional ideals. Section 6.3 contains a short but rigorous introduction to R-lattices.

It is worth emphasizing that the material from this section has many applications to the theory of separable algebras. Many of the results in Chapters 11 – 14 will be dependent on the topics of this chapter. We mention two particular instances. Example 6.5.12 is a computation which illustrates how any finite cyclic group can occur as the class group of a regular integral domain of Krull dimension two. This example plays a key role in Section 14.3.2 when we show that every finite cyclic group is a Brauer group. Example 6.3.17 describes a powerful method for constructing reflexive modules which are not projective. It is applied in Section 11.3.7 when we construct Ojanguren's example of a nontrivial locally trivial algebra.

Because Section 6.1 is comprised of some necessary background results which will be referenced in subsequent sections, some readers may prefer to start directly with Section 6.2 and return later when the need arises.

1. Background Results from Commutative Algebra

In this section we have assembled various definitions and background results that we will have occasion to apply in this chapter and in later chapters. Usually, proofs are not included, but references are provided. As general references for the material in this section we recommend [**Mat80**], [**AM69**], [**Bou89b**] and [**Rot79**].

1.1. Krull Dimension. Let R be a commutative ring. Suppose

$$P_0 \supsetneq P_1 \supsetneq \cdots \supsetneq P_n$$

is a chain of $n+1$ distinct prime ideals in $\operatorname{Spec} R$. We say this is a *prime chain* of *length* n. If $P \in \operatorname{Spec} R$, the *height* of P, denoted $\operatorname{ht}(P)$, is the supremum of the lengths of all prime chains with $P = P_0$. Let I be a proper ideal of R. The *height* of I, denoted $\operatorname{ht}(I)$, is defined to be the infimum of the heights of all prime ideals containing I. That is, $\operatorname{ht}(I) = \inf\{\operatorname{ht}(P) \mid P \in \operatorname{Spec} R, P \supseteq I\}$. The *Krull dimension* of R, denoted $\dim(R)$, is the supremum of the heights of all prime ideals in R. Thus, $\dim(R) = \sup\{\operatorname{ht}(P) \mid P \in \operatorname{Spec} R\}$. The following formula for the Krull dimension of a fiber of a homomorphism between two commutative noetherian rings will be needed.

Theorem 6.1.1. *Let $f : R \to S$ be a homomorphism of commutative noetherian rings. Let $Q \in \operatorname{Spec} S$ and $P = Q \cap R$. Then*

1. $\operatorname{ht}(Q) \le \operatorname{ht}(P) + \operatorname{ht}(Q/PS)$.
2. $\dim(S_Q) \le \dim(R_P) + \dim(S_Q \otimes_R k_P)$ *where $k_P = R_P/PR_P$ is the residue field.*
3. *If going down holds for f, then equality holds in Parts (1) and (2).*
4. *If going down holds for f and $f^\sharp : \operatorname{Spec} S \to \operatorname{Spec} R$ is surjective, then*
   (a) $\dim(S) \ge \dim(R)$, *and*
   (b) *for any ideal $I \subseteq R$, $\operatorname{ht}(I) = \operatorname{ht}(IS)$.*

Proof. See, for example, [**Mat80**, (13.B), Theorem 19]. □

We will utilize the following version of Emmy Noether's Normalization Lemma and its corollary.

Corollary 6.1.2 (Emmy Noether's Normalization Lemma). *Let k be a field and A a finitely generated commutative k-algebra. There exist $z_1, \ldots, z_m$ in A such that*

1. *the set $\{z_1, \ldots, z_m\}$ is algebraically independent over k,*
2. *A is integral over $k[z_1, \ldots, z_m]$,*
3. $\dim(A) = m$, *and*

(4) *if* A *is an integral domain with quotient field* K*, then* $\operatorname{tr.deg}_k(K) = m$.

Proof. See, for example, [**Mat80**, (14.G), Corollary 1]. □

Corollary 6.1.3. *Let* k *be a field and* A *an integral domain which is a finitely generated commutative* k*-algebra.*

(1) *If* $p \in \operatorname{Spec} A$*, then* $\dim(A/p) + \operatorname{ht}(p) = \dim(A)$.

(2) *If* p *and* q *are in* $\operatorname{Spec} A$ *such that* $p \supseteq q$*, then* $\operatorname{ht}(p/q) = \operatorname{ht}(p) - \operatorname{ht}(q)$.

Proof. See, for example, [**Mat80**, (14.H), Corollary 3]. □

1.2. The Serre Criteria for Normality. Let R be a commutative ring, M an R-module, and $a_1, \dots, a_n$ a set of elements of R. We denote by $(a_1, \dots, a_n) = Ra_1 + \dots + Ra_n$ the ideal which they generate and in the same fashion $(a_1, \dots, a_n)M = Ra_1M + \dots + Ra_nM$. We say $a_1, \dots, a_r$ is a *regular sequence for* M in case the following are satisfied.

(1) a_1 is not a zero divisor for M,

(2) for $k = 2, \dots, r$, a_k is not a zero divisor for $M/(a_1, \dots, a_{k-1})M$, and

(3) $M \neq (a_1, \dots, a_r)M$.

If this is true, and if I is an ideal of R such that $(a_1, \dots, a_r) \subseteq I$, then we say $a_1, \dots, a_r$ is a *regular sequence for* M *in* I. A regular sequence $a_1, \dots, a_r$ is *maximal in* I if there is no $b \in I$ such that $a_1, \dots, a_r, b$ is a regular sequence for M in I.

Theorem 6.1.4. *Let* R *be a commutative noetherian ring and* M *a finitely generated* R*-module. Let* I *be an ideal of* R *such that* $IM \neq M$ *and* n *a positive integer. The following are equivalent.*

(1) *There exists a regular sequence* $a_1, \dots, a_n$ *for* M *in* I.

(2) $\operatorname{Ext}^i_R(R/I, M) = (0)$ *for all* $i < n$.

Proof. See, for example, [**Mat80**, (15.C), Theorem 28]. □

Let R be a noetherian commutative ring and M a finitely generated R-module. Let I be a proper ideal in R. The I*-depth* of M, denoted $\operatorname{depth}_I(M)$, is the least element of the set $\{i \mid \operatorname{Ext}^i_R(R/I, M) \neq (0)\}$. By Theorem 6.1.4, $\operatorname{depth}_I(M)$ is equal to the length of any maximal regular sequence for M in I. If R is a local ring with maximal ideal $\mathfrak{m}$, then we sometimes write $\operatorname{depth}(M)$ instead of $\operatorname{depth}_{\mathfrak{m}}(M)$.

Assume R is a commutative noetherian local ring with maximal ideal $\mathfrak{m}$ and Krull dimension $\dim(R) = d$. If there exists a subset $\{x_1, \dots, x_d\}$ of $\mathfrak{m}$ such that $Rx_1 + \cdots + Rx_d = \mathfrak{m}$, then we say R is a *regular local ring* and in this case we call $x_1, \dots, x_d$ a *regular system of parameters.* A regular system of parameters is a regular sequence for R in $\mathfrak{m}$. A commutative ring R is a *regular ring* in case R_P is a regular local ring for each $P \in \operatorname{Spec} R$.

Theorem 6.1.5 (Serre's Criteria for Normality). *Let R be a commutative noetherian ring. Then R is normal if and only if the following two properties are satisfied.*

(R_1) *For every prime ideal P in R such that $\operatorname{ht}(P) \le 1$, R_P is a regular local ring.*

(S_2) *For every prime ideal P in R,*

$$\operatorname{depth}(R_P) \ge \begin{cases} 1 & \text{if } \operatorname{ht}(P) = 1, \\ 2 & \text{if } \operatorname{ht}(P) \ge 2. \end{cases}$$

Proof. See [**Mat80**, (17.I), Theorem 39], or [**Gro65**, Théorème (5.8.6)], for example. □

1.3. The Hilbert-Serre Criterion for Regularity. Let R be a ring, M a left R-module, and

$$\cdots \xrightarrow{d_{n+1}} P_n \xrightarrow{d_n} \cdots \xrightarrow{d_2} P_1 \xrightarrow{d_1} P_0 \xrightarrow{\epsilon} M \to 0$$

a projective resolution of M. For $n \ge 1$, let K_n be the kernel of d_n and let K_0 be the kernel of ϵ. We say K_n is the nth *syzygy* of M with respect to the projective resolution $P_\bullet \to M$. The *projective dimension* of M, written $\operatorname{proj.dim}_R M$, is the length of a shortest projective resolution for M. If $K_n = (0)$, then $\operatorname{proj.dim}_R(M) \le n$. It follows that M is projective if and only if $\operatorname{proj.dim}_R(M) = 0$. Let R be a commutative ring. The *global cohomological dimension* of R is denoted $\operatorname{coh.dim}(R)$, and is defined to be $\operatorname{coh.dim}(R) = \sup\{\operatorname{proj.dim}_R(M) \mid M \in {}_R\mathfrak{M}\}$.

Theorem 6.1.6. *Let R be a commutative noetherian ring.*

(1) *The following are equivalent.*
   (a) $\operatorname{coh.dim}(R) \le n$.
   (b) $\operatorname{proj.dim}_R(M) \le n$ *for all finitely generated R-modules M.*
   (c) $\operatorname{Ext}_R^{n+1}(M, N) = 0$ *for all finitely generated R-modules M and N.*
   (d) *For all finitely generated R-modules M, if $P_\bullet \to M$ is a projective resolution of M with syzygies $\{K_n\}$, then K_{n-1} is projective.*

(2) $\operatorname{coh.dim}(R) = \sup\{\operatorname{coh.dim}(R_P) \mid P \in \operatorname{Spec} R\}$.

(3) $\operatorname{coh.dim}(R) = \sup\{\operatorname{coh.dim}(R_{\mathfrak{m}}) \mid \mathfrak{m} \in \operatorname{Max} R\}$.

Proof. See [**Mat80**, (18.B), Lemma 5]. □

Theorem 6.1.7. *Let R be a commutative noetherian local ring with maximal ideal $\mathfrak{m}$, residue field k and Krull dimension $d = \dim(R)$. The following are equivalent*

(1) *R is regular.*

(2) $\dim_k(\mathfrak{m}/\mathfrak{m}^2) = d$.

(3) *(Hilbert-Serre) R has finite cohomological dimension.*

If any of the conditions above are satisfied, then $\operatorname{coh.dim}(R) = d$.

Proof. See [**AM69**, Theorem 11.22], and [**Mat80**, (18.G), Theorem 45]. □

Lemma 6.1.8. *Let R be a commutative noetherian ring, M a finitely generated R-module, N an R-module, and S a flat R-algebra. Then*

$$\operatorname{Ext}^n_R(M, N) \otimes_R S = \operatorname{Ext}^n_S(M \otimes_R S, N \otimes_R S)$$

for all $n \geq 0$. In particular, if $P \in \operatorname{Spec} R$, then

$$\operatorname{Ext}^n_R(M, N)_P = \operatorname{Ext}^n_{R_P}(M_P, N_P)$$

for all $n \geq 0$.

Proof. See, for example, [**Mat80**, (3.E)]. □

Theorem 6.1.9. *Let R be a commutative noetherian ring.*

(1) *R is regular if and only if $R_{\mathfrak{m}}$ is a regular local ring for each $\mathfrak{m} \in \operatorname{Max}(R)$.*

(2) *If R is a regular ring and x is an indeterminate, then $R[x]$ is a regular ring.*

(3) *Let S be a commutative noetherian faithfully flat R-algebra. If S is regular, then R is regular.*

Proof. For (1) and (2), see [**Mat80**, (18.G), Corollary] and [**Mat80**, (17.J), Theorem 40].

(3): Let $\mathfrak{p} \in \operatorname{Spec} R$. Since S is faithfully flat over R, there exists $q \in \operatorname{Spec} S$ lying over $\mathfrak{p}$. By Theorem 6.1.7, $S_{\mathfrak{q}}$ has finite cohomological dimension. Using Theorem 6.1.6 and Lemma 6.1.8, we see that $R_{\mathfrak{p}}$ also has finite cohomological dimension. By Theorem 6.1.7, $R_{\mathfrak{p}}$ is regular. □

1.4. Discrete Valuation Rings. Let R be an integral domain with quotient field K. If R and K satisfy the equivalent parts of Lemma 6.1.10, then we say R is a *valuation ring* of K.

Lemma 6.1.10. *Let R be an integral domain with quotient field K. The following are equivalent.*

(1) *For all $x \in K^*$, either $x \in R$, or $x^{-1} \in R$.*

(2) *For all a, b in R, either $a \mid b$, or $b \mid a$.*

Proof. The proof is left to the reader. □

Let G be an abelian group, written additively. We say G is an *ordered group*, if there is a partial order on G that preserves the binary operation. In other words, if $u \leq v$ and $x \leq y$, then $u+x \leq v+y$. We say G is a *totally ordered group*, if the partial order is a chain. A *valuation on a field* F is a function $\nu : F^* \to G$, for a totally ordered group G which satisfies

(1) $\nu(xy) = \nu(x) + \nu(y)$, and

(2) if $x \neq 0$, $y \neq 0$, and $x + y \neq 0$, then $\nu(x + y) \geq \min(\nu(x), \nu(y))$.

The reader should verify that $\nu(1) = 0$.

Lemma 6.1.11. *Suppose F is a field and $\nu : F^* \to G$ is a valuation on F. Let*

$$R = \{0\} \cup \{x \in F^* \mid \nu(x) \geq 0\}.$$

Then R is a valuation ring of F which we call the valuation ring associated to ν. Conversely, if R is a valuation ring of F, then there exists a valuation $v : F^* \to H$ for some totally ordered group H such that R is the valuation ring of v.

Proof. This proof is left to the reader. □

Let F be a field and $R \subseteq S$ subrings of F. Assume R and S are local rings and that the inclusion homomorphism $R \to S$ is a local homomorphism of local rings (or, equivalently, the maximal ideal of S contains the maximal ideal of R). In this case, we say S *dominates* R. The reader should verify that this defines a partial order on the set of all local subrings of F.

Lemma 6.1.12. *Let F be a field and $\nu : F^* \to G$ a valuation on F. Let R be the valuation ring of ν.*

(1) *R is a local ring with maximal ideal $\mathfrak{m}_R = \{0\} \cup \{x \in F^* \mid \nu(x) > 0\}$.*

(2) *If $R \subseteq A \subseteq F$ is a tower of local subrings of F such that A dominates R, then $R = A$. In other words, R is a maximal local subring with respect to the relation "dominates".*

(3) *R is integrally closed in F.*

Proof. Parts (1) and (2) are left to the reader.

(3): Let $x \in F$ and assume x is integral over R. We prove $x \in R$. Assume the contrary. By Lemma 6.1.10, $x^{-1} \in R$. Since x is integral over R, there are elements $r_0, \dots, r_{n-1}$ in R such that

$$x^n + r_{n-1}x^{n-1} + \cdots + r_1 x + r_0 = 0$$

where $n > 0$. Multiply by x^{1-n} and solve for x. Then

$$\begin{aligned} x &= -(x^{-1})^{n-1}(r_{n-1}x^{n-1} + \cdots + r_1 x + r_0) \\ &= -(r_{n-1} + \cdots + r_1 x^{2-n} + r_0 x^{1-n}) \end{aligned}$$

is in R, a contradiction. □

If F is a field, a *discrete valuation* on F is a valuation $\nu : F^* \to \mathbb{Z}$ such that ν is onto. The valuation ring of ν is $R = \{0\} \cup \{x \in F^* \mid \nu(x) \geq 0\}$. Then R is a valuation ring of F. In particular, Lemma 6.1.12 implies that R is a local ring with maximal ideal $\mathfrak{m} = \{0\} \cup \{x \in F^* \mid \nu(x) > 0\}$, F is the field of fractions of R, and R is integrally closed in F. Since ν is onto, we see that $\mathfrak{m} \neq (0)$, so $\dim R \geq 1$. An integral domain A is called a *discrete valuation ring*, or DVR for short, if there exists a discrete valuation on the field of fractions of A such that A is the associated valuation ring.

Lemma 6.1.13. *Let F be a field and ν a discrete valuation on F. Let R be the associated DVR, with maximal ideal $\mathfrak{m}$.*

(1) *R is a PID.*

(2) *R is noetherian.*

(3) *For any element $\pi \in R$ such that $\nu(\pi) = 1$, $\mathfrak{m} = \pi R$. A complete list of the ideals of R is $(0), R\pi, R\pi^2, \dots, R$.*

(4) $\dim R = 1$.

Proof. See, for example, [**DF04**, Proposition 16.2.5 and Corollary 16.2.6]. □

Theorem 6.1.14. *Let R be a noetherian local integral domain with field of fractions K, maximal ideal $\mathfrak{m}$ and residue field $k = R/\mathfrak{m}$. If* $\dim(R) = 1$, *then the following are equivalent.*

(1) *R is a DVR.*

(2) *R is a PID.*

(3) *R is regular.*

(4) *R is normal.*

(5) *$\mathfrak{m}$ is a principal ideal.*

(6) *There exists an element* $\pi \in R$ *such that every ideal of* R *is of the form* $R\pi^n$, *for some* $n \geq 0$. *We call* π *a* local parameter *for* R.

Proof. See, for example, [**AM69**, Proposition 9.2]. □

Theorem 6.1.15. *Let* R *be a DVR with field of fractions* K *and maximal ideal* $\mathfrak{m} = \pi R$. *Let* $\hat{R} = \varprojlim R/\mathfrak{m}^n$ *be the completion of* R *with respect to the* $\mathfrak{m}$*-adic topology.*

(1) $\hat{R}$ *is a DVR with maximal ideal* $\hat{\mathfrak{m}} = \pi\hat{R}$.
(2) K *is equal to the localization* $R[\pi^{-1}]$.
(3) *The quotient field of* $\hat{R}$ *is* $\hat{K} = \hat{R} \otimes_R K$.
(4) $\hat{K}$ *is equal to the localization* $\hat{R}[\pi^{-1}]$.
(5) $\hat{R} \cap K = R$.
(6) *Given* $a \in \hat{R}$ *and* $p > 0$ *there exists* $b \in R$ *such that* $a - b \in \mathfrak{m}^p$.
(7) *Given* $a \in \hat{K}$ *and* $p > 0$ *there exists* $b \in K$ *such that* $a - b \in \hat{\mathfrak{m}}^p$.

Proof. This proof is left to the reader. □

2. The Class Group of Weil Divisors

Let R denote a noetherian normal integral domain with quotient field K. In this section, we define the class group of R as the group of Weil divisors modulo the subgroup of principal Weil divisors. In Section 6.4, we will show that the class group can be defined as the group of all reflexive fractional ideals of R in K modulo the subgroup of all principal fractional ideals. The two approaches lead to class groups that are isomorphic to each other. One of the most useful tools for computing class groups is the exact sequence of Nagata, which we derive in Theorem 6.2.4.

By $X_1(R)$ we denote the set of all prime ideals in $\operatorname{Spec} R$ of height one. By Theorems 6.1.5 and 6.1.14, R_P is a discrete valuation ring, for each $P \in X_1(R)$. By ν_P we denote the discrete valuation on K associated to R_P. In the terminology of [**Fos73**], Theorem 6.2.1 says that R is a *Krull domain.*

Theorem 6.2.1. *Let* R *be a noetherian normal integral domain with field of fractions* K. *Let* $\alpha \in K^*$.

(1) $\nu_P(\alpha) = 0$ *for all but finitely many* $P \in X_1(R)$.
(2) $\alpha \in R$ *if and only if* $\nu_P(\alpha) \geq 0$ *for all* $P \in X_1(R)$.
(3) $\alpha \in R^*$ *if and only if* $\nu_P(\alpha) = 0$ *for all* $P \in X_1(R)$.
(4) $R = \bigcap_{P \in X_1(R)} R_P$.

Proof. See, for example, [**ZS75b**, Corollary VI.10.3, p. 42]. □

Definition 6.2.2. Let R be a noetherian normal integral domain with field of fractions K. Let $X_1(R)$ be the subset of $\operatorname{Spec} R$ consisting of those prime ideals of height one. The free $\mathbb{Z}$-module on $X_1(R)$,

$$\operatorname{Div} R = \bigoplus_{P \in X_1(R)} \mathbb{Z}P$$

is called the *group of Weil divisors* of R. According to Theorem 6.2.1, there is a homomorphism of groups $\operatorname{Div} : K^* \to \operatorname{Div}(R)$ defined by

$$\operatorname{Div}(\alpha) = \sum_{P \in X_1(R)} \nu_P(\alpha)P$$

and the kernel of $\operatorname{Div}()$ is equal to the group R^*. The *class group* of R is defined to be the cokernel of $\operatorname{Div}()$, and is denoted $\operatorname{Cl}(R)$. The sequence

$$0 \to R^* \to K^* \xrightarrow{\operatorname{Div}} \operatorname{Div}(R) \to \operatorname{Cl}(R) \to 0$$

is exact. The image of $\operatorname{Div} : K^* \to \operatorname{Div} R$ is denoted $\operatorname{Prin} R$ and is called the group of *principal Weil divisors*. In other words, $\operatorname{Cl}(R)$ is the group of Weil divisors modulo the principal Weil divisors.

As we see in Theorem 6.2.3 and Nagata's Theorem, the class group is an arithmetic invariant of a ring that in a sense measures the failure of R to be a unique factorization domain.

Theorem 6.2.3. *Let R be a noetherian integral domain. The following are equivalent.*

(1) *R is a UFD.*
(2) *Every prime ideal in R of height one is principal.*

If R is also assumed to be normal, then (1) and (2) above are equivalent to the condition that $\operatorname{Cl}(R) = (0)$.

Proof. If R is normal and $\operatorname{Cl}(R) = (0)$, then using Exercise 6.2.7 we see that every prime ideal of height one is principal. Now we prove that (2) implies (1) and leave the rest to the reader. Suppose R has the property that every height one prime is principal. Let p be an irreducible element of R. By Exercise 6.2.6, it suffices to show that the principal ideal (p) is a prime ideal. Let P be a minimal prime over-ideal of (p). By Krull's Hauptidealsatz, $\operatorname{ht}(P) = 1$. By hypothesis, $P = (\pi)$ is principal. Then π divides p and since p is irreducible, it follows that π and p are associates. This implies $P = (p)$. □

The following version of Nagata's Theorem is especially useful in many computations. Its usage is illustrated below in Examples 6.2.5 and 6.5.12, and the reader is asked to apply it in some of the exercises that appear below. For a statement of the theorem in its more traditional form, see Exercise 6.2.8.

Theorem 6.2.4 (Nagata's Theorem). *Let R denote a noetherian normal integral domain with field of fractions K. Let f be a nonzero noninvertible element of R with divisor* $\operatorname{Div}(f) = \nu_1 P_1 + \cdots + \nu_n P_n$. *The sequence of abelian groups*

$$1 \to R^* \to R[f^{-1}]^* \xrightarrow{\operatorname{Div}} \bigoplus_{i=1}^{n} \mathbb{Z}P_i \to \operatorname{Cl}(R) \to \operatorname{Cl}(R[f^{-1}]) \to 0$$

is exact.

Proof. There is a tower of subgroups $R^* \subseteq R[f^{-1}]^* \subseteq K^*$. There exists a map α such that the diagram

$$\begin{array}{ccccccccc}
1 & \longrightarrow & R^* & \longrightarrow & K^* & \xrightarrow{\operatorname{Div}} & \operatorname{Prin} R & \longrightarrow & 0 \\
 & & \downarrow \delta & & \downarrow \epsilon & & \downarrow \alpha & & \\
1 & \longrightarrow & R[f^{-1}]^* & \longrightarrow & K^* & \xrightarrow{\operatorname{Div}} & \operatorname{Prin} R[f^{-1}] & \longrightarrow & 0
\end{array}$$

is commutative, where δ is set inclusion and ϵ is set equality. Clearly, α is onto. By the Snake Lemma (Theorem 1.4.1), $\operatorname{coker} \delta \cong \ker \alpha$. Hence

$$1 \to R^* \to R[f^{-1}]^* \to \ker \alpha \to 0 \tag{6.1}$$

is exact. Using Exercise 2.2.15, $X_1(R[f^{-1}])$ is the subset of $X_1(R)$ consisting of those primes of height one in R that do not contain f. We can view $\operatorname{Div}(R[f^{-1}])$ as the free $\mathbb{Z}$-submodule of $\operatorname{Div}(R)$ generated by primes in $X_1(R[f^{-1}])$. Let β be the projection map onto this subgroup defined by $P_1 \mapsto 0, \ldots, P_n \mapsto 0$. The diagram

$$\begin{array}{ccccccccc}
0 & \longrightarrow & \operatorname{Prin}(R) & \longrightarrow & \operatorname{Div}(R) & \longrightarrow & \operatorname{Cl}(R) & \longrightarrow & 0 \\
 & & \downarrow \alpha & & \downarrow \beta & & \downarrow \gamma & & \\
0 & \longrightarrow & \operatorname{Prin}(R[f^{-1}]) & \longrightarrow & \operatorname{Div}(R[f^{-1}]) & \longrightarrow & \operatorname{Cl}(R[f^{-1}]) & \longrightarrow & 0
\end{array}$$

commutes and the rows are exact. Since β is onto, so is γ. The group $\operatorname{Div} R$ is free on $X_1(R)$. The only height one primes that contain f are $P_1, \ldots, P_n$. Therefore, the kernel of β is the free subgroup $\mathbb{Z}P_1 \oplus \cdots \oplus \mathbb{Z}P_n$. By the Snake Lemma

$$0 \to \ker \alpha \to \ker \beta \to \ker \gamma \to 0 \tag{6.2}$$

is exact. Combining sequences (6.1) and (6.2) completes the proof. □

Example 6.2.5. Let k be a field with characteristic not equal to 3. Let

$$R = \frac{k[x, y, z]}{(z^3 - y(y - x)(x + 1))}.$$

The reader should verify that R is an integrally closed noetherian integral domain. This can be done using the method outlined in Exercise 6.2.9, or by an application of Theorem 9.4.4 below. Let K be the quotient field of R. In this example we compute the class group $\operatorname{Cl}(R)$ and the group of invertible elements, R^*. In Section 11.3.7 we will revisit this example. To compute the class group $\operatorname{Cl}(R)$, we first show that there exists a localization of R which is factorial. The transformation we use is based on the blowing-up of the maximal ideal (x, y, z). For more details on this powerful technique, the interested reader is referred to [**Har77**, pp. 28–29]. Start with the equation

$$z^3 - y(y-x)(x+1)) = 0 \tag{6.3}$$

in K. Divide both sides of (6.3) by x^3 and substitute $v = y/x$ and $w = z/x$ to get

$$w^3 - v(v-1)(1+x^{-1}) = 0. \tag{6.4}$$

Solve (6.4) for x to get

$$x = \frac{v^2 - v}{w^3 - v^2 + v}. \tag{6.5}$$

Now treat v and w as indeterminates and define

$$R = \frac{k[x,y,z]}{(z^3 - y(y-x)(x+1))} \xrightarrow{\phi} k[v,w][(w^3 - v^2 + v)^{-1}] \tag{6.6}$$

by $\phi(x) = (v^2 - v)(w^3 - v^2 + v)^{-1}$, $\phi(y) = v\phi(x)$, and $\phi(z) = w\phi(x)$. The reader should verify that ϕ is a well-defined k-algebra homomorphism and that if we adjoin $(xy(y-x))^{-1}$ to R and $(v^2 - v)^{-1}$ to the ring on the right hand side of (6.6), then

$$R[x^{-1}, y^{-1}, (y-x)^{-1}] \xrightarrow{\phi} k[v,w][v^{-1}, (v-1)^{-1}, (w^3 - v^2 + v)^{-1}] \tag{6.7}$$

is a k-algebra homomorphism which is onto. Since the domain and range of ϕ are both noetherian integral domains with Krull dimension two, ϕ is an isomorphism (Corollary 6.1.3). Since $k[v,w]$ is a unique factorization domain, it follows from Theorems 6.2.3 and 6.2.4 that the group of units in the ring on the right hand side of (6.7) decomposes into the internal direct product

$$k^* \times \langle v \rangle \times \langle v - 1 \rangle \times \langle w^3 - v^2 + v \rangle. \tag{6.8}$$

Using the isomorphism (6.7) we see that the group of units in $R[x^{-1}, y^{-1}, (y-x)^{-1}]$ is generated by k^*, x, y, $y - x$. Since $z^3 - y^2$ is irreducible, $R/(x) \cong k[y,z]/(z^3 - y^2)$ is an integral domain of Krull dimension one. We also have $R/(y,z) \cong k[x]$ and $R/(y-x,z) \cong k[x]$. From this it follows that

$$\begin{aligned} \mathfrak{p}_0 &= (x) \\ \mathfrak{p}_1 &= (y,z) \\ \mathfrak{p}_2 &= (y-x,z) \end{aligned} \tag{6.9}$$

are each height one prime ideals of R. Using the identity (6.3) we see that z is a local parameter for each of the two local rings: $R_{\mathfrak{p}_1}$ and $R_{\mathfrak{p}_2}$. From this we compute the divisors:

$$\begin{aligned} \operatorname{Div}(x) &= \mathfrak{p}_0 \\ \operatorname{Div}(y) &= 3\mathfrak{p}_1 \\ \operatorname{Div}(y-x) &= 3\mathfrak{p}_2. \end{aligned} \tag{6.10}$$

Since $R[x^{-1}, y^{-1}, (y-x)^{-1}]$ is factorial, the exact sequence of Theorem 6.2.4 becomes

$$1 \to R^* \to R[(xy(y-x))^{-1}]^* \xrightarrow{\operatorname{Div}} \bigoplus_{i=0}^{2} \mathbb{Z}\mathfrak{p}_i \to \operatorname{Cl}(R) \to 0. \tag{6.11}$$

From (6.11) and (6.10), it follows that $\operatorname{Cl}(R) \cong \mathbb{Z}/3 \oplus \mathbb{Z}/3$ and is generated by the prime divisors $\mathfrak{p}_1$ and $\mathfrak{p}_2$. We remark that from (6.8) and (6.11) it follows that $R^* = k^*$.

2.1. Exercises.

Exercise 6.2.6. Let R be a noetherian integral domain. Prove that R is a UFD if and only if for every irreducible element p, the principal ideal (p) is a prime ideal.

Exercise 6.2.7. Let R be a noetherian normal integral domain with quotient field K. Assume $P \in X_1(R)$, $\alpha \in K^*$, and $\operatorname{Div} \alpha = P$. Prove that $R\alpha = P$.

Exercise 6.2.8. In this exercise, the reader is asked to prove another form of Nagata's Theorem. Let R be a noetherian normal integral domain with field of fractions K. Let $W \subseteq R^*$ be a multiplicative set. Modify the proof of Theorem 6.2.4 to show that there is an epimorphism of groups $\gamma : \operatorname{Cl}(R) \to \operatorname{Cl}(W^{-1}R)$ and that the kernel of γ is generated by the classes of those prime divisors $P \in X_1(R) - X_1(W^{-1}R)$.

Exercise 6.2.9. Let k be a field and $n \geq 2$ an integer which is invertible in k. Let $f \in k[x,y,z]$ be the polynomial $z^n - xy$ and let R be the quotient $k[x,y,z]/(f)$. In R we prefer not to use special adornment for cosets. That is, write simply x, or z for the coset represented by that element. Prove the following.

(1) R is a noetherian integral domain and $\dim(R) = 2$.

(2) If $P = (x, z)$ is the ideal in R generated by x and z, then P is a prime ideal of height one.

(3) If $I = (x)$ is the principal ideal generated by x in R, then $\operatorname{Rad} I = P$.

(4) R_P is a DVR and z generates the maximal ideal $\mathfrak{m}_P$.

(5) $\nu_P(x) = n$ and $\operatorname{Div}(x) = nP$.

(6) $R[x^{-1}] \cong k[x, z][x^{-1}]$ and $R[y^{-1}] \cong k[y, z][y^{-1}]$. $R_\mathfrak{p}$ is regular if $\mathfrak{p} \in U(x) \cup U(y)$.

(7) The only prime ideal containing both x and y is the maximal ideal $\mathfrak{m} = (x, y, z)$, which has height 2. $\operatorname{depth}(R_\mathfrak{m}) = 2$. Apply Theorem 6.1.5 to show that R is integrally closed.

(8) $\operatorname{Cl}(R[x^{-1}]) = 0$. (Hint: $R[x^{-1}]$ is a UFD.)

(9) $\operatorname{Cl}(R)$ is cyclic of order n.

See Example 13.6.8 for the computation of $\operatorname{Pic}(R)$.

Exercise 6.2.10. Let $S = \mathbb{R}[x, y]/(f)$, where $f = x^2 + y^2 - 1$. This exercise is an outline of a proof that $\operatorname{Cl}(S)$, the class group of S, is cyclic of order two. This implies S is not a UFD.

(1) Let R be the $\mathbb{R}$-subalgebra of $S[x^{-1}]$ generated by yx^{-1} and x^{-1}. Show that $R = \mathbb{R}[yx^{-1}, x^{-1}]/(1 + (yx^{-1})^2 - (x^{-1})^2)$ is a PID.

(2) Show that $R[x] = S[1/x]$ is a PID.

(3) Let $P_1 = (x, y - 1)$ and $P_2 = (x, y + 1)$. Show that S_{P_1} and S_{P_2} are local principal ideal domains. Conclude that S is normal.

(4) Show that $\operatorname{Div}(x) = P_1 + P_2$ and $\operatorname{Div}(y - 1) = 2P_1$.

(5) Use Theorems 6.2.3 and 6.2.4 to prove that that $\operatorname{Cl}(S)$ is generated by P_1 and has order two.

Exercise 6.2.11. This exercise is a continuation of Exercise 6.2.9. Let k be a field and $n \geq 2$ an integer which is invertible in k. Let $f \in k[x, y, z]$ be the polynomial $z^n - xy$ and let R be the quotient $k[x, y, z]/(f)$. Let $\mathfrak{m}$ be the maximal ideal (x, y, z) in R, and $\hat{R}$ the $\mathfrak{m}$-adic completion of R.

(1) Show that $\hat{R} \cong k[[x, y]][z]/(f)$.

(2) Follow the procedure outlined in Exercise 6.2.9 to show that $\hat{R}$ is a noetherian normal integral domain and $\operatorname{Cl}(\hat{R})$ is a cyclic group of order n generated by the class of the prime ideal $P = (x, z)$.

In Algebraic Geometry, the ring R is the affine coordinate ring of the surface $X = Z(z^n - xy)$ in $\mathbb{A}^3_k$ and the point $p = (0, 0, 0)$ is called a singular point of X. It follows from [**Dur79**, A5] and [**Lip69**] that p is a rational double point of type A_{n-1}.

Exercise 6.2.12. Let k be a field such that $\operatorname{char} k \neq 2$. For the ring

$$R = \frac{k[x, y, z]}{(z^2 - (y^2 - x^2)(x + 1))}$$

follow the method of Example 6.2.5 to prove the following.

(1) $R[x^{-1}, (y^2 - x^2)^{-1}]$ is a UFD.

(2) The group of invertible elements in $R[x^{-1}, (y^2 - x^2)^{-1}]$ is generated by $x, y - x, y + x$.

(3) $\mathfrak{q}_1 = (x, z - y)$, $\mathfrak{q}_2 = (x, z + y)$, $\mathfrak{p}_1 = (y - x, z)$, $\mathfrak{p}_2 = (y + x, z)$, are height one prime ideals in R.

(4) $\operatorname{Div}(x) = \mathfrak{q}_1 + \mathfrak{q}_2$, $\operatorname{Div}(y - x) = 2\mathfrak{p}_1$, $\operatorname{Div}(y + x) = 2\mathfrak{p}_2$.

(5) $\operatorname{Cl}(R) \cong \mathbb{Z} \oplus \mathbb{Z}/2 \oplus \mathbb{Z}/2$.

For the computation of $\operatorname{Pic}(R)$, see Exercise 10.1.20.

Exercise 6.2.13. Let k be a field and $n > 1$ an integer that is invertible in k. Assume moreover that k contains a primitive nth root of unity, say ζ. Let $a_1, \dots, a_n$ be distinct elements of k. For $1 \leq i \leq n$, define linear polynomials $\ell_i(x, y) = y - a_i x$ in $k[x, y]$, and set $f(x, y) = \ell_1(x, y) \cdots \ell_n(x, y)$. For the ring

$$R = \frac{k[x, y, z]}{(z^n - f(x, y)(x + 1))}$$

follow the method of Example 6.2.5 to prove the following.

(1) $R[x^{-1}, f(x, y)^{-1}]$ is a UFD.

(2) The group of invertible elements in the ring $R[x^{-1}, f(x, y)^{-1}]$ is generated by $x, \ell_1, \dots, \ell_n$.

(3) Let $\mathfrak{q}_i = (x, z - \zeta^i y)$, for $i = 0, \dots, n - 1$. Let $\mathfrak{p}_j = (\ell_j, z)$, for $j = 1, \dots, n$. Then $\mathfrak{q}_0, \dots, \mathfrak{q}_{n-1}$, $\mathfrak{p}_1, \dots, \mathfrak{p}_n$ are height one prime ideals in R.

(4) $\operatorname{Div}(x) = \mathfrak{q}_0 + \cdots + \mathfrak{q}_{n-1}$, and $\operatorname{Div}(\ell_j) = n\mathfrak{p}_j$, for $j = 1, \dots, n$.

(5) $\operatorname{Cl}(R) \cong (\mathbb{Z})^{(n-1)} \oplus (\mathbb{Z}/n)^{(n)}$.

Notice that for $n = 2$, this agrees with computation carried out in Exercise 6.2.12. The ring R was the focus of the article [**FH13**] where many other interesting properties of R were studied.

Exercise 6.2.14. Let k be a field and $n > 2$ an integer that is invertible in k. Let $a_1, \dots, a_{n-1}$ be distinct elements of k. For $1 \leq i \leq n - 1$, define linear polynomials $\ell_i(x, y) = y - a_i x$ in $k[x, y]$, and set $f(x, y) = \ell_1(x, y) \cdots \ell_{n-1}(x, y)$. For the ring

$$R = \frac{k[x, y, z]}{(z^n - f(x, y)(x + 1))}$$

follow the method of Example 6.2.5 to prove the following.

(1) $R[x^{-1}, f(x,y)^{-1}]$ is a UFD.

(2) The group of invertible elements in the ring $R[x^{-1}, f(x,y)^{-1}]$ is generated by $x, \ell_1, \dots, \ell_{n-1}$.

(3) Let $\mathfrak{p}_0 = (x)$, and for $i = 1, \dots, n-1$, let $\mathfrak{p}_i = (\ell_i, z)$. Then $\mathfrak{p}_0, \dots, \mathfrak{p}_{n-1}$, are height one prime ideals in R.

(4) $\operatorname{Div}(x) = \mathfrak{p}_0$, and $\operatorname{Div}(\ell_j) = n\mathfrak{p}_j$, for $j = 1, \dots, n-1$.

(5) $\operatorname{Cl}(R) \cong (\mathbb{Z}/n)^{(n-1)}$.

Notice that for $n = 3$, this agrees with computation carried out in Example 6.2.5. For the computation of $\operatorname{Pic}(R)$, see Exercise 10.1.21.

3. Lattices

We begin with a quick study of the general properties of lattices. Our emphasis is on deriving the fundamental properties of reflexive lattices that will be applied in the remaining sections of this book. As general references for this section, we recommend [**Fos73**] and [**Bou89b**].

3.1. Definition and First Properties. Let R be an integral domain with field of fractions K. Let V be a finite dimensional K-vector space. An R-submodule M of V is said to be an *R-lattice in V* if M contains a basis for V and there is a finitely generated R-submodule N in V such that $M \subseteq N$. Hence, an R-lattice is a torsion free R-module, but not necessarily finitely generated. The *rank* of M is defined to be $\dim_K V$. Listed in Example 6.3.1 and Proposition 6.3.2 are a few common examples of R-lattices that we have already encountered.

Example 6.3.1. Let R be an integral domain with field of fractions K.

(1) If M is a finitely generated torsion free R-module, the natural mapping $R \otimes_R M \to K \otimes_R M$ is one-to-one. Then M is isomorphic to the submodule $1 \otimes M$ of $K \otimes_R M$. Writing KM instead of $K \otimes_R M$, we see that M is isomorphic to a finitely generated R-lattice in KM.

(2) Let R be a noetherian integral domain and M and N finitely generated R-modules such that N is torsion free. Then $\operatorname{Hom}_R(M, N)$ is a finitely generated torsion free R-module (Exercise 2.5.11). By Proposition 2.4.10, $\operatorname{Hom}_R(M, N)$ embeds as an R-lattice in $K \otimes_R \operatorname{Hom}_R(M, N) = \operatorname{Hom}_K(K \otimes_R M, KN)$. This is a special case of Proposition 6.3.5 (3).

(3) Assume R is integrally closed in K, L/K is a finite separable field extension, and S is the integral closure of R in L. By Theorem 4.6.10, S is an R-lattice in L.

Proposition 6.3.2. *Let R be an integral domain with field of fractions K. Let V be a finite dimensional K-vector space and M an R-lattice in V.*

(1) *If R is noetherian, then M is a finitely presented R-module.*

(2) *If R is a principal ideal domain, then M is a finitely generated free R-module.*

Proof. (1): Since M is a submodule of a finitely generated module, this follows from Corollary 2.5.2.

(2): By (1) M is finitely generated torsion free. Hence M is free (see, for example, [**DF04**, Chapter 12]). □

Proposition 6.3.3. *Let R be an integral domain with field of fractions K and V a finite dimensional K-vector space. The following are equivalent for an R-submodule M of V.*

(1) *M is an R-lattice in V. That is, there is a finitely generated R-submodule N of V such that $M \subseteq N$, and if KM denotes the K-subspace of V spanned by M, then $KM = V$.*

(2) *There is a free R-submodule F in V with $\operatorname{Rank}_R(F) = \dim_K(V)$ and a nonzero element $r \in R$ such that $rF \subseteq M \subseteq F$.*

(3) *There are free R-submodules F_1, F_2 in V such that $F_1 \subseteq M \subseteq F_2$ and $\operatorname{Rank}_R(F_1) = \operatorname{Rank}_R(F_2) = \dim_K(V)$.*

(4) *There is a chain of R-submodules $L \subseteq M \subseteq N$ where $KL = V$ and N is finitely generated.*

(5) *Given any free R-submodule F in V with $\operatorname{Rank}_R(F) = \dim_K(V)$, there are nonzero elements $r, s \in R$ such that $rF \subseteq M \subseteq s^{-1}F$.*

Proof. Assume $\dim_K(V) = n$. We prove that (4) implies (5). The rest is left to the reader. Assume we are given $F = Ru_1 \oplus \cdots \oplus Ru_n$ a free R-submodule of V. Also, let $L \subseteq M \subseteq N$, where $KL = V$ and N is a finitely generated R-submodule of V. Since $KL = V$ we can pick a K-basis for V in L, say $\{\lambda_1, \dots, \lambda_n\}$. For each j there are $k_{j,i} \in K$ such that $u_j = \sum_{i=1}^n k_{j,i}\lambda_i$. Pick a nonzero $r \in R$ such that $rk_{j,i} \in R$ for all pairs j, i. Then $ru_j = \sum_{i=1}^n rk_{j,i}\lambda_i \in \sum_i R\lambda_i \subseteq L$, hence $rF = \sum_j Rru_j \subseteq L \subseteq M$. Let $\nu_1, \dots, \nu_t$ be a generating set for N. For each j there are $\kappa_{j,i} \in K$ such that $\nu_j = \sum_{i=1}^n \kappa_{j,i}u_i$. Pick a nonzero $s \in R$ such that $s\kappa_{j,i} \in R$ for all pairs j, i. Then $s\nu_j = \sum_{i=1}^n s\kappa_{j,i}u_i \in \sum_{i=1}^n Ru_i = F$. Therefore, $M \subseteq N = \sum_{j=1}^t R\nu_j \subseteq s^{-1}F$. □

Proposition 6.3.4. *Let R be an integral domain and K the field of fractions of R. In the following, U, V, $V_1, \dots, V_r$, W denote finite dimensional K-vector spaces.*

(1) *If M and N are R-lattices in V, then $M+N$ and $M\cap N$ are R-lattices in V.*

(2) *If U is a K-subspace of V, and M is an R-lattice in V, then $M\cap U$ is an R-lattice in U.*

(3) *Let $M_1,\dots,M_m$ be R-lattices in $V_1,\dots,V_m$ respectively. If $\phi : V_1\times\cdots\times V_m\to U$ is a multilinear form, then the R-module generated by $\phi(M_1\times\cdots\times M_m)$ is an R-lattice in the subspace spanned by $\phi(V_1\times\cdots\times V_m)$.*

(4) *Let L/K be an extension of fields. Let S be an R-subalgebra of L such that L is the field of fractions of S. If M is an R-lattice in V, then the image of $S\otimes_R M\to L\otimes_K V$ is an S-lattice in $L\otimes_K V$.*

Proof. (1): We apply Proposition 6.3.3 (5). Let F be a free R-submodule of V with rank $n=\dim_K(V)$. There exist nonzero elements a,b,c,d in R such that $aF\subseteq M$, $bF\subseteq N$, $M\subseteq c^{-1}F$, $N\subseteq d^{-1}F$. Then $(ab)F\subseteq M\cap N\subseteq M+N\subseteq (cd)^{-1}F$.

(2): Start with a K-basis $u_1,\dots,u_m$ for U and extend to get a K-basis $u_1,\dots,u_m,\dots,u_r$ for V. Let $E=Ru_1\oplus\cdots\oplus Ru_m$ and $F=Ru_1\oplus\cdots\oplus Ru_n$. Then $E=F\cap U$. Also, for any $\alpha\in K$, $(\alpha F)\cap U=(\sum_{i=1}^n R\alpha u_i)\cap U=\sum_{i=1}^m R\alpha u_i=\alpha E$. We apply Proposition 6.3.3 (5). Let r,s be nonzero elements in R such that $rF\subseteq M\subseteq s^{-1}F$. Then $rE\subseteq M\cap U\subseteq s^{-1}E$.

(3): For each j, M_j contains a K-spanning set for V_j. From this is follows that $\phi(M_1\times\cdots\times M_m)$ contains a spanning set for the subspace of U spanned by $\phi(V_1\times\cdots\times V_m)$. For each j, let N_j be a finitely generated R-submodule of V_j containing M_j. Then $\phi(N_1\times\cdots\times N_m)$ is contained in a finitely generated R-submodule of U.

(4): Since $K\otimes_R M=K\otimes_R V=V$, we have $L\otimes_S S\otimes_R M=L\otimes_K K\otimes_R M=L\otimes_K V$. If $M\subseteq N\subseteq V$ with N a finitely generated R-module, then the diagram of S-module homomorphisms

$$\begin{array}{ccc} S\otimes_R M & \longrightarrow & L\otimes_K V \\ & \searrow \quad \nearrow & \\ & S\otimes_R N & \end{array}$$

commutes. Therefore, the image of $S\otimes_R M$ in $L\otimes_K V$ is contained in the image of $S\otimes_R N$ which is a finitely generated S-module. $\square$

Proposition 6.3.5. *Let R be an integral domain and K the field of fractions of R. Let V and W be finite dimensional K-vector spaces. In the following, M_0,M_1,M denote R-lattices in V and N_0,N_1,N denote R-lattices in W. Using the module quotient notation, $N:M$ is defined to be*

$$N:M=\{f\in\operatorname{Hom}_K(V,W)\mid f(M)\subseteq N\}.$$

Then the following are true.

(1) *If* $M_0 \subseteq M_1$ *and* $N_0 \subseteq N_1$, *then* $N_0 : M_1 \subseteq N_1 : M_0$.

(2) *The restriction mapping* $\rho : (N : M) \to \operatorname{Hom}_R(M, N)$ *is an isomorphism of* R*-modules.*

(3) $N : M$ *is an* R*-lattice in* $\operatorname{Hom}_K(V, W)$.

(4) *Let* $Z \subseteq R - \{0\}$ *be a multiplicative set and* $Z^{-1}R$ *the localization of* R *in* K. *Then* $Z^{-1}(N : M) = Z^{-1}N : Z^{-1}M$.

Proof. (1): This proof is left to the reader.

(2): The reader should verify that restriction defines an R-module homomorphism $\rho : (N : M) \to \operatorname{Hom}_R(M, N)$. Because M contains a K-basis for V, ρ is one-to-one. Because M and N are torsion free R-modules, the maps $M \to K \otimes_R M = KM$ and $N \to K \otimes_R N = KN$ are one-to-one. If $\theta \in \operatorname{Hom}_R(M, N)$, then the diagram

$$\begin{array}{ccc} M & \xrightarrow{\theta} & N \\ \downarrow & & \downarrow \\ K \otimes_R M = V & \xrightarrow{1\otimes\theta} & K \otimes_R N = W \end{array}$$

commutes. Therefore, $1 \otimes \theta : V \to W$ is an extension of θ and belongs to $N : M$. In other words, θ is in the image of ρ.

(3): Let $E_0 \subseteq M \subseteq E_1$ be R-lattices in V with E_0 and E_1 free. Let $F_0 \subseteq N \subseteq F_1$ be R-lattices in W with F_0 and F_1 free. By (1), $F_0 : E_1 \subseteq N : M \subseteq F_1 : E_0$. By Proposition 6.3.3 (4), it suffices to prove (4) when M and N are free R-lattices. In this case, $\operatorname{Hom}_R(M, N)$ is free over R and $\operatorname{Hom}_R(M, N) \to K \otimes_R \operatorname{Hom}_R(M, N)$ is one-to-one. By Corollary 1.3.27, the assignment $\theta \mapsto 1 \otimes \theta$ embeds $\operatorname{Hom}_R(M, N)$ as an R-submodule of $\operatorname{Hom}_K(KM, KN) = \operatorname{Hom}_K(V, W)$. By (2), the image of $\operatorname{Hom}_R(M, N)$ under this embedding is equal to $N : M$. This proves $N : M$ is an R-lattice in $\operatorname{Hom}_K(V, W)$, when M and N are free R-lattices.

(4): If $f \in (N : M)$ and $z \in Z$, then $f(z^{-1}x) = z^{-1}f(x) \in z^{-1}N$ for all $x \in M$. Conversely, suppose $f \in Z^{-1}N : Z^{-1}M$. Let $y_1, \dots, y_n$ be a generating set for M. There exists $z \in Z$ such that $f(x_i) \in z^{-1}N$ for $1 \le i \le n$. Therefore, $zf \in N : M$. □

3.2. Reflexive Lattices. In the context of Proposition 6.3.5, we identify $R : M$ with the dual module $M^* = \operatorname{Hom}_R(M, R)$. By Exercise 1.3.33 the assignment $m \mapsto \varphi_m$ is an R-module homomorphism $M \to M^{**} = R : (R : M)$, where φ_m is the "evaluation at m" homomorphism. That is,

$\varphi_m(f) = f(m)$. The diagram

$$\begin{array}{ccc} M & \longrightarrow & M^{**} = R:(R:M) \\ \downarrow & & \downarrow \\ V & \longrightarrow & V^{**} \end{array} \tag{6.12}$$

commutes. By Exercise 1.3.34, the bottom horizontal arrow is an isomorphism. Since the vertical maps are one-to-one, the top horizontal arrow is one-to-one. We say M is a *reflexive R-lattice* in case $M \to R : (R : M)$ is onto. For instance, a finitely generated projective R-lattice is reflexive (Exercise 1.3.34). If M is an R-lattice, then Lemma 6.3.6 shows that $R : M$, the dual of M, is reflexive.

Lemma 6.3.6. *Let R be an integral domain with field of fractions K. Let V be a finite dimensional K-vector space and M an R-lattice in V. Then $R : M = R : (R : (R : M))$, or equivalently, $R : M$ is a reflexive R-lattice in V^*.*

Proof. By Proposition 6.3.5 (1) applied to $M \subseteq R : (R : M)$, we get the set inclusion $R : M \supseteq R : (R : (R : M))$. The reverse inclusion follows from the commutative diagram (6.12). □

Proposition 6.3.7. *Let R be an integral domain with field of fractions K. Let V be a finite dimensional K-vector space and M an R-lattice in V. Let $M \subseteq F \subseteq V$, where F is a free R-lattice (Proposition 6.3.3). Then M is a reflexive R-lattice if and only if*

$$M = \bigcap_{\alpha \in (R:M)} \left(\alpha^{-1}(R) \cap F\right).$$

Proof. It suffices to prove

$$R : (R : M) = \bigcap_{\alpha \in (R:M)} \left(\alpha^{-1}(R) \cap F\right). \tag{6.13}$$

Let $v \in V$ and assume v is in the right hand side of (6.13). Then $v \in R : (R : M)$ if and only if $\alpha(v) \in R$, for all $\alpha \in R : M$. Notice that if $\alpha \in R : M$, then $\alpha \in R : (\alpha^{-1}(R) \cap F)$. Therefore, $\alpha(v) \in R$, which shows $v \in R : (R : M)$.

For the reverse inclusion, let $\alpha \in R : M$. Then $\alpha(M) \subseteq R$, hence $M \subseteq \alpha^{-1}(R) \cap F \subseteq F$. By Proposition 6.3.3 (4), this implies $\alpha^{-1}(R) \cap F$ is an R-lattice in V. Let $v \in R : (R : (\alpha^{-1}(R) \cap F))$. Under the identification $V = V^{**}$, we identify v with a vector in V. As mentioned above, $\alpha \in R : (\alpha^{-1}(R) \cap F)$, $\alpha(v) \in R$, hence $v \in \alpha^{-1}(R)$. Since F is free, F is reflexive (Exercise 1.3.34) and we see that $R : (R : (\alpha^{-1}(R) \cap F)) \subseteq R : (R : F) = F$. Combined, this shows $R : (R : (\alpha^{-1}(R) \cap F)) \subseteq \alpha^{-1}(R) \cap F$. That is,

$\alpha^{-1}(R) \cap F$ is reflexive. This shows $R : (R : M) \subseteq \alpha^{-1}(R) \cap F$ for each α. In (6.13), the left hand side is a subset of the right hand side. □

Let R be an integral domain with field of fractions K. Let U, V, W be finite dimensional K-vector spaces. Let

$$\operatorname{Hom}_K(V, W) \otimes_K U \xrightarrow{\alpha} \operatorname{Hom}_K(\operatorname{Hom}_K(U, V), W)$$

be the isomorphism of Lemma 1.3.25 which is defined by $\alpha(f \otimes a)(h) = f(h(a))$. Let

$$\operatorname{Hom}_K(U \otimes_K V, W) \xrightarrow{\phi} \operatorname{Hom}_K(U, \operatorname{Hom}_K(V, W))$$

be the adjoint isomorphism of Theorem 1.3.24 which is defined by $\phi(\theta)(u) = \theta(u \otimes \cdot)$.

Lemma 6.3.8. *In the above context, let L, M, N be R-lattices in U, V, W respectively.*

(1) *Let $(N : M)L$ denote the image of the function $(N : M) \otimes_R L \to \operatorname{Hom}_K(V, W) \otimes_K U$. Then $\alpha\left((N : M)L\right) \subseteq N : (M : L)$.*
(2) *Let LM denote the image of the function $L \otimes_R M \to U \otimes_K V$. Then $\phi\left(N : LM\right) \subseteq (N : M) : L$, and $\phi^{-1}\left((N : M) : L\right) \subseteq N : LM$.*

Proof. (1): Let $f \in N : M$, $\ell \in L$, $h \in M : L$. Then $\alpha(f \otimes \ell)(h) = f(h(\ell)) \in N$.

(2): Suppose $\theta \in \operatorname{Hom}_K(U \otimes_K V, W)$ and $\theta(LM) \subseteq N$. For all $m \in M$ and $\ell \in L$, $\phi(\theta)(\ell)(m) = \theta(\ell \otimes m) \in N$. Therefore, $\phi(\theta)(L) \subseteq N : M$, hence $\phi(\theta) \in (N : M) : L$. For the second part, suppose $\phi(\theta)(\ell) \in N : M$ for all $\ell \in L$. Then $\phi(\theta)(\ell)(m) = \theta(\ell \otimes m) \in N$, and $\theta \in N : LM$. □

Proposition 6.3.9. *Let R be an integral domain with field of fractions K. Let N be an R-lattice in the finite dimensional K-vector space W. Let M be a reflexive R-lattice in the finite dimensional K-vector space V. Then $M : N$ is a reflexive R-lattice in $\operatorname{Hom}_K(W, V)$.*

Proof. In this context,

$$\operatorname{Hom}_K(W, V) \xrightarrow{\alpha^*} \operatorname{Hom}_K(W \otimes_K V^*, K) \xrightarrow{\phi} \operatorname{Hom}_K(W, V)$$

is the identity map. Under this identification, ϕ is the inverse of the dual of α. By Lemma 6.3.8 (2),

$$\phi(R : (R : M)N) \subseteq (R : (R : M)) : N = M : N$$

where the last equality is because M is reflexive. By Lemma 6.3.8 (1),

$$\alpha((R : M)N) \subseteq R : (M : N)$$

taking duals,

$$R : (R : (M : N)) \subseteq R : \alpha((R : M)N).$$

By the identification mentioned above, $R : (R : (M : N)) \subseteq M : N$. $\square$

By $X_1(R)$ we denote the set of all prime ideals in $\operatorname{Spec} R$ of height one.

Theorem 6.3.10. *Let R be a noetherian integrally closed integral domain with field of fractions K. Let V be a finite dimensional K-vector space and M an R-lattice in V.*

(1) *If L is another R-lattice in V, then $L_{\mathfrak{p}} = M_{\mathfrak{p}}$ for all but finitely many $\mathfrak{p} \in X_1(R)$.*

(2) *Suppose for each $\mathfrak{p} \in X_1(R)$ that $N(\mathfrak{p})$ is an $R_{\mathfrak{p}}$-lattice in V such that $N(\mathfrak{p}) = M_{\mathfrak{p}}$ for all but finitely many $\mathfrak{p} \in X_1(R)$. For $N = \bigcap_{\mathfrak{p}\in X_1(R)} N(\mathfrak{p})$, the following are true.*
   (a) *N is an R-lattice in V.*
   (b) *$N_{\mathfrak{p}} = N(\mathfrak{p})$ for all $\mathfrak{p} \in X_1(R)$.*
   (c) *If N' is an R-lattice in V such that $N'_{\mathfrak{p}} = N(\mathfrak{p})$ for all $\mathfrak{p} \in X_1(R)$, then $N' \subseteq N$.*

Proof. (1): Using Proposition 6.3.3, the reader should verify that there exist $r, s \in R$ such that $rM \subseteq L \subseteq s^{-1}M$. Let $\mathfrak{p} \in X_1(R)$ such that $\nu_{\mathfrak{p}}(r) = \nu_{\mathfrak{p}}(s) = 0$. Then $rM \otimes_R R_{\mathfrak{p}} = s^{-1}M \otimes_R R_{\mathfrak{p}}$. By Theorem 6.2.1, this proves (1).

(2): For each $\mathfrak{p} \in X_1(R)$, $R_{\mathfrak{p}}$ is a discrete valuation ring. By Proposition 6.3.2, $N(\mathfrak{p})$ is a finitely generated free $R_{\mathfrak{p}}$-module.

(a): Let F be a free R-lattice in V. By (1), $M_{\mathfrak{p}} = F_{\mathfrak{p}}$ for all but finitely many $\mathfrak{p} \in X_1(R)$. Let $\mathfrak{q}_1, \dots, \mathfrak{q}_t$ denote those height one primes in $X_1(R)$ where $F_{\mathfrak{q}_j} \neq N(\mathfrak{q}_j)$. Let $u_1, \dots, u_n$ be a free R-basis for F. Let $\{v_{j,1}, \dots, v_{j,n}\}$ be a free $R_{\mathfrak{q}_j}$-basis for $N(\mathfrak{q}_j)$. There are elements $\kappa_{k,j,i}$ in K such that $u_k = \sum_{i=1}^n \kappa_{k,j,i} v_{j,i}$. For some $r \in R - (0)$, $ru_k \in \sum_{i=1}^n Rv_{j,i} \subseteq N(\mathfrak{q}_j)$ for all k, j. For $1 \le j \le t$ this implies $rF \subseteq N(\mathfrak{q}_j)$. Also, if $F_{\mathfrak{p}} = N(\mathfrak{p})$, then $rF \subseteq rF_{\mathfrak{p}} = rN(\mathfrak{p}) \subseteq N(\mathfrak{p})$. Therefore, $rF \subseteq N = \bigcap_{\mathfrak{p}\in X_1(R)} N(\mathfrak{p})$.

There are elements $\lambda_{k,j,i}$ in K such that $v_{j,i} = \sum_{k=1}^n \lambda_{k,j,i} u_k$. For some $s \in R - (0)$, $sv_{j,i} \in \sum_{k=1}^n Ru_k = F$ for all j, i. This implies $sN(\mathfrak{q}_j) \subseteq F_{\mathfrak{q}_j}$, hence $N(\mathfrak{q}_j) \subseteq (s^{-1}F)_{\mathfrak{q}_j}$ for all j. Also, if $N(\mathfrak{p}) = F_{\mathfrak{p}}$, then $sN(\mathfrak{p}) \subseteq N(\mathfrak{p}) = F_{\mathfrak{p}}$, hence $N(\mathfrak{p}) \subseteq (s^{-1}F)_{\mathfrak{p}}$. If necessary, replace F with $s^{-1}F$, and assume $N(\mathfrak{p}) \subseteq F_{\mathfrak{p}}$ for all $\mathfrak{p} \in X_1(R)$. By taking direct sums in Theorem 6.2.1 (4) we see that $F = \bigcap_{\mathfrak{p}\in X_1(R)} F_{\mathfrak{p}}$. Then $N = \bigcap_{\mathfrak{p}\in X_1(R)} N(\mathfrak{p}) \subseteq F$. By Proposition 6.3.3, N is an R-lattice in V.

(b): By the last part of the proof of Part (a), $N(\mathfrak{p}) \subseteq F_{\mathfrak{p}}$ for all $\mathfrak{p} \in X_1(R)$ with equality for all but finitely many $\mathfrak{p} \in X_1(R)$. Let $\mathfrak{p}_1, \dots, \mathfrak{p}_w$ be those height one primes in $X_1(R)$ where $F_{\mathfrak{p}_i} \neq N(\mathfrak{p}_i)$. (Note: we do not assume this list is equal to $\mathfrak{q}_1, \dots, \mathfrak{q}_t$.) Then

$$\begin{aligned} N &= \bigcap_{\mathfrak{p} \in X_1(R)} N(\mathfrak{p}) \\ &= N(\mathfrak{p}_1) \cap \cdots \cap N(\mathfrak{p}_w) \cap \Big(\bigcap_{\mathfrak{p} \in X_1(R)} F_{\mathfrak{p}} \Big) \\ &= N(\mathfrak{p}_1) \cap \cdots \cap N(\mathfrak{p}_w) \cap F. \end{aligned}$$

It follows from the definition of localization that

$$N_{\mathfrak{p}} = N(\mathfrak{p}_1)_{\mathfrak{p}} \cap \cdots \cap N(\mathfrak{p}_w)_{\mathfrak{p}} \cap F_{\mathfrak{p}}.$$

If $\mathfrak{p}$ is not one of $\mathfrak{p}_1, \dots, \mathfrak{p}_w$, then by Lemma 6.3.11, $N(\mathfrak{p}_j)_{\mathfrak{p}} = KN(\mathfrak{p}_j) = V$, for $1 \leq j \leq w$. In this case, $N_{\mathfrak{p}} = F_{\mathfrak{p}} = N(\mathfrak{p})$. On the other hand, if $i \neq j$, then $N(\mathfrak{p}_i)_{\mathfrak{p}_j} = KN(\mathfrak{p}_i) = V$. Thus $N_{\mathfrak{p}_j} = N(\mathfrak{p}_j)_{\mathfrak{p}_j} \cap F_{\mathfrak{p}_j}$. But $N(\mathfrak{p}_j)_{\mathfrak{p}_j} = N(\mathfrak{p}_j) \subseteq F_{\mathfrak{p}_j}$, so $N_{\mathfrak{p}_j} = N(\mathfrak{p}_j)$ for $1 \leq j \leq w$.

(c): Suppose N' is an R-lattice in V such that $N'_{\mathfrak{p}} = N(\mathfrak{p})$ for all $\mathfrak{p} \in X_1(R)$. Then $N' \subseteq \bigcap_{\mathfrak{p} \in X_1(R)} N'_{\mathfrak{p}} = \bigcap_{\mathfrak{p} \in X_1(R)} N(\mathfrak{p}) = N$. □

Lemma 6.3.11. *Let R be an integral domain with field of fractions K. Let $\mathfrak{p}$, $\mathfrak{q}$ be prime ideals in R with $\mathfrak{p} \not\subseteq \mathfrak{q}$. Assume $R_{\mathfrak{p}}$ is a discrete valuation ring. Then the following are true.*

(1) $(R_{\mathfrak{p}})_{\mathfrak{q}} = K$.

(2) *If M is an $R_{\mathfrak{p}}$-module, then $M_{\mathfrak{q}} = M \otimes_R R_{\mathfrak{q}} = M \otimes_{R_{\mathfrak{p}}} K$.*

Proof. Let $a \in \mathfrak{p} - \mathfrak{q}$. Then $a \in \mathfrak{p}R_{\mathfrak{p}}$ and $a^{-1} \in R_{\mathfrak{q}}$, so the only maximal ideal in $(R_{\mathfrak{p}})_{\mathfrak{q}}$ is the zero ideal. □

Lemma 6.3.12. *Let R be an integrally closed integral domain with field of fractions K. Let V be a finite dimensional K-vector space and M an R-lattice in V. Then the following are true.*

(1) $R : M = \bigcap_{\mathfrak{p} \in X_1(R)} R_{\mathfrak{p}} : M_{\mathfrak{p}}$.

(2) *For any $\mathfrak{p} \in X_1(R)$, $(R : M)_{\mathfrak{p}} = R_{\mathfrak{p}} : M_{\mathfrak{p}}$.*

Proof. Let $F \subseteq M$ be a free R-lattice. For every $\mathfrak{p} \in X_1(R)$, the diagram

$$\begin{array}{ccc} (R : M)_{\mathfrak{p}} & \xrightarrow{\alpha} & (R : F)_{\mathfrak{p}} \\ {\scriptstyle\beta}\downarrow & & \downarrow{\scriptstyle\gamma} \\ R_{\mathfrak{p}} : M_{\mathfrak{p}} & \xrightarrow{\delta} & R_{\mathfrak{p}} : F_{\mathfrak{p}} \end{array}$$

commutes where β and γ are the natural maps induced by change of base. Since F is free, γ is an isomorphism (Corollary 1.3.27). It follows from

Proposition 6.3.5 (1) that α and δ are one-to-one. We have

$$R : M \subseteq \bigcap_{\mathfrak{p} \in X_1(R)} (R : M)_{\mathfrak{p}} \subseteq \bigcap_{\mathfrak{p} \in X_1(R)} R_{\mathfrak{p}} : M_{\mathfrak{p}}$$

where the intersection takes place in $V^* = K : V$. Let $f \in \bigcap_{\mathfrak{p} \in X_1(R)} R_{\mathfrak{p}} : M_{\mathfrak{p}}$. Then for every $\mathfrak{p} \in X_1(R)$, $f(M) \subseteq f(M_{\mathfrak{p}}) \subseteq R_{\mathfrak{p}}$. Then $f(M) \subseteq R = \bigcap_{\mathfrak{p} \in X_1(R)} R_{\mathfrak{p}}$, hence $f \in R : M$. This proves (1). Part (2) follows from Theorem 6.3.10 (2) and Part (1). □

Theorem 6.3.13. *Let R be a noetherian integrally closed integral domain with field of fractions K. Let V be a finite dimensional K-vector space and M an R-lattice in V. If we set $\tilde{M} = \bigcap_{\mathfrak{p} \in X_1(R)} M_{\mathfrak{p}}$, then the following are true.*

(1) *$R : (R : M) = \tilde{M}$.*
(2) *M is a reflexive R-lattice if and only if $M = \tilde{M}$.*
(3) *For each $\mathfrak{p} \in X_1(R)$, $\tilde{M}_{\mathfrak{p}} = M_{\mathfrak{p}}$.*
(4) *$\tilde{M}$ is a reflexive R-lattice in V containing M.*

Proof. (1): Each $M_{\mathfrak{p}}$ is a free $R_{\mathfrak{p}}$-lattice, so by Lemma 6.3.12,

$$\begin{aligned} R : (R : M) &= \bigcap_{\mathfrak{p} \in X_1(R)} R_{\mathfrak{p}} : (R : M)_{\mathfrak{p}} \\ &= \bigcap_{\mathfrak{p} \in X_1(R)} R_{\mathfrak{p}} : (R_{\mathfrak{p}} : M_{\mathfrak{p}}) \\ &= \bigcap_{\mathfrak{p} \in X_1(R)} M_{\mathfrak{p}} \\ &= \tilde{M}. \end{aligned}$$

The rest is left to the reader. □

Corollary 6.3.14. *Let R be a noetherian integrally closed integral domain with field of fractions K and let V be a finite dimensional K-vector space. Let M and N be two R-lattices in V such that N is reflexive. In order for $M \subseteq N$ it is necessary and sufficient that $M_{\mathfrak{p}} \subseteq N_{\mathfrak{p}}$ for all $\mathfrak{p} \in X_1(R)$.*

Proof. If $M \subseteq N$, then $M_{\mathfrak{p}} \subseteq N_{\mathfrak{p}}$ for all $\mathfrak{p} \in \operatorname{Spec}(R)$. Conversely, we have

$$M \subseteq R : (R : M) = \bigcap_{\mathfrak{p} \in X_1(R)} M_{\mathfrak{p}} \subseteq \bigcap_{\mathfrak{p} \in X_1(R)} N_{\mathfrak{p}} = R : (R : N) = N.$$

□

The following proposition of Auslander and Goldman ([**AG60b**]) will be applied in Chapter 11.

Proposition 6.3.15. *Let R be a noetherian integrally closed integral domain. Let M and N be finitely generated torsion free R-modules. Then there are R-module isomorphisms*

$$\operatorname{Hom}_R(M,N)^{**} \cong (N^* \otimes_R M)^* \cong \operatorname{Hom}_R(M,N^{**}) \cong \operatorname{Hom}_R(N^*,M^*)$$

where we write $(\cdot)^*$ for the dual $\operatorname{Hom}_R(\cdot,R)$. In particular,

$$\operatorname{Hom}_R(M,M)^{**} \cong \operatorname{Hom}_R(M^*,M^*) \cong \operatorname{Hom}_R(M^{**},M^{**}).$$

Proof. The homomorphism

$$N^* \otimes_R M \xrightarrow{\alpha} \operatorname{Hom}_R(M,N)^*$$

of Lemma 1.3.25 is defined by $\alpha(f \otimes x)(g) = f(g(x))$. The dual of α is

$$\operatorname{Hom}_R(M,N)^{**} \xrightarrow{\alpha^*} (N^* \otimes_R M)^*.$$

For each $\mathfrak{p} \in X_1(R)$, $R_{\mathfrak{p}}$ is a DVR and $M_{\mathfrak{p}}$ is a free $R_{\mathfrak{p}}$-module (Proposition 6.3.2). By Proposition 2.4.10 and Lemma 1.3.25,

$$N^* \otimes_R M \otimes_R R_{\mathfrak{p}} \xrightarrow{\alpha\otimes 1} \operatorname{Hom}_R(M,N)^* \otimes_R R_{\mathfrak{p}}$$

is an isomorphism. Taking duals and applying the same argument,

$$\operatorname{Hom}_R(M,N)^{**} \otimes_R R_{\mathfrak{p}} \xrightarrow{\alpha^*\otimes 1} (N^* \otimes_R M)^* \otimes_R R_{\mathfrak{p}}$$

is also an isomorphism. By Theorem 6.3.13, $\operatorname{Hom}_R(M,N)^{**}$ is a reflexive R-lattice. Without explicitly doing so, we view all of the modules as lattices in suitable vector spaces over the field of fractions of R. Applying Corollary 6.3.14, we see that α^* is an isomorphism. The second and third isomorphisms follow from the first and the adjoint isomorphisms of Theorem 1.3.24.

By the first part, $\operatorname{Hom}_R(M,M)^{**} \cong \operatorname{Hom}_R(M^*,M^*)$. Then

$$\begin{aligned}\operatorname{Hom}_R(M,M)^{**} &\cong (\operatorname{Hom}_R(M,M)^{**})^{**}\\ &\cong \operatorname{Hom}_R(M^*,M^*)^{**}\\ &\cong \operatorname{Hom}_R(M^{**},M^{**}).\end{aligned}$$

□

3.3. A Local to Global Theorem for Reflexive Lattices. Constructing examples of reflexive lattices that are not direct sums of rank one submodules and not projective is generally a difficult task. For examples of reflexive lattices that decompose into direct sums of rank one submodules, see Exercises 6.3.22 and 6.3.23. Theorem 6.3.16 provides a globalization method for constructing reflexive lattices from locally defined projective lattices. A version of Theorem 6.3.16 for sheaves of modules on a ringed space was proved by B. Auslander in [**Aus66**, Theorem VI.5]. A partial converse is [**Aus66**, Theorem VI.6]. In the language of schemes, it says that if U is an open subset of $\operatorname{Spec} R$ which contains $X_1(R)$, and M is a sheaf of $\mathcal{O}_U$-modules which is locally projective of finite rank, then M comes from a finitely generated reflexive R-module N. For an application of this method, see Proposition 11.3.26.

Before stating Theorem 6.3.16 we establish some notation. Let R be a noetherian integrally closed integral domain with quotient field K. Let $f_1, \dots, f_n$ be a set of nonzero elements of R. Let $f_0 = f_1 \cdots f_n$. Write R_i for the localization R_{f_i}, and U_i for the basic open set $U(f_i) = \operatorname{Spec} R_i = \{\mathfrak{p} \in \operatorname{Spec} R \mid f_i \notin \mathfrak{p}\}$. Then $U_0 \subseteq U_1 \cap \cdots \cap U_n$. Assume $f_1, \dots, f_n$ are chosen so that the open set $U_1 \cup \cdots \cup U_n$ contains $X_1(R)$. Let V be a finite dimensional K-vector space. Suppose for each i that M_i is a locally free R_i-lattice in V such that for each pair i, j we have $M_i \otimes_{R_i} R_{ij} = M_j \otimes_{R_j} R_{ij}$, where $R_{ij} = R_{f_i f_j}$. Let $\mathfrak{p} \in X_1(R)$. If $\mathfrak{p}$ is in U_i, then $(M_i)_\mathfrak{p}$ is an $R_\mathfrak{p}$-lattice in V. Moreover, if $\mathfrak{p}$ is in $U_i \cap U_j$, then $(M_i)_\mathfrak{p} = (M_j)_\mathfrak{p}$. Let L be a free R_0-lattice in V which contains $M_1 \otimes_{R_i} R_0 = \cdots = M_n \otimes_{R_n} R_0$. Let $v_1, \dots, v_r$ be a free R_0-basis for L. Then $F = Rv_1 + \cdots + Rv_r$ is a free R-lattice in V.

Theorem 6.3.16. *Let R, K, V, $f_1, \dots, f_n$, $M_1, \dots, M_n$, F be as above. For each $\mathfrak{p} \in X_1(R)$, define $N(\mathfrak{p})$ to be $(M_i)_\mathfrak{p}$, for any i such that $\mathfrak{p}$ is in U_i. If*

$$N = \bigcap_{\mathfrak{p} \in X_1(R)} N(\mathfrak{p}),$$

then

(1) *N is an R-lattice in V and $N_\mathfrak{p} = N(\mathfrak{p})$ for all $\mathfrak{p} \in X_1(R)$.*
(2) *N is a reflexive R-lattice in V.*
(3) *$N \otimes_R R_{f_i} = M_i$ for $1 \le i \le n$.*
(4) *$N = \bigcap_{i=1}^n M_i$.*

Proof. (1): By Krull's Hauptidealsatz (see, for example, [**AM69**, Chapter 11]), a minimal prime of f_0 has height one. By Corollary 2.5.5, f_0 is contained in only finitely many height one primes of R. Therefore, U_0 contains all but finitely many height one primes of R. By Theorem 6.3.10 (1), $(M_i)_\mathfrak{p} = F_\mathfrak{p}$ for all but finitely many $\mathfrak{p} \in X_1(R_0)$. Taken together, this implies that $N(\mathfrak{p}) = F_\mathfrak{p}$ for all but finitely many $\mathfrak{p} \in X_1(R)$. Part (1) follows from Theorem 6.3.10 (2).

Part (2) follows from Part (4) of Theorem 6.3.13.

(3): For each $\mathfrak{p} \in X_1(R_i)$, $(N \otimes_R R_i)_\mathfrak{p} = N_\mathfrak{p} = N(\mathfrak{p}) = (M_i)_\mathfrak{p}$. By Exercise 6.3.20 and Corollary 6.3.14, $N \otimes_R R_i = M_i$.

(4): We have: $N = \bigcap_{\mathfrak{p} \in X_1(R)} N(\mathfrak{p}) = \bigcap_{i=1}^n \bigcap_{\mathfrak{p} \in X_1(R_i)} (M_i)_\mathfrak{p} = \bigcap_{i=1}^n M_i$. □

Example 6.3.17. We continue to employ the notation of Section 6.3.3. Let R, K, $f_1, \dots, f_n$ be as above. Let $V = K^m$. Let $S = R_1 \oplus \cdots \oplus R_n$ and ϕ a 1-cocycle in $\mathrm{Z}^1(S/R, \mathrm{GL}_m)$ (see Definition 5.5.4). In this example we show how to associate to ϕ a set of modules $\{M_1, \dots, M_n\}$ in V satisfying the hypothesis of Theorem 6.3.16. We identify $\mathrm{GL}_m(K)$ with the set of all invertible m-by-m matrices over K. For each pair i, j, we identify $\mathrm{GL}_m(R_{ij})$ with the subgroup of $\mathrm{GL}_m(K)$ consisting of matrices which are invertible over the ring R_{ij}. As in Example 5.3.2, the Amitsur complex $C^\bullet(S/R)$ looks like

$$\bigoplus_{i \in I_n} R_i \xrightarrow{d^0} \bigoplus_{(i,j) \in I_n^2} R_{ij} \xrightarrow{d^1} \bigoplus_{(i,j,k) \in I_n^3} R_{ijk}$$

where $I_n = \{1, \dots, n\}$. Then ϕ can be viewed as a set of matrices $\{\phi_{ij} \in \mathrm{GL}_m(R_{ij}) \mid i \neq j\}$ satisfying the 1-cocycle identities

$$\phi_{ik} = \phi_{jk}\phi_{ij}$$

for all triples i, j, k. Let $B = \{b_1, \dots, b_m\}$ be the standard basis for V. Define M_1 to be the R_1-submodule of V spanned by B. For $1 \le i \le n$, define M_i to be the R_i-submodule of V spanned by the columns of ϕ_{1i}. Then M_i is spanned by vectors that are R_{1i}-linear combinations of $b_1, \dots, b_m$. Hence we see that $M_1 \otimes_{R_1} R_{1i} = M_i \otimes_{R_i} R_{1i}$. Using the 1-cocycle identity $\phi_{ij} = \phi_{1j}\phi_{1i}^{-1}$ we see that M_j is spanned by R_{ij}-linear combinations of vectors from M_i. Hence we see that $M_i \otimes_{R_i} R_{ij} = M_j \otimes_{R_j} R_{ij}$. This shows that the set of modules $M_1, \dots, M_n$ satisfies the hypothesis of Theorem 6.3.16 and

$$N = \bigcap_{i=1}^{n} M_i$$

is a reflexive R-lattice in V such that $N \otimes R_i = M_i$ for each i. In Exercise 6.3.21, the reader is asked to verify that the cohomology class of ϕ determines N up to isomorphism.

3.4. Exercises.

Exercise 6.3.18. Let R be an integral domain and M a finitely generated torsion free R-module. Let S be a submodule of M and consider $\bar{S} = KS \cap M$.

(1) Prove that $M/\bar{S}$ is a finitely generated torsion free R-module.
(2) Prove that $KS = K\bar{S}$. If S is finitely generated, prove that $\operatorname{Rank} S = \operatorname{Rank} \bar{S}$.

Exercise 6.3.19. Let R be an integral domain with field of fractions K. Let V be a finite dimensional K-vector space and M an R-lattice in V. Prove that M is a reflexive R-lattice if and only if there is an R-lattice N (in some K-vector space) such that M is isomorphic as an R-module to $R : N$.

Exercise 6.3.20. Let R be a noetherian integrally closed integral domain with field of fractions K. Let V be a finite dimensional K-vector space. Prove:

(1) If M and N are reflexive R-lattices in V, then $M \cap N$ is a reflexive R-lattice in V.

(2) If U is a K-subspace of V, and M is a reflexive R-lattice in V, then $M \cap U$ is a reflexive R-lattice in U.

(3) If M is a reflexive R-lattice in V and $Z \subseteq R - \{0\}$ is a multiplicative set, then $Z^{-1}M$ is a reflexive $Z^{-1}R$-lattice in V.

Exercise 6.3.21. In the context of Example 6.3.17, let ψ be another 1-cocycle in $\mathrm{Z}^1(S/R, \mathrm{GL}_m)$. Let N_ϕ and N_ψ be the reflexive R-lattices associated to ϕ and ψ respectively. Prove that if ψ and ϕ belong to the same cohomology class in $\mathrm{H}^1(S/R, \mathrm{GL}_m)$ (see Definition 5.5.4), then N_ϕ is isomorphic to N_ψ.

Exercise 6.3.22. In this exercise and Exercise 6.3.23 we study the class group of subrings of $k[x,y]$ generated by monomials. For more on this subject the interested reader is referred to [**And78**]. Let k be a field, and $n > 1$ an integer that is invertible in k. Assume k contains ζ, a primitive nth root of unity. Let $T = k[x,y]$ and $S = k[x^n, xy, y^n]$ (see Exercises 4.4.19 and 6.2.9). In S, let $\mathfrak{m}$ be the maximal ideal (x^n, xy, y^n). Prove:

(1) $T[x^{-n}]$ is separable over $S[x^{-n}]$ and free of rank n. The same is true for $T[y^{-n}]$ over $S[y^{-n}]$.

(2) For every prime $\mathfrak{p}$ in the open set $\operatorname{Spec}(S) - \{\mathfrak{m}\}$, $T_\mathfrak{p}$ is separable over $S_\mathfrak{p}$ and free of rank n.

(3) T is a reflexive S-module.

(4) Let $\sigma : T \to T$ be the automorphism defined by $x \mapsto \zeta x$, $y \mapsto \zeta^{-1}y$. Let $T_i = \{t \in T \mid \sigma(t) = \zeta^i t\}$. The S-module T decomposes into a direct sum $T_0 \oplus \cdots \oplus T_{n-1}$ of S-submodules where each T_i is a rank one reflexive.

Exercise 6.3.23. This exercise and Exercise 6.3.22 are examples of extensions of normal integral domains which are defined by cyclic group actions. For more on this subject, see Exercise 12.9.14. Let k be a field and $n > 1$ an integer that is invertible in k. Assume k contains ζ, a primitive nth root of unity. Let $T = k[x,y]$. Consider the subring of T defined by $S = k[x^n, x^{n-1}y, x^{n-2}y^2, \dots, xy^{n-1}, y^n]$. Let $\sigma : T \to T$ be the automorphism defined by $x \mapsto \zeta x$, $y \mapsto \zeta y$.

(1) Show that S is a normal integral domain and $\mathrm{Cl}(S) \cong \mathbb{Z}/n$ is generated by the height one prime ideal $(x^n, x^{n-1}y, x^{n-2}y^2, \dots, xy^{n-1})$.

(2) Let $T_i = \{t \in T \mid \sigma(t) = \zeta^i t\}$. Show that as an S-module T decomposes into a direct sum $T_0 \oplus \cdots \oplus T_{n-1}$ of S-submodules where each T_i is a rank one reflexive.

4. The Ideal Class Group

We show in this section that when R is a noetherian normal integral domain with quotient field K, the class group defined in Definition 6.2.2 can be identified with a group defined on classes of reflexive rank one R-lattices.

Let R be an integral domain with field of fractions K. Viewing K as a vector space over itself, a *fractional ideal* of R is an R-lattice F in K. If E and F are fractional ideals, the product EF is defined to be the R-submodule of K generated by all products of the form xy, where $x \in E$ and $y \in F$. By Proposition 6.3.4, $E+F$, $E \cap F$ and EF are fractional ideals. If F is a fractional ideal, let

$$F^{-1} = R : F = \{x \in K \mid xF \subseteq R\}.$$

where we identify $K = \operatorname{Hom}_K(K, K)$. By Proposition 6.3.5, F^{-1} is a fractional ideal of R. The reader should verify that $F^{-1}F \subseteq R$ and $F^{-1}F$ is an ideal of R. A fractional ideal F is called an *invertible ideal* of R in case $F^{-1}F = R$. An ideal I of R is a fractional ideal.

Lemma 6.4.1. *Let R be an integral domain with field of fractions K.*

(1) *If $\alpha \in K^*$, then the principal fractional ideal $I = R\alpha$ is invertible and $I^{-1} = R\alpha^{-1}$.*

(2) *If F is a nonzero R-submodule of K which is finitely generated as an R-module, then F is a fractional ideal of R.*

(3) *If F is a fractional ideal of R and $f \in \operatorname{Hom}_R(F, R)$, then for all $a, b \in F$ it is true that $af(b) = bf(a)$.*

(4) *Let F be a fractional ideal of R. For any $\alpha \in F^{-1}$, let $\ell_\alpha : F \to R$ be "left multiplication by α". The mapping $\alpha \mapsto \ell_\alpha$ is an isomorphism of R-modules $F^{-1} \to F^* = \operatorname{Hom}_R(F, R)$.*

Proof. For (3), see [**Hun80**, Lemma VIII.6.6]. The rest is left to the reader. □

Theorem 6.4.2. *Let R be an integral domain with field of fractions K and let F be a fractional ideal of R. The following are equivalent.*

(1) *F is a projective R-module.*

(2) *F is an invertible fractional ideal.*

(3) *F is an invertible R-module (that is, F is a rank one R-progenerator).*

Proof. See, for example, [**DF04**, Proposition 16.10] or [**Hun80**, Theorem VIII.6.8]. □

Lemma 6.4.3. *Let R be an integral domain with field of fractions K.*

(1) *If $F_1, \dots, F_n$ are fractional ideals of R, then $F = F_1F_2 \cdots F_n$ is invertible if and only if each F_i is invertible.*

(2) *If $P_1, \dots, P_r$ are invertible prime ideals in R, and $Q_1, \dots, Q_s$ are prime ideals in R such that $P_1P_2 \cdots P_r = Q_1Q_2 \cdots Q_s$, then $r = s$ and after relabeling, $P_i = Q_i$.*

Proof. See, for example, [**Hun80**, Lemma VIII.6.4]. □

Lemma 6.4.4. *Let R be an integral domain with field of fractions K. Let M be a finitely generated torsion free R-module. If $\dim_K(KM) = 1$, then M is isomorphic as an R-module to a fractional ideal of R in K.*

Proof. Choose any nonzero element m_0 of M and let $F = \{\alpha \in K \mid \alpha m_0 \in M\}$. Then F is an R-submodule of K. The assignment $\alpha \mapsto \alpha m_0$ defines a one-to-one R-module homomorphism $\theta : F \to M$. Since the K-vector space KM has dimension one, m_0 is a generator. Given any $m \in M$, there exists $\alpha \in K$ such that $\alpha m_0 = m$. Therefore θ is an isomorphism, and F is a finitely generated R-submodule of K. By Lemma 6.4.1 (2) we are done. □

An integral domain satisfying the equivalent conditions of Theorem 6.4.5 is called a *Dedekind domain.*

Theorem 6.4.5. *Let R be an integral domain. The following are equivalent.*

(1) *R is a noetherian normal integral domain with Krull dimension one.*

(2) *R is a noetherian integral domain and for every prime ideal P of height greater than or equal to one, the local ring R_P is a DVR.*

(3) *Every proper ideal in R has a unique representation as a product of a finite number of prime ideals.*

(4) *Every nonzero ideal in R is invertible. By Theorem 6.4.2, this is equivalent to each of the following statements.*
   (a) *Every nonzero ideal of R is R-projective.*
   (b) *Every nonzero ideal of R is an invertible R-module.*

(5) *Every fractional ideal of R is invertible. By Theorem 6.4.2, this is equivalent to each of the following statements.*
   (a) *Every fractional ideal of R is R-projective.*
   (b) *Every fractional ideal of R is an invertible R-module.*

(6) *Let $\operatorname{Frac}(R)$ denote the set of all fractional ideals of R. Then $\operatorname{Frac}(R)$ is a group under multiplication.*

Proof. See [**DF04**, Theorem 16.3.15] and [**Hun80**, Theorem VIII.6.10], for example. □

Theorem 6.4.6. *Let R be a Dedekind domain and M a finitely generated torsion free R-module of rank n. There exist fractional ideals $F_1, \dots, F_n$ of R such that $M \cong F_1 \oplus \cdots \oplus F_n$. In particular, M is an R-progenerator module.*

Proof. See, for example, [**DF04**, Theorem 16.3.22]. □

Let R be an integral domain with field of fractions K. By Lemma 6.4.1 (4), the assignment $\alpha \mapsto \ell_\alpha$ defines an isomorphism $F^{-1} \to \operatorname{Hom}_R(F, R)$. Then $F \to F^{**}$ is an isomorphism if and only if $F = (F^{-1})^{-1}$ and in this case we call F a *reflexive fractional ideal*. By Theorem 6.4.2, an invertible fractional ideal is projective, hence is reflexive. If E and F are two fractional ideals of R in K, then

$$E : F = \{\alpha \in K \mid \alpha F \subseteq E\}.$$

For any $\alpha \in E : F$, let $\ell_\alpha : F \to E$ be "left multiplication by α". By Proposition 6.3.5, the mapping $\alpha \mapsto \ell_\alpha$ defines an isomorphism of R-modules $E : F \cong \operatorname{Hom}_R(F, E)$. Notice that $F^{-1} = R : F$.

Lemma 6.4.7. *Let R be an integral domain with field of fractions K.*

(1) *If E and F are fractional ideals of R, then $E : F$ is a fractional ideal of R.*

(2) *Given fractional ideals $I_1 \subseteq I_2$ and $J_1 \subseteq J_2$, $J_1 : I_2 \subseteq J_2 : I_1$.*

(3) *If F is a fractional ideal, then*

$$F^{-1} = R : F = R : (R : (R : F)).$$

That is, F^{-1} is a reflexive fractional ideal and $F^{-1} \cong (F^{-1})^{**}$.

(4) *If F is a fractional ideal, then*

$$(F^{-1})^{-1} = \bigcap_{\alpha \in F^{-1}} \alpha^{-1} R.$$

That is, F is a reflexive fractional ideal if and only if

$$F = \bigcap_{\alpha \in F^{-1}} \alpha^{-1} R.$$

(5) *If D, E and F are fractional ideals, then*
  (a) *$D : EF = (D : E) : F$, and*
  (b) *$(D : E)F \subseteq D : (E : F)$.*

(6) *If F is a fractional ideal, then $(F^{-1}F)^{-1} = F^{-1} : F^{-1}$.*

(7) *If F is a fractional ideal and E is a reflexive fractional ideal, then $E : F$ is a reflexive ideal.*

Proof. The reader should verify that (1), (2), (3), (4), (5) and (7) are special cases of Proposition 6.3.5, Lemma 6.3.6, Proposition 6.3.7, Lemma 6.3.8, and Proposition 6.3.9. (6): By Part (5) (a), $R : F^{-1}F = (R : F) : F^{-1} = (R : F) : (R : F)$. □

Let Reflex(R) denote the set of all reflexive fractional ideals of R in K. If E and F are reflexive fractional ideals of R, then EF is not necessarily reflexive. Define a binary operation on Reflex(R) by the formula $E * F = R : (R : EF)$. By Exercise 6.4.13, this operation turns Reflex(R) into an abelian monoid with identity R. If R is a noetherian normal integral domain, then Lemma 3.3.9 (3) implies that R is completely normal and Proposition 6.4.8 implies that Reflex(R) is an abelian group.

Proposition 6.4.8. *Let R be an integral domain with field of fractions K. Then* Reflex(R) *is an abelian group if and only if R is completely normal (see Definition* 3.3.8*).*

Proof. Assume Reflex(R) is an abelian group. Let I be a fractional ideal of R in K. By Exercise 6.4.12, it is enough to show $R = I : I$. Let $J = (I^{-1})^{-1}$. By Lemma 6.4.7 (3), J is a reflexive fractional ideal. By Lemma 6.4.7 (7), $J : J$ is a reflexive fractional ideal. By Exercise 6.4.11, $J : J$ is an intermediate ring $R \subseteq J : J \subseteq K$, so $(J : J)^2 = J : J$. Then $R : (R : (J : J)^2) = R : (R : (J : J)) = J : J$ says $J : J$ is the idempotent of the group Reflex(R). That is, $R = J : J$. Again by Exercise 6.4.11,

$$R \subseteq I : I \subseteq I^{-1} : I^{-1} \subseteq J : J = R.$$

Conversely, if $I \in$ Reflex(R), then so is I^{-1} by Lemma 6.4.7 (3). It follows from Lemma 6.4.7 (6) that $R : II^{-1} = I^{-1} : I^{-1} = R$. Then $R : (R : II^{-1}) = R$, so I^{-1} is the inverse of I in Reflex(R). □

Lemma 6.4.9. *Let R be a noetherian normal integral domain with field of fractions K.*

(1) *Suppose I is an ideal in R that is maximal among all proper reflexive ideals in R. Then there exists an element $x \in K$ such that $I = R : (Rx + R)$ and I is a prime ideal.*

(2) *If P is a prime ideal of R and P is a reflexive ideal, then* $\operatorname{ht}(P) = 1$.

(3) *If $P \in X_1(R)$, then P is reflexive.*

Proof. (1): Since I is a proper reflexive ideal, $I^{-1} \neq R$. Pick $x \in I^{-1} - R$. Then $I \subseteq R : (Rx + R) \subseteq R$ and since $x \notin R$, $1 \notin R : (Rx + R)$. The ideal $R : (Rx + R)$ is reflexive, by Lemma 6.4.7 (3). By the maximality of I, $I = R : (Rx + R)$. Now suppose $a, b \in R$ and $ab \in I$. Let $A = Ra + I$ and $B = Rb + I$. Suppose $b \notin I$. Since $AB \subseteq I$, it follows that $I \subsetneq B \subseteq I : A$. Also, $I : A \subseteq I : I = R$. By Lemma 6.4.7 (7), $I : A$ is a reflexive ideal in

R. By maximality of I we conclude that $I : A = R$. Since $1 \in I : A$, we conclude that $a \in I$.

(2): Since $P \neq R$, $R \neq R : P$. Suppose $Q \in \operatorname{Spec} R$ and $(0) \subsetneq Q \subsetneq P$. Let $x \in P - Q$. Then $(R : P)x \subseteq R$, so $(R : P)xQ \subseteq Q$. But $x \notin Q$ and Q is prime, so $(R : P)Q \subseteq Q$. Thus $R : P \subseteq Q : Q$. Since R is normal, $R = Q : Q$. This is a contradiction.

(3): If $x \in P - (0)$, then Rx is free, hence reflexive. Define $\mathcal{S}$ to be the set of all $I \in \operatorname{Reflex}(R)$ such that $I \subseteq P$ and there exists $\alpha \in K^*$ such that $I = R\alpha^{-1} \cap R$. Then $\mathcal{S}$ is nonempty. Since R is noetherian, $\mathcal{S}$ has a maximal member, $M = R\alpha^{-1} \cap R$. It suffices to show that M is prime. Let a, b be elements of R such that $ab \in M$. Then $R(a\alpha)^{-1} \cap R \supseteq R\alpha^{-1} \cap R = M$. By Exercise 6.4.13, $R(a\alpha)^{-1} \cap R$ is in $\operatorname{Reflex}(R)$.

Case 1: Assume $R(a\alpha)^{-1} \cap R \subseteq P$. By the choice of M, $R(a\alpha)^{-1} \cap R = M$. Thus $ab \in R(a\alpha)^{-1} \cap R$, so there exists $r \in R$ such that $ab = r(a\alpha)^{-1} \in R$. This shows that $b = r(a\alpha)^{-1}a^{-1} \in R\alpha^{-1} \cap R = M$.

Case 2: Assume $R(a\alpha)^{-1} \cap R \not\subseteq P$. There exists $y \in R(a\alpha)^{-1} \cap R$ such that $y \notin P$. Given $w = r(y\alpha)^{-1} \in R(y\alpha)^{-1} \cap R$, $yw = r\alpha^{-1} \in M \subseteq P$. Since $y \notin P$, this proves $R(y\alpha)^{-1} \cap R \subseteq P$. We have $M = R\alpha^{-1} \cap R \subseteq R(y\alpha)^{-1} \cap R \subseteq P$. By the choice of M, this means $M = R(y\alpha)^{-1} \cap R$. Hence $a \in R(y\alpha)^{-1} \cap R = M$.

This proves that M is prime. Since $\operatorname{ht}(P) = 1$, we conclude $M = P$. Thus P is reflexive. □

Theorem 6.4.10. *Let R be a noetherian normal integral domain with field of fractions K.*

(1) *If I is an ideal in R, then I is reflexive if and only if there exist $P_1, \dots, P_n \in X_1(R)$ such that $I = R : (R : (P_1 \cdots P_n))$.*
(2) *If I is a reflexive ideal in R, then there are only finitely many $P \in X_1(R)$ such that $I \subseteq P$.*
(3) *The factorization in Part (1) is unique up to the order of the factors.*
(4) $\operatorname{Reflex}(R)$ *is a free abelian group and $X_1(R)$ is a basis.*

Proof. (1): Suppose I is a proper ideal of R and I is reflexive. If $I \in X_1(R)$, then I has the desired factorization. The proof is by contradiction. Since R is noetherian, there exists a maximal counterexample, say M. That is, M is a reflexive proper ideal in R and M does not have a factorization in the form $M = R : (R : (P_1 \cdots P_n))$, where each P_i is in $X_1(R)$. By Lemma 6.4.9, there is a maximal reflexive ideal P_1 that properly contains M. In fact, P_1 is in $X_1(R)$. Since $R \subsetneq P_1^{-1}$, it follows that $M \neq P_1^{-1} * M$, hence $M \subsetneq (R : P_1)M$. Take double duals, $M \subsetneq R : (R : (R : P_1)M)$. Also,

$M \subseteq P_1 \subseteq R$, so $(R : P_1)M \subseteq (R : P_1)P_1 \subseteq R$. That is, $R : (R : (R : P_1)M)$ is a reflexive ideal in R that properly contains M. By the choice of M, this ideal has a factorization in the desired form:

$$R : (R : (R : P_1)M) = R : (R : (P_2 \cdots P_n))$$

where $P_2, \ldots, P_n \in X_1(R)$. By Exercise 6.4.13 and Proposition 6.4.8, we see that $P_1^{-1} * M = P_2 * \cdots * P_n$ and $M = P_1 * P_2 * \cdots * P_n = R : (R : (P_1 \cdots P_n))$. The converse follows from Lemma 6.4.7 (3).

(2): Suppose $I = R : (R : (P_1 \cdots P_m))$ and each $P_i \in X_1(R)$. Then $P_1 \cdots P_m \subseteq I$. Suppose $P \in X_1(R)$ such that $I \subseteq P$. Since P is prime, there must be some i in $1, \ldots, m$ such that $P_i \subseteq P$. Since $\operatorname{ht}(P) = 1$, $P_i = P$. There are only finitely many choices for P.

(3): Suppose $I = R : (R : (P_1 \cdots P_m))$ and each $P_i \in X_1(R)$. If $m = 1$, then $I = P_1$ so the claim is trivially true. Proceed by induction on m. By Part (2), we can assume $I \subseteq P_1$. It follows that $I : P_1 \subseteq P_1 : P_1 = R$. By Lemma 6.4.7, $I : P_1$ is a reflexive ideal in R. By Exercise 6.4.14, $I : P_1 = I * P_1^{-1}$. By Exercise 6.4.13, $I : P_1 = P_2 * \cdots * P_m = R : (R : (P_2 \cdots P_m))$ and by induction we are done.

(4): By Parts (2) and (3) it suffices to show $\operatorname{Reflex}(R)$ is generated by those ideals in $X_1(R)$. Let $I \in \operatorname{Reflex}(R)$. There exists $a \in R$ such that $aI \subseteq R$. By Part (1) there are primes Q_i and P_j in $X_1(R)$ such that $aR = Q_1 * \cdots * Q_n$ and $aI = P_1 * \cdots * P_m$. Therefore, in the group $\operatorname{Reflex}(R)$ we have $I * Q_1 * \cdots * Q_n = P_1 * \cdots * P_m$. □

Let R be a noetherian normal integral domain with quotient field K. The class group of projective fractional ideals and its connection with the Picard group of R is the subject of Exercise 6.4.15. For the representation of the class group in terms of reflexive fractional ideals, see Exercise 6.4.16. Moreover, we see that the Picard group of R can be identified with the subgroup of the class group which is generated by the projective fractional ideals. Therefore, an element of the class group $\operatorname{Cl}(R)$ can either be viewed as a Weil divisor, or a reflexive fractional ideal. There are computations for which the Weil divisor representation has many advantages. This is clearly the case in Examples 6.2.5, 6.5.12, and Exercises 6.2.9 – 6.2.14. When divisors are being multiplied, then it can be advantageous to use fractional ideals. This is the case, for instance, in Example 4.2.3 and in the definition of generalized cyclic crossed products (see Section 13.5). From a computational point of view, $\operatorname{Pic}(R)$ is usually harder to compute than $\operatorname{Cl}(R)$. We will see in Theorem 6.5.11 that the two groups are equal if R is regular. More generally, the exact sequence of Exercise 6.4.17 shows that the two groups are equal if for every $\mathfrak{m} \in \operatorname{Max}(R)$, the local ring $R_{\mathfrak{m}}$ is factorial. Exercises 10.1.19 – 10.1.21 provide some more opportunities for which the exact sequence of

Exercise 6.4.17 can be applied to compute the Picard groups of some normal rings which are not regular.

4.1. Exercises.

Exercise 6.4.11. Let R be an integral domain with field of fractions K. Let F be a fractional ideal of R in K. Prove the following.

(1) $F : F$ is a ring, and $R \subseteq F : F \subseteq K$ is a tower of subrings.

(2) $F : F \subseteq F^{-1} : F^{-1} \subseteq (F^{-1})^{-1} : (F^{-1})^{-1}$.

Exercise 6.4.12. Let R be an integral domain with field of fractions K. Prove that R is completely normal if and only if $R = F : F$ for all fractional ideals F of R in K.

Exercise 6.4.13. Let R be an integral domain with field of fractions K. Let D, E, F be fractional ideals of R in K. Prove the following.

(1) $(D^{-1} : E) : F = (E^{-1} : F) : D$.

(2) $(D((EF)^{-1})^{-1})^{-1} = (((DE)^{-1})^{-1}F)^{-1} = (DEF)^{-1}$.

(3) With the binary operation $E * F = R : (R : EF) = ((EF)^{-1})^{-1}$, $\operatorname{Reflex}(R)$ is an abelian monoid.

Exercise 6.4.14. Let R be a noetherian normal integral domain with field of fractions K. Let E and F be elements of the group $\operatorname{Reflex}(R)$. Prove that $E : F = E * F^{-1}$ and $F : E = F * E^{-1}$.

Exercise 6.4.15. Let R be an integral domain with field of fractions K. Denote by $\operatorname{Invert}(R)$ the set of all invertible fractional ideals of R in K. Let $\operatorname{Prin}(R)$ denote the subset of $\operatorname{Invert}(R)$ consisting of all principal fractional ideals of R in K.

(1) Prove that $\operatorname{Invert}(R)$ is a group under multiplication and contains $\operatorname{Prin}(R)$ as a subgroup.

(2) Every invertible ideal $I \in \operatorname{Invert}(R)$ is an invertible R-module, hence I represents a class in the Picard group of R. Show that this assignment defines a homomorphism $\theta : \operatorname{Invert}(R) \to \operatorname{Pic}(R)$.

(3) Show that θ induces an isomorphism $\operatorname{Invert}(R)/\operatorname{Prin}(R) \cong \operatorname{Pic}(R)$. The group $\operatorname{Invert}(R)/\operatorname{Prin}(R)$ is called the class group of rank one projective R-modules.

Exercise 6.4.16. Let R be a noetherian normal integral domain with field of fractions K.

(1) Show that $\operatorname{Invert}(R)$ is a subgroup of $\operatorname{Reflex}(R)$.

(2) Show that $\operatorname{Prin}(R)$ is a subgroup of $\operatorname{Reflex}(R)$.

(3) The quotient $\operatorname{Reflex}(R)/\operatorname{Prin}(R)$ is called the *class group* of rank one reflexive R-modules. Show that this group is isomorphic to the class group of Weil divisors $\operatorname{Cl}(R)$ (Definition 6.2.2).

(4) Show that there is a one-to-one homomorphism

$$\operatorname{Invert}(R)/\operatorname{Prin}(R) \to \operatorname{Reflex}(R)/\operatorname{Prin}(R)$$

from the class group of rank one projectives into the class group of rank one reflexives.

(5) Show that there is a one-to-one homomorphism $\operatorname{Pic}(R) \to \operatorname{Cl}(R)$.

Exercise 6.4.17. Let R be a noetherian normal integral domain. Denote by $\operatorname{Sing}(R)$ the set of all maximal ideals $\mathfrak{m} \in \operatorname{Max}(R)$ such that $\operatorname{Cl}(R_{\mathfrak{m}}) \neq (0)$. Show that the natural maps induce an exact sequence

$$0 \to \operatorname{Pic}(R) \to \operatorname{Cl}(R) \to \prod_{\mathfrak{m} \in \operatorname{Sing}(R)} \operatorname{Cl}(R_{\mathfrak{m}})$$

of abelian groups.

Exercise 6.4.18. If R is a Dedekind domain, show that $\operatorname{Pic} R = \operatorname{Cl} R$.

5. Functorial Properties of the Class Group

Let R be a noetherian normal integral domain with field of fractions K. Let S be a noetherian normal integral domain with field of fractions L. The class group is not a functor. That is, a homomorphism $R \to S$ does not necessarily induce a homomorphism of groups $\operatorname{Cl}(R) \to \operatorname{Cl}(S)$. There are three important cases where a homomorphism on class groups does exist. The first case is when S is a localization of R in K and $K = L$. This is the context of Nagata's Theorem (Theorem 6.2.4 and Exercise 6.2.8). Secondly, if S is a flat R-algebra, we show that there is an induced homomorphism $\operatorname{Cl}(R) \to \operatorname{Cl}(S)$ (Section 6.5.1). The third scenario is when S is a faithful R-algebra which is finitely generated as an R-module (Section 6.5.2).

5.1. Flat Extensions. Now assume S/R is an extension of noetherian normal integral domains and L/K is the corresponding extension of the fields of fractions. Assume S is a flat R-algebra. Then in this context, Lemma 6.5.1 shows that there is a homomorphism of divisor groups $\operatorname{Div}(R) \to \operatorname{Div}(S)$ which induces a homomorphism of class groups $\gamma : \operatorname{Cl}(R) \to \operatorname{Cl}(S)$.

Lemma 6.5.1. *In the above context, let I be a reflexive fractional ideal of R in K.*

(1) *IS is a reflexive fractional ideal of S in L.*

(2) *$I \otimes_R S \cong IS$ by the multiplication map $a \otimes b \mapsto ab$.*

(3) *The action* $I \mapsto IS$ *induces a homomorphism* $\operatorname{Div}(R) \to \operatorname{Div}(S)$ *of abelian groups.*

(4) *The map in (3) induces a homomorphism* $\operatorname{Cl}(R) \to \operatorname{Cl}(S)$ *of abelian groups.*

(5) *If* S/R *is faithfully flat, then* I *is a projective fractional ideal if and only if* IS *is a projective fractional ideal.*

(6) *If* $\operatorname{Pic}(R) = 0$ *and* S *is a faithfully flat* R*-algebra, then* $\operatorname{Cl}(R) \to \operatorname{Cl}(S)$ *is one-to-one.*

Proof. (1): There is $\alpha \in R$ such that $\alpha I \subseteq R$. Then αI is a reflexive fractional ideal, so $\alpha I = R : (R : \alpha I)$. By Lemma 2.6.12, $\alpha IS = S : (S : \alpha IS)$. From this we get $IS = S : (S : IS)$.

(2): Since S is flat over R, $I \otimes_R S \to K \otimes_R S$ is one-to-one. But $K \otimes_R S$ is a localization of S in L, so the multiplication map $I \otimes_R S \to IS$ is an isomorphism.

Parts (3) and (4) are left to the reader.

(5): This follows from (2) and Theorem 2.6.1.

(6): If I is a reflexive fractional ideal of R in K and IS is principal, then I is an invertible fractional ideal, by (5). Since $\operatorname{Pic}(R) = 0$, I is principal. □

Corollary 6.5.2. *(Mori's Theorem) Let* R *be a commutative noetherian ring,* I *an ideal contained in the Jacobson radical of* R*, and* $\hat{R}$ *the* I*-adic completion of* R*. If* $\hat{R}$ *is an integrally closed integral domain, then* R *is an integrally closed integral domain and* $\operatorname{Cl}(R) \to \operatorname{Cl}(\hat{R})$ *is one-to-one.*

Proof. See, for example, [**Fos73**]. □

Polynomial rings are an important special case of the above. Let R be a commutative ring and x an indeterminate. The reader should verify that $R[x]$ is a faithfully flat extension of R. If R is a normal ring, then so is $R[x]$ (see, for example, [**Mat80**]).

Theorem 6.5.3. *Let* R *be a noetherian commutative ring.*

(1) *If* R *is an integrally closed integral domain, then the natural homomorphism* $\operatorname{Cl}(R) \to \operatorname{Cl}(R[x])$ *is an isomorphism.*

(2) *If* R *is a normal ring, then the natural homomorphism* $\operatorname{Pic}(R) \to \operatorname{Pic}(R[x])$ *is an isomorphism.*

Proof. (1): Let K be the quotient field of R. Since $K[x]$ is a unique factorization domain, $\operatorname{Cl}(K[x]) = 0$ (Theorem 6.2.3). By Nagata's Theorem (Exercise 6.2.8), $\operatorname{Cl}(R[x])$ is generated by the prime divisors $P \in X_1(R[x])$

such that $P \cap R \neq (0)$. Let $S = R[x]$ and $P \in X_1(S)$. Since S/R is faithfully flat, going down holds and Theorem 6.1.1 says $\operatorname{ht}(P) = \operatorname{ht}(P \cap R) + \operatorname{ht}(P/(P \cap R)S)$. If $P \cap R \neq (0)$, this means $P \cap R \in X_1(R)$, and $P = (P \cap R)S$. Therefore, $\operatorname{Cl}(R) \to \operatorname{Cl}(R[x])$ is onto. Consider the commutative diagram

$$\begin{array}{ccc} \operatorname{Pic}(R) & \xrightarrow{\alpha} & \operatorname{Pic}(R[x]) \\ \downarrow & & \downarrow \\ \operatorname{Cl}(R) & \xrightarrow{\beta} & \operatorname{Cl}(R[x]) \end{array}$$

in which β is onto and the vertical maps are one-to-one (Exercise 6.4.16). If $R[x] \to R$ is the homomorphism defined by $x \mapsto 0$, then $R \to R[x] \to R$ is an isomorphism of rings. Since $\operatorname{Pic}(\ )$ is a functor, $\operatorname{Pic}(R) \to \operatorname{Pic}(R[x]) \to \operatorname{Pic}(R)$ is an isomorphism of abelian groups, hence α is one-to-one. By Lemma 6.5.1 (5) it follows that α is onto and β is one-to-one.

(2): By the proof of (1), this is true when R is an integral domain. It follows from Lemma 3.3.10 that R is a finite direct sum of normal integral domains. By Exercise 2.6.14, the Picard group distributes across direct sums. □

Remark 6.5.4. A functor $\mathfrak{F}$ from commutative rings to abelian groups is called a *homotopy functor* if the natural map induces an isomorphism $\mathfrak{F}(R) \cong \mathfrak{F}(R[x])$. By Theorem 6.5.3, Pic is a homotopy functor on the category of normal noetherian rings. In fact, the isomorphism $\operatorname{Pic} R \cong \operatorname{Pic} R[x]$ of Theorem 6.5.3 (2) holds if R is a so-called *seminormal ring*. For the details (and more examples of homotopy functors), the interested reader is referred to [**Hoo84**]. See Corollary 13.6.6 for further results on this functor.

5.2. Finite Extensions. Now assume S/R is an extension of noetherian normal integral domains and L/K the corresponding extension of the fields of fractions. Assume S is a finitely generated R-module. In this context, we show that there is a homomorphism of divisor groups $\beta : \operatorname{Div}(R) \to \operatorname{Div}(S)$ which induces a homomorphism of class groups $\gamma : \operatorname{Cl}(R) \to \operatorname{Cl}(S)$. Let $\mathfrak{p} \in X_1(R)$. Then $R_{\mathfrak{p}}$ is a discrete valuation ring and $S_{\mathfrak{p}} = S \otimes_R R_{\mathfrak{p}}$ is a semilocal normal integral domain (Corollary 3.3.6). Let $X_1(S_{\mathfrak{p}}) = \{\mathfrak{q}_1, \dots, \mathfrak{q}_n\}$. Let $\pi_{\mathfrak{p}}$ be a local parameter for $R_{\mathfrak{p}}$. Then we view $\pi_{\mathfrak{p}}$ as an element of L, and

$$\operatorname{Div}_{S_{\mathfrak{p}}}(\pi_{\mathfrak{p}}) = \sum_{i=1}^{n} \nu_{\mathfrak{q}_i}(\pi_{\mathfrak{p}})\mathfrak{q}_i$$

is a principal divisor in $\operatorname{Div} S_{\mathfrak{p}}$. The number $\nu_{\mathfrak{q}_i}(\pi_{\mathfrak{p}})$ is called the *ramification index* of $\mathfrak{q}_i$ over $\mathfrak{p}$, and is denoted $e_{\mathfrak{q}_i}$. Notice that $e_{\mathfrak{q}_i} = 1$ if and only if $\mathfrak{q}_i$ is generated by $\pi_{\mathfrak{p}}$, and in this case we say $\mathfrak{q}_i$ is *unramified* over $\mathfrak{p}$. The homomorphism

$$\beta : \operatorname{Div} R \to \operatorname{Div} S$$

is defined by sending the generator $\mathfrak{p} \in X_1(R)$ to the divisor $\sum_{\mathfrak{q}\cap R=\mathfrak{p}} e_{\mathfrak{q}}\mathfrak{q}$, where the sum runs over the set of primes in $X_1(S)$ lying over $\mathfrak{p}$.

Proposition 6.5.5. *Let S/R be an extension of noetherian normal integral domains and L/K the corresponding extension of the fields of fractions. Assume S is a finitely generated R-module. Then there is a homomorphism $\gamma : \mathrm{Cl}(R) \to \mathrm{Cl}(S)$ which is induced on divisors by the homomorphism β defined above.*

Proof. Let $\alpha \in K^*$. Let $\mathfrak{q} \in X_1(S)$ and $\mathfrak{q} \cap R = \mathfrak{p}$. By definition of ramification index, $\nu_{\mathfrak{q}}(\alpha) = e_{\mathfrak{q}}\nu_{\mathfrak{p}}(\alpha)$. Therefore, β maps a principal divisor to a principal divisor, the diagram

$$\begin{array}{ccccccccc} 0 & \longrightarrow & \mathrm{Prin}\,(R) & \longrightarrow & \mathrm{Div}(R) & \longrightarrow & \mathrm{Cl}(R) & \longrightarrow & 0 \\ & & \downarrow & & \downarrow{\scriptstyle\beta} & & \downarrow{\scriptstyle\gamma} & & \\ 0 & \longrightarrow & \mathrm{Prin}(S) & \longrightarrow & \mathrm{Div}(S) & \longrightarrow & \mathrm{Cl}(S) & \longrightarrow & 0 \end{array}$$

commutes and the rows are exact. □

5.3. Galois Descent of Divisor Classes. References for the material in this section are [**Fos73**], [**Rim65**], and [**Sam64**]. Let R be a noetherian integrally closed integral domain with quotient field K. Let L/K be a finite field extension which is Galois with group G. The degree of the extension is denoted n. Let S be the integral closure of R in L. Then L is the quotient field of S and S is finitely generated as an R-module (Theorem 4.6.10). The reader should verify that that G acts on S as a group of R-algebra automorphisms, and $S^G = R$. In Proposition 6.5.6, we employ the notation of Section 5.1. Since G acts as a group of permutations on $X_1(S)$, and $\mathrm{Div}(S)$ is the free abelian group on $X_1(S)$, this makes $\mathrm{Div}(S)$ into a $\mathbb{Z}G$-module.

Proposition 6.5.6. *In the above context, the following are true.*

(1) *There is a monomorphism $\beta : \mathrm{Div}(R) \to \mathrm{Div}(S)^G$ of abelian groups.*

(2) *$\mathrm{Cl}(S)$ is a $\mathbb{Z}G$-module and there is a homomorphism of groups $\gamma : \mathrm{Cl}(R) \to \mathrm{Cl}(S)^G$.*

(3) *There is a natural exact sequence*

$$0 \to \ker\gamma \to \mathrm{H}^1(G, S^*) \to \mathrm{Div}(S)^G/\mathrm{Div}(R)$$

of abelian groups.

(4) *If each $\mathfrak{q} \in X_1(S)$ is unramified over $\mathfrak{q} \cap R$, then $\beta : \mathrm{Div}(R) \to \mathrm{Div}(S)^G$ is an isomorphism.*

Proof. (1): Clearly β is one-to-one. Given $\mathfrak{q} \in X_1(S)$, let $\mathfrak{p} = R \cap \mathfrak{q}$. Each $\sigma \in G$ induces a commutative diagram

$$\begin{array}{ccc} S_{\mathfrak{q}} & \xrightarrow{\sigma} & S_{\sigma(\mathfrak{q})} \\ & \searrow \quad \nearrow & \\ & R_{\mathfrak{p}} & \end{array}$$

where the top row is an isomorphism. From this we see that the ramification index of $\mathfrak{q}$ is equal to the ramification index of $\sigma(\mathfrak{q})$. Hence the image of β is fixed by σ.

(2): If $\alpha \in L^*$, then $\nu_{\mathfrak{q}}(\alpha) = \nu_{\sigma(\mathfrak{q})}(\sigma(\alpha))$, so σ maps a principal divisor to a principal divisor and $\operatorname{Cl}(S)$ is a $\mathbb{Z}G$-module. The rest follows from (1).

(3): The long exact sequence of cohomology associated to

$$1 \to S^* \to L^* \to \operatorname{Prin} S \to 0 \tag{6.14}$$

and Hilbert's Theorem 90 (Theorem 5.1.20) yield the exact sequence

$$\begin{aligned} 1 \to R^* \to K^* \to \operatorname{Prin}(S)^G &\to \mathrm{H}^1(G, S^*) \to 0 \\ &\to \mathrm{H}^1(G, \operatorname{Prin} S) \xrightarrow{\delta^1} \mathrm{H}^2(G, S^*) \xrightarrow{\epsilon} \mathrm{H}^2(G, L^*). \end{aligned} \tag{6.15}$$

By definition, $K^*/R^* = \operatorname{Prin} R$. The diagram

$$\begin{array}{ccccccccc} 0 & \longrightarrow & \operatorname{Prin} R & \longrightarrow & \operatorname{Div} R & \longrightarrow & \operatorname{Cl} R & \longrightarrow & 0 \\ & & \downarrow & & \downarrow{\scriptstyle\beta} & & \downarrow{\scriptstyle\gamma} & & \\ 0 & \longrightarrow & \operatorname{Prin}(S)^G & \longrightarrow & \operatorname{Div}(S)^G & \longrightarrow & \operatorname{Cl}(S)^G & & \end{array} \tag{6.16}$$

commutes and the rows are exact. To finish (3), combine (6.15) and (6.16) with the Snake Lemma (Theorem 1.4.1).

(4): For each $\mathfrak{p} \in X_1(R)$, let $P(\mathfrak{p}) = \{\mathfrak{q} \in X_1(S) \mid \mathfrak{q} \cap R = \mathfrak{p}\}$ be the set of those prime divisors in $X_1(S)$ lying over $\mathfrak{p}$. Then $\beta(\mathfrak{p}) = \sum_{\mathfrak{q} \in P(\mathfrak{p})} e_{\mathfrak{q}} \mathfrak{q} = \sum_{\mathfrak{q} \in P(\mathfrak{p})} \mathfrak{q}$ because each ramification index is assumed to be 1. By Theorem 3.3.5, if $\mathfrak{q} \cap R = \mathfrak{p}$, then the set $P(\mathfrak{p})$ is equal to the orbit of $\mathfrak{q}$ under the action of G on $\operatorname{Div}(S)$. Let $D = \sum_{\mathfrak{q} \in X_1(S)} a_{\mathfrak{q}} \mathfrak{q}$ be a divisor in $\operatorname{Div}(S)^G$. Since D is fixed by each $\sigma \in G$, the coefficients $a_{\mathfrak{q}}$ are constant as $\mathfrak{q}$ runs through $P(\mathfrak{p})$. If we denote this constant by $a_{\mathfrak{p}}$, then

$$D = \sum_{\mathfrak{p} \in X_1(R)} a_{\mathfrak{p}} \sum_{\mathfrak{q} \in P(\mathfrak{p})} \mathfrak{q} = \sum_{\mathfrak{p} \in X_1(R)} a_{\mathfrak{p}} \beta(\mathfrak{p})$$

which shows D is in the image of β. □

Proposition 6.5.7. *In the above context,* $\mathrm{H}^1(G, \operatorname{Div} S) = (0)$.

Proof. For each $\mathfrak{p} \in X_1(R)$ fix a prime $Q_\mathfrak{p} \in X_1(S)$ lying above $\mathfrak{p}$. Let $G_\mathfrak{p}$ be the subgroup of G fixing $Q_\mathfrak{p}$. The reader should verify that

$$\operatorname{Div}(S) = \bigoplus_{\mathfrak{p} \in X_1(R)} \mathbb{Z}G \otimes_{\mathbb{Z}G_\mathfrak{p}} \mathbb{Z}$$

as G-modules. Since G is finite, $\operatorname{Hom}_{\mathbb{Z}G_\mathfrak{p}}(\mathbb{Z}G, \mathbb{Z})$ and $\mathbb{Z}G \otimes_{\mathbb{Z}G_\mathfrak{p}} \mathbb{Z}$ are isomorphic as G-modules (Lemma 5.1.15). From Theorem 5.1.10, for each $\mathfrak{p} \in X_1(R)$ we get $\mathrm{H}^1(G, \operatorname{Hom}_{\mathbb{Z}G_\mathfrak{p}}(\mathbb{Z}G, \mathbb{Z})) = \mathrm{H}^1(G_\mathfrak{p}, \mathbb{Z})$. But $\mathbb{Z}$ is a trivial $G_\mathfrak{p}$-module and by Proposition 5.1.6 we see that $\mathrm{H}^1(G_\mathfrak{p}, \mathbb{Z}) = \operatorname{Hom}(G_\mathfrak{p}, \mathbb{Z})$. But G is finite, so the last group is 0. It follows from Exercise 5.1.25 that $\mathrm{H}^1(G, \operatorname{Div}(S)) = 0$. □

Theorem 6.5.8. *In the above context, there is an exact sequence*

$$(6.17)\quad 0 \to \operatorname{Cl}(S/R) \xrightarrow{\gamma_0} \mathrm{H}^1(G, S^*) \xrightarrow{\gamma_1} \operatorname{Div}(S)^G / \operatorname{Div}(R) \xrightarrow{\gamma_2} \operatorname{Cl}(S)^G / \operatorname{Cl}(R) \xrightarrow{\gamma_3} \mathrm{H}^2(G, S^*) \xrightarrow{\gamma_4} \mathrm{H}^2(G, L^*)$$

of abelian groups where $\operatorname{Cl}(S/R)$ *is the kernel of* $\operatorname{Cl}(R) \to \operatorname{Cl}(S)$.

Proof. The long exact sequence of cohomology associated to the short exact sequence

$$(6.18)\qquad 0 \to \operatorname{Prin} S \to \operatorname{Div} S \to \operatorname{Cl} S \to 0$$

is

$$(6.19)\quad 0 \to \operatorname{Prin}(S)^G \to \operatorname{Div}(S)^G \to \operatorname{Cl}(S)^G \xrightarrow{\delta^0} \mathrm{H}^1(G, \operatorname{Prin} S) \to \mathrm{H}^1(G, \operatorname{Div} S).$$

By Proposition 6.5.7, δ^0 is onto. Combine (6.15) with (6.19) to get

$$(6.20)\quad 0 \to \operatorname{Cl}(S/R) \xrightarrow{\gamma_0} \mathrm{H}^1(G, S^*) \xrightarrow{\gamma_1} \operatorname{Div}(S)^G \xrightarrow{\gamma_2} \operatorname{Cl}(S)^G \xrightarrow{\delta^1 \delta^0} \mathrm{H}^2(G, S^*) \xrightarrow{\gamma_4} \mathrm{H}^2(G, L^*).$$

Using Diagram (6.16) it is straightforward to derive (6.17) from (6.20). □

5.4. The Class Group of a Regular Domain. We end this section with some theorems about reflexive lattices over a regular integral domain. Probably the main theorem of this section is Theorem 6.5.10, which is due to Auslander and Goldman. There is no short proof for this important theorem, and the proofs that are known involve methods which are not used elsewhere in this text. For these reasons, we decided not to include a proof. The next two theorems will be applied in Chapter 11 in the study of the Brauer group of R.

Theorem 6.5.9. *Let R be a noetherian regular integral domain and assume the Krull dimension of R is less than or equal to two. Let M be a finitely generated R-lattice. Then M is reflexive if and only if M is projective.*

Proof. By Exercise 1.3.34, if M is projective, then M is reflexive. Assume M is a reflexive R-lattice. By Proposition 2.6.4, it suffices to show this when R is a regular local ring. If $\dim(R) = 0$, then R is a field and every R-module is projective. If $\dim(R) = 1$, then R is a DVR (Theorem 6.1.14), and M is free by Proposition 6.3.2. Assume $\dim(R) = 2$. By Proposition 6.3.5, $M^* = R : M$ is an R-lattice. Let

$$0 \to K_0 \xrightarrow{d_1} F_0 \xrightarrow{\epsilon} M^* \to 0$$

be an exact sequence, where F_0 is a finitely generated free R-module. Applying the functor $\operatorname{Hom}_R(\cdot, R)$, we get the exact sequence

$$0 \to M^{**} \xrightarrow{\epsilon^*} F_0^* \xrightarrow{d_1^*} K_0^*.$$

By hypothesis, $M = M^{**}$. Since K_0 is an R-submodule of F_0, K_0 is an R-lattice. By Proposition 6.3.5, K_0^* is an R-lattice and we can embed K_0^* in a free R-lattice F_1. If we define N to be the cokernel of $F_0^* \to F_1$, then the sequence

$$0 \to M \xrightarrow{\epsilon^*} F_0^* \xrightarrow{d_1^*} F_1 \to N \to 0 \tag{6.21}$$

is exact. Since F_0 is free, so is F_0^*. By Theorem 6.1.7, $\operatorname{coh.dim}(R) = \dim(R) = 2$. By Theorem 6.1.6, M is projective because it is the first syzygy of (6.21). □

Theorem 6.5.10. *Let R be a noetherian regular integral domain with field of fractions K. Let V be a finite dimensional K-vector space and M an R-lattice in V. If M is R-reflexive and* $\operatorname{Hom}_R(M, M)$ *is R-projective, then M is R-projective.*

Proof. The original proof of this theorem can be found in [**AG60b**]. There is also a proof by G. Garfinkel in [**OS75**] which is nearly self-contained. □

Theorem 6.5.11. *Let R be a noetherian regular integral domain with field of fractions K. Then the following are true.*

(1) $\operatorname{Pic}(R) = \operatorname{Cl}(R)$.

(2) *If R is a local ring, then* $\operatorname{Cl}(R) = (0)$ *and R is a unique factorization domain.*

Proof. (1): Let F be a reflexive fractional ideal of R in K. By Exercise 6.4.14, $F : F = R$ is free of rank one. By Theorem 6.5.10, F is projective. The equality $\operatorname{Pic}(R) = \operatorname{Cl}(R)$ follows from Exercise 6.4.16.

(2): For any local ring the Picard group is trivial. Now apply (1) and Theorem 6.2.3. □

We end this chapter with the following example, which is an application of Nagata's Theorem (Theorem 6.2.4) and Theorem 6.5.11. It shows that any finite cyclic group is the Picard group of a regular integral domain. It also forms the basis of Examples 13.6.12 and 13.6.13.

Example 6.5.12. In this example we show how to construct a regular integral domain R such that $\operatorname{Pic}(R)$ is a finite cyclic group of order n. The example comes from Algebraic Geometry and is based on the fact that if k is a field, then the class group of the projective plane $\mathbb{P}^2_k$ is an infinite cyclic group and is generated by a line. For simplicity's sake we construct our example using the projective plane. However, the same ideas apply in higher dimensions. Start with any field k and any integer $n > 1$. Let

$$S = k[x, y, z] = S_0 \oplus S_1 \oplus S_2 \oplus \cdots \oplus S_n \oplus \cdots$$

be the polynomial ring in three variables, with the usual grading. Let $f \in S_n$ be a homogeneous irreducible polynomial of degree n. The localized ring $S[f^{-1}]$ has a $\mathbb{Z}$-grading: $S[f^{-1}] = \bigoplus_{i\in\mathbb{Z}} S[f^{-1}]_i$. If $p \in S_m$ is homogeneous of degree m, then pf^{-d} is a typical homogeneous element of degree $m - dn \in S[f^{-1}]_i$. Let $R = S[f^{-1}]_0$ be the subring of homogeneous elements in $S[f^{-1}]$ of degree 0. We will show the following.

(1) R is a finitely generated k-algebra, a regular noetherian integral domain, and the Krull dimension of R is $\dim(R) = 2$.

(2) $\operatorname{Pic}(R) = \operatorname{Cl}(R) \cong \mathbb{Z}/n$.

(3) $R^* = k^*$.

A typical element of R is a fraction pf^{-d} where $p \in S_{dn}$. Since R is a subring of the field $k(x, y, z)$, R is an integral domain. Since f is irreducible and has degree $n \geq 2$, $f(0, y, z)$ is a homogeneous polynomial in $k[y, z]$ of degree n. Therefore, the homomorphism $k[x, y, z] \to k[y, z]$ defined by $x \mapsto 0$ induces

$$R = S[f^{-1}]_0 \xrightarrow{\theta} k[y, z][(f(0, y, z))^{-1}]_0.$$

Notice that θ is onto, and since the image is an integral domain, $\mathfrak{p} = \ker(\theta)$ is a prime ideal in R. Consider the local ring $R_\mathfrak{p}$. We will now show that $R_\mathfrak{p}$ is a DVR and x/y is a local parameter. If $h+i+j = dn$, then the monomial $x^h y^i z^j f^{-d}$ is in the kernel of θ if and only if $h \geq 1$. Then

$$\frac{x^h y^i z^j}{f^d} \frac{f^d}{y^{h+i} z^j} = \frac{x^h}{y^h} \tag{6.22}$$

shows $\mathfrak{p}R_\mathfrak{p}$ is generated by x/y. This also proves that $\operatorname{ht}(\mathfrak{p}) = 1$. Notice that in $S[f^{-1}]$, which is a UFD, the element $x^n f^{-1}$ belongs to the unique minimal prime ideal $(x) = (xf^{-1})$. Viewing R as a subring of $S[f^{-1}]$, we see that $x^n f^{-1}$ is irreducible in R, and $\mathfrak{p}$ is the unique minimal prime of R containing $x^n f^{-1}$. Using (6.22) we compute

$$\nu_{\mathfrak{p}}(x^n f^{-1}) = n. \tag{6.23}$$

Consider the localized ring $R[fx^{-n}]$. Given $p \in S_{dn}$ we multiply and divide by $(x^n f^{-1})^d$ to get

$$\begin{aligned} pf^{-d} &= (px^{-dn} f^d)(fx^{-n})^{-d} f^{-d} \\ &= p(1, y/x, z/x)\,(f(1, y/x, z/x))^{-d}. \end{aligned} \tag{6.24}$$

Therefore, the assignments $x \mapsto 1$, $y \mapsto u$, $z \mapsto v$ induce an isomorphism of k-algebras

$$R[fx^{-n}] \to k[u, v][(f(1, u, v))^{-1}]. \tag{6.25}$$

The homomorphism in (6.25) is usually specified by saying "dehomogenize with respect to x". Notice that the ring on the right hand side of (6.25) is a finitely generated k-algebra, a regular integral domain, and has Krull dimension two. By the same argument used in (6.25), but dehomogenizing with respect to y and z, the reader should verify that $R[fy^{-n}]$ and $R[fz^{-n}]$ are finitely generated regular integral k-algebras of Krull dimension two. For some $N > 0$, f^N is a sum of monomials of the form $x^h y^i z^j$ where at least one of h, i, j is greater than n. Therefore, $1 = f^N f^{-N}$ is in the ideal of R generated by $x^n f^{-1}, y^n f^{-1}, z^n f^{-1}$. This shows that $R[fx^{-n}] \oplus R[fy^{-n}] \oplus R[fz^{-n}]$ is a faithfully flat extension of R (Exercise 2.4.20). By Proposition 8.1.28 which is proved below, R is finitely generated as a k-algebra. For each prime ideal $P \in \operatorname{Spec} R$, the local ring R_P is regular and has dimension two. This proves (1). Since $f(x, y, z)x^{-n} = f(1, yx^{-1}, zx^{-1})$, we see that $f(1, u, v)$ is irreducible because $f(x, y, z)$ is irreducible. Applying Nagata's Theorem (Theorem 6.2.4) to the ring R, the sequence

$$1 \to R^* \to (R[fx^{-n}])^* \xrightarrow{\text{Div}} \mathbb{Z}\mathfrak{p} \to \mathrm{Cl}(R) \to \mathrm{Cl}(R[fx^{-n}]) \to 0 \tag{6.26}$$

is exact. By the isomorphism in (6.25), we see that $R[fx^{-n}]$ is a UFD. Hence $\mathrm{Cl}(R[fx^{-n}])$ is equal to (0) by Theorem 6.2.3. Using (6.25) and the fact that $k[u, v]$ is a UFD, we see that

$$(R[fx^{-n}])^* = k^* \times \langle x^n f^{-1} \rangle$$

is an internal direct sum. This and (6.23) shows that the image of Div in (6.26) is $n\mathbb{Z}\mathfrak{p}$. Therefore, $\mathrm{Cl}(R)$ is generated by $\mathfrak{p}$ and has order n. Part (2) follows from Theorem 6.5.11. Part (3) is left to the reader.

5.5. Exercises.

Exercise 6.5.13. Let R, K, L, S and G be as in Section 6.5.3.

(1) Show that the norm map $N_K^L : L^* \to K^*$ induces a homomorphism of abelian groups $N : \operatorname{Prin}(S) \to \operatorname{Prin}(R)$.

(2) Show that there is a norm map $N : \operatorname{Div}(S) \to \operatorname{Div}(R)$. (Hint: As in Proposition 6.5.7,

$$\operatorname{Div}(S) = \bigoplus_{\mathfrak{p} \in X_1(R)} \mathbb{Z}G \otimes_{\mathbb{Z}G_{\mathfrak{p}}} \mathbb{Z} = \bigoplus_{\mathfrak{p} \in X_1(R)} \operatorname{Hom}_{\mathbb{Z}G_{\mathfrak{p}}}(\mathbb{Z}G, \mathbb{Z})$$

as G-modules. Use the map $\psi : \operatorname{Hom}_{\mathbb{Z}G_{\mathfrak{p}}}(\mathbb{Z}G, \mathbb{Z}) \to \mathbb{Z}$ that is defined in Definition 5.1.12 (3).)

(3) Show that there is a norm map $N_R^S : \operatorname{Cl}(S) \to \operatorname{Cl}(R)$.

Exercise 6.5.14. Let R be a noetherian normal integral domain and x an indeterminate. Prove:

(1) $\operatorname{Cl}(R) \to \operatorname{Cl}(R[x]) \to \operatorname{Cl}(R[x, x^{-1}])$ are isomorphisms.

(2) $(R[x, x^{-1}])^* = R^* \times \langle x \rangle$.

(3) $\operatorname{Pic}(R) \to \operatorname{Pic}(R[x]) \to \operatorname{Pic}(R[x, x^{-1}])$ are isomorphisms.

Exercise 6.5.15. Give explicit definitions in terms of elements for all of the maps $\gamma_0, \ldots, \gamma_4$ in Theorem 6.5.8. Prove exactness at each term.

Exercise 6.5.16. Let k be a field and $n > 1$ an integer that is invertible in k. Assume k contains a primitive nth root of unity. Let $T = k[x, y]$ and $S = k[x^n, xy, y^n]$ (see Exercise 6.3.22). Show that the extension T/S satisfies the hypotheses of Section 6.5.3. Describe the automorphism group G which acts on T and fixes S. Compute all of the terms in the exact sequence of Theorem 6.5.8.

Chapter 7

Azumaya Algebras, I

An R-algebra A is said to be an *Azumaya* R-algebra if A is both central and separable over R. We show that the category of all Azumaya R-algebras contains every ring of the form $\operatorname{Hom}_R(P, P)$, if P is an R-progenerator module. The Commutator Theorems imply that an R-algebra A is an Azumaya algebra if and only if A is invertible in the sense that there exists an R-algebra B and an R-progenerator P such that $A \otimes_R B$ is isomorphic to $\operatorname{Hom}_R(P, P)$. Tensoring over R makes the set of isomorphism classes of Azumaya R-algebras into a monoid, and the Brauer group is defined to be the quotient modulo the submonoid of all endomorphism rings of the form $\operatorname{Hom}_R(P, P)$. Therefore, we say the Brauer group of R parametrizes the Azumaya R-algebras up to Brauer equivalence.

What we refer to as Azumaya algebras were first studied under the name central separable algebras by Auslander and Goldman in their seminal paper [**AG60a**] on the Brauer group of a commutative ring. The term Azumaya algebra was applied to these algebras by Grothendieck in his series of three papers on the Brauer group of a scheme ([**Gro68a**], [**Gro68b**], [**Gro68c**]). Prior to Auslander and Goldman's paper, and using a completely different set of terminology, Azumaya studied what we would call the Brauer group of a local ring ([**Azu51**]). What is now called the Brauer group of a field was studied by a number of twentieth century mathematicians, most notably Brauer, Hasse, Albert and E. Noether.

1. First Properties of Azumaya Algebras

As stated above, an R-algebra A is called an Azumaya R-algebra if A is both central and separable over R. The original source for the results in this section is [**AG60a**]. The following treatment is based on [**DI71**] and [**KO74b**].

Lemma 7.1.1. *If A is an Azumaya R-algebra, then $R = R{\cdot}1$ is an R-module direct summand of A.*

Proof. Let $e \in A^e$ be a separability idempotent for A. Consider the homomorphism $\varphi(e)$ in $\operatorname{Hom}_R(A, A)$, where φ is the enveloping map defined in Definition 4.1.1. Since φ is a ring homomorphism, $\varphi(e)$ is an idempotent R-module homomorphism. If B denotes the image of $\varphi(e)$ and C the kernel, then $A = B \oplus C$. But remembering that $J_{A/R}e = 0$, we see that for any $a, b \in A$,

$$a\big[\varphi(e)(b)\big] = (a \otimes 1)e \cdot b = (1 \otimes a)e \cdot b = \big[\varphi(e)(b)\big]a$$

so $\varphi(e)(b)$ is in $Z(A) = R$, the center of A. Therefore B, the image of $\varphi(e)$, is equal to R. □

Corollary 7.1.2. *Let A be an Azumaya R-algebra.*

(1) *If I is an ideal of R, then $IA \cap R = I$.*

(2) *If M is a maximal two-sided ideal of A, then there exists an ideal I of R with $IA = M$.*

Proof. (1): By Lemma 7.1.1, $A = L \oplus R$ for some R-submodule L of A. Then $IA \cap R = I(L \oplus R) \cap R = (IL \oplus I) \cap R = I$.

(2): Let $I = M \cap R$ and set $M_0 = IA$. By Proposition 4.3.6, A/M is a central separable algebra over R/I. Since M is maximal, A/M is simple and its center R/I is a field. By Part (1), $M_0 \cap R = I$. By Proposition 4.3.6, A/M_0 is a central separable algebra over the field R/I. By Corollary 4.5.4, A/M_0 is simple. Therefore M/M_0 is the zero ideal in A/M_0. This shows $M = M_0$. □

Now we show that the tensor product of two Azumaya algebras is an Azumaya algebra. We will see in Section 7.3 below that the binary operation in the Brauer group of R is induced by the tensor product.

Proposition 7.1.3. *Let A and B be Azumaya R-algebras. Then $A \otimes_R B$ is an Azumaya R-algebra.*

Proof. By Corollary 4.3.2, $A \otimes_R B$ is R-separable. Since R is an R-module direct summand of both A and B, $R \otimes_R R \cong R$ can be identified with a subring of $A \otimes_R B$. We have

$$\begin{aligned}
(A \otimes_R B)^{A \otimes_R B} &\cong \operatorname{Hom}_{(A \otimes_R B)^e}(A \otimes_R B, A \otimes_R B) && \text{(by Corollary 4.1.5)} \\
&= \operatorname{Hom}_{A^e \otimes_R B^e}(A \otimes_R B, A \otimes_R B) \\
&\cong \operatorname{Hom}_{A^e}(A, A) \otimes_R \operatorname{Hom}_{B^e}(B, B) && \text{(by Theorem 1.3.26)} \\
&\cong A^A \otimes_R B^B && \text{(by Corollary 4.1.5)} \\
&= R \otimes_R R = R.
\end{aligned}$$

Hence the center of $A \otimes_R B$ is R. □

Theorem 7.1.4. *Let A be an R-algebra. The following are equivalent.*

(1) *A is an Azumaya R-algebra.*

(2) *A is an A^e-progenerator and A is R-central.*

(3) *A is an R-progenerator and $\varphi : A^e \to \operatorname{Hom}_R(A, A)$ is an isomorphism, where φ is the enveloping map of Definition* 4.1.1.

Proof. (2) implies (1): This follows straight from Definitions 2.6.2 and 4.1.3.

(2) implies (3): By Corollary 4.1.5, $\operatorname{Hom}_{A^e}(A, A) \cong Z(A) = R$. We are in the context of the Morita Theorem (Theorem 1.5.2) with $R = A^e$ and $M = A$. By Corollary 1.5.3 (5), A is an R-progenerator because A is an A^e-progenerator. It follows from Corollary 1.5.3 (1) that $A^e \cong \operatorname{Hom}_R(A, A)$ under the enveloping map φ.

(3) implies (2): This follows by reversing the proof of (2) implies (3).

(1) implies (2): Since A is separable over R, A is projective over A^e. Since 1 generates A over A^e, it follows that A is finitely generated over A^e. It is enough to show that A is an A^e-generator module. By Lemma 1.5.1, it is enough to prove that $A^* \otimes_{\operatorname{Hom}_{A^e}(A,A)} A$ is isomorphic to A^e under the map $f \otimes a \mapsto f(a)$. By Corollary 4.1.5, $\operatorname{Hom}_{A^e}(A, A) \cong Z(A) = R$ since A is R-central. By Corollary 4.1.6, $A^* = \operatorname{Hom}_{A^e}(A, A^e)$ is isomorphic to $(0 : J_{A/R})$ under the map $f \mapsto f(1)$. It is enough to prove that $0 : J_{A/R} \otimes_R A \cong A^e$ under the map $b \otimes a \mapsto (a \otimes 1)b = (1 \otimes a)b$. The reader should verify that $A^e(0 : J_{A/R}) = (A \otimes_R 1)(0 : J_{A/R})$, so it is enough to show $A^e(0 : J_{A/R}) = A^e$. Assume $A^e(0 : J_{A/R})$ is not equal to A^e and proceed to a contradiction. If $A^e(0 : J_{A/R})$ is a proper ideal of A^e, it is contained in some maximal two-sided ideal M of A^e. By Proposition 7.1.3, A^e is a central separable R-algebra. By Corollary 7.1.2 (2), $M = IA^e$ for some proper ideal I of R, so $A^e(0 : J_{A/R}) \subseteq IA^e$. Applying the multiplication map μ, we get $A^e \cdot \mu(0 : J_{A/R}) \subseteq \mu(IA^e) = IA$. By Corollary 4.1.6, $\mu(0 : J_{A/R}) = R$. Since $A^e \cdot R = A$, it follows that $A \subseteq IA$. But $A = IA$ implies $A^e = IA^e = M$, a contradiction. □

Example 7.1.5. Let R be a commutative ring and $M_n(R)$ the ring of n-by-n matrices over R. The reader should verify that $M_n(R)$ is R-central. By Example 4.2.1, $M_n(R)$ is R-separable. Therefore, $M_n(R)$ is an Azumaya R-algebra.

If A is an Azumaya R-algebra, then we saw in the proof of Theorem 7.1.4 that there is a Morita equivalence between the category of left $A^e \cong \operatorname{Hom}_R(A, A)$-modules and the category of right R-modules. We restate this fact in Corollary 7.1.6. Under this equivalence, a right R-module M is mapped to the left A^e-module $M \otimes_R A$. Notice that multiplication by A^e is on the A factor: $a \otimes b \cdot x \otimes y = x \otimes ayb$, for $a \otimes b \in A^e$, $x \otimes y \in M \otimes_R A$. In the reverse direction, a left A^e-module N is mapped to N^A. If $x \in N^A$, then for every $a \in A$ we have $ax = xa$. For the converse to Corollary 7.1.6, see Exercise 7.4.10.

Corollary 7.1.6. *If A is an Azumaya R-algebra, then*

$$( ) \otimes_R A : \mathfrak{M}_R \to {}_{A^e}\mathfrak{M}$$

and

$$( )^A : {}_{A^e}\mathfrak{M} \to \mathfrak{M}_R$$

are inverse equivalences, establishing the equivalence of categories $\mathfrak{M}_R \sim {}_{A^e}\mathfrak{M}$. For any R-module M, $(M \otimes_R A)^A \cong M$ as R-modules under the map $m \otimes 1 \mapsto m$. For any left A^e-module N, $N^A \otimes_R A \cong N$ as left A^e-modules under the map $\sum n_i \otimes a_i \mapsto \sum n_i a_i$.

Proof. The first part follows from Theorem 7.1.4 and Exercise 1.5.6. The rest is left to the reader. □

Corollary 7.1.7. *Let A be an Azumaya R-algebra. There is a one-to-one correspondence between ideals I of R and two-sided ideals $\mathfrak{A}$ of A given by $I \mapsto IA$ and $\mathfrak{A} \mapsto \mathfrak{A} \cap R$.*

Proof. Since a two-sided ideal of A is a left A^e-module, this follows from Corollary 7.1.6. □

Corollary 7.1.8. *If Λ is an Azumaya R-algebra, then*

$$\mathrm{J}(\Lambda) = \bigcap_{\mathfrak{m} \in \operatorname{Max} R} \mathfrak{m}\Lambda = \mathrm{J}(R)\Lambda$$

where $\mathrm{J}( )$ denotes the Jacobson radical of a ring. In particular, if R is a local ring with maximal ideal $\mathfrak{m}$, then $\mathrm{J}(\Lambda) = \mathfrak{m}\Lambda$.

Proof. By Exercise 4.5.19,

$$\mathrm{J}(\Lambda) \subseteq \bigcap_{\mathfrak{m} \in \operatorname{Max} R} \mathfrak{m}\Lambda.$$

By Corollary 7.1.7,

$$\Big(\bigcap_{\mathfrak{m}\in \operatorname{Max} R} \mathfrak{m}\Lambda\Big) \cap R = \bigcap_{\mathfrak{m}\in \operatorname{Max} R} ((\mathfrak{m}\Lambda)\cap R) = \bigcap_{\mathfrak{m}\in \operatorname{Max} R} \mathfrak{m} = \mathrm{J}(R).$$

By Corollary 7.1.7 and Corollary 3.1.5,

$$\bigcap_{\mathfrak{m}\in \operatorname{Max} R} \mathfrak{m}\Lambda = \mathrm{J}(R)\Lambda \subseteq \mathrm{J}(\Lambda).$$

□

The next result shows that the property of being an Azumaya algebra is preserved under a change of base ring.

Corollary 7.1.9. *Let R be a commutative ring and S a commutative R-algebra. If A is an Azumaya R-algebra, then $A \otimes_R S$ is an Azumaya S-algebra.*

Proof. By Theorem 7.1.4, A is an R-progenerator. By Theorem 1.3.2, $A \otimes_R S$ is an S-progenerator. By Theorem 7.1.4, the enveloping map $\varphi : A^e \to \operatorname{Hom}_R(A, A)$ of Definition 4.1.1 is an isomorphism. This together with Exercise 1.3.10 and Corollary 1.3.27 give the following isomorphisms of S-algebras

$$\begin{aligned}(A \otimes_R S) \otimes_S (A \otimes_R S)^o &\cong (A \otimes_R A^o) \otimes_R S \\ &\cong \operatorname{Hom}_R(A, A) \otimes_R S \\ &\cong \operatorname{Hom}_S(A \otimes_R S, A \otimes_R S).\end{aligned}$$

The reader should verify that the composite map is φ, so Theorem 7.1.4 implies that $A \otimes_R S$ is an Azumaya S-algebra. □

Proposition 7.1.10 can be considered a generalization of Example 7.1.5.

Proposition 7.1.10. *If R is a commutative ring and E is an R-progenerator module, then $A = \operatorname{Hom}_R(E, E)$ is an Azumaya R-algebra.*

Proof. We will show that A satisfies the conditions of Theorem 7.1.4 (3). It follows from Lemma 1.5.1 (1) that $\theta_R : E^* \otimes_R E \to \operatorname{Hom}_R(E, E) = A$ is an isomorphism. By Corollary 1.5.3 (6), E^* is an R-progenerator. By Proposition 1.3.3, $A \cong E^* \otimes_R E$ is an R-progenerator. Finally we show that the map $\varphi : A^e \to \operatorname{Hom}_R(A, A)$ is an isomorphism. By Corollary 1.5.3 (4), $A^o \cong \operatorname{Hom}_R(E^*, E^*)$ and the isomorphism maps $\beta \in A^o$ to "right multiplication by β". Using Corollary 1.3.28 we have

$$\begin{aligned}A \otimes_R A^o &\cong \operatorname{Hom}_R(E, E) \otimes_R \operatorname{Hom}_R(E^*, E^*) \\ &\cong \operatorname{Hom}_R(E \otimes_R E^*, E \otimes_R E^*) \\ &\cong \operatorname{Hom}_R(A, A).\end{aligned}$$

The reader should verify that the composition of these isomorphisms is indeed φ. □

In Theorem 4.4.2 we proved the transitivity property for separable algebras. Theorem 7.1.11 is a partial converse.

Theorem 7.1.11. *An R-algebra A is separable if and only if A is an Azumaya algebra over its center and its center is a separable R-algebra.*

Proof. Let $C = Z(A)$ denote the center of A. It follows by Theorem 4.4.2 (1) that if A is separable over C and C is separable over R, then A is separable over R. Conversely, assume A is R-separable. By Theorem 4.4.2 (2), A is separable over its center C. By Theorem 7.1.4, A is a C-progenerator and by Theorem 4.4.2 (3) C is R-separable. □

We end this section with a result by Auslander and Goldman ([**AG60a**, Proposition 4.6]) on Azumaya algebras over a noetherian normal domain.

Theorem 7.1.12. *Let R be a noetherian integrally closed integral domain and Λ an R-algebra. The following are equivalent.*

(1) *Λ is an Azumaya R-algebra.*

(2) *Λ is a finitely generated projective R-module, and for every height one prime $\mathfrak{p} \in X_1(R)$, $\Lambda_\mathfrak{p} = R_\mathfrak{p} \otimes_R \Lambda$ is an Azumaya $R_\mathfrak{p}$-algebra.*

Proof. (1) implies (2): This follows from Theorem 7.1.4 and Corollary 7.1.9.

(2) implies (1): Let K be the field of fractions of R. For any minimal prime $\mathfrak{p} \in X_1(R)$, $\Lambda_\mathfrak{p} = R_\mathfrak{p} \otimes_R \Lambda$ is an Azumaya $R_\mathfrak{p}$-algebra. By Corollary 7.1.9 and Corollary 4.5.4, $A = K \otimes_R \Lambda = K \otimes_{R_\mathfrak{p}} R_\mathfrak{p} \otimes_R \Lambda$ is a central simple K-algebra. Consider the commutative diagram

$$\begin{array}{ccc} \Lambda \otimes_R \Lambda^o & \xrightarrow{\varphi_{\Lambda/R}} & \operatorname{Hom}_R(\Lambda, \Lambda) \\ \downarrow & & \downarrow \\ A \otimes_K A^o & \xrightarrow{\varphi_{A/K}} & \operatorname{Hom}_K(A, A) \end{array}$$

where $\varphi_{A/K}$ and $\varphi_{\Lambda/R}$ are the enveloping homomorphisms. Then $\varphi_{A/K}$ is an isomorphism. Since Λ is an R-progenerator, $\Lambda \to A$ is one-to-one. So Λ is an R-lattice in A, $\Lambda \otimes_R \Lambda^o$ is an R-lattice in $A \otimes_R A^o$, and $\operatorname{Hom}_R(\Lambda, \Lambda)$ is an R-lattice in $\operatorname{Hom}_K(A, A)$. For each $\mathfrak{p} \in X_1(R)$, the localization of $\varphi_{\Lambda/R}$ at $\mathfrak{p}$ is an isomorphism. By Corollary 6.3.14, $\varphi_{\Lambda/R}$ is an isomorphism. By Theorem 7.1.4, Λ is an Azumaya R-algebra. □

2. The Commutator Theorems

The commutator theorems were originally proved by Auslander and Goldman in [**AG60a**]. The treatment below is based on [**DI71**]. Consider now the following general situation. Let A be any R algebra and B any R-subalgebra of A. Then A is naturally a two-sided B/R-module and, as in Section 4.1, we let $A^B = \{a \in A \mid ab = ba \text{ for all } b \in B\}$. In this setting A^B is seen to be an R-subalgebra of A which commutes with B.

Lemma 7.2.1. *Let A be an Azumaya R-algebra. Suppose $R\cdot 1 \subseteq B \subseteq A$ is a chain of R-subalgebras of A, and $R\cdot 1 \subseteq C \subseteq A^o$ is a chain of R-subalgebras of A^o. Then*

$$(A \otimes_R A^o)^{B\otimes_R A^o} = A^B$$

and

$$(A \otimes_R A^o)^{A\otimes_R C} = (A^o)^C.$$

Proof. By Exercise 1.3.8, we have two commutative diagrams

$$(7.1)\qquad \begin{array}{ccccc} & & A\otimes_R A^o & & \\ & \nearrow & \uparrow{\scriptstyle\gamma} & \nwarrow & \\ B & \xrightarrow[\rho_1]{} & B\otimes_R A^o & \xleftarrow[\rho_2]{} & A^o \end{array} \qquad \begin{array}{ccccc} & & A\otimes_R A^o & & \\ & \nearrow & \uparrow{\scriptstyle\gamma} & \nwarrow & \\ A & \xrightarrow[\rho_1]{} & A\otimes_R C & \xleftarrow[\rho_2]{} & C \end{array}$$

of R-algebras and R-algebra homomorphisms. The reader should verify that all of the maps in (7.1) are one-to-one, so we view all of the rings as subrings of $A \otimes_R A^o$. Applying Corollary 7.1.6, we get

$$\begin{aligned} (A \otimes_R A^o)^{B\otimes_R A^o} &= \left((A \otimes_R A^o)^{A^o}\right)^B \\ &= A^B \end{aligned}$$

and

$$\begin{aligned} (A \otimes_R A^o)^{A\otimes_R C} &= \left((A \otimes_R A^o)^{A}\right)^C \\ &= (A^o)^C. \end{aligned}$$

$\square$

Theorem 7.2.2. *Let A be an Azumaya R-algebra. Suppose $R\cdot 1 \subseteq B \subseteq A$ is a chain of R-subalgebras of A.*

(1) *If B is R-separable, then $C = A^B$ is R-separable and $A^C = B$.*
(2) *If B is R-Azumaya, then $C = A^B$ is R-Azumaya and the R-algebra map $B \otimes_R C \to A$ given by $b \otimes c \mapsto bc$ is an isomorphism.*

Proof. (2): Assume that B is an Azumaya R-algebra. The specified map is an R-algebra homomorphism because B and C commute (Exercise 1.3.8). The given map is an isomorphism by Corollary 7.1.6. Since R is an R-direct summand of B, C is R-separable by Corollary 4.3.4. If x is in the center of C, it commutes with all of $B \otimes_R C \cong A$. Therefore x is in the center of A which is R, showing that C is central. By Theorem 7.1.4, C is R-Azumaya. Corollary 7.1.6 implies that $A^C = (B \otimes_R C)^C = B$.

(1): First we prove the special case where $A = \operatorname{Hom}_R(M, M)$ for some R-progenerator M. Assume B is a separable subalgebra of A containing $R \cdot 1$. Then $C = A^B = \left[\operatorname{Hom}_R(M, M)\right]^B = \operatorname{Hom}_B(M, M)$. By Theorem 7.1.11, B is central separable over its center $Z(B)$ and $Z(B)$ is separable over R. By Theorem 4.4.1, M is projective over $Z(B)$. Clearly M is A-faithful, so it is $Z(B)$-faithful. Because M is finitely generated over R, it is finitely generated over $Z(B)$. By Corollary 1.1.16, M is a $Z(B)$-progenerator. We can view B as a central separable subalgebra of $\operatorname{Hom}_{Z(B)}(M, M)$, which itself is central separable over $Z(B)$, by Proposition 7.1.10. By Part (2) applied to $Z(B) \subseteq B \subseteq \operatorname{Hom}_{Z(B)}(M, M)$,

$$C = A^B = \operatorname{Hom}_B(M, M) = \left[\operatorname{Hom}_{Z(B)}(M, M)\right]^B$$

is a central separable $Z(B)$-algebra and $\left[\operatorname{Hom}_{Z(B)}(M, M)\right]^C = B$. Theorem 4.4.2 (1) proves that C is a separable R-algebra and the fact that $Z(B) \subseteq C$ implies

$$B = \left[\operatorname{Hom}_{Z(B)}(M, M)\right]^C = \left[\operatorname{Hom}_R(M, M)\right]^C = A^C.$$

Now we prove Part (1). By Theorem 7.1.4, A is an R-progenerator and $\varphi : A \otimes_R A^o \to \operatorname{Hom}_R(A, A)$ is an isomorphism. Therefore, the special case of Part 1 applies to the subalgebras of $A \otimes_R A^o$. If B is any separable subalgebra of A containing R, then $B \otimes_R A^o$ is a separable subalgebra of $A \otimes_R A^o$ (Corollary 4.3.2). Applying the special case of Part (1) and Lemma 7.2.1, we have

$$C = A^B = \left(A \otimes_R A^o\right)^{B \otimes_R A^o}$$

is a separable R-algebra and

$$B \otimes_R A^o = \left(A \otimes_R A^o\right)^C.$$

It follows that

$$A \cap \left(B \otimes_R A^o\right) = A \cap \left(A \otimes_R A^o\right)^C = A^C.$$

The proof is complete if we show $A \cap \left(B \otimes_R A^o\right) = B$. However, since $A^o = R \oplus L$ for some R-submodule L of A^o,

$$B \otimes_R A^o \cong B \oplus (B \otimes_R L)$$

and

$$A \otimes_R A^o \cong A \oplus (A \otimes_R L),$$

so

$$(B \otimes_R L) \cap A \subseteq (A \otimes_R L) \cap A = 0.$$

Consequently any element of $B \oplus (B \otimes_R L)$ in A must be in B. □

Theorem 7.2.2 states that the Azumaya subalgebras of an Azumaya R-algebra A occur in pairs. Each member of a pair is the commutator subalgebra of the other and the tensor product of the pair is isomorphic to A. In Theorem 7.2.3 it is shown that a representation of A as a tensor product of subalgebras can only occur in this way.

Theorem 7.2.3. *Let A be an Azumaya R-algebra. Suppose B and C are subalgebras such that the map from $B \otimes_R C$ to A given by $b \otimes c \mapsto bc$ is an R-algebra isomorphism. Then B and C are Azumaya R-algebras with $A^B = C$ and $A^C = B$.*

Proof. By Proposition 2.4.6, $R \cdot 1$ is an R-module direct summand of both B and C. By Corollary 4.3.4, B and C are separable algebras. By Exercise 1.3.8, elements of B commute with elements of C. So the center of B is contained in the center of $B \otimes_R C \cong A$. Therefore B is central as well as separable. Similarly for C. Finally, by Corollary 7.1.6,

$$A^B = (B \otimes_R C)^B = C$$

and

$$A^C = (B \otimes_R C)^C = B.$$

□

Corollary 7.2.4. *Let A be an R-algebra. The following are equivalent.*

(1) *A is an Azumaya R-algebra.*
(2) *There exists an R-progenerator P, an R-algebra B and an R-algebra isomorphism $A \otimes_R B \cong \operatorname{Hom}_R(P, P)$.*
(3) *There exists an R-algebra B such that $A \otimes_R B$ is an Azumaya R-algebra.*

Proof. (1) implies (2): This follows from Theorem 7.1.4.

(2) implies (3): This follows from Proposition 7.1.10.

(3) implies (1): By Theorem 7.1.4, $A \otimes_R B$ is an R-progenerator. By Proposition 1.3.4, A and B are both progenerators over R. By Proposition 2.4.6, $A \to A \otimes_R R$ and $B \to R \otimes_R B$ are isomorphisms. By Theorem 7.2.3, we are done. □

3. The Brauer Group

In this section, the Brauer group functor is defined and its first properties are studied. The original source for the results in this section is [**AG60a**]. The approach taken below is based on [**DI71**].

Let R be a commutative ring and let $\operatorname{Az}(R)$ denote the set of isomorphism classes of Azumaya R-algebras. Although elements of $\operatorname{Az}(R)$ are isomorphism classes, we usually speak of the isomorphism class of an algebra and the algebra that represents an isomorphism class as one and the same object. Let A and B be two Azumaya R-algebras. We say that A and B are *Brauer equivalent* and write $A \sim B$ in case there exist R-progenerators P and Q such that

$$A \otimes_R \operatorname{Hom}_R(P, P) \cong B \otimes_R \operatorname{Hom}_R(Q, Q).$$

If P is an R-progenerator module, then by Proposition 7.1.10, $\operatorname{Hom}_R(P, P)$ is an Azumaya R-algebra. Let $\operatorname{Az}^0(R)$ denote the subset of $\operatorname{Az}(R)$ consisting of all isomorphism classes containing a representative of the form $\operatorname{Hom}_R(P, P)$, where P is an R-progenerator module.

Lemma 7.3.1. *Under the tensor product operation,* $\operatorname{Az}(R)$ *is an abelian monoid with identity element* R. *The set* $\operatorname{Az}^0(R)$ *is a submonoid. Brauer equivalence is an equivalence relation on* $\operatorname{Az}(R)$.

Proof. If A and B are Azumaya R-algebras, then by Proposition 7.1.3, $A \otimes_R B \cong B \otimes_R A$ is also an Azumaya R-algebra. Since R is trivially an Azumaya R-algebra, $\operatorname{Az}(R)$ is an abelian monoid under the tensor product operation. If P and Q are both R-progenerators, then by Proposition 1.3.3, so is $P \otimes_R Q$. By Corollary 1.3.28,

$$\operatorname{Hom}_R(P, P) \otimes_R \operatorname{Hom}_R(Q, Q) \to \operatorname{Hom}_R(P \otimes_R Q, P \otimes_R Q)$$

is an R-algebra isomorphism. Therefore $\operatorname{Az}^0(R)$ is a submonoid of $\operatorname{Az}(R)$. The rest is left to the reader. □

Let $\operatorname{B}(R)$ denote the set of equivalence classes of Azumaya R-algebras under the Brauer equivalence relation. We call $\operatorname{B}(R)$ the *Brauer group* of R. Lemma 7.3.2 below shows that $\operatorname{B}(R)$ is indeed an abelian group, the binary operation is induced by tensor product and the identity class is the set $\operatorname{Az}^0(R)$. If A is an Azumaya R-algebra, let $[A]$ denote the Brauer equivalence class to which A belongs.

Lemma 7.3.2. *The binary operation on* $\operatorname{Az}(R)$ *induces the structure of an abelian group on* $\operatorname{B}(R)$. *The identity element is the class* $[R]$ *and the inverse of* $[A]$ *is* $[A^o]$.

Proof. Let A and B be Azumaya R-algebras. Define a binary operation on $\mathrm{B}(R)$ by setting $[A][B] = [A \otimes_R B]$. The reader should verify that this is a well defined product and $\mathrm{B}(R)$ is an abelian monoid. By Theorem 7.1.4, A is an R-progenerator and $A \otimes_R A^o$ is isomorphic to $\mathrm{Hom}_R(A, A)$ which is in $\mathrm{Az}^0(R)$. It follows from $[A][A^o] = \left[\mathrm{Hom}_R(A, A)\right] = \left[\mathrm{Hom}_R(R, R)\right] = [R]$ that $[A]^{-1} = [A^o]$. □

Proposition 7.3.3. *The Brauer group functor, which is denoted by* $\mathrm{B}(\,)$, *is a covariant functor from the category of commutative rings (and ring homomorphisms) to the category of abelian groups (and group homomorphisms).*

Proof. Let $f : R \to S$ be a homomorphism of commutative rings. The assignment $A \mapsto A \otimes_R S$ defines a function $\mathrm{Az}(R) \to \mathrm{Az}(S)$ (Corollary 7.1.9). It follows from Theorem 1.3.2 and Corollary 1.3.27 that $\mathrm{Az}^0(R)$ is mapped to $\mathrm{Az}^0(S)$, so there is a well defined function

$$\mathrm{B}(f) : \mathrm{B}(R) \to \mathrm{B}(S)$$

under which $[A] \mapsto [A \otimes_R S]$. The rest of the proof is left to the reader. □

Proposition 7.3.4. *Let A and B be Azumaya R-algebras.*

(1) *There is a Brauer equivalence $A \sim R$ if and only if there exists an R-progenerator module M such that $A \cong \mathrm{Hom}_R(M, M)$.*

(2) *There is a Brauer equivalence $A \sim B$ if and only if there exists an R-progenerator module M such that $A \otimes_R B^o \cong \mathrm{Hom}_R(M, M)$.*

Proof. The proof of (2) is left to the reader.

(1): We prove that if $A \sim R$, then $A \in \mathrm{Az}^0(R)$. The converse is left to the reader. Assume there exist R-progenerators P and Q such that

$$A \otimes_R \mathrm{Hom}_R(P, P) \cong \mathrm{Hom}_R(Q, Q).$$

Set $B = \mathrm{Hom}_R(P, P)$. Via this isomorphism we identify A and B with subalgebras of $\mathrm{Hom}_R(Q, Q)$. By Theorem 7.2.3, $(\mathrm{Hom}_R(Q, Q))^B = A$. We view Q as a left B-module by virtue of the fact that B is a subring of $\mathrm{Hom}_R(Q, Q)$, which is a ring of endomorphisms of Q. Since $B = \mathrm{Hom}_R(P, P)$ and P is an R-progenerator, we are in a Morita context. By Theorem 1.5.2, there is an equivalence of categories ${}_R\mathfrak{M} \sim {}_B\mathfrak{M}$. Under this equivalence, the R-module $M = P^* \otimes_B Q$ corresponds to the left B-module Q. It follows from the fully faithful part of Proposition 1.1.2 that

$$\mathrm{Hom}_R(M, M) \cong \mathrm{Hom}_B(Q, Q)$$

as R-algebras. But

$$\begin{aligned}
&\operatorname{Hom}_B(Q,Q)\\
&\qquad = \{f \in \operatorname{Hom}_R(Q,Q) \mid f\bigl(b(x)\bigr) = b\bigl(f(x)\bigr), \forall b \in B \text{ and } \forall x \in Q\}\\
&\qquad = (\operatorname{Hom}_R(Q,Q))^B\\
&\qquad \cong A
\end{aligned}$$

which shows $A \cong \operatorname{Hom}_R(M,M)$. To finish the proof, by Corollary 1.1.16, it is enough to verify that M is finitely generated projective and faithful over R. Because $\operatorname{Hom}_R(M,M) \cong A$ is a faithful R-algebra, M is a faithful R-module. Because M corresponds under Morita equivalence to Q, by Corollary 1.5.5, it is enough to show that Q is finitely generated and projective over B. Since Q is R-projective, it follows from Theorem 4.4.1 that Q is B-projective. Since Q is finitely generated over R, Q is finitely generated over B. □

Despite the obvious inconsistency, we frequently use the notation of additive abelian groups and write 0 for the identity element of the group $\mathrm{B}(R)$. For instance, if $\mathrm{Az}^0(R) = \mathrm{Az}(R)$, then we write $\mathrm{B}(R) = (0)$.

Remark 7.3.5. We mention here two variations of the notions of Azumaya algebras and Brauer equivalence classes, neither of which is explored further in this book.

(1) For a commutative ring R, a "bigger Brauer group" was introduced by J. Taylor in [**Tay82**] and [**RT85**]. By construction, it classifies generalized Azumaya R-algebras which are allowed to be nonunital (in other words, algebras that do not necessarily contain 1). The interested reader is referred to [**Cae98**, Chapter 3] for more on this subject.

(2) Let R be a noetherian integrally closed integral domain. Then there is another variation of the notions of Azumaya R-algebras and Brauer equivalence classes which is based on the use of reflexive R-lattices instead of R-progenerator modules. This leads to the definitions of so-called reflexive Azumaya algebras and a reflexive Brauer group, where the binary operation is induced by a modified tensor product. For more on this subject, the interested reader is referred to [**Aus66**], [**LO82**], [**Orz82**], [**Yua74b**], and [**LB87**].

4. Splitting Rings

This section contains an introduction to the notion of a splitting ring for an Azumaya algebra. The first properties of splitting rings are studied. This important topic is revisited in Section 10.3. The main results in this section are Theorems 7.4.2 and 7.4.3, both of which are from [**AG60a**]. The proofs appearing below are based on [**KO74b**] and [**DI71**].

Let $f : R \to S$ be a homomorphism of commutative rings. The *relative Brauer group of S over R* is denoted $\mathrm{B}(S/R)$ and is defined to be the kernel of $\mathrm{B}(f) : \mathrm{B}(R) \to \mathrm{B}(S)$. If A is an Azumaya R-algebra, then by Proposition 7.3.3, $[A] \in \mathrm{B}(S/R)$ if and only if $A \otimes_R S \cong \mathrm{Hom}_S(M, M)$ for some S-progenerator M. If this is the case, then we say *S splits A* and that S is a *splitting ring* for A.

Example 7.4.1. Let $\mathfrak{m}$ be a maximal ideal of R and k an algebraically closed field containing the residue field $R/\mathfrak{m}$. Let $f : R \to k$ be the natural map. By Theorem 7.5.1 which is proved in the next section, $\mathrm{B}(k) = (0)$. Therefore, $\mathrm{B}(k/R) = \mathrm{B}(R)$.

Let R be a commutative ring and A an Azumaya R-algebra. A commutative subalgebra S of A is called a *maximal commutative subalgebra* if S is a maximal member of the set of all commutative R-subalgebras of A. By an application of Zorn's Lemma, maximal commutative subalgebras exist.

Theorem 7.4.2. *Let A be an Azumaya R-algebra and S a maximal commutative subalgebra. Then the following are true.*

(1) *There is an S-algebra isomorphism α such that the diagram*

$$\begin{array}{ccc} A \otimes_R S & \xrightarrow{\alpha} & \mathrm{Hom}_S(A, A) \\ {\scriptstyle\gamma}\downarrow & & \downarrow \\ A \otimes_R A^o & \xrightarrow{\varphi} & \mathrm{Hom}_R(A, A) \end{array}$$

commutes.

(2) *If A is a projective S-module, then A is an S-module progenerator and S is an R-module progenerator. If $\mathfrak{p}$ is a prime ideal of R such that* $\mathrm{Rank}_{S_{\mathfrak{p}}}(A_{\mathfrak{p}})$ *is defined, then*

$$\mathrm{Rank}_{R_{\mathfrak{p}}}(A_{\mathfrak{p}}) = \left(\mathrm{Rank}_{S_{\mathfrak{p}}}(A_{\mathfrak{p}})\right)^2, \text{ and}$$
$$\mathrm{Rank}_{R_{\mathfrak{p}}}(S_{\mathfrak{p}}) = \mathrm{Rank}_{S_{\mathfrak{p}}}(A_{\mathfrak{p}}).$$

(3) *If S is separable over R, then A is a projective S-module.*

(4) *If the hypothesis for either (2) or (3) is satisfied, then S is a splitting ring for A.*

Proof. (1): Tensoring $S \to A^o$ with the flat R-module A, we see that γ is one-to-one. Applying Lemma 7.2.1 and Exercise 7.4.6 we have

$$S = (A^o)^S = \left(A \otimes_R A^o\right)^{A \otimes_R S} = \left(\left(A \otimes_R A^o\right)^S\right)^A.$$

Tensoring with $A \otimes_R (\ )$ and applying Corollary 7.1.6 we get

$$A \otimes_R S = A \otimes_R \left((A \otimes_R A^o)^S \right)^A = (A \otimes_R A^o)^S.$$

By Theorem 7.1.4, φ is an isomorphism, so

$$A \otimes_R S = (A \otimes_R A^o)^S = (\operatorname{Hom}_R(A, A))^S = \operatorname{Hom}_S(A, A).$$

Therefore, we see that the image of $A \otimes_R S$ under the composite mapping $\varphi\gamma$ is equal to $\operatorname{Hom}_S(A, A)$.

(2): By Theorem 7.1.4, A is an R-progenerator. Because of the chain of subrings $R \subseteq S \subseteq A$, we see that S is a faithful R-module, and A is finitely generated and faithful over S. By Corollary 1.1.16, A is an S-progenerator. By Proposition 2.4.6, $S \cdot 1$ is an S-module direct summand of A. Since A is an R-projective, S is R-projective. By Corollary 1.1.16, S is an R-progenerator. The first rank formula comes from Part (1) and $\operatorname{Rank}_R(A) = \operatorname{Rank}_S(A \otimes_R S)$. The second formula comes from the isomorphism of S-modules: $A \otimes_R S \cong A \otimes_S (S \otimes_R S)$ and the first formula.

(3): By Theorem 4.4.3, A is a projective $A \otimes_R S$-module. But $A \otimes_R S$ is a projective S-module.

(4): A is an S-progenerator and by (1), $A \otimes_R S$ is split. □

Theorem 7.4.3. *Let A be an Azumaya R-algebra. Let S be a commutative R-algebra which is an R-module progenerator. If A is split by S, then there is an Azumaya R-algebra B which is Brauer equivalent to A such that B contains a maximal commutative subalgebra which is isomorphic to S. Moreover, B is projective as a left S-module, and for each prime $\mathfrak{p} \in \operatorname{Spec} R$, $\operatorname{Rank}_{R_\mathfrak{p}}(A_\mathfrak{p}) = \left(\operatorname{Rank}_{R_\mathfrak{p}}(S_\mathfrak{p})\right)^2$.*

Proof. Suppose S is finitely generated, projective, and faithful as an R-module, and A is an Azumaya R-algebra split by S. Since $\mathrm{B}(S/R)$ is a group, A^o is also split by S, so there exists an S-progenerator E such that

$$S \otimes_R A^o \cong \operatorname{Hom}_S(E, E). \tag{7.2}$$

Since S is an R-progenerator, it follows by Proposition 1.1.8 (3) that E is an R-progenerator. The Azumaya S-algebra $\operatorname{Hom}_S(E, E)$ is contained in the Azumaya R-algebra $\operatorname{Hom}_R(E, E)$. Both S and A^o are R-projective and hence flat. By (7.2) we can identify S and A^o with commuting R-subalgebras of $\operatorname{Hom}_R(E, E)$ such that $S \cdot A^o = \operatorname{Hom}_S(E, E)$. Set

$$B = \left[\operatorname{Hom}_R(E, E)\right]^{A^o}.$$

By Theorem 7.2.2, B is an Azumaya R-algebra with $B \otimes_R A^o \cong \operatorname{Hom}_R(E, E)$. Applying Proposition 7.3.4 (2), we conclude that $[B] = [A]$. Since S and A^o

are commuting subalgebras of $\operatorname{Hom}_R(E, E)$, we see that

$$S \subseteq \left[\operatorname{Hom}_R(E, E)\right]^{A^o} = B.$$

Therefore S is a commutative subalgebra of B. Note that

$$B^S = \left(\left(\operatorname{Hom}_R(E, E)\right)^{A^o}\right)^S \subseteq \\ \left(\operatorname{Hom}_R(E, E)\right)^S = \operatorname{Hom}_S(E, E) = S \cdot A^o \cong S \otimes_R A^o.$$

Since elements of B commute with elements of A^o, it follows that $B^S \subseteq Z(S \otimes_R A^o) = S$. Therefore S is a maximal commutative subalgebra of B, by Exercise 7.4.6. The isomorphism

$$\operatorname{Hom}_R(E, R) \otimes_R E \xrightarrow{\theta_R} \operatorname{Hom}_R(E, E)$$

of Section 1.5 is a left and right $\operatorname{Hom}_R(E, E)$-module homomorphism. Therefore, θ_R is a left S-module homomorphism. From the left, the action by the ring $\operatorname{Hom}_R(E, E)$ is on the E factor. Since E is S-projective, and $\operatorname{Hom}_R(E, R)$ is R-projective, we see that $\operatorname{Hom}_R(E, E)$ is S-projective. Using $B \otimes_R A^o \cong \operatorname{Hom}_R(E, E)$ and the fact that A is R-projective, we see that B is an S-module direct summand of $\operatorname{Hom}_R(E, E)$, hence B is S-projective. The rank formula is from Theorem 7.4.2. □

We end this section with two more variations of the Commutator Theorems. Theorem 7.4.4 is from [**KOS76**] and Theorem 7.4.5 is from [**Sal99**].

Theorem 7.4.4. *Let A be an Azumaya R-algebra and S a commutative R-subalgebra of A such that A is a projective left S-module. Then A^S is an Azumaya S-algebra. Moreover, A^S is Brauer equivalent to $A \otimes_R S$.*

Proof. The proof is similar to that of Theorem 7.4.2. The details are left to the reader. Notice that $A \otimes_R (A^o)^S = A \otimes_R \left(A \otimes_R A^o)^S\right)^A = (A \otimes_R A^o)^S$. Therefore, the image of $A \otimes_R (A^o)^S$ in $\operatorname{Hom}_R(A, A)$ is equal to $\operatorname{Hom}_S(A, A)$. □

Theorem 7.4.5. *Let S be a commutative R-algebra which is an R-module progenerator. Let A be an Azumaya R-algebra and C an Azumaya S-algebra such that $A \otimes_R S$ is Brauer equivalent to C. Then there is an Azumaya R-algebra B such that B contains a subalgebra which is isomorphic to S, $B^S = C$, B is projective as a left S-module, and B is Brauer equivalent to A.*

Proof. The proof is similar to that of Theorem 7.4.4. The details are left to the reader. There is an S-progenerator module E and an isomorphism $C \otimes_S (S \otimes_R A^o) \cong \operatorname{Hom}_S(E, E)$. Let B be the centralizer of A^o in $\operatorname{Hom}_R(E, E)$. □

4.1. Exercises.

Exercise 7.4.6. Let R be a commutative ring, A an R-algebra, and S a commutative R-subalgebra of A. Show that $S \subseteq A^S = \{x \in A \mid sx = xs, \text{ for all } s \in S\}$ with equality holding if and only if S is a maximal commutative subalgebra.

Exercise 7.4.7. Let A be an Azumaya R-algebra and S a maximal commutative subalgebra of A. Assume A is a left S-progenerator module. Prove that

$$(\ ) \otimes_S A : \mathfrak{M}_S \to {}_{S \otimes_R A^o}\mathfrak{M}$$

and

$$(\ )^S : {}_{S \otimes_R A^o}\mathfrak{M} \to \mathfrak{M}_R$$

are inverse equivalences, establishing $\mathfrak{M}_S \sim {}_{S \otimes_R A^o}\mathfrak{M}$. (Hints: Use Theorem 7.4.2, Exercise 1.5.6 and Exercise 4.1.15.)

Exercise 7.4.8. Let R_1 and R_2 be commutative rings and $R = R_1 \oplus R_2$ the direct sum. Let $\pi_i : R \to R_i$ be the projection map. Show:

(1) An Azumaya R-algebra is isomorphic to a direct sum $A_1 \oplus A_2$ where A_i is an Azumaya R_i-algebra. Moreover, $A_1 \oplus A_2$ is in $\mathrm{Az}^0(R)$ if and only if A_i is in $\mathrm{Az}^0(R_i)$ for each i.

(2) The projection maps induce an isomorphism of abelian groups $\mathrm{B}(R) \cong \mathrm{B}(R_1) \oplus \mathrm{B}(R_2)$.

Exercise 7.4.9. Let R be a commutative ring and A an Azumaya R-algebra. Show that A is Brauer equivalent to an Azumaya R-algebra B which has constant rank. (Hint: Use Exercise 7.4.8 and Corollary 2.3.6.)

Exercise 7.4.10. The point to this exercise is to prove the converse to Corollary 7.1.6. Let A be an R-algebra and assume the functors $(\ ) \otimes_R A : \mathfrak{M}_R \to {}_{A^e}\mathfrak{M}$ and $(\ )^A : {}_{A^e}\mathfrak{M} \to \mathfrak{M}_R$ define an equivalence of categories $\mathfrak{M}_R \sim {}_{A^e}\mathfrak{M}$. Show that A is an Azumaya R-algebra. (Hint: Apply Corollary 4.1.7 to show A is R-separable.)

Exercise 7.4.11 ([**DI71**, Exercise 2.8.10])**.** Prove the following generalization of Exercise 4.5.19 and Corollary 7.1.8. Let R be a commutative ring and Λ an integral separable R-algebra. Then

$$\mathrm{J}(\Lambda) = \bigcap_{\mathfrak{m} \in \operatorname{Max} R} \mathfrak{m}\Lambda.$$

5. Azumaya Algebras over a Field

Throughout this section k represents a field. By Corollary 4.5.4, being an Azumaya k-algebra is equivalent to being a central simple k-algebra. Let A be an Azumaya k-algebra. In this section we prove some powerful theorems concerning the existence of splitting subrings of A which are k-separable. First we show that the Brauer class $[A]$ contains a unique central k-division algebra D. Then we show that D contains an element α such that $k(\alpha)$ is separable over k and is a maximal subfield of D containing k. Then we show that A contains an element β such that $k(\beta)$ is separable over k and is a maximal commutative k-subalgebra of A. Notice that Theorem 7.4.2 (4) implies A is split by both $k(\alpha)$, and $k(\beta)$. Lastly, we show that A is split by a Galois field extension K/k. It is worth emphasizing that if k is a commutative ring which is not a field, then the counterparts of these results do not hold in general.

The article [**ABGV11**] is recommended as an excellent survey paper detailing many open problems on Azumaya algebras over a field in addition to providing a thorough description of the current state of Brauer group theory over a field.

Theorem 7.5.1. *Let k be a field.*

(1) *Then* $\mathrm{B}(k)$ *parametrizes the isomorphism classes of finite dimensional central division algebras over k.*
(2) *If k is algebraically closed, then* $\mathrm{B}(k)$ *is isomorphic to* (0).

Proof. (1): By Corollary 4.5.4, a k-algebra A is k-Azumaya if and only if it is k-central simple. Since k is a field, a k-progenerator module is a finite dimensional k-vector space so every class in $\mathrm{Az}^0(k)$ is represented by a matrix ring $M_n(k)$ for some positive n. Let A be an Azumaya k-algebra. By Theorem 3.2.6, A is isomorphic to $\mathrm{Hom}_D(M, M)$ where D is a finite dimensional central division algebra over k and M is a finite dimensional left D-vector space. Then A is isomorphic to the ring of matrices $M_n(D^o) = D^o \otimes_k M_n(k)$ (for example, [**Hun80**, Theorem VII.1.4]). Thus, A is Brauer equivalent to D^o, or $[A] = [D]^{-1}$ in $\mathrm{B}(k)$. If D_1 and D_2 are finite dimensional central k-division algebras and $[D_1] = [D_2]$ then there exist integers n_1 and n_2 such that $D_1 \otimes_k M_{n_1}(k) \cong D_2 \otimes_k M_{n_2}(k)$. By the uniqueness part of Theorem 3.2.6, $D_1 \cong D_2$.

Part (2) follows from Proposition 4.5.1. □

Corollary 7.5.2. *Let k be a field and A an Azumaya k-algebra with division algebra component D. If F is a maximal commutative k-subalgebra of D, then the following are true.*

(1) *F is a splitting field for D and A.*
(2) $\dim_k(F) = \dim_F(D)$ *and* $\dim_k(D) = \dim_k(F)^2$.
(3) *If K/k is a finite dimensional extension of fields and K is a splitting field for A, then* $\dim_k(F)$ *divides* $\dim_k(K)$.

Proof. Because $\dim_k(D)$ is finite, by Lemma 3.3.2 it follows that F is a field. The rest of (1) and Part (2) follow from Theorem 7.4.2.

(3): By Theorem 7.4.3, we can assume K is a maximal commutative subalgebra of $D \otimes_k M_r(k)$, for some $r \geq 1$. By Theorem 7.4.2, $\dim_k(K)^2 = r^2 \dim_k(D) = r^2 \dim_k(F)^2$. □

Let k be a field and D a central k-division algebra. The *index* of D is defined to be the square root of the dimension $\dim_k(D)$. The index of D is equal to the dimension of a maximal commutative k-subalgebra of D. For any k-central simple algebra A, the index of A is defined to be the index of the unique division algebra in the Brauer class of A.

Corollary 7.5.3. *Let A and B be Azumaya k-algebras, where k is a field. If $[B]$ is in the subgroup of $\mathrm{B}(k)$ generated by $[A]$, then the index of B divides the index of A. If $[A]$ and $[B]$ generate the same subgroup of $\mathrm{B}(k)$, then A and B have the same index.*

Proof. If K is a maximal subfield of the division algebra in $[A]$, the index of A is equal to $\dim_k(K)$. By Corollary 7.5.2 (1), K is a splitting field for A. The Brauer group is a functor, so K is a splitting field for B. By Corollary 7.5.2 (3), the index of B divides $\dim_k(K)$. □

We will apply Theorem 7.5.4 in Corollary 7.9.6 to show that the Brauer group of any finite commutative ring is trivial.

Theorem 7.5.4 (Wedderburn). *A finite division ring is a field. If k is a finite field, then $\mathrm{B}(k) = (0)$.*

Proof. Our proof is based on [**Rei75**, Theorem (7.24)]. By Theorem 7.5.1, the second statement follows from the first. Let D be a finite division ring with center k. By Corollary 7.5.2, a maximal commutative k-subalgebra of D is a field, and any two maximal subfields have the same dimension over k. If $a \in D$, then $k(a)$ is commutative, hence is contained in a maximal subfield of D. Thus D is a union of maximal subfields. Two finite fields of the same order are isomorphic (see, for example, [**DF04**, Proposition 14.3.15]), so any two maximal subfields of D are isomorphic by a k-isomorphism. Let F be a maximal subfield. By Theorem 4.5.12, any other maximal subfield F' is conjugate to F by a unit in D. Therefore,

$$D^* = \bigcup_{u \in D^*} uF^*u^{-1}.$$

If $x \in F^*$, then $(ux)F^*(ux)^{-1} = uF^*u^{-1}$, so the union can be taken over those u in a full set of left coset representatives for F^* in D^*. By Lagrange's Theorem, $[D^* : 1] = [D^* : F^*][F^* : 1]$. But the groups uF^*u^{-1} all have order $[F^* : 1]$, and they share the common element 1. This is impossible if $m > 1$. We conclude $k = F = D$. □

Theorem 7.5.5. *Let k be a field and D a k-central division algebra of degree n. Then there is an element $\alpha \in D$ such that α is separable over k and $k(\alpha)$ is a maximal subfield of D containing k.*

Proof. The idea behind the proof (which is based on [**Sal99**, Theorem 4.6]) is the fact that a nonzero polynomial with coefficients in an infinite field does not define the zero function (Exercise 1.2.20). The functions we employ are defined by characteristic polynomials and the resultant of two polynomials. Before proving the theorem we make the necessary preliminary definitions.

Let F be any field, and $F_n[x]$ the set of all monic polynomials in $F[x]$ of degree n. Then $F_n[x]$ is isomorphic as an F-vector space to F^n. Make $F_n[x]$ into a topological space by identifying it as a point set with affine n-space $\mathbb{A}^n_F$ which has the Zariski topology (see, for example, [**DF04**, Section 15.2]). Given a polynomial $f \in F_n[x]$, the resultant $\operatorname{Res}(f, f')$ is defined in Section 1.2.2. Recall that $\operatorname{Res}(f, f')$ is an element of F and is nonzero if and only if f has no repeated root. It follows from Lemma 1.2.4 that the map

$$F_n[x] \xrightarrow{f \mapsto \operatorname{Res}(f,f')} F$$

is a polynomial function on $F_n[x]$. Exercise 1.2.21 shows the set of zeros of this function is a proper closed subset of $F_n[x]$.

Let V be an F-vector space of dimension n. The characteristic polynomial of a linear transformation ϕ in $\operatorname{Hom}_F(V, V)$ is a monic polynomial of degree n, hence is in $F_n[x]$. It follows from Lemma 1.2.2, that the function

$$\operatorname{Hom}_F(V, V) \xrightarrow{\phi \mapsto \operatorname{char.poly}_F(\phi)} F_n[x]$$

is coordinate-wise a polynomial function. The composite mapping

$$\operatorname{Hom}_F(V, V) \xrightarrow{\phi \mapsto \operatorname{char.poly}_F(\phi)} F_n[x] \xrightarrow{f \mapsto \operatorname{Res}(f,f')} F \tag{7.3}$$

is a polynomial function which will be denoted P. Notice that P is not the zero function (pick a basis for V and start with the companion matrix of a separable polynomial f).

Now we return to the context of the theorem. If k is a finite field, then by Theorem 7.5.4, $k = D$ and there is nothing to prove. Assume k is infinite, and let F be any maximal commutative subfield of D that contains k. By Corollary 7.5.2, there is an F-algebra isomorphism $\theta : D \otimes_k F \cong \operatorname{Hom}_F(D, D)$. Combining θ with P from (7.3), the composite map

$$D \otimes_k F \xrightarrow{\theta} \operatorname{Hom}_F(D, D) \xrightarrow{P} F$$

is a polynomial function. As shown in Exercise 1.2.17, the characteristic polynomial of an element of D is a polynomial with coefficients in k. The restriction of $P\theta$ to $D \otimes_k k$ defines a polynomial function on D. Since k is infinite, Exercise 1.2.20 implies A contains an element $\alpha \in A$ such that the characteristic polynomial of α is separable. It follows from Exercise 1.2.16 that $\operatorname{char.poly}_k(\alpha)$ is irreducible. Therefore, $\dim_k(k(\alpha)) = n$, and by Corollary 7.5.2, $k(\alpha)$ is a maximal subfield of D. □

Corollary 7.5.6. *If k is a separably closed field, then $\mathrm{B}(k) = (0)$.*

Theorem 7.5.7. *Let k be a field and A an Azumaya k-algebra. Then there exists a commutative k-subalgebra $S \subseteq A$ satisfying the following.*

(1) *S is a maximal commutative k-subalgebra of A.*
(2) *S is separable over k.*
(3) *$S = k(\alpha)$ for some $\alpha \in A$.*
(4) *A is a free S-module.*
(5) *S is a splitting ring for A.*

Proof. By Theorem 3.2.6, A is isomorphic to $D \otimes_k M_m(k)$, for a unique central k-division algebra D.

First suppose k is finite. By Theorem 7.5.4, $A \cong M_n(k)$. There is a Galois extension $F = k(\alpha)$ of k of degree n (see, for example, [**DF04**, Proposition 14.3.15]). Let $F \to \operatorname{Hom}_k(F, F) \cong A$ be the left regular representation. By Theorem 7.4.3, we know F maps onto a maximal commutative k-subalgebra.

Now assume k is infinite. By Theorem 7.5.5 there is a maximal subfield $F = k(u)$ in D. Let $\dim_k(F) = n$. For each a in $\operatorname{Units}(k)$, $F = k(au)$. Since k is infinite, the set $\{\operatorname{Irr.poly}_k(au) \mid a \in \operatorname{Units}(k)\}$ is an infinite subset of $k[x]$. So we can find m irreducible polynomials $f_i = \operatorname{Irr.poly}_k(a_i u)$ such that $f = f_1 \cdots f_m$ has no repeated roots and $\deg(f) = nm$. In $M_m(D)$, take α to be the diagonal matrix with entries $(a_1 u, \dots, a_m u)$. The reader should verify that $\operatorname{min.poly}_k(\alpha) = f$. So $k(\alpha)$ is a commutative subalgebra and $\dim_k(\alpha) = nm$. By Exercise 4.4.11, $k(\alpha)$ is separable over k. By Theorem 4.4.1, A is a projective $k(\alpha)$-module. But $k(\alpha)$ is the internal direct sum of m copies of F, the idempotents being $e_{11}, \dots, e_{mm}$, where e_{ii} has 1 in position (i, i) and 0 elsewhere. Then $k(\alpha)e_{ii} \subseteq e_{ii}Ae_{ii}$. But $k(\alpha)e_{ii} \cong F$ is a maximal subfield in $e_{ii}Ae_{ii} \cong D$. From this it follows that $k(\alpha)$ is a maximal commutative k-subalgebra of A, and A is free of rank nm over $k(\alpha)$. □

Corollary 7.5.8. *Let k be a field and A an Azumaya k-algebra. There exists a Galois extension of fields K/k with finite group G such that K is a splitting field for A.*

Proof. Let D be the division algebra which is Brauer equivalent to A. It is enough to find a Galois splitting field for D. By Theorem 7.5.5 there is a finite dimensional separable extension of fields F/k such that F is a splitting field for D. By the Embedding Theorem (see, for example, [**Hun80**, Theorem V.3.16]), there is a finite dimensional Galois extension K/k which contains F. The Brauer group is a functor, so K also splits D. □

6. Azumaya Algebras up to Brauer Equivalence

Versions of the next theorem and proposition were proved in [**AG60a**]. The proofs given below are based on [**Sal99**] and [**DI71**] respectively. In Theorem 7.6.1, we see that in the category of Azumaya R-algebras, every homomorphism is one-to-one. In particular, an endomorphism is an automorphism.

Theorem 7.6.1. *Let A and B denote Azumaya R-algebras.*

(1) *Let $\phi : A \to B$ be a homomorphism of R-algebras. Then ϕ is one-to-one. If A and B have equal ranks (at each prime $\mathfrak{p} \in \operatorname{Spec}(R)$), then ϕ is an isomorphism.*
(2) *Every R-algebra endomorphism $f : A \to A$ is an automorphism.*

Proof. (1): By Lemma 7.1.1, $R \cdot 1$ is an R-module direct summand of both A and B. The first claim follows from Corollary 7.1.7. By Theorem 7.2.3, $B = \phi(A) \otimes_R C$, and C is an Azumaya R-subalgebra of B. By counting ranks, $\operatorname{Rank}_R(C) = 1$. Hence $R \cdot 1 = C$ and $\phi(A) = B$. Part (2) follows from (1). □

Proposition 7.6.2. *Let R be a commutative ring.*

(1) *Let P and Q be R-progenerator modules. Then $\operatorname{Hom}_R(P, P)$ and $\operatorname{Hom}_R(Q, Q)$ are isomorphic R-algebras if and only if there is a rank one R-progenerator module E such that P and $E \otimes_R Q$ are isomorphic as R-modules.*
(2) *Let A be an Azumaya R-algebra. Let M and N be left A-module progenerators. Then $\operatorname{Hom}_A(M, M)$ and $\operatorname{Hom}_A(N, N)$ are isomorphic as R-algebras if and only if there is a rank one R-progenerator module E such that N and $E \otimes_R M$ are isomorphic as left A-modules.*

Proof. First, we prove the "if" parts of (1) and (2), starting with (2). Assume E is a rank one R-progenerator module such that N and $E \otimes_R M$ are isomorphic as left A-modules. By Lemma 2.6.7, $\operatorname{Hom}_R(E, E) \cong R$.

Theorem 1.3.26 gives

$$\begin{aligned}\operatorname{Hom}_A(N,N) &\cong \operatorname{Hom}_{R\otimes_R A}(E\otimes_R M, E\otimes_R M)\\ &\cong \operatorname{Hom}_R(E,E)\otimes_R \operatorname{Hom}_A(M,M)\\ &\cong \operatorname{Hom}_A(M,M).\end{aligned}$$

Notice that (1) follows from (2) by taking $A = R$.

Now we prove the "only if" parts of (2) and (1), starting with (2). Assume that there is an R-algebra isomorphism $\operatorname{Hom}_A(M,M) \cong \operatorname{Hom}_A(N,N)$. By an application of Theorem 7.2.2, $\operatorname{Hom}_R(M,M)^A = \operatorname{Hom}_A(M,M)$, and $A\otimes_R \operatorname{Hom}_A(M,M) \cong \operatorname{Hom}_R(M,M)$. Likewise, $A\otimes_R \operatorname{Hom}_A(N,N) \cong \operatorname{Hom}_R(N,N)$. So (2) follows from (1).

Assume P and Q are R-progenerator modules, and that there is an isomorphism $\operatorname{Hom}_R(P,P) \cong \operatorname{Hom}_R(Q,Q)$ of R-algebras. Let $B = \operatorname{Hom}_R(P,P)$. By the Morita Theorems (Corollary 1.5.3), P and Q are left B-progenerator modules, and there is a right R-progenerator module E such that $Q \cong E\otimes_R P$ as left B-modules. Since B is an R-algebra, a B-module homomorphism is an R-module homomorphism. By Exercise 2.3.7, $\operatorname{Hom}_R(P,P) \cong \operatorname{Hom}_R(Q,Q)$ implies $\operatorname{Rank}_R(P) = \operatorname{Rank}_R(Q)$, and $Q \cong E\otimes_R P$ implies $\operatorname{Rank}_R(E) = 1$. □

Let R be a commutative ring. Assume A and B are Azumaya R-algebras that are Brauer equivalent. Then there exist R-progenerator modules M and N such that $A\otimes_R B^o \cong \operatorname{Hom}_R(M,M)$ and $B\otimes_R B^o \cong \operatorname{Hom}_R(N,N)$ (Proposition 7.3.4). By Theorem 7.2.2, $A \cong \operatorname{Hom}_{B^o}(M,M)$, M is a B^o-progenerator, $B^o \cong \operatorname{Hom}_B(N,N)$, and N is a B-progenerator. By two applications of Theorem 1.5.2, the categories ${}_A\mathfrak{M}$ and ${}_B\mathfrak{M}$ are naturally equivalent. We say that a projective module P over a ring is *indecomposable* if $P \neq (0)$ and there is no proper submodule of P that is a direct summand of P. The reader should verify that under the categorical equivalence ${}_A\mathfrak{M} \sim {}_B\mathfrak{M}$, the isomorphism classes of indecomposable projective A-modules are in one-to-one correspondence with the isomorphism classes of indecomposable projective B-modules.

In Theorem 7.5.1 we proved that every Brauer equivalence class over a field k contains a uniquely defined representative D that is a division ring. By Theorem 3.2.6, a central simple k-algebra D is a division ring if and only if $\operatorname{idemp}(D) = \{0,1\}$. The next theorem generalizes this result to the Brauer group of an arbitrary connected commutative ring R. Theorem 7.6.3 and its proof are based on [**DI71**, Theorem 5.1.1].

Theorem 7.6.3. *Let R be a connected commutative ring and A an Azumaya R-algebra.*

(1) *Let P be an indecomposable finitely generated projective A-module, and* $\Lambda = \operatorname{Hom}_A(P, P)$. *Then*
 (a) Λ *is an Azumaya R-algebra,*
 (b) $\operatorname{idemp}(\Lambda) = \{0, 1\}$, *and*
 (c) Λ^o *is Brauer equivalent to A.*

(2) *If B is an Azumaya R-algebra such that* $\operatorname{idemp}(B) = \{0, 1\}$ *and B is Brauer equivalent to A, then the following are true.*
 (a) *There exists an indecomposable finitely generated projective A-module P such that* $B^o \cong \operatorname{Hom}_A(P, P)$.
 (b) *The A-module P in (a) is unique in the following sense. There exists another such A-module Q as in (a), if and only if there is an invertible R-module E and an A-module isomorphism* $P \cong E \otimes_R Q$.

Proof. (1): If P is an indecomposable finitely generated projective A-module, then P is an R-progenerator. If we set $\Lambda = \operatorname{Hom}_A(P, P)$, then by Theorem 7.2.2,

$$A \otimes_R \Lambda \cong \operatorname{Hom}_R(P, P)$$

and Λ is an Azumaya R-algebra. By Proposition 7.3.4, Λ^o is Brauer equivalent to A. An idempotent in Λ defines an A-module projection on P. Since P is indecomposable, $\operatorname{idemp}(\Lambda) = \{0, 1\}$.

(2): If E is an invertible R-module and P is an indecomposable finitely generated projective A-module, then the reader should verify that $E \otimes_R P$ is indecomposable finitely generated and projective over A. The uniqueness claim in Part (b) follows from Proposition 7.6.2. We now prove Part (a). Let B be an Azumaya R-algebra such that $\operatorname{idemp}(\Lambda) = \{0, 1\}$. If A and B are Brauer equivalent, then there is an R-progenerator M with $A \otimes_R B^o \cong \operatorname{Hom}_R(M, M)$. Thus M is a finitely generated projective A-module and $B \cong \operatorname{Hom}_A(M, M)^o$. Since an A-module projection onto M corresponds to an idempotent in B, M is an indecomposable A-module. □

In [**AW14**], a connected commutative ring R is said to have the *Wedderburn property*, if for every Azumaya R-algebra A there is up to isomorphism a unique representative B in the Brauer class of A such that $\operatorname{idemp}(B) = \{0, 1\}$ and $A^o \cong \operatorname{Hom}_B(P, P)$ for some B-progenerator module P (See [**DI71**, Problem 5.3.5]). In Exercise 7.6.7, the reader is asked to prove that every local ring and every connected semilocal ring has the Wedderburn property. A class of rings which fail to have the Wedderburn property is exhibited in [**AW14**, Example 1.5]. Examples of number rings that fail to have the Wedderburn property are given in [**Bas64**, p. 46] and [**Chi69**].

6.1. Exercises.

Exercise 7.6.4 ([**Sal99**, Lemma 3.4]). Let A be an Azumaya R-algebra and e an idempotent in A such that $\operatorname{annih}_R(e) = (0)$. Following the steps outlined below, prove that eAe is an Azumaya R-algebra which is Brauer equivalent to A.

(1) Show that Ae is a left A-module direct summand of A, hence is a finitely generated projective R-module.

(2) Show that Ae is an R-progenerator, hence $\operatorname{Hom}_R(Ae, Ae)$ is an Azumaya R-algebra.

(3) Show that the left regular representation $\lambda : A \to \operatorname{Hom}_R(Ae, Ae)$ is an R-algebra monomorphism.

(4) Show that if A is identified with the image of λ, then the centralizer of A in $\operatorname{Hom}_R(Ae, Ae)$ is an Azumaya R-algebra which can be identified with $\operatorname{Hom}_A(Ae, Ae)$.

(5) Show that the map $\rho : (eAe)^o \to \operatorname{Hom}_A(Ae, Ae)$, where $\rho(eae)$ is "right multiplication by eae" is an R-algebra isomorphism. (Hint: the inverse to ρ is the map which sends f to $f(e)$.)

(6) Show that A and eAe are Brauer equivalent. (Hint: Apply Theorem 7.2.2 and Proposition 7.3.4.)

Exercise 7.6.5 ([**Sal99**, Lemma 3.4]). Let A and B be Azumaya R-algebras which are Brauer equivalent. Follow the outline below to prove that there exist matrix algebras $M_n(A)$, $M_m(B)$ together with idempotents $e \in M_n(A)$, $f \in M_m(B)$ such that $\operatorname{annih}_R(e) = (0)$, $\operatorname{annih}_R(f) = 0$, and $eM_n(A)e \cong fM_n(B)f$ as R-algebras.

(1) Let P be an R-progenerator. Assume $R^n \cong P \oplus Q$ for some $n \geq 1$, and R-module Q. Let $e \in \operatorname{Hom}_R(R^n, R^n)$ be the projection map $e : R^n \to P$. Identify the endomorphism ring $\operatorname{Hom}_R(R^n, R^n)$ with the ring of matrices $M_n(R)$. Show that $\operatorname{annih}_R(e) = (0)$.

(2) Show that as an R-algebra, $\operatorname{Hom}_R(P, P)$ can be identified with $eM_n(R)e$.

(3) Show that $A \otimes_R \operatorname{Hom}_R(P, P) \cong A \otimes_R eM_n(R)e \cong eM_n(A)e$.

(4) Prove the claim about A and B.

Exercise 7.6.6 ([**Sal99**, Lemma 3.4]). State and prove the converse of the claim in Exercise 7.6.5.

Exercise 7.6.7 ([**DI71**, Exercise 5.4.1]). Let R be a local ring with maximal ideal $\mathfrak{m}$ and A an Azumaya R-algebra. Follow the outline below to prove that R has the Wedderburn property.

(1) Show that if M and N are indecomposable finitely generated projective A-modules, then M and N are isomorphic. (Hint: Suppose

$\operatorname{Rank}_R(M) \geq \operatorname{Rank}_R(N)$. Then there is an A-module epimorphism $f : M/\mathfrak{m}M \to N/\mathfrak{m}N$. Lift f to an isomorphism $M \to N$.)

(2) Show that up to isomorphism, there is a unique Azumaya R-algebra B such that $\operatorname{idemp}(B) = \{0, 1\}$ and B is Brauer equivalent to A.

(3) Show that for some $n \geq 1$, A is isomorphic to the matrix ring $M_n(B)$.

Adapt the above proof to show that a connected semilocal ring has the Wedderburn property. This result was first proved in [**DeM69**].

7. Noetherian Reduction of Azumaya Algebras

Let R be any ring and $\theta : P \to Q$ a homomorphism between finitely generated projective R-modules. In Proposition 7.7.1 we show that θ is defined by only finitely elements of R. This result allows us to restrict θ to a homomorphism that is defined over a subring $R_0 \subseteq R$, where R_0 is a finitely generated $\mathbb{Z}$-algebra. Using this, we prove the main result of this section, Proposition 7.7.2, which states that an Azumaya R-algebra comes from an Azumaya algebra over a finitely generated subring of R.

Proposition 7.7.1. *Let R be any ring and $\theta : P \to Q$ a homomorphism between finitely generated projective R-modules. Then there is a noetherian subring $R_0 \subseteq R$ (in fact, R_0 can be taken to be a finitely generated $\mathbb{Z}$-algebra) and finitely generated projective R_0-modules $P_0 \subseteq P$, $Q_0 \subseteq Q$, such that the following are true:*

(1) *θ restricts to an R_0-module homomorphism $\theta_0 : P_0 \to Q_0$,*

(2) *$P \cong P_0 \otimes_{R_0} R$,*

(3) *$Q \cong Q_0 \otimes_{R_0} R$,*

(4) *$\theta = \theta_0 \otimes 1$,*

(5) *If θ is onto, then R_0, P_0, Q_0 can be chosen so that θ_0 is onto.*

(6) *If θ is an isomorphism, then R_0, P_0, Q_0 can be chosen so that θ_0 is an isomorphism.*

Proof. The finitely generated projective R-modules P and Q are direct summands of R^n, for some $n \geq 1$. It follows that there are idempotents a and b in $\operatorname{Hom}_R(R^n, R^n)$ such that $a^2 = a$, $b^2 = b$, $P = \operatorname{coker}(a)$, and $Q = \operatorname{coker}(b)$. All the modules are projective, so there are R-module homomorphisms ψ, ϕ such that the diagram

$$
\begin{array}{ccccccc}
R^n & \xrightarrow{a} & R^n & \xrightarrow{c} & P & \longrightarrow & 0 \\
\downarrow{\scriptstyle \psi} & & \downarrow{\scriptstyle \phi} & & \downarrow{\scriptstyle \theta} & & \\
R^n & \xrightarrow{b} & R^n & \xrightarrow{d} & Q & \longrightarrow & 0
\end{array}
\tag{7.4}
$$

commutes (for example, apply Proposition 1.1.5). Let R_0 be the subring of R generated by the entries in the matrices of a, b, ψ, ϕ with respect to the standard basis for R^n. Then the four matrices descend to define four R_0-module homomorphisms a_0, b_0, ψ_0, ϕ_0. Define $P_0 = \operatorname{coker}(a_0)$, and $Q_0 = \operatorname{coker}(b_0)$. We get a commutative diagram

$$\begin{array}{ccccccc} R_0^n & \xrightarrow{a_0} & R_0^n & \xrightarrow{c_0} & P_0 & \longrightarrow & 0 \\ \downarrow{\scriptstyle \psi_0} & & \downarrow{\scriptstyle \phi_0} & & \downarrow{\scriptstyle \theta_0} & & \\ R_0^n & \xrightarrow{b_0} & R_0^n & \xrightarrow{d_0} & Q_0 & \longrightarrow & 0 \end{array} \tag{7.5}$$

where θ_0 is induced by the rest of the diagram. Since $R_0 \subseteq R$, the identities $a_0^2 = a_0$ and $b_0^2 = b_0$ hold. Therefore, P_0 and Q_0 are finitely generated projective R_0-modules. Applying the tensor functor $(\cdot) \otimes_{R_0} R$ to Diagram (7.5) results in Diagram (7.4). Since P_0 and Q_0 are projective R_0-modules, they are flat. Tensoring $0 \to R_0 \to R$ with $P_0 \otimes_{R_0} (\cdot)$ shows $P_0 \to P$ is one-to-one. Likewise, $Q_0 \to Q$ is one-to-one. This proves (1) – (4).

(5): Assume θ is onto. The goal is to guarantee that the image of θ_0 contains a generating set for Q_0 as an R_0-module. Let $\{v_1, \dots, v_n\}$ be the standard basis for R_0^n, which we view as a subset of R^n. By (7.5) we know that Q_0 is generated as an R_0-module by $\{d_0(v_1), \dots, d_0(v_n)\}$. Since θ is onto, for each i choose $y_i \in P$ such that $\theta(y_i) = d(v_i)$. As an R-module, P is generated by $\{c(v_1), \dots, c(v_n)\}$. There are n^2 elements r_{ij} in R such that $y_i = \sum r_{ij} c(v_j)$. So if R_0 contains all of the elements r_{ij}, it follows that P_0 contains $y_1, \dots, y_n$ and the image of θ_0 contains a generating set for Q_0.

(6): By (5) we can assume $\theta_0 : P_0 \to Q_0$ is onto. Let K_0 denote the kernel of θ_0. Since P_0 and Q_0 are projective R_0-modules, the sequence of R_0-modules

$$0 \to K_0 \to P_0 \xrightarrow{\theta_0} Q_0 \to 0 \tag{7.6}$$

is split exact. Then tensoring (7.6) with $(\cdot) \otimes_{R_0} R$ yields a sequence

$$0 \to K_0 \otimes_{R_0} R \to P \xrightarrow{\theta} Q \to 0 \tag{7.7}$$

that is split exact over R. Since θ is an isomorphism, $K_0 \otimes_{R_0} R = 0$. Being a direct summand of P_0, K_0 is projective, hence flat. Then $K_0 = K_0 \otimes_{R_0} R_0 \to K_0 \otimes_{R_0} R = 0$ is one-to-one, hence $K_0 = 0$. □

Proposition 7.7.2. *Let A be an R-algebra which is a finitely generated projective R-module. Then there is a noetherian subring $R_0 \subseteq R$ (in fact, R_0 can be taken to be a finitely generated $\mathbb{Z}$-algebra) and an R_0-subalgebra A_0 of A such that the following are true:*

(1) *A_0 is a finitely generated projective R_0-module.*

(2) *$A \cong A_0 \otimes_{R_0} R$.*

(3) *If A is a separable R-algebra, then R_0 and A_0 can be chosen so that A_0 is a separable R_0-algebra.*

(4) *If A is an Azumaya R-algebra, then R_0 and A_0 can be chosen so that A_0 is an Azumaya R_0-algebra.*

Proof. The finitely generated projective R-module A is a direct summand of R^n, for some $n \geq 1$. Therefore, there is an idempotent a in $\operatorname{Hom}_R(R^n, R^n)$ and an exact sequence of R-modules

$$R^n \xrightarrow{a} R^n \xrightarrow{c} A \to 0. \tag{7.8}$$

Let $\{v_1, \dots, v_n\}$ be the standard basis for R^n. Without loss of generality, assume the image of v_1 under c is 1, the multiplicative identity of A. As in Proposition 7.7.1, the commutative diagram

$$\begin{array}{ccccccc} R^n \otimes_R R^n & \xrightarrow{a \otimes a} & R^n \otimes_R R^n & \xrightarrow{c \otimes c} & A \otimes_R A & \longrightarrow & 0 \\ \downarrow{\scriptstyle \psi} & & \downarrow{\scriptstyle \phi} & & \downarrow{\scriptstyle \mu} & & \\ R^n & \xrightarrow{a} & R^n & \xrightarrow{c} & A & \longrightarrow & 0 \end{array} \tag{7.9}$$

results from combining (7.8) with the multiplication map μ. Let R_0 be the subring of R generated by the entries in the matrices of a, ψ, and ϕ with respect to the standard basis for R^n. Then the matrices descend to define R_0-module homomorphisms a_0, ψ_0, and ϕ_0. Define A_0 to be the cokernel of a_0. We get a commutative diagram

$$\begin{array}{ccccccc} R_0^n \otimes_{R_0} R_0^n & \xrightarrow{a_0 \otimes a_0} & R_0^n \otimes_{R_0} R_0^n & \longrightarrow & A_0 \otimes_{R_0} A_0 & \longrightarrow & 0 \\ \downarrow{\scriptstyle \psi_0} & & \downarrow{\scriptstyle \phi_0} & & \downarrow{\scriptstyle \mu_0} & & \\ R_0^n & \xrightarrow{a_0} & R_0^n & \longrightarrow & A_0 & \longrightarrow & 0 \end{array} \tag{7.10}$$

where μ_0 is induced by the rest of the diagram. The proof of Proposition 7.7.1 shows that A_0 is a finitely generated projective R_0-module, $A = A_0 \otimes_{R_0} R$, and $A_0 \subseteq A$. Since (7.10) commutes, A_0 is an R_0-subalgebra of A. This proves (1) and (2).

(3): Let $e = \sum x_i \otimes y_i$ be a separability idempotent in $A \otimes_R A^0$. The diagram

$$\begin{array}{ccccc} R_0^n & \xrightarrow{c_0} & A_0 & \longrightarrow & 0 \\ \downarrow{\scriptstyle \subseteq} & & \downarrow{\scriptstyle \subseteq} & & \\ R^n & \xrightarrow{c} & A & \longrightarrow & 0 \end{array} \tag{7.11}$$

commutes and the rows are exact. In R^n find w_i, z_i such that $x_i = c(w_i)$ and $y_i = c(z_i)$. Expand each w_i and z_i in terms of the standard basis. If necessary, adjoin finitely many elements of R to R_0 so that w_i and z_i lift to

R_0^n. Therefore, the image of c_0 contains each x_i and y_i, and an idempotent $e_0 \in A_0 \otimes_{R_0} A_0^o$ exists which is a candidate for a separability idempotent. The sequence

$$0 \to J_{A_0/R_0} \to A_0 \otimes_{R_0} A_0^o \to A_0 \to 0$$

is split exact over R_0 (see Lemma 7.7.3). By (1), this implies J_{A_0/R_0} is a finitely generated projective R_0-module, hence is flat. By Lemma 7.7.3, $J_{A_0/R_0} \otimes_{R_0} R = J_{A/R}$. Therefore, $J_{A_0/R_0} \subseteq J_{A/R}$, which implies $J_{A_0/R_0} e_0 \subseteq J_{A/R} e = 0$. By Proposition 4.1.2, A_0 is separable over R_0.

(4): Assume the enveloping homomorphism

$$\varphi_{A/R} : A \otimes_R A^o \to \operatorname{Hom}_R(A, A)$$

is an R-algebra isomorphism (Theorem 7.1.4). Assume R_0 and A_0 are as in (1) and (2). Then

$$\varphi_{A_0/R_0} : A_0 \otimes_{R_0} A_0^o \to \operatorname{Hom}_{R_0}(A_0, A_0)$$

is an R_0-algebra homomorphism. Both $A_0 \otimes_{R_0} A_0^o$ and $\operatorname{Hom}_{R_0}(A_0, A_0)$ are finitely generated projective R_0-modules, and φ_{A_0/R_0} becomes the isomorphism $\varphi_{A/R}$ upon tensoring with $(\cdot) \otimes_{R_0} R$. We are in the setting of Proposition 7.7.1 (6). So there is a finitely generated extension R_1 of R_0 such that φ_{A_0/R_0} becomes an isomorphism upon tensoring with $(\cdot) \otimes_{R_0} R_1$. Then $A_0 \otimes_{R_0} R_1$ is an Azumaya R_1-algebra. □

Lemma 7.7.3. *Let R be a commutative ring, A an R-algebra and S a commutative R-algebra. Then $J_{A \otimes_R S/S} = J_{A/R} \otimes_R S$.*

Proof. The sequence

$$0 \to J_{A/R} \to A^e \xrightarrow{\mu} A \to 0$$

of A^e-modules is exact. By Exercise 1.3.8, there is an R-algebra homomorphism $\rho : A \to A \otimes_R A^o$ defined by $\rho(a) = a \otimes 1$. Since $\mu\rho = 1$, the sequence is split exact as a sequence of R-modules. Therefore, $0 \to J_{A/R} \otimes_R S \to A^e \otimes_R S \to A \otimes_R S \to 0$ is a split exact sequence of S-modules. Under the identification $A^e \otimes_R S = (A \otimes_R S) \otimes_S (A \otimes_R S)^o = (A \otimes_R S)^e$, the proof follows. □

The rest of this section contains noetherian reduction results that will arise in various contexts in the remaining chapters of the book. The proofs are based on Propositions 7.7.1 and 7.7.2, hence many details are left to the reader. There is one more result on the noetherian reduction of a Galois extension which is stated below in Proposition 12.2.11.

Proposition 7.7.4. *Let A and B be R-algebras which are finitely generated projective R-modules. Let $\theta : A \to B$ be an R-algebra homomorphism. Then there is a noetherian subring $R_0 \subseteq R$ (in fact, R_0 can be taken to be a* *finitely generated* $\mathbb{Z}$*-algebra) and* R_0*-subalgebras* $A_0 \subseteq A$, $B_0 \subseteq B$, *such that the following are true:*

(1) A_0 *and* B_0 *are finitely generated projective* R_0*-modules.*

(2) θ *restricts to an* R_0*-algebra homomorphism* $\theta_0 : A_0 \to B_0$,

(3) $A \cong A_0 \otimes_{R_0} R$,

(4) $B \cong B_0 \otimes_{R_0} R$,

(5) $\theta = \theta_0 \otimes 1$,

(6) *If* θ *is onto, then* R_0, A_0, B_0 *can be chosen so that* θ_0 *is onto.*

(7) *If* θ *is an isomorphism, then* R_0, A_0, B_0 *can be chosen so that* θ_0 *is an isomorphism.*

Proof. As in Proposition 7.7.1, there exists a finitely generated subring $R_0 \subseteq R$ such that θ_0, A_0 and B_0 satisfy Parts (1), (3), (4), (5). As in Proposition 7.7.2, we can adjoin finitely many more elements to R_0, and assume Part (2) is satisfied as well. For Parts (6) and (7) we appeal to Proposition 7.7.1. □

Proposition 7.7.5. *Let* A *and* B *be* R*-algebras which are finitely generated projective* R*-modules. Let* S *be a commutative faithful* R*-algebra and assume there exists an* S*-algebra isomorphism* $\theta : A \otimes_R S \to B \otimes_R S$. *Then there is a finitely generated* R*-subalgebra* $S_0 \subseteq S$ *such that* θ *restricts to an* S_0*-algebra isomorphism* $\theta_0 : A \otimes_R S_0 \to B \otimes_R S_0$.

Proof. Start with free resolutions for the R-modules A and B

$$R^n \xrightarrow{a} R^n \xrightarrow{c} A \longrightarrow 0$$

$$R^n \xrightarrow{b} R^n \xrightarrow{d} B \longrightarrow 0$$

where a and b are idempotents. Tensor with S, and lift θ to S-module homomorphisms ψ and ϕ such that the diagram

$$\begin{array}{ccccccc} R^n \otimes_R S & \xrightarrow{u\otimes 1} & R^n \otimes_R S & \longrightarrow & A \otimes_R S & \longrightarrow & 0 \\ \downarrow{\scriptstyle \psi} & & \downarrow{\scriptstyle \phi} & & \downarrow{\scriptstyle \theta} & & \\ R^n \otimes_R S & \xrightarrow{b\otimes 1} & R^n \otimes_R S & \longrightarrow & B \otimes_R S & \longrightarrow & 0 \end{array} \tag{7.12}$$

commutes. Let S_0 be the R-subalgebra of S generated by all of the terms in the matrices for ψ and ϕ with respect to the standard basis. Then there is a commutative diagram of S_0-modules

$$\begin{array}{ccccccc} R^n \otimes_R S_0 & \xrightarrow{a\otimes 1} & R^n \otimes_R S_0 & \longrightarrow & A \otimes_R S_0 & \longrightarrow & 0 \\ \downarrow{\scriptstyle \psi_0} & & \downarrow{\scriptstyle \phi_0} & & \downarrow{\scriptstyle \theta_0} & & \\ R^n \otimes_R S_0 & \xrightarrow{b\otimes 1} & R^n \otimes_R S_0 & \longrightarrow & B \otimes_R S_0 & \longrightarrow & 0 \end{array} \tag{7.13}$$

where θ_0 is induced by the rest of the diagram. As in Proposition 7.7.1, we can choose S_0 such that θ_0 is an S_0-module isomorphism. As in Proposition 7.7.2, we can choose S_0 such that θ_0 is an S_0-algebra isomorphism. □

Proposition 7.7.6. *Let S be a commutative R-algebra which is of finite presentation. By this we mean there is an exact sequence*

$$0 \to I \to R[x_1, \dots, x_n] \xrightarrow{\eta} S \to 0$$

where η is an R-algebra homomorphism and I, the kernel of η, is generated by a finite set of polynomials with coefficients in R. Then there is a noetherian subring $R_0 \subseteq R$ (in fact, R_0 can be taken to be a finitely generated $\mathbb{Z}$-algebra) and an R_0-algebra S_0 such that the following are true:

(1) *S_0 is a finitely presented R_0-algebra.*

(2) *$S \cong S_0 \otimes_{R_0} R$.*

Proof. Let $p_1, \dots, p_m$ be a set of generators for I. Let R_0 be the subring of R generated by the coefficients of $p_1, \dots, p_m$. Let S_0 be the factor ring $R_0[x_1, \dots, x_n]/(p_1, \dots, p_m)$. Then S_0 is an R_0-algebra satisfying (1) and (2). □

Proposition 7.7.7. *Let A be an Azumaya R-algebra. Suppose S is a commutative faithful R-algebra which is a splitting ring for A. Then there is a finitely generated faithful R-subalgebra $S_0 \subseteq S$ such that S_0 is a splitting ring for A.*

Proof. By hypothesis there exists an S-algebra isomorphism

$$\phi : A \otimes_R S \cong \operatorname{Hom}_S(P, P) \tag{7.14}$$

for some S-progenerator P. The S-module P is a direct summand of S^n, for some $n \geq 1$. It follows that there is an idempotent a in $\operatorname{Hom}_S(S^n, S^n)$ and an exact sequence of S-modules

$$S^n \xrightarrow{a} S^n \xrightarrow{c} P \to 0. \tag{7.15}$$

Let $B = \{v_1, \dots, v_n\}$ be the standard basis for S^n. Let S_1 be the R-subalgebra of S generated by the n^2 entries in the matrix of a with respect to B. As in Proposition 7.7.1, the diagram (7.15) descends to an exact sequence

$$S_1^n \xrightarrow{a_1} S_1^n \xrightarrow{c_1} P_1 \to 0 \tag{7.16}$$

and P_1 is defined to be the cokernel of a_1. Then $P \cong P_1 \otimes_{S_1} S$. The reader should verify that P_1 is faithful over S_1, hence P_1 is an S_1-progenerator. Let $\alpha_1, \dots, \alpha_m$ be a set of generators for A as an R-module. Consider $\phi(\alpha_i) \in \operatorname{Hom}_S(P, P)$. By Proposition 7.7.1, $\phi(\alpha_i)$ descends to a finitely generated subring of S. If necessary, we can adjoin a finite set of elements to the ring S_1 and assume each $\phi(\alpha_i)$ restricts to an S_1-module homomorphism in $\operatorname{Hom}_{S_1}(P_1, P_1)$. Since $\alpha_1, \dots, \alpha_m$ generate $A \otimes_R S_1$ as an S_1-module, the map ϕ restricts to

$$\phi_1 : A \otimes_R S_1 \to \operatorname{Hom}_{S_1}(P_1, P_1) \tag{7.17}$$

which is a homomorphism of S_1-Azumaya algebras. By Theorem 7.6.1, ϕ_1 is an isomorphism. □

Corollary 7.7.8. *Let R be a commutative ring. As in Exercise 2.5.16, let $\{R_\alpha\}$ be the directed system of all noetherian subrings of R such that R is equal to the direct limit $\varinjlim R_\alpha$. Then $\varinjlim \mathrm{B}(R_\alpha) \to \mathrm{B}(R)$ is an isomorphism.*

Proof. By Proposition 7.7.2, any R-Azumaya algebra A is of the form $A = A_\alpha \otimes_{R_\alpha} R$, where A_α is an R_α-Azumaya algebra. Therefore, the induced map $\varinjlim \mathrm{B}(R_\alpha) \to \mathrm{B}(R)$ is onto. If A_α is an R_α-Azumaya algebra which is split by R, then by Proposition 7.7.7, there is an intermediate ring $R_\alpha \subseteq R_\beta \subseteq R$ which is finitely generated as an R_α-algebra such that $A \otimes_{R_\alpha} R_\beta$ is split. Therefore, $\varinjlim \mathrm{B}(R_\alpha) \to \mathrm{B}(R)$ is an isomorphism. □

7.1. Exercises.

Exercise 7.7.9. In the context of Proposition 7.7.1, let R_1 be any subring of R. Show that the ring R_0 can be chosen so that it contains R_1 and is finitely generated as an R_1-algebra. In other words, instead of being a "noetherian subring which is a finitely generated $\mathbb{Z}$-algebra", it is simply a "finitely generated R_1-algebra".

Exercise 7.7.10. In the context of Proposition 7.7.2, let R_1 be any subring of R. Show that the R-algebra R_0 can be chosen so that it contains R_1 and as an R-algebra is generated by adjoining to R_1 a finite set of elements.

Exercise 7.7.11. In the context of Proposition 7.7.4, let R_1 be any subring of R. Show that the R-algebra R_0 can be chosen so that it contains R_1 and as an R-algebra is generated by adjoining to R_1 a finite set of elements.

Exercise 7.7.12. In the context of Proposition 7.7.5, let S_1 be any subring of S. Show that the R-algebra S_0 can be chosen so that it contains S_1 and as an R-algebra is generated by adjoining to S_1 a finite set of elements.

Exercise 7.7.13. In the context of Proposition 7.7.7, let S_1 be any subring of S. Show that the R-algebra S_0 can be chosen so that it contains S_1 and as an R-algebra is generated by adjoining to S_1 a finite set of elements.

Exercise 7.7.14. Let K/k be an extension of fields and A an Azumaya $k[x]$-algebra that is split by $K[x]$. Show that there is an intermediate field $k \subseteq F \subseteq K$ such that F/k is finitely generated and A is split by $F[x]$. (Hint: Since $K[x]$ is a PID, an Azumaya $K[x]$-algebra is split if and only if it is isomorphic to a ring of matrices $M_n(K[x])$.)

8. The Picard Group of Invertible Bimodules

If A is an algebra over a commutative ring R, then in this section we define the Picard group $\operatorname{Pic}_R(A)$ of invertible left $A \otimes_R A^o$-modules. When A is commutative, $\operatorname{Pic}_R(A)$ contains a subgroup isomorphic to the usual Picard group $\operatorname{Pic}(A)$. When S/R is a Galois extension of commutative rings, then in Section 13.2 we construct Kanzaki's generalized crossed product algebras using invertible left $A \otimes_R A^o$-modules associated to factor sets with values in $\operatorname{Pic}_R(S)$. If A is an Azumaya R-algebra, then we show that $\operatorname{Pic}(R)$ is isomorphic to $\operatorname{Pic}_R(A)$. Using this we prove a generalization of the Skolem-Noether Theorem which applies to any Azumaya R-algebra. General references for the results in this section are [**Bas68**, Chapter II], [**Rei75**, Section 37], and [**Frö73**].

8.1. Definition of the Picard Group.

In this section it is advantageous to employ the right hand versions of the Morita Theorem (Theorem 1.5.2). The necessary conversions are straightforward and are left to the reader.

If A is an R-algebra, then a left $A \otimes_R A^o$-module M is also viewed as a two-sided A/R-module where $a \cdot x = a \otimes 1 \cdot x$ and $x \cdot a = 1 \otimes a \cdot x$ for all $a \in A$ and $x \in M$. Conversely, if M is a two-sided A/R-module, then M can be viewed as a left $A \otimes_R A^o$-module (see Definition 4.1.1). Let ${}_AM$ denote the left A-module and M_A the right A-module.

Let M be a left $A \otimes_R A^o$-module such that the right A-module M_A is an A-progenerator. Let $B = \operatorname{Hom}_A(M_A, M_A)$. Then M is a left B-module by the product rule $bm = b(m)$. If the dual of M_A is $M^* = \operatorname{Hom}_A(M_A, A_A)$, then M^* is a left A-module with multiplication defined by $af(x) = af(x)$ (notice this uses ${}_AA$). Also, M^* is a right B-module by the rule $fb(x) = f(b(x))$. There are isomorphisms

$$\begin{array}{ccc} M \otimes_A M^* \xrightarrow{\theta_A} B & \qquad & M^* \otimes_B M \xrightarrow{\theta_B} A \\ m \otimes f \longmapsto mf(\cdot) & \qquad & f \otimes m \longmapsto f(m) \end{array}$$

where θ_A is a two-sided B/R-module isomorphism and θ_B is a two-sided A/R-module isomorphism. It follows from a right hand version of Theorem 1.5.2 (see the comment in the opening paragraph), that

$$\begin{array}{ccc} {}_A\mathfrak{M} \longrightarrow {}_B\mathfrak{M} & \qquad & {}_B\mathfrak{M} \longrightarrow {}_A\mathfrak{M} \\ N \longmapsto M \otimes_A N & & P \longmapsto M^* \otimes_B P \end{array}$$

defines a Morita equivalence between the category of left A-modules and the category of left B-modules. Since M is a two-sided A/R-module, there are natural homomorphisms of R-algebras

$$(7.18)\qquad \begin{gathered} A \xrightarrow{\lambda} B = \operatorname{Hom}_A(M_A, M_A) \\ A^o \xrightarrow{\rho} \operatorname{Hom}_A({}_AM, {}_AM) \end{gathered}$$

where $\lambda_a(x) = ax$ and $\rho_a(x) = xa$. If λ and ρ are isomorphisms of R-algebras, then we say that M is an *invertible* left $A \otimes_R A^o$-module. If this is true, then M_A induces an equivalence from the category of left A-modules to itself.

Proposition 7.8.1. *Let A be an R-algebra and M a left $A \otimes_R A^o$-module. Then M is an invertible left $A \otimes_R A^o$-module if and only if there exist a left $A \otimes_R A^o$-module N and $A \otimes_R A^o$-module isomorphisms*

$$\sigma : M \otimes_A N \xrightarrow{\cong} A \quad \text{and} \quad \tau : N \otimes_A M \xrightarrow{\cong} A$$

for which the diagrams

$$\begin{array}{ccc} M \otimes_A N \otimes_A M & \xrightarrow{\sigma\otimes 1} & A \otimes_A M \\ {\scriptstyle 1\otimes\tau}\downarrow & & \downarrow \\ M \otimes_A A & \longrightarrow & M \end{array} \qquad \text{and} \qquad \begin{array}{ccc} N \otimes_A M \otimes_A N & \xrightarrow{\tau\otimes 1} & A \otimes_A N \\ {\scriptstyle 1\otimes\sigma}\downarrow & & \downarrow \\ N \otimes_A A & \longrightarrow & N \end{array}$$

commute. In this case, $N \cong M^* = \operatorname{Hom}_A(M_A, A_A)$ as left $A \otimes_R A^o$-modules.

Proof. Assume M is invertible. Then λ is an isomorphism. Therefore, the left $\operatorname{Hom}_A(M_A, M_A)$-module structure on M and the right $\operatorname{Hom}_A(M_A, M_A)$-module structure on M^* both agree with the corresponding A-module structures. If A and $\operatorname{Hom}_A(M_A, M_A)$ are identified using λ, then $\theta_A : M \otimes_A M^* \to A$ and $\theta_B : M^* \otimes_A M \to A$ are isomorphisms of $A \otimes_R A^o$-modules. Taking $N = M^*$, the diagrams commute.

Conversely, assume the isomorphisms σ and τ are given. By Proposition 1.3.4, M is both a left and right A-module progenerator. By Theorem 1.5.2, the functor $(\ ) \otimes_A M$ induces an equivalence of the categories of right A-modules and left $\operatorname{Hom}_A({}_AM, {}_AM)$-modules. By the fully faithful part of Proposition 1.1.2,

$$\begin{aligned}A \xrightarrow{\lambda} \operatorname{Hom}_A(M_A, M_A) &\cong \operatorname{Hom}_A\left((A \otimes_A M)_A, (A \otimes_A M)_A\right)\\ &\cong \operatorname{Hom}_A\left(((M \otimes_A N) \otimes_A M)_A, ((M \otimes_A N) \otimes_A M)_A\right)\\ &\cong \operatorname{Hom}_A\left((M \otimes_A N)_A, (M \otimes_A N)_A\right)\\ &\cong \operatorname{Hom}_A(A_A, A_A)\end{aligned}$$

is an isomorphism. Likewise, the functor $M \otimes_A (\ )$ induces an equivalence of the categories of left A-modules and left $\operatorname{Hom}_A(M_A, M_A)$-modules. Then

$$\begin{aligned}A^o \xrightarrow{\rho} \operatorname{Hom}_A({}_AM, {}_AM) &\cong \operatorname{Hom}_A\left({}_A(M \otimes_A A), {}_A(M \otimes_A A)\right)\\ &\cong \operatorname{Hom}_A\left({}_A\left(M \otimes_A (N \otimes_A M)\right), {}_A\left(M \otimes_A (N \otimes_A M)\right)\right)\\ &\cong \operatorname{Hom}_A\left({}_A(N \otimes_A M), {}_A(N \otimes_A M)\right)\\ &\cong \operatorname{Hom}_A({}_AA, {}_AA)\end{aligned}$$

is an isomorphism. The reader should verify that N is isomorphic to $M^* = \operatorname{Hom}_A(M_A, A_A)$ as left $A \otimes_R A^o$-modules. □

Proposition 7.8.2. *Let A be an R-algebra. Let $\operatorname{Pic}_R(A)$ denote the set of all isomorphism classes $[P]$ of invertible left $A \otimes_R A^o$-modules P. Under the binary operation $[P][Q] = [P \otimes_A Q]$, $\operatorname{Pic}_R(A)$ is a group. The group identity is $[A]$. The inverse of a class $[P]$ is $[P]^{-1} = [P^*] = [\operatorname{Hom}_A(P_A, A_A)]$.*

Proof. Let M be an invertible left $A \otimes_R A^o$-module. Let B be the endomorphism ring $\operatorname{Hom}_A(M_A, M_A)$ and $M^* = \operatorname{Hom}_A(M_A, A_A)$. From Proposition 7.8.1, $\theta_A : M \otimes_A M^* \to A$ and $\theta_B : M^* \otimes_A M \to A$ are isomorphisms of $A \otimes_R A^o$-modules. Let P and Q be two invertible left $A \otimes_R A^o$-modules. In Proposition 7.8.1, take $N = Q^* \otimes_A P^*$ to see that $P \otimes_A Q$ is an invertible left $A \otimes_R A^o$-module. □

The group $\operatorname{Pic}_R(A)$ that was defined in Proposition 7.8.2 is called the *Picard group* of invertible left $A \otimes_R A^o$-modules. Now assume A is a commutative R-algebra. The usual Picard group of A, denoted $\operatorname{Pic}(A)$, is abelian (see Definition 2.6.8). However, in general the group $\operatorname{Pic}_R(A)$ is nonabelian. Suppose P is an A-module which is invertible in the sense of Definition 2.6.8. Since A is commutative, we can make P into a left $A \otimes_R A$-module by the rule: $a \otimes b \cdot x = (ab)x$. The reader should verify that this makes P into an invertible left $A \otimes_R A$-module. Conversely, if M is an invertible left $A \otimes_R A$-module for which the left and right A-module actions agree, then M defines an invertible A-module which represents a class in $\operatorname{Pic}(A)$. In this way, we identify $\operatorname{Pic}(A)$ with the subgroup of $\operatorname{Pic}_R(A)$ consisting of those $[M]$ for which the left and right A-module actions agree.

Let M be an invertible $A \otimes_R A$-module. Since $A = A^o$, the isomorphisms λ and ρ of (7.18) imply that for any $a \in A$, ρ_a is equal to $\lambda_{\alpha_M(a)}$, for some element $\alpha_M(a) \in A$. In other words, for all $a \in A$ there exists $\alpha_M(a) \in A$ such that $xa = \alpha_M(a)x$, for all $x \in M$.

Lemma 7.8.3. *Let A be a commutative R-algebra and M an invertible $A \otimes_R A$-module.*

(1) *The assignment $a \mapsto \alpha_M(a)$ defines an R-algebra automorphism α_M in $\operatorname{Aut}_R(A)$.*
(2) *The assignment $M \mapsto \alpha_M$ defines a homomorphism of groups $\Psi : \operatorname{Pic}_R(A) \to \operatorname{Aut}_R(A)$.*
(3) *The kernel of Ψ is $\operatorname{Pic}(A)$. In particular, $\operatorname{Pic}(A)$ is a normal subgroup of $\operatorname{Pic}_R(A)$.*

Proof. (1): Since λ and ρ are isomorphisms, the function $\alpha_M(a)$ is well defined. The reader should verify that α_M is an R-algebra automorphism of A.

(2): Let P and Q be two invertible left $A \otimes_R A$-modules. For $x \otimes y \in P \otimes_A Q$ and $a \in A$: $(x \otimes y)a = x \otimes ya = x \otimes \alpha_Q(a)y = x\alpha_Q(a) \otimes y = \alpha_P(\alpha_Q(a))x \otimes y = \alpha_P(\alpha_Q(a))(x \otimes y)$. From this it follows that $\alpha_{P \otimes_A Q} = \alpha_P \alpha_Q$.

(3): This proof is left to the reader. $\square$

Definition 7.8.4. Let M be a left $A \otimes_R A^o$-module. If α, β are in $\operatorname{Aut}_R(A)$, then we define ${}_\alpha M_\beta$ to be the left $A \otimes_R A^o$-module whose additive group is M, and whose bimodule structure is defined by $a \otimes b \cdot x = \alpha(a)x\beta(b)$ for all $a, b \in A$, and $x \in M$. The reader should verify that in this case,

$$ {}_\alpha M_\beta = {}_\alpha A_1 \otimes_A M \otimes_A {}_1 A_\beta. $$

We say ${}_\alpha M_\beta$ is the module with action *twisted* by the automorphisms α and β.

Lemma 7.8.5. *If A is an R-algebra, then as in Definition 4.1.1, A is a left $A \otimes_R A^o$-module. In the following, let $\alpha, \beta, \gamma, \delta$ represent elements of $\operatorname{Aut}_R(A)$. Then there are isomorphisms*

$$ {}_\alpha A_\beta \cong {}_{\gamma\alpha} A_{\gamma\beta} \cong {}_1 A_{\alpha^{-1}\beta} \cong {}_{\beta^{-1}\alpha} A_1 $$
$$ {}_\gamma \left({}_\alpha A_\beta\right)_\delta \cong {}_{\alpha\gamma} A_{\beta\delta} $$
$$ {}_\alpha A_\beta \otimes_A {}_\gamma A_\delta \cong {}_\alpha A_{\beta\gamma^{-1}\delta} $$
$$ {}_1 A_\alpha \otimes_A {}_1 A_\beta \cong {}_1 A_{\alpha\beta} $$
$$ {}_1 A_\alpha \otimes_A {}_\alpha A_1 \cong {}_1 A_1 $$
$$ {}_\alpha A_1 \otimes_A {}_1 A_\alpha \cong {}_\alpha A_\alpha \cong {}_1 A_1 $$

of $A \otimes_R A^o$-modules.

Proof. We prove the identity in the fourth line. The rest is left to the reader. Consider the map $\psi : {}_1A_\alpha \times {}_1A_\beta \to {}_1A_{\alpha\beta}$ defined by $x \times y \mapsto x\alpha(y)$. Then $\psi(x \cdot a \times y) = \psi(x\alpha(a) \times y) = x\alpha(a)\alpha(y) = \psi(x \times ay)$. Using this, one checks that ψ induces a homomorphism of $\mathbb{Z}$-modules $\psi : {}_1A_\alpha \otimes_A {}_1A_\beta \to {}_1A_{\alpha\beta}$. Since $\psi(a \otimes b \cdot x \otimes y) = \psi(ax \otimes y\beta(b)) = ax\alpha(y)\alpha\beta(b)) = a \otimes b \cdot (x\alpha(y)) = a \otimes b \cdot \psi(x \otimes y)$, ψ is an isomorphism of $A \otimes_R A^o$-modules. □

Lemma 7.8.6. *Let A be an R-algebra.*

(1) *For α, β in $\operatorname{Aut}_R(A)$, ${}_\alpha A_\beta$ is an invertible $A \otimes_R A^o$-module.*

(2) *The assignment $\sigma \mapsto [{}_1A_\sigma]$ defines a group homomorphism $\Phi_0 : \operatorname{Aut}_R(A) \to \operatorname{Pic}_R(A)$.*

Proof. (1): This follows from Lemma 7.8.5 and Proposition 7.8.1.

(2): That Φ_0 is a homomorphism follows from the isomorphism ${}_1A_\alpha \otimes_A {}_1A_\beta \cong {}_1A_{\alpha\beta}$, which is the identity in line four of Lemma 7.8.5. □

Lemma 7.8.7. *Let A be a commutative R-algebra.*

(1) *$\Psi\Phi_0$ is the identity map on $\operatorname{Aut}_R(A)$.*

(2) *The sequence of groups*

$$0 \to \operatorname{Pic}(A) \to \operatorname{Pic}_R(A) \xrightarrow{\Psi} \operatorname{Aut}_R(A) \to 1$$

is exact, and Ψ is split by Φ_0.

(3) *For any $[M] \in \operatorname{Pic}_R(A)$, if $\Psi([M]) = \tau$, then $M \otimes_A {}_1A_{\tau^{-1}} \cong M_{\tau^{-1}}$ is in $\operatorname{Pic}(A)$. The left and right A-module actions on M differ by a twist by an R-automorphism of A.*

(4) *The automorphism group $\operatorname{Aut}_R(A)$ acts on $\operatorname{Pic}_R(A)$ (and hence on the normal subgroup $\operatorname{Pic}(A)$) by:*

$$\tau[M] = \Phi_0(\tau)[M]\Phi_0(\tau^{-1}) = [{}_1A_\tau][M][{}_1A_\tau]^{-1} = [{}_{\tau^{-1}}M_{\tau^{-1}}]$$

for any $\tau \in \operatorname{Aut}_R(A)$ and $[M] \in \operatorname{Pic}_R(A)$.

Proof. (1): For any $a \in A$, $x \in {}_1A_\sigma$, $\rho_a(x) = x \cdot a = x\sigma(a) = \sigma(a)x = \lambda_{\sigma(a)}(x)$. This shows $\Psi([{}_1A_\sigma]) = \sigma$.

Parts (2), (3) and (4) are left to the reader. □

Theorem 7.8.8. *Let A be an R-algebra and M a left A-progenerator module. Let $B = \operatorname{Hom}_A(M, M)$. The assignment $[P] \mapsto [M^* \otimes_A P \otimes_A M]$ defines an isomorphism $\operatorname{Pic}_R(A) \to \operatorname{Pic}_R(B)$ of groups.*

Proof. The inverse map is defined by $[Q] \mapsto [M^* \otimes_B Q \otimes_B M]$. The proof follows from the same arguments used to prove Theorem 1.5.2. □

Theorem 7.8.9. *Let A be an R-algebra. Let $[P]$ and $[Q]$ be two classes in* $\operatorname{Pic}_R(A)$. *The following are equivalent:*

(1) *The left A-modules ${}_AP$ and ${}_AQ$ are isomorphic.*

(2) *$[P]$ and $[Q]$ belong to the same left coset of* $\operatorname{im}(\Phi_0)$.

(3) *For some $\tau \in \operatorname{Aut}_R(A)$, Q is isomorphic to $P \otimes_A {}_1A_\tau$ as $A \otimes_A A^o$-modules.*

(4) *For some $\tau \in \operatorname{Aut}_R(A)$, Q is isomorphic to ${}_1P_\tau$ as $A \otimes_A A^o$-modules.*

Proof. The reader should verify that (2), (3) and (4) are equivalent.

(4) implies (1): Any $A \otimes_A A^o$-module isomorphism $Q \cong {}_1P_\tau$ is an left A-module isomorphism ${}_AP \cong {}_AQ$.

(1) implies (4): Let $\theta : {}_AP \to {}_AQ$ be an isomorphism in ${}_A\mathfrak{M}$. Then θ induces an isomorphism of R-algebras

$$\mathrm{H}(\theta) : \operatorname{Hom}_A({}_AQ, {}_AQ) \to \operatorname{Hom}_A({}_AP, {}_AP), \tag{7.19}$$

where $\mathrm{H}(\theta)$ is defined by $\phi \mapsto \theta^{-1}\phi\theta$. For both P and Q, in (7.18), ρ is an isomorphism. Therefore, using $\mathrm{H}(\theta)$ we define an automorphism of the R-algebra A. Specifically, for any $b \in A$, there exists a unique $a \in A$ such that $\mathrm{H}(\theta)(\rho_b) = \rho_a$. Then

$$\theta^{-1}\rho_b\theta = \rho_a. \tag{7.20}$$

Define $\tau : A \to A$ by $\tau(b) = a$. From (7.20) one checks that τ is an R-automorphism of A. For all $x \in P$, $(\theta^{-1}\rho_b\theta)x = \rho_a x = xa$, hence $(\theta x)b = \theta(xa)$. Substituting $\tau(b) = a$, $(\theta x)b = \theta(x\tau(b))$. From this one shows θ induces an $A \otimes_R A^o$-module isomorphism $\theta : {}_1P_\tau \to Q$, so (4) is true. □

8.2. The Skolem-Noether Theorem. Let A be an R-algebra, and u an invertible element of A. The inner automorphism of A defined by u is $\sigma_u : A \to A$ where $\sigma_u(x) = uxu^{-1}$ for every $x \in A$. Let $\operatorname{Inn}(A)$ denote the set of all inner automorphisms of A. By Exercise 7.8.16, $\operatorname{Inn}(A)$ is a normal subgroup of $\operatorname{Aut}_R(A)$.

Theorem 7.8.10. *Let A be an R-algebra. The kernel of the homomorphism $\Phi_0 : \operatorname{Aut}_R(A) \to \operatorname{Pic}_R(A)$ is $\operatorname{Inn}(A)$, the group of inner automorphisms of A.*

Proof. The kernel of Φ_0 consists of those $\tau \in \operatorname{Aut}_R(A)$ such that ${}_1A_\tau$ is isomorphic to ${}_1A_1$ as $A \otimes_R A^o$-modules. A $\mathbb{Z}$-module isomorphism $\phi : {}_1A_1 \to {}_1A_\tau$ is $A \otimes_R A^o$-linear if and only if

$$\phi(axb) = a\phi(x)\tau(b) \tag{7.21}$$

for all $a, b, x \in A$. Let $u \in A^*$ be a unit in A, and let σ_u be the inner automorphism of A defined by u. Then $\sigma_u(x) = uxu^{-1}$. Define $\phi : {}_1A_1 \to {}_1A_{\sigma_u}$ by $x \mapsto xu^{-1}$. Then $\phi(axb) = (axb)u^{-1} = a(xu^{-1})(ubu^{-1}) = a\phi(x)\sigma_u(b)$. So ϕ is an $A \otimes_R A^o$-module isomorphism. Therefore, $\operatorname{Inn}(A) \subseteq \ker \Phi_0$.

Conversely, assume $\tau \in \ker \Phi_0$. Then there exists an $A \otimes_R A^o$-module isomorphism $\phi : {}_1A_1 \to {}_1A_\tau$. Let $u = \phi(1)$. Then $A\phi(1) = A$ and $A = \phi(1)A$ implies $u \in A^*$ is invertible. In (7.21) take $x = b = 1$. Then $\phi(a) = au$, for all $a \in A$. In (7.21) take $a = x = 1$. Then $\phi(b) = u\tau(b)$, for all $b \in A$. Taken together, $au = u\tau(a)$, which implies τ is the inner automorphism of A defined by u^{-1}. This shows $\operatorname{Inn}(A) \supseteq \ker \Phi_0$, and we are done. □

Theorem 7.8.11. *Let A be an Azumaya R-algebra. Then the assignment $[P] \mapsto [P \otimes_R A]$ defines an isomorphism of groups $\operatorname{Pic} R \to \operatorname{Pic}_R(A)$. The inverse mapping is defined by $[Q] \mapsto [Q^A]$.*

Proof. By Corollary 7.1.6, there is a Morita equivalence between the category of right R-modules and the category of left $A \otimes_R A^o$-modules. The functors $(\ ) \otimes_R A$ and $(\ )^A$ are inverse equivalences. The rest is left to the reader. □

Corollary 7.8.12. *Let R be a commutative ring and A an Azumaya R-algebra. There is an exact sequence*

$$1 \to \operatorname{Inn}(A) \to \operatorname{Aut}_R(A) \xrightarrow{\alpha} \operatorname{Pic}(R),$$

where α maps $\sigma \in \operatorname{Aut}_R(A)$ to $[({}_1A_\sigma)^A]$.

Proof. This follows from Theorem 7.8.10 and Theorem 7.8.11. □

Remark 7.8.13. The cokernel of the homomorphism α in Corollary 7.8.12 is described in [**RZ61**, Theorem 7]. See also [**DI71**, Theorem 2.6.5].

Corollary 7.8.14. *Let R be a commutative ring and A an Azumaya R-algebra. The group $\operatorname{Aut}_R(A)/\operatorname{Inn}(A)$ is abelian. If $\operatorname{Pic}(R) = (0)$, every R-algebra automorphism of A is inner.*

Corollary 7.8.15 (Skolem-Noether)**.** *If R is a local ring and A is an Azumaya R-algebra, then every R-algebra automorphism of A is inner. In particular, if k is a field and A is a central simple k-algebra, then every k-algebra automorphism of A is inner.*

8.3. Exercise.

Exercise 7.8.16. Let R be a commutative ring and A an R-algebra. Let u be an invertible element of A. The inner automorphism of A defined by u is $\sigma_u : A \to A$ where $\sigma_u(x) = uxu^{-1}$ for every $x \in A$. Let $\operatorname{Inn}(A)$ denote the set of all inner automorphisms of A. Show that $\operatorname{Inn}(A)$ is a normal subgroup of $\operatorname{Aut}_R(A)$, the group of all R-algebra automorphisms of A.

9. The Brauer Group Modulo an Ideal

Let R be a commutative ring and I an ideal in R. In this section we consider the question of whether an R/I-algebra can be realized as $A \otimes_R (R/I)$ for some R-algebra A. As an application of Hochschild cohomology, we derive sufficient conditions under which this lifting occurs. These results should also be compared with those of Section 9.3.6.

In the following, R always denotes a commutative ring and I an ideal in R. For any R-module, or homomorphism of R-modules, we use the bar notation to indicate the tensor product with $\bar{R} = R/I$. Theorem 7.9.1 and its proof are from [**KOS76**]. For a version that applies in the more general setting of henselian couples the interested reader is referred to [**Cip77**].

Theorem 7.9.1. *Let R be a commutative ring and I an ideal in R such that R is separated and complete in the I-adic topology (that is, $R \to \varprojlim R/I^n$ is an isomorphism). Let A be a finitely generated projective R-module. Suppose there is a multiplication $\bar{\mu} : \bar{A} \otimes_{\bar{R}} \bar{A} \to \bar{A}$ and a unit element $\bar{1} \in \bar{A}$ such that $\bar{A}$ is a $\bar{R}$-algebra. If* $\operatorname{proj.dim}_{\bar{A}^e}(\bar{A}) \leq 2$, *then $\bar{\mu}$ lifts to a multiplication $\mu : A \otimes_R A \to A$ which makes A into an R-algebra and $\bar{A} \cong A \otimes_R \bar{R}$ is a $\bar{R}$-algebra isomorphism. If* $\operatorname{proj.dim}_{\bar{A}^e}(\bar{A}) \leq 1$, *then the R-algebra A is unique up to isomorphism.*

Proof. Step 1: The theorem is true if $I^2 = (0)$. Notice that this is a special case, because if $I^2 = 0$, the inverse limit $\varprojlim R/I^n$ converges to R. Step 1 is divided into a sequence of five smaller steps.

Step 1.1: Lifting $\bar{\mu}$ to μ. Notice that $\bar{A} \otimes_R \bar{A} = \bar{A} \otimes_{\bar{R}} \bar{A}$. By Lemma 5.2.2, there is an exact sequence

$$(A \otimes_R IA) \oplus (IA \otimes_R A) \to A \otimes_R A \xrightarrow{\eta} \bar{A} \otimes_{\bar{R}} \bar{A} \to 0. \tag{7.22}$$

Since A is R-projective, $A \otimes_R IA \to A \otimes_R A$ and $IA \otimes_R A \to A \otimes_R A$ are one-to-one. Consider the diagram

$$\begin{array}{ccccccccc} 0 & \longrightarrow & \ker \eta & \longrightarrow & A \otimes_R A & \xrightarrow{\eta} & \bar{A} \otimes_{\bar{R}} \bar{A} & \longrightarrow & 0 \\ & & & & \downarrow{\scriptstyle \exists \mu} & & \downarrow{\scriptstyle \bar{\mu}} & & \\ 0 & \longrightarrow & IA & \longrightarrow & A & \longrightarrow & \bar{A} & \longrightarrow & 0 \end{array}$$

where the rows are exact. Since $A \otimes_R A$ is a projective R-module, there exists a map μ which makes the right hand square commutative (Proposition 1.1.5).

Step 1.2: Define an $\bar{A}$-module structure on IA. Since $I^2 = 0$, tensoring the exact sequence $0 \to IA \to A \to \bar{A} \to 0$ with A yields isomorphisms $A \otimes_R IA \cong \bar{A} \otimes_R IA$ and $IA \otimes_R A \cong IA \otimes_R \bar{A}$. Therefore, we can make IA

into a two-sided $\bar{A}/R$-module using

$$(\bar{A} \otimes_R IA) \oplus (IA \otimes_R \bar{A}) \xrightarrow{\nu} IA \tag{7.23}$$

where on each summand, $\nu : A \otimes_R IA \to IA$, $\nu : IA \otimes_R A \to IA$ is the restriction of μ. The multiplication map ν in (7.23) is defined on $\ker \eta$. If μ_1 is another choice for μ, then $\mu - \mu_1$ is the zero map on $\ker \eta$. This implies ν depends only on $\bar{\mu}$, and not on μ.

Step 1.3: Modify μ so it is associative. The multiplication μ may not be associative, so define

$$F(a \otimes b \otimes c) = \mu(\mu(a \otimes b) \otimes c) - \mu(a \otimes \mu(b \otimes c))$$

which is an R-module homomorphism. Consider the diagram

$$\begin{array}{ccccccccc}
0 & \longrightarrow & \ker \eta & \longrightarrow & T^3_R(A) & \xrightarrow{\eta} & T^3_{\bar{R}}(\bar{A}) & \longrightarrow & 0 \\
 & & & & \Big\downarrow F & & & & \\
0 & \longrightarrow & IA & \longrightarrow & A & \longrightarrow & \bar{A} & \longrightarrow & 0
\end{array}$$

where $T^3_R(A)$ and $T^3_{\bar{R}}(\bar{A})$ are the tensor modules (Section 5.2). Since $\bar{\mu}$ is associative, we see that the image of F is contained in IA. By Lemma 5.2.2, the sequence

$$(A \otimes_R A \otimes_R IA) \oplus (A \otimes_R IA \otimes_R A) \oplus (IA \otimes_R A \otimes_R A) \to T^3_R(A) \xrightarrow{\eta} T^3_{\bar{R}}(\bar{A}) \to 0$$

is exact. Therefore, F factors through $T^3_{\bar{R}}(\bar{A})$, and the diagram

$$\begin{array}{ccccc}
T^3_R(A) & \xrightarrow{\eta} & T^3_{\bar{R}}(\bar{A}) & \longrightarrow & 0 \\
F \Big\downarrow & \swarrow_{\bar{F}} & & & \\
IA & & & &
\end{array}$$

commutes. The reader should verify that $\delta^3(\bar{F}) = 0$, where δ^3 is the coboundary defined in (5.17). Therefore, $\bar{F} \in \mathrm{Z}^3(\bar{A}, IA)$ is a 3-cocycle for $\bar{A}$ with coefficients in IA. Since we are assuming $\operatorname{proj.dim}_{\bar{A}^e}(\bar{A}) \le 2$, this implies $\bar{F} \in \mathrm{B}^3(\bar{A}, IA)$. So there is a cochain $\bar{G} \in \operatorname{Hom}_{\bar{R}}(\bar{A} \otimes_{\bar{R}} \bar{A}, IA)$ such that

$$\bar{F}(\bar{a} \otimes \bar{b} \otimes \bar{c}) = \bar{a}\bar{G}(\bar{b} \otimes \bar{c}) - \bar{G}(\bar{a}\bar{b} \otimes \bar{c}) + \bar{G}(\bar{a} \otimes \bar{b}\bar{c}) - \bar{G}(\bar{a} \otimes \bar{b})\bar{c}.$$

Let G be defined by the commutative diagram

$$\begin{array}{ccc}
A \otimes_R A & \xrightarrow{\quad G \quad} & IA \\
 & {}_{\eta}\searrow \quad \nearrow_{\bar{G}} & \\
 & \bar{A} \otimes_{\bar{R}} \bar{A} &
\end{array}$$

Notice that G is zero on the kernel of η. The reader should verify that $\mu+G$ is associative and is a lifting of $\bar{\mu}$.

Step 1.4: Existence of a multiplicative identity element. By Step 1.3, we assume that μ defines an associative multiplication on A. We show that the multiplicative identity of $\bar{A}$ lifts to A. Let $\bar{u} = \bar{1}$. Then uA is an R-submodule of A. Since $uA \otimes_R \bar{R} = \bar{A}$, by Nakayama's Lemma (Corollary 1.1.17), $uA = A$. By Corollary 1.3.18, left multiplication by u, $x \mapsto ux$, is an R-module isomorphism from A to A. By a symmetric argument the right multiplication map, $x \mapsto xu$, is a one-to-one correspondence on A as well. Let $e \in A$ such that $ue = u$. Then $ue^2 = ue$, hence $e^2 = e$. From $ue = u$ it follows that $ueu = u^2$, hence $eu = u$. For any $a \in A$, $ua = uea$ implies $a = ea$, hence e is a left identity for A. Likewise, $au = aeu$ implies $a = ae$, so e is a right identity for A. The commutative diagram

$$\begin{array}{ccc} A \otimes_R A & \xrightarrow{\mu} & A \\ \eta \downarrow & & \downarrow \eta \\ \bar{A} \otimes_{\bar{R}} \bar{A} & \xrightarrow{\bar{\mu}} & \bar{A} \end{array} \qquad \begin{array}{ccc} e \otimes u & \longmapsto & eu = u \\ \downarrow & & \downarrow \\ \bar{e} \otimes \bar{u} = \bar{e} \otimes \bar{1} & \longmapsto & \bar{e} = \bar{1} \end{array}$$

shows that $\bar{e} = \bar{1}$.

Step 1.5: The algebra is unique, if $\operatorname{proj.dim}_{\bar{A}^e}(\bar{A}) \le 1$. Assume we have two algebra structures on A, μ_1 and μ_2, both of which agree with $\bar{\mu}$ on $\bar{A}$. Define $F : A \otimes_R A \to A$ by

$$F(a \otimes b) = \mu_1(a \otimes b) - \mu_2(a \otimes b).$$

Since $\bar{\mu}_1 = \bar{\mu}_2$, the image of F is contained in IA. Using the exact sequence (7.22), and the assumption that $I^2 = (0)$, we see that F factors through $\bar{A} \otimes_{\bar{R}} \bar{A}$, and the diagram

$$\begin{array}{ccccc} A \otimes_R A & \xrightarrow{\eta} & \bar{A} \otimes_{\bar{R}} \bar{A} & \longrightarrow & 0 \\ F \downarrow & \swarrow \bar{F} & & & \\ IA & & & & \end{array}$$

commutes. The reader should verify that $\delta^2(\bar{F}) = 0$, where δ^2 is the coboundary defined in (5.17). Therefore, $\bar{F} \in \mathrm{Z}^2(\bar{A}, IA)$ is a 2-cocycle for $\bar{A}$ with coefficients in IA. Since we are assuming $\operatorname{proj.dim}_{\bar{A}^e}(\bar{A}) \le 1$, this implies $\bar{F} \in \mathrm{B}^2(\bar{A}, IA)$. So there is a cochain $\bar{G} \in \operatorname{Hom}_{\bar{R}}(\bar{A}, IA)$ such that

$$\bar{F}(\bar{a} \otimes \bar{b}) = \bar{a}\bar{G}(\bar{b}) - \bar{G}(\bar{a}\bar{b}) + \bar{G}(\bar{a})\bar{b}.$$

Let G be defined by the commutative diagram

$$\begin{array}{ccc} A & \xrightarrow{G} & IA \\ & {\scriptstyle\eta}\searrow \quad \nearrow{\scriptstyle\bar{G}} & \\ & \bar{A} & \end{array}$$

Let $f : A \to A$ be defined by $f(x) = x + G(x)$. Notice that G is zero on the kernel of η, so f restricts to the identity map on $\bar{A}$. The reader should verify that $f(\mu_1(a \otimes b)) = \mu_2(f(a) \otimes f(b))$. Let e_1 and e_2 be the multiplicative identity elements for μ_1 and μ_2 respectively. Then $f(e_1) = f(\mu_1(e_1 \otimes e_2)) = \mu_2(f(e_1) \otimes f(e_1))$. The commutative diagram

$$\begin{array}{ccc} A & \xrightarrow{f} & A \\ {\scriptstyle\eta}\downarrow & & \downarrow{\scriptstyle\eta} \\ \bar{A} & \xrightarrow{=} & \bar{A} \end{array} \qquad \begin{array}{ccc} e_1 & \longmapsto & f(e_1) \\ \downarrow & & \downarrow \\ \bar{1} & \longmapsto & \bar{1} \end{array}$$

shows that $f(e_1)$ is a lifting for $\bar{1}$. Then $f(e_1) = e_2$ is the unique idempotent unit. This shows f is an R-algebra isomorphism.

Step 2: The general case consists of a sequence of three smaller steps.

Step 2.1: The theorem is true if I is a nilpotent ideal. Suppose the index of nilpotency is n. By Step 1, we can lift $\bar{\mu}$ to make $A \otimes_R R/I^2$ into an R/I^2-algebra. Iterate to make $A = A \otimes_R R/I^n$ into an R/I^{n-1}-algebra.

Step 2.2: The theorem is true if R is separated and complete for the I-adic topology. In this case, A is equal to $\varprojlim A \otimes_R R/I^n$ as an R-module. By Step 2, for each n, we make $A \otimes_R R/I^n$ into an R-algebra such that the natural maps $A \otimes_R R/I^{n+1} \to A \otimes_R R/I^n$ are R-algebra homomorphisms. Then $\varprojlim A \otimes_R R/I^n$ is an R-algebra which is a lifting of $\bar{A}$.

Step 2.3: If $\operatorname{proj.dim}_{\bar{A}^e}(\bar{A}) \le 1$, then the lifting is unique. This follows from the recursive construction in Step 2.1 and the uniqueness proof of Step 1.5. □

9.1. Lifting Azumaya Algebras. If R is a commutative ring, then as in Section 7.3, $\operatorname{Az}(R)$ denotes the set of isomorphism classes of Azumaya R-algebras. Theorem 7.9.2 and its corollary are from [**KOS76**].

Theorem 7.9.2. *Let R be a commutative ring and I an ideal in R such that R is separated and complete for the I-adic topology (that is, $R \to \varprojlim R/I^n$ is an isomorphism).*

(1) *The functor $A \mapsto A \otimes_R (R/I)$ from the category of R-algebras to the category of R/I-algebras defines a one-to-one correspondence between $\operatorname{Az}(R)$ and $\operatorname{Az}(R/I)$.*

(2) *The natural map* $\mathrm{B}(R) \to \mathrm{B}(R/I)$ *is an isomorphism.*

Proof. (1): Let $\bar{A}$ be an Azumaya $\bar{R} = R/I$-algebra. Then $\bar{A}$ is an $\bar{R}$-progenerator module and the enveloping map $\bar{\varphi} : \bar{A}^e \to \mathrm{Hom}_{\bar{R}}(\bar{A}, \bar{A})$ of Definition 4.1.1 is an isomorphism (Theorem 7.1.4). By Proposition 3.4.18, $\bar{A}$ lifts to an R-progenerator module A. By Proposition 4.1.2, $\bar{A}$ is projective as a left $\bar{A}^e$-module. Therefore, $\mathrm{proj.dim}_{\bar{A}^e}(\bar{A}) = 0$. By Theorem 7.9.1, there is a unique R-algebra structure on A that lifts the algebra structure on $\bar{A}$. Consider the enveloping map

$$A^e \xrightarrow{\varphi} \mathrm{Hom}_R(A, A) \tag{7.24}$$

which agrees with $\bar{\varphi}$ upon tensoring with $\bar{R}$. Since R is separated and complete for the I-adic topology, I is contained in $\mathrm{J}(R)$, the Jacobson radical of R (Proposition 3.4.8). Since A is an R-progenerator, A^e and $\mathrm{Hom}_R(A, A)$ are R-progenerators as well. Nakayama's Lemma (Corollary 1.1.17) implies φ is onto. Then φ splits, so the same argument can be applied to show φ is one-to-one. By Theorem 7.1.4, A is an Azumaya R-algebra.

Part (2) follows straight from Part (1). □

Corollary 7.9.3. *Let R be a commutative ring and I an ideal that is contained in the nil radical of R. Then the natural map* $\mathrm{B}(R) \to \mathrm{B}(R/I)$ *is an isomorphism.*

Proof. As in Exercise 2.5.16, $R = \varinjlim R_\alpha$, where $\{R_\alpha\}$ is the directed system of all noetherian subrings of R. By Corollary 7.7.8, $\varinjlim \mathrm{B}(R_\alpha) \to \mathrm{B}(R)$ is an isomorphism. For each α, $I \cap R_\alpha$ is generated by finitely many nilpotent elements. Therefore, $I \cap R_\alpha$ is a nilpotent ideal in R_α. Then $\mathrm{B}(R_\alpha) \cong \mathrm{B}(R_\alpha/(I \cap R_\alpha))$, by Theorem 7.9.2 (2). Since $\varinjlim R_\alpha/(I \cap R_\alpha) = R/I$, all of the maps in the commutative square

$$\begin{array}{ccc} \varinjlim \mathrm{B}(R_\alpha) & \longrightarrow & \mathrm{B}(R) \\ \downarrow & & \downarrow \\ \varinjlim \mathrm{B}(R_\alpha/(I \cap R_\alpha)) & \longrightarrow & \mathrm{B}(R/I) \end{array}$$

are isomorphisms. □

Corollary 7.9.4 is due to DeMeyer [**DeM75**, Corollary 4].

Corollary 7.9.4. *If R is a commutative ring, then the natural map* $\mathrm{B}(R) \to \mathrm{B}(R[[x]])$ *is an isomorphism.*

Proof. This follows from Exercise 1.4.13 and Theorem 7.9.2. □

9.2. The Brauer Group of a Commutative Artinian Ring. We end this section with three more applications of Corollary 7.9.3. First we show that the Brauer group of an artinian ring R maps isomorphically onto the Brauer group of the reduced ring $R/\operatorname{J}(R)$.

Corollary 7.9.5. *Let R be a commutative artinian ring and $R = R_1 \oplus \cdots \oplus R_n$ the unique decomposition such that each R_i is a local artinian ring (Theorem 3.2.11). If the maximal ideal of R_i is $\mathfrak{m}_i$ and the residue field is k_i, then $\operatorname{B}(R) = \operatorname{B}(k_1) \oplus \cdots \oplus \operatorname{B}(k_n)$.*

Proof. By Proposition 3.2.9, $\operatorname{Rad}_R(0) = \operatorname{J}(R)$. Then $R/\operatorname{Rad}_R(0) = k_1 \oplus \cdots \oplus k_n$. By Corollary 7.9.3 and Exercise 7.4.8, $\operatorname{B}(R) = \operatorname{B}(R/\operatorname{Rad}_R(0)) = \operatorname{B}(k_1) \oplus \cdots \oplus \operatorname{B}(k_n)$. □

In case the residue fields of R have trivial Brauer groups, we get the following two special cases.

Corollary 7.9.6. *If R is a finite commutative ring, then $\operatorname{B}(R) = (0)$.*

Proof. By Theorem 7.5.4, the Brauer group of a finite field is trivial. □

In Corollary 7.9.7 we consider algebras over a field k where k has the property that if F is a finite algebraic extension field of k, then $\operatorname{B}(F) = (0)$. For example, we will see in Corollary 11.1.8 that this condition is satisfied if k is C_1 (see Section 1.2.3). In particular, it is true for an algebraically closed field (Theorem 7.5.1) or a field of transcendence degree one over an algebraically closed field (Corollary 11.1.9). A finite field also has this property (Theorem 7.5.4).

Corollary 7.9.7. *Let k be a field with the property that if F/k is a finite algebraic extension of fields, then $\operatorname{B}(F) = (0)$ (as above, this is true if k is C_1). Let R be a commutative artinian finitely generated k-algebra. Then $\operatorname{B}(R) = 0$.*

Proof. By Exercise 3.2.17, $\dim_k(R) < \infty$. Then each of the residue fields k_i is a finite algebraic extension of k. □

Chapter 8

Derivations, Differentials and Separability

This chapter introduces two powerful methods for studying separable algebras over commutative rings. These new tools are the module of R-derivations on an R-algebra, and the module of Kähler differentials. As always, our motivation is to derive new tests for separability. Applying results about derivations allows us to prove theorems on faithfully flat descent of separability, the separability at the stalks criteria, and the residue field tests for separability. As another application of derivations we include a section on differential crossed product algebras, which is probably the most important method for constructing Azumaya algebras in characteristic p. Applying results on Kähler differentials, we derive separability criteria for commutative R-algebras. For example, the vanishing of the module of Kähler differentials leads to a separability criterion for a finitely generated commutative algebra. Lastly, differentials are applied to prove jacobian criteria for separability and regularity.

1. Derivations and Separability

This section contains an introduction to R-derivations on an R-algebra with coefficients in a two-sided module. General references for the material in this section are [**CE56**], [**KO74b**], [**Mat80**], [**Jac79**], and [**KOS76**].

1.1. The Definition and First Results. Let R be a commutative ring and A an R-algebra. The enveloping algebra is $A^e = A \otimes_R A^o$. A left right

A-bimodule M is called a two-sided A/R-module if the left and right R-actions agree (Definition 4.1.1). If M is a left A^e-module, then we can make M into a two-sided A/R-module by defining $ax = a \otimes 1 \cdot x$ and $xa = 1 \otimes a \cdot x$. In particular, A^e is a left A^e-module, hence is a two-sided A/R-module.

If M is any two-sided A/R-module, then an R-*derivation* from A to M is an R-module homomorphism $\partial : A \to M$ satisfying

$$\partial(ab) = a\partial(b) + \partial(a)b$$

for all $a, b \in A$. The set of all R-derivations from A to M is denoted $\operatorname{Der}_R(A, M)$. The reader should verify that $\operatorname{Der}_R(A, M)$ is an R-submodule of $\operatorname{Hom}_R(A, M)$ and that if ∂ is any R-derivation, then $\partial(1) = 0$.

Example 8.1.1. Let R be any ring, x an indeterminate, and $A = R[x]$ the polynomial ring. The usual derivative with respect to x is a $\mathbb{Z}$-derivation $\partial : A \to A$.

There is an exact sequence of A^e-modules

$$0 \to J_{A/R} \to A^e \xrightarrow{\mu} A \to 0 \tag{8.1}$$

(Definition 4.1.1), where μ is defined by $a \otimes b \mapsto ab$ and $J_{A/R}$ is defined to be the kernel of μ. By Definition 4.1.3, A is separable over R if and only if (8.1) is split exact as a sequence of A^e-modules. By Exercise 4.1.10, $J_{A/R}$ is generated as a left ideal in A^e by the set of all elements of the form $a \otimes 1 - 1 \otimes a$. By Exercise 1.3.8, there is an R-algebra homomorphism $\rho : A \to A \otimes_R A^o$ defined by $\rho(a) = a \otimes 1$. Using ρ we view each term in (8.1) as a left A-module. The reader should verify that $\mu\rho = 1$ and that both ρ and μ are left A-module homomorphisms. Therefore, (8.1) is always split exact as a sequence of left A-modules.

Example 8.1.2. Define an R-module homomorphism $\delta : A \to J_{A/R}$ by

$$\delta(a) = a \otimes 1 - 1 \otimes a.$$

If $a, b \in A$, then

$$\begin{aligned}
\delta(ab) &= ab \otimes 1 - 1 \otimes ab \\
&= ab \otimes 1 - a \otimes b + a \otimes b - 1 \otimes ab \\
&= (a \otimes 1)(b \otimes 1 - 1 \otimes b) + (1 \otimes b)(a \otimes 1 - 1 \otimes a) \\
&= a\delta(b) + \delta(a)b.
\end{aligned}$$

Therefore $\delta : A \to J_{A/R}$ is an R-derivation.

Lemma 8.1.3. *If $\delta : A \to J_{A/R}$ is from Example 8.1.2, then $A\delta(A) = J_{A/R}$. That is, the image of δ generates $J_{A/R}$ as a left A-module.*

Proof. A typical element of $J_{A/R}$ is $x = \sum_i x_i \otimes y_i$ such that $\sum_i x_i y_i = 0$. Then $\sum_i x_i(1 \otimes y_i - y_i \otimes 1) = \sum_i x_i \otimes y_i - \left(\sum_i x_i y_i\right) \otimes 1 = x$. □

Lemma 8.1.4. *Let R be a commutative ring and A an R-algebra.*

(1) *If A is commutative and is generated as an R-algebra by the set $X = \{x_i\}_{i \in I}$, then $J_{A/R}$ is generated as an A^e-module by the set $\delta(X) = \{x_i \otimes 1 - 1 \otimes x_i\}_{i \in I}$.*

(2) *If A is finitely generated as an R-module, then $J_{A/R}$ is finitely generated as an R-module.*

(3) *Assume either*
 (a) *A is a finitely generated R-module, or*
 (b) *A a finitely generated commutative R-algebra.*
 Then $J_{A/R}$ is a finitely generated left ideal of A^e and A is an A^e-module of finite presentation.

Proof. (1): A typical element of A can be written as a finite sum $a = \sum r_i m_i$, where $r_i \in R$ and m_i is a monomial in X. Since δ is R-linear, it is enough to show $\delta(x_1 \cdots x_n)$ is in $A^e \delta(X)$, where $x_1, \dots, x_n$ represent any elements (not necessarily distinct) of X. Because δ is an R-derivation, this follows from the generalized product rule, Exercise 8.1.9.

(2): By Proposition 1.3.3, A^e is a finitely generated R-module. The sequence (8.1) is split exact as a sequence of R-modules, hence $J_{A/R}$ is a homomorphic image of A^e.

(3): In both cases, $J_{A/R}$ is finitely generated over A^e. The exact sequence (8.1) shows that A is of finite presentation as a left A^e-module. □

Given any $f \in \operatorname{Hom}_{A^e}(J_{A/R}, M)$, let $\alpha_f : A \to M$ be defined by

$$\alpha_f(a) = f(\delta(a)).$$

The reader should verify that $\alpha_f \in \operatorname{Der}_R(A, M)$ and that there is a homomorphism of R-modules $\alpha : \operatorname{Hom}_{A^e}(J_{A/R}, M) \to \operatorname{Der}_R(A, M)$ defined by $f \mapsto \alpha_f$. Given any $m \in M$, let $\tau_m : A \to M$ be defined by

$$\tau_m(a) = am - ma.$$

The reader should verify that $\tau_m \in \operatorname{Der}_R(A, M)$ and that there is a homomorphism of R-modules $\tau : M \to \operatorname{Der}_R(A, M)$ defined by $m \mapsto \tau_m$.

Proposition 8.1.5. *In the notation developed above, there is a commutative diagram of R-modules*

$$\begin{array}{ccccccc}
0 & \longrightarrow & \operatorname{Hom}_{A^e}(A, M) & \longrightarrow & \operatorname{Hom}_{A^e}(A^e, M) & \xrightarrow{\sigma} & \operatorname{Hom}_{A^e}(J_{A/R}, M) \\
 & & \gamma \downarrow \cong & & \beta \downarrow \cong & & \alpha \downarrow \cong \\
0 & \longrightarrow & M^A & \longrightarrow & M & \xrightarrow{\tau} & \operatorname{Der}_R(A, M)
\end{array}$$

such that the three vertical maps are isomorphisms and the rows are exact.

Proof. Applying the left exact functor $\operatorname{Hom}_{A^e}(\cdot, M)$ to the exact sequence (8.1) yields the top row. Clearly the kernel of τ is M^A, so the bottom row is exact. The isomorphism β comes from Proposition 1.3.23 (1) and is defined by the action $f \mapsto f(1)$. The isomorphism γ comes from Lemma 4.1.4. We check that $\alpha\sigma = \tau\beta$. Suppose $f \in \operatorname{Hom}_{A^e}(A^e, M)$, $f(1) = m$, and $a \in A$. Then

$$\begin{aligned} \alpha(\sigma(f))(a) &= f(\delta(a)) \\ &= \delta(a)f(1) \\ &= (a \otimes 1 - 1 \otimes a)m \\ &= am - ma \\ &= \tau(\beta(f))(a). \end{aligned}$$

Next we verify that α is one-to-one. Suppose $\alpha_f = 0$. Then $f(\delta(A)) = 0$. It follows from Lemma 8.1.3 that $f(J_{A/R}) = 0$. Now we show that α is onto. Let $\partial \in \operatorname{Der}_R(A, M)$. We must show that there exists $h \in \operatorname{Hom}_{A^e}(J_{A/R}, M)$ such that $\partial = h \circ \delta$. The reader should verify that the assignment $x \otimes y \mapsto -x\partial(y)$ defines an R-module homomorphism $h : A^e \to M$ and $h(\delta(a)) = h(a \otimes 1 - 1 \otimes a) = -a\partial(1) + \partial(a) = \partial(a)$. To show that h is a homomorphism of A^e-modules, let $x = \sum_i x_i \otimes y_i$ be a typical element of $J_{A/R}$ and $a \otimes b \in A^e$. Then

$$\begin{aligned} h(a \otimes b \cdot x) &= h\Big(a \otimes b \sum_i x_i \otimes y_i\Big) \\ &= h\Big(\sum_i ax_i \otimes y_i b\Big) \\ &= -\sum_i ax_i\partial(y_i b) \\ &= -\sum_i ax_i\big(y_i\partial(b) + \partial(y_i)b\big) \\ &= -a\Big(\sum_i x_i y_i\Big)\partial(b) - a\Big(\sum_i x_i\partial(y_i)\Big)b \\ &= a \otimes b \cdot h(x) \end{aligned}$$

completes the proof. □

The image of $\tau : M \to \operatorname{Der}_R(A, M)$ is denoted $\operatorname{Inn.Der}_R(A, M)$ and is called the set of *inner derivations*. Because the diagram of Proposition 8.1.5 commutes, under the isomorphism α, the set of inner derivations corresponds to the set of $f \in \operatorname{Hom}_{A^e}(J_{A/R}, M)$ such that f extends to $A^e \to M$.

Proposition 8.1.6. *Let A and C be commutative R-algebras and*

$$u : A \to C$$

a homomorphism of R-algebras. Let I be an ideal in C such that $I^2 = 0$. Consider the map on sets

$$\beta : \mathrm{Hom}_{R\text{-}alg}(A, C) \to \mathrm{Hom}_{R\text{-}alg}(A, C/I)$$

which is induced by the natural map $\eta : C \to C/I$ on R-algebras. Let $\bar{u} = \beta(u) = \eta u$. Make I into an A-module using the homomorphism u. That is, $a \cdot x = u(a)x$.

(1) *If $D : A \to I$ is an R-derivation, then $u + D : A \to C$ is an R-algebra homomorphism in $\beta^{-1}(\bar{u})$.*

(2) *If $v : A \to C$ is in $\beta^{-1}(\bar{u})$, and $D = v - u$, then $D : A \to I$ is an R-derivation.*

(3) *The mapping $D \mapsto u + D$ defines a one-to-one correspondence*

$$\mathrm{Der}_R(A, I) \to \{v \in \mathrm{Hom}_{R\text{-}alg}(A, C) \mid \beta(v) = \beta(u)\}.$$

Proof. (1): Because

$$\begin{aligned}(u + D)(ab) &= u(ab) + D(ab)\\ &= u(a)u(b) + u(a)D(b) + u(b)D(a)\end{aligned}$$

is equal to

$$\begin{aligned}(u(a) + D(a))(u(b) + D(b)) &= u(a)u(b) + u(a)D(b) + u(b)D(a) + D(a)D(b)\\ &= u(a)u(b) + u(a)D(b) + u(b)D(a),\end{aligned}$$

$u + D$ is multiplicative. The rest is left to the reader.

(2): For $a \in A$, $D(a) = u(a) - v(a)$ is in I. The computation

$$\begin{aligned}v(ab) &= v(a)v(b)\\ &= (u(a) + D(a))(u(b) + D(b))\\ &= u(a)u(b) + u(a)D(b) + u(b)D(a) + D(a)D(b)\\ &= u(a)u(b) + u(a)D(b) + u(b)D(a),\end{aligned}$$

shows that $D(ab) = u(a)D(b) + u(b)D(a)$.

Part (3) follows from (1) and (2). □

1.2. A Noncommutative Binomial Theorem in Characteristic p.

Let R be a commutative ring of characteristic p, where p is a prime number. The next result, which is from [**Jac79**], is a pth power version of the Binomial Theorem for an R-algebra A.

Lemma 8.1.7. *Let A be an R-algebra and λ an indeterminate. Let $\tau_{a\lambda+b} : A[\lambda] \to A[\lambda]$ be the inner derivation on the polynomial ring $A[\lambda]$ defined by*

$a\lambda + b$. If $is_i(a,b)$ is the coefficient of λ^{i-1} in the polynomial

$$\tau^{p-1}_{a\lambda+b}(a) = \sum_{i=0}^{p-1} (a\lambda + b)^i a(a\lambda + b)^{p-1-i},$$

then

$$(a+b)^p = a^p + b^p + \sum_{i=1}^{p-1} s_i(a,b).$$

Proof. Write

$$(a\lambda + b)^p = (a\lambda)^p + b^p + \sum_{i=1}^{p-1} c_i(a,b)\lambda^i \tag{8.2}$$

where $c_i(a,b)$ is a polynomial in a and b of degree p. Differentiating (8.2) with respect to λ, we get

$$\sum_{i=0}^{p-1} (a\lambda + b)^i a(a\lambda + b)^{p-1-i} = \sum_{i=1}^{p-1} ic_i(a,b)\lambda^{i-1} \tag{8.3}$$

where the left hand side is from Exercise 8.1.9. By Exercise 8.1.16, the left hand side of (8.3) is $\tau^{p-1}_{a\lambda+b}(a)$. This proves $ic_i(a,b)$ is the coefficient of λ^{i-1} in $\tau^{p-1}_{a\lambda+b}(a)$. Substitution of $\lambda = 1$ in (8.2) results in the formula of the lemma. □

1.3. Extensions of Derivations. Theorem 8.1.8 is from [**KOS76**] and is a generalization of results from [**Knu70**] and [**BK71**].

Theorem 8.1.8. *Let A be a separable R-algebra and B an R-subalgebra of A such that A is projective as a left B-module. Let M be a two-sided A/R-module and $d : B \to M$ a $\mathbb{Z}$-derivation such that $d(R) \subseteq M^A$. Then d extends to a $\mathbb{Z}$-derivation $D : A \to M$.*

Proof. Let $A * M$ be the trivial extension of A by M (Exercise 8.1.14). Let $\eta : A * M \to A$ be the natural R-algebra homomorphism and $\sigma : A \to A * M$ the R-algebra homomorphism defined by $a \mapsto (a, 0)$. We identify A with the image of σ and M with the kernel of η. The reader should verify that the multiplication rule: $b(a,x) = (ba, d(b)a + bx)$ makes $A * M$ into a left B-module. Notice that η is a B-module homomorphism, but if $d \neq 0$, then σ is not. By Parts (4) and (5) of Exercise 8.1.14, there is a one-to-one correspondence between $\operatorname{Der}_{\mathbb{Z}}(A, M)$ and the set of ring homomorphisms $u : A \to A * M$ which are sections to η. Given $D \in \operatorname{Der}_{\mathbb{Z}}(A, M)$, the corresponding ring homomorphism $u : A \to A * M$ is defined by $a \mapsto (a, D(a))$. If $b \in B$, then

$$u(ba) = (ba, D(ba)) = (ba, bD(a) + D(b)a)$$

and

$$bu(a) = b(a, D(a)) = (ba, d(b)a + bD(a)).$$

From this we see that u is a B-module homomorphism if and only if $D(b) = d(b)$. Therefore, to show that d extends to A it suffices to show that there exists a section to η which is simultaneously a ring homomorphism and a B-module homomorphism.

Since $\eta : A * M \to A$ is a B-module homomorphism and A is a projective B-module, there exists $u \in \operatorname{Hom}_B(A, A * M)$ such that $\eta u = 1$. Define $v(a) = u(a) + a((1,0) - u(1))$. Then v is a B-module homomorphism, $\eta v = 1$, and $v(1) = (1,0)$. Our goal is to replace v with a homomorphism which is also multiplicative. This is where Hochschild cohomology comes into play. Define

$$f_v(x \otimes y) = v(x)v(y) - v(xy). \tag{8.4}$$

Notice that v is multiplicative if and only if f_v is the zero map. Using the fact that $\eta v(x) = x$, the reader should verify that the image of f_v is contained in M. Since $M^2 = 0$, it follows that

$$\begin{aligned} xf_v(y \otimes z) &= v(x)f_v(y \otimes z) \\ f_v(x \otimes y)z &= f_v(x \otimes y)v(z). \end{aligned} \tag{8.5}$$

From (5.17), we have

$$\delta^2 f_v(x \otimes y \otimes z) = xf_v(y \otimes z) - f(xy \otimes z) + f(x \otimes yz) - f(x \otimes y)z. \tag{8.6}$$

Using (8.4) and (8.5), the reader should verify that (8.6) reduces to $\delta^2 f_v = 0$, hence f_v is a Hochschild 2-cocycle. If $h : A \to M$ is a left B-module homomorphism such that $h(1) = 0$, and we set $w = v + h$, then it follows that w is a B-module homomorphism such that $w(1) = (1,0)$, and $\eta w = 1$. As in (8.5), $M^2 = 0$ implies $xh(y) = v(x)h(y)$ and $h(x)y = h(x)v(y)$. Using this, the reader should verify that

$$f_w(x \otimes y) = f_v(x \otimes y) + \delta^1 h(x \otimes y) \tag{8.7}$$

where δ^1 is from (5.17). If we can find an h as above such that $f_w = f_v + \delta^1 h = 0$, then $w : A \to A * M$ is the desired ring homomorphism.

We remark that the proof would be completed, if $B = R$. This is because $\mathrm{H}^2(A, M) = \operatorname{Ext}^2_{A^e}(A, M) = 0$, since A is a projective A^e-module.

In our case, we define h using a separability idempotent for A. Let $e = \sum_i x_i \otimes y_i$ be an idempotent in A^e such that the relations

$$\sum_i x_i y_i = 1 \tag{8.8}$$

and

$$\sum_i ax_i \otimes y_i = (a \otimes 1)e = (1 \otimes a)e = \sum_i x_i \otimes y_i a \tag{8.9}$$

hold for all $a \in A$. Define $h : A \to M$ by

$$h(a) = -\sum f_v(a \otimes x_i)y_i.$$

The reader should verify that h is a B-module homomorphism and $h(1) = 0$. We have

$$\begin{aligned}\delta^1 h(a \otimes b) &= ah(b) - h(ab) + h(a)b \\ &= -\sum_i \left(af_v(b \otimes x_i)y_i - f_v(ab \otimes x_i)y_i + f_v(a \otimes x_i)y_i b\right).\end{aligned}$$

And by (8.6),

$$\begin{aligned}\delta^1 h(a \otimes b) = -\sum_i \Big(&af_v(b \otimes x_i)y_i \\ &- \left(af_v(b \otimes x_i)y_i + f_v(a \otimes bx_i)y_i - f_v(a \otimes b)x_i y_i\right) + f_v(a \otimes x_i)y_i b\Big) \\ &= \sum_i \Big(f_v(a \otimes bx_i)y_i - f_v(a \otimes b)x_i y_i - f_v(a \otimes x_i)y_i b\Big).\end{aligned}$$

Again by (8.6),

$$\begin{aligned}&\sum_i f_v(a \otimes bx_i)y_i \\ &\quad= \sum_i af_v(bx_i \otimes y_i) - \sum_i f_v(abx_i \otimes y_i) + \sum_i f_v(a \otimes bx_i y_i) \\ &\qquad= af_v(b\sum_i x_i \otimes y_i) - f_v(ab\sum_i x_i \otimes y_i) + f_v(a \otimes b\sum_i x_i y_i)\end{aligned}$$

and

$$\begin{aligned}&\sum_i f_v(a \otimes x_i)y_i b \\ &\quad= \sum_i af_v(x_i \otimes y_i b) - \sum_i f_v(ax_i \otimes y_i b) + \sum_i f_v(a \otimes x_i y_i b) \\ &\qquad= af_v(\sum_i x_i \otimes y_i b) - f_v(a\sum_i x_i \otimes y_i b) + f_v(a \otimes \sum_i x_i y_i b).\end{aligned}$$

By (8.8) and (8.9) applied to the above, we get $\delta^1 h(a \otimes b) + f_v(a \otimes b) = 0$. □

1.4. Exercises.

Exercise 8.1.9. (Generalized Product and Power Rules) Suppose A is an R-algebra, M is a two-sided A/R-module, $\partial \in \operatorname{Der}_R(A, M)$ and $x, x_1, \ldots, x_n \in A$. Prove that

$$\partial(x_1 x_2 \cdots x_n) = \partial(x_1)x_2 \cdots x_n + x_1\partial(x_2)x_3 \cdots x_n + \cdots + x_1 \cdots x_{n-1}\partial(x_n)$$

and if $n \geq 1$, then $\partial(x^n) = \sum_{i=0}^{n-1} x^i \partial(x) x^{n-1-i}$.

Exercise 8.1.10. (Chain Rule) Suppose A is a commutative R-algebra and M is an A-module. Prove that if $a \in A$ and $f(x) \in R[x]$, then for any $\partial \in \operatorname{Der}_R(A, M)$, $\partial(f(a)) = f'(a)\partial(a)$.

Exercise 8.1.11. Let A be an R-algebra. Show that $M \mapsto \operatorname{Der}_R(A, M)$ defines a covariant functor from the category of two-sided A/R-modules to the category of R-modules.

Exercise 8.1.12. Suppose S is a commutative R-algebra and A is any S-algebra. Let M be a two-sided A/S-module. Show that there is an exact sequence of abelian groups

$$0 \to \operatorname{Der}_S(A, M) \xrightarrow{a} \operatorname{Der}_R(A, M) \xrightarrow{b} \operatorname{Der}_R(S, M).$$

Exercise 8.1.13. Let R be a commutative ring and S a commutative R-algebra. Let $A = S[x]$ be the polynomial ring over S in one variable and let M be any A-module. Show that $\operatorname{Der}_R(A, M) \to \operatorname{Der}_R(S, M)$ is onto. (Hint: If $\partial : S \to M$ is an R-derivation, show that the assignment $ax^i \mapsto x^i\partial(a)$ defines an R-derivation $D : A \to M$.)

Exercise 8.1.14. (The Extension of a Ring by a Module) Let A be an R-algebra and N a two-sided A/R-module (Definition 4.1.1). Define a multiplication on the two-sided A/R-module $A \oplus N$ by the formula $(a, x)(b, y) = (ab, ay + xb)$, for all a, b in A and all x, y in N.

(1) Show that the multiplication rule defined above turns the A-module $A \oplus N$ into an R-algebra with unit element $(1, 0)$. Denote this R-algebra by $A * N$.

(2) Show that the subset $\{(0, x) \mid x \in N\}$ is an ideal in $A * N$ satisfying $N^2 = 0$ and that there is a split exact sequence of two-sided A/R-modules $0 \to N \to A * N \to A \to 0$. The ring $A * N$ is called the *trivial, or split extension of A by N*.

(3) Show that the map $a \mapsto (a, 0)$ defines an R-algebra homomorphism $\sigma : A \to A * N$ which is a section to the natural map $\eta : A * N \to A$ (that is, $\eta\sigma = 1$).

(4) Let $D \in \operatorname{Der}_{\mathbb{Z}}(A, N)$. Define $u : A \to A * N$ by $u(a) = (a, D(a))$. Show that u is a ring homomorphism which is a section to the natural map $\eta : A * N \to A$.

(5) Prove the converse to (4). That is, show that if $u : A \to A * N$ is a $\mathbb{Z}$-algebra section to η, then $u(a) - \sigma(a) : A \to N$ is a $\mathbb{Z}$-derivation.

(6) Let B be a commutative R-algebra and I an ideal in B satisfying $I^2 = 0$. Let $A = B/I$. Show that there is an exact sequence of A-modules $0 \to I \to B \to A \to 0$. We say that B is an *extension of A by I*. Show that B is isomorphic to $A * I$ as R-algebras if and

only if there is an R-algebra homomorphism $\sigma : A \to B$ which is a section to the natural map $B \to A$ (in this case the extension is also said to be trivial, or split).

Exercise 8.1.15. Let A be an R-algebra and $D \in \operatorname{Der}_R(A, A)$. View $\operatorname{Der}_R(A, A)$ as an R-submodule of the ring $\operatorname{Hom}_R(A, A)$ of A-module endomorphisms of A. Then D^i denotes the composition map, which is an element of $\operatorname{Hom}_R(A, A)$ (but not necessarily an element of $\operatorname{Der}_R(A, A)$). Prove:

(1) (Leibniz Formula) For all $a, b \in A$ and $n \geq 0$,

$$D^n(ab) = \sum_{i=0}^{n} \binom{n}{i} D^i(a) D^{n-i}(b).$$

(2) If R has characteristic p, a prime number, then $D^p \in \operatorname{Der}_R(A, A)$ is an R-derivation on A.

Exercise 8.1.16. Let A be an R-algebra, $z \in A$ and $\tau_z : A \to A$ the inner derivation defined by z. That is, $\tau_z(a) = az - za$. Prove:

(1) For all $a \in A$ and $n \geq 0$, $\tau_z^n(a) = \sum_{i=0}^{n} (-1)^i \binom{n}{i} z^i a z^{n-i}$.

(2) If R has characteristic p, a prime number, then
  (a) $\tau_z^p = \tau_{z^p}$. That is, $\tau_z^p(a) = az^p - z^p a = \tau_{z^p}(a)$ for all $a \in A$.
  (b) $\tau_z^{p-1} = \sum_{i=0}^{p-1} z^i a z^{p-1-i}$.

Exercise 8.1.17. In the context of Lemma 8.1.7, derive the following.

(1) If $p = 2$, then $s_1(a, b) = \tau_b(a)$.

(2) If $p = 3$, then $s_1(a, b) = \tau_b^2(a)$ and $2s_2(a, b) = \tau_a \tau_b(a)$.

(3) If $p = 5$, then

$$s_1(a, b) = \tau_b^4(a) \qquad 2s_2(a, b) = \tau_a \tau_b^3(a) + \tau_b \tau_a \tau_b^2(a) + \tau_b^2 \tau_a \tau_b(a)$$
$$4s_4(a, b) = \tau_a^3 \tau_b(a) \qquad 3s_3(a, b) = \tau_a^2 \tau_b^2(a) + \tau_a \tau_b \tau_a \tau_b(a) + \tau_b \tau_a^2 \tau_b(a).$$

1.5. More Tests for Separability. Now we apply the above results on derivations to establish separability criteria for algebras. The main results are the vanishing of the first Hochschild cohomology criterion, the theorems on faithfully flat descent, the separability at the stalks criteria, and the residue field tests.

Let R be a commutative ring, A an R-algebra, and $A^e = A \otimes_R A^o$ the enveloping algebra. If M is a two-sided A/R-module, then the nth Hochschild cohomology group of A with coefficients in M is defined to be $\mathrm{H}^n(A, M) = \operatorname{Ext}^n_{A^e}(A, M)$ (Definition 5.4.1).

Lemma 8.1.18. *In the above context, the following are true.*

(1) $\mathrm{H}^0(A, M) = M^A = \{x \in M \mid ax = xa,$ *for all* $a \in A\}$.

(2) $\mathrm{H}^1(A, M) = \mathrm{Der}_R(A, M)/\operatorname{Inn.Der}_R(A, M)$.

Proof. The sequence of left A^e-modules

$$0 \to J_{A/R} \to A^e \xrightarrow{\mu} A \to 0$$

is exact (Definition 4.1.1). Consider the associated long exact sequence

$$0 \to \mathrm{Hom}_{A^e}(A, M) \to \mathrm{Hom}_{A^e}(A^e, M) \to \mathrm{Hom}_{A^e}(J_{A/R}, M)$$
$$\xrightarrow{\delta^0} \mathrm{Ext}^1_{A^e}(A, M) \to \mathrm{Ext}^1_{A^e}(A^e, M) \to \mathrm{Ext}^1_{A^e}(J_{A/R}, M) \to$$

of abelian groups. Since A^e is projective over A^e, $\mathrm{Ext}^1_{A^e}(A^e, M) = 0$ (see, for example, [**DF04**, Proposition 17.11] or [**Rot79**, Theorem 7.7]). The rest follows from Proposition 8.1.5. □

Theorem 8.1.19. *Let R be a commutative ring and A an R-algebra. The following are equivalent.*

(1) *A is a separable R-algebra.*

(2) *$\mathrm{H}^1(A, M) = 0$ for every two-sided A/R-module M.*

(3) *The sequence*

$$0 \to M^A \to M \xrightarrow{\tau} \mathrm{Der}_R(A, M) \to 0$$

is exact, for every two-sided A/R-module M.

Proof. (1) is equivalent to (2): Let $A^e = A \otimes_R A^o$ be the enveloping algebra (Definition 4.1.1). By Proposition 4.1.2, A is R-separable if and only if A is projective as a left A^e-module. By [**DF04**, Proposition 17.11] or [**Rot79**, Corollary 7.12], for example, A is projective as a left A^e-module if and only if $\mathrm{H}^1(A, M) = \mathrm{Ext}^1_{A^e}(A, M) = 0$ for every two-sided A/R-module M.

(2) is equivalent to (3): This follows from an application of Proposition 8.1.5 and Lemma 8.1.18. □

We now prove a faithfully flat descent theorem for separability. As a corollary we get a faithfully flat descent theorem for Azumaya algebras.

Theorem 8.1.20. *Let A be an R-algebra and S a commutative faithfully flat R-algebra. Assume $A \otimes_R S$ is separable over S and either*

(1) *A is a finitely generated R-module, or*

(2) *A a finitely generated commutative R-algebra.*

Then A is separable over R.

Proof. By Lemma 8.1.4, A is finitely presented as an A^e-module. By Proposition 2.4.11, the functors $\mathrm{Hom}_{A^e}(A, \cdot) \otimes_R S$ and $\mathrm{Hom}_{A^e \otimes_R S}(A \otimes_R S, (\cdot) \otimes_R S)$

are isomorphic. By Corollary 4.1.7, the functor $\operatorname{Hom}_{A^e \otimes_R S}(A \otimes_R S, \cdot)$ is exact. Since S is faithfully flat, it follows that $\operatorname{Hom}_{A^e}(A, \cdot)$ is exact. By Corollary 4.1.7, A is separable over R. □

Corollary 8.1.21. *Let R be a commutative ring. Let A be an R-algebra and S a commutative faithfully flat R-algebra. Then A is an Azumaya R-algebra if and only if $A \otimes_R S$ is an Azumaya S-algebra.*

Proof. Assume $A \otimes_R S$ is S-Azumaya. By Theorem 2.6.1, A is a finitely generated R-module. By Theorem 8.1.20, A is separable over R. By Proposition 2.6.3, A is R-central. For the converse, apply Corollary 7.1.9. □

The next theorem provides sufficient conditions allowing us to prove that an algebra is separable if it is separable when localized at every prime. Its corollary provides a similar test for Azumaya algebras.

Theorem 8.1.22. *Let R be a commutative ring and A an R-algebra which satisfies either*

(a) *A is a finitely generated R-module, or*

(b) *A a finitely generated commutative R-algebra.*

Then the following are equivalent.

(1) *A is a separable R-algebra.*

(2) *$A \otimes_R R_P$ is a separable R_P-algebra for every prime ideal P of R.*

(3) *$A \otimes_R R_{\mathfrak{m}}$ is a separable $R_{\mathfrak{m}}$-algebra for every maximal ideal $\mathfrak{m}$ of R.*

Proof. (1) implies (2): This follows straight from Corollary 4.3.2.

(2) implies (3): This is trivial.

(3) implies (1): To show that sequence (8.1) splits, it is enough to show that $\mu \circ () : \operatorname{Hom}_{A^e}(A, A^e) \to \operatorname{Hom}_{A^e}(A, A)$ is onto. By Lemma 8.1.4, A is of finite presentation as a left A^e-module. Let $\mathfrak{m}$ be any maximal ideal of R. Denote by $A_{\mathfrak{m}}$ the tensor product $A \otimes_R R_{\mathfrak{m}}$. Then $A^e \otimes_R R_{\mathfrak{m}} = A^e_{\mathfrak{m}}$. The diagram

$$\begin{array}{ccc}
\operatorname{Hom}_{A^e}(A, A^e) \otimes_R R_{\mathfrak{m}} & \xrightarrow{\mu\circ()\otimes 1} & \operatorname{Hom}_{A^e}(A, A) \otimes_R R_{\mathfrak{m}} \\
\downarrow & & \downarrow \\
\operatorname{Hom}_{A^e_{\mathfrak{m}}}(A_{\mathfrak{m}}, A^e_{\mathfrak{m}}) & \xrightarrow{\mu\circ()} & \operatorname{Hom}_{A^e_{\mathfrak{m}}}(A_{\mathfrak{m}}, A_{\mathfrak{m}})
\end{array}$$

commutes. The vertical maps are isomorphisms, by Proposition 2.4.11. By Corollary 4.1.7, the second horizontal map $\mu \circ ()$ is onto. Hence the top horizontal map is onto. By Exercise 2.4.12, $\mu \circ () : \operatorname{Hom}_{A^e}(A, A^e) \to \operatorname{Hom}_{A^e}(A, A)$ is onto. □

Corollary 8.1.23. *Let R be a commutative ring and A an R-algebra. The following are equivalent.*

(1) *A is an Azumaya R-algebra.*
(2) *A is finitely generated as an R-module and $A \otimes_R R_P$ is an Azumaya R_P-algebra for every prime ideal P of R.*
(3) *A is finitely generated as an R-module and $A \otimes_R R_{\mathfrak{m}}$ is an Azumaya $R_{\mathfrak{m}}$-algebra for every maximal ideal $\mathfrak{m}$ of R.*

Proof. (1) implies (2) implies (3): These are left to the reader.

(3) implies (1): By Theorem 8.1.22, A is separable over R. By Proposition 2.6.3, A is R-central. □

For an R-algebra A that is a finitely generated R-module, the next theorem and its corollaries show that separability of A over R can be reduced to the same question for certain algebras over fields. Separable algebras over fields are described by the decomposition theorems of Section 4.5.

Theorem 8.1.24. *Let R be a commutative ring and A an R-algebra which is finitely generated as an R-module. The following are equivalent.*

(1) *A is a separable R-algebra.*
(2) *$A/\mathfrak{m}A$ is a separable $R/\mathfrak{m}$-algebra for every maximal ideal $\mathfrak{m}$ of R.*

Proof. (1) implies (2): This follows straight from Corollary 4.3.2.

(2) implies (1): Let $\mathfrak{m}$ be any maximal ideal of R. Since $R_{\mathfrak{m}}/\mathfrak{m}R_{\mathfrak{m}} \cong R/\mathfrak{m}$ we have

$$\begin{aligned}(A \otimes_R R_{\mathfrak{m}})/\mathfrak{m}(A \otimes_R R_{\mathfrak{m}}) &\cong A \otimes_R (R_{\mathfrak{m}}/\mathfrak{m}R_{\mathfrak{m}}) \\ &\cong A \otimes_R (R/\mathfrak{m}) \\ &\cong A/\mathfrak{m}A.\end{aligned}$$

Since we already proved that (3) implies (1) in Theorem 8.1.22, it is enough to prove (2) implies (1) when R is a local ring.

Assume R is a local ring with maximal ideal $\mathfrak{m}$ and A is an R-algebra which is finitely generated as an R-module and such that $A/\mathfrak{m}A$ is separable over $R/\mathfrak{m}$. For the remainder of this proof, we write simply $J_{A/\mathfrak{m}A}$ instead of $J_{(A/\mathfrak{m}A)/(R/\mathfrak{m})}$ and J_A rather than $J_{A/R}$. Let $\delta : A/\mathfrak{m}A \to J_{A/\mathfrak{m}A}$ be the derivation defined by $\bar{a} \mapsto \bar{a} \otimes 1 - 1 \otimes \bar{a}$. By Theorem 8.1.19, $\delta = \tau_{\bar{z}}$ for some $\bar{z} \in J_{A/\mathfrak{m}A}$. In other words, for each $\bar{a} \in A/\mathfrak{m}A$, $\delta(\bar{a}) = \bar{a}\bar{z} - \bar{z}\bar{a} = (\bar{a} \otimes 1 - 1 \otimes \bar{a})\bar{z} = \delta(\bar{a})\bar{z}$. By Lemma 8.1.3, it follows that

$$J_{A/\mathfrak{m}A} = (A/\mathfrak{m}A)\delta(A/\mathfrak{m}A) = (A/\mathfrak{m}A)\delta(A/\mathfrak{m}A)\bar{z} = J_{A/\mathfrak{m}A}\bar{z}.$$

By Lemma 7.7.3, $J_{A/\mathfrak{m}A} = J_A/(\mathfrak{m}J_A)$. If $z \in J_A$ is a preimage of $\bar{z}$, then $J_A = J_A z + \mathfrak{m}J_A$. In Lemma 8.1.4 it was shown that J_A is finitely generated over R. By Nakayama's Lemma (Theorem 3.1.2), it follows that $J_A = J_A z$. Define a homomorphism ϕ in $\operatorname{Hom}_{A^e}(A^e, J_A)$ by $\phi(x) = xz$. Then $\phi(J_A) = J_A z = J_A$. By Corollary 1.3.19, $\phi : J_A \to J_A$ is an automorphism of A^e-modules. Therefore sequence (8.1) is split exact as A^e-modules. □

Corollary 8.1.25. *Let R be a commutative ring and A an R-algebra. The following are equivalent.*

(1) *A is an Azumaya R-algebra.*
(2) *A is faithful and finitely generated as an R-module and for every maximal ideal $\mathfrak{m}$ of R, $A/\mathfrak{m}A$ is a central simple $R/\mathfrak{m}$-algebra.*

Proof. (1) implies (2): This follows from Theorem 7.1.4 and Corollary 4.5.4.

(2) implies (1): By Theorem 8.1.24, A is a separable R-algebra. Let C denote the center of A. Then A is an Azumaya C-algebra, by Theorem 7.1.11. By Lemma 7.1.1, $C \cdot 1$ is a C-module direct summand of A. Then C is an R-module homomorphic image of a finitely generated R-module, hence is a finitely generated R-module. Because A is a faithful R-module, the algebra structure homomorphism $\theta : R \to C$ is one-to-one. For each maximal ideal, $\bar{\theta} : R/\mathfrak{m} \to C/\mathfrak{m}C$ is onto. By Corollary 1.3.19, θ is an isomorphism. □

Example 8.1.26. Let R be a commutative ring and $f \in R[x]$ a monic polynomial. We proved in Proposition 4.6.1 that $S = R[x]/(f)$ is separable over R if and only if $(f, f') = R[x]$. In this example, we apply the Residue Field Criterion to give another proof that S/R is separable if $(f, f') = R[x]$. Since f is monic, S is a free R-module of finite rank. By Theorem 8.1.24, S/R is separable if and only if $S \otimes_R k_\mathfrak{m} = k_\mathfrak{m}[x]/(f)$ is separable over $k_\mathfrak{m}$ for every maximal ideal $\mathfrak{m}$ in R, where $k_\mathfrak{m}$ denotes the residue field $R/\mathfrak{m}$. By Exercise 4.4.11, $k_\mathfrak{m}[x]/(f)$ is separable over $k_\mathfrak{m}$ if and only if (f, f') is the unit ideal in $k_\mathfrak{m}[x]$. If (f, f') is the unit ideal in $R[x]$, then for every maximal ideal $\mathfrak{m}$, (f, f') is the unit ideal in $k_\mathfrak{m}[x]$ and we are done.

Corollary 8.1.27. *Let R be a local ring with maximal ideal $\mathfrak{m}$ and residue field k. The change of base functor $(\ )\otimes_R k$ from the category of commutative separable R-algebras which are finitely generated free R-modules and the category of commutative separable k-algebras is essentially surjective.*

Proof. A commutative separable k-algebra is a direct sum $F_1 \oplus \cdots \oplus F_n$, where each F_i is a finite separable field extension of k (Corollary 4.5.8). Let F/k be a finite separable field extension. To show $(\ ) \otimes_R k$ is essentially surjective, it is enough to show that $F = S \otimes_R k$, for an appropriate extension S/R. By the Primitive Element Theorem and Corollary 4.6.2, we are done. □

1.6. Locally of Finite Type is Finitely Generated as an Algebra. The results in this section will be applied in Exercise 8.1.33 to show that an open immersion is separable. If S is a commutative R-algebra, then S is said to be *locally of finite type* in case there exist elements $f_1, \dots, f_n$ in S such that $S = Sf_1 + \cdots + Sf_n$ and for each i, $S[f_i^{-1}]$ is a finitely generated R-algebra. Proposition 8.1.28 is from [**Mum99**, Proposition 1, p. 87].

Proposition 8.1.28. *Let S be a commutative R-algebra. Then S is locally of finite type if and only if S is a finitely generated R-algebra.*

Proof. Assume S is locally of finite type and prove that S is finitely generated as an R-algebra. The converse is trivial. We are given $f_1, \dots, f_n$ in S such that $S = Sf_1 + \cdots + Sf_n$ and for each i, $S[f_i^{-1}]$ is a finitely generated R-algebra. Fix elements $u_1, \dots, u_n$ in S such that $1 = u_1 f_1 + \cdots + u_n f_n$. Fix elements $y_{i1}, \dots, y_{im}$ in $S[f_i^{-1}]$ such that $S[f_i^{-1}] = R[y_{i1}, \dots, y_{im}]$. There exist elements s_{ij} in S and nonnegative integers e_i such that $y_{ij} = s_{ij} f_i^{-e_i}$ in $S[f_i^{-1}]$. Let S_1 be the finitely generated R-subalgebra of S generated by the finite set of elements $\{s_{ij}\} \cup \{f_1, \dots, f_n\} \cup \{u_1, \dots, u_n\}$. To finish, it is enough to show that S_1 is equal to S. Let α be an arbitrary element of S and let $1 \le i \le n$. Consider $\alpha/1$ as an element of $S[f_i^{-1}]$. Since $S[f_i^{-1}]$ is generated over R by $s_{i1}, \dots, s_{im}$ and f_i^{-1}, there exists an element β_i in S_1 such that $\alpha/1 = \beta_i f_i^{-k_i}$ for some $k_i \ge 0$. For some $\ell_i \ge 0$, $f_i^{\ell_i}(\beta_i - f_i^{k_i}\alpha) = 0$ in S. For some large integer L, $f_i^L \alpha = f_i^L \beta_i$ is an element of S_1, for each i. For any positive integer N, $\alpha = 1\alpha = (u_1 f_1 + \cdots + u_n f_n)^N \alpha$. By the multinomial expansion, when N is sufficiently large, $(u_1 f_1 + \cdots + u_n f_n)^N$ is in the ideal $S_1 f_1^L + \cdots + S_1 f_n^L$. Therefore, α is in S_1. □

Corollary 8.1.29. *Let $f : R \to S$ be a homomorphism of commutative rings. If $f^\sharp : \operatorname{Spec} S \to \operatorname{Spec} R$ is an open immersion (see Exercise 2.3.9), then S is a finitely generated R-algebra.*

Proof. The proof is left to the reader. □

1.7. Exercises.

Exercise 8.1.30. Let k be a field of characteristic different from 2 and 3 and let x and y be indeterminates. Let $f = x^2 - y^3$ and $R = k[x, y]/(f)$. Define $\alpha : k[x] \to R$ by $x \mapsto x$. Show:

(1) α is one-to-one and R is a free $k[x]$-module of rank two.

(2) R is not a separable $k[x]$-algebra.

Exercise 8.1.31. This exercise is based on [**Gro71**, Proposition I.3.1, p. 2] and [**Mil80**, Proposition I.3.5] Let R be a commutative ring and S a commutative finitely generated R-algebra. Show that the following are equivalent.

(1) S is a separable R-algebra.

(2) The homomorphism of R-algebras $\mu : S^e \to S$ makes S into a flat S^e-module.

(3) For every $\mathfrak{q} \in \operatorname{Spec} S$, if $\mathfrak{p} = \mu^{-1}(\mathfrak{q})$, then $\mu : (S^e)_{\mathfrak{p}} \to S_{\mathfrak{q}}$ is an isomorphism. In the terminology of Algebraic Geometry, the diagonal morphism $\mu^\sharp : \operatorname{Spec} S \to \operatorname{Spec} S^e$ is said to be an open immersion (Exercise 2.3.9).

(Hint: Exercises 1.1.22, 1.3.15, and Proposition 2.4.8.)

Exercise 8.1.32. Let R be a commutative ring and S a commutative R-algebra. In Algebraic Geometry, the morphism $\mu^\sharp : \operatorname{Spec} S \to \operatorname{Spec} S \otimes_R S$ associated to $\mu : S \otimes_R S \to S$ is called the *diagonal morphism*.

(1) For every $\mathfrak{q} \in \operatorname{Spec} S$, show that $\mu^{-1}(\mathfrak{q})$ is the ideal $\mathfrak{q} \otimes S + S \otimes \mathfrak{q} + J_{S/R}$.

(2) Let k be an algebraically closed field. Let $\alpha \in k$ and let $\mathfrak{q}$ be the maximal ideal in $k[x]$ generated by $x - \alpha$. Show that under the diagonal map $\mu^\sharp : \operatorname{Spec} k[x] \to \operatorname{Spec} k[x] \otimes_k k[x]$, the image of $\mathfrak{q}$ is the maximal ideal in $k[x] \otimes_k k[x]$ generated by $(x - \alpha) \otimes 1$ and $1 \otimes (x - \alpha)$.

Exercise 8.1.33. (An Open Immersion is Separable) Let $f : R \to S$ be a homomorphism of commutative rings and $f^\sharp : \operatorname{Spec} S \to \operatorname{Spec} R$ the continuous map of Exercise 2.2.12. Show that if $f^\sharp$ is an open immersion (see Exercise 2.3.9), then S is separable over R. (Use Corollary 8.1.29 to show S is a finitely generated R-algebra.)

Exercise 8.1.34. Let k be a field, and define R to be $k[x,y]/(y^2 - x^2(x+1))$, where x and y are indeterminates. The algebraic set defined by $y^2 - x^2(x+1) = 0$ is an example of a nodal cubic curve.

(1) Using the fact that $x^2(x+1)$ is not a square in $k[x]$, show that R is an integral domain. Let K denote the quotient field of R.

(2) Let w represent the element $yx^{-1} \in K$. Show that x and y both belong to $k[w]$ and w is integral over R. Conclude that R is a subring of $k[w]$, that R and $k[w]$ have the same quotient field, $k[w]$ is the integral closure of R in $K = k(w)$, and $k[w]$ is a rank one R-lattice in K.

(3) Show that the rings $R[x^{-1}]$ and $k[w][x^{-1}] = k[w, (w^2 - 1)^{-1}]$ are equal.

(4) Prove that the conductor ideal from $k[w]$ to R is $\mathfrak{c} = (x, y)$.

(5) Using Theorem 8.1.24, show that $k[w]$ is separable over R.

(6) Using Corollary 2.6.6, show that $k[w]$ is not flat over R.

For more properties of the ring R, see Section 14.3.2 and Example 14.3.4.

Exercise 8.1.35. Let k be a field and $A = k[x]$ the polynomial ring over k in one variable. Set $R = k[x^2, x^3]$ (see Exercise 2.6.17). Using Theorem 8.1.24, show that A is not separable over R. Show that the rings $R[x^{-2}]$ and $A[x^{-2}]$ are equal, hence the extension $R \to A$ is separable upon localization to the nonempty basic open set $U(x^2)$.

2. Differential Crossed Product Algebras

The differential crossed product defined below is a type of cyclic crossed product (see Section 13.4) for purely inseparable radical extensions. The construction, which comes from [**KOS76**] and [**Sal76**], coincides with the "differential extensions" of [**Hoe63**], the differential crossed products of [**KO74b**] and the "regular Lie algebra extensions" of [**Yua71**] whenever the domains of definition intersect. The differential crossed products and elementary p-algebras defined in this section will be used in Section 11.4 when we study Brauer groups in characteristic p.

Let A be a ring, and $D \in \operatorname{Der}_{\mathbb{Z}}(A, A)$ a derivation on A. If u is an indeterminate, then by $A[u, D]$ we denote the *ring of differential polynomials* which is equal to the ordinary polynomial ring $A[u]$ as a left A-module, but with multiplication induced by $ua = au + D(a)$, for all $a \in A$. That this defines an associative multiplication rule follows from Lemma 8.2.1.

Lemma 8.2.1. *In the above context, $A[u, D]$ is a ring. On monomials the multiplication rule satisfies*

$$(au^m)(bu^n) = \sum_{i=0}^{m} \binom{m}{i} aD^i(b)u^{m-i+n} \tag{8.10}$$

for all $a, b \in A$.

Proof. We show that the rule $ua = au + D(a)$ induces a ring structure on the left A-module $A[u]$. The formula (8.10) follows from this and induction. Consider the ring $\operatorname{Hom}_{\mathbb{Z}}(A[u], A[u])$, of $\mathbb{Z}$-module endomorphisms of $A[u]$. The left regular representation $\lambda : A \to \operatorname{Hom}_{\mathbb{Z}}(A[u], A[u])$ sends $a \in A$ to λ_a, where $\lambda_a(bu^n) = (ab)u^n$. Under λ, A is isomorphic to a subring of $\operatorname{Hom}_{\mathbb{Z}}(A[u], A[u])$. Let $\lambda_u : A[u] \to A[u]$ be defined by $\lambda_u(bu^n) = bu^{n+1} + D(b)u^n$. Let P be the subring of $\operatorname{Hom}_{\mathbb{Z}}(A[u], A[u])$ generated by A and λ_u. Then in P the set of all elements of the form $\sum_i \lambda_{a_i}\lambda_u^i$ is a left A-module isomorphic to $A[u]$. Notice that $\lambda_u\lambda_a = \lambda_a\lambda_u + \lambda_{D(a)}$. Therefore, P is isomorphic to $A[u]$ as a left A-module. The multiplication rule on $A[u, D]$ agrees with the multiplication in the ring P. □

We fix some notation that will be in effect throughout the remainder of this section. Let p be a prime number and R a commutative ring of characteristic p. Let $r \in R$, and

$$C = \frac{R[x]}{(x^p - r)} = R[r^{1/p}].$$

Then C is a purely inseparable radical extension of R. Let v denote the image of x in C. Then C is a free R-module of rank p, with basis $1, v, \dots, v^{p-1}$. If $\partial : R[x] \to R[x]$ is the usual derivative (Example 8.1.1), then $\partial(x^p - r) = 0$. Therefore, ∂ induces a derivation $d : C \to C$ such that $d(v) = 1$ and $d(c) = 0$ if and only if $c \in R$. Thus, $d \in \operatorname{Der}_R(C, C)$ is an R-derivation of C. We call this derivation the *standard derivation on* C. Let A be an Azumaya C-algebra. By Theorem 8.1.8, d extends to a derivation $D : A \to A$. By Exercise 8.1.15, D^p is a derivation on A. If f is a polynomial in $R[x]$ of degree less than p, then $\partial^p(f) = 0$. Therefore, $D^p(c) = 0$, for any $c \in C$. That is, $D^p \in \operatorname{Der}_C(A, A)$. By Theorem 8.1.19, $D^p = \tau_u$ is the inner derivation defined by some $u \in A$. That is, $D^p(a) = au - ua$ for all $a \in A$. Notice that

$$\begin{aligned}
0 &= D(D^p(a)) - D^p(D(a)) \\
&= D(au - ua) - (D(a)u - uD(a)) \\
&= aD(u) + D(a)u - uD(a) - D(u)a - D(a)u + uD(a) \\
&= aD(u) - D(u)a
\end{aligned}$$

which shows $D(u)$ is in the center of A, which is C. We can write $D(u) = r_0 + r_1 v + \cdots + r_{p-1} v^{p-1}$, for unique $r_i \in R$. Then $0 = D^p(u) = D^{p-1}(D(u)) = (p-1)! r_{p-1}$. Since $(p-1)!$ is a unit in the prime subfield of R, this implies $r_{p-1} = 0$, hence $D(u) = r_0 + r_1 v + \cdots + r_{p-2} v^{p-2}$. Set $c_i = r_{i-1}/i$, for $i = 1, \dots, p-1$. If $c = c_1 v + \cdots + c_{p-1} v^{p-1}$, then $D(c) = D(u)$. Since $c \in C$, u and $u - c$ define the same inner derivation on A. Replace u with $u - c$, and assume from now on that $D(u) = 0$. Let z be an indeterminate and $A[z, D]$ the differential polynomial ring. Notice that $z^p - u$ is central in $A[z, D]$.

Definition 8.2.2. Let R be a commutative ring of characteristic p, $r \in R$, $C = R[x]/(x^p - r)$, and d the standard derivation on C satisfying $d(R) = 0$ and $d(v) = 1$, where v is the image of x in C. Let A be an Azumaya C-algebra, D a derivation on A which extends d, D^p the inner derivation on A determined by u, and $D(u) = 0$. In this context, by $\Delta(A, D, u)$ we denote the quotient $A[z, D]/(z^p - u)$ of the differential polynomial ring module the principal ideal generated by the central element $z^p - u$. The R-algebra $\Delta(A, D, u)$ is called a *differential crossed product*.

Lemma 8.2.3. *Let R, $C = R[x]/(x^p - r)$, $\Delta(A, D, u)$ be as in Definition 8.2.2. Then the following are true.*

(1) *As a left A-module, $\Delta(A, D, u)$ is free with basis $1, z, \dots, z^{p-1}$.*

(2) *The centralizer of* C *in* $\Delta(A, D, u)$ *is* A.

(3) $\Delta(A, D, u)$ *is an Azumaya* R*-algebra.*

Proof. (1): This proof is left to the reader.

(2): As above, v is the generator of C. Let τ_v denote the inner derivation on $\Delta(A, D, u)$ defined by v. Then $\tau_v(a) = 0$ for any $a \in A$, and $\tau_v(z) = zv - vz = vz + d(v) - vz = 1$. Let t be an element of $\Delta(A, D, u)^C$, which can be written as $t = a_0 + a_1 z + \cdots + a_{p-1} z^{p-1}$. Since t commutes with v, $0 = \tau_v^{p-1}(t) = (p-1)! a_{p-1}$. Since $(p-1)!$ is a unit in the prime subfield of R, this implies $a_{p-1} = 0$. Likewise, $0 = \tau_v^{p-2}(t) = (p-2)! a_{p-2}$ implies $a_{p-2} = 0$. Recursively, we have $a_{p-1} = a_{p-2} = \cdots = a_1 = 0$, which proves $t \in A$. Since C is equal to the center of A, this proves $A = \Delta(A, D, u)^C$.

(3): Let Z be the center of $\Delta(A, D, u)$. By (2), $Z \subseteq A$. Then Z is a subset of the center of A, which is C. If $\zeta \in C$ and $\zeta z = z\zeta$, then $D(\zeta) = d(\zeta) = 0$, which implies $\zeta \in R$. This implies $Z = R$. To finish the proof, we show $\Delta(A, D, u)$ is R-separable. By Corollary 8.1.25 and Corollary 4.5.4 it suffices to prove $\Delta(A, D, u) \otimes_R R/\mathfrak{m}$ is a central simple $R/\mathfrak{m}$-algebra, for every maximal ideal $\mathfrak{m} \in \operatorname{Max} R$. Since we already proved $\Delta(A, D, u) \otimes_R R/\mathfrak{m}$ is a central $R/\mathfrak{m}$-algebra, it suffices to prove it is simple. Without loss of generality, assume from now on that R is a field. Assume that T is a nonzero two-sided ideal in $\Delta(A, D, u)$. Let $t = a_0 + a_1 z + \cdots + a_{p-1} z^{p-1} \in T$. As in (2), let τ_v be the inner derivation defined by v. Then $\tau_v(t) \in T$, which implies $\tau_v^{p-1}(t) = (p-1)! a_{p-1} \in T$. Therefore, $a_{p-1} \in T$. Similarly, $\tau_v^{p-1}(t) = (p-2)! a_{p-2} + (p-1)! a_{p-1} z \in T$, which implies $a_{p-2} \in T$. Recursively, we find that $a_i \in T$ for each $i = 0, \ldots, p-1$. Let T_0 be the ideal in A generated by all of the a_i that appear in all of the t in T. We see that $T = T_0 + T_0 z + \cdots + T_0 z^{p-1}$. Then T_0 is a nonzero ideal in A, which is an Azumaya C-algebra. By Corollary 7.1.7, $T_0 = IA$, for some nonzero ideal I in C. Let ξ be any nonzero element of I. Since T is an ideal in $\Delta(A, D, u)$, we have $z\xi - \xi z = D(\xi) \in T$. From this it follows that $D(\xi)$ is in I. Write $\xi = \xi_0 + \xi_1 v + \cdots + \xi_j v^j$, where each ξ_i is in R, and $\xi_j \neq 0$. Now we use the fact that D is an extension of the R-derivation d. Since $d(v) = 1$, it follows that $d^{j-1}(\xi) = j! \xi_j$ is in I. But ξ_j is a unit, hence T is equal to $\Delta(A, D, u)$. □

2.1. Elementary p-Algebras. Let R be a commutative ring of characteristic p. Let a and b be two elements of R. Let $C = R[x]/(x^p - a) = R[a^{1/p}]$ and d the standard R-derivation on C such that $d(x) = 1$. If we take $A = C$ and $u = b$, then we are in the context of Definition 8.2.2. We define the *elementary p-algebra of degree p*, denoted $(a, b)_R$, to be the differential crossed product $\Delta(R[a^{1/p}], d, b)$. By Lemma 8.2.3, $(a, b)_R$ is an Azumaya R-algebra which contains C as a maximal commutative subalgebra. The elementary

p-algebras $(a,b)_R$ defined here were first studied by Jacobson [**Jac37**], in the case where R is a field. We tabulate some of the elementary properties of the algebras $(a,b)_R$ in Lemma 8.2.4, which is from [**Sal76**] and [**KOS76**]. In Proposition 13.4.7, we show that if b is invertible in R, then $(a,b)_R$ is a cyclic crossed product algebra.

Lemma 8.2.4. *Let R be a commutative ring of characteristic p. The elementary p-algebras $(a,b)_R$ satisfy the following.*

(1) *As an R-algebra, $(a,b)_R$ is generated by two elements v, w which satisfy the relations: $v^p = a$, $w^p = b$, $wv = vw + 1$.*

(2) *As a left R-module, $(a,b)_R$ is free with basis $\{v^i w^j \mid 0 \le i < p, 0 \le j < p\}$.*

(3) *If S is a commutative R-algebra, $(a,b)_R \otimes_R S \cong (a,b)_S$.*

(4) *$(0,b)_R \cong M_p(R)$. More generally, if a is a pth power in R, then $(a,b)_R \cong M_p(R)$.*

(5) *$(a,b)_R \cong (-b,a)_R$.*

(6) *The opposite algebra $(a,b)_R^o$ is isomorphic to both $(-a,b)_R$ and $(a,-b)_R$.*

(7) *Let v and w be the elements in $(a,b)_R$ mentioned in (1). If c is an element of $R[v]$ (the ring generated by R and v), then*

$$(a,b)_R \cong \left(a, b + c^p + \tau_w^{p-1}(c)\right)_R$$

where τ_w is the inner derivation defined by: $\tau_w(z) = zw - wz$. Similarly, if c is an element of $R[w]$, then

$$(a,b)_R \cong \left(a + c^p + \tau_v^{p-1}(c), b\right)_R.$$

(8) *$(a,b)_R \otimes_R (a,c)_R \cong (a,b+c)_R \otimes_R (0,b)_R$. Consequently, the algebras $(a,b)_R \otimes_R (a,c)_R$ and $(a,b+c)_R$ are Brauer equivalent.*

(9) *In the Brauer group, the class of $(a,b)_R$ is annihilated by p.*

Proof. Parts (1), (2), (3) and (9) are left to the reader.

(5): Let v and w be the elements in $(a,b)_R$ mentioned in (1). Then $v^p = a$, $w^p = b$, and $wv = vw+1$. Let x and y be the corresponding elements in $(-b,a)_R$ such that $x^p = -b$, $y^p = a$, $yx = xy+1$. The assignments $v \mapsto y$, $w \mapsto -x$ define an isomorphism of R-algebras.

(6): The isomorphism can be explicitly defined as in (5).

(4): Let v and w be the elements in $(a,b)_R$ mentioned in (1). Then $v^p = a$, $w^p = b$, and $wv = vw + 1$. Assume $a = \alpha^p$, for some $\alpha \in R$. Since $(v-\alpha)^p = 0$ and $w(v-\alpha) = (v-\alpha)w+1$, $(a,b)_R$ is isomorphic to $\Lambda = (0,b)_R$. Assume $a = 0$. Let M be the right ideal in $\Lambda = (0,b)_R$ generated by v^{p-1}.

Notice that $v^{p-1}, v^{p-1}w, \dots, v^{p-1}w^{p-1}$ is a free R-basis for M. The right regular representation

$$\Lambda^o \to \operatorname{Hom}_R(M, M)$$

is an isomorphism, by Theorem 7.6.1. (In the terminology of Section 11.1.3, v^{p-1} has rank 1 and this follows from Proposition 11.1.13.)

(7): The elementary p-algebra $(a, b)_R$ is generated as an R-algebra by v and $w + c$. We have $(w + c)v = v(w + c) + 1$. Since $\tau_w(v) = 1$, it follows from Exercise 8.1.9 that $\tau_w(R[v]) \subseteq R[v]$. If ξ is an element of $R[v]$, then

$$\begin{aligned}\tau_{c\lambda+w}(\xi) &= \xi(c\lambda + w) - (c\lambda + w)\xi \\ &= \tau_w(\xi)\end{aligned}$$

is another element of $R[v]$. Using this inductively, we get from Lemma 8.1.7 that

$$(c + w)^p = c^p + w^p + \tau_w^{p-1}(c) = b + c^p + \tau_w^{p-1}(c).$$

Consider the elementary p-algebra $(a, b + c^p + \tau_w^{p-1}(c))_R$, with generators x and y. The assignments $x \mapsto v$, $y \mapsto w + c$ define an R-algebra homomorphism

$$(a, b + c^p + \tau_w^{p-1}(c))_R \to (a, b)_R$$

which is an isomorphism because both rings are Azumaya R-algebras of the same rank (Lemma 8.2.3 and Theorem 7.6.1). The second part of (7) is similar.

(8): Let $A = (a, b)_R \otimes_R (a, c)_R$. Let v and w be the elements in $(a, b)_R$ mentioned in (1) and x and y their counterparts in $(a, c)_R$. Identify v, w, x, y with their images in A. The reader should verify the commutator identities:

$$\begin{aligned} wx &= xw \\ wy &= yw \\ vx &= xv \\ vy &= yv \\ x(v - x) &= (v - x)x \\ (w + y)(v - x) &= (v - x)(w + y) \\ (w + y)w &= w(w + y). \end{aligned} \tag{8.11}$$

Let A_1 be the R-subalgebra of A generated by $v - x$ and w. We have: $w(v - x) = wv - wx = vw + 1 - xw = (v - x)w + 1$, $w^p - b$, and $(v - x)^p = v^p - x^p = a - a = 0$. By (4), $A_1 \cong (0, b)_R$. Let A_2 be the R-subalgebra of A generated by x and $w + y$. We have $x^p = a$, $(w + y)^p = w^p + y^p = b + c$, and $(w+y)x = x(w+y)+1$. Therefore, $A_2 \cong (a, b+c)_R$. From (8.11) we see that every element of A_1 commutes with every element of A_2. By Exercise 1.3.8, there is a homomorphism of Azumaya R-algebras $A_1 \otimes_R A_2 \to A$ which is an isomorphism because the ranks agree (Theorem 7.6.1). □

3. Differentials and Separability

This section contains an introduction to the module of Kähler differentials associated to a commutative R-algebra. The module of differentials is defined and its fundamental properties are proved. Our motivation is to derive new criteria for separability. For commutative algebras these tests are found in Section 8.3.2. The results in this section are applied in Section 8.4 to study separably generated field extensions.

3.1. The Definition and Fundamental Exact Sequences. A general reference for this section is [**Mat80**]. Let A be a commutative R-algebra and $A^e = A \otimes_R A$. The sequence of A^e-modules

$$0 \to J_{A/R} \to A^e \xrightarrow{\mu} A \to 0$$

is exact (Definition 4.1.1). Using the R-algebra homomorphism $\rho : A \to A^e$ defined by $a \mapsto a \otimes 1$, we turn A^e into a left A-module. Consequently $J_{A/R}$ and $J^2_{A/R}$ are also A-modules. Let $\Omega_{A/R}$ be defined by the exact sequence

$$0 \to J^2_{A/R} \to J_{A/R} \xrightarrow{\pi} \Omega_{A/R} \to 0.$$

The left A-module $\Omega_{A/R}$ is called the *module of Kähler differentials*. As in Section 8.1.1, there is an R-derivation $\delta : A \to J_{A/R}$ defined by $a \mapsto a \otimes 1 - 1 \otimes a$. Let $d_{A/R} = \pi\delta$. The reader should verify that $d_{A/R} : A \to \Omega_{A/R}$ is an R-derivation. The derivation $d_{A/R}$, together with the module of Kähler differentials satisfies a universal mapping property. In Theorem 8.3.1, a left A-module is made into a two-sided A/R-module by making the right multiplication agree with the left multiplication. An R-module homomorphism $\partial : A \to M$ is an R-derivation of A, if $\partial(ab) = a\partial(b) + b\partial(a)$, for all $a, b \in A$.

Theorem 8.3.1. *Let A be a commutative R-algebra. For any left A-module M, if $\partial : A \to M$ is an R-derivation of A, then there exists a unique A-module homomorphism $f : \Omega_{A/R} \to M$ such that the diagram*

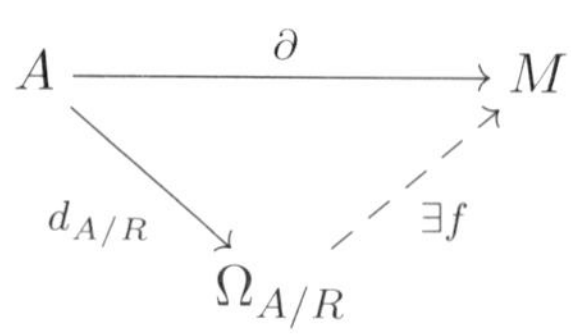

commutes. The assignment $f \mapsto f d_{A/R}$ defines an isomorphism of A-modules $\operatorname{Hom}_A(\Omega_{A/R}, M) \cong \operatorname{Der}_R(A, M)$.

Proof. The exact sequence

$$J_{A/R} \xrightarrow{\pi} \Omega_{A/R} \to 0$$

gives rise to the exact sequence

$$0 \to \operatorname{Hom}_A(\Omega_{A/R}, M) \to \operatorname{Hom}_A(J_{A/R}, M).$$

Let $f \in \operatorname{Hom}_A(\Omega_{A/R}, M)$. For any $a, b, x \in A$,

$$\begin{aligned}
(f\pi)\big((a \otimes b)(x \otimes 1 - 1 \otimes x)\big) &= (f\pi)\Big(\big(a(1 \otimes b - b \otimes 1) \\
&\qquad + ab \otimes 1\big)(x \otimes 1 - 1 \otimes x)\Big) \\
&= (f\pi)\big(a(1 \otimes b - b \otimes 1)(x \otimes 1 - 1 \otimes x) \\
&\qquad + ab(x \otimes 1 - 1 \otimes x)\big) \\
&= f\big(ab(x \otimes 1 - 1 \otimes x)\big) \\
&= abf(x \otimes 1 - 1 \otimes x).
\end{aligned}$$

This means $f\pi$ is in $\operatorname{Hom}_{A^e}(J_{A/R}, M)$, so the sequence

$$0 \to \operatorname{Hom}_A(\Omega_{A/R}, M) \xrightarrow{\zeta} \operatorname{Hom}_{A^e}(J_{A/R}, M)$$

is exact. Let $g \in \operatorname{Hom}_{A^e}(J_{A/R}, M)$. For all $a, b \in A$,

$$\begin{aligned}
g\big((a \otimes 1 - 1 \otimes a)(b \otimes 1 - 1 \otimes b)\big) &= g\big(a \otimes 1(b \otimes 1 - 1 \otimes b)\big) \\
&\quad - g\big(1 \otimes a(b \otimes 1 - 1 \otimes b)\big) \\
&= ag(b \otimes 1 - 1 \otimes b) - g(b \otimes 1 - 1 \otimes b)a \\
&= 0.
\end{aligned}$$

Since g annihilates $J^2_{A/R}$, there exists $f : \Omega_{A/R} \to M$ such that $g = f\pi$. This proves ζ is an isomorphism. Combined with Proposition 8.1.5, this shows that there is an isomorphism $\operatorname{Hom}_A(\Omega_{A/R}, M) \cong \operatorname{Der}_R(A, M)$ which is defined by $f \mapsto f\pi\delta$. Because A is commutative, the maps are A-linear. □

Proposition 8.3.2. *Let S be a commutative R-algebra which is generated as an R-algebra by the set $X = \{x_i\}_{i \in I}$. Then*

(1) *$\Omega_{S/R}$ is generated as an S-module by $d_{S/R}(X) = \{d_{S/R}x_i\}_{i \in I}$.*
(2) *If S is a polynomial ring over R (that is, if X is a set of indeterminates), then $\Omega_{S/R}$ is a free S-module with basis $d_{S/R}(X)$.*
(3) *If S is a finitely generated R-algebra, then $\Omega_{S/R}$ is a finitely generated S-module.*

Proof. Part (3) follows directly from Part (1).

(1): By Lemma 8.1.4, $J_{S/R}$ is generated as an S^e-module by the set $\delta(X) = \{x_i \otimes 1 - 1 \otimes x_i\}_{i \in I}$. Let $\pi : J_{S/R} \to J_{S/R}/J^2_{S/R}$ be the natural map.

Given any $a, b \in S$ and $x \in X$,

$$\begin{aligned}\pi\big(a \otimes b(x \otimes 1 - 1 \otimes x)\big) &= \pi\big((a(1 \otimes b - b \otimes 1) + (ab \otimes 1))(x \otimes 1 - 1 \otimes x)\big) \\ &= \pi\big(a(1 \otimes b - b \otimes 1)(x \otimes 1 - 1 \otimes x)\big) \\ &\quad + \pi\big((ab \otimes 1)(x \otimes 1 - 1 \otimes x)\big) \\ &= \pi\big((ab \otimes 1)(x \otimes 1 - 1 \otimes x)\big).\end{aligned}$$

It follows from this that $\Omega_{S/R} = J_{S/R}/J^2_{S/R}$ is generated as a left S-module by the set $\pi\delta(X) = d_{S/R}(X)$.

(2): For each $i \in I$, let $\partial_i : S \to S$ represent the "partial derivative with respect to x_i" function. By the Universal Mapping Property (Theorem 8.3.1), there exists a unique $b_i \in \operatorname{Hom}_S(\Omega_{S/R}, S)$ such that for all $j \in I$

$$b_i d_{S/R} x_j = \partial_i x_j = \begin{cases} 1 & \text{if } i = j, \\ 0 & \text{if } i \neq j. \end{cases}$$

Suppose $\sum_j s_j d_{S/R} x_j = 0$ is a finite dependence relation in $\Omega_{S/R}$ where each $s_j \in S$. Applying b_i we see that $s_i = 0$. $\square$

Now we derive the so-called fundamental exact sequences for the module of differentials.

Theorem 8.3.3 (The First Fundamental Exact Sequence). *Let S be a commutative R-algebra and A a commutative S-algebra.*

(1) *There is an exact sequence of natural homomorphisms of A-modules*

$$\Omega_{S/R} \otimes_S A \xrightarrow{a} \Omega_{A/R} \xrightarrow{b} \Omega_{A/S} \to 0.$$

(2) *There is a split-exact sequence of natural homomorphisms of A-modules*

$$0 \to \Omega_{S/R} \otimes_S A \xrightarrow{a} \Omega_{A/R} \xrightarrow{b} \Omega_{A/S} \to 0$$

if and only if given any A-module M and any R-derivation $\partial : S \to M$, there exists an R-derivation $D : A \to M$ such that the diagram

$$\begin{array}{ccc} S & \xrightarrow{\partial} & M \\ & \searrow & \nearrow_{D} \\ & A & \end{array}$$

commutes.

Proof. (1): Step 1: Define the map a. By Exercise 8.3.9, the commutative diagram of commutative rings

$$\begin{array}{ccc} R & \longrightarrow & R \\ \downarrow & & \downarrow \\ S & \longrightarrow & A \end{array}$$

induces a natural homomorphism of A-modules $a : \Omega_{S/R} \otimes_S A \to \Omega_{A/R}$.

Step 2: Define the map b. Again, by Exercise 8.3.9, the commutative diagram of commutative rings

$$\begin{array}{ccc} R & \longrightarrow & S \\ \downarrow & & \downarrow \\ A & \longrightarrow & A \end{array}$$

induces a natural homomorphism of A-modules $b : \Omega_{A/R} \to \Omega_{A/S}$.

Step 3: b is onto. A generating set for A as an R-algebra is a generating set for A as an S-algebra. It is evident that b is onto, by Proposition 8.3.2.

Step 4: The sequence is a complex. In the commutative diagram

$$\begin{array}{ccc} S & \longrightarrow & A \\ {\scriptstyle d_{S/R}}\downarrow & & \downarrow{\scriptstyle d_{A/S}} \\ \Omega_{S/R} & \xrightarrow{c} & \Omega_{A/S} \end{array}$$

c is the zero map. Therefore, $ba = 0$.

Step 5: $\ker b = \operatorname{im} a$. By Lemma 1.3.22, this is true if

(8.12)
$$0 \to \operatorname{Hom}_A(\Omega_{A/S}, M) \xrightarrow{\mathrm{H}_b} \operatorname{Hom}_A(\Omega_{A/R}, M) \xrightarrow{\mathrm{H}_a} \operatorname{Hom}_A(\Omega_{S/R} \otimes_S A, M)$$

is exact for all A-modules M. By the adjoint isomorphism of Theorem 1.3.24 and Theorem 8.3.1, (8.12) is naturally isomorphic to

$$0 \to \operatorname{Der}_S(A, M) \to \operatorname{Der}_R(A, M) \to \operatorname{Der}_R(S, M)$$

which is exact, by Exercise 8.1.12.

(2): By Exercise 1.3.31, there is a left inverse for a if and only if for all all A-modules M, the map H_a in (8.12) is onto. Equivalently, $\operatorname{Der}_R(A, M) \to \operatorname{Der}_R(S, M) \to 0$ is exact, for all A-modules M. □

Let R be a commutative ring and S a commutative R-algebra. Let I be an ideal of S and set $A = S/I$. Define a function $\gamma : I \to \Omega_{S/R} \otimes_S A$

by $x \mapsto d_{S/R}x \otimes 1$. If $x, y \in I$, then $\gamma(xy) = xd_{S/R}y \otimes 1 + yd_{S/R}x \otimes 1 = d_{S/R}y \otimes x + d_{S/R} \otimes y = 0$. Therefore, γ factors through I^2 and we have the A-module homomorphism (also denoted by γ)

$$\gamma : I/I^2 \to \Omega_{S/R} \otimes_S A.$$

Theorem 8.3.4 (The Second Fundamental Exact Sequence). *Let S be a commutative R-algebra, I an ideal in S, and $A = S/I$. The sequence of A-modules*

$$I/I^2 \xrightarrow{\gamma} \Omega_{S/R} \otimes_S A \xrightarrow{a} \Omega_{A/R} \to 0$$

is exact.

Proof. Step 1: a is onto and the sequence is a complex. By Exercise 8.3.9, the diagram

$$\begin{array}{ccc} S & \xrightarrow{\theta} & A \\ {\scriptstyle d_{S/R}}\downarrow & & \downarrow{\scriptstyle d_{A/R}} \\ \Omega_{S/R} & \xrightarrow{a} & \Omega_{A/R} \end{array}$$

commutes. Since θ is onto and the vertical maps are onto, a is onto. If $x \in I$, then $d_{A/R}\theta(x) = 0$, hence $\operatorname{im}\gamma \subseteq \ker a$.

Step 2: $\operatorname{im}\gamma = \ker a$. As in the proof of Theorem 8.3.3, it suffices to prove

$$0 \to \operatorname{Hom}_A(\Omega_{A/R}, M) \xrightarrow{\mathrm{H}_a} \operatorname{Hom}_A(\Omega_{S/R} \otimes_S A, M) \xrightarrow{\mathrm{H}_\gamma} \operatorname{Hom}_A(I/I^2, M)$$

is exact, for every A-module M. By the adjoint isomorphism of Theorem 1.3.24 and Theorem 8.3.1, this last sequence is isomorphic to

$$0 \to \operatorname{Der}_R(A, M) \to \operatorname{Der}_R(S, M) \to \operatorname{Hom}_S(I, M).$$

The reader should verify that this last sequence is exact. □

3.2. More Tests for Separability. In this section we apply the ideas from Section 8.3 to derive separability criteria for commutative R-algebras. For example, the vanishing of the module of Kähler differentials is equivalent to being formally unramified (Theorem 8.3.5), and for a finitely generated algebra, this is equivalent to being separable (Theorem 8.3.6). As an application, we prove the Jacobian Criterion for Separability (Proposition 8.3.8). General references for the material in this section are [**KO74b**] and [**Ray70**].

Theorem 8.3.5. *Let S be a commutative R-algebra. The following are equivalent:*

(1) *S is formally unramified over R (see Definition 4.7.1).*

(2) $\Omega_{S/R} = 0$.

(3) $\operatorname{Der}_R(S, M) = 0$ *for every left S-module M.*

Proof. (2) implies (3): This follows from Theorem 8.3.1.

(3) implies (2): If $\operatorname{Der}_R(S, \Omega_{S/R}) = 0$, then $\operatorname{Hom}_S(\Omega_{S/R}, \Omega_{S/R}) = 0$, by Theorem 8.3.1. From this we conclude that $\Omega_{S/R} = 0$.

(2) implies (1): Suppose A is a commutative R-algebra, I is a nilpotent ideal in A, and $g : S \to A/I$ is an R-algebra homomorphism. Assume there exists an R-algebra homomorphism $\gamma : S \to A$ such that the diagram

$$\begin{array}{ccc} S & \xrightarrow{g} & A/I \\ \uparrow & \searrow^{\gamma} & \uparrow \eta \\ R & \longrightarrow & A \end{array}$$

commutes. We must show that γ is unique. We are given that $I^n = 0$, for some $n \geq 1$ and by a finite induction argument on n, it is enough to prove this when $n = 2$. By Proposition 8.1.6, the set of all R-algebra homomorphisms $v : S \to A$ such that $\eta v = \eta\gamma$ is in one-to-one correspondence with $\operatorname{Der}_R(S, I)$. By Theorem 8.3.1, $\operatorname{Der}_R(S, I) \cong \operatorname{Hom}_S(\Omega_{S/R}, I)$, which is the singleton set $\{0\}$. That is, γ is the unique lifting of g.

(1) implies (2): As in Exercise 8.1.14, let $C = S * \Omega_{S/R}$ be the trivial extension of S by $\Omega_{S/R}$. The natural map $\eta : C \to S$ is defined by $\eta(a, x) = a$. There is an R-algebra homomorphism $u : S \to S * \Omega_{S/R}$ defined by $u(a) = (a, 0)$ and u is a section to η. That is, $\bar{u} = \eta u : S \to S$ is the identity map. Since S/R is formally unramified, u is the unique lifting of $\bar{u}$ in $\operatorname{Hom}_{R\text{-alg}}(S, C)$. By Proposition 8.1.6, this implies $\operatorname{Der}_R(S, \Omega_{S/R}) = (0)$. By Theorem 8.3.1, $\operatorname{Hom}_S(\Omega_{S/R}, \Omega_{S/R}) = (0)$, hence $\Omega_{S/R} = 0$. $\square$

When S is a finitely generated R-algebra, (2) implies (1) in Theorem 8.3.6 is a converse to Corollary 4.7.2.

Theorem 8.3.6. *Let S be a commutative finitely generated R-algebra. The following are equivalent.*

(1) *S is a separable R-algebra.*

(2) *S is formally unramified over R (see Definition 4.7.1).*

(3) $\operatorname{Der}_R(S, M) = 0$ *for every left S-module M.*

(4) $\Omega_{S/R} = 0$.

Proof. By Theorem 8.3.5, (2), (3) and (4) are equivalent.

(1) implies (4): By Proposition 4.1.2, $J_{S/R}$ is an idempotent generated ideal in S^e. Therefore, $J^2_{S/R} = J_{S/R}$.

(4) implies (1): This is the only part of the proof where we need to assume S is finitely generated. By Lemma 8.1.4, $J_{S/R}$ is a finitely generated

ideal of S^e. We are given that $J_{S/R}^2 = J_{S/R}$. It follows from Exercise 1.1.24 and Proposition 4.1.2 that S/R is separable. □

Theorem 8.3.7. *Let S be a commutative finitely generated R-algebra with structure homomorphism $\theta : R \to S$. The following are equivalent.*

(1) *S is a separable R-algebra.*

(2) *For every $\mathfrak{p} \in \operatorname{Spec} R$, if $k_\mathfrak{p} = R_\mathfrak{p}/(\mathfrak{p}R_\mathfrak{p})$, then $S \otimes_R k_\mathfrak{p}$ is a separable $k_\mathfrak{p}$-algebra.*

(3) *For every $\mathfrak{p} \in \operatorname{Spec} R$, and every $\mathfrak{q} \in \operatorname{Spec} S$ such that $\mathfrak{p} = \theta^{-1}(\mathfrak{q})$, $\mathfrak{p}S_\mathfrak{q} = \mathfrak{q}S_\mathfrak{q}$, and $k_\mathfrak{q} = S_\mathfrak{q}/(\mathfrak{q}S_\mathfrak{q})$ is a finite separable extension of the field $k_\mathfrak{p} = R_\mathfrak{p}/(\mathfrak{p}R_\mathfrak{p})$.*

(4) *For every algebraically closed field F and homomorphism of rings $\phi : R \to F$, $S \otimes_R F$ is a separable F-algebra.*

Proof. (1) implies (2), and (1) implies (4): Follow straight from Corollary 4.3.2.

(4) implies (2): Let $\mathfrak{p} \in \operatorname{Spec} R$. Let F be the algebraic closure of $k_\mathfrak{p} = R_\mathfrak{p}/(\mathfrak{p}R_\mathfrak{p})$ and $\phi : R \to F$ the natural map. By assumption, $S \otimes_R F$ is separable over F. Corollary 4.3.5 implies $S \otimes_R k_\mathfrak{p}$ is separable over $k_\mathfrak{p}$.

(2) implies (1): By Proposition 8.3.2, $\Omega_{S/R}$ is a finitely generated S-module. By Theorem 8.3.6, to finish the proof it is enough to show $\Omega_{S/R} = 0$. By Proposition 2.1.2, it is enough to show $\Omega_{S/R} \otimes_S S_\mathfrak{q} = 0$ for every $\mathfrak{q} \in \operatorname{Spec} S$. Fix $\mathfrak{q} \in \operatorname{Spec} S$ and let $\mathfrak{p} = \mathfrak{p} \cap R$. Since $\left(\Omega_{S/R}\right)_\mathfrak{q} = \Omega_{S/R} \otimes_S S_\mathfrak{q}$ is finitely generated over $S_\mathfrak{q}$ and $\mathfrak{m}_\mathfrak{p} \subseteq \mathfrak{m}_\mathfrak{q}$, by Theorem 3.1.2 (Nakayama's Lemma), it is enough to show $\left(\Omega_{S/R}\right)_\mathfrak{q}/\mathfrak{m}_\mathfrak{p}\left(\Omega_{S/R}\right)_\mathfrak{q} = 0$. By Exercise 8.3.10, $\Omega_{S/R} \otimes_R k_\mathfrak{p} = \Omega_{S \otimes_R k_\mathfrak{p}/k_\mathfrak{p}}$, and by Theorem 8.3.6, $\Omega_{S \otimes_R k_\mathfrak{p}/k_\mathfrak{p}} = 0$. The reader should verify that

$$\begin{aligned}
\left(\Omega_{S/R}\right)_\mathfrak{q}/\mathfrak{m}_\mathfrak{p}\left(\Omega_{S/R}\right)_\mathfrak{q} &= \left(\Omega_{S/R}\right)_\mathfrak{q} \otimes_{R_\mathfrak{p}} R_\mathfrak{p}/\mathfrak{m}_\mathfrak{p} \\
&= \left(\Omega_{S/R}\right)_\mathfrak{q} \otimes_R k_\mathfrak{p} \\
&\cong S_\mathfrak{q} \otimes_S \Omega_{S/R} \otimes_R k_\mathfrak{p} \\
&\cong S_\mathfrak{q} \otimes_S \Omega_{S \otimes_R k_\mathfrak{p}/k_\mathfrak{p}} \\
&= 0.
\end{aligned}$$

(1) implies (3): Assume S is R-separable, $\mathfrak{q} \in \operatorname{Spec} S$ and $\mathfrak{p} = \mathfrak{q} \cap R$. By Exercise 4.4.9, $S_\mathfrak{q}$ is separable over $R_\mathfrak{p}$. By Exercise 4.5.16, $\mathfrak{m}_\mathfrak{p} S_\mathfrak{q} = \mathfrak{m}_\mathfrak{q}$ and $k_\mathfrak{q} = S_\mathfrak{q} \otimes_R k_\mathfrak{p}$ is a separable field extension of $k_\mathfrak{p}$.

(3) implies (2): Fix $\mathfrak{p} \in \operatorname{Spec} R$ such that there exists some $\mathfrak{q} \in \operatorname{Spec} S$ and $\mathfrak{p} = \mathfrak{q} \cap R$. By Exercise 2.1.12, $\mathfrak{q}_\mathfrak{p} = \mathfrak{q} \otimes_R R_\mathfrak{p}$ is a prime ideal of $S_\mathfrak{p} = S \otimes_R R_\mathfrak{p}$ and the local ring of $S_\mathfrak{p}$ at $\mathfrak{q}_\mathfrak{p}$ is $S_\mathfrak{q}$. By Exercise 2.1.9, $S_\mathfrak{p}/\mathfrak{q}_\mathfrak{p}$

is an integral domain with quotient field $k_{\mathfrak{q}} = S_{\mathfrak{q}}/\mathfrak{m}_{\mathfrak{q}}$. The diagram

$$\begin{array}{ccc} S_{\mathfrak{p}}/\mathfrak{q}_{\mathfrak{p}} & \longrightarrow & k_{\mathfrak{q}} \\ \uparrow & & \uparrow \\ R_{\mathfrak{p}}/\mathfrak{m}_{\mathfrak{p}} & \xrightarrow{=} & k_{\mathfrak{p}} \end{array}$$

commutes. By hypothesis, $k_{\mathfrak{q}}/k_{\mathfrak{p}}$ is a finite dimensional field extension. It follows from Lemma 3.3.2 that $S_{\mathfrak{p}}/\mathfrak{q}_{\mathfrak{p}}$ is a field. That is, $\mathfrak{q}_{\mathfrak{p}}$ is a maximal ideal in $S_{\mathfrak{p}}$. It follows from Exercise 2.3.8 that every prime ideal in $S \otimes_R k_{\mathfrak{p}}$ is a maximal ideal, and moreover each maximal ideal is of the form $\mathfrak{q} \otimes_R k_{\mathfrak{p}}$ for some $\mathfrak{q}$ lying over $\mathfrak{p}$. Because $S \otimes_R k_{\mathfrak{p}}$ is finitely generated as a $k_{\mathfrak{p}}$-algebra, $S \otimes_R k_{\mathfrak{p}}$ is noetherian by the Hilbert Basis Theorem (see, for example, [**Hun80**, Theorem VIII.4.9]). By Proposition 3.2.10, $S \otimes_R k_{\mathfrak{p}}$ is artinian. By Theorem 3.2.11, if $\operatorname{Max}(S \otimes_R k_{\mathfrak{p}}) = \{\mathfrak{n}_1, \dots, \mathfrak{n}_n\}$, then $S \otimes_R k_{\mathfrak{p}} = (S \otimes_R k_{\mathfrak{p}})_{\mathfrak{n}_1} \oplus \cdots \oplus (S \otimes_R k_{\mathfrak{p}})_{\mathfrak{n}_n}$. Suppose $\mathfrak{n}_i = \mathfrak{q}_i \otimes_R k_{\mathfrak{p}}$ is an arbitrary maximal ideal of $S \otimes_R k_{\mathfrak{p}}$. By Exercise 2.3.8,

$$(S \otimes_R k_{\mathfrak{p}})_{\mathfrak{n}_i} = (S \otimes_R k_{\mathfrak{p}})_{\mathfrak{q}_i \otimes_R k_{\mathfrak{p}}} = S_{\mathfrak{q}_i}/\mathfrak{m}_{\mathfrak{p}} S_{\mathfrak{q}_i} = S_{\mathfrak{q}_i}/\mathfrak{m}_{\mathfrak{q}_i} = k_{\mathfrak{q}_i}.$$

This proves that $S \otimes_R k_{\mathfrak{p}} \cong k_{\mathfrak{q}_1} \oplus \cdots \oplus k_{\mathfrak{q}_n}$ and by Corollary 4.5.8, we are done. $\square$

We conclude this section with a proof of a jacobian criterion for separability. For computations it turns out to be one of the most useful tests for separability.

Proposition 8.3.8. *Let R be a commutative ring. Let $I = (f_1, \dots, f_n)$ be an ideal in $S = R[x_1, \dots, x_n]$ generated by a set of n polynomials in n indeterminates. Then S/I is separable over R if and only if the determinant of the jacobian matrix $(\partial f_i/\partial x_j)$ maps to a unit in S/I.*

Proof. Let $A = S/I$. We use the notation of Theorem 8.3.4. The sequence

$$I/I^2 \xrightarrow{\gamma} \Omega_{S/R} \otimes_S A \xrightarrow{a} \Omega_{A/R} \to 0$$

is exact. By Theorem 8.3.6, A/R is separable if and only if γ is onto. By Proposition 8.3.2, $\Omega_{S/R} \otimes_S A$ is a free A-module on the basis $\{dx_1, \dots, dx_n\}$. For each i,

$$\gamma(f_i) = \sum_{j=1}^{n} \frac{\partial f_i}{\partial x_j} dx_j.$$

Thus, $\Omega_{A/R}$ is isomorphic to the cokernel of the A-module homomorphism

$$A^{(n)} \xrightarrow{J} A^{(n)}$$

where J denotes multiplication by the jacobian matrix $(\partial f_i/\partial x_j)$. The result follows from properties of the determinant (see [**Hun80**, Theorem VII.3.5], for example). □

The following definition will be used in some of the exercises below. Let M be a module over the commutative ring R. If $P \in \operatorname{Spec} R$, then the *stalk* of M at P is the localization M_P of M with respect to the multiplicative set $R - P$. The *support* of M is the set of all points in $\operatorname{Spec} R$ for which the stalk of M is nontrivial,

$$\operatorname{Supp}_R(M) = \{P \in \operatorname{Spec} R \mid M_P \neq 0\}.$$

If R is understood, we write simply $\operatorname{Supp}(M)$.

3.3. Exercises.

Exercise 8.3.9. Let

$$\begin{array}{ccc} R & \longrightarrow & S \\ \downarrow & & \downarrow \\ A & \xrightarrow{\theta} & B \end{array}$$

be a commutative diagram of commutative rings. Show that there exists a unique homomorphism ψ such that the diagram

$$\begin{array}{ccc} A & \xrightarrow{\theta} & B \\ {\scriptstyle d_{A/R}}\downarrow & & \downarrow{\scriptstyle d_{B/S}} \\ \Omega_{A/R} & \overset{\exists\psi}{\dashrightarrow} & \Omega_{B/S} \end{array}$$

of A-modules commutes. Show that ψ induces a homomorphism $\Omega_{A/R} \otimes_A B \to \Omega_{B/S}$ of B-modules.

Exercise 8.3.10. Suppose A and S are commutative R-algebras. Show that there exists a unique isomorphism ϕ such that the diagram

$$\begin{array}{ccc} A \otimes_R S & \xrightarrow{d_{A/R} \otimes 1} & \Omega_{A/R} \otimes_R S \\ & {\scriptstyle d_{A \otimes_R S/S}} \searrow & \nearrow {\scriptstyle \exists\phi} \\ & \Omega_{A \otimes_R S/S} & \end{array}$$

of S-modules commutes. (Hint: The inverse of ϕ is constructed in Exercise 8.3.9.)

Exercise 8.3.11. Let A be a commutative R-algebra and $W \subseteq A$ a multiplicative set. Let A_W denote the localization $W^{-1}A$. Show that there exists an isomorphism of A_W-modules $\Omega_{A_W/R} \cong \Omega_{A/R} \otimes_A A_W = W^{-1}\Omega_{A/R}$. (Hint: Construct $\Omega_{A/R} \otimes_A A_W \to \Omega_{A_W/R}$ using Exercise 8.3.9.)

Exercise 8.3.12. Let R be a commutative ring, f a monic polynomial in $R[x]$ and $S = R[x]/(f)$. Let $t \in S$. Show that $S[t^{-1}]$ is separable over R if and only if the image of f' is invertible in $S[t^{-1}]$. (Hint: $S = R[x,y]/(f, yg-1)$, where g is a preimage of t in $R[x]$.)

Exercise 8.3.13. Let R be a commutative ring and S a commutative R-algebra. Let $A = S[x_1, \dots, x_n]$ be the polynomial ring over S in n variables. Show that the sequence

$$0 \to \Omega_{S/R} \otimes_S A \xrightarrow{a} \Omega_{A/R} \xrightarrow{b} \Omega_{A/S} \to 0$$

is split-exact.

Exercise 8.3.14. Let R be a commutative ring and I an ideal in R. Let $P \in \operatorname{Spec} R$. Prove that $(R/I)_P \neq 0$ if and only if $I \subseteq P$. Conclude that $\operatorname{Supp}(R/I)$ is equal to $V(I)$. In particular, $\operatorname{Supp}(R) = \operatorname{Spec} R$.

Exercise 8.3.15. Let R be a commutative ring, M an R-module and N a submodule. Prove that $\operatorname{Supp}(M) = \operatorname{Supp}(N) \cup \operatorname{Supp}(M/N)$. (Hint: Localize the exact sequence $0 \to N \to M \to M/N \to 0$.)

Exercise 8.3.16. Let R be a commutative ring and M a finitely generated R-module. Prove that $\operatorname{Supp}(M) = V\big(\operatorname{annih}(M)\big)$. Conclude that $\operatorname{Supp}(M)$ is a closed subset of $\operatorname{Spec} R$.

Exercise 8.3.17. Let S be a finitely generated commutative R-algebra. Let U be the set of all points P in $\operatorname{Spec} S$ such that S_P is a separable R-algebra. Prove that U is an open (possibly empty) subset of $\operatorname{Spec} S$. (Hint: Apply Exercise 8.3.16 to $\Omega_{S/R}$.)

4. Separably Generated Extension Fields

This section contains an introduction to the notion of separably generated field extensions. We prove a strong version of the Noether Normalization Lemma (Theorem 8.4.6) and apply it to prove the theorem on the finiteness of the integral closure of a finitely generated k-algebra (Theorem 8.4.7). General references for this section are [**Mat80**] and [**ZS75a**].

Lemma 8.4.1. *Let $k \subseteq K \subseteq F$ be a tower of field extensions. If $F = K(\alpha)$ is a simple algebraic extension of K, then*

$$\dim_K \Omega_{K/k} \leq \dim_F \Omega_{F/k} \leq 1 + \dim_K \Omega_{K/k}.$$

Proof. Let $f \in K[x]$ be the irreducible polynomial of α. Let I be the principal ideal in $K[x]$ generated by f. By Theorem 8.3.4,

$$I/I^2 \xrightarrow{\gamma} \Omega_{K[x]/k} \otimes_{K[x]} F \xrightarrow{a} \Omega_{F/k} \to 0$$

is an exact sequence of F-vector spaces. By Exercise 8.3.13 and Proposition 8.3.2, $\Omega_{K[x]/k}$ is a free $K[x]$-module of rank $1 + \dim_K \Omega_{K/k}$. The image of γ is generated over F by $\gamma(f)$, hence has dimension less than or equal to one. □

Let F/k be a finitely generated extension of fields. Let $\Xi \subseteq F$ be a transcendence base for F/k. We say Ξ is a *separating transcendence base* of F/k in case F is a separable algebraic extension of $k(\Xi)$. We say F/k is *separably generated* if there exists a separating transcendence base for F/k.

Theorem 8.4.2. *Let F be a finitely generated extension field of k.*

(1) $\dim_F \Omega_{F/k} \geq \operatorname{tr.deg}_k F$.

(2) $\dim_F \Omega_{F/k} = \operatorname{tr.deg}_k F$ *if and only if F/k is separably generated.*

(3) $\Omega_{F/k} = 0$ *if and only if F is separable over k.*

Proof. (3): This follows from Theorem 8.3.6, Corollary 4.5.4, and Theorem 4.5.6.

(1): A transcendence base $\xi_1, \dots, \xi_n$ exists for F/k (see, for example, [**Hun80**, Corollary VI.1.7]). If we set $K = k(\xi_1, \dots, \xi_n)$, then F/K is finite dimensional. Applying Lemma 8.4.1 iteratively, we get $\dim_F \Omega_{F/k} \geq \dim_K \Omega_{K/k}$. Note that K is the quotient field of $k[\xi_1, \dots, \xi_n]$. By Proposition 8.3.2 and Exercise 8.3.11, $\dim_K \Omega_{K/k} = n = \operatorname{tr.deg}_k F$.

(2): Assume $\xi_1, \dots, \xi_n$ is a transcendence base and $K = k(\xi_1, \dots, \xi_n)$. If F/K is separable, then $\Omega_{F/K} = 0$, by Theorem 8.3.6. Theorem 8.3.3 implies

$$\Omega_{K/k} \otimes_K F \xrightarrow{a} \Omega_{F/k} \to 0$$

is exact. Therefore, equality holds in Part (1). Conversely, suppose in Part (1) that equality holds. Let $n = \operatorname{tr.deg}_k F$ and choose $\xi_1, \dots, \xi_n$ in F such that the set $d_{F/k}(\xi_1), \dots, d_{F/k}(\xi_n)$ is a basis for the F-vector space $\Omega_{F/k}$. Let $K = k(\xi_1, \dots, \xi_n)$. The diagram

$$\begin{array}{ccc} K & \longrightarrow & F \\ {\scriptstyle d_{K/k}}\downarrow & & \downarrow{\scriptstyle d_{F/k}} \\ \Omega_{K/k} & \xrightarrow{\psi} & \Omega_{F/k} \end{array}$$

commutes. The image of ψ contains a generating set for $\Omega_{F/k}$, hence $a : \Omega_{K/k} \otimes_R F \to \Omega_{F/k}$ is onto. By Theorem 8.3.3, $\Omega_{F/K} = 0$. By Part (3), F/K is separable and finite dimensional. Since $n = \operatorname{tr.deg}_k F$, this implies that the set $\xi_1, \dots, \xi_n$ is a transcendence base for F/k (see, for example, [**Hun80**, Section VI.1]). □

Proposition 8.4.3 (S. MacLane). *Let k be a field and $F = k(a_1, \dots, a_n)$ a finitely generated extension field of k. If F/k is separably generated, then* *there exists a subset of* $\{a_1, \ldots, a_n\}$ *which is a separating transcendence base for* F/k.

Proof. Let $r = \operatorname{tr.deg}_k(F)$. Let $S = k[x_1, \ldots, x_n]$ be the polynomial ring over k in n indeterminates. Define $\phi : S \to F$ by $x_i \mapsto a_i$. Since the image of ϕ is $k[a_1, \ldots, a_n]$, an integral domain, the kernel of ϕ is a prime ideal P of S. The ideal P is finitely generated, hence we can write $P = (f_1, \ldots, f_m)$. Let $A = S/P$. Then F is the quotient field of A. The sequence

$$P/P^2 \xrightarrow{\gamma} \Omega_{S/k} \otimes_S A \xrightarrow{a} \Omega_{A/k} \to 0 \tag{8.13}$$

of Theorem 8.3.4 is exact, $\Omega_{S/k} \otimes_S A$ is a free A-module, and $\{dx_1, \ldots, dx_n\}$ is a free basis. For each i,

$$\gamma(f_i) = \sum_{j=1}^{n} \frac{\partial f_i}{\partial x_j} dx_j.$$

Tensor (8.13) with $(\ ) \otimes_A F$. The sequence

$$F^{(m)} \xrightarrow{J} F^{(n)} \to \Omega_{F/k} \to 0$$

is exact, where J is multiplication by the jacobian matrix $J = (\partial f_i/\partial x_j)$. Since F/k is separably generated, by Theorem 8.4.2, the rank of J is $n - r$. This implies there exists an $(n-r)$-by-$(n-r)$ submatrix of J which also has rank $n-r$. Relabel the x_i if necessary and assume the rank of the submatrix

$$(\partial f_i/\partial x_j \mid 1 \le i \le n - r, r + 1 \le j \le n)$$

is $n - r$. The proof of Theorem 8.4.2 shows the set $\{d_{F/k}(a_1), \ldots, d_{F/k}(a_r)\}$ is a basis for $\Omega_{F/k}$ over F and $a_1, \ldots, a_r$ is a separating transcendence base for F/k. □

Lemma 8.4.4. *Let k be a field and $F = k(a_1, \ldots, a_n)$ a finitely generated extension field of k. If $\operatorname{tr.deg}_k F = r$ and F/k is not separably generated, then upon relabeling the a_i, the field $k(a_1, \ldots, a_{r+1})$ is of transcendence degree r over k, and is not separably generated over k.*

Proof. The proof is by induction on n. If $n = r + 1$, then there is nothing to prove. Assume $n > r + 1$ and that the result is true for $n - 1$. Relabel the a_i and assume a_1 is algebraically dependent on $a_2, \ldots, a_n$ over k. Then $k(a_2, \ldots, a_n)$ has transcendence degree r over k. If $k(a_2, \ldots, a_n)$ is not separably generated over k, then by induction we are done. Assume $k(a_2, \ldots, a_n)$ is separably generated over k. By Proposition 8.4.3, we can relabel the a_i and assume $a_2, \ldots, a_{r+1}$ is a separating transcendence base for $k(a_2, \ldots, a_n)$ over k. Then $k(a_2, \ldots, a_n)$ is separable and finite dimensional over $k(a_2, \ldots, a_{r+1})$. It follows that $k(a_1, a_2, \ldots, a_n)$ is separable and finite dimensional over $k(a_1, a_2, \ldots, a_{r+1})$. By the transitive property of separable field extensions (see, for example, [**Hun80**, Corollary V.6.8]), it follows that $k(a_1, a_2, \ldots, a_{r+1})$ is not separably generated over k. □

Theorem 8.4.5. *Let k be a perfect field, and F/k a finitely generated extension of fields.*

(1) *(F. K. Schmidt) F/k is separably generated.*
(2) *(Primitive Element Theorem) If $r = \operatorname{tr.deg}_k F$, then there exists a transcendence base $\Xi = \{\xi_1, \ldots, \xi_r\}$ for F/k, an element $u \in F$ which is separable over $k(\Xi)$, and $F = k(\Xi)[u]$.*

Proof. (1): Let $r = \operatorname{tr.deg}_k F$ and assume $F = k(a_1, \ldots, a_n)$. For contradiction's sake, assume F/k is not separably generated. Let $p = \operatorname{char} k$. By Lemma 8.4.4, we reduce to the case where $n = r + 1$. Let $S = k[x_1, \ldots, x_n]$ be the polynomial ring over k in n indeterminates. Define $\phi : S \to F$ by $x_i \mapsto a_i$. Since the image of ϕ is $k[a_1, \ldots, a_n]$, an integral domain, the kernel of ϕ is a prime ideal P of S. By Noether's Normalization Lemma (Corollary 6.1.2 and Corollary 6.1.3), P has height one. Since S is a unique factorization domain, there exists an irreducible polynomial f in S such that $P = (f)$. View $f(a_1, \ldots, a_r, x_{r+1})$ as an element of $k(a_1, \ldots, a_r)[x_{r+1}]$. Since a_{r+1} is not separable over $k(a_1, \ldots, a_r)$, it follows that f is a polynomial in $k[x_1, \ldots, x_r][x_{r+1}^p]$. Iterate this argument $r + 1$ times. Then $f \in k[x_1^p, \ldots, x_r^p, x_{r+1}^p]$. Since k is perfect, $f = g^p$ for some $g \in S$, a contradiction.

(2): This follows from Part (1), Theorem 8.4.3, and the Primitive Element Theorem (for example, [**DF04**, Theorem 14.4.25]). □

4.1. Emmy Noether's Normalization Lemma.

We prove a second version of Emmy Noether's Normalization Lemma (Corollary 6.1.2). It requires the ground field to be infinite. The advantage of this version is that it allows us to construct the underlying polynomial ring in such a way that it contains a separating transcendence base.

Theorem 8.4.6 (Emmy Noether's Normalization Lemma). *Let k be an infinite field and A a finitely generated commutative k-algebra. Assume A is an integral domain with field of fractions K. Then there exist $z_1, \ldots, z_m$ in A such that*

(1) *the set $\{z_1, \ldots, z_m\}$ is algebraically independent over k,*
(2) *A is integral over $k[z_1, \ldots, z_m]$,*
(3) $\dim(A) = m$,
(4) $\operatorname{tr.deg}_k(K) = m$, *and*

(5) *if A is generated as a k-algebra by $x_1, \ldots, x_n$, then there are elements a_{ij} in k such that $z_i = \sum_{j=1}^n a_{ij} x_j$,*

(6) *if K is separably generated over k, then $\{z_1, \ldots, z_m\}$ can be chosen in such a way that K is separable over $k(z_1, \ldots, z_m)$.*

Proof. We prove (6). The other cases are left to the reader. Our proof is based on [**ZS75a**, I, Chapter V, Theorem 8, p. 266]. Let $x_1, \ldots, x_n$ be a generating set for A as a k-algebra. By Proposition 8.4.3, resort the list and assume $\{x_1, \ldots, x_m\}$ is a separating transcendence base for K over k. Proceed by induction on n. If $m = n$, then take $z_i = x_i$, for $1 \le i \le m$, and stop. Otherwise, assume $n > m$ and assume the claim is true for any algebra on $n-1$ generators. Then each of $x_{m+1}, \ldots, x_n$ is algebraic over $k(x_1, \ldots, x_m)$.

Let $A' = k[x_1, \ldots, x_{n-1}]$, and K' the field of fractions of A'. By assumption, x_n is separable over K'. Starting with the minimum polynomial for x_n over K', we can find a polynomial P in $k[X_1, \ldots, X_n]$ such that $P(x_1, \ldots, x_{n-1}, X_n)$ is a separable polynomial in $K'[X_n]$ and such that $P(x_1, \ldots, x_{n-1}, x_n) = 0$. Write P as a sum

$$P(X_1, \ldots, X_n) = \sum_{i=0}^{q} P_i(X_1, \ldots, X_n) \tag{8.14}$$

where $P_i(X_1, \ldots, X_n)$ is a homogeneous polynomial of degree i in the polynomial ring $k[X_1, \ldots, X_n]$, and $P_q \neq 0$. Introduce new indeterminates $Z_1, \ldots, Z_{n-1}, \Lambda_1, \ldots, \Lambda_{n-1}$ and define an embedding of k-algebras

$$\begin{aligned} \theta : k[X_1, \ldots, X_n] &\to k[Z_1, \ldots, Z_{n-1}, \Lambda_1, \ldots, \Lambda_{n-1}, X_n] \\ X_1 &\mapsto Z_1 + \Lambda_1 X_n \\ &\vdots \\ X_{n-1} &\mapsto Z_{n-1} + \Lambda_{n-1} X_n. \end{aligned}$$

If we denote by F the image of P under θ, then

$$\begin{aligned} F &= F(Z_1, \ldots, Z_{n-1}, \Lambda_1, \ldots, \Lambda_{n-1}, X_n) \\ &= P(Z_1 + \Lambda_1 X_n, \ldots, Z_{n-1} + \Lambda_{n-1} X_n, X_n) \\ &= \sum_{i=0}^{q} P_i(Z_1 + \Lambda_1 X_n, \ldots, Z_{n-1} + \Lambda_{n-1} X_n, X_n). \end{aligned} \tag{8.15}$$

Because each P_i is homogeneous of degree i, if we expand F as a polynomial in X_n, the highest degree term is

$$X_n^q P_q(\Lambda_1, \ldots, \Lambda_{n-1}, 1). \tag{8.16}$$

By $\mathbb{A}_k^{n-1}$ we denote affine $n-1$-space over k with the Zariski topology (see, for example, [**DF04**, Section 15.2]). The zero set of $P_q(\Lambda_1, \ldots, \Lambda_{n-1}, 1)$ in

$\mathbb{A}_k^{n-1}$ is a closed subset, call it V_1. Because $P_q(\Lambda_1, \ldots, \Lambda_{n-1}, 1)$ is a nonzero polynomial and k is infinite, $V_1 \neq \mathbb{A}_k^{n-1}$. There exists $(\lambda_1, \ldots, \lambda_{n-1}) \in \mathbb{A}_k^{n-1}$ such that if we set $z_1 = x_1 - \lambda_1 x_n, z_{n-1} = x_{n-1} - \lambda_{n-1} x_n$, then

$$F(z_1, \ldots, z_{n-1}, \lambda_1, \ldots, \lambda_{n-1}, X_n) \tag{8.17}$$

is a polynomial of degree q in $k[z_1, \ldots, z_{n-1}][X_n]$ and the leading coefficient is a nonzero element of k. Since $F(z_1, \ldots, z_{n-1}, \lambda_1, \ldots, \lambda_{n-1}, x_n) = P(x_1, \ldots, x_n) = 0$, this shows x_n is integral over $k[z_1, \ldots, z_{n-1}]$. To finish the proof, we show that there exists a choice for $(\lambda_1, \ldots, \lambda_{n-1})$ such that x_n is a simple root of the polynomial in (8.17). In (8.15), compute the derivative of F with respect to X_n:

$$\frac{\partial F}{\partial X_n} = \sum_{i=1}^{n-1} \Lambda_i \frac{\partial P}{\partial X_i} + \frac{\partial P}{\partial X_n}. \tag{8.18}$$

Substituting $X_1 = x_1, \ldots, X_n = x_n$, we have

$$\frac{\partial F}{\partial X_n}(x_1, \ldots, x_n) = \sum_{i=1}^{n-1} \Lambda_i \frac{\partial P}{\partial X_i}(x_1, \ldots, x_n) + \frac{\partial P}{\partial X_n}(x_1, \ldots, x_n). \tag{8.19}$$

which is a linear polynomial in $k[\Lambda_1, \ldots, \Lambda_{n-1}]$. The polynomial (8.19) is not identically zero, because for $\Lambda_1 = 0, \ldots, \Lambda_{n-1} = 0$ it evaluates to $\partial P / \partial X_n(x_1, \ldots, x_n)$ which is nonzero since x_n is separable over K'. The zero set of (8.19) in $\mathbb{A}_k^{n-1}$ is a proper closed subset, call it V_2. Since $V_1 \cup V_2$ is the zero set of a nonzero polynomial in $k[\Lambda_1, \ldots, \Lambda_{n-1}]$, it is a proper closed subset. Therefore, there is a point $(\lambda_1, \ldots, \lambda_{n-1})$ such that (8.19) is nonzero and x_n is a simple root of the polynomial (8.17). □

As an application, we get the following finiteness theorem for the integral closure of an integral domain in an extension of its quotient field. Theorem 8.4.7, which requires A to be a finitely generated algebra over a field, is a strong version of Theorem 4.6.10.

Theorem 8.4.7. *Let A be an integral domain which is a finitely generated algebra over a field k. Let K be the quotient field of A, and let L be a finitely generated algebraic extension of K. If S is the integral closure of A in L, then S is a finitely generated A-module, and is also a finitely generated k-algebra.*

Proof. Our proof is based on [**ZS75a**, I, Chapter V, Theorem 9, p. 267]. By the proof of Theorem 4.6.10, there are elements $\lambda_1, \ldots, \lambda_n$ in S which generate L as a vector space over K. Let B be the A-subalgebra of L generated by $\lambda_1, \ldots, \lambda_n$. Then B is finitely generated as an A-module, finitely generated as a k-algebra, L is the field of fractions of B, and S is the integral closure of B in L. After replacing A with B and K with L, we assume S is the integral closure of A in K. It is enough to show S is finitely generated as an A-module.

Let Ω be an algebraically closed field containing K. For the remainder of this proof, every k-algebra is tacitly assumed to be a subring of Ω. Assume A is generated as a k-algebra by $x_1, \dots, x_n$. Let $\bar{k}$ be the algebraic closure of k, and $\bar{A}$ the $\bar{k}$-algebra generated by $x_1, \dots, x_n$. Let $\bar{K}$ be the field of fractions of $\bar{A}$. By Theorem 8.4.5, $\bar{K}$ is separably generated over $\bar{k}$. By Theorem 8.4.6, there are elements $z_1, \dots, z_m$ in $\bar{A}$ which satisfy:

(a) $\bar{k}[z_1, \dots, z_m]$ is a polynomial subring of $\bar{A}$,

(b) $\bar{A}$ is integral over $\bar{k}[z_1, \dots, z_m]$,

(c) there are elements a_{ij} in $\bar{k}$ such that $z_i = \sum_{j=1}^{n} a_{ij}x_j$, for $1 \le i \le m$,

(d) $\bar{K}$ is separable over $\bar{k}(z_1, \dots, z_m)$.

Let P_j be the minimum polynomial for x_j over $\bar{k}(z_1, \dots, z_m)$. By Theorem 3.3.4, P_j is a polynomial with coefficients in $\bar{k}[z_1, \dots, z_m]$. Let F be the subfield of $\bar{k}$ generated by adjoining to k all of the elements a_{ij} of (c), and all of the $\bar{k}$-coefficients that appear in $P_1, \dots, P_n$. Let A' be the F-algebra generated by $x_1, \dots, x_n$ and let K' be the field of fractions of A'. By construction, we have:

(e) $F[z_1, \dots, z_m]$ is a polynomial subring of A',

(f) A' is integral over $F[z_1, \dots, z_m]$, and

(g) K' is separable over $F(z_1, \dots, z_m)$.

Let T be the integral closure of $F[z_1, \dots, z_m]$ in K'. By Theorem 4.6.10, T is a finitely generated $F[z_1, \dots, z_m]$-module. By (f), T contains A', hence T is a finitely generated A'-module. Since $\dim_k(F)$ is finite, A' is a finitely generated A-module. Therefore, T is a finitely generated A-module. Since $S = T \cap K$, S is an A-submodule of T. Since A is noetherian, S is a finitely generated A-module (Corollary 2.5.2). □

4.2. Algebraic Curves. As an application of Theorem 8.4.7, we prove the algebraic analog of the fact that a nonsingular projective algebraic curve over a field has only finitely many "points at infinity". Consider a Dedekind domain A which is finitely generated as an algebra over a field k. By Noether Normalization (Corollary 6.1.2) there is an element $\alpha \in A$ which is transcendental over k and A is an integral extension of $k[\alpha]$. The field of fractions of A is a finitely generated algebraic extension of $k(\alpha)$. In the following, $X_1(K)$ denotes the set of all discrete valuation rings R in K containing k as a subring. The following lemmas will be applied in Section 14.3.4.

Lemma 8.4.8. *Let k be any field, α an indeterminate, and K a finitely generated algebraic extension field of $k(\alpha)$. Let $x \in K$. Then the set of all $R \in X_1(K)$ such that x is not in R is a finite set.*

Proof. The idea for this proof comes from [**Har77**, §I.6]. Given $R \in X_1(K)$, let $\mathfrak{m}_R$ denote the maximal ideal of R. Then $x \notin R$ if and only if $x^{-1} \in \mathfrak{m}_R$. Letting $y = x^{-1}$, it is equivalent to show that the set $\{R \in X_1(K) \mid y \in \mathfrak{m}_R\}$ is finite. For every $R \in X_1(K)$, R is integrally closed in K, so R contains $\bar{k}$, the algebraic closure of k in K. If y is algebraic over k, then $y \in \bar{k} \subseteq R$, hence y is a unit in R, and $y \notin \mathfrak{m}_R$. Now assume y is transcendental over k. Since K has transcendence degree one over k, it follows that K is finitely generated and algebraic over $k(y)$. Let A be the integral closure of $k[y]$ in K. Then A is finitely generated as a $k[y]$-module and as a k-algebra (Theorem 8.4.7). Then A is an integrally closed noetherian integral domain with Krull dimension one, hence A is a Dedekind domain. By Theorem 6.2.1, y belongs to only finitely many maximal ideals in A. We show that there is a one-to-one correspondence between maximal ideals of A containing y and the set of $R \in X_1(K)$ such that $y \in \mathfrak{m}_R$. Let $R \in X_1(K)$ such that $y \in \mathfrak{m}_R$. Then $k[y] \subseteq R$ and since R is integrally closed in K, $A \subseteq R$. Let $\mathfrak{n} = \mathfrak{m}_R \cap A$. Then $\mathfrak{n}$ is a prime ideal of A containing y, so $\mathfrak{n}$ is a maximal ideal. Localizing, $A_\mathfrak{n} \subseteq R$ and $A_\mathfrak{n}$ is dominated by R. By Lemma 6.1.12, $A_\mathfrak{n} = R$. □

Lemma 8.4.9. *Let k be any field, α an indeterminate, and K a finitely generated algebraic extension field of $k(\alpha)$. Let A be the integral closure of $k[\alpha]$ in K. Then the set of all $R \in X_1(K)$ which do not contain A is a finite set.*

Proof. As in the proof of Lemma 8.4.8, A is a Dedekind domain with field of fractions K and finitely generated as a k-algebra. Let $U = \{A_\mathfrak{p} \mid \mathfrak{p} \in X_1(A)\}$ be the set of localizations of A at its maximal ideals. Then U is a subset of $X_1(K)$. We need to prove that $X_1(K) - U$ is a finite set.

Let $R \in X_1(K)$ with maximal ideal $\mathfrak{m}_R$. Assume $A \subseteq R$. Consider $\mathfrak{m}_R \cap A = \mathfrak{n}$. Then $\mathfrak{n}$ is a prime ideal of A. Since A is a Dedekind domain, $\operatorname{ht}(\mathfrak{n})$ is 0 or 1. If $x \in A - \mathfrak{n}$, then we see that x is an element of R, since $A \subseteq R$. Then $x \notin \mathfrak{m}_R$, so $x^{-1} \in R$. This shows the localization of A at the prime ideal $\mathfrak{n}$ is a subring of R. In other words, $\mathfrak{n} \neq (0)$. Hence $\mathfrak{n}$ is a maximal ideal of A and $A_\mathfrak{n} = R$. So there is a one-to-one correspondence between maximal ideals of A and $R \in X_1(K)$ such that $A \subseteq R$. Therefore, $U = \{R \in X_1(K) \mid A \subseteq R\}$. Let $a_1, \dots, a_n$ be a generating set for A as a k-algebra. Given $R \in X_1(K)$, we have $A \subseteq R$ if and only if $\{a_1, \dots, a_n\} \subseteq R$. Let $U_i = \{R \in X_1(K) \mid a_i \in R\}$. By Lemma 8.4.8, $X_1(K) - U_i$ is a finite set. Since $U = U_1 \cap \cdots \cap U_n$, it follows that $X_1(K) - U$ is a finite set. □

5. Tests for Regularity

In this section, all rings are commutative. Suppose R is a local ring with maximal ideal $\mathfrak{m}$ and residue field $k = R/\mathfrak{m}$. If R is noetherian and has Krull dimension $\dim(R) = d$, then R is a regular local ring if and only if $\dim_k(\mathfrak{m}/\mathfrak{m}^2) = d$ (see, for example, [**AM69**, Theorem 11.22]). The results in this section will be applied in Section 9.2.3.

5.1. A Differential Criterion for Regularity.

As above, let R be a local ring with maximal ideal $\mathfrak{m}$. A *coefficient field* of R is a subfield k of R which is mapped onto $R/\mathfrak{m}$ under the natural map $R \to R/\mathfrak{m}$. In this case, R is a k-algebra, and $k \to R/\mathfrak{m}$ is a k-algebra isomorphism. The reader should verify that every $x \in R$ has a unique representation in the form $x = y + z$, where $y \in k$ and $z \in \mathfrak{m}$.

Proposition 8.5.1. *Let R be a local ring with maximal ideal $\mathfrak{m}$ and assume R contains a coefficient field k. Then the k-linear map*

$$\mathfrak{m}/\mathfrak{m}^2 \xrightarrow{\gamma} \Omega_{R/k} \otimes_R k$$

of Theorem 8.3.4 is an isomorphism.

Proof. The cokernel of γ is $\Omega_{k/k}$ which is 0, so γ is onto. To show γ is one-to-one, it is enough to apply the exact functor $\operatorname{Hom}_k(\cdot, k)$ and show that

$$\operatorname{Hom}_k(\Omega_{R/k} \otimes_R k, k) \xrightarrow{\mathrm{H}_\gamma} \operatorname{Hom}_k(\mathfrak{m}/\mathfrak{m}^2, k)$$

is onto. As in the proof of Theorem 8.3.4, the map H_γ is isomorphic to

$$\operatorname{Der}_k(R, k) \xrightarrow{\rho} \operatorname{Hom}_R(\mathfrak{m}, k)$$

where ρ is defined by $\partial \mapsto \partial|_{\mathfrak{m}}$. It suffices to show ρ is onto. Let $h \in \operatorname{Hom}_R(\mathfrak{m}, k)$. Given $x \in R$, write $x = y + z$, where $y \in k$ and $z \in \mathfrak{m}$. This representation is unique. Define $\partial : R \to k$ by $\partial(x) = h(z)$. It is easy to see that ∂ is a well defined function that extends h, and $\partial(k) = 0$. The reader should verify that ∂ is a k-derivation on R. □

Theorem 8.5.2. *Let R be a local ring with maximal ideal $\mathfrak{m}$ and assume R contains a coefficient field k which is a perfect field. Assume R is a localization of a finitely generated k-algebra. The following are equivalent.*

(1) *R is regular.*
(2) *$\Omega_{R/k}$ is a free R-module of rank $d = \dim(R)$.*

Proof. We know R is noetherian and of finite Krull dimension (see, for example, [**AM69**, Chapter 11]).

(2) implies (1): By Proposition 8.5.1, $\dim_k(\mathfrak{m}/\mathfrak{m}^2) = d$ and R is regular.

(1) implies (2): Assume $\dim_k(\mathfrak{m}/\mathfrak{m}^2) = d$. By Proposition 8.5.1, it follows that $\dim_k(\Omega_{R/k} \otimes_R k) = d$. Since R is regular, R is a normal integral domain (see, for example, [**Mat80**, (17.D), Theorem 34 and (17.E), Theorem 35]). Let K be the quotient field of R. By Exercise 8.3.11, $\Omega_{R/k} \otimes_R K = \Omega_{K/k}$. By Theorem 8.4.5 and Theorem 8.4.2, $\dim_K(\Omega_{K/k}) = \operatorname{tr.deg}_k(K)$. By Noether's Normalization Lemma (Corollary 6.1.2), $d = \operatorname{tr.deg}_k(K)$. By Proposition 8.3.2 and Exercise 8.3.11, $\Omega_{R/k}$ is a finitely generated R-module. By Corollary 2.6.5, $\Omega_{R/k}$ is a free R-module of rank d. □

5.2. A Jacobian Criterion for Regularity. Throughout this section, k is an algebraically closed field, and all rings are commutative. From a utilitarian point of view, the jacobian criterion of Theorem 8.5.3 is one of the most useful and powerful methods for showing that a finitely generated k-algebra R is regular. Before stating the theorem, it is first necessary to understand the relationship between the maximal ideals of the ring R and the points on an affine algebraic set corresponding to R.

Define *affine n-space over k* to be $\mathbb{A}^n = \{(a_1, \dots, a_n) \mid a_i \in k\}$. For any subset $Y \subseteq \mathbb{A}^n$, the ideal of Y in $A = k[x_1, \dots, x_n]$ is defined by

$$I(Y) = \{f \in A \mid f(P) = 0, \text{ for all } P \in Y\}.$$

This is an ideal, as is easily checked. One can check that $I(Y)$ is equal to its own nil radical, hence $I(Y)$ is a radical ideal (see Section 2.2). If $T \subseteq A$, then the *zero set* of T is

$$Z(T) = \{P \in \mathbb{A}^n \mid f(P) = 0, \text{ for all } f \in T\}.$$

If I is the ideal generated by T in A, then $Z(I) = Z(T)$. An *affine algebraic set* is any $Y \subseteq \mathbb{A}^n$ which is equal to $Z(T)$ for some set $T \subseteq A$. Hilbert's Nullstellensatz (see, for instance, [**DF04**, Theorem 15.32.]) says that the assignments $Y \mapsto I(Y)$ and $I \mapsto Z(I)$ define a one-to-one correspondence between the algebraic sets in $\mathbb{A}^n$ and the radical ideals in A.

If $Y \subseteq \mathbb{A}^n$ is a affine algebraic set, then the *affine coordinate ring* of Y is $\mathcal{O}(Y) = A/I(Y)$. Now assume I is a radical ideal in A, and $Y = Z(I)$ is the associated affine algebraic set. Then $I = I(Y)$ and $\mathcal{O}(Y) = A/I$. A corollary to Hilbert's Nullstellensatz says that the maximal ideals in $\mathcal{O}(Y) = A/I$ are in one-to-one correspondence with the points $P \in Y$. A point $P = (a_1, \dots, a_n) \in Y$, corresponds to the maximal ideal $\mathfrak{m}$ in $\mathcal{O}(Y)$ generated by $x_1 - a_1, \dots, x_n - a_n$. The localization of $\mathcal{O}(Y)$ at the maximal ideal $\mathfrak{m}$ is called the *local ring* at P on Y and is denoted $\mathcal{O}_{P,Y}$. Theorem 8.5.3 is a jacobian criterion for $\mathcal{O}_{P,Y}$ to be a regular local ring. By Theorem 6.1.9, $\mathcal{O}(Y)$ is a regular ring if $\mathcal{O}_{P,Y}$ is a regular local ring for each $P \in Y$. For an elementary proof of the following theorem, see [**Har77**, Theorem I.5.1].

Theorem 8.5.3. *Let $Y \subseteq \mathbb{A}^n$ be an affine algebraic set and $f_1, \dots, f_t$ a set of generators for $I(Y)$. Let $P \in Y$ and assume the Krull dimension of the local ring $\mathcal{O}_{P,Y}$ is d. Then the jacobian matrix*

$$J = \left(\frac{\partial f_i}{\partial x_j}(P) \right)$$

has rank $n - d$ if and only if $\mathcal{O}_{P,Y}$ is a regular local ring.

Example 8.5.4. In the above context, let $F = Z(f)$ be an algebraic set in $\mathbb{A}^n$ defined by a square-free polynomial f in $A = k[x_1, \dots, x_n]$. Then $\dim(\mathcal{O}(F)) = n - 1$. Let $d \geq 2$ be an integer that is invertible in k. An application of the Eisenstein Criterion shows that $z^d - f$ is irreducible in $k[x_1, \dots, x_n, z]$. Let $Y = Z(z^d - f)$ be the corresponding algebraic set in $\mathbb{A}^{n+1}$. Then $\mathcal{O}(Y) = A[z]/(z^d - f)$ is an integral domain with Krull dimension n. If $\pi : \mathbb{A}^{n+1} \to \mathbb{A}^n$ is the projection along the z-axis defined by $(a_1, \dots, a_n, b) \mapsto (a_1, \dots, a_n)$, then $\pi^{-1}(F)$ is the algebraic subset of Y equal to $Y \cap Z(z)$. The jacobian of $z^d - f$ is $(f_{x_1}, \dots, f_{x_n}, dz^{d-1})$. By $\operatorname{Sing}(Y)$ (or $\operatorname{Sing}(F)$) we denote the set of points where the local ring on Y (or F) is not a regular local ring. From Theorem 8.5.3, we see at once that $P \in \operatorname{Sing}(Y)$ if and only if $P = (a_1, \dots, a_n, 0)$ and $\pi(P) = (a_1, \dots, a_n)$ is in $\operatorname{Sing}(F)$. Using Theorem 6.1.9 we see that $\mathcal{O}(Y)$ is a regular integral domain if and only if $\mathcal{O}(F)$ is a regular ring.

Example 8.5.5. Although the field in Theorem 8.5.3 is required to be algebraically closed, it is sometimes possible to work around this obstacle. In this paragraph, one such method is presented. Let k be a field and in this example do not assume k is algebraically closed. Let $\bar{k}$ be an algebraic closure of k. Let I be an ideal in $k[x_1, \dots, x_n]$ and $T = k[x_1, \dots, x_n]/I$. If $\bar{T} = T \otimes_k \bar{k}$, then the natural map $T \to \bar{T}$ is faithfully flat (Exercise 2.4.14). By Theorem 6.1.9 (3), if $\bar{T}$ is regular, then T is regular. By Lemma 3.3.11, if $\bar{T}$ is an integrally closed integral domain, then T is an integrally closed integral domain. By Lemma 3.3.11, if $\bar{T}$ is a normal ring, then T is a normal ring.

Example 8.5.6. Let k be a field and $\bar{k}$ an algebraic closure of k. Let $\lambda_1, \dots, \lambda_v$ be distinct elements of k, where $v \geq 2$, and for each i, let $\ell_i = x - \lambda_i y$. Then $f(x, y) = \ell_1 \cdots \ell_v \in k[x, y]$ is a homogeneous polynomial of degree v in $k[x, y]$. Let $d \geq 2$ be an integer that is invertible in k, and consider the noetherian ring $T = k[x, y, z]/(z^d - f(x, y))$. Since f is square-free and $\bar{k}[x, y]$ is a UFD, by Eisenstein's Criterion, $\bar{T} = T \otimes_k \bar{k}$ is an integral domain. Using Theorem 8.5.3, the reader should verify that there is only one maximal ideal in $\operatorname{Max}(\bar{T})$ for which the local ring is not regular, namely, the ideal $\mathfrak{m} = (x, y, z)$. In particular, $\bar{T}_{\mathfrak{p}}$ is regular for every height one prime ideal $\mathfrak{p}$. By an application of Serre's Criteria (Theorem 6.1.5), the reader should verify that $\bar{T}$ is an integrally closed integral domain. As mentioned in Example 8.5.5, this implies T is an integrally closed integral domain.

Chapter 9

Étale Algebras

This chapter contains results that will be applied in Chapters 10 through 14. A commutative finitely presented R-algebra S is defined to be étale if S is separable and flat (see Definition 9.2.3 below). Our major focus in this chapter is to prove the fundamental properties of étale algebras. An additional motivation is to show that if an algebra is formally étale (see Definition 4.7.1), then it is étale. Hence, this chapter is a continuation of the study of formally smooth and formally étale algebras that was begun in Section 4.7. On this topic the main result (Theorem 9.2.19) demonstrates that if R is noetherian and S is finitely generated, then S is étale if and only if S is formally étale. Section 9.3 contains the important theorems of permanence. These theorems outline those properties of a ring R that any étale R-algebra also possesses. As an application of the theorems of permanence, we include in Section 9.4 a useful technique for constructing an extension of a noetherian normal domain which is itself a noetherian normal domain.

1. Complete Noetherian Rings

The goal of this section is to derive sufficient conditions on a directed system of noetherian local rings in order to guarantee that the direct limit is again a noetherian local ring. This result, which is stated below as Corollary 9.1.6, will have many important applications in Chapter 10. It will be applied for instance, to prove that the henselization of a noetherian local ring is a noetherian local ring. For this reason, and for lack of a reference with a suitably accessible proof, we include one here. The proof we give is a compilation of results from all of the following sources: [**Mat80**], [**Bou89b**], [**Ray70**], and [**Gro61**].

Let R be a commutative ring and I an ideal of R. Let M be an R-module. In Example 3.4.10 and Example 3.4.11 we defined the associated graded ring

$$\operatorname{gr}_I(R) = \bigoplus_{n \geq 0} I^n / I^{n+1}$$

and the associated graded module

$$\operatorname{gr}_I(M) = \bigoplus_{n=0}^{\infty} I^n M / I^{n+1} M.$$

Then $\operatorname{gr}_I(M)$ is a graded $\operatorname{gr}_I(R)$-module. For the following, set $R_0 = \operatorname{gr}_I(R)_0 = R/I$ and $M_0 = \operatorname{gr}_I(M)_0 = M/I$. The ring $\operatorname{gr}_I(R)$ is an R_0-algebra, and M_0 is an R_0-module. For all $n \geq 0$, the multiplication map

$$\mu_{n0} : \frac{I^n}{I^{n+1}} \otimes_{R_0} M_0 \to \frac{I^n M}{I^{n+1} M}$$

is onto. Taking the direct sum, there is a surjective degree-preserving homomorphism

$$\mu : \operatorname{gr}_I(R) \otimes_{R_0} M_0 \to \operatorname{gr}_I(M)$$

of R_0-modules. We say that M is *ideal-wise separated for I* if for each finitely generated ideal J of R, the R-module $J \otimes_R M$ is separated in the I-adic topology.

Example 9.1.1. Some examples of modules that are ideal-wise separated are listed here.

(1) Let S be a commutative R-algebra and M a finitely generated S-module. Suppose S is noetherian and I is an ideal of R such that $IS \subseteq \mathrm{J}(S)$. Let L be any ideal of R. The reader should verify that the I-adic topology on $L \otimes_R M$ is equal to the $I \otimes_R S$-adic topology, which is equal to the IS-adic topology. Since $L \otimes_R M$ is a finitely generated S-module, the Krull Intersection Theorem (see, for example, [**Mat80**, (11.D), Corollaries 1, 2 and 3]) says $J \otimes_R M$ is separated in the I-adic topology. Therefore M is ideal-wise separated for I.

(2) Let R be a commutative ring and M a flat R-module. If J is an ideal of R, then $0 \to J \otimes_R M \to M \to M/JM \to 0$ is exact. That is, $J \otimes_R M = JM$. If I is an ideal of R and M is separated for the I-adic topology, then $I^n JM \subseteq I^n M$ so JM is separated for the I-adic topology. Therefore M is ideal-wise separated for I.

(3) Let R be a principal ideal domain. Let I and J be ideals of R and M an R-module. If $w \in I^n(J \otimes_R M)$, then w can be written in the form $1 \otimes z$ where $z \in I^n M$. If M is separated in the I-adic topology, then M is ideal-wise separated for I.

The first main theorem of this section is Theorem 9.1.2, which is a local criteria for flatness. Its proof can be found in [**Mat80**, (20.C), Theorem 49].

Theorem 9.1.2. *Let R be a commutative ring, I an ideal of R, and M an R-module. Let $\operatorname{gr}_I(M)$ be the associated graded $\operatorname{gr}_I(R)$-module. Set $R_0 = R/I$ and $M_0 = M/I$. Assume*

(A) *I is nilpotent, or*

(B) *R is noetherian and M is ideal-wise separated for I.*

Then the following are equivalent.

(1) *M is a flat R-module.*

(2) *$\operatorname{Tor}_1^R(N, M) = 0$ for all R_0-modules N.*

(3) *M_0 is a flat R_0-module and $0 \to I \otimes_R M \to IM$ is an exact sequence.*

(4) *M_0 is a flat R_0-module and $\operatorname{Tor}_1^R(R_0, M) = 0$.*

(5) *M_0 is a flat R_0-module and the multiplication maps*

$$\mu_{n0} : \frac{I^n}{I^{n+1}} \otimes_{R_0} M_0 \to \frac{I^n M}{I^{n+1} M}$$

are isomorphisms for all $n \geq 0$.

(6) *$M_n = M/I^{n+1}M$ is a flat $R_n = R/I^{n+1}$-module for each $n \geq 0$.*

Now we prove two finiteness theorems for the inverse limit of modules.

Proposition 9.1.3. *Let $\{A_i, \phi_i^j\}$ be an inverse system of discrete commutative rings for the index set $\{0, 1, 2, \ldots\}$. Let $\{M_i, \psi_i^j\}$ be an inverse system of modules over the inverse system of rings $\{A_i, \phi_i^j\}$. For each $0 \leq i \leq j$, define $\mathfrak{n}_j$ to be the kernel of $\phi_0^j : A_j \to A_0$, assume $\phi_i^i : A_i \to A_i$ is the identity mapping, and*

$$0 \to \mathfrak{n}_j^{i+1} \to A_j \xrightarrow{\phi_i^j} A_i \to 0$$

and

$$0 \to \mathfrak{n}_j^{i+1} M_j \to M_j \xrightarrow{\psi_i^j} M_i \to 0$$

are exact sequences. If $A = \varprojlim A_i$ and $M = \varprojlim M_i$, then the following are true.

(1) *A is a separated and complete topological ring, M is a separated and complete topological A-module, and the natural maps $\alpha_j : A \to A_j$, $\beta_j : M \to M_j$, are onto.*

(2) *If M_0 is a finitely generated A_0-module, then M is a finitely generated A-module. More specifically, if S is a finite subset of M and $\beta_0(S)$ is a generating set for M_0, then S is a generating set for M.*

Proof. (1): This follows from Proposition 3.4.4, Corollary 3.4.7, and the definition of inverse limit.

(2): For all $\ell \le k$, the diagram

$$\begin{array}{ccccccccc} 0 & \longrightarrow & \mathfrak{n}^{i+1}_{i+\ell} & \longrightarrow & A_{i+\ell} & \xrightarrow{\phi^{i+\ell}_i} & A_i & \longrightarrow & 0 \\ & & \downarrow & & \downarrow \phi^{i+\ell}_{i+k} & & \downarrow \phi^i_i & & \\ 0 & \longrightarrow & \mathfrak{n}^{i+1}_{i+k} & \longrightarrow & A_{i+k} & \xrightarrow{\phi^{i+k}_i} & A_i & \longrightarrow & 0 \end{array}$$

commutes and the vertical arrows are onto. By Proposition 1.4.8, if we define $\mathfrak{m}_{i+1}$ to be the kernel of $\alpha_i : A \to A_i$, then

$$\mathfrak{m}_{i+1} = \varprojlim_k \mathfrak{n}^{i+1}_{i+k}.$$

Similarly, if we set N_{i+1} to be the kernel of $\beta_i : M \to M_i$, then

$$N_{i+1} = \varprojlim_k \mathfrak{n}^{i+1}_{i+k} M_{i+k}.$$

It follows from the commutative diagram

$$\begin{array}{ccccccccc} 0 & \longrightarrow & \mathfrak{m}_{i+k+1} & \longrightarrow & A & \xrightarrow{\alpha_{i+k}} & A_{i+k} & \longrightarrow & 0 \\ & & \downarrow & & \downarrow = & & \downarrow \phi^{i+k}_i & & \\ 0 & \longrightarrow & \mathfrak{m}_{i+1} & \longrightarrow & A & \xrightarrow{\alpha_i} & A_i & \longrightarrow & 0 \end{array}$$

that

$$\alpha_{i+k}(\mathfrak{m}_{i+1}) = \ker \phi^{i+k}_i = \mathfrak{n}^{i+1}_{i+k}. \tag{9.1}$$

Likewise,

$$\beta_{i+k}(N_{i+1}) = \mathfrak{n}^{i+1}_{i+k} M_{i+k}. \tag{9.2}$$

For $i \ge 1$ and $j \ge 1$,

$$\begin{aligned} \beta_{i+j-1}(\mathfrak{m}_i N_j) &= \alpha_{i+j-1}(\mathfrak{m}_i)\beta_{i+j-1}(N_j) \\ &= \mathfrak{n}^i_{i+j-1}\mathfrak{n}^j_{i+j-1} M_{i+j-1} \\ &= \mathfrak{n}^{i+j}_{i+j-1} M_{i+j-1} \\ &= 0 \end{aligned}$$

since $\mathfrak{n}^{i+j}_{i+j-1}$ is the kernel of α^{i+j}_{i+j}. This shows that $\mathfrak{m}_i N_j \subseteq \ker \beta_{i+j-1} = N_{i+j}$. Similarly, one checks that $\mathfrak{m}_i \mathfrak{m}_j \subseteq \mathfrak{m}_{i+j}$. Defining $\mathfrak{m}_0 = A$, and $N_0 = M$, $\{\mathfrak{m}_i\}$ is a filtration on A and $\{N_i\}$ is a compatible filtration on M. The reader should verify that the topologies on A and M are those defined by the filtrations $\{\mathfrak{m}_i\}$ and $\{N_i\}$.

Let S be a finite subset of M and assume $\beta_0(S)$ is a generating set for M_0. Let M' be the submodule of M generated by S. Let $\mathfrak{a}$ be an ideal in A such that $\alpha_1(\mathfrak{a}) = \mathfrak{n}_1$. We are going to prove

$$N_i = \mathfrak{a}^i M' + N_{i+1} \tag{9.3}$$

for all $i \geq 0$. Define $\mathfrak{a}_i = \alpha_i(\mathfrak{a})$ and $M'_i = \beta_i(M')$. Since $N_{i+1} = \ker \beta_i$, to prove (9.3) it suffices to prove

$$\beta_i(N_i) = \beta_i(\mathfrak{a}^i M') = \alpha_i(\mathfrak{a}^i)\beta_i(M') = \mathfrak{a}_i^i M'_i. \tag{9.4}$$

Since $\beta_0(N_0) = \beta_0(M) = M_0$ is equal to $M'_0 = \beta_0(M') = M_0$, we see that (9.4) is satisfied for $i = 0$. For $i \geq 1$, the diagram

$$\begin{array}{ccccccccc}
0 & \longrightarrow & \mathfrak{n}_i & \longrightarrow & A_i & \xrightarrow{\phi_0^i} & A_0 & \longrightarrow & 0 \\
 & & \downarrow & & \downarrow{\scriptstyle \phi_1^i} & & \downarrow{\scriptstyle =} & & \\
0 & \longrightarrow & \mathfrak{n}_1 & \longrightarrow & A_1 & \xrightarrow{\phi_0^1} & A_0 & \longrightarrow & 0
\end{array}$$

commutes and the vertical arrows are onto. Therefore, $\phi_1^i(\mathfrak{n}_i) = \mathfrak{n}_1$. Since the diagram

$$\begin{array}{ccc}
A & \xrightarrow{\alpha_1} & A_1 \\
 & {\scriptstyle \alpha_i}\searrow & \uparrow{\scriptstyle \phi_1^i} \\
 & & A_i
\end{array}$$

commutes, $\phi_1^i(\mathfrak{n}_i) = \mathfrak{n}_1 = \alpha_1(\mathfrak{a}) = \phi_1^i \alpha_i(\mathfrak{a}) = \phi_1^i(\mathfrak{a}_i)$. Since $\mathfrak{n}_i^2 = \ker \phi_1^i$, it follows that $\mathfrak{n}_i = \mathfrak{a}_i + \mathfrak{n}_i^2$. For $i \geq 1$ the diagram

$$\begin{array}{ccc}
M & \xrightarrow{\beta_0} & M_0 \\
 & {\scriptstyle \beta_i}\searrow & \uparrow{\scriptstyle \psi_0^i} \\
 & & M_i
\end{array}$$

commutes and ψ_0^i is onto. Therefore, $\psi_0^i(M'_i) = \psi_0^i \beta_i(M') = \beta_0(M') = M_0 = \psi_0^i(M_i)$. Since $\mathfrak{n}_i M_i = \ker \psi_0^i$, it follows that $M_i = M'_i + \mathfrak{n}_i M_i$. Combining these results, we have

$$\mathfrak{n}_i^i M_i = (\mathfrak{a}_i + \mathfrak{n}_i^2)^i (M'_i + \mathfrak{n}_i M_i). \tag{9.5}$$

For $0 \leq k \leq i$ we have $\mathfrak{a}_i^k \mathfrak{n}_i^{i+1-k} \subseteq \mathfrak{n}_i^{i+1} = 0$. From this and (9.2), we see that (9.5) collapses to

$$\beta_i(N_i) = \mathfrak{n}_i^i M_i = \mathfrak{a}_i^i M'_i.$$

Together with (9.4), this proves (9.3).

From (9.1), $\mathfrak{m}_1 = \alpha_1^{-1}(\mathfrak{n}_1)$. Therefore, $\mathfrak{a} \subseteq \mathfrak{m}_1$, and $\mathfrak{a}^i \subseteq \mathfrak{m}_1^i \subseteq \mathfrak{m}_i$. From (9.3), this shows $N_i \subseteq m_i M' + N_{i+1}$. On the other hand, $m_i M \subseteq N_i$, from which it follows that

$$N_i = m_i M' + N_{i+1}.$$

It follows from Corollary 3.4.15 that $M' = M$. □

Corollary 9.1.4. *In the context of Proposition 9.1.3, assume M_0 is a finitely generated A_0-module and that the ideal $\mathfrak{n}_1$ of A_1 is finitely generated. Let $\mathfrak{m}_1$ be the kernel of $\alpha_0 : A \to A_0$. Then the following are true.*

(1) *The topologies on A and M are the $\mathfrak{m}_1$-adic topologies.*

(2) *For all $i \geq 0$, the sequences*

$$0 \to \mathfrak{m}_1^{i+1} \to A \xrightarrow{\alpha_i} A_i \to 0$$

and

$$0 \to \mathfrak{m}_1^{i+1} M \to M \xrightarrow{\beta_i} M_i \to 0$$

are exact.

(3) *$\mathfrak{m}_1/\mathfrak{m}_1^2$ is a finitely generated A-module.*

Proof. We retain the notation established in the proof of Proposition 9.1.3. Since $\mathfrak{n}_1$ is a finitely generated ideal in A_1, we assume $\mathfrak{a}$ is a finitely generated ideal in A such that $\alpha_1(\mathfrak{a}) = \mathfrak{n}_1$. Let $i \geq 0$ be any integer. Since $\mathfrak{a}$ and M are finitely generated A-modules, so is $\mathfrak{a}^i M$. For all $j \geq 0$, it follows from (9.3) that

$$N_{i+j} = \mathfrak{a}^j(\mathfrak{a}^i M) + N_{i+j+1} \subseteq \mathfrak{m}_j(\mathfrak{a}^i M) + N_{i+j+1}.$$

On the other hand, $\mathfrak{m}_j(\mathfrak{a}^i M) \subseteq \mathfrak{m}_j \mathfrak{m}_i M \subseteq \mathfrak{m}_{i+j} M \subseteq N_{i+j}$. This shows

$$N_{i+j} = \mathfrak{m}_j(\mathfrak{a}^i M) + N_{i+j+1}.$$

Define a filtration $\{N_{ij}\}_{j \in \mathbb{Z}}$ on N_i by

$$N_{ij} = \begin{cases} N_i & \text{if } j < 0, \\ N_{i+j} & \text{if } j \geq 0. \end{cases}$$

Applying Corollary 3.4.15, we obtain $N_i = \mathfrak{a}^i M$. Since $\mathfrak{a}^i \subseteq \mathfrak{m}_1^i \subseteq \mathfrak{m}_i$, we have $N_i \subseteq \mathfrak{m}_1^i M \subseteq \mathfrak{m}_i M \subseteq N_i$. Hence, $N_i = \mathfrak{m}_1^i M$. If we take $M_i = A_i$, this shows $\mathfrak{m}_i = \mathfrak{m}_1^i$, and the proof of (1) is complete. Part (2) follows from (1) and the definitions for $\mathfrak{m}_i$ and N_i. By (9.3), $\mathfrak{m}_1 = \mathfrak{a} + \mathfrak{m}_1^2$, which proves (3). □

Example 9.1.5. Let R be a commutative ring and I an ideal in R such that I/I^2 is a finitely generated R/I-module. Let $\hat{R} = \varprojlim_n R/I^n$ be the I-adic completion of R. With respect to the filtration $\{\widehat{I^n}\}$, $\hat{R}$ is separated and complete (Corollary 3.4.7). The reader should verify that the inverse system of rings $\{R/I^n\}$ satisfies the hypotheses of Corollary 9.1.4, hence the topology on $\hat{R}$ is the $\hat{I}$-adic topology. Moreover, $\hat{I}/\hat{I}^2 \cong I/I^2$ is finitely generated over $\hat{R}/\hat{I}$.

As an application of the above results, we derive sufficient conditions such that the limit of a directed system of noetherian rings is noetherian.

Corollary 9.1.6. *Let $\{A_i, \phi^i_j\}$ be a directed system of commutative local rings for a directed index set I. Let $\mathfrak{m}_i$ denote the maximal ideal of A_i. For each $i \le j$, assume $\phi^i_j : A_i \to A_j$ is a local homomorphism of local rings. If $A = \varinjlim A_i$, then the following are true.*

(1) *A is a local ring with maximal ideal $\mathfrak{m} = \varinjlim_i \mathfrak{m}_i$, each homomorphism $\alpha_i : A_i \to A$ is a local homomorphism of local rings, and the residue field of A is $\varinjlim_i A_i/\mathfrak{m}_i$.*

(2) *If $\mathfrak{m}_j = \mathfrak{m}_i A_j$ for each $i \le j$, then $\mathfrak{m}_i A = \mathfrak{m}$.*

(3) *For each $i \le j$, assume $\mathfrak{m}_j = \mathfrak{m}_i A_j$ and A_j is a faithfully flat A_i-module. If each A_i is noetherian, then A is noetherian.*

Proof. (1): Let $\mathfrak{m} = \bigcup_i \alpha_i(\mathfrak{m}_i)$. The reader should verify that $\mathfrak{m}$ is the unique maximal ideal of A. Take the direct limit of the exact sequences

$$0 \to \mathfrak{m}_i \to A_i \to A_i/\mathfrak{m}_i \to 0$$

and apply Theorem 1.4.3 to get the exact sequence

$$0 \to \mathfrak{m} \to A \to A/\mathfrak{m} \to 0.$$

(2): The sequence $\mathfrak{m}_i \otimes_{A_i} A_j \to \mathfrak{m}_j \to 0$ is exact. The functor $\varinjlim_j(\ )$ is exact (Theorem 1.4.3) and commutes with tensor products (Proposition 1.4.4). Hence the sequence $\mathfrak{m}_i \otimes_{A_i} A \to \mathfrak{m} \to 0$ is exact.

(3): By Exercise 1.4.14 and Exercise 2.4.19, A is faithfully flat over each A_i. Therefore, $0 \to \mathfrak{m}_i^n \otimes_{A_i} A \to A_i \otimes_{A_i} A$ is exact, and $\mathfrak{m}_i^n \otimes_{A_i} A \to \mathfrak{m}_i^n A = \mathfrak{m}^n$ is an isomorphism. It follows that

$$\begin{aligned} \mathfrak{m}^n/\mathfrak{m}^{n+1} &\cong (\mathfrak{m}_i^n A) \,/\, (\mathfrak{m}_i^{n+1} A) \\ &\cong \left(\mathfrak{m}_i^n/\mathfrak{m}_i^{n+1}\right) \otimes_{A_i} A \\ &\cong \left(\mathfrak{m}_i^n/\mathfrak{m}_i^{n+1}\right) \otimes_{A_i/\mathfrak{m}_i} (A_i/\mathfrak{m}_i \otimes_{A_i} A) \\ &\cong \left(\mathfrak{m}_i^n/\mathfrak{m}_i^{n+1}\right) \otimes_{A_i/\mathfrak{m}_i} A/\mathfrak{m} \end{aligned}$$

are isomorphisms of $A/\mathfrak{m}$-vector spaces. Since A_i is noetherian, $\mathfrak{m}_i^n/\mathfrak{m}_i^{n+1}$ is a finite dimensional $A_i/\mathfrak{m}_i$-vector space. Therefore, $\mathfrak{m}^n/\mathfrak{m}^{n+1}$ is finite dimensional over $A/\mathfrak{m}$. Let $\hat{A} = \varprojlim A/\mathfrak{m}^n$. By (2), $\hat{A} = \varprojlim A/\mathfrak{m}_i^n A$, for each i. By Example 9.1.5 and Proposition 3.4.9, $\hat{A}$ is noetherian.

The maximal ideal of $\hat{A}$ is $\hat{\mathfrak{m}}$. By Proposition 3.4.12, we have $\hat{\mathfrak{m}} = \mathfrak{m}\hat{A} = \mathfrak{m}_i\hat{A}$, for each i. Because A is flat over A_i, $(A_i/\mathfrak{m}_i{}^n)\otimes_{A_i} A$ is flat over $A_i/\mathfrak{m}_i^n$. Therefore,

$$\hat{A}/\mathfrak{m}_i{}^n\hat{A} = A/\mathfrak{m}_i{}^n A = A_i/\mathfrak{m}_i^n \otimes_{A_i} A$$

is flat over $A_i/\mathfrak{m}_i^n$. In the terminology of Example 9.1.1 (1), the A_i-module $\hat{A}$ is ideal-wise separated for $\mathfrak{m}_i$. By (6) implies (1) of Theorem 9.1.2, it follows that $\hat{A}$ is flat over A_i. By Exercise 2.4.19, $\hat{A}$ is faithfully flat over A_i. By Exercise 1.4.15, $\hat{A}$ is faithfully flat over A. By Exercise 2.5.10, A is noetherian. □

2. Étale and Smooth Algebras

In this section we introduce the notion of an étale R-algebra. First the necessary definitions and proofs of the basic properties are presented. After that, the emphasis is on the problem of showing that an algebra which is formally étale (see Definition 4.7.1) is étale. This result is proved in Theorem 9.2.19 and in Section 9.2.4 we apply it to Raynaud's construction of an étale R-algebra A_{et} associated to an R-algebra A which is a finitely generated free R-module. Our study of étale algebras continues in Section 9.3. General references for this section are [**Mil80**], [**Gro64**], and [**MCPA15**].

2.1. Étale Algebras.

Definition 9.2.1. Let S be a commutative R-algebra. We say S is a *finitely presented R-algebra*, or an *R-algebra of finite presentation*, if there is a polynomial ring in $n \geq 1$ variables over R and a surjective R-algebra homomorphism $\theta : R[x_1, \dots, x_n] \to S$ such that the kernel of θ is a finitely generated ideal. An R-algebra A is said to be *locally of finite presentation*, or *essentially of finite presentation*, if A is isomorphic to $S_\mathfrak{q}$ where S is an R-algebra of finite presentation, and $\mathfrak{q} \in \operatorname{Spec} S$.

Example 9.2.2. Let R, S and T be commutative rings. Here are some examples of finitely presented R-algebras.

(1) If R is a noetherian ring, then by the Hilbert Basis Theorem, $R[x_1, \dots, x_n]$ is noetherian. An R-algebra S is finitely presented if and only if it is finitely generated.

(2) Let S be an R-algebra which is finitely generated and projective as an R-module. The reader should verify that S is a finitely presented R-algebra.

(3) Let f be an element of R which is not in the nil radical $\operatorname{Rad}_R(0)$. Let S be the localization $R[f^{-1}]$. Define $\theta : R[x] \to S$ by $x \mapsto f^{-1}$. The reader should verify that the kernel of θ is the principal ideal

generated by the polynomial $xf - 1$. Then S is a finitely presented R-algebra.

(4) (Finitely Presented over Finitely Presented is Finitely Presented) If S is finitely presented over R and T is finitely presented over S, then the reader should verify that T is finitely presented over R.

(5) (Finitely Presented is Preserved by a Change of Base) If S is a finitely presented R-algebra, and T is an R-algebra, then the reader should verify that $S \otimes_R T$ is a finitely presented T-algebra.

(6) (A Product of Finitely Presented is Finitely Presented) If S and T are finitely presented R-algebras, then the reader should verify that $S \otimes_R T$ is a finitely presented R-algebra.

Definition 9.2.3. Let S be a commutative R-algebra. We say that S is *étale* over R in case S is a flat separable finitely presented R-algebra. If R is noetherian, then S is étale over R if and only if S is a commutative R-algebra which is finitely generated, flat and separable.

Example 9.2.4. Listed here are some examples of algebras that are étale and algebras that are not étale. In the following, R is a commutative ring.

(1) Let S be a commutative separable R-algebra which is finitely generated and projective as an R-module. The reader should verify that S is étale.

(2) Let f be an element of R which is not in the nil radical $\operatorname{Rad}_R(0)$. If S is equal to the localization $R[f^{-1}]$, then S is étale over R.

(3) (An Open Immersion is Étale) Let $f : R \to S$ be a homomorphism of commutative rings. If $f^\sharp : \operatorname{Spec} S \to \operatorname{Spec} R$ is an open immersion (see Exercise 2.3.9), then by Exercise 8.1.33, S is a finitely generated separable R-algebra. From Proposition 2.4.8, it follows that S is flat over R. If R is noetherian, then S is étale over R. Although we will not prove it here, it is true that even when R is not noetherian, S is a finitely presented R algebra, hence is étale over R (see [**Gro64**, Proposition I.4.6, p. 328]).

(4) Let I be a finitely generated ideal in R and assume I is not idempotent generated. Then $S = R/I$ is finitely presented. By Exercise 4.1.9, S is separable over R. By Exercise 1.1.22 and Exercise 1.3.15, S is not flat over R. Therefore, S is not étale over R.

(5) Let k be a field and $R = k[x, y]/(xy)$ the affine coordinate ring of the plane curve $Y = Z(xy)$ corresponding to the union of the two coordinate axes. Let $A = R/(x) \oplus R/(y)$ and $\phi : R \to A$ the natural map. Then A is the affine coordinate ring of the disjoint union of

the two irreducible components of Y. In the sense of the Zariski topology on Y, $\phi : R \to A$ corresponds to a closed cover, but not an open cover. By Exercise 4.4.21, A is separable over R, but not flat over R. In particular, A is separable but not étale over R.

(6) Let k be a field and $R = k[x,y]/(y^2 - x^2(x+1))$ the affine coordinate ring of a nodal cubic curve. By Exercise 8.1.34, if we set $w = yx^{-1}$, then the quotient field of R is $K = k(w)$, the integral closure of R in K is $k[w]$, $k[w]$ is a rank one R-lattice in K, the R-algebra $k[w]$ is separable over R, but not flat. Therefore, the extension $R \to k[w]$ is separable but not étale.

(7) Let x be an indeterminate and $n > 1$. Then the extension $R[x^n] \subseteq R[x]$ is free of rank n, hence is flat. By Exercise 4.6.20, it is not separable, hence not étale.

Proposition 9.2.5. *Let R, S and T be commutative rings.*

(1) *(Étale over Étale is Étale) If S is étale over R and T is étale over S, then T is étale over R.*

(2) *(Étale is Preserved by a Change of Base) If S is étale over R and T is any R-algebra, then $S \otimes_R T$ is étale over T.*

(3) *(A Product of Étale extensions is Étale) If S and T are both étale over R, then $S \otimes_R T$ is étale over R.*

Proof. The proof is left to the reader. □

Corollary 9.2.6. *Let S be a commutative R-algebra and T a commutative S-algebra. If S/R is separable and T/R is étale, then T/S is étale.*

Proof. It follows from Theorem 4.4.3, that T/S is flat and separable and the kernel of $\gamma : S \otimes_R T \to T$ is idempotent generated. Thus, T is a finitely presented $S \otimes_R T$-algebra. By Example 9.2.2 (4) and (5), T is a finitely presented S-algebra. □

Corollary 9.2.7. *Let R be a commutative noetherian ring and S a finitely generated commutative R-algebra. Let U be the set of all points P in* Spec S *such that S_P is an étale $R_{P\cap R}$-algebra. Then U is an open (possibly empty) subset of* Spec S.

Proof. This follows from the Theorem of Generic Flatness (see, for example, [**Mat80**, (22.B), Theorem 53]) and Exercise 8.3.17. □

Corollary 9.2.8. *Let $f : S \to T$ and $g : S \to T$ be R-algebra homomorphisms, where S and T are commutative R-algebras. Assume S is separable over R and T is a connected ring. If there exists $\mathfrak{q} \in$* Spec T *such that*

$\mathfrak{p} = f^{-1}(\mathfrak{q}) = g^{-1}(\mathfrak{q})$ *and the maps induced by* f *and* g *agree on the residue fields* $k(\mathfrak{p}) \to k(\mathfrak{q})$, *then the maps* f *and* g *are equal.*

Proof. By Corollary 4.3.2, $S \otimes_R T$ is separable over T. By Exercise 1.3.8 (2), f and g induce R-algebra homomorphisms γ_f and γ_g such that the two diagrams

commute. To show $f = g$, it suffices to prove $\gamma_f = \gamma_g$. Since γ_f and γ_g are left inverses for ρ_2, by Corollary 4.4.8 there are idempotents e_f and e_g in $S \otimes_R T$ such that γ_f factors through $(S \otimes_R T)e_f$ and γ_g factors through $(S \otimes_R T)e_g$. Therefore, it suffices to prove $e_f = e_g$. By Corollary 2.2.11, since T is connected, it suffices to prove $\gamma_f^{-1}(\mathfrak{q})$ and $\gamma_g^{-1}(\mathfrak{q})$ belong to the same connected component of $\operatorname{Spec}(S \otimes_R T)$. The diagrams

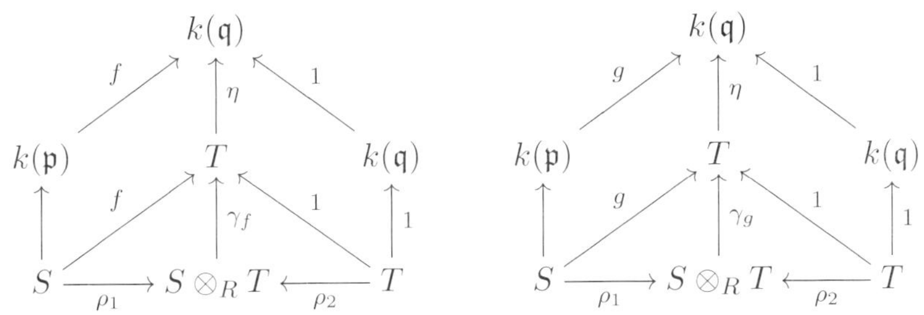

commute. By hypothesis, f and g induce the same map on the fields $k(\mathfrak{p}) \to k(\mathfrak{q})$. By Exercise 1.3.8 (2), the R-algebra homomorphisms $\eta\gamma_f$ and $\eta\gamma_g$ are equal. This implies $\gamma_f^{-1}(\mathfrak{q}) = \gamma_g^{-1}(\mathfrak{q})$, hence the idempotents e_f, e_g are equal. □

Proposition 9.2.9. *Let* S *be a noetherian étale* R*-algebra with structure homomorphism* $\theta : R \to S$. *Then* $\theta^\sharp : \operatorname{Spec} S \to \operatorname{Spec} R$ *is an open map.*

Proof. Since S is flat over R, going-down holds (see, for example, [**Mat80**, (5.D), Theorem 4]) and this follows from [**Mat80**, (6.I), Theorem 8]. □

2.2. Formally Smooth Algebras. Let S be a commutative R-algebra. In Definition 4.7.1 we defined S to be formally smooth over R if the map on sets

$$\beta : \operatorname{Hom}_{R\text{-alg}}(S, A) \to \operatorname{Hom}_{R\text{-alg}}(S, A/I)$$

is onto, for all commutative R-algebras A and for all nilpotent ideals I in A. We already proved in Corollary 4.7.3 that a commutative separable flat R-algebra S is formally smooth (in fact, S is formally étale). The main result of this section is a proof that over a noetherian ring R a finitely generated formally smooth algebra is flat. Additionally, we derive the standard properties of formal smoothness and important tests for formal smoothness. Among them are the criteria based on the stalks over the local rings, faithfully flat descent, and the jacobian. The results of this section were originally proved in [**Gro64**]. The presentation below is based on [**MCPA15**, Chapter 17].

Example 9.2.10. Let R be a commutative ring, I an index set, $X = \{x_i\}_{i \in I}$ a set of indeterminates indexed by I, and $S = R[X]$ be the polynomial algebra over R (Exercise 1.4.10). Let A be a commutative R-algebra and I an ideal in A. Let $\eta : A \to A/I$ be the natural map. Let $\bar{\theta} : S \to A/I$ be an R-algebra homomorphism. By the Axiom of Choice, for each i, let $a_i \in A$ such that $\bar{\theta}(x_i) = \eta(a_i)$. If we define $\theta : S \to A$ by $\theta(x_i) = a_i$, then $\beta(\theta) = \bar{\theta}$. This shows S is formally smooth over R.

We begin with a proof that every finitely generated formally smooth algebra over a noetherian ring R is flat.

Theorem 9.2.11. *Let R be a commutative noetherian ring and S a commutative R-algebra which is formally smooth and of finite presentation. Then S is flat over R.*

Proof. This proof, which is due to J. Conde-Lago, is from [**CL17**]. Since S is a finitely generated R-algebra, there exists an exact sequence of R-modules

$$0 \to I \to R[x_1, \ldots, x_n] \xrightarrow{\phi} S \to 0$$

where ϕ is an R-algebra homomorphism and I is the kernel of ϕ. Let A denote the polynomial ring $R[x_1, \ldots, x_n]$ and let $\hat{A}$ be the I-adic completion of A. Then A is noetherian by the Hilbert Basis Theorem, and $\hat{A}$ is a flat A-module, by Proposition 3.4.12. Since A is a free R-module, A is flat over R. By Exercise 2.4.16, $\hat{A}$ is flat over R. Let $\alpha : \hat{A} \to \hat{A}/\hat{I} = A/I = S$ be the natural map on R-algebras. To show that S is flat over R, by Exercise 2.4.15, it suffices to show that there is an R-algebra homomorphism $\beta : S \to \hat{A}$ such that $\alpha\beta$ is the identity map on S. By the universal mapping property, it suffices to construct R-algebra homomorphisms $\beta_k : S \to A/I^k$ such that for all $k \geq 1$ the diagram

$$\begin{array}{ccc} S & \xrightarrow{\beta_{k+1}} & A/I^{k+1} \\ & \searrow^{\beta_k} & \downarrow \\ & & A/I^k \end{array}$$

commutes. The map $\beta_1 : S \to A/I$ is the identity map. Inductively, assume $k \geq 1$ and β_k exists. Since I^k/I^{k+1} is a nilpotent ideal in A/I^{k+1}, β_{k+1} exists because S is formally smooth. $\square$

The rest of this section is devoted to proving the standard properties of formal smoothness.

Lemma 9.2.12. *Let S be a formally smooth R-algebra.*

(1) *(Formally Smooth is Preserved by Base Extension) If T is a commutative R-algebra, then $S \otimes_R T$ is formally smooth over T.*
(2) *(Formally Smooth over Formally Smooth Is Formally Smooth) If T is a formally smooth S-algebra, then T is a formally smooth R-algebra.*
(3) *If $W \subseteq R$ is a multiplicative set, then $W^{-1}R$ is formally smooth over R.*

Proof. Parts (1) and (2) are left to the reader.

(3): Let A be a commutative R-algebra with structure homomorphism $\theta : R \to A$. Let I be a nilpotent ideal in A and $g : W^{-1}R \to A/I$. By Exercise 3.1.6, $\theta(W) \subseteq A^*$. Therefore, there exists a unique $\gamma : W^{-1}R \to A$ such that the diagram

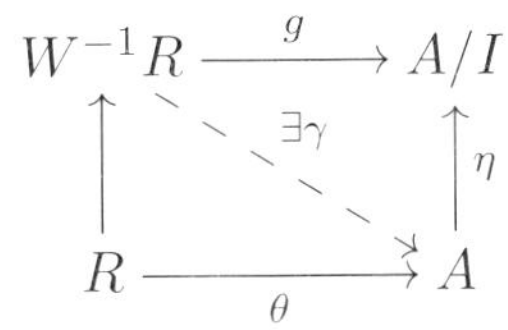

commutes. □

Theorem 9.2.13. *Let S be a formally smooth R-algebra and I an ideal in S. Then $A = S/I$ is a formally smooth R-algebra if and only if the sequence of A-modules (Theorem 8.3.4)*

$$0 \to I/I^2 \xrightarrow{\gamma} \Omega_{S/R} \otimes_S A \xrightarrow{\alpha} \Omega_{A/R} \to 0$$

is split exact.

Proof. By Exercise 1.3.31 and Theorem 8.3.4, the sequence in the theorem is split exact if and only if for every A-module M, the sequence

$$\operatorname{Hom}_A(\Omega_{S/R} \otimes_S A, M) \xrightarrow{H_\gamma} \operatorname{Hom}_A(I/I^2, M) \to 0 \tag{9.6}$$

is exact. By the adjoint isomorphism of Theorem 1.3.24, (9.6) is isomorphic to

$$\operatorname{Hom}_S(\Omega_{S/R}, M) \xrightarrow{H_\gamma} \operatorname{Hom}_S(I, M) \to 0 \tag{9.7}$$

which by Theorem 8.3.1 is isomorphic to

$$\operatorname{Der}_R(S, M) \to \operatorname{Hom}_S(I, M) \to 0. \tag{9.8}$$

To complete the proof, we show $A = S/I$ is formally smooth over R if and only if (9.8) is exact for an arbitrary A-module M.

First assume $A = S/I$ is formally smooth over R. Then there exists an R-algebra homomorphism γ such that the diagram

$$\begin{array}{ccc} S/I & \xrightarrow{=} & S/I \\ \uparrow & \searrow^{\exists\gamma} & \uparrow{\scriptstyle\eta} \\ R & \longrightarrow & S/I^2 \end{array}$$

commutes. In this case, γ is a section to the natural map η

$$0 \to I/I^2 \xrightarrow{\subseteq} S/I^2 \xrightarrow{\eta} S/I \to 0$$

and by Exercise 8.1.14, S/I^2 is a split extension of S/I by I/I^2. Let M be an S/I-module and $\phi \in \operatorname{Hom}_{S/I}(I/I^2, M)$. To show that ϕ comes from an R-derivation in $\operatorname{Der}_R(S, M)$, define $\delta : S/I^2 \to M$ by $\delta(x) = \phi(x - \gamma\eta(x))$. The reader should verify that δ is an R-derivation in $\operatorname{Der}_R(S/I^2, M)$. By Exercise 8.1.12, δ lifts to an R-derivation in $\operatorname{Der}_R(S, M)$ such that the diagram

$$\begin{array}{ccccccc} 0 & \longrightarrow & I & \xrightarrow{\subseteq} & S & \xrightarrow{\subseteq} & S/I \\ & & \downarrow & & \downarrow & & \downarrow{\scriptstyle =} \\ 0 & \longrightarrow & I/I^2 & \xrightarrow{\subseteq} & S/I^2 & \longrightarrow & S/I \\ & & \downarrow{\scriptstyle\phi} & \swarrow_{\delta} & & & \\ & & M & & & & \end{array}$$

commutes. From this it follows that (9.8) is exact.

Conversely, assume (9.8) is exact for every A-module M. Let B be an R-algebra and J an ideal in B such that $J^n = 0$, for some $n \geq 1$. By a finite induction argument on n, it suffices to assume $J^2 = 0$. Assume $a : A \to B/J$ is an R-algebra homomorphism. Consider the commutative diagram

$$\begin{array}{ccccc} S & \longrightarrow & S/I & \xrightarrow{a} & B/J \\ \uparrow & \searrow^{\exists\alpha} & & \searrow^{\exists f} & \uparrow \\ R & \longrightarrow & & \longrightarrow & B \end{array}$$

Since S/R is formally smooth, an R-algebra homomorphism α exists which commutes with the rest of the diagram. To complete, the proof, we need to show the existence of the R-algebra homomorphism f. Since the diagram commutes, $\alpha(I) \subseteq J$. Since $J^2 = 0$, J is a B/J-module. Using $a : A \to B/J$, we also make J into an A-module. Since $\alpha(I^2) \subseteq J^2 = 0$, there is induced a homomorphism $I/I^2 \to J$ of A-modules. Since (9.8) is exact, there is an R-derivation $\delta : S \to J$ which agrees with α on I. By Proposition 8.1.6,

$\alpha - \delta : S \to B$ is an R-algebra homomorphism which lifts $S \to S/I \to B/J$. By construction, $\alpha - \delta$ factors through S/I, so f exists. □

Corollary 9.2.14. *Let S be a formally smooth R-algebra. Then the following are true.*

(1) *$\Omega_{S/R}$ is a projective S-module.*

(2) *If I is an ideal in S and $A = S/I$, then A is a formally smooth R-algebra if and only if $\gamma : I/I^2 \to \Omega_{S/R} \otimes_S A$ is one-to-one and $\Omega_{A/R}$ is a projective A-module.*

Proof. (1): Write S as the homomorphic image of a polynomial ring over R: $R[X] \to S \to 0$, where $X = \{x_i\}_{i \in I}$ is a set of indeterminates. By Example 9.2.10, $R[X]$ is formally smooth over R. By Theorem 9.2.13, there is a split exact sequence

$$0 \to I/I^2 \to \Omega_{R[X]/R} \otimes_{R[X]} S \to \Omega_{S/R} \to 0$$

of S-modules. By Proposition 8.3.2, $\Omega_{R[X]/R} \otimes_{R[X]} S$ is a free S-module.

(2): In the sequence

$$0 \to I/I^2 \xrightarrow{\gamma} \Omega_{S/R} \otimes_S A \xrightarrow{a} \Omega_{A/R} \to 0$$

of Theorem 9.2.13, a is always onto, and by (1), the center term is a projective A-module. Therefore, the sequence is split exact if and only if $\Omega_{A/R}$ is A-projective and γ is one-to-one. □

The next result is a local ring criterion for formal smoothness.

Corollary 9.2.15 (Formally Smooth if and only if Locally Formally Smooth). *If S is a commutative finitely presented R-algebra, then the following are equivalent.*

(1) *S is formally smooth over R.*

(2) *For every $\mathfrak{q} \in \operatorname{Spec} S$, if $\mathfrak{p} = \mathfrak{q} \cap R$, then $S_{\mathfrak{q}}$ is formally smooth over $R_{\mathfrak{p}}$.*

Proof. (1) implies (2): This follows from Lemma 9.2.12 and does not require the finitely presented hypothesis.

(2) implies (1): By assumption, S is a homomorphic image of $A = R[x_1, \dots, x_n]$ for some n, and the kernel is a finitely generated ideal $I \subseteq A$. In the exact sequence

$$I/I^2 \xrightarrow{\gamma} \Omega_{A/R} \otimes_A S \to \Omega_{S/R} \to 0 \tag{9.9}$$

of S-modules, $\Omega_{A/R} \otimes_A S$ is free of finite rank (Proposition 8.3.2). Since I is a finitely generated A-module, I/I^2 is a finitely generated S-module. Therefore, $\Omega_{S/R}$ is an S-module of finite presentation. By Corollary 9.2.14, to complete the proof it suffices to show γ is one-to-one, and $\Omega_{S/R}$ is S-projective.

Let $\mathfrak{q} \in \operatorname{Spec} S$, $\mathfrak{p} = \mathfrak{q} \cap R$, and $Q = \mathfrak{q} \cap A$. Using Exercise 8.3.10 and Exercise 8.3.11, one verifies the following

$$\Omega_{S/R} \otimes_S S_{\mathfrak{p}} = \Omega_{S_{\mathfrak{p}}/R_{\mathfrak{p}}}$$
$$\Omega_{S/R} \otimes_S S_{\mathfrak{q}} = \Omega_{S_{\mathfrak{q}}/R_{\mathfrak{p}}}$$
$$\Omega_{A/R} \otimes_A S_{\mathfrak{q}} = \Omega_{A_Q/R_{\mathfrak{p}}} \otimes_{A_Q} S_{\mathfrak{q}}.$$

By Proposition 2.6.4, $\Omega_{S/R}$ is S-projective if and only if $\Omega_{S_{\mathfrak{q}}/R_{\mathfrak{p}}}$ is $S_{\mathfrak{q}}$-projective for all $\mathfrak{q}$. By Exercise 2.1.10, $S_{\mathfrak{q}} = S_Q = S \otimes_A A_Q$, hence $(I/I^2) \otimes_S S_{\mathfrak{q}} = (IA_Q)/(I^2 A_Q)$. By Lemma 9.2.12, A_Q is formally smooth over $R_{\mathfrak{p}}$, so Theorem 9.2.13 applies to $S_{\mathfrak{q}} = A_Q/IA_Q$. Tensoring (9.9) with $(\ ) \otimes_S S_{\mathfrak{q}}$ yields the split exact sequence

$$0 \to (IA_Q)/(I^2 A_Q) \to \Omega_{A_Q/R_{\mathfrak{p}}} \otimes_{A_Q} S_{\mathfrak{q}} \to \Omega_{S_{\mathfrak{q}}/R_{\mathfrak{p}}} \to 0 \tag{9.10}$$

of $S_{\mathfrak{q}}$-modules. From Exercise 2.4.12, it follows that the map γ in (9.9) is one-to-one. Since (9.10) is split exact, and the center term is $S_{\mathfrak{q}}$-free, $\Omega_{S_{\mathfrak{q}}/R_{\mathfrak{p}}}$ is $S_{\mathfrak{q}}$-projective. As mentioned above, this proves $\Omega_{S/R}$ is projective over S, and the proof is complete. □

Now we can prove a faithfully flat descent theorem for formally smooth algebras.

Corollary 9.2.16 (Faithfully Flat Descent of a Formally Smooth Algebra). *Let T be a commutative faithfully flat R-algebra. If S is a commutative finitely generated R-algebra, then S is formally smooth over R if and only if $S \otimes_R T$ is formally smooth over T.*

Proof. For some n, if we set $A = R[x_1, \dots, x_n]$, then there is an R-algebra epimorphism $\sigma : A \to S$. Let I be the kernel of σ. In the sequence of S-modules

$$0 \to I/I^2 \xrightarrow{\gamma} \Omega_{A/R} \otimes_A S \to \Omega_{S/R} \to 0 \tag{9.11}$$

the middle term is free of finite rank. Therefore, $\Omega_{S/R}$ is finitely generated as an S-module. Since $S \otimes_R T$ is faithfully flat over S, (9.11) is exact if and only if

$$0 \to (I/I^2) \otimes_R T \xrightarrow{\gamma \otimes 1} \Omega_{A/R} \otimes_A S \otimes_R T \to \Omega_{S/R} \otimes_R T \to 0 \tag{9.12}$$

is an exact sequence of $S \otimes_R T$-modules. Using Exercise 8.3.10, one verifies the identities

$$\Omega_{S \otimes_R T/T} = \Omega_{S/R} \otimes_R T$$
$$\Omega_{A/R} \otimes_A S \otimes_R T = \Omega_{A \otimes_R T/T} \otimes_{A \otimes_R T} (S \otimes_R T).$$

By Theorem 2.6.1, $\Omega_{S\otimes_R T/T}$ is $S\otimes_R T$-projective if and only if $\Omega_{S/R}$ is S-projective. Since T/R is flat,

$$0 \to I^2 \otimes_R T \to I \otimes_R T \to (I/I^2) \otimes_R T \to 0$$

is exact. By Lemma 9.2.12, $A \otimes_R T$ is formally smooth over T. Therefore, (9.12) is isomorphic to the counterpart of (9.11) for $(A \otimes_R T)/(I \otimes_R T) = S \otimes_R T$. The proof now follows from Corollary 9.2.14 (2). □

In the following theorem and its corollary we prove the jacobian criterion for an algebra over a local ring R to be formally smooth.

Theorem 9.2.17. *Let R be a commutative local ring, I a finitely generated ideal in the polynomial ring $R[x_1, \dots, x_n]$, and $A = R[x_1, \dots, x_n]/I$. Let $\mathfrak{q} \in \operatorname{Spec} A$, and $S = A_{\mathfrak{q}}$. If S is formally smooth over R, then there is a set of polynomials $f_1, \dots, f_r$ in I such that*

(1) *$f_1, \dots, f_r$ map to a free basis for the S-module $(I/I^2)_{\mathfrak{q}}$, and*

(2) *the jacobian matrix $(\partial f_i/\partial x_j)$ has rank r over the field $S/\mathfrak{q}S$.*

Proof. Let $Q = \mathfrak{q} \cap R[x_1, \dots, x_n]$ and $B = R[x_1, \dots, x_n]_Q$. Then by Exercise 2.1.10, $S = B/IB$. By Lemma 9.2.12 and Example 9.2.10, B/R is formally smooth. By Proposition 8.3.2 and Exercise 8.3.11, $\Omega_{B/R}$ is a free B-module of rank n. By Theorem 9.2.13, the sequence of S-modules

$$0 \to I/I^2 \xrightarrow{\gamma} \Omega_{B/R} \otimes_B S$$

is split exact, where we write I for IB. Then I/I^2 is a finitely generated free S-module (Proposition 2.3.2). Let $k = S/\mathfrak{q}S$. By Theorem 8.3.4, the map

$$I/I^2 \otimes_B k \xrightarrow{\gamma\otimes 1} \Omega_{B/R} \otimes_B k$$

sends a polynomial in I to its gradient (reduced modulo $\mathfrak{q}S$). Using Proposition 2.3.2 we pick a set of polynomials $\{f_1, \dots, f_r\}$ in I that maps to a free S-basis for $(I/I^2)_{\mathfrak{q}}$. Then $\{f_1, \dots, f_r\}$ reduces modulo $\mathfrak{q}S$ to a k-basis for $I/I^2 \otimes_B k$, and the jacobian matrix $(\partial f_i/\partial x_j)$ has rank r over k. □

A converse of Theorem 9.2.17 is proved in Corollary 9.2.18.

Corollary 9.2.18 (Jacobian Criterion). *Let R be a commutative local ring, $I = (f_1, \dots, f_r)$ an ideal in the polynomial ring $R[x_1, \dots, x_n]$ generated by r elements, and $A = R[x_1, \dots, x_n]/I$. Let $\mathfrak{q} \in \operatorname{Spec} A$, and $S = A_{\mathfrak{q}}$. If the jacobian matrix $(\partial f_i/\partial x_j)$ has rank r over the field $S/\mathfrak{q}S$, then S is formally smooth over R.*

Proof. Retain the notation established in the proof of Theorem 9.2.17. By Theorem 9.2.13, it suffices to show that the sequence of S-modules

$$0 \to I/I^2 \xrightarrow{\gamma} \Omega_{B/R} \otimes_B S$$

is split exact, where we write I for IB. Let $k = S/\mathfrak{q}S$. Since

$$I/I^2 \otimes_B k \xrightarrow{\gamma \otimes 1} \Omega_{B/R} \otimes_B k$$

is one-to-one, this follows from Corollary 2.3.3. □

2.3. Formally Étale is Étale. In this short section we apply the results of the preceding section to prove two important theorems on étale algebras over a noetherian ring R. First we show that a finitely generated formally étale algebra over R is étale. Secondly, we derive a useful jacobian criterion for a finitely generated R-algebra to be étale. Theorem 9.2.19 is a converse to Corollary 4.7.3 when R is noetherian and S is finitely generated.

Theorem 9.2.19. *Let R be a commutative noetherian ring and S a commutative finitely generated R-algebra. The following are equivalent:*

(1) *S is formally étale over R (see Definition 4.7.1).*

(2) *S is étale over R (see Definition 9.2.3).*

Proof. (1) implies (2): By definition, formally étale is equivalent to formally unramified and formally smooth. By Theorem 8.3.6, S is separable over R, and by Theorem 9.2.11, S is flat.

(2) implies (1): This is Corollary 4.7.3. □

Corollary 9.2.20 (Jacobian Criterion). *Let R be a commutative noetherian ring and $A = R[x_1, \dots, x_n]$ the polynomial ring in n indeterminates. Let $I = (f_1, \dots, f_n)$ be an ideal in A generated by a set of n polynomials. Then $S = A/I$ is an étale R-algebra if and only if the determinant of the jacobian matrix, $\det(\partial f_i/\partial x_j)$, is mapped to a unit in S.*

Proof. By Proposition 8.3.8, S is separable over R if and only if the determinant of the jacobian is mapped to a unit in S. To finish the proof, we show that S is flat over R, if the determinant of the jacobian is mapped to a unit in S. Let $\mathfrak{q} \in \operatorname{Spec} S$ and $\mathfrak{p} = \mathfrak{q} \cap R$. Then $A_\mathfrak{p} = R_\mathfrak{p}[x_1, \dots, x_n]$. If $I_\mathfrak{p}$ is the ideal in $A_\mathfrak{p}$ generated by $f_1, \dots, f_n$, then $S_\mathfrak{p} = A_\mathfrak{p}/I_\mathfrak{p}$. By Corollary 9.2.18, $S_\mathfrak{q}$ is formally smooth over $R_\mathfrak{p}$. By Corollary 9.2.15, S is formally smooth over R. By Theorem 9.2.11, S is a flat R-algebra. □

2.4. An Example of Raynaud. Let R be a local ring and A a commutative R-algebra that is finitely generated and free of rank n as an R-module. In this section we outline a construction from [**Ray70**, Chap. I, §4] which shows that there is an étale R-algebra associated to A. Let $\{a_1, \dots, a_n\}$ be a free R-basis for A. Additionally, by Proposition 2.4.6 we assume $a_1 = 1$.

For each pair of indices (i, j), the multiplication operation on A defines

$$a_i a_j = \sum_{k=1}^{n} a_{i,j,k} a_k \tag{9.13}$$

where the coefficients $\{a_{i,j,k}\}$ are in R. Note that if i is fixed, then $(a_{i,j,k})$ is the matrix associated to "left multiplication by a_i". Also, because A is commutative, if k is fixed, then we have $a_{i,j,k} = a_{j,i,k}$. For an arbitrary $\alpha = \sum_{k=1}^{n} r_k a_k$ in A, we have

$$\alpha^2 = \left(\sum_{k=1}^{n} r_k a_k\right)^2 = \sum_{k=1}^{n} \left(\sum_{i=1}^{n}\sum_{j=1}^{n} r_i r_j a_{i,j,k}\right) a_k. \tag{9.14}$$

Let $x_1, \dots, x_n$ be indeterminates and for $1 \le k \le n$ define quadratic polynomials

$$f_k = x_k - \sum_{i=1}^{n}\sum_{j=1}^{n} x_i x_j a_{i,j,k} \tag{9.15}$$

in the polynomial ring $R[x_1, \dots, x_n]$. It follows from (9.14) and (9.15) that $\alpha = \sum_{k=1}^{n} r_k a_k$ is an idempotent in A if and only if $f_k(r_1, \dots, r_n) = 0$ for $1 \le k \le n$.

Let A_{et} be the R-algebra defined by $R[x_1, \dots, x_n]/(f_1, \dots, f_n)$. Without using any extra adornment, we will denote by x_i both the monomial and the coset it represents in A_{et}.

Let S be another commutative R-algebra. Then $a_1 \otimes 1, \dots, a_n \otimes 1$ is a free S-module basis for $A \otimes_R S$. The arguments given above show that a typical element $\beta = \sum_{k=1}^{n} a_k \otimes s_k$ in $A \otimes_R S$ is idempotent if and only if $f_k(s_1, \dots, s_n) = 0$ for $1 \le k \le n$. In particular, the element $e = \sum_{k=1}^{n} a_k \otimes x_k$ in $A \otimes_R A_{\mathrm{et}}$ is idempotent. In the language of Category Theory, Proposition 9.2.21 shows that the functor $\mathrm{idemp}(A \otimes_R (\cdot))$ from the category of R-algebras to the category of sets is represented by the R-algebra A_{et} and the idempotent e.

Proposition 9.2.21. *In the above context, if S is any commutative R-algebra, then there is a one-to-one correspondence*

$$\mathrm{Hom}_{R\text{-}alg}(A_{\mathrm{et}}, S) \to \mathrm{idemp}(A \otimes_R S)$$

which maps an R-algebra homomorphism θ to $(1 \otimes \theta)(e) = \sum_{k=1}^{n} a_k \otimes \theta(x_k)$.

Proof. The proof is left to the reader. □

For an application of Proposition 9.2.22, the reader is referred to the proof of Proposition 10.1.7 given below.

Proposition 9.2.22. *Let R be a local ring and A a commutative R-algebra which is a finitely generated free R-module of rank n. Let $a_1, \dots, a_n$ be a free R-module basis for A with $a_1 = 1$. Let $f_1, \dots, f_n$ be defined as in (9.15), and let A_{et} be the R-algebra $R[x_1, \dots, x_n]/(f_1, \dots, f_n)$. Then A_{et} is an étale R-algebra.*

Proof. Step 1: Show that A_{et} is formally étale over R. Let S be an arbitrary R-algebra and assume I is a nilpotent ideal in S. By Corollary 3.4.16, $\operatorname{idemp}(A \otimes_R S) \to \operatorname{idemp}(A \otimes_R (S/I))$ is a one-to-one correspondence. Proposition 9.2.21 implies $\operatorname{Hom}_{R\text{-alg}}(A_{\mathrm{et}}, S) \to \operatorname{Hom}_{R\text{-alg}}(A_{\mathrm{et}}, S/I)$ is a one-to-one correspondence.

Step 2: Show that A_{et} is étale over R. Let $\mathfrak{m}$ be the maximal ideal in R. Let R_0 be the subring of R generated by all of the elements $\{a_{i,j,k}\}$ appearing in (9.13). Let R_1 be the localization of R_0 at the prime ideal $\mathfrak{m} \cap R_0$. Then R_1 is a noetherian local subring of R and the coefficients of the polynomials $f_1, \dots, f_n$ are in R_1. Let A_1 be the R_1-submodule of A generated by $a_1, \dots, a_n$. Using the equations (9.13), we see that A_1 is an R_1-algebra. Define $(A_1)_{\mathrm{et}}$ to be $R_1[x_1, \dots, x_n]/(f_1, \dots, f_n)$. By Step 1 and Theorem 9.2.19, $(A_1)_{\mathrm{et}}$ is étale over R_1. By Proposition 9.2.5, $A_{\mathrm{et}} = (A_1)_{\mathrm{et}} \otimes_{R_1} R$, is étale over R. □

3. More Properties of Étale Algebras

In this section we show that locally every étale R-algebra S is isomorphic to a standard étale algebra. This means that for any prime ideal $\mathfrak{q}$ in $\operatorname{Spec} S$, if $\mathfrak{p} = \mathfrak{q} \cap R$, then there is a basic open neighborhood $U(f)$ containing $\mathfrak{q}$, a basic open neighborhood $U(h)$ containing $\mathfrak{p}$, and a polynomial $P \in R[h^{-1}][x]$ such that $S[f^{-1}]$ is isomorphic to a localization of $R[h^{-1}][x]/(P)$. This means that locally, S is generated by a primitive element and upon restriction to the residue field of $R_{\mathfrak{p}}$, the polynomial P is separable. This section contains the theorems of permanence for étale R-algebras as well as the proof that étale algebras can be lifted modulo an ideal generated by nilpotent elements. The theorems of permanence allow us to define the étale neighborhood of a local ring. General references for the material in this section are [**Ray70**], [**Mil80**], SGA 1 ([**Gro71**], [**Gro03**]), and EGA IV ([**Gro64**] [**Gro65**] [**Gro66**] [**Gro67**]).

3.1. Quasi-finite Algebras. This section contains background results (the most important being Zariski's Main Theorem) that will be applied when we construct standard étale algebras in Section 9.3.3.

Proposition 9.3.1. *Let k be a field, B a finitely generated commutative k-algebra, and $q \in \operatorname{Spec} B$. The following are equivalent.*

(1) *q is an isolated point in* $\operatorname{Spec} B$.

(2) *B_q is a finite dimensional k-algebra.*

Proof. (1) implies (2): If the point q is isolated in the Zariski topology, then it is an open set. There exists $f \in B$ such that $q = \operatorname{Spec} B - V(f) = \operatorname{Spec} B_f$. Since B_f is noetherian and has only one prime ideal, B_f is artinian by Proposition 3.2.10. Since B_f has only one prime ideal, B_f is local with maximal ideal qB_f. By Exercise 3.2.17, B_f is finite dimensional over k. Since B_f is local, $B_f = (B_f)_q = B_q$, which shows B_q is finite dimensional over k.

(2) implies (1): Suppose B_q is finite dimensional over k. Let K and C be the kernel and cokernel of the localization map $B \to B_q$. Consider the sequence of B-modules

$$0 \to K \to B \to B_q \to C \to 0.$$

Then $K_q = C_q = 0$. Since B is noetherian, K is finitely generated over B. Since B_q is finite dimensional over k, C is finite dimensional over k hence finitely generated over B. By Lemma 2.1.3, there exists $f \in B - q$ such that $K_f = C_f = 0$. Therefore $B_f = B_q$. But B_q is local and finite dimensional over k, hence is artinian. So $\operatorname{Spec} B_q = q = \operatorname{Spec} B_f$. So q is isolated. □

In the following, if p is a prime ideal in A, and A_p is the local ring, then the residue field of A_p is denoted k_p.

Proposition 9.3.2. *Let B be a finitely generated commutative A-algebra. If q is in* $\operatorname{Spec} B$ *and $p = q \cap A$, then following are equivalent.*

(1) *q is an isolated point in* $\operatorname{Spec}(B \otimes_A k_p) = \operatorname{Spec}(B \otimes_A (A_p/pA_p))$.

(2) *B_q/pB_q is finite dimensional over k_p.*

Proof. By k_p we denote the residue field of A at the prime p. That is, $k_p = A_p/pA_p$. Then $B \otimes_A k_p = B \otimes_A A_p \otimes_{A_p} k_p = B_p \otimes_{A_p} k_p$. Also, $B_q = (B_p)_q$, from which we get $B_q/pB_q = (B_p)_q/p(B_p)_q$. It is enough to prove the proposition when A is a local ring with maximal ideal p. In this case, $B/pB = B \otimes_A k_p$ is a finitely generated algebra over the field $A/p = k_p$ and $(B/pB)_q = B_q/pB_q$. The proof follows from Proposition 9.3.1 applied to the algebra B/pB over the field k_p. □

If A and B are as in Proposition 9.3.2 and either (1) or (2) is satisfied, then we say B is *quasi-finite* over A at q. If this is true for all $q \in \operatorname{Spec} B$, then we say B is *quasi-finite* over A.

Let $A \to B \to C$ be homomorphisms of rings. Assume B is finitely generated as an A-module, C is finitely generated as a B-algebra and $\operatorname{Spec} C \to \operatorname{Spec} B$ is an open immersion (Exercise 2.3.9). Then C is quasi-finite over

A. The form of Zariski's Main Theorem stated below says every quasi-finite homomorphism factors in this way.

Theorem 9.3.3 (Zariski's Main Theorem). *Let B be a commutative A-algebra which is finitely generated as an A-algebra and which is quasi-finite over A. If $\tilde{A}$ is the integral closure of A in B, then*

(1) $\operatorname{Spec} B \to \operatorname{Spec} \tilde{A}$ *is an open immersion and*

(2) *there exists an A-subalgebra R of $\tilde{A}$ such that R is finitely generated as an A-module and* $\operatorname{Spec} B \to \operatorname{Spec} R$ *is an open immersion.*

Proof. See [**Ray70**, Corollaire 2, p. 42]. □

3.2. Exercises.

Exercise 9.3.4. (Quasi-finite over Quasi-finite is Quasi-finite) Prove that if B is quasi-finite over A and C is quasi-finite over B, then C is quasi-finite over A.

Exercise 9.3.5. Prove that if S is a commutative finitely generated separable R-algebra, then S is quasi-finite over R.

Exercise 9.3.6. Let A be a commutative ring. Prove:

(1) If B is a commutative A-algebra that is finitely generated as an A-module, then B is quasi-finite over A.

(2) If $f \in A$, then the localization A_f is quasi-finite over A.

3.3. Standard Étale Algebras. Let R be a commutative ring. A *standard étale R-algebra* is an étale R-algebra of the form $S[g^{-1}]$, where $f, g \in R[x]$, and $S = R[x]/(f)$ are as in the next proposition. We will prove in Theorem 9.3.9 that if R is noetherian, then locally every étale R-algebra is isomorphic to a standard étale R-algebra.

Proposition 9.3.7. *Let R be a commutative ring, f and g two polynomials in $R[x]$, and $S = R[x]/(f)$. Assume f', the derivative of f, restricts to a unit in the localization $S[g^{-1}]$. Then $S[g^{-1}]$ is an étale R-algebra.*

Proof. Case 1: Assume R is noetherian. Consider $I = (f(x), g(x)y - 1)$, which is an ideal in $A = R[x, y]$. Then A/I can be identified with $S[g^{-1}]$. The jacobian matrix corresponding to I is

$$J = \begin{bmatrix} f'(x) & g'(x)y \\ 0 & g(x) \end{bmatrix}.$$

The determinant of J is $f'(x)g(x)$, which is invertible in A/I. By Corollary 9.2.20, $S[g^{-1}]$ is étale over R.

Case 2: Reduce to the noetherian case. Localizing the natural map $R[x] \to S$ gives rise to a surjection $R[x][g^{-1}] \to S[g^{-1}]$. There is a polynomial $h \in R[x]$ such that hg^{-r} maps to the inverse of f' in $S[g^{-1}]$. Let $R_0[x]$ be a noetherian subring of $R[x]$ such that h, f, and g are in $R_0[x]$. Let $S_0 = R_0[x]/(f)$. Then f' restricts to a unit in the localization $S_0[g^{-1}]$. By Case 1, $S_0[g^{-1}]$ is étale over R_0. By Proposition 9.2.5 (2), $S[g^{-1}] = S_0[g^{-1}] \otimes_{R_0} R$ is étale over R. □

Remark 9.3.8. If S is a standard étale R-algebra, then we can present another proof of Corollary 4.7.3. By a finite induction argument applied to the sequence $A \to A/I^{m-1} \to \cdots \to A/I^2 \to A/I$, it suffices to prove the proposition when $I^2 = 0$. Assume we are given a standard étale R-algebra $S = (R[X]/P)[b^{-1}]$. Write x for the element of S represented by the polynomial X. Then $g(x) = a + I$, for some $a \in A$. Suppose $P(X) = \sum_{i=0}^{n} r_i x^i$. If $h \in I$, then $h^2 = 0$ and for $i > 0$, $(a+h)^i = a^i + iha^{i-1}$. In the ring A, we have

$$P(a+h) = \sum_{i=0}^{n} r_i (a+h)^i = P(a) + hP'(a).$$

Since $P'(x)$ is invertible in S, it follows that $g(P'(x)) = P'(a) + I$ is invertible in A/I. Since I is contained in the nil radical of A, $P'(a)$ is invertible in A. In A we set $h = -P(a)(P'(a))^{-1}$ and $\alpha = a + h$. Then $P(\alpha) = 0$, and the R-algebra homomorphism $R[X]/P \to A$ defined by $X \mapsto \alpha$ is well defined. To show that this map extends to S, suppose b is defined by a polynomial $Q \in R[X]$. Then $g(b) = Q(a)$ is invertible in A/I. But $Q(\alpha) + I = Q(a) + I$ shows $Q(\alpha)$ is invertible in A. Then $X \mapsto \alpha$ induces the desired R-algebra homomorphism γ.

Let S be a commutative R-algebra and $\mathfrak{q} \in \operatorname{Spec} S$. We say S is *unramified over R in a neighborhood of* $\mathfrak{q}$, if there exists $f \in S - \mathfrak{q}$ such that $S[f^{-1}]$ is a finitely generated separable R-algebra. We say S is *étale over R in a neighborhood of* $\mathfrak{q}$, if there exists $f \in S - \mathfrak{q}$ such that $S[f^{-1}]$ is étale over R.

Theorem 9.3.9. *Let R be a commutative noetherian ring, S a commutative R-algebra, $\mathfrak{q}$ a prime ideal in S, and $\mathfrak{p} = \mathfrak{q} \cap R$. If S is étale over R in a neighborhood of $\mathfrak{q}$, then there exist $f \in S - \mathfrak{q}$ and $h \in R - \mathfrak{p}$ such that $S[f^{-1}]$ is isomorphic to a standard étale $R[h^{-1}]$-algebra.*

Proof. The proof is divided into four steps.

Step 1: Reduction to the case where S is a finitely generated R-module. After replacing S by a localization, we assume S is separable and finitely generated as an R algebra. By Exercise 9.3.5, S is quasi-finite over R. By Theorem 9.3.3, there exists a factorization $R \to T \to S$ such that T is a finitely generated R-module and there exists $f \in T - \mathfrak{q} \cap T$ such that

$T[f^{-1}] = S[f^{-1}]$. The reader should verify that T is étale over R in a neighborhood of $\mathfrak{q} \cap \tilde{R}$, and that if the theorem is true for T, it is true for S.

Step 2: Reduction to the case where R is local. As in Step 1, we assume S is separable over R and finitely generated as an R-algebra. Assume the theorem is true for $R_\mathfrak{p}$. So there exist polynomials $f, g \in R_\mathfrak{p}[x]$ as in Proposition 9.3.7, $c \in S_\mathfrak{p} - \mathfrak{q}S_\mathfrak{p}$, and an $R_\mathfrak{p}$-algebra isomorphism

$$\left(\frac{R_\mathfrak{p}[x]}{(f)}\right)[g^{-1}] \xrightarrow{\phi} S_\mathfrak{p}[c^{-1}].$$

The reader should verify that there is an element $\alpha \in R - \mathfrak{p}$ such that f and g come from $R[\alpha^{-1}][x]$, c comes from $S[\alpha^{-1}]$, $\phi(x)$ comes from $S[\alpha^{-1}][c^{-1}]$, and $\phi(g^{-1})$ comes from a unit in $S[\alpha^{-1}][c^{-1}]$. Therefore, there is a commutative diagram

$$\begin{array}{ccc} \left(\frac{R[\alpha^{-1}][x]}{(f)}\right)[g^{-1}] & \xrightarrow{\psi} & S[\alpha^{-1}][c^{-1}] \\ \downarrow & & \downarrow \\ \left(\frac{R_\mathfrak{p}[x]}{(f)}\right)[g^{-1}] & \xrightarrow{\phi} & S_\mathfrak{p}[c^{-1}] \end{array}$$

where the vertical maps are induced by tensoring with $R_\mathfrak{p}$, and ψ is an $R[\alpha^{-1}]$-algebra homomorphism. Since R is noetherian, the kernel of ψ is a finitely generated ideal. Also, $S[\alpha^{-1}][c^{-1}]$ is a finitely generated $R[\alpha^{-1}]$-algebra. We can pass from the local to the global, by Lemma 2.1.5. That is, there exists an element $\beta \in R - \mathfrak{p}$ such that

$$\left(\frac{R[\beta^{-1}][x]}{(f)}\right)[g^{-1}] \xrightarrow{\phi} S[\beta^{-1}][c^{-1}]$$

is an isomorphism of $R[\beta^{-1}]$-algebras. So if the theorem is true for $R_\mathfrak{p}$, it is true for R.

Step 3: Reduction to the case where $S = R[x]$ is a cyclic extension. By the previous steps, we assume R is local, S is separable over R, and finitely generated as an R-module. By Theorem 8.3.7, $\mathfrak{p}$ generates the maximal ideal of $S_\mathfrak{q}$, and the extension of residue fields $k_\mathfrak{p} \to k_\mathfrak{q}$ is separable. Let $\bar{S} = S \otimes_R k_\mathfrak{p}$. Then $\bar{S}$ is artinian and the prime ideals of $\bar{S}$ correspond to those primes of S lying over $\mathfrak{p}$. Let $\bar{\mathfrak{q}}$ be the maximal ideal of $\bar{S}$ corresponding to $\mathfrak{q}$. Then $(\bar{S})_{\bar{\mathfrak{q}}} = k_\mathfrak{q}$. By the Primitive Element Theorem, $k_\mathfrak{q} = k_\mathfrak{p}(t)$ is a simple extension. By Theorem 3.2.11, there is $x \in S$ which maps to t in $(\bar{S})_{\bar{\mathfrak{q}}}$, and such that x restricts to 0 for all other prime ideals of $\bar{S}$. Let $C = R[x]$, and $\mathfrak{r} = \mathfrak{q} \cap C$. Then S is a finitely generated C-module and $\mathfrak{q}$ is the only prime ideal of S lying over $\mathfrak{r}$. Thus $S_\mathfrak{r} = S_\mathfrak{q}$ is a finitely generated $C_\mathfrak{r}$-module. The composite map

$$k_\mathfrak{p}(t) \to C_\mathfrak{r}/\mathfrak{r}C_\mathfrak{r} \to S_\mathfrak{q}/\mathfrak{q}C_\mathfrak{q} = k_\mathfrak{q}$$

is onto, hence $C_{\mathfrak{r}} \to S_{\mathfrak{q}}$ is onto (Exercise 1.1.18). By Lemma 2.1.4, there is $f \in C - \mathfrak{r}$ such that $C[f^{-1}] = S[f^{-1}]$.

Step 4: Assume R is local, $S = R[x]$ is finitely generated as an R-module, $\mathfrak{q} \in \operatorname{Spec} S$ is the only prime lying above the maximal ideal $\mathfrak{p} \in \operatorname{Spec} R$, and S is separable and flat over R. So $S/\mathfrak{p}S$ is an artinian local ring, and is separable over the field $k_{\mathfrak{p}}$. It follows from Corollary 4.5.8 that $S/\mathfrak{p}S$ is equal to the field $k_{\mathfrak{q}}$. Let n be the dimension of $k_{\mathfrak{q}}$ over $k_{\mathfrak{p}}$. Then $1, x, \dots, x^{n-1}$ generate $S/\mathfrak{p}S = k_{\mathfrak{q}}$ over $R/\mathfrak{p} = k_{\mathfrak{p}}$. By Corollary 1.1.17 (Nakayama's Lemma), the R-module S is generated by $1, x, \dots, x^{n-1}$. There is a monic polynomial $P \in R[X]$ such that $h : R[X]/(P) \to S$ is onto. The natural map $\bar{h} : (R/\mathfrak{p})[X]/(P) \to S/\mathfrak{p}S$ is an isomorphism. The image of P in $k_{\mathfrak{p}}[X]$ is therefore the irreducible polynomial for t over $k_{\mathfrak{p}}$. Since $k_{\mathfrak{q}}$ is separable over $k_{\mathfrak{p}}$, (P, P') is the unit ideal in $k_{\mathfrak{p}}[X]$. Let $\mathfrak{q}' = h^{-1}(\mathfrak{q})$. By Theorem 8.1.24, $(R[X]/(P))_{\mathfrak{q}'}$ is separable over R. By Proposition 4.6.1, (P, P') is the unit ideal in $(R[X]/(P))_{\mathfrak{q}'}$. By Lemma 2.1.3, there is some $b \in (R[X]/(P)) - \mathfrak{q}'$ such that $(R[X]/(P))[b^{-1}]$ is a standard étale R-algebra. If we set $c = h(b)$, then $h : (R[X]/(P))[b^{-1}] \to S[c^{-1}]$ is onto. By Corollary 9.2.6, $h : (R[X]/(P))[b^{-1}] \to S[c^{-1}]$ is étale. By Exercise 1.1.22, h is a localization. □

Remark 9.3.10. As in [**Ray70**, Théorème V.1], Theorem 9.3.9 can be proved without assuming R is noetherian. In the above proof, the noetherian hypothesis was used only in Step 2 to simplify the local to global argument. If R is not noetherian, it is necessary to assume S is finitely presented over R. The interested reader is referred to [**Gro64**, §I.4] for the necessary background results on homomorphisms between finitely presented algebras.

Theorem 9.3.11. *Let S a commutative R-algebra, $\mathfrak{q}$ a prime ideal in S, and $\mathfrak{p} = \mathfrak{q} \cap R$. If S is unramified over R in a neighborhood of $\mathfrak{q}$, then there exist $f \in S - \mathfrak{q}$ and $h \in R - \mathfrak{p}$ such that $S[f^{-1}]$ is the homomorphic image of a standard étale $R[h^{-1}]$-algebra.*

Proof. Mimic the proof of Theorem 9.3.9. In Step 2 assume only that ϕ is onto. To pass from the local to the global, it is enough to assume S is finitely generated as an R-algebra. The only place in the proof of Theorem 9.3.9 where it was necessary to assume S/R is flat was at the end of Step 4, where it was used to prove that h is flat. □

3.4. Theorems of Permanence. We investigate some properties of R that are inherited by an étale R-algebra.

Theorem 9.3.12. *Let S be a noetherian étale R-algebra.*

(1) *For every* $\mathfrak{q} \in \operatorname{Spec} S$, *if* $\mathfrak{p} = \mathfrak{q} \cap R$, *then* $\dim(S_{\mathfrak{q}}) = \dim(R_{\mathfrak{p}})$.

(2) *If R is regular, then S is regular. If S is regular, and is a faithfully flat R-algebra, then R is regular.*

(3) *If R is a normal ring, then S is a normal ring. If S is a normal ring which is a faithfully flat R-algebra, then R is normal.*

(4) *If* $\operatorname{Rad}_R(0) = (0)$, *then* $\operatorname{Rad}_S(0) = (0)$. *If S is an integral domain which is a faithfully flat R-algebra, then R is an integral domain.*

Proof. Let $\mathfrak{q} \in \operatorname{Spec} S$, and $\mathfrak{p} = \mathfrak{q} \cap R$. Let $\mathfrak{m}_\mathfrak{p} = \mathfrak{p}R_\mathfrak{p}$, and $\mathfrak{m}_\mathfrak{q} = \mathfrak{q}S_\mathfrak{q}$. By Theorem 8.3.7 (3), $\mathfrak{m}_\mathfrak{q}$ is generated by $\mathfrak{m}_\mathfrak{p}$.

(1): Let $k_\mathfrak{p} = R_\mathfrak{p}/\mathfrak{m}_\mathfrak{p}$ denote the residue field at $\mathfrak{p}$. By Theorem 8.3.7 (3), $S_\mathfrak{q} \otimes_R k_\mathfrak{p}$ is a finite dimensional separable extension field of $k_\mathfrak{p}$, hence has Krull dimension zero. Because S/R is flat, the rest follows from Theorem 6.1.1.

(2): Since $R_\mathfrak{p}$ is regular, $\mathfrak{m}_\mathfrak{p}$ is generated by $\dim(R_\mathfrak{p})$ elements. Hence $\mathfrak{m}_\mathfrak{q} = \mathfrak{m}_\mathfrak{p} S_\mathfrak{q}$ is generated by $\dim(R_\mathfrak{p})$ elements. Therefore, $S_\mathfrak{q}$ has a regular system of parameters and is regular. The second statement follows from Theorem 6.1.9 (3).

(3): We apply the Serre Criteria, Theorem 6.1.5. By Part (2), if $\operatorname{ht}(\mathfrak{q}) = 1$, then $S_\mathfrak{q}$ is regular and (R_1) is satisfied. Assume $\operatorname{ht}(\mathfrak{q}) = 2$. By Part (1), $\dim(R_\mathfrak{p}) = 2$. Since R is normal, $\operatorname{depth}(R_\mathfrak{p}) = 2$. Since S/R is flat, it follows from Lemma 6.1.8 and Theorem 6.1.4 that

$$\operatorname{Ext}^i_{R_\mathfrak{p}}(k_\mathfrak{p}, R_\mathfrak{p}) \otimes_{R_\mathfrak{p}} S_\mathfrak{q} = \operatorname{Ext}^i_{S_\mathfrak{q}}(k_\mathfrak{q}, S_\mathfrak{q})$$

is zero if $i < 2$, and nonzero if $i = 2$. Therefore, $\operatorname{depth}(S_\mathfrak{q}) = 2$ and (S_2) is satisfied. The second statement is Lemma 3.3.11.

(4): Let $\mathfrak{p}_1, \dots, \mathfrak{p}_n$ be the distinct minimal primes of R. The kernel of the natural map

$$R \xrightarrow{\phi} \bigoplus_{i=1}^{n} R/\mathfrak{p}_i$$

is $\operatorname{Rad}_R(0) = (0)$. Since S/R is flat,

$$S \xrightarrow{1 \otimes \phi} \bigoplus_{i=1}^{n} S \otimes_R R/\mathfrak{p}_i$$

is one-to-one. By Proposition 9.2.5, $S \otimes_R R/\mathfrak{p}_i$ is an étale $R/\mathfrak{p}_i$-algebra. If the nil radical of each of the rings $S \otimes_R R/\mathfrak{p}_i$ is trivial, then we are done. Therefore, it is enough to prove the claim when R is an integral domain. Suppose R is an integral domain with quotient field K. Since S/R is flat, $S \to S \otimes_R K$ is one-to-one. By Corollary 4.3.2, $S \otimes_R K$ is a separable K-algebra. By Corollary 4.5.8, the nil radical of $S \otimes_R K$ is trivial. The second statement follows from Lemma 2.4.4. □

3.5. Étale Algebras over a Normal Ring. We study étale algebras over a commutative noetherian normal ring R. With this extra hypothesis on the base ring, we can strengthen some of results of the previous section. Our treatment of this subject is based on [**Mil80**, § I.3].

Proposition 9.3.13. *Let R be a commutative noetherian normal ring, S a commutative R-algebra, $\mathfrak{q}$ a prime ideal in S, and $\mathfrak{p} = \mathfrak{q} \cap R$. If S is étale over R in a neighborhood of $\mathfrak{q}$, then there exist $f \in S - \mathfrak{q}$ and $h \in R - \mathfrak{p}$ such that $S[f^{-1}]$ is isomorphic to a standard étale $R[h^{-1}]$-algebra of the form $(R[h^{-1}][x]/(P))[b^{-1}]$ where $R[h^{-1}]$ is an integral domain and P is irreducible over the quotient field of $R[h^{-1}]$.*

Proof. After localizing, we can assume S is étale over R. By Theorem 9.3.12, S is a normal ring. Using Lemma 3.3.10, we can localize to appropriate basic open neighborhoods of $\mathfrak{p}$ and $\mathfrak{q}$ and assume R and S are normal domains. By Theorem 9.3.9, we can localize again, and assume S is R-isomorphic to a standard étale R-algebra $(R[X]/(P))[g^{-1}]$. To finish the proof, we need to show that $P \in R[X]$ can be replaced with a polynomial that is irreducible over the quotient field K of R. As in Step 2 of the proof of Theorem 9.3.9, it suffices to assume R is local. By Step 4 of the proof of Theorem 9.3.9, assume $S = (R[X]/(P))[g^{-1}]$, and $\mathfrak{q}$ is the only prime ideal lying over $\mathfrak{p}$. Then $S \otimes_R k_{\mathfrak{p}} = k_{\mathfrak{q}}$ is separable over the field $k_{\mathfrak{p}}$, and the restriction of P to $k_{\mathfrak{p}}[X]$ has no repeated roots. Assume $P = P_1 Q$ is a factorization in $K[X]$ where P_1 is monic, irreducible and such that the restriction of P_1 to $S \otimes_R k_{\mathfrak{p}}$ is zero. By Lemma 3.3.3 (Gauss' Lemma), $P_1 \in R[X]$. Consider the ideal $I = (P_1, Q)$ in $R[x]$. Since P_1 and Q are relatively prime over $k_{\mathfrak{p}}$, $(R[x]/I) \otimes_R k_{\mathfrak{p}} = (0)$. By Nakayama's Lemma, Corollary 1.1.14, $I = R[x]$. By the Chinese Remainder Theorem,

$$R[X]/(P) \xrightarrow{\phi} R[X]/(P_1) \oplus R[X]/(Q)$$

is an isomorphism. Then S is isomorphic to $\big(R[X]/(P_1) \oplus R[X]/(Q)\big)[\phi(g)^{-1}]$. Let g_1 be the projection of $\phi(g)$ onto the first summand. The projection $S \to (R[X]/(P_1))[g_1^{-1}]$ is étale by Proposition 9.2.5 (1). □

Theorem 9.3.14. *Let R be a commutative noetherian normal ring and S an R-algebra.*

(1) *If S is a flat R-algebra, then for every $\mathfrak{q} \in \operatorname{Spec} S$, the natural map $R_{\mathfrak{p}} \to S_{\mathfrak{q}}$ is one-to-one, where $\mathfrak{p} = R \cap \mathfrak{q}$.*

(2) *If S is unramified over R and for every $\mathfrak{q} \in \operatorname{Spec} S$ the natural map $R_{\mathfrak{p}} \to S_{\mathfrak{q}}$ is one-to-one, where $\mathfrak{p} = R \cap \mathfrak{q}$, then S is étale over R.*

Proof. (1): The proof is left to the reader.

(2): We need to prove S is a flat R-algebra. It suffices to prove $S_{\mathfrak{q}}$ is a flat $R_{\mathfrak{p}}$-algebra (Proposition 2.4.8). We can assume R is an integrally closed local integral domain with quotient field K. By Theorem 9.3.9, we can assume S is the homomorphic image of a standard étale R-algebra of the form $(R[X]/(P))[g^{-1}]$. By Proposition 9.3.13, we can assume P is irreducible in $K[X]$. Write $C = R[X]/(P)$, and let $\mathfrak{q}' = \mathfrak{q} \cap C$. Consider the commutative diagram

$$\begin{array}{ccccc} R & \longrightarrow & C_{\mathfrak{q}'} & \xrightarrow{\phi} & S_{\mathfrak{q}} \\ \downarrow & & \downarrow \alpha & & \downarrow \beta \\ K & \longrightarrow & C_{\mathfrak{q}'} \otimes_R K & \xrightarrow{\phi \otimes 1} & S_{\mathfrak{q}} \otimes_R K \end{array}$$

of R-algebras. We are given that $R \to S_{\mathfrak{q}}$ is one-to-one. Tensoring with $(\ ) \otimes_R K$ is exact, so $K \to S_{\mathfrak{q}} \otimes_R K$ is one-to-one. This implies $\phi \otimes 1$ is not the zero map. Notice that $C_{\mathfrak{q}'} \otimes_R K = K[X]/(P)$ is an extension field of K, hence $\phi \otimes 1$ is one-to-one. Because $C_{\mathfrak{q}'}$ is étale over R, the map α is one-to-one. It follows that ϕ is one-to-one. Since $C[g^{-1}] \to S$ is onto, ϕ is onto. This proves ϕ is an isomorphism, so $S_{\mathfrak{q}}$ is étale over R. □

Corollary 9.3.15. *Let R be a noetherian integrally closed integral domain with quotient field K.*

(1) *Let L be a finite separable field extension of K, and $\tilde{R}$ the integral closure of R in L.*
   (a) *The set* $\operatorname{Spec} \tilde{R} - \operatorname{Supp}(\Omega_{\tilde{R}/R})$ *is a nonempty open.*
   (b) *If $\alpha \in \tilde{R}$ and $V(\alpha) \supseteq \operatorname{Supp}(\Omega_{\tilde{R}/R})$, then $R \to \tilde{R}[\alpha^{-1}]$ is étale.*

(2) *If $R \to S$ is étale, then $S = S_1 \oplus \cdots S_n$ where each S_i is an integrally closed integral domain.*

(3) *For each i, if L_i is the quotient field of S_i and $\tilde{R}_i$ is the integral closure of R in L_i, then* $\operatorname{Spec} S_i \to \operatorname{Spec} \tilde{R}_i$ *is an open immersion.*

Proof. (2): By Theorem 9.3.12, S is a normal ring. By Lemma 3.3.10, S decomposes into a direct sum of a finite number of integrally closed integral domains.

(1): By Proposition 8.3.2 and Exercise 8.3.16, $\operatorname{Supp}(\Omega_{\tilde{R}/R})$ is a closed set. By Theorem 8.4.2, $\Omega_{\tilde{R}/R}$ is zero at the generic point of $\operatorname{Spec} \tilde{R}$. This proves Part (a). If α is as in Part (b), then $R \to \tilde{R}[\alpha^{-1}]$ is unramified by Theorem 8.3.6 and étale by Theorem 9.3.14.

(3): Since $S \to S_i$ is a localization, it follows from Proposition 9.2.5 that $R \to S_i$ is étale. Therefore, S_i is quasi-finite over R and the claim follows from Theorem 9.3.3. □

3.6. Topological Invariance of Étale Coverings. If R is a commutative ring and I is an ideal contained in the nil radical of R, then $\operatorname{Spec} R \to \operatorname{Spec} R/I$ is a homeomorphism. The next two theorems give sufficient conditions for the unique lifting of étale coverings of R/I to étale coverings of R. By comparison, Section 7.9 considers the similar problem of lifting Azumaya R/I-algebras.

Theorem 9.3.16. *Let R be a commutative noetherian ring and I an ideal contained in* $\operatorname{Rad}_R(0)$, *the nil radical of R.*

(1) *If S is an étale R-algebra, then $S \otimes_R R/I$ is an étale R/I-algebra.*
(2) *If T is a faithfully flat noetherian étale R/I-algebra, then there exists a faithfully flat noetherian étale R-algebra S, such that T is isomorphic to $S \otimes_R R/I$ as R/I-algebras.*
(3) *If T is a noetherian étale R/I-algebra, then there exists a noetherian étale R-algebra S, such that T is isomorphic to $S \otimes_R R/I$ as R/I-algebras.*
(4) *The change of base functor $(\ ) \otimes_R R/I$ from the category of noetherian étale R-algebras to the category of noetherian étale R/I-algebras is essentially surjective.*

Proof. Part (1) follows from Proposition 9.2.5 (3).

(2): The proof is divided into 3 steps.

Step 1: A standard étale R/I-algebra lifts to a standard étale R-algebra. Let $\mathfrak{q}$ be a prime ideal of T, and $\mathfrak{p} = \mathfrak{q} \cap (R/I)$. By Theorem 9.3.9, find $f \in T - \mathfrak{q}$ and $h \in R - \mathfrak{p}$ such that $T[f^{-1}]$ is isomorphic to a standard étale $(R/I)[h^{-1}]$-algebra. By Theorem 2.5.6 we can assume $T[f^{-1}]$ is connected. Using Proposition 9.3.7, the reader should verify that there exists a standard étale $R[h^{-1}]$-algebra A such that $A \otimes_R (R/I) \cong T[f^{-1}]$, as $(R/I)[h^{-1}]$-algebras. To complete the proof, we need to glue the various A together to produce an étale R-algebra.

Step 2: Construct algebras on each open set of a finite covering of $\operatorname{Spec} R$. We are given that $\operatorname{Spec} T \to \operatorname{Spec} R/I$ is onto (Lemma 2.4.4). So as $\mathfrak{q}$ ranges over $\operatorname{Spec} T$, if $f \in T$ and $h \in R/I$ are chosen as in Step 1, then the basic open sets $U(f)$ and $U(h)$ make up open coverings of $\operatorname{Spec} T$ and $\operatorname{Spec} R$. By Exercise 2.2.16, there is a finite set of such elements $f_1, \dots, f_n$ in T and $h_1, \dots, h_n$ in R such that $\operatorname{Spec} T = U(f_1) \cup \dots \cup U(f_n)$ and $\operatorname{Spec} R = U(h_1) \cup \dots \cup U(h_n)$.

Step 3: The desired R-algebra S is obtained by gluing together the algebras defined on the open cover. For each i, let A_i be a standard étale $R[h_i^{-1}]$-algebra such that $A_i \otimes_R (R/I) \cong T[f_i^{-1}]$, as $(R/I)[h_i^{-1}]$-algebras. Upon tensoring with R/I, $R[h_i^{-1}] \otimes_R A_j$ and $A_i \otimes_R R[h_j^{-1}]$ both become isomorphic to $T[(f_if_j)^{-1}]$ as $(R/I)[(h_ih_j)^{-1}]$-algebras. By Corollary 4.7.3, there are $R[(h_ih_j)^{-1}]$-algebra homomorphisms

$$R[h_i^{-1}] \otimes_R A_j \xrightarrow{g_{ij}} A_i \otimes_R R[h_j^{-1}]$$

Using Corollary 9.2.8 one verifies that the maps g_{ij} are isomorphisms and $g_{ik} = g_{ij}g_{jk}$, for all triples (i, j, k). By Example 5.3.9 and Theorem 5.3.10 there exists an R-algebra S such that $S \otimes_R R[h_i^{-1}] \cong A_i$, for each i. It follows from Theorem 8.3.7 that S is separable over R. Since S is locally flat, it is flat. This shows that S is étale over R. That $S \otimes_R R/I$ is isomorphic to T follows from Proposition 5.3.1 and the fact that they agree on the open cover $U(h_1), \dots, U(h_n)$.

(3): We apply (2) to the faithfully flat étale R/I-algebra $T \oplus R/I$. There is an R-algebra S such that $S \otimes_R R/I \cong T \oplus R/I$. Let f be the idempotent corresponding to the direct summand T and e the unique idempotent of S that lifts f (Corollary 3.4.16). It follows from Exercises 2.4.15 and 4.4.10 that Se is an étale R-algebra. Since $Se \otimes_R R/I \cong T$, we are done.

Part (4) follows from Part (3). □

Theorem 9.3.17. *Let R be a commutative ring and I a nilpotent ideal in R. The functor $S \mapsto S \otimes_R R/I$ establishes an equivalence between the category of commutative separable flat R-algebras and the category of commutative separable flat R/I-algebras.*

Proof. The functor is essentially surjective, by Theorem 9.3.16. It is enough to show that the functor is also fully faithful, by Proposition 1.1.2. Let S and T be commutative separable flat R-algebras. We show that the map

$$\operatorname{Hom}(S, T) \to \operatorname{Hom}(S \otimes_R R/I, T \otimes_R R/I)$$

is a one-to-one correspondence. By Exercise 1.3.8, to give an R-algebra homomorphism $\alpha : S \to T$ is equivalent to giving an R-algebra homomorphism $\gamma : S \otimes_R T \to T$ such that the diagram

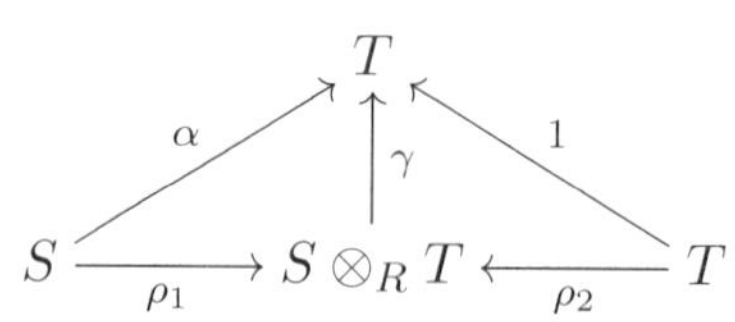

commutes. But $S \otimes_R T$ is separable and flat over T (Corollary 4.3.2 and Theorem 1.3.2). By Corollary 4.4.8, the homomorphisms γ correspond bijectively to idempotents e in $S \otimes_R T$ such that the composite mapping $T \to (1 \otimes T)e \to (S \otimes_R T)e$ is an isomorphism. Since I is nilpotent, Corollary 3.4.16 implies $\operatorname{idemp}(S \otimes_R T) \to \operatorname{idemp}(S \otimes_R T \otimes_R R/I)$ is a one-to-one correspondence. Clearly the existence of γ gives rise to a splitting $g : S \otimes_R T \otimes_R R/I \to T \otimes_R R/I$. Conversely, by Corollary 4.7.3, a T-algebra homomorphism g gives rise to a T-algebra homomorphism γ such that the diagram

$$\begin{array}{ccccc} S \otimes_R T & \longrightarrow & S \otimes_R T \otimes_R R/I & \xrightarrow{g} & T \otimes_R R/I \\ \uparrow{\scriptstyle \rho_2} & & \searrow^{\gamma} & & \uparrow \\ T & & \xrightarrow{=} & & T \end{array}$$

commutes. □

3.7. Étale Neighborhood of a Local Ring. Let R be a local ring with maximal ideal $\mathfrak{m}$ and residue field k. An *étale neighborhood* of R is a pair $(A, \mathfrak{p})$ where A is an étale R-algebra, $\mathfrak{p} \in \operatorname{Spec} A$, $\mathfrak{p} \cap R = \mathfrak{m}$, and $k \to k(\mathfrak{p})$ is an isomorphism. If $\operatorname{Spec} A$ is connected, then we say $(A, \mathfrak{p})$ is a *connected étale neighborhood* of R. The results of this section will have applications in the next chapter.

Lemma 9.3.18. *Let R be a noetherian local ring with maximal ideal $\mathfrak{m}$ and residue field k. Let $(A, \mathfrak{p})$ and $(B, \mathfrak{q})$ be two étale neighborhoods of R.*

(1) *If B is connected, then there is at most one R-algebra homomorphism $f : A \to B$ such that $f^{-1}(\mathfrak{q}) = \mathfrak{p}$.*
(2) *There exists a connected étale neighborhood $(C, \mathfrak{r})$ of R, together with R-algebra homomorphisms $f : A \to C$, $g : B \to C$ such that $f^{-1}(\mathfrak{r}) = \mathfrak{p}$ and $g^{-1}(\mathfrak{r}) = \mathfrak{q}$.*
(3) *The collection $\mathfrak{N}$ of all connected étale neighborhoods of R is a directed system of R-algebras.*

Proof. Part (1) follows straight from Corollary 9.2.8.

(2): The homomorphisms $A \to k(\mathfrak{p}) = k$ and $B \to k(\mathfrak{q}) = k$ induce a homomorphism $h : A \otimes_R B \to k$ (Exercise 1.3.8). Let $\mathfrak{r}$ be the kernel of h. Let C be the connected component of $A \otimes_R B$ which contains the prime ideal $\mathfrak{r}$ (Theorem 2.5.6).

(3): Let $\mathfrak{N} = \{(A, \mathfrak{p})\}$ be the collection of all connected étale neighborhoods of R. Elements of $\mathfrak{N}$ are unique up to isomorphism. The binary relation on $\mathfrak{N}$ is $(A, \mathfrak{p}) \le (B, \mathfrak{q})$ if there exists an R-algebra homomorphism $f : A \to B$ such that $f^{-1}(\mathfrak{q}) = \mathfrak{p}$. By Parts (1) and (2) $\mathfrak{N}$ is a directed system of R-algebras. □

Lemma 9.3.19. *Let $(R, \mathfrak{m})$ be a local ring with residue field k and let $(A, \mathfrak{p})$ be an étale neighborhood of R. If $(S, \mathfrak{n})$ is a local ring and $\theta : R \to S$ a local homomorphism of local rings, then there is a maximal ideal $\mathfrak{r}$ in $A \otimes_R S$ such that $(A \otimes_R S, \mathfrak{r})$ is an étale neighborhood of S.*

Proof. Let $k(\mathfrak{n})$ be the residue field of S. Then θ induces a homomorphism of fields $k \to k(\mathfrak{n})$. The homomorphisms $A \to k(\mathfrak{p}) = k$ and $S \to k(\mathfrak{n})$ induce a homomorphism $h : A \otimes_R S \to k(\mathfrak{n})$ (Exercise 1.3.8). Let $\mathfrak{r}$ be the kernel of h. Since h is onto, $\mathfrak{r}$ is maximal and the diagram

$$\begin{array}{ccc} A \otimes_R S & \xrightarrow{h} & k(\mathfrak{n}) \\ \uparrow & \nearrow & \\ S & & \end{array}$$

commutes. By Proposition 9.2.5, $A \otimes_R S$ is étale over S. □

Definition 9.3.20. Let A be a local ring with maximal ideal $\mathfrak{m}$ and B a commutative A-algebra. We say that B is a *local étale* A-algebra if there is an étale A-algebra C and a prime ideal $\mathfrak{n}$ in $\operatorname{Spec} C$ such that $\mathfrak{m} = \mathfrak{n} \cap A$ and B is isomorphic as an A-algebra to $C_{\mathfrak{n}}$. Notice that in this case $A \to B$ is a local homomorphism of local rings. If R is noetherian, then by the Hilbert Basis Theorem (see, for example, [**Hun80**, Theorem VIII.4.9]), so are C and A. In the terminology of [**Gro67**, (18.6.1)], a local étale algebra is called *essentiellement étale*. Let I be a directed index set and $\{A_i, \phi^i_j\}$ a directed system of local étale A-algebras such that each $\phi^i_j : A_i \to A_j$ is a local homomorphism of local rings. The direct limit $\varinjlim A_i$ is said to be a *local ind-étale* A-algebra.

Theorem 9.3.21. *If R is a local ring and A a local ind-étale R-algebra, then the following are true.*

(1) *A is a local ring and a faithfully flat R-algebra.*

(2) *If $\mathfrak{m}$ is the maximal ideal of R and $\mathfrak{n}$ the maximal ideal of A, then $\mathfrak{m}^q A = \mathfrak{n}^q$ for any $q \geq 1$.*

(3) *For every $\mathfrak{p} \in \operatorname{Spec} R$, $A \otimes_R k(\mathfrak{p})$ is an integral $k(\mathfrak{p})$-algebra.*

(4) *A is reduced if and only if R is reduced.*

(5) *A is normal if and only if R is normal.*

(6) *A is noetherian if and only if R is noetherian.*

Proof. Let $A = \varinjlim A_i$, where $\{A_i, \phi^i_j\}$ is a directed system of local étale R-algebras for the directed index set I such that for all $i \leq j$, $\phi^i_j : A_i \to A_j$ is a local homomorphism of local rings. By Theorem 4.4.3, $\phi^i_j : A_i \to A_j$ is separable and faithfully flat. Let $\mathfrak{m}_i$ be the maximal ideal of A_i. By Theorem 8.3.7, $\mathfrak{m}_i A_j = \mathfrak{m}_j$. By Corollary 9.1.6, A is a local ring with maximal ideal $\mathfrak{m} = \varinjlim \mathfrak{m}_i$. By Exercise 1.4.15, A is faithfully flat over R. If R is noetherian, then so is each A_i (Hilbert Basis Theorem) and by Corollary 9.1.6, A is noetherian. The rest is left to the reader. □

4. Ramified Radical Extensions

As another application of Theorem 9.3.12, we study the important class of finite extensions of commutative rings defined by adjoining an nth root of an element. Let R be a commutative ring, $n \geq 2$, $a \in R$, and set $S = R[x]/(x^n - a)$. We say S/R is a *radical extension of degree n*. In this section, the emphasis is on radical extensions which are not separable over R. Such an extension is also said to be a *ramified extension*. Our goal is to derive necessary and sufficient conditions on n and a such that if R is a noetherian normal integral domain, then so is S. Necessary conditions are provided by Lemma 9.4.2 (2). Sufficient conditions are stated in Lemma 9.4.3 and Theorem 9.4.4. The results of this section will be applied below in Section 12.9.1, when we study cyclic Galois extensions. For reference, we state sufficient conditions for S to be a separable R-algebra.

Lemma 9.4.1. *Let R be a commutative ring, $n \geq 2$, and $a \in R$. Then the following are true for the radical extension $S = R[x]/(x^n - a)$.*

(1) *S is an R-algebra which is a finitely generated free R-module of rank n with basis $1, x, \dots, x^{n-1}$.*

(2) *If a and n are both invertible in R, then S is separable and hence étale over R.*

Proof. Part (1) is left to the reader. Part (2) follows from Part (1) and Exercise 4.6.20. □

Lemma 9.4.2. *Let R be a commutative ring and a an element of R that is not a zero divisor. If $n \geq 2$ and $e \geq 1$, then the following are true for the radical extension $S = R[x]/(x^n - a^e)$.*

(1) *a and x are not zero divisors in S.*

(2) *If a is not a unit in R and $e \geq 2$, then S is not integrally closed in $Q(S)$, the total ring of quotients of S.*

Proof. (1): Since S is a free R-module (Lemma 9.4.1), a is not a zero divisor of S. Suppose $a_0, \dots, a_{n-1}$ are elements of R and $(a_0 + a_1 x + \cdots + a_{n-1}x^{n-1})x = 0$. Then $a_0 x + a_1 x^2 + \cdots + a_{n-2}x^{n-1} + a_{n-1}a = 0$ implies $0 = a_0 = \cdots = a_{n-1}$.

(2): Let $w = ax^{-1}$ and $v = xa^{-1}$, which are elements of $Q(S)$ by Lemma 9.4.1. If $n \geq e$, then $w^n = a^n(x^n)^{-1} = a^{n-e} \in S$. Therefore, w is integral over S. For contradiction's sake, assume $w \in S$. Then there are elements a_i of R such that $a_0 + a_1 x + \cdots + a_{n-1}x^{n-1} = ax^{-1}$. Then $a_0 x + a_1 x^2 + \cdots + a_{n-2}x^{n-1} + a_{n-1}x^n = a$, which implies $0 = a_0 = \cdots = a_{n-2}$, and $a_{n-1}a^e = a$. This is a contradiction, since a is not a zero divisor and not invertible. If $n < e$, then by a similar argument one shows v is integral over S, and $v \notin S$. $\square$

Now we derive sufficient conditions for a radical extension of a noetherian normal integral domain R to be a noetherian normal integral domain. Let a be a nonzero element of R and assume the divisor of a is

$$\operatorname{Div}(a) = P_1 + \cdots + P_v$$

where $P_1, \dots, P_v$ are the distinct minimal primes of a in $X_1(R)$. In this case we say that $\operatorname{Div}(a)$ is a *reduced effective divisor.*

Lemma 9.4.3. *Let R be a DVR with maximal ideal $\mathfrak{m} = (\pi)$. Let $S = R[x]/(x^n - \pi)$, where $n \geq 2$. Then S is a DVR with maximal ideal $M = (x)$.*

Proof. Since R is a UFD, so is $R[x]$. By Eisenstein's Criterion, $x^n - \pi$ is irreducible in $R[x]$. Therefore, S is an integral domain. By the Hilbert Basis Theorem, S is noetherian. Since $S/(x) = R/(\pi)$ is a field, $M = (x)$ is a maximal ideal in S. By Theorem 3.3.5 every maximal ideal of S contains π. Since $x^n = \pi$, this implies M is the unique maximal ideal, so S is a local ring. By Krull's Hauptidealsatz, $\operatorname{ht}(M) = 1$. Therefore, $\dim(S) = 1$ and by Theorem 6.1.14, S is a DVR. $\square$

Theorem 9.4.4. *Let R be a noetherian normal integral domain with quotient field K. Let a be a nonzero element of R and assume $\operatorname{Div}(a)$ is a reduced effective divisor and $n \geq 2$ is invertible in R. If $S = R[x]/(x^n - a)$ and $L = K[x]/(x^n - a)$, then the following are true.*

(1) *L is a field.*
(2) *S is a noetherian integral domain.*
(3) *L is the quotient field of S.*
(4) *S is a noetherian normal integral domain.*
(5) *S is the integral closure of R in L.*

Proof. (1): For each $P \in X_1(R)$, R_P is a DVR with valuation ν_P. Let $\operatorname{Div}(a) = P_1 + \cdots + P_v$, where $P_1, \dots, P_v$ are the distinct minimal primes of a in $X_1(R)$. For each i, $\nu_{P_i}(a) = 1$, so a is a local parameter for R_{P_i}. By Lemma 9.4.3, $x^n - a$ is irreducible in $R_{P_i}[x]$. By Gauss' Lemma, $x^n - a$ is irreducible in $K[x]$. Therefore, L is a field.

(2): By Lemma 9.4.1, S is a free R-module of rank n and $1, x, \dots, x^{n-1}$ is a basis. The natural mapping $S = S \otimes_R R \to S \otimes_R K = L$ is one-to-one since S is a flat R-module. Hence S is a subring of L and consequently an integral domain. By the Hilbert Basis Theorem, S is noetherian.

(3): The natural mapping $S \to L$ is a localization of S.

(4): By Proposition 3.3.7 it suffices to show S_Q is a normal integral domain for each $Q \in \operatorname{Spec} S$. Let $P = Q \cap R$. If $a \notin P$, then by Lemma 9.4.1, $S_P = S \otimes_R R_P$ is étale over R_P. By Theorem 9.3.12, S_P is normal. By Proposition 3.3.7 S_Q is normal.

Now we assume $a \in P = Q \cap R$ and show S_Q is normal. We apply the Serre Criteria, Theorem 6.1.5. First we show S is regular in codimension one. Since $a \in Q$ and Q is prime, it follows that $x \in Q$. We have

$$\frac{S}{(x)} = \frac{R[x]}{(x^n - a, x)} = \frac{R}{(a)} \tag{9.16}$$

so the prime ideals of S containing x correspond bijectively with the prime ideals of R containing a. A prime ideal $Q \in \operatorname{Spec} S$ corresponds to $P = Q \cap R$. A prime ideal $P \in \operatorname{Spec} R$ corresponds to $Q = PS + (x)$. In particular, for $1 \le i \le v$, if we set $Q_i = P_i S + (x)$, then $Q_i \cap R = P_i$. Since a is a local parameter for R_{P_i}, Lemma 9.4.3 shows that S_{Q_i} is a DVR with local parameter x. We have shown that S is regular in codimension one. Now suppose Q is a prime ideal in S containing x and $\operatorname{ht}(Q) \ge 2$. We show $\operatorname{depth}(S_Q) \ge 2$. Let $P = Q \cap R$. Since R_P is normal, the associated primes of aR_P are $P_1 R_P, \dots, P_v R_P$, the height one primes containing a (see, for example, [**Mat80**, (17.H), Theorem 38]). By (9.16), $S_Q/(x) = R_P/(a)$, so the associated primes of xS_Q are $Q_1 S_Q, \dots, Q_v S_Q$, the height one primes containing x. By Theorem 6.1.5, $\operatorname{depth}(R_P) \ge 2$, so there exists $b \in PR_P$ such that $b \notin P_i R_P$ for each i. Since $P_i = Q_i \cap R$, it follows that $b \notin Q_i S_Q$ for each i. Therefore, b is not a zero divisor for xS_Q. This proves x, b is a regular sequence for S_Q in QS_Q, hence $\operatorname{depth}(S_Q) \ge 2$.

(5): S is integral over R and S is integrally closed in L. □

Corollary 9.4.5. *In the context of Theorem* 9.4.4, *assume* $\operatorname{Div}(a) = P_1 + \cdots + P_v$. *Then the following are true.*

(1) *There are unique primes* $Q_1, \dots, Q_v$ *in* $X_1(S)$ *such that* $P_i = Q_i \cap R$ *and* $\operatorname{Div}(x) = Q_1 + \cdots + Q_v$.

(2) *For each* i, *the ramification index* e_{Q_i} *is equal to* n.

Proof. This follows from the proof of Part (4) of Theorem 9.4.4. □

Corollary 9.4.6. *In the context of Theorem* 9.4.4, *assume moreover that* R *is a* $\mathbb{Z}[n^{-1}, \zeta]$*-algebra, where* ζ *is a primitive* n*th root of unity in* $\mathbb{C}$. *Then the following are true.*

(1) L/K *is a cyclic Galois extension with group* $G = \langle \sigma \rangle$, *and* $\sigma(x) = \zeta x$.

(2) *If* $\operatorname{Div}(x) = Q_1 + \cdots + Q_v$, *then in the context of Theorem* 6.5.8,

$$\bigoplus_{i=1}^{v} (\mathbb{Z}/n)Q_i = \operatorname{Div}(S)^G / \operatorname{Div}(R)$$

is a free $\mathbb{Z}/n$*-module of rank* v.

(3) *There is an is an exact sequence*

$$0 \to \operatorname{Cl}(S/R) \xrightarrow{\gamma_0} \mathrm{H}^1(G, S^*) \xrightarrow{\gamma_1} \bigoplus_{i=1}^{v} (\mathbb{Z}/n)Q_i \xrightarrow{\gamma_2} \operatorname{Cl}(S)^G / \operatorname{Cl}(R) \xrightarrow{\gamma_3} \mathrm{H}^2(G, S^*) \xrightarrow{\gamma_4} \mathrm{H}^2(G, L^*)$$

of abelian groups.

Proof. The action by G on $\operatorname{Div}(S)$ fixes each prime divisor Q_i. The diagram

$$\begin{array}{ccccccccc} 0 & \longrightarrow & \bigoplus_{i=1}^{v} \mathbb{Z}P_i & \longrightarrow & \operatorname{Div}(R) & \longrightarrow & \operatorname{Div}(R[a^{-1}]) & \longrightarrow & 0 \\ & & \downarrow \phi & & \downarrow \psi & & \downarrow \tau & & \\ 0 & \longrightarrow & \bigoplus_{i=1}^{v} \mathbb{Z}Q_i & \longrightarrow & \operatorname{Div}(S)^G & \longrightarrow & \operatorname{Div}(S[x^{-1}])^G & \longrightarrow & 0 \end{array}$$

commutes, and the rows are split exact sequences of $\mathbb{Z}$-modules. By Proposition 6.5.6, the map ϕ sends P_i to nQ_i, the vertical maps are one-to-one, and τ is an isomorphism. Part (2) follows from Theorem 1.4.1 applied to this diagram. Part (3) is the exact sequence of Theorem 6.5.8 with the isomorphism of (2) inserted. Part (1) is left to the reader. □

4.1. Exercises.

Exercise 9.4.7. In this exercise we study a ramified quadratic extension of $R = k[x]$, where k is assumed to be an algebraically closed field of characteristic different from 2. The associated double cover of $\mathbb{A}^1$ is sometimes called an affine hyperelliptic curve. Let $p(x)$ be a monic polynomial in $k[x]$ of degree $d \geq 2$. Let $p(x) = \ell_1^{e_1} \cdots \ell_v^{e_v}$ be the unique factorization of $p(x)$. Assume $\ell_1, \ldots, \ell_v$ are distinct and of the form $\ell_i = x - \lambda_i$. Let $f = y^2 - p(x)$. Let $D = \gcd(e_1, \ldots, e_v)$ and for each i, write $e_i = 2q_i + r_i$, where $0 \leq r_i < 2$. Set $q(x) = \ell_1^{q_1} \cdots \ell_v^{q_v}$ and $r(x) = \ell_1^{r_1} \cdots \ell_v^{r_v}$. In the terminology of Section 8.5.2, let F be the affine curve in $\mathbb{A}^2$ defined by $f = 0$. Let $\mathcal{O}(F) = k[x, y]/(f)$ be the affine coordinate ring of F. Prove the following.

(1) F is irreducible if and only if D is odd.
(2) $\mathcal{O}(F)$ is a regular domain if and only if each $e_i = 1$.
(3) For $1 \leq i \leq v$ let Q_i be the point on F where $y = 0$ and $\ell_i = 0$. If $P \in F$, then $\mathcal{O}_{P,Y}$ is a regular local ring if and only if P is not in the set $\{Q_i \mid e_i \geq 2\}$.

(4) If D is even, then $F = F_1 \cup F_2$, where F_1 and F_2 are the affine curves defined by $y - q(x) = 0$ and $y + q(x) = 0$.
   (a) The natural map $\mathcal{O}(F) \to \mathcal{O}(F_1) \oplus \mathcal{O}(F_2)$ is one-to-one.
   (b) The conductor ideal from $\mathcal{O}(F_1) \oplus \mathcal{O}(F_2)$ to $\mathcal{O}(F)$ is $(y, q(x))$.
   (c) For each i, $\mathcal{O}(F_i)$ is isomorphic to $k[x]$.
   (d) $\mathcal{O}(F_1) \oplus \mathcal{O}(F_2)$ is a separable $\mathcal{O}(F)$-algebra.
   (e) $\mathcal{O}(F_1) \oplus \mathcal{O}(F_2)$ is not a flat $\mathcal{O}(F)$-algebra.

(5) Assume D is odd. Let w be another variable and $\tilde{F}$ the affine curve defined by $w^2 - r(x) = 0$. Let $\mathcal{O}(\tilde{F}) = k[x, w]/(w^2 - r(x))$ be the affine coordinate ring of $\tilde{F}$. Define $\phi : \mathcal{O}(F) \to \mathcal{O}(\tilde{F})$ by $x \mapsto x$, $y \mapsto wq(x)$.
   (a) ϕ is a k-algebra homomorphism and ϕ induces an isomorphism on the quotient fields $K(F) \cong K(\tilde{F})$.
   (b) If both rings are identified with subrings of $K(\tilde{F})$, then the integral closure of $\mathcal{O}(F)$ is equal to $\mathcal{O}(\tilde{F})$.
   (c) The conductor ideal from $\mathcal{O}(\tilde{F})$ to $\mathcal{O}(F)$ is equal to $(y, q(x))$. As an ideal in $\mathcal{O}(\tilde{F})$, the conductor is the principal ideal (q).
   (d) Let $\mathfrak{m}_i$ be the maximal ideal in $\mathcal{O}(F)$ generated by y and ℓ_i. Then $\mathfrak{m}_i$ is the ideal corresponding to Q_i. Consider $\phi^\sharp : \operatorname{Spec} \mathcal{O}(\tilde{F}) \to \operatorname{Spec} \mathcal{O}(F)$. If e_i is odd, then $(\phi^\sharp)^{-1}(\mathfrak{m}_i)$ is a singleton set, or equivalently, there is exactly one maximal ideal lying over $\mathfrak{m}_i$. If e_i is even, then $(\phi^\sharp)^{-1}(\mathfrak{m}_i)$ consists of two points and there are two maximal ideals in $\mathcal{O}(\tilde{F})$ lying over $\mathfrak{m}_i$.
   (e) Determine necessary and sufficient conditions on $e_1, \dots, e_v$ such that $\mathcal{O}(\tilde{F})$ is a separable $\mathcal{O}(F)$-algebra. (Hint: Exercises 8.1.34 and 8.1.35.)

Exercise 9.4.8. This exercise is an n-dimensional version of Exercise 9.4.7. Because of the similarities, we will be less verbose. Let k be a field and p a polynomial in $A = k[x_1, \dots, x_n]$ of degree $d \geq 2$. Assume p is not a square. Let $p = p_1^{e_1} \cdots p_s^{e_s}$ be the unique factorization of p into irreducibles. Write $e_i = 2q_i + r_i$, where $0 \leq r_i < 2$. Set $r = p_1^{r_1} \cdots p_s^{r_s}$ and $q = p_1^{q_1} \cdots p_s^{q_s}$. Then $p = rq^2$. Let X be the affine hypersurface in $\mathbb{A}_k^{n+1}$ defined by $z^2 - rq^2 = 0$. The affine coordinate ring of X is $\mathcal{O}(X) = A[z]/(z^2 - rq^2)$. Let $\tilde{X}$ be the affine hypersurface in $\mathbb{A}_k^{n+1}$ defined by $w^2 - r = 0$. The affine coordinate ring of $\tilde{X}$ is $\mathcal{O}(\tilde{X}) = A[w]/(w^2 - r)$. Define an A-algebra homomorphism $\phi : \mathcal{O}(X) \to \mathcal{O}(\tilde{X})$ by $\alpha \mapsto \alpha$ for $\alpha \in A$ and $z \mapsto wq$. Identify both rings with subrings of the quotient field of $\mathcal{O}(X)$ and show that $\mathcal{O}(\tilde{X})$ is the integral closure of $\mathcal{O}(X)$. Show that the conductor ideal from $\mathcal{O}(\tilde{X})$ to $\mathcal{O}(X)$ is $(z, q) \subseteq \mathcal{O}(X)$. As an ideal in $\mathcal{O}(\tilde{X})$, the conductor is the principal ideal (q).

Exercise 9.4.9. Let $n \geq 2$ be an integer and k a field in which $2n$ is invertible. Also assume k contains a primitive $2n$th root of unity, ζ. For $T = k[x, y, z]/(z^n - x^{n-1}y + 1)$, prove the following.

(1) Show that T is regular. (Hint: Example 8.5.5.)

(2) For $i = 1, \dots, n$, the ideal $Q_i = (x, z + \zeta^{2i-1})$ is a height one prime ideal of T.

(3) The divisor of x is $\operatorname{Div} x = Q_1 + \cdots + Q_n$.

(4) $\operatorname{Pic}(T) = \operatorname{Cl}(T)$ is generated by $Q_1, \dots, Q_n$, and is a free $\mathbb{Z}$-module of rank $n - 1$.

(5) Let σ be the $k[x, y]$-algebra automorphism of T defined by $z \mapsto \zeta^2 z$. Let $G = \langle \sigma \rangle$. Compute: $\mathrm{H}^j(G, \operatorname{Cl}(T))$ for $j \geq 0$.

Exercise 9.4.10. Let $n \geq 2$ be an integer and k a field in which $2n$ is invertible. Also assume k contains a primitive $2n$th root of unity, ζ. For $T = k[x, y, z]/(z^n - x^{n-1} + y^n)$, prove the following.

(1) $\operatorname{Cl}(T)$ is a free $\mathbb{Z}/(n-1)$ module of rank $n - 1$, and is generated by the height one prime ideals of T that contain x.

(2) Let σ be the $k[x, y]$-algebra automorphism of T defined by $z \mapsto \zeta^2 z$. Then $\mathrm{H}^j(\langle \sigma \rangle, \operatorname{Cl}(T)) = (0)$ for all $j \geq 0$.

See Example 13.6.8 for a computation of $\operatorname{Pic}(T)$.

Chapter 10

Henselization and Splitting Rings

Let A be a local ring with maximal ideal $\mathfrak{m}$ and residue field k. There are a number of equivalent conditions that can be taken as the definition of henselian, but they all are based on the ability to lift idempotents modulo $\mathfrak{m}$. By definition, we say A is *henselian* if for every commutative A-algebra B which is finitely generated as an A-module, idempotents in the ring $B/\mathfrak{m}B$ can be lifted to idempotents in B. For instance, if S is a faithfully flat étale A-algebra, then the closed fiber $k \to S/\mathfrak{m}S$ is an étale k-algebra and Corollary 4.5.8 implies $S/\mathfrak{m}S$ is a finite direct sum of separable extension fields of k. We will show that A is henselian if and only if for every étale A-algebra S, if the closed fiber $k \to S/\mathfrak{m}S$ has a section, say $\bar{\sigma} : S/\mathfrak{m}S \to k$ (or equivalently, $S/\mathfrak{m}S$ has a direct summand isomorphic to k), then $\bar{\sigma}$ lifts to a section $\sigma : S \to A$ and A is isomorphic to a direct summand of S. A henselian local ring is said to be strictly henselian if the residue field k is separably closed. In this case, if S is a faithfully flat étale A-algebra, then there is always a direct summand isomorphic to k, and consequently A is isomorphic to a direct summand of S. For any henselian local ring A, the category of étale A-algebras that are finitely generated as A-modules is equivalent to the category of commutative separable k-algebras. Therefore, a henselian local ring A is strictly henselian if and only if for any faithfully flat étale A-algebra S, the closed fiber $k \to S/\mathfrak{m}S$ has a section. In this book we do not define sites or Grothendieck topologies, so we will not go any further on this subject than to say that it is because of this attribute that for the étale topology the counterpart of the local ring at a prime $\mathfrak{p}$ is the strict henselization $R^{sh}_{\mathfrak{p}}$. Notice that for the Zariski topology this property is practically taken for granted. If $\mathfrak{p}$ is a prime ideal of a commutative ring R and $U(\alpha)$ a basic open neighborhood of $\mathfrak{p}$, then the existence of a section is trivial because the local rings $R_\mathfrak{p}$ and $(R_\alpha)_\mathfrak{p}$ are equal.

Every Azumaya algebra A over a commutative ring R is shown to have a faithfully flat étale splitting ring S. The construction allows us to prove even more. We show that if A has constant rank n^2, then there is a faithfully flat étale R-algebra S such that $A \otimes_R S$ is isomorphic to the ring of n-by-n matrices over S. In other words, A is a twisted form of matrices for S.

In Section 10.4 we define the Čech cohomology groups for both the flat and étale topologies. If the units functor is denoted $\mathbb{G}_m$, then for the étale topology we show that there is a natural isomorphism $\operatorname{Pic}(R) = \check{\mathrm{H}}^1_{\mathrm{et}}(R, \mathbb{G}_m)$, and we construct a natural embedding of the Brauer group $\beta : \mathrm{B}(R) \to \check{\mathrm{H}}^2_{\mathrm{et}}(R, \mathbb{G}_m)$. When R is noetherian, the image of β is equal to the subgroup of all torsion elements in the cohomology group.

1. Henselian Local Rings

Throughout this section, all rings are commutative. Henselian local rings were first studied by Azumaya ([**Azu51**]). As we will see in this section, there are many equivalent ways to define henselian local rings. In fact, in the following we show that there are more than a half dozen equivalent conditions that could be used. The property which we find most convenient to take as the definition is based on the ability to lift idempotents modulo the maximal ideal. Let A be a local ring with maximal ideal $\mathfrak{m}$. As mentioned in the introduction above, A is defined to be henselian if for every commutative A-algebra B which is finitely generated as an A-module, idempotents in the ring $B/\mathfrak{m}B$ can be lifted to idempotents in B. We show that this definition is equivalent to the more-or-less traditional definition involving the lifting of a factorization of a polynomial modulo $\mathfrak{m}$. The presentation given here is based on [**Ray70**], [**Mil80**], and [**Gro71**] (see also [**Gro03**]).

1.1. The Definition.

Lemma 10.1.1. *Let A be a local ring with maximal ideal $\mathfrak{m}$. Let B be a commutative A-algebra which is finitely generated as an A-module. Then*

(1) *B has only finitely many maximal ideals, hence (by Exercise 3.1.8) is semilocal, and*

(2) *if $\mathfrak{n}_1, \dots, \mathfrak{n}_m$ are the maximal ideals of B, then $\bar{B} = B/\mathfrak{m}B \cong \bar{B}_{\bar{\mathfrak{n}}_1} \oplus \cdots \oplus \bar{B}_{\bar{\mathfrak{n}}_m}$ is a direct sum of artinian local rings.*

Proof. If $\tau : A \to B$ is the structure homomorphism, then τ factors into

$$A \xrightarrow{\tau_1} A_0 \xrightarrow{\tau_0} B$$

where τ_0 is one-to-one and τ_1 is onto. Since the kernel of τ_1 is contained in $\mathfrak{m}$, A_0 is a local ring. Let $\mathfrak{m}_0$ be the maximal ideal of A_0. By Theorem 3.3.5, the maximal ideals of B are those prime ideals $P \in \operatorname{Spec} B$ such that $P \cap A_0 = \mathfrak{m}_0$.

For simplicity's sake, assume that τ is one-to-one and that $A = A_0$. Let M be a maximal ideal of B. Then $\mathfrak{m} = M \cap A$, so $\mathfrak{m}B \subseteq M$. The maximal ideals of B correspond to the maximal ideals of $\bar{B} = B/\mathfrak{m}B$. Since $\bar{B}$ is a finite dimensional $A/\mathfrak{m}$-vector space, by Exercise 2.5.15, $\bar{B}$ is artinian. By Proposition 3.2.9, $\bar{B}$ has only finitely many maximal ideals. This proves (1).

Let $\mathfrak{n}_1, \dots \mathfrak{n}_m$ be the complete list of maximal ideals in $\operatorname{Spec} B$. If the image of an ideal in $\bar{B}$ is represented by a bar, then $\bar{\mathfrak{n}}_1, \dots \bar{\mathfrak{n}}_m$ is the complete list of prime ideals in $\operatorname{Spec} \bar{B}$. According to Theorem 3.2.11, the natural homomorphism $\bar{B} \to \bar{B}_{\bar{\mathfrak{n}}_1} \oplus \cdots \oplus \bar{B}_{\bar{\mathfrak{n}}_m}$ is an isomorphism. This proves Part (2). □

Lemma 10.1.2. *Let A be a local ring with maximal ideal $\mathfrak{m}$. Let B be a commutative A-algebra which is finitely generated as an A-module. Let $\mathfrak{n}_1, \dots, \mathfrak{n}_m$ be the complete list of maximal ideals of B. The following are equivalent.*

(1) *B is a direct sum of local rings.*
(2) *The natural map $B \to B_{\mathfrak{n}_1} \oplus \cdots \oplus B_{\mathfrak{n}_m}$ is an isomorphism.*
(3) *The decomposition $\bar{B}/\mathfrak{m}\bar{B} \cong \bar{B}_{\bar{\mathfrak{n}}_1} \oplus \cdots \oplus \bar{B}_{\bar{\mathfrak{n}}_m}$ lifts to a decomposition of B into a direct sum of local rings.*

Proof. (2) implies (3): This is immediate. (3) implies (1): This is immediate. (1) implies (2): This is Exercise 2.1.8. □

Let $\operatorname{idemp}(R)$ denote the set of idempotents of the commutative ring R. The homomorphic image of an idempotent is an idempotent, so given a homomorphism of rings $A \to B$, there is a function $\operatorname{idemp}(A) \to \operatorname{idemp}(B)$.

Lemma 10.1.3. *Let A be a local ring with maximal ideal $\mathfrak{m}$. Let B be a commutative A-algebra which is finitely generated as an A-module.*

(1) *The map* $\operatorname{idemp}(B) \to \operatorname{idemp}(B/\mathfrak{m}B)$ *is one-to-one.*
(2) *The map* $\operatorname{idemp}(B) \to \operatorname{idemp}(B/\mathfrak{m}B)$ *is onto if and only if B is a direct sum of local rings.*

Proof. (1): This follows from Corollary 3.4.16 since $\mathfrak{m}B \subseteq \operatorname{J}(B)$.

(2): As above, write $\bar{B}$ for $B/\mathfrak{m}B$. By Lemma 10.1.1, $\bar{B} = \bar{B}f_1 \oplus \cdots \oplus \bar{B}f_m$ and each $\bar{B}f_i$ is a local ring. The idempotents of $\bar{B}f_i$ are 0 and 1. Therefore the idempotents of $\bar{B}$ are of the form $a_1f_1 + \cdots + a_mf_m$ where each a_i is either 0 or 1. If B is a direct sum of local rings, then by Lemma 10.1.2, the map $\operatorname{idemp}(B) \to \operatorname{idemp}(\bar{B})$ is onto. Conversely, assume $\operatorname{idemp}(B) \to \operatorname{idemp}(\bar{B})$ is onto. Let $\eta : B \to \bar{B}$ be the natural map. For each i find $e_i \in B$ such that $\eta e_i = f_i$, and $e_i^2 = e_i$. If $i \neq j$, then $\eta(e_i e_j) = \eta(e_i)\eta(e_j) = f_i f_j = 0$. But $(e_i e_j)^2 = e_i e_j$ and by Part (1) it follows that $e_i e_j = 0$. If $e = e_1 + \cdots + e_m$, then $e^2 = e$ and $\eta(e) = 1$ and Part (1) says $e = 1$. The set of orthogonal idempotents $\{e_1, \dots, e_m\}$ gives a decomposition $B = Be_1 \oplus \cdots \oplus Be_m$. By construction, $Be_j \to \bar{B} f_j$ is a surjection. Then Be_j is local and its unique maximal ideal contains $\mathfrak{m}Be_j$. □

Let A be a local ring with maximal ideal $\mathfrak{m}$. In the introductory paragraph of Section 10, we defined A to be henselian if for every commutative A-algebra B which is finitely generated as an A-module, the map $\operatorname{idemp}(B) \to \operatorname{idemp}(B/\mathfrak{m}B)$ is onto.

Theorem 10.1.4. *Let A be a local ring with maximal ideal $\mathfrak{m}$ and residue field $k = A/\mathfrak{m}$. The following are equivalent.*

(1) *A is henselian.*

(2) *If B is a commutative A-algebra which is finitely generated and free as an A-module, then* $\operatorname{idemp}(B) \to \operatorname{idemp}(B/\mathfrak{m}B)$ *is onto.*

(3) *For every monic polynomial $p \in A[x]$,*

$$\operatorname{idemp}(A[x]/(p)) \to \operatorname{idemp}((A/\mathfrak{m})[x]/(\bar{p}))$$

is onto, where $\bar{p}$ is the image of p in $(A/\mathfrak{m})[x]$.

(4) *If there exist polynomials $f, g_0, h_0 \in A[x]$ such that*
  (a) *f, g_0 and h_0 are monic,*
  (b) *$f - g_0 h_0 \in \mathfrak{m}A[x]$, and*
  (c) *$A[x] = g_0 A[x] + h_0 A[x] + \mathfrak{m}A[x]$,*
  then there exist polynomials $g, h \in A[x]$ such that
  (d) *g and h are monic,*
  (e) *$g - g_0 \in \mathfrak{m}A[x]$,*
  (f) *$h - h_0 \in \mathfrak{m}A[x]$, and*
  (g) *$f = gh \in A[x]$.*

(5) *If B is a commutative A-algebra which is finitely generated as an A-module, then B is a direct sum of local rings.*

Proof. It follows from Lemma 10.1.3 that (1) and (5) are equivalent. It is immediate that (1) implies (2), and that (2) implies (3).

(3) implies (4): Write $\bar{A}$ for $A/\mathfrak{m}$ and let $\bar{f}, \bar{g}_0, \bar{h}_0$ denote the images of the polynomials in $\bar{A}[x]$. By (b) we have $\bar{f} = \bar{g}_0 \bar{h}_0$ and by (c), $1 = (\bar{g}_0, \bar{h}_0)$. If we set $B = A[x]/(f)$, then B is a finitely generated free A-module and the

rank of B is equal to $\deg f = \deg g_0 + \deg h_0$. Write $\bar{B}$ for $B/\mathfrak{m}B = B \otimes_A \bar{A}$. By the Chinese Remainder Theorem,

$$\bar{B} = \frac{\bar{A}[x]}{(\bar{f})} = \frac{\bar{A}[x]}{(\bar{g}_0)} \oplus \frac{\bar{A}[x]}{(\bar{h}_0)}.$$

By (3), the decomposition of $\bar{B}$ lifts to a decomposition $B = A[x]/(f) = B_1 \oplus B_2$. Consider the projection

$$B_1 \to \frac{B_1}{\mathfrak{m}B_1} \cong \frac{\bar{A}[x]}{(\bar{g}_0)}.$$

Let $\bar{x}$ denote the coset $x + (\bar{g}_0)$ in $\bar{A}[x]/(\bar{g}_0)$. Lift $\bar{x}$ to get $y_1 \in B_1$. Let n_0 be the degree of g_0. In B_1 consider the A-submodule $S_0 = A \cdot 1 + Ay_1 + \dots + Ay_1^{n_0-1}$. Since $B_1 = B_0 + \mathfrak{m}B_1$, Nakayama's Lemma (Corollary 1.1.14) says that $B_0 = B_1$. So $y_1^{n_0} \in B_0$ and there is a monic polynomial $g \in A[x]$ of degree n_0 such that $g(y_1) = 0$. Restricting to $\bar{A}[x]$ shows $\bar{g}$ is in the ideal $(\bar{g}_0)$. That is, $\bar{g}_0$ divides $\bar{g}$. Since both are monic of degree n_0, we see that $\bar{g}_0 = \bar{g}$. The map $x \mapsto y_1$ defines $\theta_1 : A[x] \to B_1$ and $\theta_1 g = 0$. Likewise there is a monic polynomial $h \in A[x]$ such that the degree of h is equal to the degree of h_0, $\bar{h} = \bar{h}_0$, and $A[x]/(h) \cong B_2$. Since gh is in the kernel of the map $A[x] \to A[x]/(f) = B = B_1 \oplus B_2$, it follows that f divides gh. Since gh and f are both monic of the same degree, it follows that $f = gh$.

(4) implies (3): Let p be a monic polynomial in $A[x]$ and write $\bar{p}$ as the image of p in $k[x]$. Suppose $\bar{p}_1, \dots, \bar{p}_m$ are the distinct irreducible divisors of $\bar{p}$. Factor $\bar{p}$ in $k[x]$

$$\bar{p} = (\bar{p}_1)^{e_1} \cdots (\bar{p}_m)^{e_m}$$

and assume each p_i is a monic polynomial in $A[x]$ and each $e_i > 0$. To simplify notation, let q_i denote $(p_i)^{e_i}$. Then $\bar{q}_1, \dots, \bar{q}_m$ are pairwise relatively prime. Applying (4) iteratively, we may lift the factorization $\bar{p} = \bar{q}_1 \cdots \bar{q}_m$ to a factorization $p = q_1 \cdots q_m$ in $A[x]$. The diagram

$$\begin{array}{ccc}
\frac{A[x]}{(p)} & \xrightarrow{\phi} & \frac{A[x]}{(q_1)} \oplus \cdots \oplus \frac{A[x]}{(q_m)} \\
\downarrow & & \downarrow \\
\frac{k[x]}{(\bar{p})} & \longrightarrow & \frac{k[x]}{(\bar{q}_1)} \oplus \cdots \oplus \frac{k[x]}{(\bar{q}_m)}
\end{array}$$

commutes. By the Chinese Remainder Theorem, the second horizontal map is an isomorphism. By Corollary 1.3.19, ϕ is onto. Since ϕ is an A-module homomorphism between finitely generated free modules of the same rank, ϕ is an isomorphism. Each $k[x]/(\bar{q}_i)$ is a local ring and (3) follows from Lemma 10.1.3.

(3) implies (5): Let B be a commutative A-algebra which is finitely generated as an A-module. We show that B is a direct sum of local rings.

Write $\bar{B}$ for $B/\mathfrak{m}B$ and let $\eta : B \to \bar{B}$ be the natural map. By Lemma 10.1.3, it is enough to show that idempotents of $\bar{B}$ lift to idempotents of B. That is, it suffices to show $\operatorname{idemp} B \to \operatorname{idemp} \bar{B}$ is onto. Let $\mathfrak{n}_1, \dots, \mathfrak{n}_m$ be the maximal ideals of B. Let $\bar{e}_i$ be the primitive idempotent of $\bar{B}$ corresponding to the maximal ideal $\mathfrak{n}_i$. We find an idempotent $e \in \operatorname{idemp} B$ such that $\eta(e) = \bar{e}_i$. Pick $b \in B$ such that $\eta(b) = \bar{e}_i$. Of course b is not necessarily an idempotent, but if $\tau_j : B \to B_{\mathfrak{n}_j}$ is the localization at $\mathfrak{n}_j$, then we have

$$\tau_j(b) = \begin{cases} 0 & \text{if } j \neq i, \\ 1 & \text{if } j = i. \end{cases}$$

Since B is finitely generated as an A-module, there is a monic polynomial $f \in A[x]$ such that $f(b) = 0$. Let $A[x] \to B$ be the A-algebra homomorphism defined by $x \mapsto b$. Set $S = A[x]/(f)$ and $\phi : S \to B$ the induced A-algebra homomorphism. For each j, let $Q_j = \phi^{-1}(\mathfrak{n}_j)$. Each Q_j is a prime ideal of S lying over $\mathfrak{m}$, hence by Lemma 10.1.1 is a maximal ideal. Let $\bar{x} \in S$ denote the image of x. Now $\phi(\bar{x}) = b \in \mathfrak{n}_j$ if and only if $j \neq i$. Therefore $\bar{x} \in Q_j$ if and only if $j \neq i$. By (3), S decomposes into a direct sum of local rings. Let f be the primitive idempotent of S corresponding to the maximal ideal Q_i. Then $e = \phi(f)$ is an idempotent of B. Since $f \in Q_j$ if and only if $j \neq i$, we get $e \in \mathfrak{n}_j$ if and only if $j \neq i$. Therefore e is a primitive idempotent in B and $\eta(e) = \bar{e}_i$. □

Proposition 10.1.5. *Let $\{A_i, \phi^i_j\}$ be a directed system of commutative local rings for a directed index set I. Assume $\phi^i_j : A_i \to A_j$ is a local homomorphism of local rings, for each $i \leq j$. Let $A = \varinjlim A_i$. If each A_i is a henselian local ring, then A is a henselian local ring.*

Proof. Let $\mathfrak{m}_i$ be the maximal ideal of A_i. Let $\mathfrak{m} = \bigcup_i \alpha_i(\mathfrak{m}_i)$. By Corollary 9.1.6, $\mathfrak{m}$ is the unique maximal ideal of A.

Let $p \in A[x]$ be a monic polynomial. By Theorem 10.1.4 (3) it is enough to show $A[x]/(p)$ is a direct sum of local rings. For some $i_0 \in I$, if $i \geq i_0$, all of the coefficients of p are in A_i. Let p_0 be a polynomial in $A_{i_0}[x]$ which maps onto p. Write p_i for the image of p_0 under $\phi^{i_0}_i$. Write $\bar{p}$ and $\bar{p}_i$ for the restriction of a polynomial modulo $\mathfrak{m}$ and $\mathfrak{m}_i$. The diagram of sets

$$\begin{array}{ccc} \operatorname{idemp}(A_i[x]/(p_i)) & \longrightarrow & \operatorname{idemp}((A_i/\mathfrak{m}_i)[x]/(\bar{p}_i)) \\ \downarrow & & \downarrow \\ \operatorname{idemp}(A[x]/(p)) & \longrightarrow & \operatorname{idemp}((A/\mathfrak{m})[x]/(\bar{p})) \end{array}$$

commutes. Since A_i is henselian, the top horizontal arrow is onto. Since $A/\mathfrak{m} = \varinjlim A_i/\mathfrak{m}_i$, there exists $i_1 \geq i_0$ such that if $i \geq i_1$, then the irreducible factors of $\bar{p} \in (A/\mathfrak{m})[x]$ are in $(A_i/\mathfrak{m}_i)[x]$. The right vertical arrow is onto, for

$i \geq i_1$. Therefore, the bottom horizontal arrow is onto. By Lemma 10.1.3, $A[x]/(p)$ is a direct sum of local rings. □

Proposition 10.1.6. *Let R be a henselian local ring with maximal ideal $\mathfrak{m}$ and residue field k. Let A be a commutative R-algebra. Assume A is integral over R, and $\mathfrak{n}$ is a maximal ideal in A.*

(1) *The map* $\operatorname{idemp}(A) \to \operatorname{idemp}(A \otimes_R k)$ *is a one-to-one correspondence.*

(2) *$A_{\mathfrak{n}}$ is integral over R.*

(3) *$A_{\mathfrak{n}}$ is henselian.*

Proof. (1): As in Exercise 1.4.12, $A = \varinjlim A_\alpha$, where A_α runs over the set of all R-subalgebras of A such that A_α is finitely generated as an R-module. By Proposition 1.4.4, $A \otimes_R k = \varinjlim (A_\alpha \otimes_R k)$. By Lemma 10.1.3, $\operatorname{idemp}(A_\alpha) = \operatorname{idemp}(A_\alpha \otimes_R k)$. The reader should verify that

$$\begin{aligned} \operatorname{idemp}(A) &= \varinjlim \operatorname{idemp}(A_\alpha) \\ &= \varinjlim \operatorname{idemp}(A_\alpha \otimes_R k) \\ &= \operatorname{idemp}(A \otimes_R k). \end{aligned}$$

(2): Let $\mathfrak{n}_\alpha = \mathfrak{m} \cap A_\alpha$. By Theorem 3.3.5, $\mathfrak{n}_\alpha$ is a maximal ideal of A_α. Since R is henselian, Lemma 10.1.2 implies $(A_\alpha)_{\mathfrak{n}_\alpha}$ is a direct summand of A_α, hence is a finitely generated R-module. The reader should verify that $\varinjlim (A_\alpha)_{\mathfrak{n}_\alpha} = A_{\mathfrak{n}}$. Therefore, every element of $A_{\mathfrak{n}}$ is integral over R.

(3): This follows by applying Exercise 10.1.17 and Proposition 10.1.5 to the direct limit $\varinjlim (A_\alpha)_{\mathfrak{n}_\alpha} = A_{\mathfrak{n}}$ of Part (2). □

Proposition 10.1.7. *Let R be a local ring with maximal ideal $\mathfrak{m}$ and residue field k. The following are equivalent.*

(1) *R is henselian.*

(2) *If S is a commutative quasi-finite R-algebra and $\mathfrak{m}S$ is not the unit ideal, then for every maximal ideal $\mathfrak{n}$ of S lying over $\mathfrak{m}$, $S_{\mathfrak{n}}$ is finitely generated as an R-module and is a direct summand of S.*

(3) *Suppose S is an étale R-algebra and there exists $\mathfrak{n} \in \operatorname{Spec} S$ such that $\mathfrak{n} \cap R = \mathfrak{m}$ and the induced map on residue fields $k \to k(\mathfrak{n})$ is an isomorphism. Then $R \to S_{\mathfrak{n}}$ is an isomorphism, there exists an R-algebra homomorphism $\sigma : S \to R$, and $S_{\mathfrak{n}}$ is a direct summand of S.*

Proof. (1) implies (2): Let $\theta : R \to S$ be the structure homomorphism. By Theorem 9.3.3, θ factors

$$\begin{array}{ccc} R & \xrightarrow{\theta} & S \\ & \searrow^{\phi} \quad \nearrow_{\psi} & \\ & A & \end{array}$$

where A is finitely generated as an R-module and $\psi^\sharp : \operatorname{Spec} S \to \operatorname{Spec} A$ is an open immersion. Let $\mathfrak{m}_1, \dots, \mathfrak{m}_n$ be the complete list of maximal ideals in A (Lemma 10.1.1). By Theorem 3.3.5, $\mathfrak{m}_1, \dots, \mathfrak{m}_n$ are precisely those prime ideals in A lying over $\mathfrak{m}$. Let A_i denote the local ring $A_{\mathfrak{m}_i}$. Since R is henselian, the natural map $A \to A_1 \oplus \cdots \oplus A_n$ is an isomorphism (Lemma 10.1.2). If necessary, relabel so that $\mathfrak{m}_1, \dots, \mathfrak{m}_s$ are those maximal ideals in A that are in the image of $\psi^\sharp : \operatorname{Spec} S \to \operatorname{Spec} A$. By hypothesis, $\mathfrak{m}S$ is not the unit ideal, hence $s \geq 1$. Since $\psi^\sharp$ is an open immersion, for each $i = 1, \dots, s$, there is a unique maximal ideal of S, say $\mathfrak{n}_i$, which lies over $\mathfrak{m}_i$ and $S_{\mathfrak{n}_i} \cong A_i$. The diagram

$$\begin{array}{ccc} \operatorname{Spec} S_{\mathfrak{n}_i} & \xrightarrow{\cong} & \operatorname{Spec} A_i \\ \downarrow & & \downarrow \\ \operatorname{Spec} S & \xrightarrow{\psi^\sharp} & \operatorname{Spec} A \end{array}$$

commutes. The vertical maps are one-to-one, by Exercise 2.2.15. The image of $\operatorname{Spec} A_i \to \operatorname{Spec} A$ is both open and closed, and is contained in the image of $\psi^\sharp$. Being an open immersion, $\psi^\sharp$ is one-to-one. Therefore, the disjoint union $\operatorname{Spec} S_{\mathfrak{n}_1} \cup \cdots \cup \operatorname{Spec} S_{\mathfrak{n}_s}$ identifies with a subset of $\operatorname{Spec} S$ that is both open and closed. By Theorem 2.2.9, S decomposes into a direct sum which is either of the form $S = S_{\mathfrak{n}_1} \oplus \cdots \oplus S_{\mathfrak{n}_s}$, or $S = S_0 \oplus S_{\mathfrak{n}_1} \oplus \cdots \oplus S_{\mathfrak{n}_s}$. By construction, in the second case, $\mathfrak{m}S_0$ is the unit ideal.

(2) implies (3): Since S is an étale R-algebra, by Exercise 9.3.5, S is quasi-finite. By (2), $S_{\mathfrak{n}}$ is finitely generated over R and is a direct summand of S. Then $S_{\mathfrak{n}}$ is étale over R and faithfully flat by Lemma 2.4.5. Lemma 2.4.4 implies $R \to S_{\mathfrak{n}}$ is one-to-one. By Proposition 8.3.7, $\mathfrak{m}S_{\mathfrak{n}} = \mathfrak{n}S_{\mathfrak{n}}$. Since the homomorphism on residue fields $k \to k(\mathfrak{n})$ is onto, Exercise 1.1.18 implies $R \to S_{\mathfrak{n}}$ is onto.

(3) implies (1): Let A be a commutative R-algebra that is finitely generated and free as an R-module. To show R is henselian, we must show $\operatorname{idemp}(A) \to \operatorname{idemp}(A \otimes_R k)$ is onto (Lemma 10.1.3). Let A_{et} be the étale R-algebra that represents the idempotents of A (Proposition 9.2.22). There is a one-to-one correspondence $\operatorname{Hom}_{R\text{-alg}}(A_{\mathrm{et}}, k) \to \operatorname{idemp}(A \otimes_R k)$ (Proposition 9.2.21). Let $\bar{e}$ be an idempotent in $A \otimes_R k$. We show $\bar{e}$ comes from

an idempotent e in A. Let $\bar{\theta} : A_{\mathrm{et}} \to k$ be the R-algebra homomorphism corresponding to $\bar{e}$. Since $R \to k$ is onto, so is $\bar{\theta}$. Let $\mathfrak{n}$ be the kernel of $\bar{\theta}$. Then $\mathfrak{n}$ is a maximal ideal, $\bar{\theta}$ factors through the local ring $(A_{\mathrm{et}})_{\mathfrak{n}}$, and there is a commutative diagram

$$\begin{array}{ccc} A_{\mathrm{et}} & \xrightarrow{\ \bar{\theta}\ } & k \\ & \searrow^{\eta} \qquad \nearrow_{\bar{\tau}} & \\ & (A_{\mathrm{et}})_{\mathfrak{n}} & \end{array}$$

of R-algebras. By (3), $R \to (A_{\mathrm{et}})_{\mathfrak{n}}$ is an isomorphism, there is an R-algebra homomorphism $\sigma : A_{\mathrm{et}} \to R$, and $(A_{\mathrm{et}})_{\mathfrak{n}}$ is a direct summand of A_{et}. Corollary 4.4.8 shows that the kernel of σ is idempotent generated. □

Proposition 10.1.8. *Let R be a henselian local ring with maximal ideal $\mathfrak{m}$ and residue field k. The change of base functor $(\ ) \otimes_R k$ establishes an equivalence between the category of étale R-algebras which are finitely generated R-modules and the category of commutative separable k-algebras.*

Proof. The functor is essentially surjective, by Corollary 8.1.27. It is enough to show that the functor is also fully faithful, by Proposition 1.1.2. Let S and T be two étale R-algebras that are finitely generated R-modules. We show that the map

$$\mathrm{Hom}_{R\text{-alg}}(S,T) \to \mathrm{Hom}_{k\text{-alg}}(S \otimes_R k, T \otimes_R k)$$

is a one-to-one correspondence. Using Theorem 10.1.4, it suffices to prove this if S and T are local rings. By Exercise 10.1.17, S and T are henselian. The one-to-one property follows from Corollary 9.2.8. By Theorem 8.3.7 (3), $S \otimes_R k$ and $T \otimes_R k$ are the residue fields of S and T, respectively. We must show that a k-algebra homomorphism $\phi : S \otimes_R k \to T \otimes_R k$ comes from an R-algebra homomorphism $\alpha : S \to T$. By Exercise 1.3.8, to give α is equivalent to giving a T-algebra homomorphism $\gamma : S \otimes_R T \to T$ such that the diagram

$$\begin{array}{ccccc} & & T & & \\ & \nearrow^{\alpha} & \uparrow{\scriptstyle \gamma} & \nwarrow^{1} & \\ S & \xrightarrow[\rho_1]{} & S \otimes_R T & \xleftarrow[\rho_2]{} & T \end{array}$$

commutes. Consider the diagram

$$\begin{array}{ccccc} & & T \otimes_R k & & \\ & \nearrow^{a} & \uparrow{\scriptstyle \psi} & \nwarrow^{b} & \\ S & \xrightarrow[\rho_1]{} & S \otimes_R T & \xleftarrow[\rho_2]{} & T \end{array}$$

where $a(x) = \phi(x \otimes 1)$ and $b(y) = y \otimes 1$. By Exercise 1.3.8, there is a unique R-algebra homomorphism ψ such that the diagram commutes. But $S \otimes_R T$ is étale and finitely generated as a module over T (Proposition 9.2.5). The diagram

$$\begin{array}{ccc} S \otimes_R T & \xrightarrow{\psi} & T \otimes_R k \\ {\scriptstyle \rho_2}\uparrow & & \uparrow{\scriptstyle \cong} \\ T & \longrightarrow & T \otimes_R k \end{array}$$

commutes. By Proposition 10.1.7 (3), there is a section $\gamma : S \otimes_R T \to T$ to ρ_2, which is the desired map. $\square$

Lemma 10.1.9, which is due to Azumaya [**Azu51**], shows that for henselian local rings, simple roots have unique liftings modulo the maximal ideal.

Lemma 10.1.9. *Let R be a henselian local ring with maximal ideal $\mathfrak{m}$ and residue field k. For a polynomial f in $R[x]$, let $\bar{f}$ denote the restriction of f to $k[x]$. Let $f \in R[x]$ be a monic polynomial and $a \in R$. Assume $\bar{a} \in k$ is a simple root of $\bar{f}$. Then in the coset $a + \mathfrak{m}$, there exists one and only one root of f.*

Proof. There exists a monic polynomial $g_0 \in R[x]$ such that $\bar{f} = (x - \bar{a})\bar{g}_0$ in $k[x]$. Since $\bar{a}$ is a simple root, $\bar{g}_0(\bar{a}) \neq 0$. Therefore, $x - \bar{a}$ and $\bar{g}_0$ generate the unit ideal in $k[x]$. By Theorem 10.1.4 (4), there are $b \in R$, $g \in R[x]$ such that $f = (x - b)g$, $b - a \in \mathfrak{m}$ and $\bar{g} = \bar{g}_0$. This shows $f(b) = 0$. Now suppose $c - a \in \mathfrak{m}$ and $f(c) = 0$. Then $(c - b)g(c) = 0$. But $g(c) \notin \mathfrak{m}$ because $\bar{c} = \bar{a}$ is not a root of $\bar{g}$. This implies $c = b$, so b is unique. $\square$

Theorem 10.1.10 (Hensel's Lemma). *Let $(R, \mathfrak{m})$ be a local ring which is separated and complete for the $\mathfrak{m}$-adic topology. Then R is henselian.*

Proof. Let B be a commutative R-algebra that is finitely generated and free as an R-module. By Corollary 3.4.17, $\operatorname{idemp}(B) \to \operatorname{idemp}(B \otimes_R (R/\mathfrak{m}))$ is onto. By Lemma 10.1.2 and Theorem 10.1.4, R is henselian. $\square$

Remark 10.1.11. Let $(R, \mathfrak{m})$ be a local ring which is separated and complete with respect to the $\mathfrak{m}$-adic topology. Using Proposition 4.7.4 we see that R satisfies Proposition 10.1.7 (3). Hence R is henselian, which gives another proof of Theorem 10.1.10. For a more-or-less traditional proof of Hensel's Lemma, see for example, [**Nag62**].

1.2. Henselian Noetherian Local Rings. For a noetherian local ring, we have the following stronger form of Theorem 10.1.4.

Theorem 10.1.12. *Let R be a commutative noetherian local ring with maximal ideal $\mathfrak{m}$ and residue field $k = k(\mathfrak{m})$. For any polynomial f with coefficients in R, $\bar{f}$ will denote the polynomial with coefficients in k. The following are equivalent.*

(1) *R is henselian.*

(2) *Let $I = (f_1, \dots, f_n)$ be an ideal in $A = R[x_1, \dots, x_n]$ generated by n polynomials and set $S = A/I$. Assume $P = (a_1, \dots, a_n)$ is a k-rational point of $S \otimes_R k$ in $\mathbb{A}^n_k$. Assume the determinant of the jacobian matrix evaluated at P, $\det(\partial \bar{f}_i/\partial x_j)(P)$, is a unit in k. Then P lifts to an R-rational point of S in $\mathbb{A}^n_R$.*

(3) *If there exist polynomials $f, g_0, h_0 \in R[x]$ such that*
   (a) *$\bar{g}_0$ is monic,*
   (b) *$\bar{f} = \bar{g}_0 \bar{h}_0$, and*
   (c) *$\bar{g}_0$ and $\bar{h}_0$ generate the unit ideal in $k[x]$,*

   then there exist polynomials $g, h \in R[x]$ such that
   (d) *g is monic,*
   (e) *$\bar{g} = \bar{g}_0$,*
   (f) *$\bar{h} = \bar{h}_0$, and*
   (g) *$f = gh \in R[x]$.*

Proof. The proof is based on [**Mil80**, Theorem I.4.2]. (1) implies (2): Let $S = R[x_1, \dots, x_n]/I$. Let J denote the determinant of the jacobian matrix $(\partial f_i/\partial x_j)$. Then $J \in R[x_1, \dots, x_n]$. Let $\mathfrak{p}$ denote the maximal ideal in $S \otimes_R k$ generated by $x_1 - a_1, \dots, x_n - a_n$. The diagram

$$\begin{array}{ccccc} S & \longrightarrow & S \otimes_R k & \longrightarrow & \frac{S \otimes_R k}{\mathfrak{p}} \\ \uparrow & & \uparrow & \nearrow \cong & \\ R & \longrightarrow & \frac{R}{\mathfrak{m}} = k & & \end{array}$$

commutes. Then $\mathfrak{p}$ lifts to a prime ideal $\mathfrak{n}$ in S that lies over $\mathfrak{m}$. Since J restricts to a unit in the residue field $k(\mathfrak{n})$, J represents a unit in the local ring $S_\mathfrak{n}$. That means J represents a unit in a neighborhood of $\mathfrak{n}$. In other words, there is a polynomial $g \in R[x_1, \dots, x_n]$ such that g is not in $\mathfrak{n}$, and J restricts to a unit in $S[g^{-1}]$. Let $\mathfrak{a}$ denote the ideal $(f_1, \dots, f_n, gy - 1)$ in $R[x_1, \dots, x_n, y]$. Then $S[g^{-1}]$ is isomorphic to $R[x_1, \dots, x_n, y]/\mathfrak{a}$. The jacobian matrix looks like

$$J_\mathfrak{a} = \begin{bmatrix} (\partial f_i/\partial x_j) & * \\ 0 & g \end{bmatrix}$$

So the determinant of $J_\mathfrak{a}$ is invertible in the ring $S[g^{-1}] \cong R[x_1, \dots, x_n, y]/\mathfrak{a}$. By Corollary 9.2.20, $S[g^{-1}]$ is a finitely generated separable and flat over R

(this is where we use the fact that R is noetherian). By Proposition 10.1.7 (3) there is a section $\sigma : S[g^{-1}] \to R$, which corresponds to an R-rational point of S.

(2) implies (3): Let $f = f_{m+n}x^{m+n} + \cdots + f_1 x + f_0$, $g_0 = x^m + a_{m-1}x^{m-1} + \cdots + a_1 x + a_0$, and $h_0 = b_n x^n + b_{n-1}x^{n-1} + \cdots + b_1 x + b_0$. Let K denote the polynomial ring $K = R[u_{m-1}, \dots, u_1, u_0, v_n, \dots, v_1, v_0]$ in $m+n+1$ variables over R. In $K[x]$, consider $g = x^m + u_{m-1}x^{m-1} + \cdots + u_1 x + u_0$, and $h = v_n x^n + v_{n-1}x^{n-1} + \cdots + v_1 x + v_0$. Equating coefficients of $x^{m+n}, \dots, x^2, x, 1$ in the product $f = gh$, we get the $m + n + 1$ equations

$$\begin{aligned} f_{m+n} &= v_n \\ f_{m+n-1} &= u_{m-1}v_n + v_{n-1} \\ &\vdots \\ f_2 &= u_2 v_0 + u_1 v_1 + u_0 v_2 \\ f_1 &= u_1 v_0 + u_0 v_1 \\ f_0 &= u_0 v_0. \end{aligned} \tag{10.1}$$

Upon restricting the coefficients to k, the equations (10.1) define an algebraic subset Z of $\mathbb{A}_k^{m+n+1}$. Moreover, we are given that the point

$$P = (\bar{a}_{m-1}, \dots, \bar{a}_1, \bar{a}_0, \bar{b}_n, \dots, \bar{b}_1, \bar{b}_0)$$

is in Z. To prove (3) it suffices to show that P lifts to a solution in $\mathbb{A}_R^{m+n+1}$. By (2) it suffices to show that the determinant of the jacobian of the system of equations (10.1) is nonzero when evaluated at P. The reader should verify that the jacobian matrix of (10.1) is

$$J = \begin{bmatrix} 0 & & & & & 1 & & & & \\ v_n & 0 & & & & u_{m-1} & 1 & & & \\ v_{n-1} & v_n & 0 & & & u_{m-2} & u_{m-1} & 1 & & \\ v_{n-2} & v_{n-1} & v_n & & & u_{m-3} & u_{m-2} & u_{m-1} & & \\ \vdots & \vdots & \vdots & & & \vdots & \vdots & \vdots & & \\ v_0 & v_1 & v_2 & \cdots & & u_0 & u_1 & u_2 & \cdots & \\ & v_0 & v_1 & \cdots & & & u_0 & u_1 & \cdots & \\ & & & \vdots & \vdots & & & & \vdots & \vdots \\ & & & v_1 & v_2 & & & & u_1 & u_2 \\ & & & v_0 & v_1 & & & & u_0 & u_1 \\ & & & & v_0 & & & & & u_0 \end{bmatrix}$$

which is of size $(m+n+1)$-by-$(m+n+1)$. The transpose of J is the Sylvester matrix of the two polynomials $0x^{n+1} + v_n x^n + v_{n-1}x^{n-1} + \cdots + v_1 x + v_0$ and $x^m + u_{m-1}x^{m-1} + \cdots + u_1 x + u_0$ (see Section 1.2). The determinant, $\det(J)$, is an element of the ring K. Upon evaluating $\det(J)$ at P we get an element of k that (up to multiplying by ± 1) is equal to the resultant $\operatorname{Res}(h_0, g_0)$ (Lemma 1.2.5). Since g_0 and h_0 generate the unit ideal in $k[x]$, $\operatorname{Res}(g_0, h_0) \neq 0$, by Corollary 1.2.8.

(3) implies (1) because Part (4) of Theorem 10.1.4 is immediate. □

1.3. Exercises.

Exercise 10.1.13. Let k be a field.

(1) Let $R = k[x]/(f^r)$, where $r \geq 1$ and f is an irreducible monic polynomial in $k[x]$. Show that R is a local ring.

(2) Prove that k is henselian. (Hint: use Part (3) of Theorem 10.1.4.)

Exercise 10.1.14. Let R be a henselian local ring with maximal ideal $\mathfrak{m}$, and I a proper ideal of R. Show that R/I is a henselian local ring with maximal ideal $\mathfrak{m}/I$.

Exercise 10.1.15. Let R be a commutative local ring and $I = \operatorname{Rad}_R(0)$ the nil radical of R. Show that R is henselian if and only if R/I is henselian.

Exercise 10.1.16. Let A be a local ring with maximal ideal $\mathfrak{m}$. Let B be a commutative A-algebra which is finitely generated as an A-module. Let J be an ideal in B. Show that if B is a direct sum of local rings, then $\operatorname{idemp}(B) \to \operatorname{idemp}(B/J)$ is onto. That is, idempotents of B/J lift to idempotents of B.

Exercise 10.1.17. Let R be a henselian local ring and S a commutative R-algebra which is finitely generated as an R-module. Show that if S is a local ring, then S is henselian.

Exercise 10.1.18. Let F be a separably closed field and n an integer that is invertible in F. Let g be a polynomial in $F[x, y]$ with a nonzero constant term. Let $S = F[x, y][z]/(z^n - xyg)$, $\mathfrak{n} = (x, y, z)$ the maximal ideal in S generated by x, y, z, and $\hat{S}$ the $\mathfrak{n}$-adic completion of S.

(1) Using Theorem 10.1.10, show that there exists an invertible element $u \in F[[x, y]]$ satisfying $u^n = g$.

(2) Let $A = F[x, y, z]/(z^n - xy)$, $\mathfrak{m}$ the maximal ideal in A generated by x, y, z, and $\hat{A}$ the $\mathfrak{m}$-adic completion of A. Prove that $\hat{S}$ is isomorphic to $\hat{A} = F[[x, y]][z]/(z^n - xy)$. We say that $S_{\mathfrak{n}}$ and $A_{\mathfrak{m}}$ are *analytically isomorphic*. This also proves that the singularity of S at $\mathfrak{n}$ is a rational double point of type A_{n-1} (see Exercise 6.2.11).

(3) Prove that $\hat{S}$ is a noetherian normal integral domain, $\operatorname{Cl}(\hat{S})$ is cyclic of order n and $\operatorname{Cl}(\hat{S})$ is generated by the prime divisor $P = (x, z)$.

(4) Prove that the local ring $S_\mathfrak{n}$ is a noetherian normal integral domain, $\mathrm{Cl}(S_\mathfrak{n})$ is cyclic of order n, and $\mathrm{Cl}(S_\mathfrak{n})$ is generated by the prime divisor $P = (x, z)$. (Hint: Corollary 6.5.2.)

(5) Let k be a subfield of F such that $g \in k[x, y]$. Define R to be $k[x, y][z]/(z^n - xyg)$ and let $\mathfrak{m} = (x, y, z)$ be the maximal ideal in R generated by x, y, z. Prove that the local ring $R_\mathfrak{m}$ is a noetherian normal integral domain, $\mathrm{Cl}(R_\mathfrak{m})$ is cyclic of order n, and $\mathrm{Cl}(R_\mathfrak{m})$ is generated by the prime divisor $P = (x, z)$.

Exercise 10.1.19. Let F be an algebraically closed field with characteristic not equal to 3, and

$$R = \frac{F[x, y, z]}{(z^3 - y(y - x)(x + 1))}$$

the ring of Example 6.2.5. Let $\mathrm{Sing}(R)$ denote the set of all maximal ideals $\mathfrak{m} \in \mathrm{Max}(R)$ such that $R_\mathfrak{m}$ is not regular. Prove the following.

(1) $\mathrm{Sing}(R) = \{\mathfrak{m}_0 = (z, y, x), \mathfrak{m}_1 = (z, y, x+1), \mathfrak{m}_2 = (z, y-x, x+1)\}$. (Hint: Theorem 8.5.3.)

(2) $\mathrm{Pic}(R) = (0)$. (Hint: Exercise 6.4.17.)

(3) If k is any subfield of F, then $\mathrm{Pic}\big(k[x, y, z]/(z^3 - y(y-x)(x+1))\big) = (0)$.

Exercise 10.1.20. Let R be the ring of Exercise 6.2.12. Using the method of Exercise 10.1.19, prove that $\mathrm{Pic}(R) \cong \mathbb{Z}$.

Exercise 10.1.21. Let R be the ring of Exercise 6.2.14. Using the method of Exercise 10.1.19, prove that $\mathrm{Pic}(R) = (0)$.

Exercise 10.1.22. Let R denote a finite commutative local ring. Using Proposition 10.1.8, Theorem 4.5.7, and Corollary 7.9.6, state and prove a classification theorem for separable R-algebras which are finitely generated and projective R-modules.

2. Henselization of a Local Ring

First we show that a henselization exists for any noetherian local ring (Corollary 10.2.6). This is the method used, for example, in [**Mil80**]. The henselization of an arbitrary local ring is then obtained by a direct limit over the directed system of noetherian subrings. See Remark 10.2.7 for another approach. In Section 10.2.3 we show that a noetherian local ring has a strict henselization.

2.1. Henselization of a Noetherian Local Ring.

Definition 10.2.1. Let R be a commutative local ring with maximal ideal $\mathfrak{m}$ and residue field k. A commutative local ring R^h with maximal ideal m^h is said to be a *henselization* of R if the following are satisfied.

(1) There is a local homomorphism of local rings $i : R \to R^h$.

(2) The residue field $R^h/\mathfrak{m}^h$ is isomorphic to k.

(3) R^h is henselian.

(4) (Universal Mapping Property) If $\theta : R \to S$ is a local homomorphism of commutative local rings and S is henselian, then there is a unique R-algebra homomorphism ϕ such that the diagram

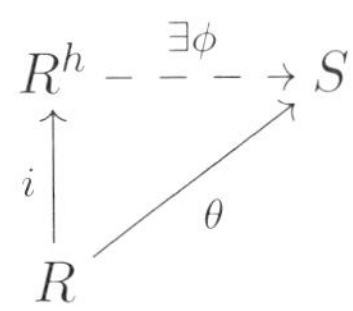

commutes.

Definition 10.2.2. Let R be a noetherian local ring with maximal ideal $\mathfrak{m}$ and residue field k. Let R^h be the direct limit $\varinjlim A$ over $\mathfrak{N}$, the directed system of connected étale neighborhoods of R (see Lemma 9.3.18). The ring R^h is called the *henselization* of R. In Corollary 10.2.6 we show that the properties (1) – (4) of Definition 10.2.1 are satisfied by R^h and that these properties uniquely determine R^h.

Lemma 10.2.3. *In the context of Definition 10.2.2, let $\mathfrak{L} = \{A_\mathfrak{p} \mid (A, \mathfrak{p}) \in \mathfrak{N}\}$. Then $\mathfrak{L}$ is a directed system of noetherian local étale R-algebras such that for any comparable pair $(A, \mathfrak{p}) \le (B, \mathfrak{q})$ in $\mathfrak{N}$, the R-algebra homomorphism $\phi : A_\mathfrak{p} \to B_\mathfrak{q}$ is a local homomorphism of local rings. In this case $B_\mathfrak{q}$ is faithfully flat and separable over $A_\mathfrak{p}$. The maximal ideal $\mathfrak{q}B_\mathfrak{q}$ is equal to $\mathfrak{p}B_\mathfrak{q}$.*

Proof. The proof follows from Lemma 9.3.18, Theorem 4.4.3 and Theorem 8.3.7. □

Lemma 10.2.4. *Let R be a noetherian local ring with maximal ideal $\mathfrak{m}$ and residue field k. Let R^h be the direct limit $\varinjlim A$ over $\mathfrak{N}$, the directed system of connected étale neighborhoods of R. Let $(A, \mathfrak{p})$ be a connected étale neighborhood of R in $\mathfrak{N}$. Then the following are true.*

(1) *For each $n \ge 1$, $\mathfrak{m}^n A_\mathfrak{p} = (\mathfrak{p}A_\mathfrak{p})^n$.*

(2) *For each $n \ge 1$, $\mathfrak{m}^n/\mathfrak{m}^{n+1} \cong (\mathfrak{p}A_\mathfrak{p})^n / (\mathfrak{p}A_\mathfrak{p})^{n+1}$.*

(3) *For each $n \ge 1$, $R/\mathfrak{m}^n \cong A_\mathfrak{p}/ (\mathfrak{p}A_\mathfrak{p})^n$.*

(4) $\hat{A}_\mathfrak{p} = \varprojlim A_\mathfrak{p} / (\mathfrak{p}A_\mathfrak{p})^n$ *is isomorphic to* $\hat{R} = \varprojlim R/\mathfrak{m}^n$.

Proof. Part (1) follows from Theorem 8.3.7.

(2): Tensor the short exact sequence

$$0 \to \mathfrak{m}^{n+1} \to \mathfrak{m}^n \to \mathfrak{m}^n/\mathfrak{m}^{n+1} \to 0$$

with the flat R-algebra $A_\mathfrak{p}$. By flatness and (1), $\mathfrak{m}^n \otimes_R A_\mathfrak{p} = \mathfrak{m}^n A_\mathfrak{p} = (\mathfrak{p}A_\mathfrak{p})^n$. From

$$\begin{aligned}(\mathfrak{p}A_\mathfrak{p})^n / (\mathfrak{p}A_\mathfrak{p})^{n+1} &\cong \mathfrak{m}^n/\mathfrak{m}^{n+1} \otimes_R A_\mathfrak{p} \\ &\cong \mathfrak{m}^n/\mathfrak{m}^{n+1} \otimes_{R/\mathfrak{m}} R/\mathfrak{m} \otimes_R A_\mathfrak{p} \\ &\cong \mathfrak{m}^n/\mathfrak{m}^{n+1}\end{aligned}$$

we get (2).

(3): The $n = 1$ case follows from the definition of étale neighborhood. For $n > 1$, apply induction, the Snake Lemma (Theorem 1.4.1), and (2) to the commutative diagram:

$$\begin{array}{ccccccccc} 0 & \longrightarrow & \frac{\mathfrak{m}^n}{\mathfrak{m}^{n+1}} & \longrightarrow & \frac{R}{\mathfrak{m}^{n+1}} & \longrightarrow & \frac{R}{\mathfrak{m}^n} & \longrightarrow & 0 \\ & & \downarrow & & \downarrow & & \downarrow & & \\ 0 & \longrightarrow & \frac{\mathfrak{m}^n A_\mathfrak{p}}{\mathfrak{m}^{n+1} A_\mathfrak{p}} & \longrightarrow & \frac{A_\mathfrak{p}}{\mathfrak{m}^{n+1} A_\mathfrak{p}} & \longrightarrow & \frac{A_\mathfrak{p}}{\mathfrak{m}^n A_\mathfrak{p}} & \longrightarrow & 0. \end{array}$$

Part (4) follows from Part (3). □

Theorem 10.2.5. *Let R be a noetherian local ring with maximal ideal $\mathfrak{m}$ and residue field k. Let R^h be the direct limit $\varinjlim A$ over $\mathfrak{N}$, the directed system of connected étale neighborhoods of R. Let $i : R \to R^h$ be the natural map. As in Lemma 10.2.3, let $\mathfrak{L} = \{A_\mathfrak{p} \mid (A, \mathfrak{p}) \in \mathfrak{N}\}$. Then the following are true.*

(1) $R^h = \varinjlim_{\mathfrak{N}} A = \varinjlim_{\mathfrak{L}} A_\mathfrak{p}$.

(2) *R^h is a local ring, with maximal ideal $\mathfrak{m}^h = \varinjlim_{\mathfrak{L}} \mathfrak{p}A_\mathfrak{p}$ and residue field $R^h/\mathfrak{m}^h = R/\mathfrak{m}$.*

(3) *R^h is noetherian.*

(4) *R^h is a faithfully flat R-algebra.*

(5) *The completions $\hat{R} = \varprojlim R/\mathfrak{m}^n$ and $\widehat{R^h} = \varprojlim R^h/\mathfrak{m}^n$ are isomorphic.*

Proof. (1): The localization maps $A \to A_\mathfrak{p}$ induce a natural map $\beta : \varinjlim_{\mathfrak{N}} A \to \varinjlim_{\mathfrak{L}} A_\mathfrak{p}$. If $(A, \mathfrak{p})$ in $\mathfrak{N}$, then for any $\alpha \in A - \mathfrak{p}$, $(A[\alpha^{-1}], \mathfrak{p})$ is another connected étale neighborhood of R. It follows that $A \to \varinjlim_{\mathfrak{N}} A$ factors through the local ring $A_{\mathfrak{p}}$. This induces a homomorphism $\varinjlim_{\mathfrak{L}} A_{\mathfrak{p}} \to \varinjlim_{\mathfrak{N}} A$, which is the inverse to β.

(2), (3), (4): By Lemma 10.2.3 and Corollary 9.1.6, $R^h = \varinjlim_{\mathfrak{L}} A_{\mathfrak{p}}$ is a noetherian local ring with maximal ideal $\mathfrak{m}^h = \varinjlim_{\mathfrak{L}} \mathfrak{p}A_{\mathfrak{p}}$ and is faithfully flat over R. Moreover, $\mathfrak{m}^h = \mathfrak{m}R^h$ and the residue field of R^h is $\varinjlim_{\mathfrak{L}} A_{\mathfrak{p}}/\mathfrak{p}A_{\mathfrak{p}} = R/\mathfrak{m} = k$.

(5): From Lemma 10.2.4 (3), $R/\mathfrak{m}^n \cong A_{\mathfrak{p}}/\mathfrak{m}^n A_{\mathfrak{p}}$. The direct limit functor $\varinjlim_{\mathfrak{L}}()$ is exact (Theorem 1.4.3). Applying it to the exact sequence

$$0 \to \mathfrak{m}^n A_{\mathfrak{p}} \to A_{\mathfrak{p}} \to A_{\mathfrak{p}}/\mathfrak{m}^n A_{\mathfrak{p}} \to 0$$

shows that $R/\mathfrak{m}^n \to R^h/(\mathfrak{m}^h)^n$ is an isomorphism. □

Corollary 10.2.6. *Let R be a noetherian local ring with maximal ideal $\mathfrak{m}$ and residue field k. Let R^h be the direct limit $\varinjlim A$ over $\mathfrak{N}$, the directed system of connected étale neighborhoods of R. Let $i : R \to R^h$ be the natural map. Then R^h is the henselization of R. Moreover, R^h is unique up to R-algebra isomorphism.*

Proof. By Theorem 10.2.5 (2), properties (1) and (2) of Definition 10.2.1 are satisfied by R^h. We show that (3) and (4) are satisfied as well. First we prove that $R \to R^h$ satisfies the universal mapping property. Let S be a henselian local ring, and $\theta : R \to S$ a local homomorphism of local rings. We want to show that θ factors through $R \to R^h$. Let $(A, \mathfrak{p})$ be a connected étale neighborhood of R. By Lemma 9.3.19, there is a maximal ideal $\mathfrak{r}$ in $S \otimes_R A$ such that $(S \otimes_R A, \mathfrak{r})$ is an étale neighborhood of S. Applying Proposition 10.1.7 (3), there is a section $\sigma : S \otimes_R A \to S$. If ϕ is defined to be the composite map $A \to S \otimes_R A \to S$, then by Corollary 9.2.8, ϕ is uniquely determined by A and θ. The universal mapping property of the direct limit gives the desired map $R^h \to S$.

Now we show that R^h is henselian. By Theorem 10.2.5, R^h is a noetherian local ring with maximal ideal $\mathfrak{m}^h$. Let S be an étale R^h-algebra. Assume there is a prime ideal $\mathfrak{n}$ in S such that $\mathfrak{n} \cap R^h = \mathfrak{m}^h$ and the map on residue fields $k \to k(\mathfrak{n})$ is an isomorphism. By Proposition 10.1.7, it suffices to show that $R^h \to S_{\mathfrak{n}}$ is an isomorphism and $S_{\mathfrak{n}}$ is a direct summand of S. By Theorem 9.3.9, we can assume S is a standard étale R^h-algebra, $S = (R^h[x]/(f))[g^{-1}]$. By Proposition 1.4.2 and Proposition 9.3.7 there is a connected étale neighborhood $(A, \mathfrak{p})$ in the directed system $\mathfrak{N}$ such that f and g come from polynomials in $A[x]$, and $B = (A[x]/(f))[g^{-1}]$ is a standard étale A-algebra. Let $\mathfrak{q} = \mathfrak{n} \cap B$. On the residue fields, the maps $k \to k(\mathfrak{p}) \to k(\mathfrak{q}) \to k(\mathfrak{n})$ are isomorphisms. Then $(B, \mathfrak{q})$ is an étale neighborhood of R. By the direct limit, there is a natural map $\beta : B \to R^h$. So β

induces $\sigma : S = R^h \otimes_R B \to R^h$, which is the desired section. Corollary 4.4.8 shows the kernel of σ is idempotent generated.

The proof that the henselization is unique is left to the reader. $\square$

Remark 10.2.7. As in [**Ray70**, Théorème VIII.1], Corollary 10.2.6 can be proved without assuming R is noetherian. In the above proof, the noetherian hypothesis was only used to apply Theorem 9.3.9. As mentioned in Remark 9.3.10, with some extra effort, this assumption can be avoided.

2.2. Henselization of an Arbitrary Local Ring.

Proposition 10.2.8. *Let $\{A_i, \phi^i_j\}$ be a directed system of commutative noetherian henselian local rings for a directed index set I. Assume $\phi^i_j : A_i \to A_j$ is a local homomorphism of local rings, for each $i \leq j$. Let $A = \varinjlim A_i$. Then A is a local ring and $\varinjlim A^h_i$ is a henselization of A.*

Proof. By Proposition 10.1.5, A is a local ring and $\varinjlim A^h_i$ is a henselian local ring. The reader should verify that $A \to \varinjlim A^h_i$ satisfies (1) – (4) of Definition 10.2.1, hence is a henselization of A. $\square$

Theorem 10.2.9. *If R is a local ring with maximal ideal $\mathfrak{m}$ and residue field k, then R^h, the henselization of R, exists. The map $R \to R^h$ is faithfully flat.*

Proof. As in Exercise 2.5.17, let $\{R_\alpha\}$ be a directed system of noetherian local subrings of R satisfying the following:

(1) The maximal ideal of R_α is $\mathfrak{m}_\alpha = R_\alpha \cap \mathfrak{m}$.
(2) $R = \varinjlim R_\alpha$.
(3) $\mathfrak{m} = \varinjlim \mathfrak{m}_\alpha$.
(4) $R/\mathfrak{m} = \varinjlim (R_\alpha/\mathfrak{m}_\alpha)$.

The reader should verify that $R^h = \varinjlim (R_\alpha)^h$ is a henselization of R. To show R^h is faithfully flat, apply Exercise 1.4.15 and Proposition 10.1.5. $\square$

Corollary 10.2.10. *If $\theta : R \to S$ is a local homomorphism of commutative local rings, then there is a unique local homomorphism of local rings θ^h such that the diagram*

$$\begin{array}{ccc} R & \xrightarrow{\theta} & S \\ \downarrow & & \downarrow \\ R^h & \xrightarrow{\theta^h} & S^h \end{array}$$

commutes.

Proof. The proof is left to the reader. $\square$

2.3. Strict Henselization of a Noetherian Local Ring. First we establish some notation that will be in effect for the remainder of this section. Let R be a noetherian local ring with maximal ideal $\mathfrak{m}$ and residue field k. Let k_s be a fixed separable closure of k.

Denote by $\mathfrak{E}$ the set of all couples (A, α) where A is a local étale R-algebra (Definition 9.3.20) with maximal ideal $\mathfrak{m}(A)$, residue field $k(A)$, and $\alpha : A \to k_s$ is a homomorphism of R-algebras which factors through $k(A)$ such that the diagram

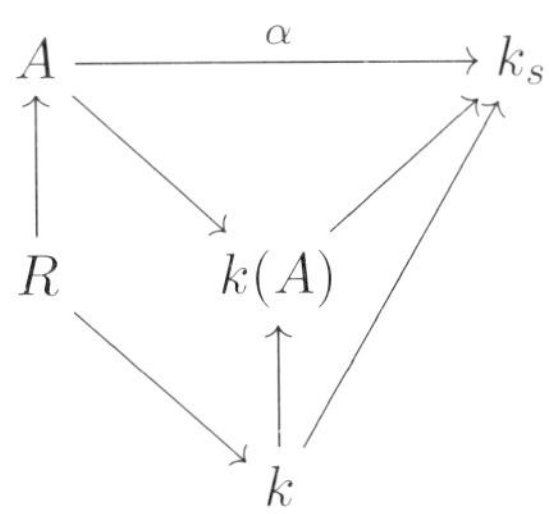

commutes. Let (A, α) and (B, β) be two pairs in $\mathfrak{E}$. We say $(A, \alpha) \leq (B, \beta)$ in case there is an R-algebra homomorphism $\theta : A \to B$ such that

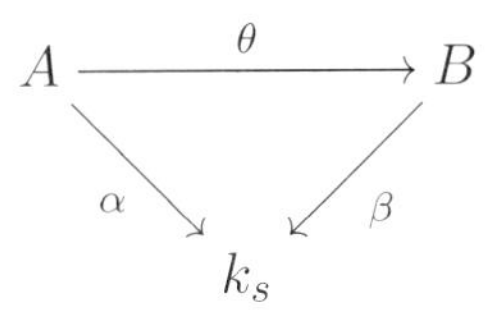

commutes and θ is a local homomorphism of local rings.

Lemma 10.2.11. *Let R be a noetherian local ring and $\mathfrak{E}$ the collection of all couples (A, α) of local étale R-algebras and local homomorphisms $\alpha : A \to k_s$. Let (A, α) and (B, β) be two couples in $\mathfrak{E}$.*

(1) *There is at most one local R-algebra homomorphism of local rings $\theta : A \to B$ such that $\alpha = \beta\theta$.*
(2) *There exists a local étale R-algebra (C, γ) such that $(A, \alpha) \leq (C, \gamma)$ and $(B, \beta) \leq (C, \gamma)$.*
(3) *The collection $\mathfrak{E}$ is a directed system of R-algebras.*
(4) *If $(A, \alpha) \leq (B, \beta)$, then B/A is faithfully flat, separable, and we have $\mathfrak{m}(B) = \mathfrak{m}(A)B$.*

Proof. Part (1) follows straight from Corollary 9.2.8.

(2): The homomorphisms $\alpha : A \to k_s$ and $\beta : B \to k_s$ induce a homomorphism $h : A \otimes_R B \to k_s$ (Exercise 1.3.8). Let $\mathfrak{p}$ be the kernel of h. Then $\mathfrak{p}$ is a prime ideal of $A \otimes_R B$ lying over $\mathfrak{m}$. The reader should verify that $C = (A \otimes_R B)_{\mathfrak{p}}$ is a local étale R-algebra. Let $\gamma : C \to k_s$ be the localization of h.

Part (3) is left to the reader.

Part (4) follows from Theorem 4.4.3 and Theorem 8.3.7. □

A henselian local ring is said to be *strictly henselian* if it has a separably closed residue field. The next theorem shows that a noetherian local ring can be embedded in a strictly henselian local ring.

Theorem 10.2.12. *In the notation of Lemma 10.2.11, let R^{sh} be the R-algebra direct limit over all A such that $(A, \alpha) \in \mathfrak{C}$, and $R^{sh} \to k_s$ the corresponding R-algebra homomorphism. Then the following are true.*

(1) *R^{sh} is a strictly henselian local ring, noetherian, is faithfully flat over R, and the residue field is mapped onto k_s by the natural map $R^{sh} \to k_s$.*

(2) *Given a strictly henselian local ring B, a local homomorphism of local rings $f : R \to B$, an extension of fields F/k_s, and a local homomorphism $\beta : B \to F$, then there exists a unique homomorphism $f^{sh} : R^{sh} \to B$ such that the diagram*

$$\begin{array}{ccccc} R & \longrightarrow & R^{sh} & \longrightarrow & k_s \\ & \searrow^{f} & \Big\downarrow{\scriptstyle \exists f^{sh}} & & \Big\downarrow{\scriptstyle \subseteq} \\ & & B & \xrightarrow{\beta} & F \end{array}$$

commutes.

Proof. (1): By Lemma 10.2.11 and Corollary 9.1.6, $R^{sh} = \varinjlim_{\mathfrak{C}} A$ is a noetherian local ring with maximal ideal $\mathfrak{m}^{sh} = \varinjlim_{\mathfrak{C}} \mathfrak{m}(A)$, and is a faithfully flat R-algebra. Moreover, $\mathfrak{m}^{sh} = \mathfrak{m}(A)R^{sh}$ and the residue field of R^{sh} is $\varinjlim_{\mathfrak{C}} k(A)$. By applying Corollary 4.6.2, the reader should verify that the residue field of R^{sh} is equal to k_s. By mimicking the proof of Corollary 10.2.6, the reader should verify that R^{sh} is henselian.

(2): Let $(A, \alpha) \in \mathfrak{C}$. Using Exercise 4.1.9 we see that $k(A)$ is separable over R. Then $k(A) \otimes_R k(B)$ is separable over $k(B)$. By Exercise 1.3.8, there is a homomorphism of R-algebras $h : k(A) \otimes_R k(B) \to F$ and the image of h is separable over $k(B)$. Since $k(B)$ is separably closed, the image of h is equal to $k(B)$. There is an étale R-algebra C such that A is the localization of C at a prime ideal lying over $\mathfrak{m}$. The maps on residue fields $C \to A \to k(A)$, $B \to k(B)$ induce an R-algebra homomorphism $\theta : C \otimes_R B \to F$ which factors through h. Therefore, the image of θ is equal to $k(B)$. By Proposition 10.1.7 (3), there is a section $\sigma : C \otimes_R B \to B$ to $B \to C \otimes_R B$. The image of the composite map $C \to C \otimes_R B \to B \to k(B)$ is a subring of $k(A)$. Therefore, $C \to C \otimes_R B \to B$ factors into $C \to A \to B$.

By Corollary 9.2.8, the homomorphism $\phi : A \to B$ just constructed is uniquely determined by f, A, and α. The universal mapping property of the direct limit gives the desired homomorphism $f^{sh} : R^{sh} \to B$. □

2.4. Exercises.

Exercise 10.2.13. Let R be a commutative noetherian local ring and R^h the henselization of R. Prove the following.

(1) If R is reduced, then R^h is reduced.
(2) $\dim(R) = \dim(R^h)$.
(3) If R is regular, then R^h is regular.
(4) If R is normal, then R^h is normal.

Exercise 10.2.14. Let R be a commutative noetherian local ring and R^{sh} the strict henselization of R. Prove the following.

(1) If R is reduced, then R^{sh} is reduced.
(2) $\dim(R) = \dim(R^{sh})$.
(3) If R is regular, then R^{sh} is regular.
(4) If R is normal, then R^{sh} is normal.

3. Splitting Rings for Azumaya Algebras

This section is a continuation of the study of splitting rings that was begun in Section 7.4. Our goal is to show that any Azumaya algebra has a faithfully flat étale splitting ring. The construction proceeds in three steps. First we consider the problem over a local ring. We have already seen that an Azumaya algebra over a field k has a splitting ring which is a maximal commutative separable k-subalgebra that is generated by a primitive element. Theorem 10.3.5 shows that an Azumaya algebra over a local ring has a splitting subring with the analogous properties. Then we show how to extend the results from a local ring at a prime ideal $\mathfrak{p}$ in $\operatorname{Spec} R$ to to a basic open neighborhood of $\mathfrak{p}$ (Section 10.3.2). Lastly, we prove the existence of faithfully flat étale splitting rings in the general case (Theorem 10.3.9). General references for the results in this section are [**KO74b**, §III.6] and [**OS75**, Chapter 11].

3.1. Existence of Splitting Rings (Local Version). The following theorem of Azumaya is from [**Azu51**].

Theorem 10.3.1. *Let R be a henselian local ring and A an R-algebra which is integral over R. If I is a two-sided ideal in A, $\eta : A \to A/I$ the natural map, and $\bar{e}$ an idempotent in A/I, then there is an idempotent $e \in A$ such that $\eta(e) = \bar{e}$.*

Proof. Let $\theta : R \to A$ be the structure homomorphism. Let $R_0 = R \cdot 1$ be the image of θ. By Exercise 10.1.14, R_0 is a henselian local ring. Also A is an R_0-algebra which is integral over R_0. Without loss of generality, assume $R = R_0$. Pick $a \in A$ such that $\eta(a) = \bar{e}$. Then $B = R[a]$ is finitely generated as an R-module. If $J = B \cap I$, then the image of $B = R[a]$ under η is isomorphic to B/J. We see that $\eta(B)$ is a commutative R-subalgebra of A/I containing $\bar{e}$ and is finitely generated as an R-module. Since R is henselian, both B and B/J are direct sums of local rings. By Exercise 10.1.16, the idempotent $\bar{e}$ lifts to an idempotent $e \in R[a]$. □

Theorem 10.3.2. *Let R be a henselian local ring with maximal ideal $\mathfrak{m}$ and residue field k.*

(1) *The map on Brauer groups* $\mathrm{B}(R) \to B(k)$ *is one-to-one.*
(2) *If k is a separably closed field, then* $\mathrm{B}(R) = (0)$.
(3) *If A is an Azumaya R-algebra, then there exists a commutative R-algebra S satisfying the following.*
    (a) *$A \otimes_R S$ is split.*
    (b) *S is separable over R.*
    (c) *S is a finitely generated free R-module.*
    (d) *S is a henselian local ring and $R \to S$ is a local homomorphism.*

Proof. (1): Let A be an Azumaya R-algebra and assume $A \otimes_R k \cong M_n(k)$. By Proposition 2.3.2, $\mathrm{Rank}_R(A) = n^2$. Let M_{11} be the idempotent matrix over k with 1 in position $(1,1)$ and 0 elsewhere. By Theorem 10.3.1, lift M_{11} to an idempotent $e \in A$. Then $A = Ae \oplus A(1-e)$, so Ae is a finitely generated projective R-module. Since $Ae \otimes_R k \cong k^n$, Proposition 2.3.2 shows $\mathrm{Rank}_R(Ae) = n$. The left regular representation $\varphi : A \to \mathrm{Hom}_R(Ae, Ae)$ is an R-algebra homomorphism. By Theorem 7.6.1, φ is an isomorphism. Therefore, A is split.

Part (2) follows from Part (1) and Corollary 7.5.6.

(3): Since $A \otimes_R k$ is an Azumaya k-algebra, by Theorem 7.5.5 there exists a separable extension field $k(\alpha)/k$ such that $A \otimes_R k(\alpha)$ is split. Let $f \in k[x]$ be the irreducible polynomial for α. Lift f to a monic polynomial $p \in R[x]$. By Corollary 8.1.27, $S = R[x]/(p)$ is a separable R-algebra which is finitely generated and free. Also, $S \otimes_R k = S/\mathfrak{m}S = k(\alpha)$. So $\mathfrak{m}S$ is a maximal ideal of S. By Theorem 3.3.5, $\mathfrak{m}S \subseteq J(S)$. Therefore, $J(S) = \mathfrak{m}S$

and S is a local ring. By Exercise 10.1.17, S is henselian. The diagram

$$\begin{array}{ccc} \mathrm{B}(R) & \longrightarrow & \mathrm{B}(S) \\ \downarrow & & \downarrow \\ \mathrm{B}(k) & \longrightarrow & \mathrm{B}(k(\alpha)) \end{array}$$

commutes, A is split by $k(\alpha)$, and by Part (1), the vertical maps are one-to-one. Therefore, A is split by S. □

Remark 10.3.3. In Corollary 12.6.4 we show that the splitting ring S constructed in Theorem 10.3.2 (3) can be embedded in a Galois extension. Hence A has a splitting ring which is Galois over R.

Corollary 10.3.4. *Let R be a local ring and A an Azumaya R-algebra of rank* $\mathrm{Rank}_R(A) = n^2$.

(1) *There exists a faithfully flat R-algebra which is a splitting ring for A.*
(2) *(Cayley-Hamilton, Local Ring Version) Every element $\alpha \in A$ is a root of its characteristic polynomial, which is a monic polynomial of degree n in $R[x]$.*

Proof. (1): By Theorem 10.3.2, there exists a faithfully flat étale R^h-algebra S which splits $A \otimes_R R^h$. Then S is faithfully flat over R, and splits A.

(2): By (1), let S be a faithfully flat commutative R-algebra such that $A \otimes_R S$ is isomorphic to $\mathrm{Hom}_S(P, P)$, for an S-progenerator. Then $A \to A \otimes_R S$ is one-to-one because A is faithfully flat over R. By Example 5.3.3, the characteristic polynomial $p_S(x) = \mathrm{char.\,poly}_S(\alpha) \in S[x]$ exists and $p_S(\alpha) = 0$. The degree of $p_S(x)$ is n. Since S is faithfully flat, and taking determinants commutes with change of base ring, the polynomial $p_S(x)$ is in the kernel of d^0 in the Amitsur complex

$$0 \to R[x] \xrightarrow{\theta} S[x] \xrightarrow{d^0} S \otimes_R S[x] \xrightarrow{d^1} \cdots$$

which is an exact sequence. That is, $\mathrm{char.\,poly}_S(\alpha)$ comes from $R[x]$. □

Theorem 10.3.5. *Let R be a local ring and A an Azumaya R-algebra. Then there exists an R subalgebra $S \subseteq A$ which satisfies:*

(1) *$S = R[\beta]$ for some $\beta \in A$.*
(2) *S is a finitely generated free R-module.*
(3) *S is separable over R.*
(4) *A is a free S-module.*
(5) *S is a maximal commutative R-subalgebra of A.*

(6) *The S-algebra homomorphism $A \otimes_R S \to \operatorname{Hom}_S(A, A)$ is an isomorphism. Hence, S is a splitting ring for A.*

Proof. Let $\mathfrak{m}$ be the maximal ideal of R, and $k = R/\mathfrak{m}$ the residue field. Then $A/\mathfrak{m}A$ is an Azumaya k-algebra. By Theorem 7.5.7 there is a maximal commutative k-subalgebra $k(\alpha) \subseteq A/\mathfrak{m}A$ such that $A/\mathfrak{m}A$ is a free $k(\alpha)$-module, and $k(\alpha)$ is a direct sum of the form $F \oplus \cdots \oplus F$, where F/k is an extension of fields. Let $\beta \in A$ be a preimage of α. If n is the dimension $\dim_k k(\alpha)$, then by Theorem 7.4.2, $n^2 = \dim_k(A/\mathfrak{m}A)$. We can find a k-basis for $A/\mathfrak{m}A$ of the form $1, \alpha, \dots, \alpha^{n-1}, \gamma_1, \dots, \gamma_{n^2-n}$. Lift $\gamma_1, \gamma_2, \dots$ to preimages $y_1, y_2, \dots$ in A. Since A is R-projective, Proposition 2.6.4 shows that $1, \beta, \dots, \beta^{n-1}, y_1, \dots, y_{n^2-n}$ is a free R-basis for A. Let S be the R-submodule of A spanned by $1, \beta, \dots, \beta^{n-1}$. By Corollary 10.3.4, $\beta^n \in S$, which implies S is an R-subalgebra of A. By construction, S is a free R-module. Since $S/\mathfrak{m}S = k(\alpha)$ is separable over k, it follows from Theorem 8.1.24 that S is separable over R. By Proposition 4.4.1, A is a projective S-module. Since $S/\mathfrak{m}S = k(\alpha)$ is a finite direct sum of copies of a field F, S has only finitely many maximal ideals. For every maximal ideal $\mathfrak{n}$ of S, $A/\mathfrak{n}A$ is free over $S/\mathfrak{n}$ of rank n. We see that A is an S-module of constant rank. Since S is semilocal, A is a free S-module (Exercise 3.1.8). We have shown (1) – (4). By Theorem 7.4.2, (6) follows from (5).

(5): By Theorem 7.4.2, it is enough to prove that S is a maximal commutative subalgebra of A. By Exercise 7.4.6, it suffices to prove that $S = A^S$. Since $S/\mathfrak{m}S = k(\alpha)$ is a maximal commutative subalgebra of $A/\mathfrak{m}A$, the homomorphism

$$S/\mathfrak{m}S \to (A/\mathfrak{m}A)^{S/\mathfrak{m}S} \tag{10.2}$$

is onto. Hence, under the natural map $\eta : A \to A/\mathfrak{m}A$, the preimage of $(A/\mathfrak{m}A)^{S/\mathfrak{m}S}$ is $S + \mathfrak{m}A$. Since $\eta(A^S)$ is in the right hand side of (10.2), $A^S \subseteq S + \mathfrak{m}A$. Intersecting both sides with A^S, we get

$$A^S = S + A^S \cap \mathfrak{m}A. \tag{10.3}$$

The ring A is a left $A \otimes_R A^o$-module. Mapping $x \otimes y$ to the "left multiplication" map defines $\varphi : A^e \to \operatorname{Hom}_R(A, A)$. By Theorem 7.1.4, φ is an isomorphism. Since elements of $(A^o)^S$ commute with elements of S, the same action induces λ and β such that the diagram of ring homomorphisms

$$\begin{array}{ccc}
A \otimes_R S & \xrightarrow{\lambda} & \operatorname{Hom}_S(A, A) \\
\downarrow & \nearrow_{\beta} & \\
A \otimes_R (A^o)^S & & \Big\downarrow \gamma \\
\downarrow & & \\
A \otimes_R A^o & \xrightarrow{\varphi} & \operatorname{Hom}_R(A, A)
\end{array}$$

commutes. The map γ is one-to-one by Exercise 1.3.35. Since A is flat over R, the two left hand vertical maps are one-to-one. Therefore, λ and β are one-to-one. Applying $(\ )^{A \otimes_R S}$, and using Lemma 7.2.1 we get

$$\operatorname{Hom}_S(A, A)^{A \otimes_R S} \subseteq \operatorname{Hom}_R(A, A)^{A \otimes_R S} = (A \otimes_R A^o)^{A \otimes_R S} = (A^o)^S.$$

Applying the same argument to the image of β, we get

$$1 \otimes (A^o)^S \subseteq \left(A \otimes_R (A^o)^S\right)^{A \otimes_R S} \subseteq \operatorname{Hom}_S(A, A)^{A \otimes_R S}.$$

Taken together, this shows that $\operatorname{Hom}_S(A, A)^{A \otimes_R S} = (A^o)^S$. By Corollary 7.1.9 and Proposition 7.1.10, it follows that $A \otimes_R S$ and $\operatorname{Hom}_S(A, A)$ are Azumaya S-algebras. By Theorem 7.2.2 (The Commutator Theorem), $(A^o)^S$ is an S-Azumaya algebra. Hence A^S is an Azumaya algebra.

By (10.3), $A^S \cap \mathfrak{m}A$ is a two-sided ideal in A^S. By Corollary 7.1.7 there is an ideal I in S such that $IA^S = A^S \cap \mathfrak{m}A$ and $IA^S \cap S = I$. Therefore, $I = A^S \cap \mathfrak{m}A \cap S = \mathfrak{m}A \cap S$. But $S/\mathfrak{m}S \to A/\mathfrak{m}A$ is one-to-one, so $I = \mathfrak{m}A \cap S = \mathfrak{m}S$. This leads to $A^S = S + A^S \cap \mathfrak{m}A = S + \mathfrak{m}SA^S = S + \mathfrak{m}A^S$. The modules A^S and S are finitely generated over R, so Corollary 1.1.17 (Nakayama's Lemma), implies $A^S = S$. □

Remark 10.3.6. In Corollary 12.6.4 we show that the splitting ring S of Theorem 10.3.5 can be embedded in a Galois extension. Hence A has a splitting ring which is Galois over R.

3.2. Local to Global Lemmas. Lemma 10.3.7 shows that an algebra which is Azumaya over the local ring at a prime $\mathfrak{p} \in \operatorname{Spec} R$ can be extended to an Azumaya algebra on a basic open neighborhood of $\mathfrak{p}$, provided that A is a finitely presented R-module.

Lemma 10.3.7. *Let R be a commutative ring, $\mathfrak{p} \in \operatorname{Spec} R$, and A an R-algebra.*

(1) *If A is a finitely presented R-module and $A_{\mathfrak{p}}$ is a free $R_{\mathfrak{p}}$-module, then there exists $\alpha \in R - \mathfrak{p}$ such that A_α is a free R_α-module.*

(2) *If A is a finitely presented R-module and $A_{\mathfrak{p}}$ is an Azumaya $R_{\mathfrak{p}}$-algebra, then there exists $\alpha \in R - \mathfrak{p}$ such that A_α is an R_α-Azumaya algebra.*

(3) *Assume either*
   (a) *A is a finitely generated R-module, or*
   (b) *A a finitely generated commutative R-algebra.*
   If $A_{\mathfrak{p}}$ is a separable $R_{\mathfrak{p}}$-algebra, then there exists $\alpha \in R - \mathfrak{p}$ such that A_α is a separable R_α-algebra.

Proof. (1): This comes from Lemma 2.1.6.

(2): By Theorem 7.1.4, $A_{\mathfrak{p}}$ is a finitely generated free $R_{\mathfrak{p}}$-module, and the enveloping homomorphism

$$\varphi_{A_{\mathfrak{p}}/R_{\mathfrak{p}}} : A_{\mathfrak{p}} \otimes_{R_{\mathfrak{p}}} A_{\mathfrak{p}}^o \to \operatorname{Hom}_{R_{\mathfrak{p}}}(A_{\mathfrak{p}}, A_{\mathfrak{p}})$$

is an $R_{\mathfrak{p}}$-algebra isomorphism. Assume R_α and A_α are chosen as in (1). Then

$$\varphi_{A_\alpha/R_\alpha} : A_\alpha \otimes_{R_\alpha} A_\alpha^o \to \operatorname{Hom}_{R_\alpha}(A_\alpha, A_\alpha)$$

is an R_α-algebra homomorphism. Both $A_\alpha \otimes_{R_\alpha} A_\alpha^o$ and $\operatorname{Hom}_{R_\alpha}(A_\alpha, A_\alpha)$ are finitely generated free R_α-modules, and $\varphi_{A_\alpha/R_\alpha}$ becomes the isomorphism $\varphi_{A_{\mathfrak{p}}/R_{\mathfrak{p}}}$ upon tensoring with $(\cdot) \otimes_{R_\alpha} R_{\mathfrak{p}}$. We are in the setting of Exercise 2.3.10. For some $\beta \in R - \mathfrak{p}$, $\varphi_{A_\beta/R_\beta}$ is an isomorphism. Then A_β is an R_β-Azumaya algebra.

(3): Let $e \in A_{\mathfrak{p}} \otimes_{R_{\mathfrak{p}}} A_{\mathfrak{p}}^o$ be a separability idempotent for $A_{\mathfrak{p}}$. By Exercise 2.1.13 there exists $\alpha \in R - \mathfrak{p}$ such that e comes from an idempotent in $A_\alpha \otimes_{R_\alpha} A_\alpha^o$. By Lemma 8.1.4, J_{A_α/R_α} is generated by a finite set $x_1, \dots, x_n$ as a left ideal in $A_\alpha \otimes_{R_\alpha} A_\alpha^o$. Since $(R_{\mathfrak{p}}x_1 + \cdots + R_{\mathfrak{p}}x_n)e = 0$, by Lemma 2.1.3, we can replace α by an appropriate element of $R - \mathfrak{p}$ and assume $(R_\alpha x_1 + \cdots + R_\alpha x_n)e = 0$. Therefore, $J_{A_\alpha/R_\alpha}e$ is the zero ideal of $A_\alpha \otimes_{R_\alpha} A_\alpha^o$, and by Proposition 4.1.2, A_α is separable over R_α. □

Lemma 10.3.8 shows that a splitting ring for an Azumaya $R_{\mathfrak{p}}$-algebra extends to a splitting ring on a basic open neighborhood of the prime $\mathfrak{p}$. Note that after extending the splitting ring to a neighborhood of the prime, the splitting ring we construct is not generated by a primitive element. It does however retain all of the other properties of the splitting ring at the local ring.

Lemma 10.3.8. *Let R be a commutative ring and $\mathfrak{p} \in \operatorname{Spec} R$. Let A be an R-algebra which is an R-module of finite presentation. Assume $A_{\mathfrak{p}}$ contains a commutative $R_{\mathfrak{p}}$-subalgebra S which satisfies all of the following.*

(A) *$A_{\mathfrak{p}}$ is a free S-module.*

(B) *S is a separable $R_{\mathfrak{p}}$-algebra and a finitely generated free $R_{\mathfrak{p}}$-module.*

(C) *The S-algebra homomorphism*

$$A_{\mathfrak{p}} \otimes_{R_{\mathfrak{p}}} S \to \operatorname{Hom}_S(A_{\mathfrak{p}}, A_{\mathfrak{p}})$$

(see Theorem 7.4.2) is an isomorphism.

Then there exists $\alpha \in R - \mathfrak{p}$ such that the following are true:

(1) *A_α is an R_α-Azumaya algebra which is a finitely generated free R_α-module.*

(2) *A_α contains a commutative R_α-subalgebra S_α which satisfies:*
   (a) *$S = S_\alpha \otimes_{R_\alpha} R_{\mathfrak{p}}$.*
   (b) *S_α is a free R_α-module.*
   (c) *S_α is separable over R_α.*
   (d) *A_α is a free S_α-module.*
   (e) *The homomorphism $A_\alpha \otimes_{R_\alpha} S_\alpha \to \operatorname{Hom}_{S_\alpha}(A_\alpha, A_\alpha)$ is an isomorphism of S_α-algebras. So S_α splits A_α.*

Proof. By Proposition 7.1.10, $A_{\mathfrak{p}} \otimes_{R_{\mathfrak{p}}} S \cong \operatorname{Hom}_S(A_{\mathfrak{p}}, A_{\mathfrak{p}})$ is S-Azumaya. By Proposition 2.4.6 and Corollary 4.3.5, $A_{\mathfrak{p}}$ is $R_{\mathfrak{p}}$-Azumaya.

The local to global proof is a series of five steps. At each step it may be necessary to modify the localization by "inverting another element". Because this is a finite process, in the end we only invert one element. For notational simplicity, we prefer to call the new element and the old element by the same name, α.

Step 1. Start with a free basis $\{b_1, \dots, b_m\}$ for S over $R_{\mathfrak{p}}$, where we assume $b_1 = 1$. Let $\{c_1, \dots, c_n\}$ be a free basis for $A_{\mathfrak{p}}$ over $R_{\mathfrak{p}}$. Then $\{b_i c_j\}$ is a free basis for $A_{\mathfrak{p}}$ over $R_{\mathfrak{p}}$ (Exercise 1.1.9). By Lemma 10.3.7 we can find $\alpha \in R - \mathfrak{p}$ such that A_α is an R_α-Azumaya algebra which is finitely generated free as an R_α-module. By this construction, we can assume A_α contains $\{b_1, \dots, b_m\}$ and $\{c_1, \dots, c_n\}$.

Step 2. Let $R_{\mathfrak{p}}^m \to S$ be the $R_{\mathfrak{p}}$-module isomorphism defined by mapping the standard basis to the free basis $\{b_1, \dots, b_m\}$. The multiplication map μ lifts to a homomorphism ϕ such that the diagram

$$\begin{array}{ccc} R_{\mathfrak{p}}^m \otimes_{R_{\mathfrak{p}}} R_{\mathfrak{p}}^m & \longrightarrow & S \otimes_{R_{\mathfrak{p}}} S \\ \downarrow{\scriptstyle \phi} & & \downarrow{\scriptstyle \mu} \\ R_{\mathfrak{p}} & \longrightarrow & S \end{array}$$

commutes. The matrix for ϕ with respect to the standard basis involves only finitely many elements of $R_{\mathfrak{p}}$. So we can modify the α from Step 1 such that ϕ is defined over the ring R_α. Define S_α to be the R_α-submodule of A_α generated by $\{b_1, \dots, b_m\}$. Then the multiplication $S_\alpha \otimes_{R_\alpha} S_\alpha \to S_\alpha$ is defined. Therefore, S_α is an R_α subalgebra of A_α which is a free R_α-module.

Step 3. Applying the argument of Lemma 10.3.7 (3), we can modify α from Step 2 and assume there exists a separability idempotent for S_α. Therefore, S_α is separable over R_α.

Step 4. By the construction in Step 1, there is an S_α-module homomorphism $\theta : S_\alpha^n \to A_\alpha$ defined by $\theta(a_1, \dots, a_n) = \sum a_i c_i$. Then θ is a homomorphism of projective R_α-modules which becomes an isomorphism upon tensoring with $(\ ) \otimes_{R_\alpha} R_\mathfrak{p}$. We are in the setting of Exercise 2.3.10. Modify α from Step 3, and assume A_α is a free S_α-module.

Step 5. There is a homomorphism $\varphi_{A_\alpha/R_\alpha} : A_\alpha \otimes_{R_\alpha} S_\alpha \to \operatorname{Hom}_{S_\alpha}(A_\alpha, A_\alpha)$ of S_α-algebras which becomes an S-algebra isomorphism upon tensoring with $(\ ) \otimes_{R_\alpha} R_\mathfrak{p}$. We are in the setting of Exercise 2.3.10. Modify α from Step 4, and assume $\varphi_{A_\alpha/R_\alpha}$ is an isomorphism. The proof is complete. □

3.3. Splitting Rings for Azumaya Algebras. To construct a faithfully flat splitting ring for an Azumaya R-algebra, we apply the results of Lemma 10.3.8 one prime at at time. Since $\operatorname{Spec} R$ is quasi-compact, there exists a faithfully flat R-algebra that splits A. Note that the globalized splitting ring is not a subring of the algebra A.

Theorem 10.3.9. *Let A be an R-algebra. The following are equivalent.*

(1) *A is an Azumaya R-algebra.*

(2) *For every prime ideal $\mathfrak{p}$ in $\operatorname{Spec} R$ there is an element $\alpha \in R - \mathfrak{p}$ such that A_α contains a commutative R_α-subalgebra S_α which satisfies:*
   - (a) *S_α is a finitely generated free R_α-module.*
   - (b) *S_α is separable over R_α.*
   - (c) *A_α is a finitely generated free S_α-module.*
   - (d) *There is an S_α-algebra isomorphism $A_\alpha \otimes_{R_\alpha} S_\alpha \cong M_n(S_\alpha)$ for some $n \geq 1$.*

(3) *There exists a faithfully flat étale R-algebra S, an S-progenerator P, and an S-algebra isomorphism $A \otimes_R S \cong \operatorname{Hom}_S(P, P)$.*

(4) *There exists a commutative faithfully flat R-algebra S, an S-module progenerator P, and an isomorphism $A \otimes_R S \cong \operatorname{Hom}_S(P, P)$ of S-algebras.*

Proof. (1) implies (2): This follows directly from Corollary 8.1.23, and Theorem 10.3.5 and Lemma 10.3.8.

(2) implies (3): By Exercise 2.2.16 there is an open cover $\operatorname{Spec} R = U(\alpha_1) \cup \cdots \cup U(\alpha_n)$ where each α_i is as in (2). Then each S_{α_i} is an étale R-algebra, and $S = \bigoplus_i S_{\alpha_i}$ is a faithfully flat étale R-algebra. Then $A \otimes_R S = \bigoplus_i A_{\alpha_i} \otimes_{R_{\alpha_i}} S_{\alpha_i} \cong \bigoplus_i M_{n_i}(S_{\alpha_i})$. Take P to be the S-module $\bigoplus_i S_{\alpha_i}^{n_i}$.

(3) implies (4): This is immediate.

(4) implies (1): This follows from Corollary 8.1.21. □

In general, the faithfully flat étale splitting ring S of Theorem 10.3.9 (3) is not an R-progenerator module. For an example, see Exercise 11.3.31 below.

Corollary 10.3.10. *Let A be an Azumaya R-algebra. Assume A has constant rank. Then the following are true.*

(1) $\text{Rank}_R(A) = n^2$ *for some* $n \geq 1$.

(2) *There exists a commutative faithfully flat étale R-algebra S such that $A \otimes_R S$ is isomorphic to $M_n(S)$, for some $n \geq 1$.*

Proof. The proof is left to the reader. □

Corollary 10.3.10 shows that an Azumaya algebra A/R of constant rank has a faithfully flat splitting ring S such that upon extension of the ring of scalars to S, $A \otimes_R S$ is isomorphic to a ring of matrices over S. In other words, A is a twisted form of matrices for S (see Remark 5.5.11). In particular, every central simple algebra, and every division algebra is a form of matrices.

4. Cech Cohomology

One of the most powerful tools for computing Picard groups and Brauer groups is étale cohomology. This section contains an introduction to the theory using the Čech cohomology. From here, and with a suitable background in schemes, a student of this subject could then proceed on to the derived functor cohomology which is treated in the references [**Mil80**], [**AGV72a**], [**AGV72b**], [**AGV73**] and [**Del77**]. As pointed out below (Remark 10.4.5), Čech étale cohomology and derived functor cohomology agree under quite general conditions.

The Čech étale cohomology groups of a commutative ring R are defined as direct limits of Amitsur cohomology groups, where the limit is taken over the category of all faithfully flat étale R-algebras S. For example, we already know that the Amitsur cohomology group $\mathrm{H}^1(S/R, \mathbb{G}_m)$ classifies the twisted forms of a free R-module of rank one (Corollary 5.5.8) for S/R. From this we show that the Picard group of R is isomorphic to the Čech cohomology group $\check{\mathrm{H}}^1_{\mathrm{et}}(R, \mathbb{G}_m)$.

As we saw in Corollary 10.3.10 of the previous section, every Azumaya R-algebra A of constant rank n^2 is a twisted form of the matrix algebra $M_n(R)$ for some S/R. Using Artin's Refinement Theorem (Theorem 10.4.6) and the general version of the Skolem-Noether Theorem (Corollary 7.8.12), we show that up to isomorphism the Azumaya R-algebras of constant rank

n^2 are classified by the pointed set of Čech cohomology $\check{\mathrm{H}}^1_{\mathrm{et}}(R, \mathrm{PGL}_n)$. Since $\mathrm{PGL}_n(S)$ is the quotient of $\mathrm{GL}_n(S)$ by the normal subgroup $\mathbb{G}_m(S)$, there is an associated exact sequence of cohomology, with connecting map ∂ : $\check{\mathrm{H}}^1_{\mathrm{et}}(R, \mathrm{PGL}_n) \to \check{\mathrm{H}}^2_{\mathrm{et}}(R, \mathbb{G}_m)$. Given an Azumaya algebra of constant rank, we show that this map is constant on the Brauer class of A. This leads to the main result of this section, Theorem 10.4.11, which proves that there is a natural embedding of the Brauer group $\mathrm{B}(R)$ into the second Čech étale cohomology group with coefficients in the group of units.

4.1. The Definition. In this section we define Čech cohomology for both the étale and faithfully flat topologies. The definition is based on the Amitsur cohomology of Section 5.5. When the coefficients are in an abelian group, we get Čech cohomology groups for all nonnegative degrees. In the nonabelian case, the cohomology in degrees zero and one are pointed sets, but not groups. The functorial properties of Čech cohomology are derived.

Definition 10.4.1. Let R be a commutative ring and $\mathfrak{F}$ a functor from the category of commutative R-algebras to the category of groups. Let $\mathfrak{C}_{\mathrm{et}}(R)$ be the category whose objects are the isomorphism classes of faithfully flat étale R-algebras and whose morphisms are R-algebra homomorphisms. Define a partial ordering on the objects of $\mathfrak{C}_{\mathrm{et}}(R)$ by specifying $S \leq T$ if and only if there is an R-algebra homomorphism $S \to T$. By Proposition 9.2.5, this defines a directed system on the objects of $\mathfrak{C}_{\mathrm{et}}(R)$. By Theorem 5.5.5, Amitsur cohomology $\mathrm{H}^n(\cdot/R, \mathfrak{F})$ defines a functor from $\mathfrak{C}_{\mathrm{et}}(R)$ to the category of groups if $\mathfrak{F}$ is abelian, otherwise to the category of sets. Then

$$\check{\mathrm{H}}^n_{\mathrm{et}}(R, \mathfrak{F}) = \varinjlim_{S \in \mathfrak{C}_{\mathrm{et}}(R)} \mathrm{H}^n(S/R, \mathfrak{F})$$

exists and is called the *nth Čech cohomology of R with coefficients in $\mathfrak{F}$ with respect to the étale topology.* If $\mathfrak{F}$ is abelian, $\check{\mathrm{H}}^n_{\mathrm{et}}(R, \mathfrak{F})$ is defined for all $n \geq 0$ and is a group. If $\mathfrak{F}$ is nonabelian, then the Čech cohomology is defined for $0 \leq n \leq 1$ and is a pointed set.

Let $\mathfrak{C}_{\mathrm{fl}}(R)$ be the category whose objects are the isomorphism classes of faithfully flat R-algebras and whose morphisms are R-algebra homomorphisms. As for the étale category, $\mathfrak{C}_{\mathrm{fl}}(R)$ can also be made into a directed system of R-algebras. Then

$$\check{\mathrm{H}}^n_{\mathrm{fl}}(R, \mathfrak{F}) = \varinjlim_{S \in \mathfrak{C}_{\mathrm{fl}}(R)} \mathrm{H}^n(S/R, \mathfrak{F})$$

is the *nth Čech cohomology of R with coefficients in $\mathfrak{F}$ with respect to the flat topology.*

Proposition 10.4.2. *In the context of Definition* 10.4.1, $\check{\mathrm{H}}^n_{\mathrm{et}}(\cdot, \mathfrak{F})$ *is a functor from the category of commutative rings to the category of abelian groups* *if $\mathfrak{F}$ is abelian and to the category of pointed sets otherwise. The same result holds for the flat cohomology $\check{\mathrm{H}}^n_{\mathrm{fl}}(\cdot, \mathfrak{F})$.*

Proof. Let $\phi : R \to R'$ be a homomorphism of commutative rings. Let S be a faithfully flat étale R-algebra. Then $S' = S \otimes_R R'$ is a faithfully flat étale R' algebra (Proposition 9.2.5 and Exercise 2.4.14). The diagram

$$\begin{array}{ccc} S & \xrightarrow{\rho_1} & S \otimes_R R' = S' \\ \uparrow{\scriptstyle\theta} & & \uparrow{\scriptstyle\rho_2} \\ R & \xrightarrow{\phi} & R' \end{array}$$

R-algebra homomorphisms commutes (Exercise 1.3.8). By Theorem 5.5.5, there is an induced map $\mathrm{H}^n(S/R, \mathfrak{F}) \to \mathrm{H}^n(S'/R', \mathfrak{F})$ which depends only on ϕ, and which induces a map on the direct limits $\phi^* : \check{\mathrm{H}}^n_{\mathrm{et}}(R, \mathfrak{F}) \to \check{\mathrm{H}}^n_{\mathrm{et}}(R', \mathfrak{F})$. The same proof applies to the flat cohomology. □

In the above context, let $\mathfrak{F}$ and $\mathfrak{G}$ be two covariant functors on the category of commutative R-algebras to the category of groups. Let $\varphi : \mathfrak{F} \to \mathfrak{G}$ be a morphism of functors. That is, for each R-algebra S there is a homomorphism of groups $\varphi(S) : \mathfrak{F}(S) \to \mathfrak{G}(S)$ such that for any homomorphism of R-algebras $f : S \to T$, the diagram of group homomorphisms

$$(10.4)\qquad \begin{array}{ccc} \mathfrak{F}(S) & \xrightarrow{\varphi(S)} & \mathfrak{G}(S) \\ \downarrow{\scriptstyle\mathfrak{F}(f)} & & \downarrow{\scriptstyle\mathfrak{G}(f)} \\ \mathfrak{F}(T) & \xrightarrow{\varphi(T)} & \mathfrak{G}(T) \end{array}$$

commutes. In the language of [**Mil80**, Chapter II], $\mathfrak{F}$ and $\mathfrak{G}$ are called presheaves and φ a morphism of presheaves.

Proposition 10.4.3. *In the above context, let $\varphi : \mathfrak{F} \to \mathfrak{G}$ be a morphism of functors on $\mathfrak{C}_{\mathrm{fl}}$. For any $S \in \mathfrak{C}_{\mathrm{fl}}$, there is an induced map $\varphi_* : \mathrm{H}^i(S/R, \mathfrak{F}) \to \mathrm{H}^i(S/R, \mathfrak{G})$. If $\mathfrak{F}$ and $\mathfrak{G}$ are abelian, then φ_* is a homomorphism for all $i \geq 0$. Otherwise, φ_* is a morphism of pointed sets for $0 \leq i \leq 1$. The same result is true with étale cohomology replacing faithfully flat cohomology.*

Proof. In the abelian case, φ induces a morphism of cochain complexes which in turn induces homomorphisms on the cohomology groups (see, for example, [**Rot79**, Theorem 6.1]). The nonabelian case follows from (10.4) and the definitions (Definition 5.5.4). We leave the details to the reader. □

Proposition 10.4.4. *In the above context, let $\varphi_1 : \mathfrak{F}_1 \to \mathfrak{F}_2$ and $\varphi_2 : \mathfrak{F}_2 \to \mathfrak{F}_3$ be morphisms where each $\mathfrak{F}_i$ takes values in the category of abelian groups. Assume for any $T \in \mathfrak{C}_{\mathrm{fl}}$ that*

$$1 \to \mathfrak{F}_1(T) \xrightarrow{\varphi_1} \mathfrak{F}_2(T) \xrightarrow{\varphi_1} \mathfrak{F}_3(T) \to 1$$

is an exact sequence. For any $S \in \mathfrak{C}_{\mathrm{fl}}$, there is a long exact sequence

$$\cdots \to \mathrm{H}^n(S/R, \mathfrak{F}_1) \xrightarrow{\varphi_1} \mathrm{H}^n(S/R, \mathfrak{F}_2) \xrightarrow{\varphi_2} \mathrm{H}^n(S/R, \mathfrak{F}_3) \xrightarrow{\delta^n} \mathrm{H}^{n+1}(S/R, \mathfrak{F}_1) \to \cdots$$

of Amitsur cohomology groups. Taking the limit over S yields a long exact sequence

$$\cdots \to \check{\mathrm{H}}^n_{\mathrm{fl}}(R, \mathfrak{F}_1) \xrightarrow{\varphi_1} \check{\mathrm{H}}^n_{\mathrm{fl}}(R, \mathfrak{F}_2) \xrightarrow{\varphi_2} \check{\mathrm{H}}^n_{\mathrm{fl}}(R, \mathfrak{F}_3) \xrightarrow{\delta^n} \check{\mathrm{H}}^{n+1}_{\mathrm{fl}}(R, \mathfrak{F}_1) \to \cdots$$

of Čech cohomology groups. The same result is true with étale cohomology replacing faithfully flat cohomology.

Proof. A short exact sequence of complexes gives rise to a long exact sequence of abelian groups (see, for example, [**Rot79**, Theorem 6.3]). For the second sequence, taking direct limits preserves exactness (Theorem 1.4.3). We leave the details to the reader. □

Remark 10.4.5. The étale cohomology studied in [**Mil80**] is based on derived functors. If $\mathfrak{F}$ takes values in the category of abelian groups, then for each $q \geq 0$ there is a natural homomorphism $\check{\mathrm{H}}^q_{\mathrm{et}}(R, \mathfrak{F}) \to \mathrm{H}^q_{\mathrm{et}}(R, \mathfrak{F})$ from the Čech cohomology group that we are using here to the derived functor cohomology group. In [**Art71**, Corollary 4.2], M. Artin proved that if R is noetherian, then these maps are isomorphisms for all q (see also [**Mil80**, Theorem III.2.17]).

4.2. The Brauer group and Amitsur Cohomology. In this section we show that the Brauer group of a commutative noetherian ring R can be embedded as a subgroup of the second cohomology group $\check{\mathrm{H}}^2_{\mathrm{et}}(R, \mathbb{G}_m)$, where we write $\mathbb{G}_m$ for the functor which maps a commutative R-algebra to its group of units. The approach used here is based on that of L. Childs [**OS75**, Theorem 13.12] and J. Milne [**Mil80**, Theorem IV.2.5].

Let R be a commutative ring and S a commutative R-algebra. For $n \geq 1$, consider the ring of matrices $M_n(S)$. The group of invertible elements in $M_n(S)$ is denoted $\mathrm{GL}_n(S)$. The center of $M_n(S)$ is the set of constant diagonal matrices, which we identify with S. The reader should verify that the center of $\mathrm{GL}_n(S)$ is identified with $\mathbb{G}_m(S)$, the group of units of S. The group of inner automorphisms of $M_n(S)$, denoted $\mathrm{Inn}(M_n(S))$, is therefore the quotient group $\mathrm{GL}_n(S)/\mathbb{G}_m(S)$, which is also called the *projective general linear group* and denoted $\mathrm{PGL}_n(S)$. Therefore, the sequence of pointed sets

$$1 \to \mathbb{G}_m(S) \to \mathrm{GL}_n(S) \to \mathrm{PGL}_n(S) \to 1 \tag{10.5}$$

is exact, and we view $\mathrm{PGL}_n(S)$ as $\mathrm{Inn}(M_n(S))$, the group of inner automorphisms of $M_n(S)$.

Theorem 10.4.6. *(M. Artin) Let R be a noetherian commutative ring and S an étale R-algebra. If W is a faithfully flat $S^{\otimes n}$-algebra for $n \geq 1$, then there is a faithfully flat étale S-algebra T and a homomorphism $\beta : W \to T^{\otimes n} = T \otimes_R T \otimes \cdots \otimes_R T$ such that the diagram*

$$\begin{array}{ccc} S^{\otimes n} & \xrightarrow{\eta} & T^{\otimes n} \\ & \searrow \quad \nearrow_{\beta} & \\ & W & \end{array}$$

of R-algebra homomorphisms commutes, where η is the natural map induced by the structure homomorphism $S \to T$.

Proof. See [**Art71**, Theorem 4.1] or [**Mil80**, Lemma III.2.19]. □

Proposition 10.4.7. *Let R be a commutative noetherian ring. The set of isomorphism classes of Azumaya R-algebras of rank n^2 is classified by $\check{\mathrm{H}}^1_{\mathrm{et}}(R, \mathrm{PGL}_n)$.*

Proof. Let $S \in \mathfrak{C}_{\mathrm{et}}$ be a faithfully flat étale R-algebra. Let θ be a 1-cocycle in $\mathrm{Z}^1(S/R, \mathrm{PGL}_n)$. Then θ is an inner automorphism of $M_n(S \otimes_R S)$. By Proposition 5.5.7, there is an R-algebra B and an S-algebra isomorphism $\beta : B \otimes_R S \to M_n(S)$. By Theorem 10.3.9 and Remark 5.5.11, B is an Azumaya R-algebra of rank n^2. This shows that every 1-cocycle in $\check{\mathrm{H}}^1_{\mathrm{et}}(R, \mathrm{PGL}_n)$ comes from an Azumaya R-algebra.

Conversely, let A be an Azumaya R-algebra of constant rank n^2. By Corollary 10.3.10, there is a faithfully flat étale R-algebra S such that $A \otimes_R S \cong M_n(S)$. So A is a twisted form of $M_n(R)$ for S/R. By Proposition 5.5.7, there is associated to A a 1-cocycle θ in $\mathrm{Aut}_{S \otimes_R S}(M_n(S \otimes_R S))$. By Corollary 7.8.12 there is an exact sequence of pointed sets

$$1 \to \mathrm{PGL}_n(S \otimes_R S) \to \mathrm{Aut}_{S \otimes_R S}(M_n(S \otimes_R S)) \xrightarrow{\alpha} \mathrm{Pic}(S \otimes_R S).$$

Then $\alpha(\theta) = I$ is a rank one projective $S \otimes_R S$-module. By Exercise 2.6.16, there is a faithfully flat $S \otimes_R S$-algebra, call it W, such that $I \otimes_{S \otimes_R S} W$ is isomorphic to W. In fact, W is a direct sum of localizations, hence is étale over $S \otimes_R S$, by Example 9.2.4 (2). Denote by θ_W the image of θ in $\mathrm{Aut}_W(M_n(W))$. By the commutative diagram

$$\begin{array}{ccccc} \mathrm{PGL}_n(S \otimes_R S) & \longrightarrow & \mathrm{Aut}_{S \otimes_R S}(M_n(S \otimes_R S)) & \xrightarrow{\alpha} & \mathrm{Pic}(S \otimes_R S) \\ \downarrow & & \downarrow & & \downarrow \\ \mathrm{PGL}_n(W) & \longrightarrow & \mathrm{Aut}_W(M_n(W)) & \xrightarrow{\alpha} & \mathrm{Pic}(W) \end{array}$$

θ_W is in the image of $\mathrm{PGL}_n(W)$, hence is an inner automorphism. By Theorem 10.4.6, there exists a faithfully flat étale S-algebra T and a homomorphism $\beta : W \to T \otimes_R T$ such that the natural map $S \otimes_R S \to T \otimes_R T$ factors through W. This implies the image of θ in $\mathrm{Aut}_{T\otimes_R T}(M_n(T \otimes_R T))$ is in $\mathrm{PGL}_n(T \otimes_R T)$. By Proposition 9.2.5, T is a faithfully flat étale R-algebra. Since $A \otimes_R T \cong M_n(T)$, A is a twisted form of $M_n(R)$ for T/R. The automorphism θ defines a cohomology class in $\mathrm{H}^1(T/R, \mathrm{PGL}_n)$, and by Definition 10.4.1, a cohomology class in $\check{\mathrm{H}}^1_{\mathrm{et}}(R, \mathrm{PGL}_n)$. The rest is similar to Proposition 5.5.7 and is left to the reader. □

Proposition 10.4.8. *Let R be a commutative noetherian ring. The set of isomorphism classes of finitely generated projective R-modules of rank n is equal to $\check{\mathrm{H}}^1_{\mathrm{et}}(R, \mathrm{GL}_n)$. In particular,* $\mathrm{Pic}(R) = \check{\mathrm{H}}^1_{\mathrm{et}}(R, \mathbb{G}_m)$.

Proof. By Corollary 5.5.8, every 1-cocycle in $\check{\mathrm{H}}^1_{\mathrm{et}}(R, \mathrm{GL}_n)$ comes from a finitely generated projective R-module of constant rank n.

Let M be a finitely generated projective R-module of constant rank n. By Exercise 2.6.16, there is a faithfully flat R-algebra S such that $M \otimes_R S \cong S^n$. In fact, S is a direct sum $R_{f_1} \oplus \cdots \oplus R_{f_m}$ of localizations of R, hence is étale over R. Thus M is a twisted form of R^n for S/R. By Proposition 5.5.7, associated to M is a cohomology class in $\mathrm{H}^1(S/R, \mathrm{GL}_n)$, hence a cohomology class in $\check{\mathrm{H}}^1_{\mathrm{et}}(R, \mathrm{GL}_n)$. The rest is similar to Proposition 5.5.7 and is left to the reader. □

Proposition 10.4.9. *Let R be a commutative noetherian ring. There is an exact sequence of pointed sets*

$$1 \to \check{\mathrm{H}}^1_{\mathrm{et}}(R, \mathbb{G}_m) \xrightarrow{\rho} \check{\mathrm{H}}^1_{\mathrm{et}}(R, \mathrm{GL}_n) \xrightarrow{\chi} \check{\mathrm{H}}^1_{\mathrm{et}}(R, \mathrm{PGL}_n) \xrightarrow{\partial} \check{\mathrm{H}}^2_{\mathrm{et}}(R, \mathbb{G}_m)$$

and the maps satisfy the following.

(1) *ρ and χ are induced by the exact sequence* (10.5).

(2) *Under ρ, a rank one projective module I is mapped to the rank n projective module $I^n = I \oplus \cdots \oplus I$.*

(3) *Under χ, a projective R-module M is mapped to the Azumaya R-algebra* $\mathrm{Hom}_R(M, M)$.

(4) *The sequence is natural in R. That is, if $R \to R'$ is a homomorphism of commutative noetherian rings, then the diagram*

$$\begin{array}{ccccccc}
\check{\mathrm{H}}^1_{\mathrm{et}}(R, \mathbb{G}_m) & \xrightarrow{\rho} & \check{\mathrm{H}}^1_{\mathrm{et}}(R, \mathrm{GL}_n) & \xrightarrow{\chi} & \check{\mathrm{H}}^1_{\mathrm{et}}(R, \mathrm{PGL}_n) & \xrightarrow{\partial} & \check{\mathrm{H}}^2_{\mathrm{et}}(R, \mathbb{G}_m) \\
\downarrow & & \downarrow & & \downarrow & & \downarrow \\
\check{\mathrm{H}}^1_{\mathrm{et}}(R', \mathbb{G}_m) & \xrightarrow{\rho} & \check{\mathrm{H}}^1_{\mathrm{et}}(R', \mathrm{GL}_n) & \xrightarrow{\chi} & \check{\mathrm{H}}^1_{\mathrm{et}}(R', \mathrm{PGL}_n) & \xrightarrow{\partial} & \check{\mathrm{H}}^2_{\mathrm{et}}(R', \mathbb{G}_m)
\end{array}$$

commutes where the vertical maps are from Proposition 10.4.2.

Proof. The maps ρ and χ follow from Proposition 10.4.3. The proof that ρ is one-to-one is left to the reader.

(2): The proof is left to the reader.

(3): The map $\chi_S : \mathrm{H}^1(S/R, \mathrm{GL}_n) \to \mathrm{H}^1(S/R, \mathrm{PGL}_n)$ is induced by sending the 1-cocycle represented by ϕ to the 1-cocycle represented by "conjugation by ϕ". Let M be a projective R-module of constant rank n. As in Exercise 2.6.16, let $S = R_{f_1} \oplus \cdots \oplus R_{f_m}$ be such that there is an S-module isomorphism $\beta : M \otimes_R S \cong S^n$. By Proposition 5.5.7, the 1-cocycle in $\mathrm{Z}^1(S/R, \mathrm{GL}_n)$ corresponding to M is $\phi = \beta_1\beta_0^{-1}$. Let $A = \mathrm{Hom}_R(M, M)$. There is an $S \otimes_R S$-algebra isomorphism $\gamma : A \otimes_R S \cong \mathrm{Hom}_S(S^n, S^n) \cong M_n(S)$ given by $\gamma(h) = \beta h \beta^{-1}$, which shows A is a twisted form of $M_n(R)$. The 1-cocycle in $\mathrm{Z}^1(S/R, \mathrm{Aut}(M_n))$ associated to A is $\theta = \gamma_1\gamma_0^{-1}$. Notice that for $a \in M_n(S \otimes_R S)$, $\theta(a) = \beta_1\beta_0^{-1}a\beta_0\beta_1^{-1} = \phi a \phi^{-1}$. Thus θ is the inner automorphism corresponding to conjugation by ϕ. Since we identify $\mathrm{PGL}_n(S \otimes_R S)$ with $\mathrm{Inn}(M_n(S \otimes_R S))$, this proves (3).

Definition of ∂: Let θ be a 1-cocycle in $\mathrm{Z}^1(S/R, \mathrm{PGL}_n)$, where $S \in \mathfrak{C}_{\mathrm{et}}(R)$. Then $\theta \in \mathrm{PGL}_n(S \otimes_R S)$ is conjugation by some $f \in \mathrm{GL}_n(S \otimes_R S)$. Let $x \in (S \otimes_R S)^n$. The relation $\theta_1(x) = \theta_2\theta_0(x)$ yields $f_1 x f_1^{-1} = f_2 f_0 x f_0^{-1} f_2^{-1}$, hence x is equal to $f_1^{-1} f_2 f_0 x f_0^{-1} f_2^{-1} f_1$. Since x was arbitrary, this implies $f_1^{-1} f_2 f_0$ is central. That is, $f_1^{-1} f_2 f_0 \in \mathbb{G}_m(S \otimes_R S \otimes_R S)$. The image of $f_1^{-1} f_2 f_0$ under the coboundary map $d^2 : \mathbb{G}_m(S^{\otimes 3}) \to \mathbb{G}_m(S^{\otimes 4})$ is given by

$$(f_1^{-1} f_2 f_0)_0 (f_1^{-1} f_2 f_0)_1^{-1} (f_1^{-1} f_2 f_0)_2 (f_1^{-1} f_2 f_0)_3^{-1}.$$

The reader should verify that $(f_i)_j = (f_{j-1})_i$ if $i < j$. Using this and the fact that $\mathbb{G}_m(S^{\otimes 4})$ is an abelian group, we see that $f_1^{-1} f_2 f_0$ is a 2-cocycle in $\mathrm{Z}^2(S/R, \mathbb{G}_m)$. The map

$$\mathrm{H}^1(S/R, \mathrm{PGL}_n) \xrightarrow{\partial_S} \mathrm{H}^2(S/R, \mathbb{G}_m)$$

is defined by $\theta \mapsto f_1^{-1} f_2 f_0$. If θ and ϕ represent the same cohomology class in $\mathrm{H}^1(S/R, \mathrm{PGL}_n)$, then by a routine calculation one shows θ and ϕ give rise to the same 2-cocycle in $\mathrm{Z}^2(S/R, \mathbb{G}_m)$. Let S and T be two faithfully flat étale R-algebras, and $\alpha : S \to T$ an R-algebra homomorphism. The reader should verify that the diagram

$$\begin{array}{ccc}
\mathrm{H}^1(S/R, \mathrm{PGL}_n) & \xrightarrow{\partial_S} & \mathrm{H}^2(S/R, \mathbb{G}_m) \\
\downarrow{\scriptstyle \alpha^*} & & \downarrow{\scriptstyle \alpha^*} \\
\mathrm{H}^1(T/R, \mathrm{PGL}_n) & \xrightarrow{\partial_T} & \mathrm{H}^2(T/R, \mathbb{G}_m)
\end{array}$$

commutes. The map ∂ is defined by taking the direct limit of ∂_S, over all $S \in \mathfrak{C}_{\mathrm{et}}$.

Exactness at $\check{\mathrm{H}}^1_{\mathrm{et}}(R, \mathrm{GL}_n)$: This follows from Proposition 7.6.2.

Exactness at $\check{\mathrm{H}}^1_{\mathrm{et}}(R, \mathrm{PGL}_n)$: Assume $\partial([\theta])$ is a 2-coboundary. In the notation from above, assume θ is conjugation by $f \in \mathrm{GL}_n(S \otimes_R S)$, and $f_1^{-1} f_2 f_0$ represents the image of θ under ∂. Let $v \in \mathbb{G}_m(S \otimes_R S)$ and assume $d^1(v) = v_0 v_1^{-1} v_2 = f_1^{-1} f_2 f_0$. Set $g = fv^{-1}$. Using the fact that v is central, it follows that $g_1^{-1} g_2 g_0 = v_1 f_1^{-1} f_2 v_2^{-1} f_0 v_0^{-1} = 1$. This implies g is a 1-cocycle in $\mathrm{Z}^1(S/R, \mathrm{GL}_n)$. By (3), χ maps the cohomology class of g to the inner automorphism defined by g. Since v is central, conjugation by g is equal to conjugation by f. Therefore, $[\theta]$ is in the image of χ. Now assume $\phi \in \mathrm{Z}^1(S/R, \mathrm{GL}_n)$. By (3), χ maps the cohomology class $[\phi]$ to the 1-cocycle $\theta \in \mathrm{Z}^1(S/R, \mathrm{PGL}_n(S \otimes_R S))$ corresponding to conjugation by ϕ. Then $\partial([\theta])$ is the unit $\phi_1^{-1} \phi_2 \phi_0 \in \mathbb{G}_m(S \otimes_R S \otimes_R S)$. But ϕ is already a 1-cocycle, so $\phi_1^{-1} \phi_2 \phi_0 = 1$.

(4): The proof is left to the reader. □

Theorem 10.4.10. *Let R be a commutative noetherian ring. There is a natural one-to-one homomorphism $\beta : \mathrm{B}(R) \to \check{\mathrm{H}}^2_{\mathrm{et}}(R, \mathbb{G}_m)$ of abelian groups.*

Proof. By Exercise 7.4.9, $\mathrm{B}(R)$ is generated by Azumaya R-algebras of constant rank. Let $\mathrm{Az}(R)$ denote the set of isomorphism classes of Azumaya R-algebras of constant rank. Let A and B be two Azumaya R-algebras, where $\mathrm{Rank}_R(A) = m^2$ and $\mathrm{Rank}_R(B) = n^2$. By Proposition 10.4.7, there are unique cohomology classes $c(A) \in \check{\mathrm{H}}^1_{\mathrm{et}}(R, \mathrm{PGL}_m)$ and $c(B) \in \check{\mathrm{H}}^1_{\mathrm{et}}(R, \mathrm{PGL}_n)$ corresponding to A and B respectively. The map

$$\beta_0 : \mathrm{Az}(R) \to \mathrm{H}^2_{\mathrm{et}}(R, \mathbb{G}_m)$$

is defined by sending A to $\partial(c(A))$, where ∂ is the connecting map defined in Proposition 10.4.9.

Suppose $A = \mathrm{Hom}_R(M, M)$ is split, for some R-progenerator module M of constant rank m. By Proposition 10.4.8, corresponding to M is a unique cohomology class $b(M) \in \check{\mathrm{H}}^1_{\mathrm{et}}(R, \mathrm{GL}_m)$. By Proposition 10.4.9, $\chi(b(M)) = c(A)$, which implies $\beta_0(A) = 1$.

Let $S \in \mathfrak{C}_{\mathrm{et}}(R)$. By Corollary 1.3.28, we have an isomorphism $\psi : M_m(S) \otimes_S M_n(S) \cong M_{mn}(S)$ of S-algebras. Let ψ_m and ψ_n be defined by the commutative diagram

$$\begin{array}{ccccc} & & M_{mn}(S) & & \\ & \overset{\psi_m}{\nearrow} & \uparrow{\scriptstyle \psi} & \overset{\psi_n}{\nwarrow} & \\ M_m(S) & \longrightarrow & M_m(S) \otimes_S M_n(S) & \longleftarrow & M_n(S) \end{array}$$

of R-algebras. Then ψ_m, and ψ_n induce group homomorphisms for which the diagram

$$\begin{array}{ccccc} \mathrm{GL}_m(S) & \xrightarrow{\psi_m} & \mathrm{GL}_{mn}(S) & \xleftarrow{\psi_n} & \mathrm{GL}_n(S) \\ \downarrow & & \downarrow & & \downarrow \\ \mathrm{PGL}_m(S) & \xrightarrow{\psi_m} & \mathrm{PGL}_{mn}(S) & \xleftarrow{\psi_n} & \mathrm{PGL}_n(S) \end{array}$$

commutes. Using the definition of ∂ in Proposition 10.4.9, the reader should verify that the diagram

$$\begin{array}{ccccc} \mathrm{H}^1(S/R, \mathrm{PGL}_m) & \xrightarrow{\partial} & \check{\mathrm{H}}^2_{\mathrm{et}}(R, \mathbb{G}_m) & \xleftarrow{\partial} & \mathrm{H}^1(S/R, \mathrm{PGL}_n) \\ \downarrow{\scriptstyle (\psi_m)_*} & & \downarrow{\scriptstyle =} & & \downarrow{\scriptstyle (\psi_n)_*} \\ \mathrm{H}^1(S/R, \mathrm{PGL}_{mn}) & \xrightarrow{\partial} & \check{\mathrm{H}}^2_{\mathrm{et}}(R, \mathbb{G}_m) & \xleftarrow{\partial} & \mathrm{H}^1(S/R, \mathrm{PGL}_{mn}) \end{array}$$

commutes. For some $S \in \mathfrak{C}_{\mathrm{et}}$, $c(A)$ comes from $\mathrm{H}^1(S/R, \mathrm{PGL}_m)$ and $c(B)$ comes from $\mathrm{H}^1(S/R, \mathrm{PGL}_n)$. If $c(A)$ is conjugation by $u(A) \in \mathrm{GL}_m(S \otimes_R S)$, then it follows that $(\psi_m)_*(c(A))$ is conjugation by $u(A) \otimes 1 \in \mathrm{GL}_{mn}(S \otimes_R S)$. Likewise, $(\psi_m)_*(c(B))$ is conjugation by $1 \otimes u(B) \in \mathrm{GL}_{mn}(S \otimes_R S)$. Starting with the commutative diagram

$$\begin{array}{ccc} (A \otimes_R S) \otimes_S (B \otimes_R S) & \longrightarrow & M_m(S) \otimes_S M_n(S) \\ \downarrow & & \downarrow{\scriptstyle \psi} \\ A \otimes_R B \otimes_R S & \longrightarrow & M_{mn}(S) \end{array}$$

the reader should verify that $c(A \otimes_R B)$, the 1-cocycle in $\mathrm{H}^1(S/R, \mathrm{PGL}_{mn})$ corresponding to $A \otimes_R B$, is conjugation by $u(A) \otimes u(B)$. Therefore, $\partial(c(A \otimes_R B)) = \partial(c(A))\partial(c(B))$. It follows from this that β_0 factors into a group homomorphism $\beta : \mathrm{B}(R) \to \check{\mathrm{H}}^2_{\mathrm{et}}(R, \mathbb{G}_m)$. By exactness of the sequence in Proposition 10.4.9, β is one-to-one. □

Theorem 10.4.11. *Let R be a commutative ring. There is a natural one-to-one homomorphism $\beta : \mathrm{B}(R) \to \check{\mathrm{H}}^2_{\mathrm{et}}(R, \mathbb{G}_m)$ of abelian groups.*

Proof. Applying Corollary 7.7.8, we write $\mathrm{B}(R) = \varinjlim \mathrm{B}(R_\alpha)$, where $\{R_\alpha\}$ is the directed system of noetherian subrings of R. By Theorem 10.4.10 and Proposition 10.4.2, there are homomorphisms $\beta_\alpha : \mathrm{B}(R_\alpha) \to \check{\mathrm{H}}^2_{\mathrm{et}}(R_\alpha, \mathbb{G}_m) \to \check{\mathrm{H}}^2_{\mathrm{et}}(R, \mathbb{G}_m)$ for each α. The map β is the direct limit of the various β_α. Any 2-coboundary in $\check{\mathrm{H}}^2_{\mathrm{et}}(R, \mathbb{G}_m)$ comes from a 2-coboundary in $\mathrm{B}^2(S_\alpha/R_\alpha, \mathbb{G}_m)$, for some α. It follows from this that β is one-to-one. The details are left to the reader. □

Remark 10.4.12. In general, the map β of Theorem 10.4.11 is not onto. This is because $\mathrm{B}(R)$ is a torsion group (see Corollary 11.2.5), but the cohomology group $\check{\mathrm{H}}^2_{\mathrm{et}}(R, \mathbb{G}_m)$ is in general not. Some examples are listed here.

(1) In [**Gro68b**, Remarques 1.11 (b), p. 75] Grothendieck adapted an example of Mumford to show that $\check{\mathrm{H}}^2_{\mathrm{et}}(X, \mathbb{G}_m)$ can contain a nontrivial torsion free subgroup when X is a complex algebraic surface.

(2) Let $f(x, y, z) \in \mathbb{C}[x, y, z]$ be a nonconstant homogeneous polynomial of degree $d > 4$ defining a nonsingular projective curve $C = Z(f) \subseteq \mathbb{P}^2$. The *cone* over C is the affine algebraic surface $X = Z(f) \subseteq \mathbb{A}^3$. If $R = k[x, y, z]/(f)$ is the affine coordinate ring of X, then in [**Chi76**] Childs applies results of Danilov to show that $\check{\mathrm{H}}^2_{\mathrm{et}}(R, \mathbb{G}_m)$ is nontrivial and torsion free.

(3) Let k be an algebraically closed field of characteristic zero, $n \geq 3$, and $a_1, \dots, a_n$ distinct elements of k. Set $f = z^n - (y - a_1 x) \cdots (y - a_n x)(x - 1)$ and $X = Z(f) \subseteq \mathbb{A}^3$. Then X is the algebraic surface studied in [**FH13**], where it was shown that X has a nonrational singularity at the point $p = (0, 0, 0)$. The desingularization of X at p is resolved by one blowing-up and the exceptional curve is a nonsingular hyperelliptic curve E of genus $g = (n-1)(n-2)/2$. Moreover, if $\mathcal{O}_p$ is the local ring on X at p, then it was shown that $\check{\mathrm{H}}^2_{\mathrm{et}}(\mathcal{O}_p, \mathbb{G}_m)$ has a subgroup isomorphic to the image of the jacobian variety $\mathrm{Pic}^0(E)$, modulo a finitely generated subgroup. By [**Mum70**, p. 64], $\mathrm{Pic}^0(E)$ is an abelian variety with dimension equal to g, the genus of E. By [**Mum70**, (iv), p. 42], $\mathrm{Pic}^0(E)$ is a divisible group. The theory of abelian varieties implies that the subgroup of torsion elements in $\mathrm{Pic}^0(E)$ is a direct sum of $2g$ copies of $\mathbb{Q}/\mathbb{Z}$ (see [**Mil80**, pp. 126–127]), hence is countable. Assuming k is uncountable, this implies $\check{\mathrm{H}}^2_{\mathrm{et}}(\mathcal{O}_p, \mathbb{G}_m)$ has a nontrivial torsion free subgroup.

Theorem 10.4.13 (O. Gabber). *If R is a noetherian commutative ring, the image of $\beta : \mathrm{B}(R) \to \check{\mathrm{H}}^2_{\mathrm{et}}(R, \mathbb{G}_m)$ is equal to the subgroup of all torsion elements.*

Proof. A Theorem of Gabber ([**Gab81**, Theorem II.1, p. 163], [**Hoo82**, Theorem 7], see also [**Cae98**, Theorem 6.7.4] and [**KO81**]) states that the image of β is equal to the subgroup of torsion elements of the derived functor cohomology group $\mathrm{H}^2_{\mathrm{et}}(R, \mathbb{G}_m)$. When R is noetherian, the Čech cohomology that we are using here agrees with the derived functor cohomology (Remark 10.4.5). □

Proposition 10.4.14. *Let R be a commutative ring and S an étale R-algebra which is an R-module progenerator. Consider the diagram of abelian groups*

$$\begin{array}{ccc} & & \mathrm{H}^2(S/R, \mathbb{G}_m) \\ & & \big\downarrow \phi_{S/R} \\ \mathrm{B}(R) & \xrightarrow{\beta} & \check{\mathrm{H}}^2_{\mathrm{et}}(R, \mathbb{G}_m) \end{array}$$

with natural maps. Let $\gamma \in \check{\mathrm{H}}^2_{\mathrm{et}}(R, \mathbb{G}_m)$. If γ is in the image of $\phi_{S/R}$, then γ is in the image of β.

Proof. Let $u = \sum_i a_i \otimes b_i \otimes c_i$ be a unit in $(S \otimes_R S \otimes_R S)^*$ which is a 2-cocycle in $\mathrm{Z}^2(S/R, \mathbb{G}_m)$ representing a preimage of γ. Let $P = S \otimes_R S$. We make P into an S-progenerator module by letting S act on the first factor. Define $f : S \otimes_R P \to P \otimes_R S$ by $f(x \otimes y \otimes z) = \sum_i a_i x \otimes c_i z \otimes b_i y$. The reader should verify that f is an $S \otimes_R S$-module isomorphism and $f_2^{-1} f_3 f_1 : S \otimes_R S \otimes_R P \to S \otimes_R S \otimes_R P$ is multiplication by $u_1 u_2^{-1} u_3 = u_4$. Define

$$\mathrm{Hom}_{S \otimes_R S}(S \otimes_R P, S \otimes_R P) \xrightarrow{\theta} \mathrm{Hom}_{S \otimes_R S}(P \otimes_R S, P \otimes_R S)$$

by $\theta(g) = f g f^{-1}$. Then $\theta_3 \theta_1 = \theta_2$, so θ is a descent datum for $\mathrm{Hom}_S(P, P)$. By Theorem 5.3.10, there is an R-algebra A such that $A \otimes_R S \cong \mathrm{Hom}_S(P, P)$. The reader should verify that under β, the Brauer class of A is mapped to γ. □

Remark 10.4.15. Proposition 10.4.14 is a special case of [**Hoo72**, Proposition 3.1]. See also [**Gab81**, Lemma II.4, p. 165], [**Hoo82**, Proposition 3] and [**Mil80**, Remark IV.2.14 (b)].

Proposition 10.4.16. [**KO74b**, Proposition V.2.1] *Let R be a commutative ring and S a commutative faithfully flat R-algebra.*

(1) *$\mathrm{H}^0(S/R, \mathbb{G}_m)$ is equal to R^*, the group of units of R.*
(2) *$\mathrm{H}^1(S/R, \mathbb{G}_m)$ is equal to the kernel of the natural map* $\mathrm{Pic}(R) \to \mathrm{Pic}(S)$.
(3) *If* $\mathrm{Pic}(S) = \mathrm{Pic}(S \otimes_R S) = (0)$, *there is a natural one-to-one homomorphism of abelian groups $\beta : \mathrm{B}(S/R) \to \mathrm{H}^2(S/R, \mathbb{G}_m)$. If S is an R-module progenerator, β is an isomorphism.*

Proof. Part (1) follows from Proposition 5.3.1 and Part (2) from Corollary 5.5.9.

(3): Let A be an Azumaya R-algebra and assume there is an isomorphism of S-algebras $\phi : A \otimes_R S \cong \operatorname{Hom}_S(P, P)$. Define $\theta : \operatorname{Hom}_{S\otimes_R S}(S \otimes_R P, S \otimes_R P) \to \operatorname{Hom}_{S\otimes_R S}(P \otimes_R S, P \otimes_R S)$ by commutativity of the diagram

$$\begin{array}{ccc} S \otimes_R A \otimes_R S & \xrightarrow{1\otimes\phi} & \operatorname{Hom}_{S\otimes_R S}(S \otimes_R P, S \otimes_R P) \\ \downarrow{\scriptstyle \tau\otimes 1} & & \downarrow{\scriptstyle \theta} \\ A \otimes_R S \otimes_R S & \xrightarrow{\phi\otimes 1} & \operatorname{Hom}_{S\otimes_R S}(P \otimes_R S, P \otimes_R S). \end{array}$$

Then $\theta_1 = \theta_2\theta_0$, so θ is a descent datum, which corresponds to A (Theorem 5.3.10). By Corollary 7.8.12, θ is conjugation by an $S \otimes_R S$-module isomorphism $f : S \otimes_R P \to P \otimes_R S$. The cocycle identity $\theta_1 = \theta_2\theta_0$ implies $f_1^{-1} f_2 f_0$ is an element of $\mathbb{G}_m(S \otimes_R S \otimes_R S)$. The rest, which is similar to the proofs of Proposition 10.4.9 and Theorem 10.4.10, is left to the reader. The condition for β to be onto follows from Proposition 10.4.14. □

Chapter 11

Azumaya Algebras, II

The first section below is concerned with properties belonging to individual elements in an Azumaya algebra A over a commutative ring R. This section is an application of faithfully flat descent (Section 5.3) and the main results of Section 10.3. Using the fact that if A has constant rank, then A is a form of matrices for a faithfully flat splitting ring, we proceed to show that an element α in A has many of the same attributes as a matrix. The second section below contains a proof due to D. Saltman that associated to A is an integer $n > 0$ such that $A^{\otimes n}$ is split. In other words, the Brauer group of R is a torsion abelian group. The third section contains an introduction to the theory of maximal orders. When R is an integral domain with field of fractions K, we study maximal R-orders Λ in central simple K-algebras A. Our emphasis is on those R-orders that are Azumaya R-algebras. We describe the kernel of the natural map $\mathrm{B}(R) \to \mathrm{B}(K)$ in terms of Azumaya R-algebras that are Brauer equivalent to $\mathrm{Hom}_R(M, M)$ for some reflexive R-lattice M. We give a description due to B. Auslander for the subgroup of $\mathrm{B}(R)$ represented by Azumaya algebras that are split by the local ring $R_{\mathfrak{p}}$, for all $\mathfrak{p} \in \operatorname{Spec} R$. There is an explicit construction of Ojanguren's example of an Azumaya algebra that is locally split, but which has order three in the Brauer group. The fourth section contains a description due to Knus, Ojanguren and Saltman, of the p-torsion of the Brauer group $\mathrm{B}(R)$, when R is a commutative ring of characteristic p.

1. Invariants Attached to Elements in Azumaya Algebras

An Azumaya algebra A of constant rank over a commutative ring R is a twisted form of matrices (Corollary 10.3.10). In this section we exploit this feature to assign to elements of A a number of invariants that are typically associated to matrices over R. A reference for the material in this section is [**Sal99**, Chapter 4].

1.1. The Characteristic Polynomial. Let A be an Azumaya R-algebra. Let S be a faithfully flat commutative R-algebra, P an S-progenerator and $\sigma : A \otimes_R S \cong \operatorname{Hom}_S(P, P)$ an isomorphism of S-algebras (S, P and σ exist, by Theorem 10.3.9). Let $R[x]$ be the ring of polynomials in the variable x. Then $S[x] = R[x] \otimes_R S$ is a faithfully flat extension of $R[x]$.

Let $\alpha \in A$. By virtue of σ, we view $\alpha \otimes 1 \in A \otimes_R S$ as an element of $\operatorname{Hom}_S(P, P)$. In Example 5.3.3 we defined the characteristic polynomial $p(x) = \operatorname{char.poly}_S(\alpha \otimes 1)$. Then $p(x)$ is an element of the ring $S[x]$, and the Cayley-Hamilton theorem applies. Namely, $p(\alpha \otimes 1) = 0$. Now we show that the characteristic polynomial actually comes from a polynomial in $R[x]$, and does not depend on the choice of the splitting ring S. The proof will utilize Lemma 11.1.1.

Lemma 11.1.1. *Let R be a commutative ring. Let P and Q be finitely generated projective R-modules. If $\varphi : \operatorname{Hom}_R(P, P) \to \operatorname{Hom}_R(Q, Q)$ is an R-algebra isomorphism, then $\det(\phi) = \det(\varphi(\phi))$, for every $\phi \in \operatorname{Hom}_R(P, P)$.*

Proof. The determinant construction commutes with extension of scalars. If we prove that the determinants agree at every local ring $R_{\mathfrak{p}}$, then they agree in R, by Proposition 2.1.2. We have therefore reduced to the case where R is a local ring. Then P and Q are free R-modules and φ corresponds to an automorphism of $M_n(R^n)$ for some $n \geq 0$. By Skolem-Noether (Corollary 7.8.15), φ is an inner automorphism, hence ϕ and $\varphi(\phi)$ have the same determinant. □

Lemma 11.1.2. *Let A be an Azumaya R-algebra and α an element of A. For any faithfully flat splitting ring S for A, the characteristic polynomial* $\operatorname{char.poly}_S(\alpha \otimes 1)$ *comes from $R[x]$ and does not depend on the choice of S.*

Proof. By hypothesis there exist a commutative faithfully flat R-algebra S, an S-progenerator P, and an isomorphism $\sigma : A \otimes_R S \cong \operatorname{Hom}_S(P, P)$ of S-algebras. Let T be another faithfully flat commutative R-algebra, Q a T-progenerator, and $\rho : A \otimes_R T \cong \operatorname{Hom}_T(Q, Q)$ an isomorphism of T-algebras.

We prove the result for determinants. By replacing R with $R[x]$ and S with $S[t]$, the reader should verify that the result holds for all of the coefficients of the characteristic polynomial.

Consider the commutative diagram of $S \otimes_R T$-algebras

$$\begin{array}{ccc} A \otimes_R S \otimes_R T & \xrightarrow{\sigma \otimes 1} & \operatorname{Hom}_{S \otimes_R T}(P \otimes_R T, P \otimes_R T) \\ \downarrow{\scriptstyle \tau \otimes 1} & & \downarrow{\scriptstyle \varphi} \\ S \otimes_R A \otimes_R T & \xrightarrow{1 \otimes \rho} & \operatorname{Hom}_{S \otimes_R T}(S \otimes_R Q, S \otimes_R Q) \end{array}$$

where $\tau(x \otimes y) = y \otimes x$ and φ is defined by the rest of the diagram. Lemma 11.1.1 and the fact that the determinant commutes with extension of scalars imply that

$$\det_S(\sigma(\alpha \otimes 1_S)) \otimes 1_T = 1_S \otimes \det_T(\rho(\alpha \otimes 1_T)). \tag{11.1}$$

The Amitsur complex

$$0 \to R \xrightarrow{\theta} S \xrightarrow{d^0} S \otimes_R S \xrightarrow{d^1} \cdots$$

is exact (Proposition 5.3.1). Putting $S = T$ in (11.1) shows that $\det_S(\sigma(\alpha \otimes 1))$ is in the kernel of d^0, hence belongs to R. Then (11.1) shows that $1_S \otimes \det_S(\sigma(\alpha \otimes 1))$ and $1_S \otimes \det_T(\rho(\alpha \otimes 1))$ are equal in $S \otimes_R T$. The Amitsur complex for the faithfully flat cover $S \otimes_R T$ shows that $\det_S(\sigma(\alpha \otimes 1))$ and $\det_T(\rho(\alpha \otimes 1))$ descend to the same element in R. □

Corollary 11.1.3 (Cayley-Hamilton Theorem). *Let A be an Azumaya R-algebra and α an element of A. If $p(x) = \operatorname{char.poly}_R(\alpha) \in R[x]$ is the characteristic polynomial of α, then $p(\alpha) = 0$.*

Proof. Let S be a faithfully flat commutative R-algebra which splits A. Let $\sigma : A \otimes_R S \to \operatorname{Hom}_S(P, P)$ be an S-algebra isomorphism. Then $A \to A \otimes_R S$ is one-to-one, because S is flat and $R \to A$ is one-to-one. As shown in Example 5.3.3, $p(\alpha \otimes 1_S) = 0$. □

Theorem 11.1.4, which is a generalization of Lemma 1.2.2, is due to Saltman ([**Sal99**]). It can be proved using the methods from above.

Theorem 11.1.4. *Let R be a commutative ring, A an Azumaya R-algebra of constant rank n^2, and $R[A^*]$ the ring of polynomial functions on A (see Section 1.2.1). There are $s_1, \dots, s_n$ in $R[A^*]$ satisfying the following properties:*

(1) *If $A = M_n(R)$ is the ring of matrices, then for any $\alpha \in A$, $x^n - s_1(\alpha)x^{n-1} + s_2(\alpha)x^{n-2} - \cdots + (-1)^n s_n(\alpha)$ is the characteristic polynomial of α.*

(2) *For any $r \in R$, $s_i(r\alpha) = r^i s_i(\alpha)$. That is, s_i is homogeneous of degree i.*

(3) *Let* $\theta : R \to S$ *be a homomorphism of commutative rings and* $\theta : R[A^*] \to R[(A \otimes_R S)^*]$ *the induced map on polynomial rings. Let* $t_1, \dots, t_n$ *be the corresponding polynomial functions in* $R[(A \otimes_R S)^*]$. *Then* $\theta(s_i) = t_i$.

Proof. The proof is left to the reader. □

Let A be an Azumaya R-algebra with constant rank n^2. In Theorem 11.1.4, the function $s_n : A \to R$ is called the *norm*, and we write $N_R^A(\alpha) = s_n(\alpha)$. The function $s_1 : A \to R$ is called the *trace*, and we write $T_R^A(\alpha) = s_1(\alpha)$.

The norm and trace we have just defined are sometimes called the *reduced norm* and *reduced trace* (for example, see [**Rei75**, §9]). This is to distinguish them from the norm and trace that are defined in Exercise 1.2.14 and Example 1.2.3. The difference is in how we view A as a ring of endomorphisms. In the present setting, A is an Azumaya algebra, hence is a form of matrices. In Lemma 11.1.2, it is this attribute of A that allows us to define the characteristic polynomial of an element of A. In Exercise 1.2.14 and Example 1.2.3, the algebra A was not assumed to be an Azumaya algebra. Hence, to define the characteristic polynomial of an element of A, it was necessary to embed A in $\operatorname{Hom}_R(A, A)$ using the left regular representation. The left regular embedding has the effect of raising the degree of the characteristic polynomial, and therefore the degree of the norm polynomial.

Remark 11.1.5. The reader should verify that the usual formulas hold:

$$\begin{aligned} N_R^A(\alpha\beta) &= N_R^A(\alpha)N_R^A(\beta) \\ N_R^A(r) &= r^n \\ R_R^A(\alpha+\beta) &= T_R^A(\alpha) + R_R^A(\beta) \\ T_R^A(\alpha\beta) &= T_R^A(\beta\alpha) \\ T_R^A(r\alpha) &= rT_R^A(\alpha) \end{aligned}$$

for all $\alpha, \beta \in A$, and $r \in R$. Moreover, α is invertible in A if and only if $N_R^A(\alpha)$ is invertible in R.

Corollary 11.1.6. *Let A be an Azumaya R-algebra of constant rank n^2.*

(1) *The trace map $T_R^A : A \to R$ is onto. There is an element in A with trace 1.*

(2) *Let $\tau : A \to A^o$ be the anti-isomorphism between A and the opposite algebra. Then $s_i(\tau(\alpha)) = s_i(\alpha)$, for each $i = 1, \dots, n$.*

(3) *The trace pairing $\tau : A \times A \to R$ defined by $\tau(\alpha, \beta) = T_R^A(\alpha\beta)$ is a symmetric nondegenerate R-bilinear form.*

Proof. Let S be a faithfully flat splitting ring for A/R, which exists by Corollary 10.3.10. For each i, the diagram

$$\begin{array}{ccc} A & \xrightarrow{s_i} & R \\ \subseteq\downarrow & & \downarrow\subseteq \\ M_n(S) & \xrightarrow{s_i} & S \end{array}$$

commutes.

(1): For $i = 1$, the trace map in the bottom row is obviously onto. By faithful flatness, the trace in the top row is onto.

(2): On the ring of matrices τ is the transpose operator. Therefore, this follows from the usual formulas for determinants of matrices.

(3): τ is bilinear by the usual properties of the trace. Since the trace pairing is nondegenerate upon restriction of scalars to S (Exercise 1.2.19) it is nondegenerate. □

Let k be a field and A a central simple k-algebra with $\dim_k(A) = n^2$. Fix a k-basis $\alpha_1, \dots, \alpha_{n^2}$ for A. With respect to this basis, we identify A with affine n^2-space over k. That is, an element $a_1\alpha_1 + \dots + a_{n^2}\alpha_{n^2} \in A$ corresponds to the point $a_1, \dots, a_{n^2}) \in \mathbb{A}_k^{n^2}$. With this identification, the norm $N_k^A : A \to k$ corresponds to a homogeneous polynomial in $k[x_1, \dots, x_{n^2}]$ of degree n.

Corollary 11.1.7. *Let k be a field, and A a central simple k-algebra with $\dim_k(A) = n^2$. Fix a k-basis for A and let $N_k^A : A \to k$ be the norm polynomial, viewed as a homogeneous polynomial of degree n in n^2 variables. Then A is a division algebra if and only if N_k^A has no nontrivial zero.*

Proof. See Remark 11.1.5. □

Corollary 11.1.8. *Let k be a field which is C_1. If D is a k-central division algebra, then $D = k$. In particular,* $\mathrm{B}(k) = (0)$.

Proof. Assume A is a central simple k-algebra and $\dim_k(A) = n^2 > 1$. The norm polynomial (with respect to a basis for A over k) has degree n in $k[x_1, \dots, x_{n^2}]$. Since k is C_1, the norm has nontrivial zeros, hence A contains noninvertible elements. □

Corollary 11.1.9 (Tsen's Theorem). *Let k be an algebraically closed field and F an algebraic extension field of $k(x)$. Then* $\mathrm{B}(F) = (0)$.

Proof. By Proposition 1.2.10 and Theorem 1.2.13, F is C_1. □

1.2. Exercises.

Exercise 11.1.10. Let K be a field and A a central simple K-algebra. Prove a noncommutative version of Exercise 4.6.16.

(1) Show that the trace T_K^A is a generator for $\operatorname{Hom}_K(A, K)$ as a right A-module.

(2) Let $\lambda_1, \ldots, \lambda_r$ be a K-basis for A. Show that there exists another K-basis $\mu_1, \ldots, \mu_r$ for A such that $T_K^A(\mu_i \lambda_j) = \delta_{i,j}$ (the Kronecker delta function).

Exercise 11.1.11. In this exercise we consider an example from [**AG60a**, Remark, p. 388]. Let $R = \mathbb{R}[x, y]/(x^2 + y^2)$. The following steps outline a proof that R is a nonnormal integral domain. If the quotient field of R is denoted K, then the integral closure of R in K is computed and it is shown that the natural map $\mathrm{B}(R) \to \mathrm{B}(K)$ is not one-to-one.

(1) Let $\mathfrak{m}$ be the maximal ideal in R generated by x and y. Show that the composite map $\mathrm{B}(\mathbb{R}) \to \mathrm{B}(R) \to \mathrm{B}(R/\mathfrak{m})$ is an isomorphism, hence $0 \to \mathrm{B}(\mathbb{R}) \to \mathrm{B}(R)$ is split exact. Using the fact that the ring of real quaternions $\mathbb{H}$ is an $\mathbb{R}$-central division algebra, conclude that $\mathrm{B}(R)$ is nontrivial.

(2) Show that R is an integral domain. Let K denote the quotient field of R and $\bar{R} = R[yx^{-1}]$ the extension of R obtained by adjoining yx^{-1}. Show that $\bar{R}$ is isomorphic to the polynomial ring $\mathbb{C}[x]$.

(3) Show that $\mathrm{B}(K) = (0)$ and conclude that $\mathrm{B}(K/R) = \mathrm{B}(R)$.

(4) Show that $\bar{R}$ is equal to the integral closure of R in K and that the conductor ideal from $\bar{R}$ to R is $\mathfrak{m}$.

(5) Show that the two rings $R[x^{-1}]$ and $\bar{R}[x^{-1}]$ are equal.

(6) Use Theorem 8.1.24 to show that $\bar{R}$ is separable over R.

(7) Use Corollary 2.6.6 to show that $\bar{R}$ is not flat over R. By Part (5), it is flat on the nonempty basic open subset $U(x) \subseteq \operatorname{Spec} R$.

For a continuation of this example, including computations of $\operatorname{Pic}(R)$ and $\mathrm{B}(R)$, see Exercise 14.2.18.

1.3. The Rank of an Element. Let Λ be an Azumaya R-algebra. Assume Λ has constant rank n^2. Let $\alpha \in \Lambda$ and consider the right ideal $\alpha\Lambda$. We say α has *rank* r if $\alpha\Lambda$ is a Λ-module direct summand of Λ and the R-module $\alpha\Lambda$ has constant rank nr.

Lemma 11.1.12. *Let R be a commutative ring, P an R-module progenerator of constant rank n and $\Lambda = \operatorname{Hom}_R(P, P)$. If $\alpha \in \Lambda$, then the following are equivalent.*

(1) $\alpha\Lambda$ *is a* Λ*-module direct summand of* Λ *and* $\operatorname{Rank}_R(\alpha\Lambda) = nr$.

(2) $\alpha(P)$ *is an* R*-module direct summand of* P *and* $\operatorname{Rank}_R(\alpha(P)) = r$.

Proof. Consider the exact sequence of R-modules

$$0 \to \ker\alpha \to P \xrightarrow{\alpha} P \to \operatorname{coker}\alpha \to 0. \tag{11.2}$$

Applying the exact functor $\operatorname{Hom}_R(P, \cdot)$ to (11.2) yields

$$\Lambda \xrightarrow{\mathrm{H}_\alpha} \Lambda \to \operatorname{Hom}_R(P, \operatorname{coker}\alpha) \to 0 \tag{11.3}$$

where $\mathrm{H}_\alpha(\lambda) = \alpha\lambda$ (Proposition 1.3.20). Hence (11.3) shows that $\Lambda/\alpha\Lambda$ is isomorphic to $\operatorname{Hom}_R(P, \operatorname{coker}\alpha)$ as an R-module.

(2) implies (1): If the image of α is a direct summand of P, then (11.3) is split as a sequence of R-modules. By Theorem 4.4.1, (11.3) splits as a sequence of right Λ-modules.

(1) implies (2): By Lemma 1.1.20, $\alpha\Lambda = e\Lambda$ for some idempotent $e \in \Lambda$. Using e one verifies that $\alpha(P)$ is a direct summand of P. □

Proposition 11.1.13. *Let* Λ *be an Azumaya* R*-algebra of constant rank* n^2.

(1) *If there exists an element* α *in* Λ *of rank* 1, *then* Λ *is split.*

(2) *If there exists an idempotent* e *in* Λ *of rank* r, *then* $\operatorname{Rank}_R(eAe) = r^2$.

Proof. (1): Since α has rank 1, the right ideal $\alpha\Lambda$ is an R-module progenerator of rank n. The right regular representation $\Lambda^o \to \operatorname{Hom}_R(\alpha\Lambda, \alpha\Lambda)$ is an isomorphism, by Theorem 7.6.1. Therefore, Λ^o is split. Because $\mathrm{B}(R)$ is a group, Λ is split.

(2): By Exercise 7.6.4, $\Lambda \otimes_R e\Lambda e \cong \operatorname{Hom}_R(e\Lambda, e\Lambda)$. Hence $\operatorname{Rank}_R(eAe) = r^2$. □

Proposition 11.1.14. *Let* k *be a field,* A *an Azumaya* k*-algebra and* $\alpha \in A$. *If* $n^2 = \dim_k(A)$, *then the following are true.*

(1) *The index of* A *divides the rank of* α.

(2) α *is invertible in* A *if and only if the rank of* α *is equal to* n.

Proof. (1): By a right handed version of Theorem 3.2.6, $A \cong \operatorname{Hom}_D(M, M)$, where M is the unique simple right A-module and $D = \operatorname{Hom}_A(M, M)$ is a k-central division algebra. The index of A is m, where $\dim_k(D) = m^2$. If r is the rank of α and $\dim_D(M) = s$, then $\dim_k(\alpha A) = rn = rsm$. The right ideal αA is isomorphic to a direct sum of copies of M, hence $\alpha A \cong M^t \cong D^{st}$, for some $t \geq 1$. We have $\dim_k(\alpha A) = rsm = stm^2$, so m divides r.

(2): Note that $\alpha A = A$ if and only if $\dim_k(\alpha A) = n^2$. □

2. The Brauer Group is Torsion

Let Λ be an Azumaya R-algebra of constant rank n^2. In this section we present a proof due to David Saltman ([**Sal81**], [**Sal99**, Chapter 5]) that the Brauer class of Λ is annihilated by n. As in Section 5.3.1, by $\Lambda^{\otimes r}$ we denote $\Lambda \otimes_R \cdots \otimes_R \Lambda$, the tensor product of r copies of Λ. Let S_r denote the symmetric group, which acts as a group of R-algebra automorphisms of $\Lambda^{\otimes r}$ by permuting the factors. If $\sigma \in S_r$, then on a basic element λ of the tensor product, we have

$$\sigma(\lambda) = \sigma(\lambda_1 \otimes \cdots \otimes \lambda_r) = \lambda_{\sigma^{-1}(1)} \otimes \cdots \otimes \lambda_{\sigma^{-1}(r)}. \tag{11.4}$$

Our goal is to show that the automorphism σ is an inner automorphism on $\Lambda^{\otimes r}$. Notice that if R is a field, then this is true by the Skolem-Noether Theorem (Corollary 4.5.13). Since S_r is generated by transpositions, we first consider the $r = 2$ case.

Proposition 11.2.1. *Let Λ be an Azumaya R-algebra of constant rank n^2. There is a unique invertible $\alpha \in \Lambda^{\otimes 2}$ such that the following are true:*

(1) $\alpha(x \otimes y)\alpha^{-1} = y \otimes x$.

(2) *If S is a commutative R-algebra which is a splitting ring for Λ and $\Lambda \otimes_R S \cong \operatorname{Hom}_S(P, P)$, for an S-progenerator module P, then the image of $\alpha \otimes 1$ under the composite mapping*

$$\begin{aligned}\Lambda^{\otimes 2} \otimes_R S &\cong (\Lambda \otimes_R S) \otimes_S (\Lambda \otimes_R S)\\ &\cong \operatorname{Hom}_S(P, P) \otimes_S \operatorname{Hom}_S(P, P)\\ &\cong \operatorname{Hom}_S(P^{\otimes 2}, P^{\otimes 2})\end{aligned}$$

is the map $P^{\otimes 2} \to P^{\otimes 2}$ defined by $a \otimes b \mapsto b \otimes a$.

Proof. Step 1: We prove the proposition when $\Lambda = M_n(R)$ is a matrix algebra over R. Let $e_1, \dots, e_n$ denote the standard basis for R^n. In $M_n(R)$, let e_{ij} denote the elementary matrix. Then for all i, j, k, ℓ, s,

$$\begin{aligned} e_{ij}e_s &= \begin{cases} e_i & \text{if } j = s,\\ 0 & \text{otherwise, and}\end{cases}\\ e_{ij}e_{k\ell} &= \begin{cases} e_{i\ell} & \text{if } j = k,\\ 0 & \text{otherwise.}\end{cases}\end{aligned} \tag{11.5}$$

For any matrix $M \in M_n(R)$,

$$\sum_{i=1}^{n}\sum_{j=1}^{n} e_{ij}Me_{ji} = T_R^{\Lambda}(M) \tag{11.6}$$

where $T_R^\Lambda : \Lambda = M_n(R) \to R$ is the trace (Exercise 1.2.18). In $\Lambda \otimes_R \Lambda$, let

$$\alpha = \sum_{i=1}^{n} \sum_{j=1}^{n} e_{ij} \otimes e_{ji}. \tag{11.7}$$

Direct computations show that

$$\alpha(e_r \otimes e_s) = e_s \otimes e_r \tag{11.8}$$

for all basis vectors e_r, e_s in R^n, and

$$\alpha(e_{rs} \otimes e_{uv}) = e_{us} \otimes e_{rv} = (e_{uv} \otimes e_{rs})\alpha \tag{11.9}$$

for arbitrary elementary matrices e_{rs}, e_{uv}. Therefore,

$$\alpha(u \otimes v) = v \otimes u \tag{11.10}$$

for all u, v in R^n and

$$\alpha(x \otimes y) = (x \otimes y)\alpha \tag{11.11}$$

for all x, y in $\Lambda = M_n(R)$. From (11.11) we see that the right ideal $\alpha\Lambda^{\otimes 2}$ is equal to the left ideal $\Lambda^{\otimes 2}\alpha$. By (11.9), both ideals contain a generating set for $\Lambda^{\otimes 2}$. This implies α is invertible in $\Lambda^{\otimes 2}$. This proves (1). It follows from (11.10) that (2) is satisfied. Lastly, the equations (11.10) uniquely determine α as an element in $\Lambda^{\otimes 2}$.

Step 2: Let Λ be an Azumaya R-algebra of constant rank n^2. Since $0 \to R \to \Lambda$ is exact as R-modules, $0 \to \operatorname{Hom}_R(\Lambda, R) \to \operatorname{Hom}_R(\Lambda, \Lambda)$ is exact as well. By Theorem 7.1.4, there is an isomorphism $\Lambda \otimes_R \Lambda^o \to \operatorname{Hom}_R(\Lambda, \Lambda)$ which sends γ to "left multiplication by γ". In particular, we can view the trace $T_R^\Lambda : \Lambda \to R$ as an element of $\operatorname{Hom}_R(\Lambda, \Lambda)$ and there is an element $\sum_i x_i \otimes y_i \in \Lambda \otimes_R \Lambda^o$ such that $\sum_i x_i \lambda y_i = T_R^\Lambda(\lambda)$ for all $\lambda \in \Lambda$. Identifying the R-modules $\Lambda \otimes_R \Lambda^o$ and $\Lambda \otimes_R \Lambda$, let $\alpha = \sum_i x_i \otimes y_i \in \Lambda \otimes_R \Lambda$. If $\theta : R \to S$ is a homomorphism of commutative rings, then $T_S^{\Lambda \otimes_R S}$ is given by left multiplication by the element $\sum_i x_i \otimes y_i \otimes 1$ of $\Lambda \otimes_R \Lambda^o \otimes_R S$ (Theorem 11.1.4). Now let S be a faithfully flat R-algebra such that $\Lambda \otimes_R S$ is isomorphic to $M_n(S)$ (Corollary 10.3.10). Since S is faithfully flat over R, $\Lambda^{\otimes 2} \to \Lambda^{\otimes 2} \otimes_R S \cong M_n(S)$ is one-to-one. From Step 1, the image of α in $M_n(S)$ is given by (11.7). Therefore, α is unique. From this we also see that (11.11) holds for α. Because S is faithfully flat over R, it follows that $\alpha\Lambda^{\otimes 2} = \Lambda^{\otimes 2} = \Lambda^{\otimes 2}\alpha$ and α is invertible in $\Lambda^{\otimes 2}$. This proves (1). Let P be as in (2). Using Step 1 and the fact that P is locally free, we see that (2) holds. □

As mentioned in [**KO74b**, Remarque IV.4.2], the first part of Proposition 11.2.1 is due to Oscar Goldman and we will refer to the element $\alpha \in \Lambda^{\otimes 2}$ as the *Goldman element*. Since the symmetric group S_r is generated by transpositions, in Theorem 11.2.2 we show that there are Goldman-like elements in $\Lambda^{\otimes r}$ for $r \geq 2$.

Theorem 11.2.2. *Let Λ be an Azumaya R-algebra of constant rank n^2, $r > 1$, and S_r the symmetric group which acts on $\Lambda^{\otimes r}$ according to (11.4). For each $\sigma \in S_r$ there is an invertible element u_σ in $\Lambda^{\otimes r}$ such that the following are true:*

(1) *$\sigma(\lambda) = u_\sigma \lambda u_\sigma^{-1}$ for all $\lambda \in \Lambda^{\otimes r}$.*

(2) *If S is a commutative R-algebra which is a splitting ring for Λ and $\Lambda \otimes_R S \cong \operatorname{Hom}_S(P, P)$, for an S-progenerator module P, then the image of $\alpha \otimes 1$ under the composite mapping*
$$\Lambda^{\otimes r} \otimes_R S \cong (\Lambda \otimes_R S)^{\otimes r} \cong \operatorname{Hom}_S(P^{\otimes r}, P^{\otimes r})$$
is the map $P^{\otimes r} \to P^{\otimes r}$ defined by $a_1 \otimes \cdots \otimes a_r \mapsto a_{\sigma^{-1}(1)} \otimes \cdots \otimes a_{\sigma^{-1}(r)}$.

(3) *The mapping $\sigma \mapsto u_\sigma$ defines a homomorphism of groups $S_r \to (\Lambda^{\otimes r})^*$, hence there is a homomorphism of R-algebras from the group ring $R(S_r)$ to $\Lambda^{\otimes r}$.*

Proof. Notice that (3) is a direct consequence of (2) and Corollary 10.3.10. For each pair (i, j) such that $i < j$, let $\tau_{ij} \in S_r$ be the corresponding transposition. Let $\theta_{ij} : \Lambda^{\otimes 2} \to \Lambda^{\otimes r}$ the R-algebra monomorphism defined by $\theta_{ij}(a \otimes b) = \lambda_1 \otimes \cdots \otimes \lambda_r$, where
$$\lambda_k = \begin{cases} a & \text{if } k = i, \\ b & \text{if } k = j, \\ 1 & \text{otherwise.} \end{cases}$$
Let $\alpha \in \Lambda^{\otimes 2}$ be the Goldman element from Proposition 11.2.1 and define $u_{\tau_{ij}} = \theta_{ij}(\alpha)$. Then $u_{\tau_{ij}}$ clearly satisfies (1) and (2). Now let S be a faithfully flat R-algebra such that $\Lambda \otimes_R S$ is isomorphic to $M_n(S)$ (Corollary 10.3.10). If we replace Λ with $M_n(S)$, then the rest of the proof follows from the fact that the elements $u_{\tau_{ij}}$ generate a subgroup of $(\Lambda^{\otimes r})^*$ which is isomorphic to S_r. □

Proposition 11.2.3. *Let Λ be an Azumaya R-algebra of constant rank n^2. In the notation of Theorem 11.2.2, let*
$$\beta_r = \sum_{\sigma \in S_r} \operatorname{sign}(\sigma) u_\sigma$$
in $\Lambda^{\otimes r}$. Then β_r has rank equal to the binomial coefficient $\binom{n}{r}$.

Proof. By Corollary 10.3.10, it is enough to show this for $\Lambda = \operatorname{Hom}_R(P, P)$, where $P = Rv_1 \oplus \cdots Rv_n$ is the free R-module of rank n. Then each u_σ acts on the tensor module $T^r(P)$ by permuting the factors. Let $J = \{1, \dots, n\}$. By taking the lexicographical order on $J^r = J \times \cdots \times J$, the cartesian product of r copies of J, we linearly order the free basis $B = \{v_{i_1} \otimes \cdots \otimes v_{i_r} \mid (i_1, \dots, i_r) \in J^r\}$ of $P^{\otimes r}$. The five steps below outline a proof that the image of β_r is a free R-module direct summand of $P^{\otimes r}$ with rank $\binom{n}{r}$. The details are left to the reader.

(1): Let $(i_1, \dots, i_r) \in J^r$. If there is a pair $j < k$ such that $i_j = i_k$, then $\beta_r(v_{i_1} \otimes \cdots \otimes v_{i_r}) = 0$. Therefore, $\beta_r : P^{\otimes r} \to P^{\otimes r}$ is a multilinear alternating map. The proof is similar to the proof that the determinant function on the ring of matrices is alternating.

(2): Let $x_1 \otimes \cdots \otimes x_r$ be a basic element in $P^{\otimes r}$. For each $\tau \in S_r$, $\beta_r(\tau(x_1 \otimes \cdots \otimes x_r)) = \operatorname{sign}(\tau)\beta_r(x_1 \otimes \cdots \otimes x_r)$. Therefore, β_r is skew symmetric.

(3): Let $B_0 = \{(v_{i_1} \otimes \cdots \otimes v_{i_r} \mid (i_1, \dots, i_r) \in J^r,\ i_1 < i_2 < \cdots < i_r\}$. Then B_0 is a subset of the basis B for $P^{\otimes r}$. The R-module spanned by B_0 is a direct summand of $P^{\otimes r}$ of rank $\binom{n}{r}$.

(4): The image of β_r is spanned by $\{\beta_r(b) \mid b \in B_0\}$, where B_0 is as in (3). This follows from (1) and (2).

(5): If we restrict β_r to the submodule spanned by B_0, then the matrix of β_r with respect to the basis B_0 is lower triangular with diagonal $(1, 1, \dots, 1)$. Therefore, the image of β_r is a direct summand of $P^{\otimes r}$ of rank $\binom{n}{r}$. □

Corollary 11.2.4. *Let Λ be an Azumaya R-algebra of constant rank n^2. Then $\Lambda^{\otimes n}$ is split. Equivalently, the Brauer class represented by Λ is annihilated by n.*

Proof. This follows by Proposition 11.1.13 and Proposition 11.2.3 because β_n is an element in $\Lambda^{\otimes n}$ with rank 1. □

Corollary 11.2.5. *If R is a commutative ring, then the Brauer group of R is torsion.*

Proof. This follows from Corollary 11.2.4 because Exercise 7.4.9 shows that any Azumaya R-algebra is Brauer equivalent to an Azumaya R-algebra of constant rank. □

2.1. Applications to Division Algebras. Let k be a field. If A is a central simple k-algebra, then by Corollary 11.2.4, the Brauer class of A is annihilated by n, where $n^2 = \dim_k(A)$. The *order* of A is defined to be the order of $[A]$ in $\mathrm{B}(k)$. The following factorization theorem (Theorem 11.2.8) and its proof are based on [**Sal99**, Theorem 5.7].

Lemma 11.2.6. *Let k be a field and D a k-central division algebra of index d.*

(1) *If* $1 \le r \le d$, *then the index of* $D^{\otimes r}$ *divides* $\gcd\left(\binom{d}{r}, d\right)$.

(2) *If* p *is a prime divisor of* d, *then the index of* $D^{\otimes p}$ *divides* d/p.

(3) *If* s *is a divisor of* d, *then the index of* $D^{\otimes s}$ *divides* d/s.

Proof. (1): By Proposition 11.2.3, $D^{\otimes r}$ contains an element β_r of rank $\binom{d}{r}$. By Proposition 11.1.14, the index of $D^{\otimes r}$ divides $\binom{d}{r}$. By Corollary 7.5.3, the index of $D^{\otimes r}$ divides d.

(2): Let p^e be the highest power of p that divides d. The reader should verify that the highest power of p that divides $\binom{d}{p}$ is p^{e-1}. Therefore, (2) follows from (1).

Part (3) follows from Part (2) and induction on the number of prime factors of d. □

Theorem 11.2.7. *Let* k *be a field,* D *a* k*-central division algebra of index* d, *and* $1 \le r \le d$.

(1) *The order of* $D^{\otimes r}$ *divides* d.

(2) *If* r *and* d *are relatively prime, then the index of* $D^{\otimes r}$ *is equal to* d.

(3) *The index of* $D^{\otimes r}$ *divides* $d/\gcd(r, d)$.

Proof. Part (1) follows straight from Corollary 11.2.4.

(2): By (1), D and $D^{\otimes r}$ generate the same subgroup of $\mathrm{B}(k)$. By Corollary 7.5.3, the index of D is equal to the index of $D^{\otimes r}$.

(3): Let $g = \gcd(r, d)$. In the Brauer group we have $[D^{\otimes r}] = [D^{\otimes g}]^{r/g}$. By Corollary 7.5.3, the index of $D^{\otimes r}$ divides the index of $D^{\otimes g}$. By Lemma 11.2.6, the index of $D^{\otimes g}$ divides d/g. □

Theorem 11.2.8. *Let* k *be a field and* D *a* k*-central division algebra such that* $\dim_k(D) = d^2$.

(1) *If* $d = ab$, *with* $\gcd(a, b) = 1$, *then there exist unique* k*-central simple algebras* A, B *satisfying*
   (a) $D \cong A \otimes_k B$,
   (b) A *and* B *are division algebras,*
   (c) $\dim_k(A) = a^2$, *and*
   (d) $\dim_k(B) = b^2$.

(2) *If* $d = p_1^{e_1} \cdots p_r^{e_r}$ *is the unique factorization of* d, *where* $p_1, \dots, p_r$ *are distinct prime numbers, then there exist unique* k*-central division algebras* $D_1, \dots, D_r$ *satisfying*
   (a) $D \cong D_1 \otimes_k \cdots \otimes_k D_r$, *and*
   (b) $\dim_k(D_i) = (p_i^{e_i})^2$ *for* $i = 1, \dots, r$.

Proof. (1): Step 1: A factorization exists. By the Bézout Identity for $\mathbb{Z}$, we can write $1 = au + bv$. Let A be the division algebra component of $D^{\otimes bv}$, and B the division algebra component of $D^{\otimes au}$. Then $[D] = [D^{\otimes au}][D^{\otimes bv}] = [A][B] = [A \otimes_k B]$ implies D is the division algebra component of $A \otimes_k B$. In particular, d^2 divides $\dim_k(A)\dim_k(B)$. By Theorem 11.2.7, the index of $D^{\otimes b}$ divides a. By Corollary 7.5.3, the index of A divides a. Likewise, the index of B divides b. We have shown that $\dim_k(D) = \dim_k(A \otimes_k B)$. By Theorem 3.2.6, this implies $A \otimes_k B$ is isomorphic to D.

Step 2: The factorization is unique. Suppose A_1 and B_1 also satisfy (a) – (d). Then $[A_1]^a = [k] = [B_1]^b$ implies $[D]^{bv} = [A]$ is in the subgroup generated by $[A_1]$. Likewise, $[D]^{au} = [B]$ is in the subgroup generated by $[B_1]$. Since the order of A_1 is relatively prime to the order of B_1, the subgroup $\langle [A_1], [B_1] \rangle$ is a direct product $\langle [A_1] \rangle \oplus \langle [B_1] \rangle$. Then $[D] = [A][B] = [A_1][B_1]$ implies that $[A] = [A_1]$ and $[B] = [B_1]$. Since the algebras are division algebras, $A \cong A_1$ and $B \cong B_1$.

(2): This follows from Part (1) and a finite induction argument. □

Remark 11.2.9. In [**AW14**], B. Antieau and B. Williams construct a ring R and an Azumaya R-algebra A such that the rank of A is of the form $(2n)^2$ with n odd, and such that there is no factorization $A \cong A_2 \otimes_R A_n$ into Azumaya algebras of ranks 2^2 and n^2, respectively.

3. Maximal Orders

Throughout this section R is an integral domain with field of fractions K. Usually R is assumed to be integrally closed in K. General references for the results presented in this section are [**Rei75**] and [**AG60b**].

3.1. Definition, First Properties.

Let R be an integral domain with field of fractions K and V a finite dimensional K-vector space. Recall that an R-submodule M of V is an R-lattice in V if there is a finitely generated R-submodule N in V such that $M \subseteq N$ and $KM = V$ (Section 6.3.1). Let A be a K-algebra which is finite dimensional as a K-vector space. An *R-order in A* is an R-subalgebra of A which is also an R-lattice in A. A *maximal order* in A is an order that is maximal with respect to set inclusion.

Let R be a noetherian integral domain with field of fractions K. Let A be a finite dimensional K-algebra. Assume $\{x_1, \dots, x_m\} \subseteq A$ is a spanning set for A. Then $M = \sum_i Rx_i$ is an R-lattice in A. The set

$$\mathfrak{O}_l(M) = \{\alpha \in A \mid \alpha M \subseteq M\} \tag{11.12}$$

is called the *left order of M*. The set

$$\mathfrak{O}_r(M) = \{\alpha \in A \mid M\alpha \subseteq M\} \tag{11.13}$$

is called the *right order of* M. The reader should verify that $\mathfrak{O}_l(M)$ and $\mathfrak{O}_r(M)$ are R-subalgebras of A. We prove that $\mathfrak{O}_l(M)$ is an R-lattice in A, hence is an R-order. By a similar argument, $\mathfrak{O}_r(M)$ is an R-order in A. Let $y \in A = KM$. There are $k_i \in K$ such that $y = k_1x_1 + \cdots + k_mx_m$. Let r be a nonzero element of R such that $rk_i \in R$ for all i. Then $ry = rk_1x_1 + \cdots + rk_mx_m \in M$. In particular, $1 \in A$ and there is a nonzero $s \in R$ such that $s \cdot 1 \in M$. Given $\alpha \in \mathfrak{O}_l(M)$, $\alpha s \cdot 1 \in M$. Therefore, $\mathfrak{O}_l(M) \subseteq s^{-1}M$. Since R is noetherian, this implies $\mathfrak{O}_l(M)$ is a finitely generated R-module. Let y be an arbitrary element of A. Pick a nonzero element r of R such that $ryx_j \in M$ for all j. Then $ryM = Rryx_1 + \cdots + Rryx_m \subseteq M$. This proves $ry \in \mathfrak{O}_l(M)$. From this it follows that $A = K\mathfrak{O}_l(M)$. This proves that the left order of M is an R-order in A.

Lemma 11.3.1. *Let S/R be an extension of noetherian integral domains and K/L the associated extension of the fields of fractions. Let A be a central simple L-algebra and M an R-lattice in A. If S is a flat R-module, then there is an isomorphism of S-algebras $S \otimes_R \mathfrak{O}_r(M) \cong \mathfrak{O}_r(S \otimes_R M)$.*

Proof. Let Γ be any R-order in A such that $\Gamma M \subseteq M$. Then M is a left Γ-module. As in Exercise 1.3.32, there is an isomorphism of rings $A^o \cong \operatorname{Hom}_A(A, A)$ which is defined by mapping $\alpha \in A$ to ρ_α, where $\rho_\alpha(x) = x\alpha$. Given $\alpha \in A$, right multiplication by α is in $\operatorname{Hom}_\Gamma(M, M)$ if and only if $\alpha \in \mathfrak{O}_r(M)$. The diagram

$$\begin{array}{ccc} \mathfrak{O}_r(M)^o & \xrightarrow{\sigma} & \operatorname{Hom}_\Gamma(M, M) \\ \downarrow & & \downarrow \gamma \\ A^o & \xrightarrow{\rho} & \operatorname{Hom}_A(A, A) \end{array} \tag{11.14}$$

of R-algebras commutes and σ is an isomorphism. Notice that $\operatorname{Hom}_\Gamma(M, M)$ does not depend on the choice of Γ. By Proposition 6.3.5, $S \otimes_R M$ is an S-lattice in $K \otimes_L A$. By Proposition 2.4.10, $S \otimes_R \operatorname{Hom}_\Gamma(M, M) \cong \operatorname{Hom}_{S\otimes_R\Gamma}(S \otimes_R M, S \otimes_R M)$. From the isomorphism σ of (11.14), $S \otimes_R \mathfrak{O}_r(M) \cong \mathfrak{O}_r(S \otimes_R M)$. □

Lemma 11.3.2. *Let R be a noetherian integrally closed integral domain with field of fractions K. Let A be a central simple K-algebra and Λ an R-subalgebra of A such that $K\Lambda = A$. The following are equivalent.*

(1) *Λ is an R-order in A.*
(2) *Every $\alpha \in \Lambda$ is integral over R.*
(3) *For every $\alpha \in \Lambda$, the coefficients of the characteristic polynomial of α are in R.*
(4) *For every $\alpha \in \Lambda$, $T_K^A(\alpha) \in R$.*

Proof. (1) implies (2): This follows from Proposition 3.3.1.

(2) implies (3): Let F be an extension field of K such that $A \otimes_K F \cong M_n(F)$. Let S be the integral closure of R in F. Let $\alpha \in \Lambda$. By Theorem 3.3.4, $\text{min. poly}_F(\alpha) \in S[x]$. By Gauss' Lemma (Lemma 3.3.3), each irreducible factor of $\text{char. poly}_F(\alpha)$ is in $S[x]$. Therefore, $\text{char. poly}_F(\alpha) \in S[x]$. By Lemma 11.1.2, the coefficients of the characteristic polynomial of α are in $S \cap K$. Since R is integrally closed in K, $S \cap K = R$.

(3) implies (4): If the characteristic polynomial of α is $x^m - s_1 x^{m-1} + \dots$, then s_1 is the trace of α.

(4) implies (1): Since $K\Lambda = A$, there exists a K-basis for A, say $\{\lambda_1, \dots, \lambda_m\}$, which is contained in Λ. By Exercise 11.1.10 there exists another K-basis $\mu_1, \dots, \mu_r$ for A such that $T_K^A(\mu_i\lambda_j) = \delta_{i,j}$ (the Kronecker delta function). Let $\alpha = \sum_{i=1}^m k_i\mu_i$ be a typical element of A. Then

$$\begin{aligned} T_K^A(\alpha\lambda_j) &= T_K^A\Big(\sum_{i=1}^m k_i\mu_i\lambda_j\Big) \\ &= \sum_{i=1}^m k_i T_K^A(\mu_i\lambda_j) \\ &= k_j. \end{aligned}$$

If $\alpha \in \Lambda$, then by (4), $T_K^A(\alpha\lambda_j) = k_j$ is in R. Thus, $\Lambda \subseteq \sum_{i=1}^m R\mu_i$. Since R is noetherian, Λ is a finitely generated R-module, which proves (1). The reader should notice the similarity between this proof and that for Theorem 4.6.10. □

Theorem 11.3.3. *Let R be a noetherian integrally closed integral domain with field of fractions K. Let A be a central simple K-algebra and Λ_0 an R-order in A. Then there exists a maximal R-order Λ in A such that $\Lambda_0 \subseteq \Lambda \subseteq A$.*

Proof. Let $\mathcal{S}$ be the set of all R-orders in A containing Λ_0, ordered by set inclusion. Let $\{\Lambda_i \mid i \in I\}$ be a chain in $\mathcal{S}$. Let $\Lambda = \bigcup_{i\in I}\Lambda_i$. Then Λ is an R-subalgebra of A containing Λ_0 and $K\Lambda = A$. If $\alpha \in \Lambda$, then for some $i \in I$, $\alpha \in \Lambda_i$. By Lemma 11.3.2, α is integral over R and Λ is an R-order, hence belongs to $\mathcal{S}$. Zorn's Lemma implies $\mathcal{S}$ contains a maximal member. □

Lemma 11.3.4. *Let R be an integral domain with field of fractions K and A a central simple K-algebra. Let Λ and Γ be R-orders in A. Define*

$$\mathfrak{F} = \{x \in A \mid x\Lambda \subseteq \Gamma\}.$$

Then the following are true.

(1) *$\mathfrak{F}$ is a left ideal in Γ.*

(2) $\mathfrak{F}$ *is a right* Λ*-module.*

(3) $\mathfrak{F}$ *contains every right* Λ*-module in* Γ.

(4) $\mathfrak{F} \cap R \neq (0)$.

Proof. By Proposition 6.3.3 there exists a nonzero $r \in R$ such that $r\Lambda \subseteq \Gamma$. This proves (4). The rest is left to the reader. □

In Lemma 11.3.4 the ideal $\mathfrak{F}$ is called the *conductor of* Λ *to* Γ *in* A.

Lemma 11.3.5. *Let R be an integrally closed integral domain with field of fractions K. Let A be an R-algebra that is generated by the single element, a. Assume the R-module A is torsion free and finitely generated. Viewing A as an R-order in $KA = K \otimes_R A$, let $f = \operatorname{min.poly}_K(a)$. Then $f \in R[x]$, and A is isomorphic to $R[x]/(f)$. It follows that A is a finitely generated free R-module.*

Proof. Consider the commutative diagram

$$\begin{array}{ccccccccc} 0 & \longrightarrow & \ker\sigma & \longrightarrow & R[x] & \xrightarrow{\sigma} & A & \longrightarrow & 0 \\ & & \downarrow{\subseteq} & & \downarrow{\subseteq} & & \downarrow{\subseteq} & & \\ 0 & \longrightarrow & (f) & \longrightarrow & K[x] & \xrightarrow{\tau} & K \otimes_R A & \longrightarrow & 0 \end{array} \tag{11.15}$$

with exact rows, where the maps σ and τ are defined by $x \mapsto a$. The vertical maps in (11.15) are one-to-one because A is torsion free. By Theorem 3.3.4, $f \in R[x]$. The rest is left to the reader. □

3.2. Localization and Completion of Maximal Orders.

Theorem 11.3.6. *Let R be a noetherian integral domain with field of fractions K. Let A be a central simple K-algebra and Λ an R-order in A. Let W be a multiplicative subset of R.*

(1) $W^{-1}\Lambda$ *is a* $W^{-1}R$*-order in* A.

(2) *If* Λ *is a maximal order, then* $W^{-1}\Lambda$ *is a maximal order.*

Proof. (1): The proof is left to the reader.

(2): Suppose Γ is a $W^{-1}R$-order in A and $W^{-1}\Lambda \subseteq \Gamma$. By Proposition 6.3.3, there exists some nonzero $g \in R$ such that $g\Gamma \subseteq W^{-1}\Lambda$. Let $\Delta = g\Gamma \cap \Lambda$. The reader should verify that Δ is an R-lattice in A. Since $\Lambda\Delta \subseteq \Lambda(g\Gamma) \cap \Lambda \subseteq g\Gamma \cap \Lambda = \Delta$, we see that $\Lambda \subseteq \mathfrak{O}_l(\Delta)$. Since Λ is maximal, this implies $\Lambda = \mathfrak{O}_l(\Delta)$. Let $\gamma \in \Gamma$. Then $\gamma\Delta \subseteq \gamma(g\Gamma \cap \Lambda) \subseteq \gamma g\Gamma \subseteq g\Gamma \subseteq W^{-1}\Lambda$. Since Δ is a finitely generated R-module, there exists some $t \in W$ such that $t\gamma\Delta \subseteq \Lambda$. Moreover, $t\gamma\Delta \subseteq t\gamma g\Gamma \subseteq g\Gamma$, hence $t\gamma\Delta \subseteq g\Gamma \cap \Lambda = \Delta$. Therefore, $t\gamma \in \mathfrak{O}_l(\Delta) = \Lambda$, and $\gamma \in W^{-1}\Lambda$. □

Corollary 11.3.7. *Let R be a noetherian integrally closed integral domain with field of fractions K. Let A be a central simple K-algebra and Λ and R-order in A. Then Λ is maximal if and only if $\Lambda_{\mathfrak{p}}$ is a maximal $R_{\mathfrak{p}}$-order for every prime ideal $\mathfrak{p} \in \operatorname{Spec}(R)$.*

Proof. The proof is left to the reader. □

Theorem 11.3.8. *Let R be a noetherian integrally closed integral domain with field of fractions K. Let A be a central simple K-algebra and Λ and R-order in A. Then Λ is maximal if and only if Λ is a reflexive R-module and $\Lambda_{\mathfrak{p}}$ is a maximal $R_{\mathfrak{p}}$-order for every height one prime ideal $\mathfrak{p} \in X_1(R)$.*

Proof. Assume Λ is a maximal R-order in A. By Theorem 11.3.6, $\Lambda_{\mathfrak{p}}$ is a maximal $R_{\mathfrak{p}}$-order for every $\mathfrak{p} \in X_1(R)$. By Theorem 6.3.13, $R : (R : \Lambda) = \bigcap_{\mathfrak{p} \in X_1(R)} \Lambda_{\mathfrak{p}}$ is a reflexive R-lattice in A containing Λ. Since each $\Lambda_{\mathfrak{p}}$ is an R-algebra, so is $R : (R : \Lambda)$. Since Λ is maximal, $\Lambda = R : (R : \Lambda)$, so Λ is reflexive.

Conversely, assume Λ is a reflexive R-module and $\Lambda_{\mathfrak{p}}$ is a maximal $R_{\mathfrak{p}}$-order for every height one prime ideal $\mathfrak{p} \in X_1(R)$. Let Γ be another R-order in A such that $\Lambda \subseteq \Gamma$. Then $\Lambda_{\mathfrak{p}} \subseteq \Gamma_{\mathfrak{p}}$, and by maximality, $\Lambda_{\mathfrak{p}} = \Gamma_{\mathfrak{p}}$. By Theorem 6.3.13, $\Gamma \subseteq \bigcap_{\mathfrak{p} \in X_1(R)} \Gamma_{\mathfrak{p}} = \bigcap_{\mathfrak{p} \in X_1(R)} \Lambda_{\mathfrak{p}} = \Lambda$. Therefore, Λ is maximal. □

Theorem 11.3.9. *Let R be a DVR with field of fractions K. Let $\hat{R}$ be the completion of R and $\hat{K}$ the completion of K. Let V be a finite dimensional K-vector space and $\hat{V} = \hat{K} \otimes_K V$. The assignments $M \mapsto \hat{R} \otimes_R M$ and $T \mapsto T \cap V$ define a one-to-one order preserving correspondence between the set of R-lattices in V and the set of $\hat{R}$-lattices in $\hat{V}$.*

Proof. This proof is from [**Rei75**, Theorem (5.2)].

Step 1: If M is an R-lattice in V, we show $\hat{R} \otimes_R M$ is a $\hat{R}$-lattice in $\hat{V}$. By Proposition 6.3.2, a lattice over a DVR is a finitely generated free module. Let $\mu_1, \dots, \mu_n$ be a free R-basis for M. Then $\mu_1, \dots, \mu_n$ is also a K-basis for V, a $\hat{K}$-basis for $\hat{K} \otimes_K V$, and a free $\hat{R}$-basis for $\hat{R} \otimes_R M$.

Step 2: The correspondence is one-to-one. By Theorem 6.1.15, $\hat{R} \cap K = R$, so $\hat{R} \otimes_R M \cap V = \bigoplus_{i=1}^{n} (\hat{R} \cap K)\mu_i = \bigoplus_{i=1}^{n} R\mu_i = M$.

Step 3: If T is a $\hat{R}$-lattice in $\hat{V}$, we show $T \cap V$ is an R-lattice in V. Let $X = \{x_1, \dots, x_n\}$ be a free $\hat{R}$-basis for T. Let $Y = \{y_1, \dots, y_n\}$ be a K-basis for V. Then X and Y are both $\hat{K}$-bases for $\hat{V}$. Let $S = (s_{ij})$ be the change of basis matrix in $M_n(\hat{K})$ such that $y_j = \sum_i s_{ij} x_i$.

This part of the proof uses Theorem 6.1.15. Let $\pi \in R$ such that $\nu(\pi) = 1$. Let p be a positive integer such that $\pi^p S \in M_n(\hat{R})$. Let $Q = (q_{ij}) \in M_n(K)$ such that $Q - S^{-1} \in M_n(\hat{\mathfrak{m}}^{p+1})$. Then $(\pi^p S)(Q - S^{-1}) \in M_n(\hat{\mathfrak{m}}^{p+1})$, which implies $S(Q-S^{-1}) = \pi^{-p}(\pi^p S)(Q-S^{-1}) \in M_n(\hat{\mathfrak{m}})$. By Exercise 3.1.9 and Nakayama's Lemma (Theorem 3.1.2), SQ is an invertible matrix in $M_n(\hat{R})$. Define z_k by the following.

$$\begin{aligned} z_k &= \sum_i q_{ik} y_i \\ &= \sum_i q_{ik} \Big(\sum_j s_{ji} x_j \Big) \\ &= \sum_i \sum_j q_{ik} s_{ji} x_j \\ &= \sum_j \Big(\sum_i s_{ji} q_{ik} \Big) x_j . \end{aligned}$$

Since SQ is a unit in $M_n(\hat{R})$, $Z = \{z_1, \ldots, z_n\}$ is a $\hat{R}$-basis for T in $\hat{V}$, and $T = \sum \hat{R} x_i = \sum \hat{R} z_i$. Since $Q \in M_n(K)$ and $Y \subseteq V$, we have $Z \subseteq V$. Therefore, Z is a K-basis for V and $T \cap V = \sum (\hat{R} \cap K) z_i = \sum R z_i$ is an R-lattice in V.

Step 4: $\hat{R}(T \cap V) = \sum \hat{R} z_i = T$, which shows the correspondence is onto. □

Theorem 11.3.10. *Let R be a DVR with maximal ideal $\mathfrak{m}$ and field of fractions K. Let $\hat{R}$ be the completion of R and $\hat{K}$ the field of fractions of $\hat{R}$. If A is a central simple K-algebra and Λ an R-order in A, then Λ is a maximal order if and only if $\hat{\Lambda} = \hat{R} \otimes_R \Lambda$ is a maximal $\hat{R}$-order in $\hat{A} = \hat{K} \otimes_K A$.*

Proof. By Theorem 6.1.15, $\hat{\Lambda} \to \hat{A}$ is one-to-one. By Proposition 6.3.4, $\hat{\Lambda}$ is an $\hat{R}$-lattice in $\hat{A}$. Consequently, $\hat{\Lambda}$ is an $\hat{R}$-order in $\hat{A}$. Let Δ be an $\hat{R}$-order in $\hat{\Lambda}$ such that $\hat{\Lambda} \subseteq \Delta$. By Theorem 11.3.9, $\Delta \cap A$ is an R-order in A, and $\hat{R}(\Delta \cap A) = \Delta$. Since $\Lambda \subseteq \Delta \cap A$, if Λ is maximal, then $\hat{\Lambda} = \Delta$, so $\hat{\Lambda}$ is maximal.

Let Γ be an R-order in A such that $\Lambda \subseteq \Gamma$. Then $\hat{\Lambda} \subseteq \hat{\Gamma}$ and if we assume $\hat{\Lambda}$ is maximal, then $\hat{\Lambda} = \hat{\Gamma}$. By Theorem 11.3.9, $\Lambda = \hat{\Lambda} \cap A = \hat{\Gamma} \cap A = \Gamma$. □

3.3. When is a Maximal Order an Azumaya Algebra? We derive necessary and sufficient conditions for a maximal order over a noetherian integrally closed integral domain to be an Azumaya algebra. First we show that an Azumaya algebra over a normal domain is always a maximal order.

Theorem 11.3.11. *Let R be an integrally closed integral domain with field of fractions K. Let Λ be an Azumaya R-algebra. Then Λ is a maximal R-order in $K \otimes_R \Lambda$.*

Proof. By Corollary 4.5.4, $A = K \otimes_R \Lambda$ is a central simple K-algebra. Since Λ is an R-progenerator module (Theorem 7.1.4), $\Lambda \to A$ is one-to-one. Identifying Λ with its image in A, Λ is an R-order. Suppose Δ is an R-order in A and $\Lambda \subseteq \Delta$. Then Δ is a two-sided Λ/R-module and we are in the Morita context of Corollary 7.1.6. If $C = \Delta^\Lambda$ denotes the centralizer of Λ in Δ, then $C \otimes_R \Lambda \cong \Delta$ (by the multiplication map). Since $K\Lambda = A$, C centralizes A. Therefore, $C \subseteq Z(A) = K$. But $C \subseteq \Delta$, Δ is integral over R and R is integrally closed in K. This implies $C = R$, and consequently, $\Lambda = \Delta$. □

Lemma 11.3.12. *Let R be a Dedekind domain with field of fractions K. Let B be a central simple K-algebra such that the Brauer class $[B]$ is in the image of the homomorphism $\mathrm{B}(R) \to \mathrm{B}(K)$. If Λ is a maximal R-order in B, then Λ is an Azumaya R-algebra.*

Proof. This proof is from [**Sal99**, Lemma 9.5]. Let A be an Azumaya R-algebra such that $K \otimes_R A$ is Brauer equivalent to B. For some $r \geq 1$, there is a K-algebra isomorphism $K \otimes_R A \cong B \otimes_K M_r(K)$. Then $\Lambda \otimes_R M_r(R)$ embeds as an R-lattice in $B \otimes_K M_r(K)$, and A embeds as an R-lattice in $K \otimes_R A$. Let M be the image of $(\Lambda \otimes_R M_r(R)) \otimes_R A^o \to K \otimes_R A \cong B \otimes_K M_r(K)$. Then M is an R-lattice in $K \otimes_R A \cong B \otimes_K M_r(K)$ (Proposition 6.3.4). Since R is a Dedekind domain, M is an R-progenerator (Theorem 6.4.6). The ring $(\Lambda \otimes_R M_r(R)) \otimes_R A^o$ acts on M from the left by $(\alpha \otimes \beta)(x) = \alpha x \beta$. Using the left regular representation

$$(\Lambda \otimes_R M_r(R)) \otimes_R A^o \to \mathrm{Hom}_R(M, M)$$

we can map Λ, $M_r(R)$ and A^o onto R-subalgebras of $\mathrm{Hom}_R(M, M)$ which commute with each other. The map $\psi : M_r(R) \otimes_R A^o \to \mathrm{Hom}_R(M, M)$ is a homomorphism between Azumaya R-algebras, hence is one-to-one (Theorem 7.6.1). Consider the commutative diagram

$$\begin{array}{ccc}
M_r(R) \otimes_R A^o & \xrightarrow{\psi} & \mathrm{Hom}_R(M, M) \\
\downarrow & & \\
B \otimes_K M_r(K) \otimes_K M_r(B)^o & & \downarrow \\
\cong \downarrow & & \\
M_r(B) \otimes_K M_r(B)^o & \xrightarrow{\varphi} & \mathrm{Hom}_K(M_r(B), M_r(B)).
\end{array} \tag{11.16}$$

The enveloping homomorphism φ is an isomorphism (Proposition 4.5.3). The centralizer of $M_r(K) \otimes_K M_r(B)^o$ in $\mathrm{Hom}_K(M_r(B), M_r(B))$ is B (Theorem 7.2.3), where the individual rings are identified with their images under the isomorphism φ. Let Γ denote the centralizer of $M_r(R) \otimes_R A^o$ in

$\operatorname{Hom}_R(M, M)$. By Theorem 7.2.2, Γ is R-Azumaya and $\Gamma \otimes_R M_r(R) \otimes_R A^o \cong \operatorname{Hom}_R(M, M)$. Since (11.16) commutes, we have the chain of R-subalgebras $\Lambda \subseteq \Gamma \subseteq B$. Since Λ is a maximal order in B, we conclude that $\Lambda = \Gamma$. Therefore, Λ is R-Azumaya. □

Proposition 11.3.13. *Let R be a noetherian integrally closed integral domain with field of fractions K. Let A be a central simple K-algebra such that the Brauer class $[A]$ is in the image of the homomorphism $\mathrm{B}(R) \to \mathrm{B}(K)$. Let Λ be a maximal R-order in A. If Λ is a projective R-module, then Λ is an Azumaya R-algebra.*

Proof. If $\mathfrak{p}$ is a prime ideal of R with height one, then $R_{\mathfrak{p}}$ is a DVR, and in particular, a Dedekind domain. The proposition follows directly from Theorem 11.3.6, Lemma 11.3.12, and Theorem 7.1.12. □

Theorem 11.3.14. *Let R be a noetherian regular integral domain with Krull dimension less than or equal to two. Let K be the field of fractions of R and A a central simple K algebra such that the Brauer class $[A]$ is in the image of the homomorphism $\mathrm{B}(R) \to \mathrm{B}(K)$. If Λ is a maximal R-order in A, then Λ is an Azumaya R-algebra.*

Proof. By Theorem 11.3.8, Λ is a reflexive R-module. By Theorem 6.5.9, Λ is a projective R-module. By Proposition 11.3.13, Λ is an Azumaya R-algebra. □

Remark 11.3.15. Theorem 11.3.14 is due to Auslander and Goldman ([**AG60a**]). In [**AC16**], B. Antieau and K. Chan prove that the converse is true.

Remark 11.3.16. In [**AW14**, Corollary 1.2], B. Antieau and B. Williams construct a noetherian regular integral domain R with Krull dimension six such that there exists a Brauer class in $\mathrm{B}(R)$ defining a central K-division algebra D and such that no maximal R-order for D is R-Azumaya.

3.4. Azumaya Algebras at the Generic Point. In this section R is a noetherian integrally closed integral domain with field of fractions K. We consider Azumaya R-algebras that represent Brauer classes that are in the image of the natural map $\mathrm{B}(R) \to \mathrm{B}(K)$. The next proposition is from [**AG60b**].

Proposition 11.3.17. *Let R be a noetherian integrally closed integral domain with field of fractions K. Let V be a finite dimensional K-vector space and Λ an R-order in $\operatorname{Hom}_K(V, V)$. Then Λ is maximal if and only if there is a reflexive R-lattice M in V such that $\Lambda = \operatorname{Hom}_R(M, M)$.*

Proof. Let M be a reflexive R-lattice in V such that $\Lambda = \operatorname{Hom}_R(M, M)$. By Proposition 6.3.15, $\Lambda^{**} = \operatorname{Hom}_R(M^{**}, M^{**}) = \operatorname{Hom}_R(M, M) = \Lambda$. Therefore, Λ is reflexive. Let $\mathfrak{p} \in X_1(R)$. By Proposition 2.4.10, $\Lambda_{\mathfrak{p}} = \Lambda \otimes_R R_{\mathfrak{p}} = \operatorname{Hom}_R(M, M) \otimes_R R_{\mathfrak{p}} = \operatorname{Hom}_{R_{\mathfrak{p}}}(M_{\mathfrak{p}}, M_{\mathfrak{p}})$. Since $R_{\mathfrak{p}}$ is a DVR, $M_{\mathfrak{p}}$ is a free $R_{\mathfrak{p}}$-module (Proposition 6.3.2). Therefore, $\Lambda_{\mathfrak{p}}$ is an Azumaya $R_{\mathfrak{p}}$-algebra (Proposition 7.1.10) and a maximal $R_{\mathfrak{p}}$-order (Theorem 11.3.11). By Theorem 11.3.8, Λ is a maximal order.

Assume Λ is an R-order in $A = \operatorname{Hom}_K(V, V)$. Let $v \in V - (0)$. Let $\psi : A \to V$ be the "evaluation at v" homomorphism which is defined by $\psi(f) = f(v)$. Then ψ is a left A-module epimorphism. If we set $L = \psi(\Lambda)$, then $\psi : \Lambda \to L$ is a left Λ-module epimorphism. The reader should verify that L is an R-lattice in V. The diagram

$$\begin{array}{ccc} \Lambda & \longrightarrow & \operatorname{Hom}_R(L, L) \\ \downarrow & & \downarrow \\ A & \xrightarrow{=} & \operatorname{Hom}_K(V, V) \end{array} \tag{11.17}$$

commutes, the top horizontal arrow being the left regular representation of Λ in $\operatorname{Hom}_R(L, L)$. All the arrows in (11.17) are one-to-one. By Proposition 6.3.15 we have $\Lambda \subseteq \operatorname{Hom}_R(L, L) \subseteq \operatorname{Hom}_R(L, L)^{**} = \operatorname{Hom}_R(L^{**}, L^{**})$. If we set $M = L^{**}$, then M is a reflexive R-lattice in V and $\operatorname{Hom}_R(M, M)$ is an R-order in $\operatorname{Hom}_K(V, V)$ that contains Λ. Assuming Λ is a maximal order, $\Lambda = \operatorname{Hom}_R(M, M)$. □

Theorem 11.3.18. *If R is a noetherian regular integral domain with field of fractions K, then the natural homomorphism $\mathrm{B}(R) \to \mathrm{B}(K)$ is one-to-one.*

Proof. Let Λ be an Azumaya R-algebra which is split by K. Then $\Lambda \otimes_R K$ is isomorphic to a ring of matrices $M_r(K)$, for some $r \geq 1$. By Theorem 11.3.11, Λ is a maximal order in $M_r(K)$. By Proposition 11.3.17, Λ is isomorphic to $\operatorname{Hom}_R(M, M)$ where M is a reflexive R-lattice in a vector space over K of dimension r. Since Λ is a projective R-module (Theorem 7.1.4), it follows from Theorem 6.5.10 that M is projective. Hence Λ is split (Section 7.4). □

Theorem 11.3.19. *Let R be a noetherian regular integral domain with field of fractions K. Assume*

(1) *the Krull dimension of R is less than or equal to two, or*
(2) *R is a finitely generated algebra over a field k and $R \otimes_k \bar{k}$ is regular, where $\bar{k}$ is an algebraic closure of k.*

Then

$$\mathrm{B}(R) = \bigcap_{\mathfrak{p} \in X_1(R)} \mathrm{B}(R_{\mathfrak{p}})$$

where the intersection takes place inside $\mathrm{B}(K)$.

Proof. When R has dimension one or two this result is due to Auslander and Goldman [**AG60a**, Theorem 7.4]. This is the only part which we prove here. The statement of the theorem which places no restriction on the Krull dimension of R is due to Ray Hoobler. The interested reader is referred to [**Hoo80**, Corollary 2] for the proof.

Assume $\dim(R) \leq 2$. By Theorem 11.3.18, the diagram

$$\begin{array}{ccc} \mathrm{B}(R) & \longrightarrow & \mathrm{B}(K) \\ & \searrow \quad \nearrow & \\ & \mathrm{B}(R_{\mathfrak{p}}) & \end{array}$$

commutes and all of the arrows are one-to-one. Let A be a central simple K-algebra and assume the Brauer class $[A]$ is in $\mathrm{B}(R_{\mathfrak{p}})$ for all $\mathfrak{p} \in X_1(R)$. Let Λ be a maximal R-order in A. By Theorem 11.3.8 and Theorem 6.5.9, Λ is a projective R-module. By Proposition 11.3.13, Λ is an Azumaya R-algebra. $\square$

Remark 11.3.20. Let R be a noetherian integral domain with field of fractions K. Then in general the natural map $\mathrm{B}(R) \to \mathrm{B}(K)$ is not one-to-one. For example, if R is the ring $\mathbb{R}[x, y]/(x^2 + y^2)$ of Exercise 11.1.11, then $\mathrm{B}(K/R)$ contains an element of order two. If R is a noetherian normal local ring, then in [**Gro68b**] Grothendieck showed that the group $\mathrm{B}(K/R)$ is related to the group of torsion in the cokernel of the natural map $\mathrm{Cl}(R) \to \mathrm{Cl}(R^{sh})$. In other words, the Brauer group of R detects the existence of nonprincipal minimal prime ideals in the strict henselization of R which do not come from R. For examples and more results on this subject, the interested reader is referred to [**DFM93b**] and [**BFH13**].

3.5. Azumaya Algebras over a DVR. In this section, R denotes a discrete valuation ring (a DVR) with maximal ideal $\mathfrak{m}$, residue field k, and field of fractions K. Let $\hat{R}$ be the $\mathfrak{m}$-adic completion of R. By Theorem 6.1.15, $\hat{R}$ is a DVR with maximal ideal $\hat{\mathfrak{m}} = \mathfrak{m}\hat{R}$, residue field $\hat{R}/\mathfrak{m}\hat{R} = R/\mathfrak{m} = k$, and field of fractions $\hat{K} = \hat{R} \otimes_R K$.

Proposition 11.3.21. *Let R be a DVR with maximal ideal $\mathfrak{m}$, residue field k, and field of fractions K. Let $\hat{R}$ be the $\mathfrak{m}$-adic completion of R and $\hat{K}$ the field of fractions of $\hat{R}$. If $\mathrm{B}(k) = (0)$, then the sequence of abelian groups*

$$0 \to \mathrm{B}(R) \to \mathrm{B}(K) \to \mathrm{B}(\hat{K})$$

is exact.

Proof. The diagram

$$\begin{array}{ccc} \mathrm{B}(R) & \xrightarrow{\beta} & \mathrm{B}(\hat{R}) \\ {\scriptstyle\alpha}\downarrow & & \downarrow{\scriptstyle\gamma} \\ \mathrm{B}(K) & \xrightarrow{\delta} & \mathrm{B}(\hat{K}) \end{array} \tag{11.18}$$

of abelian groups is commutative. By Theorem 11.3.18, α is one-to-one. Since we are assuming $\mathrm{B}(k) = (0)$, it follows from Theorem 10.3.2 that $\mathrm{B}(\hat{R}) = 0$. Therefore, $\delta\alpha = 0$. To finish the proof, let A be a central simple K-algebra which is split by $\hat{K}$. We show $[A]$ is in the image of α. Let Λ be a maximal R-order in A. Since $A = K \otimes_R \Lambda$, it suffices to show Λ is an Azumaya R-algebra. By Theorem 11.3.10, $\hat{\Lambda} = \hat{R} \otimes_R \Lambda$ is a maximal $\hat{R}$-order in $\hat{A} = \hat{K} \otimes_K A$. By Proposition 6.3.2, Λ is finitely generated and free as an R-module and $\hat{\Lambda}$ is finitely generated and free as an $\hat{R}$-module. Since $\Lambda/\mathfrak{m}\Lambda \cong \hat{\Lambda}/\hat{\mathfrak{m}}\hat{\Lambda}$, by Corollary 8.1.25, $\hat{\Lambda}$ is an Azumaya $\hat{R}$-algebra if and only if Λ is an Azumaya R-algebra. Since $\hat{A}$ is split, by Lemma 11.3.12, we are done. □

Lemma 11.3.22. *Let R be a DVR with maximal ideal $\mathfrak{m}$ and residue field $k = R/\mathfrak{m}$. Let Λ be an Azumaya R-algebra and I a left ideal in Λ. Then I is principal.*

Proof. By Corollary 7.1.8, $\mathrm{J}(\Lambda) = \mathfrak{m}\Lambda$. Let I be a left ideal in Λ. By Theorem 7.1.4, Λ is an R-progenerator module. Since R is a DVR, I and Λ are finitely generated free R-modules and $\mathrm{Rank}_R(I) \le \mathrm{Rank}_R(\Lambda)$. Therefore,

$$\mathrm{Rank}_{R/\mathfrak{m}}(I/\mathfrak{m}I) \le \mathrm{Rank}_{R/\mathfrak{m}}(\Lambda/\mathfrak{m}\Lambda). \tag{11.19}$$

By Corollary 4.5.4, $(R/\mathfrak{m}) \otimes_R \Lambda = \Lambda/\mathfrak{m}\Lambda = \Lambda/\mathrm{J}(\Lambda)$ is a central simple algebra over the field $R/\mathfrak{m} = k$. By Theorem 3.2.5, up to isomorphism $\Lambda/\mathrm{J}(\Lambda)$ has a unique simple left module, say L. By Theorem 3.2.1, for some $t > 0$, $\Lambda/\mathrm{J}(\Lambda)$ is isomorphic to a direct sum of t copies of L and for some $s \ge 0$, $I/\mathfrak{m}I$ is isomorphic to a direct sum of s copies of L. By (11.19), $s \le t$. Therefore, $I/\mathfrak{m}I$ is isomorphic to a module direct summand of $\Lambda/\mathrm{J}(\Lambda)$. By Lemma 1.1.20, $I/\mathfrak{m}I$ is a principal left $\Lambda/\mathrm{J}(\Lambda)$-module. Let $x \in I$ and assume $\Lambda x + \mathrm{J}(\Lambda)I = I$. By Nakayama's Lemma (Theorem 3.1.2), $\Lambda x = I$. □

Proposition 11.3.23. *Let R be a DVR with maximal ideal $\mathfrak{m}$, residue field k, and field of fractions K. Let Λ be an Azumaya R-algebra. If Γ is a maximal R-order in $\Lambda \otimes_R K$, then Λ and Γ are isomorphic under an inner automorphism of $\Lambda \otimes_R K$.*

Proof. Let $\mathfrak{F} = \{x \in A \mid x\Gamma \subseteq \Lambda\}$ be the conductor of Γ to Λ in A. By Lemma 11.3.4, $\mathfrak{F}$ is a left ideal in Λ. By Lemma 11.3.22, $\mathfrak{F} = \Lambda t$ for some

$t \in \Lambda$. By Lemma 11.3.4, $\mathfrak{F} \cap R \neq (0)$. The left ideal generated by t in $A = \Lambda \otimes_R K$ contains k, hence is equal to A. Therefore, t is a unit in A. Now $\mathfrak{F} = \Lambda t$ is a right Γ-module, so $t\Gamma \subseteq \Lambda t$, and it follows that $\Gamma \subseteq t^{-1}\Lambda t$. The reader should verify that $t^{-1}\Lambda t$ is an R-order in A. Since Γ is a maximal order, $\Gamma = t^{-1}\Lambda t$. □

3.6. Locally Trivial Azumaya Algebras. In this section R is an integrally closed integral domain with field of fractions K. We study Azumaya R-algebras that are split by every local ring $R_\mathfrak{p}$ for $\mathfrak{p} \in \operatorname{Spec} R$. Lemmas 11.3.24 and 11.3.25 are due to Bernice Auslander ([**Aus71**]). Additional results along this line are presented in the exercises (Section 11.3.8).

Lemma 11.3.24. *Let R be a noetherian integrally closed integral domain with field of fractions K. Let M be a reflexive R-lattice of rank n such that $\Lambda = \operatorname{Hom}_R(M, M)$ is projective. Then the following are true.*

(1) *Λ is an Azumaya R-algebra.*
(2) *Λ is split if and only if there is an R-progenerator P and a reflexive fractional ideal I of R in K such that $M \cong I \otimes_R P$.*
(3) *In (2), if R is a local ring, then there is a reflexive fractional ideal I of R in K such that M is isomorphic to the direct sum of n copies of I.*

Proof. Part (1) follows from Propositions 11.3.17 and 11.3.13.

(2): Assume $\Lambda = \operatorname{Hom}_R(M, M) \cong \operatorname{Hom}_R(P, P)$ for some R-progenerator module P. Under the Morita correspondence $\mathfrak{M}_R \sim {}_\Lambda\mathfrak{M}$, M is isomorphic to the tensor product $\operatorname{Hom}_\Lambda(P, M) \otimes_R P$ (Exercise 1.5.6). Moreover, one can check that the isomorphism $\operatorname{Hom}_\Lambda(P, M) \otimes_R P \to M$ is given by $f \otimes x \mapsto f(x)$. If we set $A = \Lambda \otimes_R K = \operatorname{Hom}_K(KM, KM) = \operatorname{Hom}_K(KP, KP)$, then the diagram

$$
\begin{array}{ccc}
\operatorname{Hom}_\Lambda(P, M) & \longrightarrow & \operatorname{Hom}_R(P, M) \\
\downarrow & & \downarrow \\
\operatorname{Hom}_A(KP, KM) & \longrightarrow & \operatorname{Hom}_K(KP, KM)
\end{array}
\tag{11.20}
$$

commutes and all of the arrows are one-to-one. By Proposition 6.3.9, we know that $\operatorname{Hom}_R(P, M)$ is a reflexive R-lattice in $\operatorname{Hom}_K(KP, KM)$. By Exercise 6.3.20, $\operatorname{Hom}_\Lambda(P, M) = \operatorname{Hom}_R(P, M) \cap \operatorname{Hom}_A(KP, KM)$ is a reflexive R-lattice in $\operatorname{Hom}_A(KP, KM)$. But $\operatorname{Hom}_A(KP, KM)$ is a K-vector space of dimension one. Therefore, $\operatorname{Hom}_\Lambda(P, M)$ is a rank one reflexive R-lattice, and is isomorphic to a reflexive fractional ideal (Lemma 6.4.4).

Conversely, assume P is an R-progenerator of rank n and I is a reflexive fractional ideal of R in K. For each $\mathfrak{p} \in \operatorname{Spec} R$, $(I \otimes_R P)_\mathfrak{p}$ is isomorphic to a direct sum of n copies of $I_{\mathfrak{p}}$. Therefore, $\operatorname{Hom}_R(I \otimes_R P, I \otimes_R P) \otimes_R R_{\mathfrak{p}}$ is isomorphic to a direct sum of n^2 copies of $\operatorname{Hom}_{R_{\mathfrak{p}}}(I_{\mathfrak{p}}, I_{\mathfrak{p}})$. By Exercise 6.4.12, $\operatorname{Hom}_R(I, I) \cong I : I = R$. Therefore, $\operatorname{Hom}_R(I \otimes_R P, I \otimes_R P)$ is locally free of rank n^2, hence projective. By Exercise 1.3.34, $I \otimes_R P$ is reflexive. By Part (1), $\operatorname{Hom}_R(I \otimes_R P, I \otimes_R P)$ is R-Azumaya. The assignment $\alpha \mapsto 1 \otimes \alpha$ defines a homomorphism of Azumaya R-algebras $\operatorname{Hom}_R(P, P) \to \operatorname{Hom}_R(I \otimes_R P, I \otimes_R P)$ which is an isomorphism, by Theorem 7.6.1.

(3): If R is local, then the projective module P in (2) is free of rank n (Proposition 2.3.2). □

Lemma 11.3.25. *Let R be a noetherian integrally closed integral domain with field of fractions K. Let Λ be an Azumaya R-algebra which represents a Brauer class in the kernel of the natural homomorphism*

$$\mathrm{B}(R) \to \prod_{\mathfrak{p} \in \operatorname{Spec} R} \mathrm{B}(R_{\mathfrak{p}}).$$

Then there is a reflexive R-lattice M of rank n satisfying the following.

(1) $\Lambda \cong \operatorname{Hom}_R(M, M)$.

(2) *For each $\mathfrak{p} \in \operatorname{Spec} R$, there is a reflexive fractional ideal $I_{\mathfrak{p}}$ of $R_{\mathfrak{p}}$ in K such that $M_{\mathfrak{p}}$ is isomorphic to the direct sum of n copies of $I_{\mathfrak{p}}$.*

Proof. By Theorem 11.3.11, Λ is a maximal R-order in $\Lambda \otimes_R K$. Since (0) is a prime ideal in R, Λ is split by K. By Proposition 11.3.17, $\Lambda \cong \operatorname{Hom}_R(M, M)$ for some reflexive R-lattice M of rank $n = \dim_K(KM)$. This proves (1). Part (2) is a special case of Lemma 11.3.24. □

3.7. An Example of Ojanguren. The purpose of this section is to present an example due to M. Ojanguren [**Oja74**] of a noetherian integrally closed integral domain R with the property that in Lemma 11.3.25, the kernel of the natural homomorphism $\mathrm{B}(R) \to \prod_{\mathfrak{p} \in \operatorname{Spec} R} \mathrm{B}(R_{\mathfrak{p}})$ contains an element of order three. In other words, R has a nontrivial locally trivial Azumaya algebra.

Let k be a field with characteristic not equal to 3. Start with the three linear polynomials $\ell_1 = y$, $\ell_2 = y - x$, $\ell_3 = x + 1$ in the polynomial ring $k[x, y]$ and set

$$R = \frac{k[x, y, z]}{(z^3 - \ell_1 \ell_2 \ell_3)}.$$

The ring R was the subject of Example 6.2.5 and we already know for instance that R is a noetherian normal integral domain with Krull dimension 2. We leave it to the interested reader to verify that R is the affine coordinate ring of an algebraic surface with three rational singularities each of type A_2.

In K, the quotient field of R, the relation $z^3 = \ell_1\ell_2\ell_3$ is satisfied. It follows that

$$\begin{aligned}\mathfrak{p}_1 &= (\ell_1, z)\\ \mathfrak{p}_2 &= (\ell_2, z)\\ \mathfrak{p}_3 &= (\ell_3, z)\end{aligned} \tag{11.21}$$

are the only minimal primes containing z. In Example 6.2.5 we showed that $\mathrm{Cl}(R)$ is isomorphic to $\mathbb{Z}/3 \oplus \mathbb{Z}/3$ and is generated by $\mathfrak{p}_1$ and $\mathfrak{p}_2$. We also proved that $R^* = k^*$. One can compute the principal divisors

$$\begin{aligned}\mathrm{Div}(z) &= \mathfrak{p}_1 + \mathfrak{p}_2 + \mathfrak{p}_3\\ \mathrm{Div}(\ell_1) &= 3\mathfrak{p}_1\\ \mathrm{Div}(\ell_2) &= 3\mathfrak{p}_2\\ \mathrm{Div}(\ell_3) &= 3\mathfrak{p}_3.\end{aligned} \tag{11.22}$$

Nagata's sequence (Theorem 6.2.4) becomes

$$1 \to R^* \to R[z^{-1}]^* \xrightarrow{\mathrm{Div}} \bigoplus_{i=1}^{3} \mathbb{Z}\mathfrak{p}_i \to \mathrm{Cl}(R) \to \mathrm{Cl}(R[z^{-1}]) \to 0. \tag{11.23}$$

Therefore, $\mathrm{Cl}(R[z^{-1}]) = 0$ and the image of Div is a free $\mathbb{Z}$-module of rank 3. From (11.22) we conclude that $\mathrm{Cl}(R)$ is generated by the divisor classes of any two of $\mathfrak{p}_1$, $\mathfrak{p}_2$, $\mathfrak{p}_3$. We also conclude that

$$R[z^{-1}]^* = k^* \times \langle z\rangle \times \langle \ell_1\rangle \times \langle \ell_2\rangle \tag{11.24}$$

is an internal direct product of abelian groups and $\ell_3 = z^3\ell_1^{-1}\ell_2^{-1}$. Notice that $R_z = R[z^{-1}] = R[(\ell_1\ell_2\ell_3)^{-1}]$. Since ℓ_3 belongs to only one minimal prime, the Nagata sequence yields

$$1 \to R^* \to R^*_{\ell_3} \xrightarrow{\mathrm{Div}} \mathbb{Z}\mathfrak{p}_3 \to \mathrm{Cl}(R) \to \mathrm{Cl}(R_{\ell_3}) \to 0. \tag{11.25}$$

We know from above that $\mathfrak{p}_3$ generates a direct summand of $\mathrm{Cl}(R)$. We conclude that $\mathrm{Cl}(R_{\ell_3}) \cong \mathbb{Z}/3$ is generated by the divisor class of $\mathfrak{p}_1$ or $\mathfrak{p}_2$. If we set $S = R_{\ell_1\ell_3} \oplus R_{\ell_2\ell_3}$, then S is a flat extension of R_{ℓ_3}. Notice that the maximal ideal (ℓ_1, ℓ_2, z) generates the unit ideal of S. Therefore, $\operatorname{Spec} S \to \operatorname{Spec} R_{\ell_3}$ is not onto and S is not faithfully flat over R_{ℓ_3}. Nevertheless, the image of $\operatorname{Spec} S$ does contain every prime of height one in R_{ℓ_3}. Therefore, we employ the method of Example 6.3.17 to show how the reflexive ideal $\mathfrak{p}_1$ corresponds to a 1-cocycle in $\mathrm{Z}^1(S/R_{\ell_3}, \mathrm{GL}_1)$. The cochain complex is obtained by applying the units functor $\mathrm{GL}_1(\cdot)$ to the Amitsur complex $C^\bullet(S/R_{\ell_3})$. The key portion is

$$R^*_{\ell_1\ell_3} \times R^*_{\ell_2\ell_3} \xrightarrow{d^0} R^*_z \xrightarrow{d^1} \langle 1\rangle \tag{11.26}$$

where $d^0(\alpha, \beta) = \alpha\beta^{-1}$. Using (11.24) we see that the homology of (11.26) is cyclic of order 3 and is generated by the 1-cocycle $\{z\}$. Following Example 6.3.17, if we set

$$N = R_{\ell_1\ell_3} \cap R_{\ell_2\ell_3} z,$$

then N is a reflexive R_{ℓ_3}-lattice in K such that $N \otimes R_{\ell_1\ell_3} = R_{\ell_1\ell_3}$ and $N \otimes R_{\ell_2\ell_3} = R_{\ell_2\ell_3} z$. The ideal $\mathfrak{p}_1 = (\ell_1, z)$ has these properties, hence $N = \mathfrak{p}_1$ (Corollary 6.3.14).

Proposition 11.3.26. *In the above context, there exists a reflexive R-lattice M satisfying the following:*

(1) *M has rank 3.*

(2) *$\Lambda = \operatorname{Hom}_R(M, M)$ is an Azumaya R-algebra.*

(3) *The Brauer class $[\Lambda]$ has order 3 in $\mathrm{B}(R)$.*

(4) *Λ is locally trivial. That is, $\Lambda \otimes R_{\mathfrak{p}}$ is split, for every $\mathfrak{p} \in \operatorname{Spec} R$.*

Proof. The proof we give is based on [**For95**, Example 2.1]. In the notation of Theorem 6.3.16, let $f_1 = \ell_2\ell_3$, $f_2 = \ell_1\ell_3$, $f_3 = \ell_1\ell_2$. One can check that the only primes containing $\{f_1, f_2, f_3\}$ are the three maximal ideals $\mathfrak{m}_1 = (\ell_2, \ell_3, z)$, $\mathfrak{m}_2 = (\ell_1, \ell_3, z)$, and $\mathfrak{m}_3 = (\ell_1, \ell_2, z)$. Set $U_1 = U(f_1) = \operatorname{Spec} R_{f_1}$, $U_2 = U(f_2) = \operatorname{Spec} R_{f_2}$, $U_3 = U(f_3) = \operatorname{Spec} R_{f_3}$. Then $U_1 \cup U_2 \cup U_3 = \operatorname{Spec} R - \{\mathfrak{m}_1, \mathfrak{m}_2, \mathfrak{m}_3\}$. Notice that $U_1 \cap U_2 = U_1 \cap U_3 = U_2 \cap U_3 = U(\ell_1\ell_2\ell_3) = U(z) = \operatorname{Spec} R_z$. Let $S = R_{f_1} \oplus R_{f_2} \oplus R_{f_3}$. Then we are in the context of Example 6.3.17. To construct M, it suffices to find a 1-cocycle $\{\phi_{12}, \phi_{32}, \phi_{13}\}$ in $\mathrm{Z}^1(S/R, \mathrm{GL}_3)$. Let

$$\phi_{12} = \begin{bmatrix} z & 0 & 0 \\ l_1 - z & z & 0 \\ l_1 & -l_1 & z \end{bmatrix} = D \begin{bmatrix} z & 0 & 0 \\ l_1 - z & z & 0 \\ 1 & -1 & z \end{bmatrix} D^{-1},$$

$$\phi_{32} = \begin{bmatrix} z & z & l_3 \\ l_1 - z & l_1 & 0 \\ l_1 & 0 & 0 \end{bmatrix} = D \begin{bmatrix} z & z & l_3 \\ l_1 - z & l_1 & 0 \\ 1 & 0 & 0 \end{bmatrix},$$

$$\phi_{13} = \begin{bmatrix} 1 & -1 & z/l_1 \\ 0 & 1 & z(z/l_1 - 1)/l_1 \\ 0 & 0 & -l_2/l_1 \end{bmatrix} = \begin{bmatrix} 1 & -1 & z \\ 0 & 1 & z(z/l_1 - 1) \\ 0 & 0 & -l_2 \end{bmatrix} D^{-1},$$

where

$$D = \begin{bmatrix} 1 & 0 & 0 \\ 0 & 1 & 0 \\ 0 & 0 & l_1 \end{bmatrix}.$$

It is easy to check that each ϕ_{ij} is in $\mathrm{GL}_3(R_z)$ and the 1-cocycle identity $\phi_{12} = \phi_{32}\phi_{13}$ is satisfied. As in Example 6.3.17, associated to the 1-cocycle $\phi = \{\phi_{ij}\}$ is a reflexive R-lattice M in K^3. It is easy to see that ϕ_{32} is in

$\mathrm{GL}_3(R_{f_2})$. Using the argument of the paragraph preceding this proposition, this means $M \otimes_R R_{\ell_1}$ is a free R_{ℓ_1}-module. Likewise, we see that ϕ_{13} is in $\mathrm{GL}_3(R_{f_3})$ and for the same reason, $M \otimes_R R_{\ell_2}$ is a free R_{ℓ_2}-module. Now we prove that $M \otimes_R R_{\ell_3}$ is the direct sum of three copies of the ideal $\mathfrak{p}_1$. By the computation in the paragraph preceding this proposition, the 3-by-3 scalar matrix $\psi = z$ defines a 1-cocycle for the extension $R_{f_1} \oplus R_{f_2}$ of R_{f_3} which corresponds to the reflexive lattice $\mathfrak{p}_1 \oplus \mathfrak{p}_1 \oplus \mathfrak{p}_1$. Now

$$\psi^{-1}\phi_{12} = \begin{bmatrix} 1 & 0 & 0 \\ \ell_1/z - 1 & 1 & 0 \\ \ell_1/z & -\ell_1/z & 1 \end{bmatrix} = \begin{bmatrix} 1 & 0 & 0 \\ \frac{z^2}{f_1} - 1 & 1 & 0 \\ \frac{z^2}{f_1} & -\frac{z^2}{f_1} & 1 \end{bmatrix}.$$

Thus $\psi^{-1}\phi_{12}$ is in $\mathrm{GL}_3(R_{f_1})$. This means $\psi^{-1}\phi_{12}$ is a 1-coboundary, hence ϕ_{12} is cohomologous to ψ. Therefore, $M \otimes_R R_{\ell_3} \cong \mathfrak{p}_1 \oplus \mathfrak{p}_1 \oplus \mathfrak{p}_1$. We have constructed the reflexive R-lattice M. Now we verify that M satisfies properties (1) – (4).

(1): This is clear from Theorem 6.3.16.

(2): By Lemma 11.3.24 (2), we have $\Lambda \otimes R_{\ell_i} = \mathrm{Hom}_{R_{\ell_i}}(M \otimes R_{\ell_i}, M \otimes R_{\ell_i})$ is a split Azumaya R_{ℓ_i}-algebra. Since $R_{\ell_1} \oplus R_{\ell_2} \oplus R_{\ell_3}$ is faithfully flat, by Corollary 8.1.21 we see that Λ is an Azumaya R-algebra.

(4): This follows from (2) since $R_{\ell_1} \oplus R_{\ell_2} \oplus R_{\ell_3}$ is faithfully flat.

(3): Consider the homomorphism of abelian groups

$$\mathrm{Cl}(R) \xrightarrow{\theta} \bigoplus_{i=1}^{3} \mathrm{Cl}(R_{\ell_i}).$$

As in (11.25) we know $\mathrm{Cl}(R_{\ell_i}) \cong \mathbb{Z}/3$ for each i. Since $\mathrm{Cl}(R)$ is generated by $\mathfrak{p}_1, \mathfrak{p}_2, \mathfrak{p}_3$, the image of θ is generated by the three vectors

$$\begin{bmatrix} 1 \\ 1 \\ 0 \end{bmatrix}, \begin{bmatrix} 1 \\ 0 \\ 1 \end{bmatrix}, \begin{bmatrix} 0 \\ 1 \\ 1 \end{bmatrix}.$$

There is no rank one reflexive R-module I such that $I \otimes R_{\ell_1}$ is R_{ℓ_1}-projective, $I \otimes R_{\ell_2}$ is R_{ℓ_2}-projective, and $I \otimes R_{\ell_1} \cong \mathfrak{p}_1$. By Lemma 11.3.24 (2), Λ is not split. The fact that $[\Lambda]$ has order three follows from the exact sequence of Exercise 11.3.28. □

3.8. Exercises. In Exercises 11.3.27 and 11.3.28, we outline a series of steps that lead to two exact sequences that were originally derived by B. Auslander in [**Aus66**] and [**Aus71**] (see also [**Tre78**], [**LO82**], [**CGO75**], [**Orz82**]).

Exercise 11.3.27. Let R be a noetherian integrally closed integral domain with field of fractions K. Let $\operatorname{BReflex}(R)$ denote the set of all finitely generated reflexive R-modules M such that $\operatorname{Hom}_R(M, M)$ is projective.

(1) Define a relation on $\operatorname{BReflex}(R)$ by specifying that two modules M and N are related if there exist R-progenerator modules P and Q such that $(M \otimes_R P)^{**} \cong (N \otimes Q)^{**}$. Show that this defines an equivalence relation. Denote by $\operatorname{BCl}(R)$ the set of equivalence classes $[M]$.

(2) Show that there is a binary operation on $\operatorname{BCl}(R)$ defined by tensoring and taking the double dual, that is, $[M][N] = [(M \otimes_R N)^{**}]$. Show that this which makes $\operatorname{BCl}(R)$ into an abelian group where the group identity is the class $[R]$ and the inverse class of $[M]$ is $[M^*]$.

(3) Show that the assignment $[M] \mapsto \operatorname{Hom}_R(M, M)$ induces a homomorphism of groups $\operatorname{BCl}(R) \to \operatorname{B}(R)$.

(4) Show that there is an exact sequence of abelian groups

$$0 \to \operatorname{Cl}(R)/\operatorname{Pic}(R) \to \operatorname{BCl}(R) \to \operatorname{B}(R) \to \operatorname{B}(K)$$

describing the subgroup of the Brauer group of R consisting of classes of Azumaya algebras that are split by K.

Exercise 11.3.28. Let R be a noetherian integrally closed integral domain with field of fractions K. Let $\operatorname{BCl}(\mathbf{P}, \mathbf{I}_R)$ be the subgroup of $\operatorname{BCl}(T)$ consisting of classes of reflexive R-modules M with the additional property that for each $\mathfrak{p} \in \operatorname{Spec} R$ there is a reflexive fractional ideal $I_{\mathfrak{p}}$ in $R_{\mathfrak{p}}$ such that $M \otimes_R R_{\mathfrak{p}} \cong F \otimes_{R_{\mathfrak{p}}} I_{\mathfrak{p}}$, where F is a free $R_{\mathfrak{p}}$-module. Show that there is an exact sequence of abelian groups

$$0 \to \operatorname{Cl}(R)/\operatorname{Pic}(R) \to \operatorname{BCl}(\mathbf{P}, \mathbf{I}_R) \to \operatorname{B}(R) \to \prod_{\mathfrak{p} \in \operatorname{Spec} R} \operatorname{B}(R_{\mathfrak{p}})$$

describing the subgroup of the Brauer group of R consisting of classes of Azumaya algebras that are split by the local ring at every prime ideal of R.

Exercise 11.3.29. Let $n > 1$ and assume k is a field in which n is invertible. As in the example of Section 11.3.7, set

$$R = \frac{k[x, y, z]}{(z^n - \ell_1 \ell_2 \ell_3)}$$

where $\ell_1 = y$, $\ell_2 = y - x$, $\ell_3 = x + 1$. For a few low values of n (for example, $2, 3, 4, 5$), show that there exists a reflexive R-lattice M satisfying:

(1) M has rank n.

(2) $\Lambda = \operatorname{Hom}_R(M, M)$ is an Azumaya R-algebra.

(3) Λ is locally trivial. That is, $\Lambda \otimes_R R_{\mathfrak{p}}$ is split, for all $\mathfrak{p} \in \operatorname{Spec} R$.

Use the method of Proposition 11.3.26, where the $n = 3$ case is proved. As a challenge, try to find a general construction that works for all $n > 1$. (As this book is being written, I have gone up to $n = 8$ and I am unaware of a general solution. T.F.)

Exercise 11.3.30. Let k be a field such that $\operatorname{char} k \neq 2$. In the notation of Exercise 11.3.29, let $R = k[x, y, z](z^2 - \ell_1\ell_2\ell_3)$. Then R is isomorphic to the ring of Exercises 6.2.12 and 10.1.20. Prove that the natural map

$$\mathrm{B}(R) \to \prod_{\mathfrak{p}\in\operatorname{Spec} R} \mathrm{B}(R_{\mathfrak{p}})$$

is one-to-one.

Exercise 11.3.31. This exercise is based on [**KO74b**, Remarque III.6.5]. Let $R = \mathbb{Z}[\sqrt{2}]$ and $F = \mathbb{Q}[\sqrt{2}]$, where we view both rings as subrings of $\mathbb{R}$.

(1) Show that R is a PID. (Hint: See for example, [**DF04**, Exercise 8.1.9].)
(2) Let $\Lambda = F \oplus Fi \oplus Fj \oplus Fij$ be the ring of quaternions over F (see Example 1.1.1). Show that Λ is an F-central division algebra.
(3) In Λ, let

$$\alpha = \frac{1+i}{\sqrt{2}}, \quad \beta = \frac{1+j}{\sqrt{2}}$$

and let A be the R-submodule of Λ defined by $A = R + R\alpha + R\beta + R\alpha\beta$. Show that A is a maximal R-order in Λ, hence is an Azumaya R-algebra and defines a nontrivial element in $\mathrm{B}(R)$. Using Theorem 14.3.8 below, one can verify that $\mathrm{B}(R)$ is a cyclic group of order two.
(4) It follows from Algebraic Number Theory that R is separably closed (for a proof, see [**KO74b**, Corollaire III.4.11]). In other words, a commutative separable R-algebra which is an R-progenerator is isomorphic to the trivial extension R^n, for some n. Conclude that any faithfully flat étale splitting ring S of A (as in Theorem 10.3.9 (3)) is not a finitely generated R-module.

4. Brauer Groups in Characteristic p

Most of the material in this section comes from [**KOS76**] and [**Sal76**]. Throughout this section p is a prime number and R is a commutative ring of characteristic p. If S is a commutative faithful R-algebra, then we say S is *purely inseparable* over R if for every $x \in S$ there exists $e \geq 0$ such that $x^{p^e} \in R$. Let $q = p^e$ be a power of p. The qth power map $x \mapsto x^q$

defines a homomorphism $R \to R$. By R^q we denote the image of this map. If $a_1, \dots, a_n$ are elements of R, then the factor ring

$$R[x_1, \dots, x_n]/(x_1^q - a_1, \dots, x_n^q - a_n)$$

is denoted by $R[a_1^{1/q}, \dots, a_n^{1/q}]$. If $S = R[a_1^{1/q}, \dots, a_n^{1/q}]$, then by Exercise 11.4.14, $S^q \subseteq R$, hence S is purely inseparable over R. In the literature, an extension of the form $S = R[a_1^{1/p^{e_1}}, \dots, a_n^{1/p^{e_n}}]$ is sometimes said to be a *Galois purely inseparable extension* (for example, see [**Sal76**], [**Yua74a**]).

4.1. The Brauer Group is p-divisible.

Theorem 11.4.1. *Let R be a commutative ring of characteristic p and $q = p^a$, for some $a \geq 1$. Let $\varphi : R \to R$ be the homomorphism defined by $\varphi(x) = x^q$. Then the induced map $\varphi^* : \mathrm{B}(R) \to \mathrm{B}(R)$ is the qth power map defined by $\varphi^*(\alpha) = \alpha^q$.*

Proof. If S is a faithfully flat étale R-algebra, then S also has characteristic p. Then $x \mapsto x^q$ defines a ring homomorphism $\varphi : S \to S$ and the diagram

$$\begin{array}{ccc} S & \xrightarrow{\varphi} & S \\ \uparrow & & \uparrow \\ R & \xrightarrow{\varphi} & R \end{array}$$

commutes. By Theorem 5.5.5, $\varphi^* : \mathrm{H}^2(S/R, \mathbb{G}_m) \to \mathrm{H}^2(S/R, \mathbb{G}_m)$ maps a 2-cocycle $u \in S^{\otimes 3}$ to u^q. This proves φ^* is the qth power map. The rest follows from Theorem 10.4.11. □

Corollary 11.4.2. *Let R be a commutative ring of characteristic p and let q be a power of p. Let S be a faithful commutative R-algebra such that $S^q = R$. That is, for every $s \in S$, $s^q \in R$. Then the relative Brauer group $\mathrm{B}(S/R)$ is equal to the subgroup of $\mathrm{B}(R)$ annihilated by q.*

Proof. Let $\theta : R \to S$ be the structure homomorphism, which is one-to-one by assumption. We identify R with the image of θ. Let $\varphi : S \to S$ be the qth power homomorphism, $\varphi(x) = x^q$. Let I be the kernel of φ. Then $\bar{\varphi} : S/I \to R$ is an isomorphism. The diagram

$$\begin{array}{ccc} \mathrm{B}(R) & \xrightarrow{\varphi^*} & \mathrm{B}(R) \\ {\scriptstyle\theta^*}\downarrow & & \uparrow{\scriptstyle\bar{\varphi}^*} \\ \mathrm{B}(S) & \xrightarrow{\eta^*} & \mathrm{B}(S/I) \end{array}$$

commutes. Since I is contained in the nil radical of S, $\eta^* : \mathrm{B}(S) \to \mathrm{B}(S/I)$ is an isomorphism (Corollary 7.9.3). The kernel of θ^* is equal to the kernel of φ^*, which by Theorem 11.4.1 is the subgroup annihilated by q. □

When R is a field, Theorem 11.4.3 is due to G. Hochschild [**Hoc55**, Theorem 5].

Theorem 11.4.3. *Let R be a commutative ring of characteristic p. Let S be a commutative faithful R-algebra such that for every $y \in S$, there exists $q = p^a$, where $a \geq 0$, such that $y^q \in R$. Then the natural map $\mathrm{B}(R) \to \mathrm{B}(S)$ is onto.*

Proof. Let Λ be an Azumaya S-algebra. By Exercise 7.7.10 there is a finitely generated R-subalgebra $S_0 \subseteq S$ and an Azumaya S_0-algebra $\Lambda_0 \subseteq \Lambda$ such that $\Lambda_0 \otimes_{S_0} S \cong \Lambda$. Suppose $\{v_1, \dots, v_n\} \subseteq S_0$ is a set of generators for S_0 as an R-algebra. For some $a > 0$, if $q = p^a$, then $v_i^q \in R$. Let $a_i = v_i^q$. Under the Frobenius homomorphism $S_0 \to S_0$ defined by $x \mapsto x^q$, the image of S_0 is contained in $R[a_1, \dots, a_n] = R$. Let $x_1, \dots, x_n$ be indeterminates and define $f : R[x_1, \dots, x_n] \to S_0$ by mapping $x_i \mapsto v_i$. Then f is onto, and $f(x^q - a_i) = 0$ for each i. So f factors through the natural map $\eta : R[x_1, \dots, x_n] \to R[x_1, \dots, x_n]/(x_1^q - a_1, \dots, x_n^q - a_n)$ and the diagram

$$\begin{array}{ccc} R[x_1, \dots, x_n] & \xrightarrow{\quad\theta\quad} & S_0 \\ & {\scriptstyle\eta}\searrow \qquad \nearrow{\scriptstyle\bar{\theta}} & \\ & S_1 = R[a_1^{1/q}, \dots, a_n^{1/q}] & \end{array}$$

commutes. Let I denote the kernel of $\bar{\theta} : S_1 \to S_0$. Under the homomorphism $x \mapsto x^q$, the image of $S_1 \to S_1$ is contained in $R[a_1, \dots, a_n] = R$ (Exercise 11.4.14). Because $\bar{\theta}$ is an R-algebra homomorphism, $I^q = 0$. By Theorem 7.9.2, there is an Azumaya S_1-algebra, say Λ_1, such that $\Lambda_1 \otimes_{S_1} S_0 \cong \Lambda_0$. To prove the theorem, it suffices to show $\mathrm{B}(R) \to \mathrm{B}(S_1)$ is onto. We can view S_1 as the nth term of a sequence of radical extensions, each of the form $R_i = R_{i-1}[x]/(x^q - a_i)$. Therefore, it suffices to show that $\mathrm{B}(R) \to B(C)$ is onto, if C is a purely inseparable radical extension of the form $C = R[a^{1/q}]$, for some $a \in R$. As in Section 8.2, let A be an Azumaya C-algebra and $\Delta(A, D, u)$ any differential crossed product algebra (Definition 8.2.2). Then $\Delta(A, D, u)$ is an Azumaya R-algebra, a finitely generated free left A-module, and the centralizer of C in $\Delta(A, D, u)$ is A. Since A is a C-Azumaya algebra, A is projective as a left C-module. By Theorem 7.4.4, $A = \Delta(A, D, u)^C$ is Brauer equivalent to $\Delta(A, D, u) \otimes_R C$. Since A was an arbitrary C-Azumaya algebra, this shows $\mathrm{B}(R) \to \mathrm{B}(C)$ is onto. $\square$

Corollary 11.4.4. *If R is a commutative ring of characteristic p, then $\mathrm{B}(R)$ is p-divisible.*

Proof. Let $\theta : R \to R$ be the Frobenius homomorphism, $\theta(x) = x^p$. If R^p denotes the image of θ, then θ factors

$$\begin{array}{ccc} R & \xrightarrow{\theta} & R \\ & {\scriptstyle\bar\theta}\searrow \quad \nearrow{\scriptstyle\iota} & \\ & R^p & \end{array}$$

where ι is the set inclusion homomorphism. The diagram of abelian groups

$$\begin{array}{ccc} \mathrm{B}(R) & \xrightarrow{\theta^*} & \mathrm{B}(R) \\ & {\scriptstyle\bar\theta^*}\searrow \quad \nearrow{\scriptstyle\iota^*} & \\ & \mathrm{B}(R^p) & \end{array}$$

commutes. By Theorem 11.4.1, θ^* is the pth power map. By Theorem 11.4.3, $\iota^* : \mathrm{B}(R^p) \to \mathrm{B}(R)$ is onto. The kernel of θ is contained in the nil radical of R, so $\bar\theta^*$ is an isomorphism, by Corollary 7.9.3. This shows θ^* is onto. □

Corollary 11.4.5. *Let k be a perfect field of characteristic p. The pth power map $\mathrm{B}(k) \to \mathrm{B}(k)$ is an isomorphism. The Brauer group of k contains no element of order p.*

Proof. The Frobenius homomorphism $\theta : k \to k$ is an automorphism, hence induces an automorphism of abelian groups $\theta^* : \mathrm{B}(k) \to \mathrm{B}(k)$. □

4.2. Generators for the Subgroup Annihilated by p. If A is an Azumaya R-algebra, the *order of A* is the order of $[A]$ in the Brauer group $\mathrm{B}(R)$. In the literature, this is also called the *exponent*, or *period* of A.

Proposition 11.4.6. *Let R be a commutative ring of characteristic p. Let A be an Azumaya R-algebra of order $q = p^a$. Then there are $a_1, \dots, a_n \in R$ such that A is split by $R[a_1^{1/q}, \dots, a_n^{1/q}]$.*

Proof. Let $\{a_i \mid i \in I\}$ be a set of generators for R as an algebra over the subring R^q, where I is some index set. Let $S = \varinjlim R[a_{i_1}^{1/q}, \dots, a_{i_{n_j}}^{1/q}]$, where $\{i_1, \dots, i_{n_j}\}$ runs through the directed system of all finite subsets of I. Given $y \in R$, there is a polynomial $g \in R^q[x_1, \dots, x_m]$, for some m and $y = g(a_{i_1}, \dots, a_{i_m})$, for some finite subset $\{i_1, \dots, i_m\}$ of I. Viewing y as an element of S, each coefficient of g, and each of $a_{i_1}, \dots, a_{i_m}$ is a qth power in S. Therefore, y is a qth power in S. This shows $R \subseteq S^q$. Conversely, qth power of $R[a_{i_1}^{1/q}, \dots, a_{i_{n_j}}^{1/q}]$ is in R (Exercise 11.4.14), hence $R \supseteq S^q$.

By Corollary 11.4.2, the kernel of the natural map $\mathrm{B}(R) \to \mathrm{B}(S)$ is equal to the subgroup annihilated by q. Therefore, A is split by S. By Proposition 7.7.7, there is a finitely generated R-subalgebra $S_0 \subseteq S$ such that A is split by S_0. There is a finite subset $\{i_1, \dots, i_{n_j}\} \subseteq I$ such that the generators of S_0 come from $S_1 = R[a_{i_1}^{1/q}, \dots, a_{i_{n_j}}^{1/q}]$. Then S_1 splits A, and is of the desired form. □

Lemma 11.4.7. *Let R be a commutative ring of characteristic p and A an Azumaya R-algebra. Let q be a power of p, $r \in R$, and suppose A contains an R-subalgebra $C \cong R[r^{1/q}]$ such that A is projective as a left C-module. Let $B = A^C$. Then A is isomorphic to a differential crossed product $\Delta(B, D, u)$ for some derivation D on B and some element $u \in B$.*

Proof. By Theorem 7.4.4, B is an Azumaya C-algebra and B is Brauer equivalent to $A \otimes_R C$. Write v for $r^{1/q}$, the generator of C over R. Let d be the R-derivation on C such that $d(v) = 1$. By Theorem 8.1.8, d extends to a derivation D on A. Using the fact that $D(C) \subseteq C$, the reader should verify that $D(A^C) \subseteq A^C$. By Theorem 8.1.19, D is an inner derivation. Let $z \in A$ be such that $D(a) = za - az$, for all $a \in A$. It follows from Exercise 8.1.16 that $D^p(a) = z^p a - az^p$. Let $u = z^p$. If $f \in R[x]$ has degree less than p, then $d^p(f(v)) = 0$. Therefore, $D^p(C) = 0$, which implies u centralizes C. That is, $u \in A^C = B$. Notice that $D(u) = uz - zu = z^{p+1} - z^{p+1} = 0$, so we are in the setting of Definition 8.2.2. The differential crossed product $\Delta(B, D, u)$ exists, and is equal to the quotient $B[t, D]/(t^p - r)$ of the differential polynomial ring modulo the principal ideal generated by $t^p - u$. Define $g : B[t, D] \to A$ by $\sum_i b_i t^i = \sum_i b_i z^i$. Then g is a left B-module homomorphism. Since $g(tb) = g(bt + D(b)) = bz + D(b) = zb = g(z)b$, it follows from Lemma 8.2.1 that g is a ring homomorphism. Since $g(t^p - u) = 0$, g induces a homomorphism of rings $\bar{g} : \Delta(B, D, u) \to A$. By Lemma 8.2.3 and Theorem 7.6.1, $\bar{g}$ is one-to-one. Let A_1 denote the image of $\bar{g}$. Then $A^{A_1} \subseteq A^B$. Since $C \subseteq B$, $A^B \subseteq A^C = B$. Since $B^B = C$, it follows that $A^B = C$. This implies A^{A_1} is commutative. Since A_1 is R-Azumaya, by Theorem 7.2.2, A^{A_1} is R-Azumaya. Therefore, $A^{A_1} = R$ and $A_1 = A$. □

Corollary 11.4.8. *Let R be a commutative ring of characteristic p, A an Azumaya R-algebra, and $a \in R$. Suppose A contains a maximal commutative R-subalgebra C which is isomorphic to $R[a^{1/p}]$ and that A is projective as a left C-module. Then there exists $b \in R$ such that A is isomorphic to the elementary p-algebra $(a, b)_R$ (Section 8.2.1).*

Proof. In Lemma 11.4.7, if C is a maximal commutative R-subalgebra of A, then $B = A^C = C$. If D is an extension of d to A, then for some $z \in A$, $D(a) = za - az$. If $u = z^p$, then $u \in C$, and $D(u) = d(u) = 0$, which implies $u \in R$. Lemma 11.4.7 shows that A is isomorphic to $\Delta(C, d, u)$, which is the elementary p-algebra $(a, u)_R$. □

Lemma 11.4.9. *Let R be a commutative ring of characteristic p and let $a_1,\dots,a_n$ be elements of R. If A is an Azumaya R-algebra which is split by $R[a_1^{1/p},\dots,a_n^{1/p}]$, then A is Brauer equivalent to a tensor product of the form $A_1\otimes_R A_2\otimes_R\cdots\otimes_R A_n$, where each A_i is an elementary p-algebra of the form $(a_i,b_i)_R$.*

Proof. The proof is by induction on n. If $n=1$, then by Theorem 7.4.3 and Corollary 11.4.8, A is Brauer equivalent to $(a_1,b_1)_R$, for some $b_1\in R$. Assume $n>1$ and the result is true for $n-1$. Let C denote the ring $R[a_1^{1/p},\dots,a_{n-1}^{1/p}]$. Then $A\otimes_R C$ is split by $C[a_n^{1/p}]$. By Corollary 11.4.8, $A\otimes_R C$ is Brauer equivalent to $(a_n,c)_C$, for some $c\in C$. Let v and w be the generators of $(a_n,c)_C$, which satisfy the relations: $v^p=a_n$, $w^p=c$, $wv=vw+1$. Consider $f=cv^{p-1}\in(a_n,c)_C$. Since c is a central element, $f^p=v^{p(p-1)}c^p=a_n^{p-1}c^p$. Since $\tau_w(v)=wv-vw=1$, using Exercise 8.1.9, we see that $\tau_w(cv^{p-1})=(p-1)cv^{p-2}$. Recursively, we find that $\tau_w^{p-1}(f)=(p-1)!c=-c$. Hence $f^p+\tau_w^{p-1}(f)=a_n^{p-1}c^p-c$. By Lemma 8.2.4 (7), $(a_n,c)_C\cong(a_n,c+f^p+\tau_w^{p-1}(f))_C=(a_n,c+a_n^{p-1}c^p-c)_C=(a_n,a_n^{p-1}c^p)_C$. Since $a_n\in R$ and $c^p\in C^p\subseteq R$ (Exercise 11.4.14), we have shown that $A\otimes_R C$ is Brauer equivalent to $(a_n,b)_C$ for some $b\in R$. By Lemma 8.2.4, $A\otimes_R(a_n,-b)_R\otimes_R C\cong A\otimes_R(a_n,-b)_C$ is Brauer equivalent to $(a_n,b)_C\otimes_C(a_n,b)_C$, which is split. By the induction hypothesis, $A\otimes_R(a_n,-b)_R$ is Brauer equivalent to $A_1\otimes_R\cdots\otimes_R A_{n-1}$. If we set $A_n=(a_n,b)_R$, then we are done. □

Corollary 11.4.10. *Let R be a commutative ring of characteristic p. The subgroup of* $\mathrm{B}(R)$ *annihilated by p is generated by the elementary p-algebras $(a,b)_R$.*

Proof. By Lemma 8.2.4, the Brauer class of an elementary p-algebra $(a,b)_R$ is annihilated by p. If an Azumaya R-algebra A has order p, then by Proposition 11.4.6 and Lemma 11.4.9, A is Brauer equivalent to a product of elementary p-algebras. □

Lemma 11.4.11. *Let R be a commutative ring of characteristic p and I an ideal in R. Then the natural homomorphism* $\mathrm{B}(R)\to\mathrm{B}(R/I)$ *is onto when restricted to elements of order p.*

Proof. Let a and b be arbitrary elements of R. If we use the bar notation to designate "reduction modulo I", then by Lemma 8.2.4 we have $(\bar a,\bar b)_{R/I}\cong(a,b)_R\otimes_R(R/I)$. By Corollary 11.4.10, we are done. □

We require some facts about abelian p-groups contained in the next lemma.

Lemma 11.4.12. *Let $f : A \to B$ be a homomorphism of abelian p-groups. If A is p-divisible and the image of f contains every element in B annihilated by p, then the following are true.*

(1) *f is onto.*

(2) *If every y in B with exponent p is the image of some x in A of exponent p, then the kernel of f is p-divisible.*

Proof. (1): The proof is left to the reader.

(2): Let K denote the kernel of f. If x is a nonzero element in K, then there exists $y \in A$ such that $py = x$. If $y \in K$, then we are done. Assume $f(y) \neq 0$. Then $pf(y) = f(x) = 0$, so $f(y)$ has order p. For some $k \in K$, $k + y$ has order p. Then $0 = p(k + y) = pk + x$. That is, $x = p(-k)$. □

By ${}_p\mathrm{B}(R)$ we denote the subgroup of $\mathrm{B}(R)$ consisting of those elements which are annihilated by some power of p.

Theorem 11.4.13. *Let R be a commutative ring of characteristic p and I an ideal in R. Then the following are true.*

(1) *The sequence*

$$0 \to K \to {}_p\mathrm{B}(R) \xrightarrow{\theta} {}_p\mathrm{B}(R/I) \to 0$$

is split exact, where θ is the natural homomorphism, and K is the kernel of θ.

(2) *The kernel of the natural homomorphism $\eta : \mathrm{B}(R) \to \mathrm{B}(R/I)$ is p-divisible.*

Proof. (1): By Corollary 11.4.4, ${}_p\mathrm{B}(R)$ is p-divisible. By Lemmas 11.4.11 and 11.4.12, θ is onto and the kernel is divisible.

(2): The Brauer group is torsion (Corollary 11.2.5), so the kernel of η decomposes into a direct sum $\ker \eta = {}_p\ker(\eta) \oplus Q$, where every element of Q has order that is relatively prime to p. Let $\pi : \ker(\eta) \to \ker(\eta)$ be the "multiplication by p" homomorphism. By (1) $\pi : {}_p\ker(\eta) \to {}_p\ker(\eta)$ is onto. The reader should verify that $\pi : Q \to Q$ is an isomorphism, hence π is onto. □

4.3. Exercises.

Exercise 11.4.14. Let R be a commutative ring of characteristic p, where p is a prime number. Let $q = p^a$, for some $a \geq 1$. Show that if $a_1, \dots, a_n$ are elements of R and $S = R[a_1^{1/q}, \dots, a_n^{1/q}]$, then $S^q \subseteq R$.

Exercise 11.4.15. Let R be a commutative ring of characteristic p. Let S be a faithful commutative R-algebra such that for every $y \in S$ there exists

$e > 0$ such that $y^{p^e} \in R$. Show that for any commutative S-algebra T, the sequence

$$0 \to \mathrm{B}(S/R) \to \mathrm{B}(T/R) \to \mathrm{B}(T/S) \to 0$$

of abelian groups is exact. When R and S are fields, a cohomological version of this sequence was derived in [**Yua67**, Theorem 3.4].

Chapter 12

Galois Extensions of Commutative Rings

This chapter contains an introduction to the Galois theory of commutative rings. The starting point is a commutative ring R, a faithful commutative R-algebra S, and a finite group G of R-algebra automorphisms of S. The topics in this chapter are closely tied to those of Chapter 7, the connection being the notion of crossed product algebras. Given the extension S/R, the group G, and a 2-cocycle $f \in Z^2(G, S^*)$, we define the crossed product $\Delta(S/R, G, f)$, an R-algebra that in general is noncommutative. By definition, we say that S/R is a Galois extension with group G, if the trivial crossed product $\Delta(S/R, G, 1)$ is an Azumaya R-algebra. The crossed product is probably the most important tool for the construction of Azumaya algebras and will be revisited in Chapter 13. There is a Morita equivalence between the category of left $\Delta(S/R, G, 1)$-modules and the category of right R-modules. A strong form of the Fundamental Theorem of Galois Theory is proved when S is a connected ring, and a weaker version which applies when S has nontrivial idempotents. A commutative separable R-algebra A which is an R-progenerator module of constant rank can be embedded as a subalgebra of a Galois extension T/R with finite group G in such a way that $A = T^H$, for some subgroup H. There is an entire section devoted to cyclic Galois extensions and Kummer Theory.

1. Crossed Product Algebras, the Definition

Let S be a commutative ring and G a finite group of automorphisms of S. Let $1 \in G$ denote the group identity element. By Corollary 5.1.7, a normalized 2-cocycle in $\mathrm{Z}^2(G, S^*)$ is a function $f : G \times G \to S^*$ satisfying the identities

$$f(\rho, \sigma) f(\rho\sigma, \tau) = \rho f(\sigma, \tau) f(\rho, \sigma\tau) \tag{12.1}$$

$$f(\sigma, 1) = f(1, \tau) = 1 \tag{12.2}$$

for all $(\rho, \sigma, \tau) \in G^3$. Recall that f is a 2-coboundary if there exists $g : G \to S^*$ and

$$f(\sigma, \tau) = \sigma g(\tau) g(\sigma) g(\sigma\tau)^{-1} \tag{12.3}$$

for all $(\sigma, \tau) \in G^2$. Let $\Delta = \bigoplus_{\sigma \in G} S u_\sigma$ be the free left S-module on the set $\{u_\sigma \mid \sigma \in G\}$. Using G and f, we make S into a ring. On the generators the multiplication rule is

$$s u_\sigma t u_\tau = s \sigma(t) f(\sigma, \tau) u_{\sigma\tau} \tag{12.4}$$

for all $\sigma, \tau \in G$, and $s, t \in S$.

Lemma 12.1.1. *As above, let S be a commutative ring, G a finite group of automorphisms of S, and f a normalized 2-cocycle in $\mathrm{Z}^2(G, S^*)$. Then the following are true.*

(1) *The multiplication formula (12.4) makes the left S-module $\Delta = \bigoplus_{\sigma \in G} S u_\sigma$ into a ring. The multiplicative identity for Δ is u_1.*
(2) *$S u_1$ is a subring isomorphic to S, and $S^G u_1$ is a subring of the center of Δ.*
(3) *Each basis element u_σ is an invertible element of Δ.*

Proof. The reader should verify that u_1 is a multiplicative identity for Δ. Since f satisfies the 2-cocycle identity (12.1),

$$\begin{aligned}(r u_\rho s u_\sigma)(t u_\tau) &= (r \rho(s) f(\rho, \sigma) u_{\rho\sigma})(t u_\tau)\\ &= r \rho(s) f(\rho, \sigma) \rho\sigma(t) f(\rho\sigma, \tau) u_{\rho\sigma\tau}\end{aligned}$$

is equal to

$$\begin{aligned}(r u_\rho)(s u_\sigma t u_\tau) &= (r u_\rho)(s \sigma(t) f(\sigma, \tau) u_{\sigma\tau})\\ &= r \rho(s) \rho\sigma(t) \rho f(\sigma, \tau) f(\rho, \sigma\tau) u_{\rho\sigma\tau}\end{aligned}$$

hence, the associative law for multiplication holds. (3) follows from the fact that $u_\sigma u_{\sigma^{-1}} = f(\sigma, \sigma^{-1}) u_1$ is in $S^* u_1$. The rest is left to the reader. □

Definition 12.1.2. If R is a subring of S^G, then the ring defined on $\bigoplus_{\sigma \in G} S u_\sigma$ by the formula (12.4) is denoted $\Delta(S/R, G, f)$ and is called a *crossed product algebra*. The function f is called the *factor set*. By Lemma 12.1.1 (2), $\Delta(S/R, G, f)$ is an R-algebra. Notice that we have assumed f is a normalized factor set. It follows from Exercise 12.5.5 together with Lemma 12.1.3 that this assumption does not result in a loss of generality. Sometimes we will write 1 instead of u_1 for the multiplicative identity of the crossed product algebra.

Lemma 12.1.3. *In the above context, let f, h be 2 cocycles in $Z^2(G, S^*)$ and assume h is a 2-coboundary. Then $\Delta(S/R, G, f)$ and $\Delta(S/R, G, fh)$ are isomorphic as R-algebras.*

Proof. By (12.3) there is $g : G \to S^*$ such that $h(\sigma, \tau) = \sigma g(\tau)g(\sigma)g(\sigma\tau)^{-1}$. Let $\Delta(S/R, G, f) = \bigoplus_{i=1}^n Su_\sigma$ and $\Delta(S/R, G, fh) = \bigoplus_{i=1}^n Sv_\sigma$. Define

$$\phi : \Delta(S/R, G, f) \to \Delta(S/R, G, fh)$$

by sending the basis element u_σ to $g(\sigma)^{-1}v_\sigma$. The reader should verify that this extends to an R-algebra isomorphism. □

Definition 12.1.4. Notice that the constant function $f(\sigma, \tau) = 1$ satisfies (12.1) and (12.2). We call $\Delta(S/R, G, 1)$ the *trivial crossed product.* There is an R-algebra homomorphism

$$\phi_{S/R} : \Delta(S/R, G, 1) \to \operatorname{Hom}_R(S, S)$$

which maps $s \in S$ to "left multiplication by s", and which maps u_σ to σ.

2. Galois Extension, the Definition

General references for this section are [**DI71**], [**Sal99**], and [**CHR65**].

Definition 12.2.1. Let S be a commutative ring and G a finite set of automorphisms of S. Let R be a subring of S^G. We say S is a *Galois extension of R* if the trivial crossed product $\Delta(S/R, G, 1)$ is an Azumaya R-algebra.

Lemma 12.2.2. *Let S/R be a Galois extension of commutative rings with finite group G. Then the following hold.*

(1) *$S^G = R$.*
(2) *S is an R-progenerator module. The natural map $R \to S$ splits and $R \cdot 1$ is an R-module direct summand of S.*
(3) *G acts faithfully on S.*
(4) *S is a maximal commutative R-subalgebra of $\Delta(S/R, G, 1)$.*
(5) *S/R has constant rank* $\operatorname{Rank}_R(S) = [G : 1]$.
(6) *The map $\phi_{S/R} : \Delta(S/R, G, 1) \to \operatorname{Hom}_R(S, S)$ is an R-algebra isomorphism.*
(7) *If I is an ideal of S which is an $R(G)$-submodule, then there is an ideal J in R such that $I = JS$.*

Proof. (1): If $x \in S^G$, then x is in the center of $\Delta(S/R, G, 1)$, which is R.

(2): By Theorem 7.1.4, $\Delta(S/R, G, 1)$ is an R-progenerator. By definition, $S \cdot 1$ is a direct summand of $\Delta(S/R, G, 1)$, hence S is finitely generated and projective over R. By Corollary 1.1.16, S is an R-progenerator. The second statement is true by Proposition 2.4.6.

(3): By Proposition 7.1.10 and Theorem 7.6.1, $\phi_{S/R}$ is one-to-one.

(4): By (3), $\phi_{S/R}$ is one-to-one. But $\mathrm{Hom}_R(S,S)^S = \mathrm{Hom}_S(S,S) = S$. By Exercise 7.4.6, S is a maximal commutative subalgebra of $\mathrm{Hom}_R(S,S)$.

(5): As a left S-module, $\Delta(S/R, G, 1)$ is free of rank $[G:1]$. By (4) and Theorem 7.4.2, the rank of $\Delta(S/R, G, 1)$ over R is equal to $\mathrm{Rank}_R(S)^2$.

(6): By (5) and Theorem 7.6.1, $\phi_{S/R}$ is an isomorphism.

(7): In $\Delta(S/R, G, 1)$ let $\mathfrak{A} = \sum_\sigma I u_\sigma$. Then $\mathfrak{A}$ is a two-sided ideal. Let $J = \mathfrak{A} \cap R$. By Corollary 7.1.7, $\mathfrak{A} = J\Delta(S/R, G, 1) = \bigoplus_\sigma JSu_\sigma$. Then $Iu_1 = JSu_1$, hence $I = JS$. □

Remark 12.2.3. If S/R is a Galois extension of commutative rings with finite group G, then by Lemma 12.2.2 (6), $\Delta(S/R, G, 1) \cong \mathrm{Hom}_R(S,S)$ is an isomorphism of R-algebras. Therefore, there is a Morita equivalence (Section 1.5) between the category of right R-modules and the category of left $\Delta(S/R, G, 1)$-modules. In Proposition 12.4.2 we show that under this equivalence a right R-module N corresponds to $N \otimes_R S$. Under the reverse correspondence, a left $\Delta(S/R, G, 1)$-module M corresponds to $M^G = \{x \in M \mid u_\sigma(x) = x, \text{ for all } \sigma \in G\}$.

Example 12.2.4. Let G be a group of order $[G:1] = n$ which acts transitively on $\{1, \dots, n\}$. Let R be any commutative ring, and $S = \bigoplus_{i=1}^n R$ the ring direct sum of n copies of R. Then we define an $R(G)$-module action on S which makes S into a Galois extension of R. Let $e_1, \dots, e_n$ be the idempotents in S corresponding to the direct summands. Given $\sigma \in G$, let $\sigma(e_i) = e_{\sigma(i)}$. Extend σ by linearity to an R-module homomorphism of S. Since the idempotents are orthogonal, it is clear that σ is an R-algebra automorphism of S. For each $i \neq j$ there is some $\sigma \in G$ such that $\sigma(i) = j$. Then $\phi_{S/R}(e_j u_\sigma) = e_j\sigma$, and

$$e_j\sigma(e_k) = \begin{cases} e_j & \text{if } i = k, \\ 0 & \text{otherwise.} \end{cases}$$

Thus $e_j\sigma$ corresponds to the linear map defined by the elementary matrix e_{ji}. Since the image of $\phi_{S/R}$ contains a generating set for the R-module $\mathrm{Hom}_R(S,S)$, $\phi_{S/R}$ is onto. Since $\mathrm{Hom}_R(S,S)$ and $\Delta(S/R, G, 1)$ are free R-modules of the same rank, $\phi_{S/R}$ is one-to-one (Exercise 2.4.13). This proves $\Delta(S/R, G, 1)$ is an Azumaya R-algebra, so S/R is Galois.

Example 12.2.5. Let G be a finite group acting on itself by left translations. As in Example 12.2.4, make $\bigoplus_{\sigma \in G} Re_\sigma$ into a Galois extension of R with group G. The multiplication is defined by $\sigma e_\tau = e_{\sigma\tau}$, for $\sigma, \tau \in G$. This is the so-called *split*, or *trivial* G-Galois extension of R. Following the notation used by [**CHR65**], [**Har65**], [**Mag74a**] and [**Mag74b**], we sometimes denote the trivial G-Galois extension of R by $E_G(R)$.

Example 12.2.6. Let S be a commutative ring, G a finite set of automorphisms of S, and R a subring of S^G. We define an S-algebra, denoted $\nabla(S, G)$, in the following way. The underlying set is the product $\prod_{\sigma \in G} S$ of $[G:1]$ copies of S indexed by the set G. That is, $\nabla(S, G)$ is the set of all functions $v : G \to S$. Because S is a commutative ring, we make $\nabla(S, G)$ into a commutative ring by point-wise multiplication and addition of functions: $(v_1 + v_2)(\tau) = v_1(\tau) + v_2(\tau)$, $(v_1 v_2)(\tau) = v_1(\tau) v_2(\tau)$. The constant functions make up a subring of $\nabla(S, G)$ which is isomorphic to S. In this way, we view $\nabla(S, G)$ as an S-algebra. For each $\sigma \in G$, let $e_\sigma \in \nabla(S, G)$ be the function defined by

$$e_\sigma(\tau) = \delta_{\sigma,\tau} = \begin{cases} 1 & \text{if } \sigma = \tau, \\ 0 & \text{otherwise.} \end{cases}$$

Then $\{e_\sigma \mid \sigma \in G\}$ is a set of orthogonal idempotents in $\nabla(S, G)$ and $\sum_\sigma e_\sigma$ is the constant function 1. Hence, $\nabla(S, G) = \bigoplus_{\sigma \in G} S e_\sigma$. Regard $S \otimes_R S$ as an S-algebra using the left factor. The reader should verify that the map

$$\ell : S \otimes_R S \to \nabla(S, G)$$

defined by $\ell(a \otimes b)(\tau) = a\tau(b)$ is an S-algebra homomorphism. Notice that in terms of the idempotents, $\ell(a \otimes b) = \sum_{\sigma \in G} a\sigma(b) e_\sigma$.

Lemma 12.2.7. *Let S be a commutative ring, G a finite group of automorphisms of S, and R a subring of S^G. Let T be a commutative R-algebra.*

(1) *(Galois is Preserved by Change of Base) If S/R is a Galois extension with group G, then $S \otimes_R T$ is a Galois extension of T with group G.*
(2) *(Faithfully Flat Descent of Galois) Assume T is a faithfully flat R-algebra. Then S/R is a Galois extension with group G if and only if $S \otimes_R T$ is a Galois extension of T with group G.*

Proof. By Exercise 12.5.6, $\Delta(S/R, G, 1) \otimes_R T$ is isomorphic to $\Delta(S \otimes_R T/T, G, 1)$ as T-algebras.

(1): Since S is an R-progenerator, $S \otimes_R T$ is a T-progenerator (Theorem 1.3.2). By Proposition 2.4.6, $T \to S \otimes_R T$ is one-to-one. The rest follows from Corollary 7.1.9.

(2): By Corollary 8.1.21, $\Delta(S/R, G, 1) \otimes_R T$ is an Azumaya T-algebra if and only if $\Delta(S/R, G, 1)$ is an Azumaya R-algebra. □

Proposition 12.2.8 is the counterpart of Corollaries 4.6.8 and 4.6.9 for Galois extensions.

Proposition 12.2.8. *Let S/R be a Galois extension of commutative rings with finite group G. Consider the element $t = \sum_{\sigma \in G} u_\sigma$ in $\Delta(S/R, G, 1)$.*

(1) $\phi_{S/R}(t) \in \operatorname{Hom}_R(S, R)$.

(2) *As a right S-module,* $\operatorname{Hom}_R(S, R)$ *is free of rank one and is generated by $\phi_{S/R}(t)$.*

(3) *There exists $c \in S$ such that $T_R^S(c) = 1$. Moreover, $R \cdot c$ is an R-module direct summand of S.*

Proof. (1): For any $x \in S$, $\phi_{S/R}(t)(x) = \sum_{\sigma \in G} \sigma(x)$ is fixed by G.

(2): By Lemma 1.3.17, $\operatorname{Hom}_R(S, R)$ is a right S-module by the action $(fa)(x) = f(ax)$, for all $a \in S$ and $x \in S$. Suppose $f \in \operatorname{Hom}_R(S, R)$. By Lemma 12.2.2 (5), $f = \phi_{S/R}(w)$, for some $w = \sum_{\sigma \in G} a_\sigma u_\sigma$. Then for all $x \in S$, $\sum_{\sigma \in G} a_\sigma \sigma(x) \in R$. For any $\rho \in G$ we have

$$\begin{aligned}\rho\Big(\sum_{\sigma \in G} a_\sigma \sigma(x)\Big) &= \sum_{\tau \in G} \rho(a_{\rho^{-1}\tau})\tau(x) \\ &= \sum_{\tau \in G} a_\tau \tau(x)\end{aligned}$$

where $\tau = \rho\sigma$. Since $\phi_{S/R}$ is an isomorphism, it follows that

$$\sum_{\tau \in G} \rho(a_{\rho^{-1}\tau}) u_\tau = \sum_{\tau \in G} a_\tau u_\tau .$$

Therefore, $\rho(a_{\rho^{-1}\tau}) = a_\tau$, or equivalently, $a_\sigma = \sigma(a_1)$ for all $\sigma \in G$. Thus $w = \sum_{\sigma \in G} \sigma(a_1) u_\sigma = (\sum_{\sigma \in G} u_\sigma) a_1 = t a_1$.

Part (3) follows straight from Corollary 4.6.9. □

Theorem 12.2.9. *Let S be a commutative ring, G be a finite group of automorphisms of S, and R a subring of S^G. Then the following statements are equivalent:*

(1) *The trivial crossed product $\Delta(S/R, G, 1)$ is an R-Azumaya algebra, or equivalently, S is a Galois extension of R with group G (see Definition 12.2.1).*

(2) (a) $S^G = R$,
   (b) *for each nonzero idempotent $e \in S$ and each pair $\sigma \neq \tau$ in G there is an element $x \in S$ with $\sigma(x)e \neq \tau(x)e$, and*
   (c) *S is a separable R-algebra.*

(3) (a) $S^G = R$, *and*
(b) *there exist* $x_1, \dots, x_n, y_1, \dots, y_n$ *in* S *such that*

$$\sum_{j=1}^{n} x_j \sigma(y_j) = \begin{cases} 1 & \textit{if } \sigma = 1, \\ 0 & \textit{otherwise}, \end{cases}$$

for all $\sigma \in G$.

(4) (a) S *is a finitely generated projective* R*-module, and*
(b) $\phi_{S/R} : \Delta(S/R, G, 1) \to \operatorname{Hom}_R(S, S)$ *is an isomorphism (see Definition* 12.1.2*).*

(5) (a) $S^G = R$, *and*
(b) $\ell : S \otimes_R S \to \nabla(S, G)$ *is an isomorphism (see Example* 12.2.6*).*

(6) (a) $S^G = R$, *and*
(b) *for each maximal ideal* $\mathfrak{m}$ *of* S *and for each* $1 \neq \sigma \in G$*, there is an* $x \in S$ *such that* $\sigma(x) - x$ *is not in* $\mathfrak{m}$.

Proof. (1) implies (4): This is Lemma 12.2.2 (4).

(4) implies (1): By Corollary 1.1.16, S is an R-progenerator. By Proposition 7.1.10 $\operatorname{Hom}_R(S, S)$ is an Azumaya R-algebra.

(2) implies (3): Let $e = \sum_{j=1}^{n} x_j \otimes y_j \in S \otimes_R S$ be a separability idempotent for S over R (Proposition 4.1.2). Then $\sum_{j=1}^{n} x_j y_j = 1$ and $(1 \otimes x - x \otimes 1)e = 0$ for all $x \in S$. If σ is an R-automorphism of S, then $1 \otimes \sigma$ is an S-automorphism of $S \otimes_R S$. Since S is commutative, the map $\mu : S \otimes_R S \to S$ of Definition 4.1.1 is an R-algebra homomorphism as well as an $S \otimes_R S$-module homomorphism. Because e is an idempotent in $S \otimes_R S$, for each $\sigma \in G$, $e_\sigma = \mu((1 \otimes \sigma)(e))$ is an idempotent in S. For any $x \in S$,

$$\begin{aligned} xe_\sigma &= x\mu((1 \otimes \sigma)(e)) \\ &= \mu((x \otimes 1)(1 \otimes \sigma)(e)) \\ &= \mu((1 \otimes \sigma)((x \otimes 1)e)) \\ &= \mu((1 \otimes \sigma)((1 \otimes x)e)) \\ &= \mu(1 \otimes \sigma(x))\mu((1 \otimes \sigma)e)) \\ &= \sigma(x)e_\sigma. \end{aligned}$$

Therefore, $xe_\sigma = \sigma(x)e_\sigma$ for all $x \in S$. By (2) (b), either $e_\sigma = 0$ or $\sigma = 1$. Because $e_1 = 1$, this leads to

$$e_\sigma = \mu((1 \otimes \sigma)(e)) = \sum_{j=1}^{n} x_j \sigma(y_j) = \begin{cases} 1 & \text{if } \sigma = 1, \\ 0 & \text{otherwise}, \end{cases}$$

for all $\sigma \in G$.

(3) implies (4): Let $x_1, \dots, x_n, y_1, \dots, y_n$ be the elements of S given in (3). Define $f_j \in \operatorname{Hom}_R(S, R)$ by $f_j(x) = \sum_{\sigma \in G} \sigma(xy_j)$. Then for all $x \in S$

we have

$$\begin{aligned}\sum_{j=1}^{n} f_j(x)x_j &= \sum_{j=1}^{n}\sum_{\sigma\in G}\sigma(x)\sigma(y_j)x_j\\ &= \sum_{\sigma\in G}\sigma(x)\sum_{j=1}^{n}x_j\sigma(y_j)\\ &= x.\end{aligned}$$

The set $f_1,\ldots,f_n$, $x_1,\ldots,x_n$ is therefore a dual basis for S over R. This proves S is a finitely generated projective R-module. To prove $\phi_{S/R}$ is onto, let $h\in \operatorname{Hom}_R(S,S)$. Consider

$$w=\sum_{\sigma\in G}\sum_{j=1}^{n}h(x_j)\sigma(y_j)u_\sigma$$

which is an element of $\Delta(S,G,1)$. For any $x\in S$ we have

$$\begin{aligned}\phi_{S/R}(w)(x) &= \sum_{\sigma\in G}\sum_{j=1}^{n}h(x_j)\sigma(y_j)\sigma(x)\\ &= \sum_{j=1}^{n}h(x_j)\sum_{\sigma\in G}\sigma(y_jx)\\ &= h\Big(\sum_{\sigma\in G}\sum_{j=1}^{n}x_j\sigma(y_j)\sigma(x)\Big)\\ &= h(x).\end{aligned}$$

Thus $\phi_{S/R}(w)=h$. To prove $\phi_{S/R}$ is one-to-one, let $w=\sum_{\sigma\in G}a_\sigma u_\sigma$ be an element of $\Delta(S,G,1)$ and assume $\phi_{S/R}(w)=0$. Then

$$\begin{aligned}0 &= \sum_{j=1}^{n}\sum_{\sigma\in G}\phi_{S/R}(w)(x_j)\sigma(y_j)u_\sigma\\ &= \sum_{\sigma\in G}\sum_{\tau\in G}\sum_{j=1}^{n}a_\tau\tau(x_j)\sigma(y_j)u_\sigma\\ &= \sum_{\sigma\in G}\sum_{\tau\in G}a_\tau\tau\Big(\sum_{j=1}^{n}x_j\tau^{-1}\sigma(y_j)\Big)u_\sigma\\ &= \sum_{\sigma\in G}a_\sigma u_\sigma\\ &= w.\end{aligned}$$

(1) plus (4) implies (5): By Lemma 12.2.2, $S^G = R$. Consider the diagram of R-module homomorphisms

$$\begin{array}{ccccc} S \otimes_R S & \xrightarrow{\ell} & \nabla(S,G) & \xrightarrow{f_3} & \Delta(S/R, G, 1) \\ \downarrow f_1 & & & & \downarrow \phi_{S/R} \\ S \otimes_R \operatorname{Hom}_R(S,R) & & \xrightarrow{f_2} & & \operatorname{Hom}_R(S,S) \end{array}$$

where the maps f_1, f_2, and f_3 are to be defined. The S-algebra map ℓ comes from Example 12.2.6 and is defined by $x \otimes y \mapsto \sum_\sigma x\sigma(y)e_\sigma$. The S-module isomorphism f_3 is defined on basis elements by $e_\sigma \mapsto u_\sigma$. Therefore,

$$\phi_{S/R}\left(f_3\left(\ell(x \otimes y)\right)\right) = \sum_\sigma x\sigma(y)\sigma.$$

The map f_1 is induced by the isomorphism $S \cong \operatorname{Hom}_R(S,R)$ of Proposition 12.2.8. That is, $f_1(x \otimes y) = x \otimes \phi_{S/R}(ty)$. The isomorphism f_2 is from Lemma 1.5.1. Then

$$f_2\left(f_1(x \otimes y)\right) = f_2(x \otimes \phi_{S/R}(ty)) = \sum_\sigma x\sigma(y)\sigma.$$

Therefore, the diagram commutes, and ℓ is an isomorphism.

(5) implies (3): Let $\ell^{-1}(e_1) = \sum_{j=1}^n x_j \otimes y_j$. The definition of ℓ implies

$$\sum_{j=1}^n x_j\sigma(y_j) = \begin{cases} 1 & \text{if } \sigma = 1, \\ 0 & \text{otherwise,} \end{cases}$$

for all $\sigma \in G$.

(3) implies (2): Let $x_1, \dots, x_n, y_1, \dots, y_n$ be the elements of S described in (3). Let $T_R^S \in \operatorname{Hom}_R(S,R)$ be defined by $T_R^S(x) = \sum_{\sigma \in G} \sigma(x)$. For any $x \in S$

$$\begin{aligned} \sum_{j=1}^n x_j T_R^S(xy_j) &= \sum_{k=1}^n x_j \sum_{\sigma \in G} \sigma(x)\sigma(y_j) \\ &= \sum_{\sigma \in G} \sigma(x) \sum_{j=1}^n x_j\sigma(y_j) \\ &= x. \end{aligned}$$

By Theorem 4.6.7, S is separable over R (moreover, S is an R-progenerator).

To prove (2) (b), let e be a nonzero idempotent in S. Assume σ, τ are in G, and $\sigma(x)e = \tau(x)e$ for all $x \in S$. Then $x\sigma^{-1}(e) = \sigma^{-1}\tau(x)\sigma^{-1}(e)$ for all $x \in S$ so

$$\begin{aligned}\sigma^{-1}(e) &= \sum_{j=1}^{n} x_j y_j \sigma^{-1}(e) \\ &= \sum_{j=1}^{n} x_j \sigma^{-1}\tau(y_j)\sigma^{-1}(e) \\ &= \begin{cases} \sigma^{-1}(e) & \text{if } \sigma = \tau, \\ 0 & \text{otherwise.} \end{cases}\end{aligned}$$

Since $\sigma^{-1}(e) \neq 0$, we must have $\sigma = \tau$.

(3) implies (6): Assume there exists $\sigma \neq 1$ in G and a maximal ideal $\mathfrak{m}$ of S such that $(1-\sigma)S \subseteq \mathfrak{m}$. Then by (3) we have that

$$1 = \sum_{j=1}^{n} x_j(y_j - \sigma(y_j))$$

is in $\mathfrak{m}$, a contradiction.

(6) implies (3): Let $\sigma \neq 1$ be in G. By (6) the ideal generated by elements of the form $\sum_{j=1}^{n} x_j(1-\sigma)y_j$ is all of S. Thus there are elements $x_1, \dots, x_n, y_1, \dots, y_n$ in S with

$$\sum_{j=1}^{n} x_j(1-\sigma)y_j = 1.$$

Denote by $\delta_{\sigma,\tau}$ the Kronecker delta function, which is equal to 1 if $\sigma = \tau$, and 0 otherwise. For all τ in the set $\{1, \sigma\}$ we have

$$\sum_{j=1}^{n} x_j(1-\tau)y_j = 1 - \delta_{\tau,1}.$$

Let $x_{n+1} = 1 - \sum_{j=1}^{n} x_j y_j$ and $y_{n+1} = 1$. For all τ in the set $\{1, \sigma\}$ we have

$$\sum_{j=1}^{n+1} x_j \tau(y_j) = \delta_{\tau,1}.$$

Next let V and V' be any two subsets of G containing the identity of G and for which there are elements $x_1, \dots, x_n, y_1, \dots, y_n, x'_1, \dots, x'_m, y'_1, \dots, y'_m$ in S so that for all $\sigma \in V$, $\sigma' \in V'$ the following hold:

$$\sum_{j=1}^{n} x_j \sigma(y_j) = \delta_{\sigma,1}$$

and

$$\sum_{k=1}^{m} x'_k \sigma'(y'_k) = \delta_{\sigma',1}.$$

Then for any $\tau \in V \cup V'$ we have

$$\sum_{(j,k)=(1,1)}^{(n,m)} x_j x'_k \tau(y'_k y_j) = \delta_{\tau,1}.$$

Since $G = \bigcup_{\sigma\neq 1}\{1,\sigma\}$, the proof is complete. □

In Theorem 12.2.10 we see that S is a projective module over the group ring $R(G)$. We show in Proposition 12.4.2 (2) that S is a progenerator module over the trivial crossed product $\Delta(S/R, G, 1)$.

Theorem 12.2.10. *Let S/R be a Galois extension of commutative rings with finite group G.*

(1) *The action $(s\sigma)(x\otimes y) = sx\otimes\sigma(y)$ makes $S\otimes_R S$ into a left $S(G)$-module.*
(2) *There are $S(G)$-module isomorphisms $S\otimes_R S \cong S(G) \cong S\otimes_R R(G)$.*
(3) *S is a finitely generated projective $R(G)$-module.*
(4) *For all $n \geq 1$, $\mathrm{H}^n(G, S) = 0$.*
(5) *The G-Galois extension S/R is a twisted form of $E_G(R)$, the trivial G-Galois extension of R (see Example 12.2.5).*
(6) *S is a faithfully flat étale R-algebra.*

Proof. Parts (1) and (6) are left to the reader.

(2): Let $\nabla(S,G)$ be the S-algebra defined in Example 12.2.6. Let $S(G)$ act from the left on $\nabla(S,G)$ by $(s\sigma v)(\tau) = sv(\tau\sigma)$. Since

$$((\rho\sigma)v)(\tau) = v(\tau\rho\sigma) = (\sigma v)(\tau\rho) = (\rho(\sigma v))(\tau)$$

one checks that this makes $\nabla(S,G)$ into a left $S(G)$-module and the function $\ell : S\otimes_R S \to \nabla(S,G)$ from Example 12.2.6 is a left $S(G)$-module isomorphism. Define

$$\nabla(S,G) \xrightarrow{\psi} S(G)$$

by $v \mapsto \sum_{\tau\in G} v(\tau)\tau^{-1}$. One checks that ψ is S-linear. The idempotent e_σ is mapped to σ^{-1}, hence ψ is an S-module isomorphism. For $\sigma\in G$,

$$\begin{aligned}\psi(\sigma v) = \sum_{\tau\in G}(\sigma v)(\tau)\tau^{-1} &= \sum_{\tau\in G} v(\tau\sigma)\tau^{-1}\\ &= \sum_{\rho\in G} v(\rho)\sigma\rho^{-1} = \sigma\sum_{\rho\in G} v(\rho)\rho^{-1} = \sigma\psi(v)\end{aligned}$$

so ψ is an $S(G)$-module isomorphism. Combining ℓ, ψ with the isomorphism of rings $S(G)\cong S\otimes_R R(G)$ proves (2).

(3): By Lemma 12.2.2 (2), S is an R-progenerator. Therefore, $S(G)$ is an $R(G)$-progenerator. By Lemma 12.2.2 (2) the natural map $\eta : R\to S$ has a left inverse. Suppose $\lambda : S \to R$ and $\lambda\eta(x) = x$ for all $x \in R$. Thus $\lambda \otimes 1$ is a left inverse to $\eta \otimes 1 : R \otimes_R S \to S \otimes_R S$. Because G acts on the right factor, $\eta \otimes 1$ and $\lambda \otimes 1$ are $R(G)$-module homomorphisms, which proves (3).

(4): For some $m > 0$, S is an $R(G)$-module direct summand of $R(G)^m = (\mathbb{Z}G \otimes_{\mathbb{Z}} R)^m = \mathbb{Z}G \otimes_{\mathbb{Z}} R^m$. This follows from Exercise 5.1.24.

(5): By Theorem 12.2.9, $\ell : S \otimes_R S \to \nabla(S, G)$ is an isomorphism of S-algebras. By the proof of (2), ℓ is also an isomorphism of left $S(G)$-modules. In terms of the idempotent decomposition $\nabla(S, G) = \bigoplus_{\sigma \in G} Se_\sigma$, the left G-action is $\rho e_\sigma = e_{\sigma\rho^{-1}}$. The left G-action on $E_G(S) = \bigoplus_{\sigma \in G} Se_\sigma$, the trivial G-Galois extension of S, is given by $\rho e_\sigma = e_{\rho\sigma}$. Define $f : E_G(S) \to \nabla(S, G)$ by $e_\sigma \mapsto e_{\sigma^{-1}}$. Then $f(\rho e_\sigma) = f(e_{\rho\sigma}) = e_{(\rho\sigma)^{-1}} = e_{\sigma^{-1}\rho^{-1}} = \rho f(e_\sigma)$. Therefore, f is an isomorphism of left $S(G)$-modules and S-algebras. □

2.1. Noetherian Reduction of a Galois Extension. The next proposition can be proved using the methods of Section 7.7. The proof is left to the reader.

Proposition 12.2.11. *Let S/R be a Galois extension of commutative rings with finite group G (see Definition* 12.2.1*). Then there is a noetherian subring $R_0 \subseteq R$ (in fact, R_0 can be taken to be a finitely generated $\mathbb{Z}$-algebra) and an R_0-subalgebra $S_0 \subseteq S$ such that the following are true:*

(1) *S_0 is a Galois extension of R_0 with group G.*
(2) *Each R_0-automorphism of S_0 is the restriction of an R-automorphism of S.*
(3) *$S \cong S_0 \otimes_{R_0} R$, the isomorphism preserving G-actions.*
(4) *If $\Delta(S/R, G, f)$ is a crossed product (see Definition* 12.1.2*), then S_0 can be chosen so that there is a crossed product $\Delta(S_0/R_0, G, f_0)$ and $\Delta(S/R, G, f) \cong \Delta(S_0/R_0, G, f_0) \otimes_{R_0} R$.*

3. Induced Galois Extensions and Galois Extensions of Fields

In Corollary 12.3.1 and Theorem 12.3.4 we show that if k is a field and S/k is a Galois extension with finite group G, then S is a so-called induced extension that is obtained from an extension of fields F/k which is Galois with group H, and H is a subgroup of G. First we show that a Galois extension of fields in the usual sense is a Galois extension in the sense of Definition 12.2.1. A general reference for the material in this section is [**Sal99**].

Corollary 12.3.1. *Let F/k be an extension of fields and let G be a finite group of automorphisms of F with $F^G = k$. Then F is a Galois extension of k.*

Proof. The verification of Theorem 12.2.9 (6) is immediate. □

Assume T/R is a Galois extension of commutative rings with finite group H. Additionally, assume G is a finite group which contains H as a subgroup. Using T and H, The goal is to define a G-Galois extension S/R which is induced by the H-Galois extension T/R. The underlying R-module of S is the induced $\mathbb{Z}G$-module $\operatorname{Hom}_{\mathbb{Z}H}(\mathbb{Z}G, T)$ (see Definition 5.1.9). Therefore, we view elements of S as functions with values in T. Recall that for $v \in \operatorname{Hom}_{\mathbb{Z}H}(\mathbb{Z}G, T)$ and $\sigma \in G$, $(\sigma v)(x) = v(x\sigma)$. Therefore, S is a left $\mathbb{Z}G$-module. To make S into a ring, we pick a free basis for $\mathbb{Z}G$ as a $\mathbb{Z}H$-module. In other words, we pick a set of right coset representatives $\sigma_1, \dots, \sigma_r$ for H in G. Then

$$\operatorname{Hom}_{\mathbb{Z}H}(\mathbb{Z}G, T) \cong \bigoplus_{i=1}^{r} \operatorname{Hom}_{\mathbb{Z}H}(\mathbb{Z}H\sigma_i, T) \cong \bigoplus_{i=1}^{r} T$$

is defined by mapping $f \mapsto \sum_{i=1}^r f(\sigma_i)$. Under this identification, the ring structure on S is defined to be that of $\bigoplus_{i=1}^r T$. We emphasize that the multiplication operation depends on the basis choice. By Lemma 5.1.15, there is an isomorphism of $\mathbb{Z}G$-modules $\operatorname{Hom}_{\mathbb{Z}H}(\mathbb{Z}G, T) \cong \mathbb{Z}G \otimes_{\mathbb{Z}H} T$ defined by $f \mapsto \sum_{i=1}^r \sigma_i^{-1} \otimes f(\sigma_i)$, which is independent of the basis choice. The representation is therefore

$$\operatorname{Hom}_{\mathbb{Z}H}(\mathbb{Z}G, T) \cong \bigoplus_{i=1}^{r} \operatorname{Hom}_{\mathbb{Z}H}(\mathbb{Z}H\sigma_i, T) \cong \bigoplus_{i=1}^{r} \sigma_i^{-1} \otimes_{\mathbb{Z}H} T \cong \bigoplus_{i=1}^{r} T$$

and the elements $\sigma_i^{-1} \otimes 1$ are idempotents in the ring structure on S. Let $\operatorname{Ind}_H^G(T)$ denote the ring which we have just defined on the $\mathbb{Z}G$-module $\operatorname{Hom}_{\mathbb{Z}H}(\mathbb{Z}G, T)$.

Lemma 12.3.2. *Let F/k be a Galois extension of fields with finite group H. Let G be a finite group containing H as a subgroup, $\{\sigma_1, \dots, \sigma_r\}$ a set of right coset representatives for H in G, and $\operatorname{Ind}_H^G(F)$ the ring defined above. Then $\operatorname{Ind}_H^G(F)$ is a Galois extension of k with group G.*

Proof. Assume $\sigma_1 = 1$. The subset of $\mathbb{Z}G \otimes_{\mathbb{Z}H} F$ fixed by G is $1 \otimes_{\mathbb{Z}H} F^H$, which is equal to k. The k-algebra $\operatorname{Ind}_H^G(F)$ is a direct sum of copies of the separable k-algebra F. By Corollary 4.5.8, $\operatorname{Ind}_H^G(F)$ is separable. By Theorem 12.2.9 (2), to finish the proof, it suffices to show that for every primitive idempotent $\sigma_i^{-1} \otimes 1$ and every pair $\alpha \neq \beta$ in G, there is $x \in \operatorname{Ind}_H^G(F)$ such that the $\sigma_i^{-1} \otimes 1$-components of $\alpha(x)$ and $\beta(x)$ are different.

Left multiplication by α and β define permutations of the left cosets of H. For each i, there are unique h_i, k_i in H, and α_i, β_i in $\{1, \dots, r\}$, such that $\alpha x_i^{-1} = x_{\alpha_i}^{-1} h_i$ and $\beta x_i^{-1} = x_{\beta_i}^{-1} k_i$. Consider a typical $x = \sum_{i=1}^r \sigma_i^{-1} \otimes_{\mathbb{Z}H} t_i \in \mathbb{Z}G \otimes_{\mathbb{Z}H} F$. Then

$$\begin{aligned}\alpha(x) &= \sum_{i=1}^r \alpha x_i^{-1} \otimes t_i \\ &= \sum_{i=1}^r x_{\alpha_i}^{-1} \otimes h_i(t_i).\end{aligned}$$

If $\alpha_i^{-1} \neq \beta_i^{-1}$, then we pick x so that $t_{\alpha_i^{-1}} = 0$ and $t_{\beta_i^{-1}} = 1$. Then the $x_i^{-1} \otimes 1$- component of $\alpha(x)$ is 0, but for β it is $x_i^{-1} \otimes 1$. If $\alpha_i^{-1} = \beta_i^{-1}$, then h_i and k_i are two different k-automorphisms of F. So there is some $t_{\alpha_i^{-1}} \in T$ for which the functions have different values. □

Theorem 12.3.3. *Let T/R be a Galois extension of commutative rings with finite group H. Let G be a finite group containing H as a subgroup, $\{\sigma_1, \dots, \sigma_r\}$ a set of right coset representatives for H in G, and $\operatorname{Ind}_H^G(T)$ the ring defined above. Then $\operatorname{Ind}_H^G(T)$ is a Galois extension of R with group G.*

Proof. Let $S = \operatorname{Ind}_H^G(T)$. It is enough to show $\Delta(S/R, G, 1)$ is an Azumaya R-algebra (Definition 12.2.1). Since S is a finite direct sum of copies of T, S is an R-progenerator module (Lemma 12.2.2). Let $\mathfrak{m}$ be a maximal ideal in R. By Lemma 12.2.7, $T/(\mathfrak{m}T)$ is Galois over $R/\mathfrak{m}$ with group G. The reader should verify that $\operatorname{Ind}_H^G(T) \otimes_R (R/\mathfrak{m}) = \operatorname{Ind}_H^G(T/\mathfrak{m}T)$. It follows from Lemma 12.3.2 that $\Delta((S/\mathfrak{m}S)/(R/\mathfrak{m}), G, 1)$ is an Azumaya $R/\mathfrak{m}$-algebra. Since $\Delta(S/R, G, 1) \otimes_R (R/\mathfrak{m}) \cong \Delta((S/\mathfrak{m}S)/(R/\mathfrak{m}), G, 1)$, the proof follows from Corollary 8.1.25. □

Now we show that every Galois extension of a field k is induced from a Galois extension of fields F/k.

Theorem 12.3.4. *Let k be a field and S a Galois extension of k with finite group G. Then there exists a subgroup $H \subseteq G$ and a Galois extension of fields F/k with group H such that $S \cong \operatorname{Ind}_H^G(F)$.*

Proof. By Theorem 12.2.9 (2), S/k is separable. By Corollary 4.5.8, S is a finite direct sum $K_1 \oplus \dots \oplus K_r$ where each K_i is a finite separable extension field of k. Let e be a primitive idempotent in S corresponding to one of the direct summands. If we set $F = Se$, then F is equal to one of the fields K_i, hence is a separable extension field of k. Let H be the subgroup of G that fixes e. For each $\sigma \in G$, $\sigma : Se \to S\sigma(e)$ is an isomorphism of fields, the inverse being σ^{-1}. For any σ, τ, since the field Se has only two idempotents,

$\sigma(e)\tau(e)$ is either 0 or $\sigma(e)$. If $Ge = \{\sigma(e) \mid \sigma \in G\}$ is the orbit of e, then Ge has length $[G : H]$ and consists of pairwise orthogonal primitive idempotents. Let $\epsilon = \sum_{x \in Ge} x$. Then ϵ is a nonzero idempotent of $S^G = k$. So $\epsilon = 1$, and $S\epsilon = \bigoplus_{x \in Ge} Sx$. We have shown that each K_i is isomorphic to F, and $r = [G : H]$. Since $[G : 1] = \dim_k(S)$, this implies $[H : 1] = \dim_k(F)$. We know from Theorem 12.2.9 (2) that distinct elements of H define distinct functions on $F = Se$. Therefore, $H \subseteq \operatorname{Aut}_k(F)$. Since F/F^H is Galois with group H (Theorem 12.2.9 (6)), it follows that $\dim_{F^H}(F) = [H : 1]$. This implies $k = F^H$, and F/k is Galois with group H.

Let $\sigma_1, \dots, \sigma_r$ be a full set of right coset representatives for H in G. Let $\operatorname{Ind}_H^G(F)$ be the induced Galois extension of k with group G. From the definition of induced Galois extension, the mapping $f \mapsto \sum_{i=1}^r f(x_i)x_i(e)$ defines a homomorphism of k-algebras $\gamma : \operatorname{Ind}_H^G(F) \to S$ which also respects the $\mathbb{Z}G$-module structures. Both rings are direct sums of fields and γ is one-to-one on primitive idempotents, hence γ is one-to-one. Counting dimensions over k shows that γ is onto. □

4. Galois Descent of Modules and Algebras

General references for this section are [**DI71**], [**Sal99**], [**CHR65**], [**KO74b**], and [**Kan68**].

Let S be a commutative ring, G a finite group of automorphisms of S, and R a subring of S^G. Let $\Delta = \Delta(S/R, G, 1)$ be the trivial crossed product. Let M be a left Δ-module. The ring homomorphism $S \to \Delta$ turns M into a left S-module. There is also a ring homomorphism $\theta : \mathbb{Z}G \to \Delta$, from the group ring to Δ, which is defined by $\theta(\sigma) = u_\sigma$. So M is also a left $\mathbb{Z}G$-module.

Conversely, suppose M is a left S-module as well as a left $\mathbb{Z}G$-module. If the two module actions satisfy the property

$$\sigma(sx) = \sigma(s)\sigma(x)$$

for all $s \in S$, $\sigma \in G$, $x \in M$, then we say the G-module action is *semi-linear*. In this case, the reader should verify that the action defined by: $(su_\sigma)x = s(\sigma(x))$ makes M into a left Δ-module. Let $M^G = \{x \in M \mid \sigma(x) = x, \text{ for all } \sigma \in G\}$.

Example 12.4.1. Let S be a commutative ring, G a finite set of automorphisms of S, and R a subring of S^G. Let $\nabla(S, G)$ be the S-algebra defined in Example 12.2.6. In this example, we show how to make $\nabla(S, G)$ into a left $\Delta(S/R, G, 1)$-module and show that $\ell : 1 \otimes_R S \to \nabla(S, G)^G$ is a surjective homomorphism of R-algebras.

(1) For any $v \in \nabla(S, G)$ and $\sigma \in G$, let σv be the function from G into S defined by

$$(\sigma v)(\tau) = \sigma(v(\sigma^{-1}\tau)). \tag{12.5}$$

If $\sigma_1, \sigma_2 \in G$, then

$$\begin{aligned}(\sigma_1(\sigma_2 v))(\tau) &= \sigma_1(\sigma_2 v)(\sigma_1^{-1}\tau)\\ &= \sigma_1(\sigma_2(v(\sigma_2^{-1}\sigma_1^{-1}\tau)))\\ &= (\sigma_1\sigma_2)v((\sigma_1\sigma_2)^{-1}\tau)\\ &= ((\sigma_1\sigma_2)v)(\tau),\end{aligned}$$

which shows $\sigma_1(\sigma_2 v) = (\sigma_1\sigma_2)v$. If $s \in S$, then

$$\begin{aligned}(\sigma(sv))(\tau) &= \sigma(sv)(\sigma^{-1}\tau)\\ &= \sigma(sv)(\sigma^{-1}\tau))\\ &= \sigma(s)\sigma(v(\sigma^{-1}\tau))\\ &= \sigma(s)\sigma(v)(\tau),\end{aligned}$$

which shows $\sigma(sv) = \sigma(s)\sigma(v)$. We have shown that $\nabla(G, S)$ has the structure of a left $\Delta(S/R, G, 1)$-module.

(2) By Example 12.2.6, $\nabla(S, G) = \bigoplus_{\sigma\in G} Se_\sigma$. The reader should verify that the group action defined in (1) permutes the idempotents according to the formula

$$\sigma e_\tau = e_{\sigma\tau}. \tag{12.6}$$

(3) In the above notation, $\nabla(G, S)^G$ is the set of all functions $v : G \to S$ such that $\sigma v = v$ for all $\sigma \in G$. From (12.5), this means $\sigma(v(\sigma^{-1}\tau)) = v(\tau)$ for all $\sigma, \tau \in G$. In particular, taking $\sigma = \tau$, $\sigma(v(1)) = v(\sigma)$. This says v is the G-orbit of $v(1)$. We see that $\nabla(G, S)^G$ can be identified with the set of all G-module homomorphisms from G into S. By Proposition 1.3.23 (1), $\operatorname{Hom}_{\mathbb{Z}G}(\mathbb{Z}G, S)$ is isomorphic to S as left $\mathbb{Z}G$-modules. Because of the possible confusion, it is helpful to recall that by Lemma 1.3.17 (1), we make $\operatorname{Hom}_{\mathbb{Z}G}(\mathbb{Z}G, S)$ into a left $\mathbb{Z}G$-module by the action $(\sigma v)(\tau) = v(\tau\sigma) = (\tau\sigma)(v(1)) = \tau(\sigma(v(1)))$. Consider the commutative diagram

$$\begin{array}{ccccc} & & \nabla(G,S) & & \\ & \overset{\theta_1}{\nearrow} & \uparrow \ell & \overset{\theta_2}{\nwarrow} & \\ S & \xrightarrow[\rho_1]{} & S\otimes_R S & \xleftarrow[\rho_2]{} & S \end{array} \tag{12.7}$$

where ρ_1 and ρ_2 are the R-algebra homomorphisms of Exercise 1.3.8. By Example 12.2.6, $S \otimes_R S$ is viewed as an S-algebra by ρ_1. Therefore, the arrows in the triangle on the left hand side of (12.7) are S-algebra homomorphisms.

On the right hand side, the arrows are R-algebra homomorphisms. The composite mapping $\theta_2 : S \to \nabla(G, S)$ is defined by $\theta(s)(\sigma) = \sigma(s)$. The reader should verify that the image of θ_2 is equal to $\nabla(G, S)^G$.

Proposition 12.4.2. *Let S/R be a Galois extension of commutative rings with finite group G. Let $\Delta = \Delta(S/R, G, 1)$ be the trivial crossed product. There is a Morita equivalence between the categories of left Δ-modules and right R-modules.*

(1) *A right R-module N corresponds to the left Δ-module $N \otimes_R S$.*

(2) *S is a left Δ-progenerator module.*

(3) *Under this correspondence, a left Δ-module M maps to the right R-module $\operatorname{Hom}_\Delta(S, M) \cong M^G$.*

Proof. As mentioned in Remark 12.2.3, S is an R-progenerator and there is an isomorphism of R-algebras $\Delta(S/R, G, 1) \cong \operatorname{Hom}_R(S, S)$. Most of the claims follow straight from Theorem 1.5.2. The reader should verify that the assignment $f \mapsto f(1)$ defines an isomorphism of R-modules: $\operatorname{Hom}_\Delta(S, M) \cong M^G$. This and Theorem 1.3.29 can be used to show that a Δ-module M maps to the R-module $\operatorname{Hom}_\Delta(S, \Delta) \otimes_\Delta M \cong \operatorname{Hom}_\Delta(S, M) \cong M^G$. □

Theorem 12.4.3. *Let S be a commutative ring, G a finite group of automorphisms of S, and $R = S^G$. Let $\Delta = \Delta(S/R, G, 1)$ be the trivial crossed product. The following are equivalent*

(1) *S/R is a Galois extension with group G.*

(2) *For any left Δ-module M, the mapping $s \otimes x \mapsto sx$ defines an isomorphism of S-modules $S \otimes_R M^G \cong M$.*

Proof. (1) implies (2): By the categorical equivalence of Proposition 12.4.2, the R-module R corresponds to the left Δ-module S and the left Δ-module M is isomorphic to $N \otimes_R S$ for some right R-module N. By the categorical equivalence, there are isomorphisms

$$M^G \cong \operatorname{Hom}_\Delta(S, M) \cong \operatorname{Hom}_R(R, N) \cong N$$

of R-modules. Hence $M^G \otimes_R S \cong N \otimes_R S \cong M$ as left Δ-modules. Because R is commutative and S is faithfully flat over R, $1 \otimes_R M^G \to S \otimes_R M^G = S(1 \otimes_R M^G)$ is one-to-one (Lemma 2.4.4), and the map is as claimed.

(2) implies (1): Let $\nabla(G, S)$ be the left $\Delta(S/R, G, 1)$-module of Example 12.4.1. The R-algebra epimorphism $\theta_2 : S \to \nabla(G, S)^G$ of (12.7) is defined by $\theta_2(s)(\sigma) = \sigma(s)$. By hypothesis (2), the composite map

$$S \otimes_R S \xrightarrow{1 \otimes \theta_2} S \otimes_R \nabla(G, S)^G \to \nabla(G, S)$$

is an isomorphism. On elements, the mapping is $a \otimes b \mapsto a \otimes \theta(b) \mapsto a\theta(b)$ which we recognize as the function ℓ from Example 12.2.6. By Theorem 12.2.9 (5), S/R is Galois with group G. □

Corollary 12.4.4 (Descent of Azumaya Algebras). *Let S/R be a Galois extension with finite group G. Let A be an S-algebra. Assume G acts on A as a group of R-algebra automorphisms and that the action is S-semilinear. That is, $\sigma(sx) = \sigma(s)\sigma(x)$ for all $\sigma \in G$, $s \in S$, $x \in A$. Then A^G is an R-algebra and $S \otimes_R A^G \cong A$. Moreover, A is S-Azumaya if and only if A^G is R-Azumaya.*

Proof. The reader should verify that A^G is an R-algebra. The last claim follows from Corollary 8.1.21. □

Corollary 12.4.5 (Descent of Invertible Modules). *Let S/R be a Galois extension with finite group G. Let P be an invertible left S-module. By the left regular representation, we view $S \subseteq \operatorname{Hom}_R(P, P)$ as a subring. Then the following are equivalent:*

(1) *$\operatorname{Hom}_R(P, P)$ is isomorphic to $\Delta(S/R, G, 1)$ by an R-algebra isomorphism which is also an $S \otimes_R S$-module isomorphism.*
(2) *There is an invertible R-module N such that $N \otimes_R S \cong P$ as S-modules.*

Proof. (1) implies (2): Since P is a left $\operatorname{Hom}_R(P, P)$-module, we can make P into a left $\Delta(S/R, G, 1)$-module. By Proposition 12.4.2, $P \cong P^G \otimes_R S$ is an S-module isomorphism. By Corollary 1.5.5, P^G is an R-progenerator of rank one.

(2) implies (1): This follows from the fact that the R-algebra isomorphisms

$$\begin{aligned}\operatorname{Hom}_R(P, P) &\cong \operatorname{Hom}_R(N \otimes_R S, N \otimes_R S)\\ &\cong \operatorname{Hom}_R(N, N) \otimes_R \operatorname{Hom}_R(S, S) \cong \Delta(S/R, G, 1)\end{aligned}$$

are also $S \otimes_R S$-module isomorphisms. □

5. The Fundamental Theorem of Galois Theory

First we prove a general result which is from [**CHR65**] and applies when the Galois extension contains nontrivial idempotents (Theorem 12.5.1). A stronger form of the theorem (which is from [**DI71**]) is proved when the Galois extension is connected (Theorem 12.5.4).

Theorem 12.5.1 (Fundamental Theorem of Galois Theory). *Let S/R be a Galois extension of commutative rings with finite group G. If H is a* *subgroup of G, then S/S^H is a Galois extension with group H. Moreover, S^H is a separable R-algebra and H is equal to the subgroup of G that leaves S^H point-wise fixed. If H is a normal subgroup of G, then S^H/R is a Galois extension with group G/H.*

Proof. Let $x_1, \dots, x_n, y_1, \dots, y_n$ be the elements of S which satisfy Theorem 12.2.9 (3). The same theorem together with the same set of elements show S/S^H is a Galois extension with group H. By Lemma 12.2.2, S is an S^H-progenerator module. Theorem 4.4.2 implies S^H is separable over R. Let K be the subgroup of G that fixes S^H point-wise. Then $H \subseteq K \subseteq G$, and $S^H = S^K$. By the first part of the proof, S/S^H is a Galois extension with group K. Lemma 12.2.2 shows $[H : 1] = [K : 1]$, hence $H = K$.

Now suppose H is a normal subgroup of G. Then G/H acts as a group of R-algebra automorphisms of S^H by $(\sigma H)(x) = \sigma(x)$. Since H is the subgroup of G that fixes S^H point-wise, G/H can be identified with a subgroup of $\operatorname{Aut}_R(S^H)$. To show that S^H is a Galois extension of R, we prove that Theorem 12.2.9 (3) is satisfied. First, we have $(S^H)^{G/H} = S^G = R$. As above, let $x_1, \dots, x_n, y_1, \dots, y_n$ be the elements in S satisfying Theorem 12.2.9 (3) for the group G. As in Proposition 12.2.8, let $T^S_{S^H} : S \to S^H$ be the trace map. Then there is $c \in S$ such that $T^S_{S^H}(c) = \sum_{\rho \in H} \rho(c) = 1$. Set $a_j = T^S_{S^H}(x_j)$ and $b_j = T^S_{S^H}(y_j c)$. For every $\rho \in G$, we have

$$\begin{aligned}
\sum_{j=1}^{n} a_j \rho(b_j) &= \sum_{\sigma \in H} \sum_{\tau \in H} \sum_{j=1}^{n} \sigma(x_j) \rho\tau(y_j c) \\
&= \sum_{\sigma \in H} \sigma \Bigl(\sum_{\tau \in H} \Bigl(\sum_{j=1}^{n} x_j \sigma^{-1} \rho\tau(y_j) \Bigr) \sigma^{-1} \rho\tau(c) \Bigr)
\end{aligned}$$

If $\rho \notin H$, then $\sigma^{-1}\rho\tau \neq 1$ and the entire summation collapses to 0. If $\rho \in H$, then the inner sum is 1 whenever $\tau = \rho^{-1}\sigma$, otherwise it is 0. Therefore, the summation reduces to $\sum_{\sigma \in H} \sigma(c) = 1$. □

5.1. Fundamental Theorem for a Connected Galois Extension.

Theorem 12.5.2. *Let S and T be commutative R-algebras. Assume T is a connected ring and S is a separable R-algebra which is finitely generated as an R-module.*

(1) *There are only finitely many R-algebra homomorphisms from S to T.*

(2) *If S is a finitely generated projective R-module of constant rank, then there are at most $\operatorname{Rank}_R(S)$ homomorphisms from S to T.*

(3) *If S/R is a Galois extension with finite group G and if f and g are two R-algebra homomorphisms from S to T, then there is a unique $\sigma \in G$ such that $g = f\sigma$.*

Proof. (1): Let $\sigma_1, \dots, \sigma_n$ be distinct R-algebra homomorphisms from S to T. By Exercise 1.3.8, the homomorphisms $h_1, \dots, h_n$ defined by $h_i(x \otimes y) = x\sigma_i(y)$ are distinct T-algebra homomorphisms from $T \otimes_R S$ to T. By Corollary 4.4.8, there are orthogonal idempotents $e_1, \dots, e_n$ in $T \otimes_R S$ such that h_i factors through $(T \otimes_R S)e_i$ and the composite mapping $T \to (T \otimes_R 1)e_i \to (T \otimes_R S)e_i$ is one-to-one and onto. As a ring, $T \otimes_R S$ decomposes

$$T \otimes_R S = (T \otimes_R S)e_1 \oplus \cdots \oplus (T \otimes_R S)e_n \oplus (T \otimes_R S)(1 - e_1 - \cdots - e_n).$$

As a T-module, we have

$$T \otimes_R S \cong T^n \oplus (T \otimes_R S)(1 - e_1 - \cdots - e_n).$$

If $\mathfrak{m}$ is a maximal ideal of T, then $(T/\mathfrak{m}) \otimes_R S$ contains a subspace of dimension n. If the R-module S is generated by k elements, then $k \geq \dim_{T/\mathfrak{m}}((T/\mathfrak{m}) \otimes_R S) \geq n$.

(2): If S is projective of constant rank, then

$$\operatorname{Rank}_R(S) = \dim_{T/\mathfrak{m}}((T/\mathfrak{m}) \otimes_R S) \geq n.$$

(3): If S/R is Galois with finite group G, then $[G : 1] = \operatorname{Rank}_R(S) \geq n$, by Lemma 12.2.2. Let $\sigma \in G - \langle 1 \rangle$. By Theorem 12.2.9 (6), the set $\{\sigma(x) - x \mid x \in S\}$ generates the unit ideal. Suppose f is an R-algebra homomorphism from S to T. For some $x \in S$, $\sigma(x) - x$ is not in the kernel of f. Therefore, $f\sigma \neq f$. The set $\{f\sigma \mid \sigma \in G\}$ contains $[G : 1]$ distinct elements. □

Corollary 12.5.3. *Let S/R be a Galois extension with finite group G. Assume S is a connected ring. Any R-algebra homomorphism from S to S is an automorphism in G.*

Proof. The proof is left to the reader. □

Theorem 12.5.4 (Fundamental Theorem of Galois Theory). *Let S be a commutative connected ring, R a subring of S, and assume S is separable over R and finitely generated as an R module. Let $G = \operatorname{Aut}_R(S)$ be the group of all R-algebra automorphisms of S and assume $R = S^G$. Then the following are true.*

(1) *G is finite of order $[G : 1] = \operatorname{Rank}_R(S)$.*

(2) *There is a one-to-one order inverting correspondence between the subgroups of G and the subrings of S which contain R and are separable over R. The subgroup $H \subseteq G$ corresponds to the subring S^H, the subring $R \subseteq T \subseteq S$ corresponds to the subgroup $\{\sigma \in G \mid \sigma(x) = x$ for all $x \in T\}$.*

(3) *A subgroup* $H \subseteq G$ *is normal if and only if the corresponding subring* $T = S^H$ *is a Galois extension of* R *and in this case the Galois group is* G/H.

Proof. It follows from Theorem 12.5.2 that $[G : 1]$ is finite. By Theorem 12.2.9 (2), S/R is Galois. By Lemma 12.2.2, $[G : 1] = \operatorname{Rank}_R(S)$. Theorem 12.5.1 shows that S/S^H is Galois with group H, and if H is normal in G, then S^H is Galois over R with group G/H. Suppose H and K are subgroups of G such that $S^H = S^K$. Applying Corollary 12.5.3 to the Galois extensions S/S^H and S/S^K, we conclude that $K \subseteq H \subseteq K$.

Now let T be a subring of S containing R and separable over R. Let $H = \{\sigma \in G \mid \sigma(x) = x \text{ for all } x \in T\}$. Then H is a subgroup of G and $T \subseteq S^H$. Let $\sigma_1, \dots, \sigma_n$ be a full set of distinct left coset representatives of H in G and assume $\sigma_1 = 1$. By Exercise 1.3.8, the homomorphisms $h_1, \dots, h_n$ defined by $h_i(x \otimes y) = x\sigma_i(y)$ are distinct S-algebra homomorphisms from $S \otimes_R T$ to S. By Corollary 4.3.2, $S \otimes_R T$ is a separable S-algebra. Since S is connected, Corollary 4.4.8 implies there is an idempotent e_1 in $S \otimes_R T$ with $h_1(e_1) = 1$, $h_1(w)e_1 = we_1$ for all $w \in S \otimes_R T$, and $h_i(e_1) = 0$ if $i \neq 1$. Let $e_1 = \sum_{j=1}^m s_j \otimes t_j$ be a representation of e_1. Then

$$h_i(e_1) = \sum_{j=1}^n s_j \sigma_i(t_j) = \begin{cases} 0 & \text{if } i \neq 1, \\ 1 & \text{if } i = 1. \end{cases}$$

It follows that

$$\sum_{j=1}^n s_j \sigma(t_j) = \begin{cases} 1 & \text{if } \sigma \in H, \\ 0 & \text{if } \sigma \notin H, \end{cases}$$

for every $\sigma \in G$. Applying σ^{-1} to both sides of this equation and relabeling we have

$$\sum_{j=1}^n \tau(s_j) t_j = \begin{cases} 1 & \text{if } \tau \in H, \\ 0 & \text{if } \tau \notin H. \end{cases}$$

By Proposition 12.2.8, there exists $c \in S$ with $\sum_{\rho \in H} \rho(c) = 1$. Then for any a in S^H,

$$\begin{aligned} a &= \sum_{\sigma \in H} \sigma(ca) \\ &= \sum_{\sigma \in G} \sum_{j=1}^n \sigma(s_j ca) t_j \\ &= \sum_{j=1}^n \Big(\sum_{\sigma \in G} \sigma(s_j ca) \Big) t_j. \end{aligned}$$

Since $\sum_{\sigma \in G} \sigma(s_j ca)$ is in R, and and $t_j \in T$, we see that $a \in T$. Hence $T = S^H$.

Finally let T be a separable extension of R in S such that T/R is Galois. We have already shown that $T = S^H$ for some subgroup H of G. The reader should verify that $\sigma H \sigma^{-1} = H$ for all $\sigma \in G$ if and only if $\sigma(T) = T$ for all $\sigma \in G$. There are $\operatorname{Rank}_R(T)$ automorphisms of T since T is a Galois extension of R. By Theorem 12.5.2, any homomorphism from T to S is an automorphism of T. In particular $\sigma(T) = T$ for all $\sigma \in G$. This completes the proof. □

5.2. Exercises.

Exercise 12.5.5. In this exercise we outline a proof that every crossed product is isomorphic to one defined by a factor set which is normalized. This implies that there was no loss of generality in Definition 12.1.2 when we started with a normalized 2-cocycle. Let S be a commutative ring and G a finite group of automorphisms of S. Let Δ denote the left S-module $\bigoplus_{\sigma \in G} S u_\sigma$. Let $f \in Z^2(G, S^*)$ be a 2-cocycle which is not necessarily normalized.

(1) Let $1 \in G$ denote the group identity element. Show that the multiplication formula (12.4) makes the left S-module Δ into a ring with multiplicative identity $f(1,1)^{-1} u_1$. (Hint: Use the 2-cocycle identity (12.1) to show $f(\sigma, 1) = \sigma f(1,1)$ and $f(1,1) = f(1,\sigma)$.)

(2) Let $g : G \to S^*$ be defined by $g(\sigma) = (\sigma f(1,1))^{-1}$. Define a 2-coboundary h by $h(\sigma, \tau) = \sigma g(\tau) g(\sigma) g(\sigma\tau)^{-1}$. Show that $f_1 = fh$ is a normalized 2-cocycle. That is, $f_1(\sigma, 1) = f_1(1, \tau) = 1$.

Exercise 12.5.6. Let S be a commutative ring, G a finite group of automorphisms of S, R a subring of S^G, f a normalized 2-cocycle in $Z^2(G, S^*)$, and T a commutative R-algebra. Prove the following.

(1) Every $\sigma \in G$ defines a T-algebra automorphism $\sigma \otimes 1 : S \otimes_R T \to S \otimes_R T$.

(2) By composition with the map $S \to S \otimes_R R \to S \otimes_R T$, f defines a 2-cocycle $f : G \times G \to S^* \to (S \otimes_R T)^*$ in $Z^2(G, (S \otimes_R T)^*)$.

(3) $\Delta(S/R, G, f) \otimes_R T$ is isomorphic to $\Delta(S \otimes_R T/T, G, f \otimes 1)$ by the map which sends $au_\sigma \otimes x$ to $(a \otimes x)u_\sigma$.

Exercise 12.5.7. This exercise is based on [**Sal82**, Lemma 0.1]. Suppose $R \subseteq S \subseteq T$ are extensions of commutative rings such that S/R is Galois with finite group H and T/S is Galois with finite group K. Then we have a chain of subgroups $K \subseteq \operatorname{Aut}_S(T) \subseteq \operatorname{Aut}_R(T)$. For each $\sigma \in H$ assume that

σ extends to an automorphism $z_\sigma \in \operatorname{Aut}_R(T)$. Moreover, assume that the following two properties hold.

(a) For all $\sigma, \tau \in H$, $z_\sigma z_\tau \in z_{\sigma\tau} K$.

(b) For all $\sigma \in H$, $z_\sigma K z_\sigma^{-1} \subseteq K$.

Prove the following.

(1) $G = \{z_\sigma k \mid \sigma \in H, k \in K\}$ is a subgroup of $\operatorname{Aut}_R(T)$.

(2) T/R is a Galois extension with group G.

(Hint: For (2) use Theorem 12.2.9 (6). The set $\{z_\sigma k(x) - x \mid x \in T\}$ contains the set $\{z_\sigma(x) - x \mid x \in S\}$.)

Exercise 12.5.8. Let R be a henselian local ring with maximal ideal $\mathfrak{m}$ and residue field k. Use Proposition 10.1.8 to prove the following.

(1) Let S be a connected Galois extension of R with finite group G. show that $S/\mathfrak{m}S$ is a Galois extension field of k, and the Galois group is isomorphic to G.

(2) Let F/k be a Galois extension of fields with finite group G. Show that there is a Galois extension S/R such that S is a local ring, $F \cong S \otimes_R k$ and $\operatorname{Aut}_R(S) \cong \operatorname{Aut}_k(F)$.

Exercise 12.5.9. Let S/R be a commutative Galois extension with finite group G and T/R a commutative Galois extension with finite group H. Show that $S \otimes_R T$ is a Galois extension of R with group $G \times H$. (Hint: Show that $\Delta(S/R, G, 1) \otimes_R \Delta(T/R, H, 1) \cong \Delta((S \otimes_R T)/R, G \times H, 1)$.)

Exercise 12.5.10. In the context of Proposition 12.2.11, let R_1 be any subring of R. Show that the R-algebra R_0 can be chosen so that it contains R_1 and as a ring is generated by adjoining to R_1 a finite set of elements.

Exercise 12.5.11. Let S/R be a Galois extension of commutative rings with finite group G. Let $\Delta = \Delta(S/R, G, 1) = \bigoplus_{\sigma \in G} S u_\sigma$ be the trivial crossed product.

(1) If $Z(G)$ denotes the center of G, then show that there is a monomorphism of groups $Z(G) \to \operatorname{Aut}_R(\Delta)$ defined by sending $\tau \in Z(G)$ to the inner automorphism "conjugation by u_τ".

(2) Assume G is abelian. By (1), every $\tau \in \operatorname{Aut}_R(S)$ extends to an R-algebra automorphism of Δ. In this way, view G as a subgroup of $\operatorname{Aut}_R(\Delta)$, and show that Δ^G is the commutative ring $\bigoplus_{\sigma \in G} R u_\sigma$.

(3) Assume G is abelian. As in (2), show that the ring $\bigoplus_{\sigma \in G} R u_\sigma$ is isomorphic to the group ring $R(G)$ and is a maximal commutative subring of Δ.

Exercise 12.5.12. Let S/R and S'/R be two Galois extensions of commutative rings with the same finite group G. Let $f : S \to S'$ be an R-algebra homomorphism which is also an $R(G)$-module homomorphism.

(1) Show that f extends to an R-algebra homomorphism

$$\Delta(S/R, G, 1) \xrightarrow{\bar{f}} \Delta(S'/R, G, 1)$$

which is an isomorphism. (Hint: Theorem 7.6.1.)

(2) Prove that f is an isomorphism.

6. The Embedding Theorem

The Embedding Theorem says that a commutative separable R-algebra A that is an R-module progenerator can be embedded as a R-subalgebra of a Galois extension T/R in such a way that $A \cong T^H$ for some subgroup H of the Galois group G. We prove two versions of this theorem. The first, Theorem 12.6.1, applies when A is not necessarily connected and in general it yields a Galois extension that is not connected. When the ring A is connected, it is sometimes desirable to embed A into a Galois extension T which is itself connected. If F/k is a finite dimensional separable extension of fields, then there is a finite dimensional Galois extension K/k which contains F as an intermediate field. In fact, K can be taken to be the splitting field of the set of irreducible polynomials of a finite generating set for F/k. The counterpart of this theorem for connected rings is proved in Theorem 12.6.3.

6.1. Embedding a Separable Algebra.

Let $J = \{1, \dots, n\}$ and $J^n = J \times \cdots \times J$ (n times). We view the symmetric group S_n as the group of permutations of J.

Theorem 12.6.1. *(Embedding Theorem) Let A be a commutative separable R-algebra that as a module is an R-progenerator of constant rank n. Then there exists a Galois extension T/R with finite group G, a subgroup $H \subseteq G$, and an R-algebra isomorphism $A \to T^H$. The group G can be taken to be isomorphic to the symmetric group S_n, and H isomorphic to S_{n-1}.*

Proof. The following proof is from [**Sal99**]. Let $T = A^{\otimes n} = A \otimes_R \cdots \otimes_R A$ be the tensor product of n copies of A. Let $G = S_n$ be the symmetric group. Then G acts on T as a group of R-algebra automorphisms by permuting the factors. If $\sigma \in G$, then on a basic element of the tensor product, we have

$$\sigma(a_1 \otimes \cdots \otimes a_n) = a_{\sigma^{-1}(1)} \otimes \cdots \otimes a_{\sigma^{-1}(n)}.$$

For $i \in J$, let $A_i = R \otimes_R \cdots \otimes_R A \otimes_R \cdots \otimes_R R$ denote the R-subalgebra of T corresponding to the ith factor of T. Because A is an R-progenerator, A is faithfully flat over R (Proposition 2.4.6). Therefore, the assignment

$x \mapsto 1 \otimes \cdots \otimes x \otimes \cdots \otimes 1$ defines an R-algebra isomorphism $A \to A_i$. For every pair $i \neq j$, let A_{ij} denote the R-subalgebra of T generated by A_i and A_j. Then $A_{ij} = R \otimes_R \cdots \otimes_R A \otimes_R \cdots \otimes_R A \otimes_R \cdots \otimes_R R$ is isomorphic to $A \otimes_R A$. Because A is a commutative separable R-algebra, there is a unique separability idempotent $e \in A \otimes_R A$ (Exercise 4.1.8). Let $e_{ij} \in A_{ij}$ be the separability idempotent. Let

$$f = \prod_{i \neq j} (1 - e_{ij})$$

where the product is over all $(i, j) \in J^2$ such that $i \neq j$. Then f is an idempotent in T. For any $\sigma \in G$, $\sigma(A_{\sigma^{-1}(i)\sigma^{-1}(j)}) = A_{ij}$, hence $\sigma(f) = f$. We make Tf into an R-algebra by $R \to Rf$. Therefore G acts as a group of R-algebra automorphisms on Tf.

To show that Tf is a Galois extension of R with group G, it suffices to show that it is true after a faithfully flat change of base (Lemma 12.2.7). Proposition 4.6.11 shows there is a faithfully flat R-algebra B such that $A \otimes_R B$ is isomorphic to B^n as B-algebras. Therefore, we assume $A = R^n$ is the trivial extension. Let $e_1, \ldots, e_n$ be the standard orthogonal idempotents in $A = R^n$. Given a sequence $(i_1, \ldots, i_n) \in J^n$ there is a corresponding idempotent $e_{i_1} \otimes \cdots \otimes e_{i_n}$ in $T = A^{\otimes n}$. By Exercise 4.1.11 the separability idempotent in $A \otimes_R A = R^n \otimes_R R^n$ is $e = \sum_{k=1}^{n} e_k \otimes e_k$. Using this, one verifies that

$$(e_i \otimes e_j)(1 - e) = \begin{cases} e_i \otimes e_j & \text{if } i \neq j, \\ 0 & \text{otherwise.} \end{cases}$$

Therefore, in $T = A^{\otimes n}$ it is true that

$$(e_{i_1} \otimes \cdots \otimes e_{i_n})f = \begin{cases} e_{i_1} \otimes \cdots \otimes e_{i_n} & \text{if } (i_1, \ldots, i_n) \text{ is a permutation in } S_n, \\ 0 & \text{otherwise.} \end{cases}$$

The set $\{e_{i_1} \otimes \cdots \otimes e_{i_n} \mid (i_1, \ldots, i_n) \in J^n\}$ is a free R-module basis for T. We have seen that the R-module Tf is generated by the set $\{e_{i_1} \otimes \cdots \otimes e_{i_n} \mid (i_1, \ldots, i_n) \in S_n\}$. So Tf is a free R-module of rank $n!$. As shown in Example 12.2.4, Tf is a Galois extension of R with group $G = S_n$.

The injection $A \to A_1 \subseteq T$ followed by the projection onto Tf defines the surjection $\phi : A \to A_1 f$. Let $H \cong S_{n-1}$ be the subgroup of $G = S_n$ that fixes 1. Then $A_1 f \subseteq (Tf)^H$. By Theorem 12.5.1, Tf is a Galois extension of $(Tf)^H$ with group H. By Corollary 4.4.4, we know $(Tf)^H$ is an R-progenerator module of rank n. Since Tf has constant rank $(n-1)!$ over $(Tf)^H$, by Corollary 3.2.13 we know $(Tf)^H$ has constant rank n over R.

To see that $\phi : A \to (Tf)^H$ is an isomorphism, it suffices to show it is an isomorphism after a faithfully flat base change. As before, we assume that

$A = R^n$. For $j \in J$, the image of ϕ contains

$$\begin{aligned}\xi_j = (e_j \otimes 1 \otimes \cdots \otimes 1)f &= (e_j \otimes \sum_{k=1}^{n} e_k \otimes \cdots \otimes \sum_{k=1}^{n} e_k)f \\ &= \sum_{\sigma \in S_n, \sigma(j)=j} e_j \otimes e_{\sigma(2)} \otimes \cdots \otimes e_{\sigma(n)}\end{aligned}$$

where the last summation is over all permutations $\sigma \in S_n$ that fix j. The reader should verify:

(1) $\xi_1, \ldots, \xi_n$ are independent over R, hence ϕ is one-to-one.

(2) $\xi_1, \ldots, \xi_n$ can be extended to a free basis for Tf, thus the image of ϕ is a direct summand of Tf.

From (1) and (2) it follows that $\phi : A \to (Tf)^H$ has a left inverse. Since $\operatorname{Rank}_R(A) = \operatorname{Rank}_R((Tf)^H) = n$, ϕ is an isomorphism. $\square$

6.2. Embedding a Connected Separable Algebra.

Lemma 12.6.2. *Let R be a commutative connected ring, S a commutative separable R-algebra which is an R-progenerator, and T a commutative connected faithful R-algebra. Let $m = \operatorname{Rank}_R(S)$. Then the following are equivalent.*

(1) *There exist $\operatorname{Rank}_R(S)$ distinct R-algebra homomorphisms from S to T.*

(2) *There is a T-algebra isomorphism $S \otimes_R T \cong T^m$.*

Proof. (1) implies (2): Let $\sigma_1, \ldots, \sigma_m$ be distinct R-algebra homomorphisms from S to T. Since S is a flat R-module, $S \to S \otimes_R R \to S \otimes_R T$ is one-to-one. Let $\pi_i : S \otimes_R T \to T$ be defined by $\pi_i(s \otimes t) = \sigma_i(s)t$. By Exercise 1.3.8, the π_i are m distinct T-algebra homomorphisms from $S \otimes_R T$ to T. By Corollary 4.4.8, there are orthogonal idempotents $e_1, \ldots, e_m$ in $S \otimes_R T$ with $(S \otimes_R T)e_i \cong T$ and

$$S \otimes_R T = (S \otimes_R T)e_1 \oplus \cdots \oplus (S \otimes_R T)e_m \oplus (S \otimes_R T)(1 - e_1 - e_2 - \cdots - e_m).$$

Because $S \otimes_R T$ has constant rank m over T, $1 - e_1 - e_2 - \cdots - e_m = 0$.

(2) implies (1): Let $\pi_i : S \otimes_R T \to T$ be the T-algebra projection onto the i-th component of the given direct sum decomposition. Let σ_i be the restriction of π_i to $S \otimes 1$. Then one quickly checks that $\sigma_1, \ldots, \sigma_m$ are the required R-algebra homomorphisms. $\square$

Theorem 12.6.3. *(Embedding Theorem) Let S be a commutative connected separable R-algebra which is an R-module progenerator. Then there is a connected Galois extension N/R with finite group G, a subgroup $H \subseteq G$, and an R-algebra isomorphism $S \to N^H$.*

Proof. This proof is from [**DI71**]. Let $n = \operatorname{Rank}_R(S)$. By Exercise 4.6.19, there is a commutative connected separable R-algebra T which is an R-progenerator and $T \otimes_R S \cong T^n$. By Proposition 2.4.6, $R \cdot 1$ is an R-module direct summand of both S and T so we can identify S with $1 \otimes S$ and T with $T \otimes 1$ in $T \otimes_R S$. If π_i is the T-algebra projection onto the ith coordinate in the decomposition of $T \otimes S$, then π_i restricted to $1 \otimes S$ is an R-algebra homomorphism from S to T which we denote σ_i.

$$\begin{array}{ccccc} 1 \otimes S & \xrightarrow{\subseteq} & T \otimes_R S & \xrightarrow{\cong} & T^n \\ {\scriptstyle =}\uparrow & & & & \downarrow{\scriptstyle \pi_i} \\ S & & \xrightarrow{\quad\sigma_i\quad} & & T. \end{array}$$

Since π_i is a T-algebra homomorphism, by Exercise 1.3.8, π_i is completely determined by σ_i. By Corollary 4.4.7, each σ_i is a monomorphism. By Theorem 12.5.2, $\sigma_1, \ldots, \sigma_n$ are all the distinct monomorphisms from S to T. By Exercise 1.3.8, the R-algebra homomorphisms $\sigma_1, \ldots, \sigma_n$ determine a unique R-algebra homomorphism $h : S^{\otimes n} \to T$, where $S^{\otimes n}$ is the tensor product over R of n copies of S. Then $h(x_1 \otimes \cdots x_m) = \sigma_1(x_1) \cdots \sigma_n(x_n)$. Let

$$N = \sigma_1(S) \cdots \sigma_n(S)$$

be the image of h. By Proposition 4.3.6 and Corollary 4.4.6, N is separable over R and an R-progenerator.

Now we show that N is a Galois extension of R. By Theorem 12.2.9 (2), it is enough to find a group G of R-automorphisms of N such that $R = N^G$. Since N is an R-progenerator, by Proposition 2.4.6 we can identify S with $1 \otimes S \subseteq N \otimes_R S$. Let $h_i : N \otimes_R S \to N$ be defined by $h_i(x \otimes y) = x\sigma_i(y)$. By Exercise 1.3.8, $h_1, \ldots, h_n$ are distinct N-algebra homomorphisms. By Lemma 12.6.2, there is an N-algebra isomorphism $N \otimes_R S \cong N^n$. The homomorphism $1 \otimes h : N \otimes_R (S^{\otimes n}) \to N \otimes_R N$ is an N-algebra epimorphism. By Corollary 4.4.7, the kernel of $1 \otimes h$ is generated by an idempotent in

$$N \otimes_R \left(S^{\otimes n}\right) \cong \bigoplus_{i=1}^{n^n} N.$$

Since N has no idempotents other than 0 and 1, the only idempotents in the ring $\bigoplus_{i=1}^{n^n} N$ are sums of the standard idempotents. Therefore, $N \otimes_R N$ is a free N-module. Thus

$$N \otimes_R N = N \oplus \cdots \oplus N$$

where there are $m = \operatorname{Rank}_R(N)$ copies of N on the right hand side. Let ρ_i be the N-algebra projection of $N \otimes_R N$ onto the ith coordinate in the direct sum decomposition of $N \otimes_R N$. Define an R-algebra homomorphism

τ_i from N to N by letting τ_i be the restriction of ρ_i to $1 \otimes N$. By Corollary 4.4.7, τ_i is a monomorphism, and $\tau_i(N)$ is separable over R and an R-progenerator. Therefore by Corollary 4.4.4, N is a separable extension of $\tau_i(N)$ and finitely generated projective as a $\tau_i(N)$-module. Because $\operatorname{Rank}_R(N) = \operatorname{Rank}_R(\tau_i(N))$, Corollary 3.2.13 implies $\operatorname{Rank}_{\tau_i(N)}(N) = 1$. Then Proposition 2.4.6 implies $\tau_i(N) = N$. We have constructed a set $\tau_1, \dots, \tau_m$ of automorphisms of N. By Theorem 12.5.2, there are no more than m distinct R-automorphisms of N. The N-algebra homomorphisms ρ_i are completely determined by the R-algebra homomorphisms τ_i. Consequently, $G = \{\tau_1, \dots, \tau_m\}$ is the group of all R-automorphisms of N. By Theorem 4.4.2, N is a separable extension of N^G. By Theorem 12.5.4 N/N^G is Galois with group G, and

$$\operatorname{Rank}_{N^G}(N) = [G : 1] = \operatorname{Rank}_R(N).$$

Clearly $R \subseteq N^G$. By Corollary 4.4.4, N^G is separable over R and is an R-module progenerator. Therefore, Corollary 3.2.13 implies $\operatorname{Rank}_R(N^G) = 1$. Hence $R = N^G$, by Proposition 2.4.6. By Theorem 12.5.4, the image of the R-algebra monomorphism $\sigma_1 : S \to N$ corresponds to a fixed subring N^H. This completes the proof. □

Corollary 12.6.4. *Let R be a local ring and A an Azumaya R-algebra.*

(1) *There exists a Galois extension T/R which is a splitting ring for A.*

(2) *In (1), T can be assumed to be connected.*

(3) *If R is a henselian local ring, then in (1), T can be assumed to be a local ring and $R \to T$ a local homomorphism of local rings.*

Proof. (1): By Theorem 10.3.5 there is a separable R-algebra S which is finitely generated and free as an R-module and which is a splitting ring for A. By Theorem 12.6.1, we can embed S as a subalgebra of a Galois extension T/R with finite group G. Since the Brauer group is a functor, T is a splitting ring for A.

(2): Start with S as in (1). By Exercise 2.3.12, $\operatorname{Spec} S$ has only a finite number of connected components. Let $S = S_0 \oplus S_1$ where S_0 is connected. The R-algebra homomorphism $S \to S_0$ makes S_0 into a separable R-algebra which is finitely generated and projective as an R-module. Since S splits A, so does S_0. By Theorem 12.6.3, we can embed S_0 as a subalgebra of a connected Galois extension T/R with finite group G.

(3): Let $\mathfrak{m}$ be the maximal ideal of R and $k = R/\mathfrak{m}$ the residue field. By Corollary 7.5.8 there exists a Galois extension of fields K/k with finite group G such that $A \otimes_R k$ is split by K. By the Primitive Element Theorem (see, for example, [**DF04**, Theorem 14.4.25]), $K = k(\alpha)$. Let $f \in k[x]$ be the irreducible polynomial for α. Lift f to a monic polynomial $p \in R[x]$. As in Theorem 10.3.2, $S = R[x]/(p)$ is a separable R-algebra which is finitely generated and free. Also, $S \otimes_R k = S/\mathfrak{m}S = k(\alpha)$, $R \to S$ is a local homomorphism of local rings, and A is split by S. By Proposition 10.1.8, $\operatorname{Aut}_R(S) \cong G = \operatorname{Aut}_k(K)$. By Exercise 12.5.6, $\Delta(S/R, G, 1) \otimes_R k$ is isomorphic to $\Delta(K/k, G, 1)$. By Corollary 8.1.25 and Definition 12.2.1, S/R is Galois with group G. □

7. Separable Polynomials

Most of the material in this section was originally published in [**Jan66**].

Let R be a commutative ring. As in Section 4.6.1, a polynomial $f \in R[x]$ is said to be separable in case f is monic and $R[x]/(f)$ is a separable R-algebra. If S is a commutative R-algebra, then an element $b \in S$ is separable if b is the root of a separable polynomial $f \in R[x]$. By $R[b]$ we denote the R-subalgebra of S generated by b.

Proposition 12.7.1. *Let S be a commutative R algebra and b an element of S.*

(1) *If b is a separable element in S, then $R[b]$ is a separable R-subalgebra of S.*
(2) *Assume S is a connected, separable R-algebra and an R-module progenerator.*
   (a) *If $R[b]$ is a separable extension of R in S, then b is separable. Moreover, there is a separable polynomial $f \in R[x]$ and $x \mapsto b$ defines an R-algebra isomorphism $R[x]/(f(x)) \to R[b]$.*
   (b) *If b is separable, then $R[b]$ is a free R-module and $R[b]$ is a separable extension of R in S.*

Proof. (1): If b is a separable element in S, then there is a separable polynomial $f(x) \in R[x]$ with $f(b) = 0$. Thus $R[b]$ is a homomorphic image of $R[x]/(f(x))$. By Proposition 4.3.6, $R[b]$ is separable over R.

(2): By the Embedding Theorem (Theorem 12.6.3), we assume S is a Galois extension of R with finite Galois group G. By Theorem 12.5.4, there is a subgroup H of G with $S^H = R[b]$ and S is Galois over S^H with group H. By Corollary 4.4.4, S^H is an R-progenerator. By Corollary 3.2.13, we have $\operatorname{Rank}_R(S^H) = [G : H]$. Let $\sigma_1, \dots, \sigma_n$ be a full set of left coset representatives of H in G and let

$$f(x) = \prod_{i=1}^{n} (x - \sigma_i(b)).$$

Since $R[b]$ is fixed by elements of H, $f(x)$ is a monic polynomial in $R[x]$ of degree n and $f(b) = 0$. To finish the proof, we show that $f(x)$ is separable. Since f is monic, $R[x]/(f(x))$ is a free R-module of rank n. Let $\eta : R[x]/(f(x)) \to R[b]$ be the R-algebra epimorphism induced by $x \mapsto b$. Because $R[b]$ is R-projective, the sequence of R-modules $R[x]/(f(x)) \to R[b] \to 0$ splits. But $R[b]$ and $R[x]/(f(x))$ have the same rank, so η is an isomorphism. This proves f is separable.

Part (b) follows from (1) and the proof of (a). □

For a connected ring R, Corollary 12.7.2 classifies homomorphic images of $R[x]$ that are separable over R and R-module progenerators.

Corollary 12.7.2. *Let R be a commutative connected ring and I an ideal in $R[x]$ such that $A = R[x]/I$ is a separable R-algebra that is an R-module progenerator. Then I is a principal ideal that is generated by a separable polynomial.*

Proof. By Exercise 2.3.12, there are primitive idempotents $e_1, \dots, e_n$ in A such that $A = Ae_1 \oplus \cdots \oplus Ae_n$. Then each Ae_i is generated as an R-algebra by xe_i, is a connected separable R-algebra, and an R-progenerator. By Proposition 12.7.1, there is a separable polynomial $g_i(x) \in R[x]$ such that $R[x]/(g_i) \to R[xe_i] = Ae_i$ is an isomorphism. By the Chinese Remainder Theorem, the ideals (g_i) are pairwise comaximal and $I = (g_1 \cdots g_n)$ is principal. □

Corollary 12.7.3. *Let R be a local ring with maximal ideal $\mathfrak{m}$. Let S be a commutative separable R-algebra which is an R-module progenerator. Assume either*

(1) *S is a local ring and the structure homomorphism $\theta : R \to S$ is a local homomorphism of local rings, or*

(2) *the residue field $R/\mathfrak{m}$ is infinite.*

Then there is a separable element $a \in S$ such that $S = R[a]$.

Proof. By Corollary 4.6.3 if (1) is satisfied or Corollary 4.6.4 if (2) is satisfied, there is a primitive element $a \in S$ such that $S = R[a]$. By Corollary 12.7.2, a is a separable element. □

The next result is due to T. McKenzie ([**McK98**]).

Corollary 12.7.4. *Let R be a local ring with maximal ideal $\mathfrak{m}$ and assume the residue field $k = R/\mathfrak{m}$ is finite. Let S be a commutative connected separable R-algebra which is an R-progenerator module. Then there exist*

(1) *a monic separable polynomial $f \in R[x]$ such that $R[x]/(f)$ is connected, and*

(2) *an R-algebra monomorphism $S \to R[x]/(f)$.*

Proof. By the Embedding Theorem (Theorem 12.6.3), it suffices to prove the corollary when S is a Galois extension of R. Then $S \otimes_R k$ is a Galois extension of k. By Theorem 12.3.4, $S \otimes_R k \cong F \oplus \cdots \oplus F = F^n$, for some finite field extension F/k of degree d. Therefore, S has exactly n maximal ideals, say $\mathfrak{q}_1, \ldots, \mathfrak{q}_n$. For each i, $\mathfrak{q}_i \cap R = \mathfrak{m}$ and $S/\mathfrak{q}_i \cong F$. By Exercise 4.5.18, there is some integer $m \geq 1$ which is relatively prime to d such that if E/k is a field extension of degree md, then $E^n = E \oplus \cdots \oplus E$ is a homomorphic image of $k[x]$. Let $g_0 \in k[x]$ be a monic irreducible polynomial of degree m. Lift g_0 to a monic polynomial $g \in R[x]$. Since $R[x]/(g) \otimes_R k$ is separable over k, g is a separable polynomial. Set $T = S[x]/(g)$. Then T is separable over S and a free S-module of rank m. Since $\gcd(d, m) = 1$, g_0 is irreducible in $(S/\mathfrak{q})[x] \cong F[x]$ (for example, Corollary 13.2.22). Using the proof of Corollary 12.7.2, we see that T is connected. We have $T \otimes_R k = (S[x]/(g)) \otimes_R k = (S/\mathfrak{m})[x]/(g) = F[x]/(g_0) \oplus \cdots \oplus F[x]/(g_0) \cong E \oplus \cdots \oplus E$ is a homomorphic image of $k[x]$. Pick an element $\alpha \in T$ such that α restricts to a generator for $T \otimes_R k$ over k. Define $\theta : R[x] \to T$ by $\theta(x) = \alpha$. Then T is generated as an R-module by the image of θ and $\mathfrak{m}T$. By Nakayama's Lemma (Corollary 1.1.17), θ is onto. By Corollary 12.7.2, the kernel of θ is generated by a separable polynomial. □

Example 12.7.5. Separable polynomials do not play the central role in the Galois theory of commutative rings as they do in the field theory. Examples exist of separable extensions S/R for which the only elements $x \in S$ that are separable over R are those $x \in R$. For one such example due to Dedekind, see [**Wei63**, p. 170]. In this example, R is a local principal ideal domain and S a connected finitely generated projective separable extension of R.

Example 12.7.6. Let R be a commutative ring and $f(x) = x^2 - x \in R[x]$. If R has at least two elements, then $0 \neq 1$ and by the Chinese Remainder Theorem $R[x]/(f(x)) \cong R \oplus R$. By Exercise 4.1.11, $f(x)$ is a separable polynomial. Every idempotent in R is a root of $f(x)$. This shows that in Lemma 12.7.7 it is necessary to assume that S is connected.

Lemma 12.7.7. *Let R be a commutative ring and $f \in R[x]$ a separable polynomial. Let S be a commutative connected R-algebra and $a_1, a_2 \in S$. If $f(a_1) = f(a_2) = 0$ and $a_1 - a_2 \neq 0$, then $a_1 - a_2$ is a unit in S.*

Proof. The reader should verify that $a_2 - a_1$ is a unit in S if and only if the principal ideals $(x - a_1)$, $(x - a_2)$ are comaximal in $S[x]$. Notice that $S[x]/(f(x)) = R[x]/(f(x)) \otimes_R S$ is separable over S by Corollary 4.3.2.

Consider the commutative diagram

$$\begin{array}{ccc} S[x] & \xrightarrow{\quad\psi\quad} & \frac{S[x]}{(x-a_1)} \oplus \frac{S[x]}{(x-a_2)} \\ & {\scriptstyle\eta}\searrow \qquad \nearrow{\scriptstyle \phi=(\phi_1,\phi_2)} & \\ & \frac{S[x]}{(f(x))} & \end{array}$$

where all of the maps are the natural homomorphisms. Then ψ is onto if and only if ϕ is onto. Using Corollary 4.4.8 we find that for each i, the kernel of ϕ_i is idempotent generated. More specifically, since S is connected, there are idempotents e_1, e_2 in $S[x]/(f(x))$ such that $e_1e_2 = 0$, $xe_i = a_ie_i$, and the kernel of ϕ_i is equal to the principal ideal generated by $1 - e_i$. Since $e_2(1 - e_1) + (1 - e_2) = e_2 + 1 - e_2 = 1$, the Chinese Remainder Theorem shows ϕ is onto. Therefore ψ is onto. Applying the Chinese Remainder Theorem once again, we conclude that the principal ideals $(x - a_1)$, $(x - a_2)$ are comaximal in $S[x]$. □

Theorem 12.7.8. *Let R be a commutative connected ring and $f(x)$ a monic polynomial of degree n in $R[x]$. Then the following statements are equivalent.*

(1) *$f(x)$ is a separable polynomial.*

(2) *The ideal in $R[x]$ generated by $f(x)$ and $f'(x)$ is the unit ideal.*

(3) *There is a commutative connected separable R-algebra S which is an R-module progenerator such that S contains elements $a_1, \dots, a_n$ satisfying:*
   (a) *$a_i - a_j$ is invertible in S when $i \neq j$, and*
   (b) *$f(x) = (x - a_1) \cdots (x - a_n)$ in $S[x]$.*

(4) *Let $S = R[x]/(f(x))$ and $T_R^S : S \to R$ the trace map. Write x (without any extra adornment) for the image of x in $R[x]/(f(x))$. If M is the n-by-n matrix in $M_n(R)$ whose entry in position (i, j) is $T_R^S(x^{i-1}x^{j-1})$, then the determinant of M is invertible in R.*

Proof. (1) is equivalent to (2): This is Proposition 4.6.1.

(2) implies (3): By Exercise 4.6.19, there is a connected separable R-algebra S which is an R-progenerator and an S-algebra isomorphism

$$S \otimes_R R[x]/(f(x)) \cong S \oplus \cdots \oplus S$$

where there are $n = \deg f(x)$ copies of S on the right hand side. Since $S \otimes_R R[x]/(f(x)) = S[x]/(f(x))$, we have

$$S[x]/(f(x)) = Se_1 \oplus \cdots \oplus Se_n$$

for orthogonal idempotents $e_1, \dots, e_n$ in $S[x]/(f(x))$. Let $a_1, \dots, a_n \in S$

such that $xe_i = a_ie_i$. Then $0 = f(x) = f(a_1e_1+\cdots+a_ne_n) = f(a_1)e_1+\cdots+f(a_n)e_n$ implies $f(a_i) = 0$ for each i. The correspondences $x \mapsto a_i$ provide distinct S-algebra homomorphisms $h_i : S[x]/(f(x)) \to S$. Therefore, if $i \neq j$, then $a_i \neq a_j$. This shows $f(x)$ has n roots in S. By Lemma 12.7.7, $a_i - a_j$ is invertible in S, if $i \neq j$. By Synthetic Division there is $p_1(x) \in S[x]$ such that $f(x) = (x - a_1)p_1(x)$. Since $a_1 - a_2$ is invertible and $f(a_2) = 0$, a_2 is a root of $p_1(x)$, and by the same argument $f(x) = (x - a_1)(x - a_2)p_2(x)$, for some $p_2(x) \in S[x]$. Recursively, we see that $f(x) = (x - a_1)(x - a_2)\cdots(x - a_n)$.

(3) implies (2): Assume S has the properties stated in (3). Then the ideals $(x-a_i)$ are pairwise comaximal in $S[x]$ for $i = 1,\dots,n$ since the ideal containing $x-a_i$ and $x-a_j$ contains the unit a_i-a_j. By Exercise 2.2.17, the intersection of the ideals $(x-a_i)$ is the principal ideal $(f(x))$. Since $S[x]/(x-a_i) \cong S$ by the mapping $x \mapsto a_i$, the Chinese Remainder Theorem shows that $S[x]/(f(x))$ is isomorphic to S^n. By Exercise 4.4.10 $S \otimes_R R[x]/(f(x))$ is separable over S. so by Proposition 2.4.6 and Corollary 4.3.5, $R[x]/(f(x))$ is separable over R.

(1) implies (4): If $n = \deg f(x)$, then $S = R[x]/(f(x))$ is a free R-module with basis $\{1, x, \dots, x^{n-1}\}$. Let π_i be the R-module projection of $R[x]/(f(x))$ onto the coefficient of x^i. By Exercise 4.6.16, the trace map is given by

$$T_R^S(z) = \sum_{i=0}^{n-1} \pi_i(zx^i)$$

for $z \in S$ and S/R is separable if and only if there are elements $\mu_i \in S$ with $T_R^S \cdot \mu_i = \pi_i$. Assume S is separable over R and $\mu_1, \dots, \mu_n$ are given. Then $T_R^S(\mu_ix^j) = \pi_i(x^j) = \delta_{ij}$ (the Kronecker delta function). Write

$$\mu_i = \sum_{j=0}^{n-1} a_{ik}x^k$$

where $a_{ij} \in R$. Then

$$\delta_{i,j} = T_R^S(\mu_ix^j) = \sum_{k=0}^{n-1} a_{ik}T_R^S(x^kx^j)$$

which shows the matrices (a_{ij}) and $(T_R^S(x^ix^j))$ are inverses of each other in $M_n(R)$. Hence the determinant of $(T_R^S(x^ix^j))$ is invertible in R.

(4) implies (1): Reverse the proof of (1) implies (4). □

Theorem 12.7.9. *Let S be a connected Galois extension of R with finite group G. Let $f(x)$ be a separable polynomial in $R[x]$. If $b_1, \dots, b_m$ are all the roots of $f(x)$ in S, then $R[b_1, \dots, b_m]$ is a Galois extension of R in S and $m \leq \deg(f(x))$.*

Proof. Since $R[b_1, \dots, b_m]$ is a homomorphic image of $R[b_1]\otimes_R \cdots \otimes_R R[b_m]$ and each $R[b_i]$ is separable over R, by Proposition 12.7.1, we have that $R[b_1, \dots, b_m]$ is separable over R. Let G be the Galois group of S over R. By Theorem 12.5.4, there is a subgroup H of G with $R[b_1, \dots, b_m] = S^H$. The reader should verify that H is a normal subgroup of G if and only if $\sigma(R[b_1, \dots, b_m]) \subseteq R[b_1, \dots, b_m]$ for all $\sigma \in G$. If $\sigma \in G$, then $f(\sigma(b_i)) = \sigma(f(b_i)) = 0$, hence σ is a permutation on the set $\{b_1, \dots, b_m\}$. Therefore $\sigma(R[b_1, \dots, b_m]) \subseteq R[b_1, \dots, b_m]$ for all $\sigma \in G$. By Theorem 12.5.4 (3), $R[b_1, \dots, b_m]$ is a Galois extension of R. By Theorem 12.5.2, there are at most n distinct algebra homomorphisms from $R[x]/(f(x))$ to S where $n = \operatorname{Rank}_R\big(R[x]/(f(x))\big) = \deg(f)$. The correspondences $x \mapsto b_i$ define a set of m distinct algebra homomorphisms from $R[x]/(f(x))$ to S. Thus $m \le n$. □

Definition 12.7.10. Let R be a connected commutative ring and S a connected commutative separable R-algebra which is an R-module progenerator. Let $f \in R[x]$ be a separable polynomial. Then S is called a *splitting ring* for f if there are elements $a_1, \dots, a_n$ in S such that $f(x) = (x-a_1)\cdots(x-a_n)$ and $S = R[a_1, \dots, a_n]$. If $P \subseteq R[x]$ is a set of separable polynomials, then S is called a *splitting ring* for P if each $f \in P$ factors into a product of linear polynomials over S, and if S is generated as an R-algebra by roots of polynomials in P.

Corollary 12.7.11. *Let R be a connected commutative ring and $f \in R[x]$ a separable polynomial. Then a splitting ring for f exists.*

Proof. By Theorem 12.7.8 there is a commutative connected separable R-algebra A which is an R-module progenerator and elements $a_1, \dots, a_n$ in A such that $f(x) = (x - a_1)\cdots(x - a_n)$. By Theorem 12.6.3 we can assume A is a Galois extension of R. By Theorem 12.7.9, $S = R[a_1, \dots, a_n]$ is a splitting ring for f. □

7.1. Exercise.

Exercise 12.7.12. This exercise is based on a result of [**HN70**]. Assume S is a connected, separable R-algebra which is an R-module progenerator. Assume $S \neq R$ and S is not a field. Show that S contains an element b such that $R[b]$ is not separable over R.

8. Separable Closure and Infinite Galois Theory

8.1. The Separable Closure. Let R be a connected commutative ring and Ω a connected commutative extension ring of R. We call Ω *locally separable* in case any finite subset of Ω is contained in a separable extension of R in Ω which is an R-progenerator. We call Ω *separably closed* in case any separable Ω-algebra which is an Ω-progenerator is isomorphic to the trivial extension Ω^n for some $n \geq 1$. We call Ω the *separable closure* of R in case Ω is a locally separable separably closed extension of R. In Theorem 12.8.4 we prove that a separable closure exists and is unique. The proof we give is due to David Harrison, and originally appeared in [**Jan66**]. It utilizes the arithmetic of ordinal numbers and because this subject will not be discussed in this book, the reader is referred to a book on set theory, for example [**Sto79**], or [**Kel75**] for the fundamental results. The method of Harrison is to construct the separable closure from the bottom up. In [**Mag74b**, Chapter III], A. Magid constructs the separable closure using a top down approach.

A topological space X is said to be *Hausdorff* if for any two distinct points $x, y \in X$, there are neighborhoods $x \in U$ and $y \in V$ such that $U \cap V = \emptyset$. We say X is *compact* if for any open cover $\{U_i \mid i \in D\}$ of X, there is a finite subset $J \subseteq D$ such that $\{U_i \mid i \in D\}$ is an open cover of X. Let $\{X_i \mid i \in D\}$ be a family of topological spaces indexed by a set D. The *product topology* on $\prod_{i \in D} X_i$ is defined to be the finest topology such that all of the projection maps $\pi_i : \prod_{i \in D} X_i \to X_i$ are continuous. For proofs of Theorems 12.8.1 and 12.8.2, the reader is referred to a book on Point Set Topology, for example [**Kel75**].

Theorem 12.8.1. *(Tychonoff Product Theorem) If $\{X_i \mid i \in D\}$ is a family of compact topological spaces indexed by a set D, then with the product topology, $\prod_{i \in D} X_i$ is a compact topological space.*

Theorem 12.8.2. *If $\{X_i \mid i \in D\}$ is a family of Hausdorff topological spaces indexed by a set D, then with the product topology, $\prod_{i \in D} X_i$ is a Hausdorff topological space.*

We require Lemma 12.8.3, which is a corollary to the Tychonoff Product Theorem.

Lemma 12.8.3. *Let $\{X_i, \phi_i^j\}$ be an inverse system of nonempty compact Hausdorff topological spaces for a directed index set I. Then $\varprojlim_{i \in I} X_i$ is a nonempty compact Hausdorff space in the topology inherited from $\prod_{i \in I} X_i$.*

Proof. The product $\prod_{k \in I} X_k$ is a nonempty compact Hausdorff space, by the Axiom of Choice and Theorems 12.8.1 and 12.8.2. By Proposition 1.4.5,

$$\varprojlim_{i \in I} X_i = \bigcap_{j \in I} W(j)$$

where we define

$$W(j) = \{x \in \prod_{k \in I} X_k \mid \phi_i^j(x_j) = x_i, \text{ for all } i \leq j\}.$$

First we show that $W(j)$ is nonempty and closed. Let $y \in \prod_{k \in I} X_k - W(j)$. Then there is some $i \in I$ such that $i \leq j$ and $\phi_i^j(y_j) \neq y_i$. Since X_i is Hausdorff we can find disjoint open sets U_i and V_i in X_i such that $y_i \in U_i$, $\phi_i^j(y_j) \in V_i$. Let $P_j = (\phi_i^j)^{-1}(V_i)$. Then P_j is an open subset of X_j with $\phi_i^j(P_j) \cap U_i = \emptyset$. Let

$$U = \{x \in \prod_{k \in I} X_k \mid x_i \in U_i,\, x_j \in P_j\}.$$

Since U is an open neighborhood of y in $\prod_{k \in I} X_k$ and $U \cap W(j) = \emptyset$, this shows $W(j)$ is closed. The reader should verify that

$$W(j) = \bigcup_{x_j \in X_j} \left(\bigcap_{i \leq j} \pi_i^{-1}\left(\phi_i^j(x_j)\right)\right)$$

is nonempty.

To show that $\bigcap_{j \in I} W(j) \neq \emptyset$, by Exercise 12.8.12 it suffices to show the family $\{W(j) \mid j \in I\}$ has the finite intersection property. Since I is a directed set, if $J \subseteq I$ is a finite subset, then there is an upper bound $u \in I$ for J. Therefore, if $y \in W(u)$, then $\phi_i^u(y_u) = y_i$ for all $i \leq u$. If $i \leq j \leq u$, then $\phi_i^j(y_j) = \phi_i^j \phi_j^u(y_u) = \phi_i^u(y_u) = y_i$. Thus $y \in W(j)$, for all $j \in J$. Since $\bigcap_{j \in J} W(j) \supseteq W(k) \neq \emptyset$, the family $\{W(j) \mid j \in I\}$ has the finite intersection property. This completes the proof. □

Theorem 12.8.4. *Let R be a commutative connected ring.*

(1) *A separable closure Ω of R exists.*

(2) *For any commutative separable R-algebra S which is also an R-progenerator, there are* $\operatorname{Rank}_R(S)$ *algebra homomorphisms from S to Ω. If S is connected, then the homomorphisms are monomorphisms.*

(3) *The separable closure is unique up to isomorphism.*

Proof. (1): Suppose for contradiction's sake that a separable closure does not exist. Let α be an ordinal. If $\alpha = 0$, then set $\Omega_\alpha = R$. Now suppose $\alpha > 0$. Inductively assume that for each ordinal β less than α that Ω_β is a locally separable R-algebra and that the algebras are nested. That is, if $\beta < \gamma$, then $\Omega_\beta \subseteq \Omega_\gamma$.

First suppose α is not a limit ordinal. Since $\Omega_{\alpha-1}$ is not a separable closure, there exists a connected separable algebra Ω_α over $\Omega_{\alpha-1}$ which is a progenerator module and $\Omega_\alpha \neq \Omega_{\alpha-1}$. Now we show that Ω_α is locally separable over R. Let $z_1, \dots, z_m$ be a subset of Ω_α. If $T : \Omega_\alpha \to \Omega_{\alpha-1}$ is the trace map, then by Theorem 4.6.7 there are elements $x_1, \dots, x_n, y_1, \dots, y_n$

in Ω_α such that

$$
\begin{aligned}
z_j &= \sum_{i=1}^{n} x_i T(y_i z_j) \\
x_i x_j &= \sum_{k=1}^{n} x_k T(y_k x_i x_j) \\
y_j &= \sum_{i=1}^{n} x_i T(y_i y_j).
\end{aligned}
$$

Let A be the finite subset of $\Omega_{\alpha-1}$ consisting of all of the elements $T(y_k x_i x_j)$, $T(y_i y_j)$, $T(x_j)$, $T(y_i z_j)$ where i, j, k range over all possible values. Since $\Omega_{\alpha-1}$ is locally separable over R there is an R-subalgebra $S \subseteq \Omega_{\alpha-1}$ which contains A and such that S/R is separable and S is an R-module progenerator. Let C be the S-subalgebra of Ω_α generated by $x_1, \dots, x_n$. Let T' be the restriction of T to C. Observe that $T' \in \operatorname{Hom}_S(C, S)$ and $y_1, \dots, y_n$, $z_1, \dots, z_m$ are elements of C. Applying Theorem 4.6.7 we see C/S is separable and C is an S-module progenerator. By Corollary 4.4.4, C is a separable R-algebra and an R-progenerator module. This proves Ω_α is a locally separable R-algebra.

Now suppose α is a limit ordinal. By the induction hypothesis, $\{\Omega_\beta\}_{\beta<\alpha}$ is a directed system. The reader should verify that

$$
\begin{aligned}
\Omega_\alpha &= \varinjlim_{\beta<\alpha} \Omega_\beta \\
&= \bigcup_{\beta<\alpha} \Omega_\beta
\end{aligned}
$$

is a connected locally separable extension of R and that for all $\beta < \alpha$, $\Omega_\beta \neq \Omega_\alpha$.

Let $S(R)$ be the collection of all isomorphism classes of commutative connected separable R-algebras which are R-progenerator modules. We note that $S(R)$ is a set because every $S \in S(R)$ is an R-algebra of finite presentation, and the class of all such algebras is a set. Let λ be the ordinal whose cardinality is greater than

$$
\sum_{S \in S(R)} \operatorname{Rank}_R(S)|S|
$$

where $|S|$ denotes the cardinality of S. By Theorem 12.5.2 there are at most $\operatorname{Rank}_R(S)$ isomorphisms of a separable R-algebra S which is an R-progenerator into Ω_λ. However since Ω_λ is locally separable we have

$$
\Omega_\lambda = \bigcup \{S \in S(R) \mid S \subseteq \Omega_\lambda\}.
$$

Now for each S there are at most $\operatorname{Rank}_R(S)$ copies of S in the union, hence

$$\lambda > \sum_{S \in S(R)} \operatorname{Rank}_R(S)|S| \geq |\Omega_\lambda| \geq \lambda$$

which is a contradiction. Therefore a separable closure of R must exist.

(2): Let Ω be a separable closure of R. Let S be a commutative separable R-algebra which is an R-progenerator of rank $n = \operatorname{Rank}_R(S)$. Since $\Omega \otimes_R S$ is a separable Ω-algebra that is a progenerator module, $\Omega \otimes_R S$ is isomorphic to Ω^n as an Ω-algebra. Let $\pi_1, \dots, \pi_n$ be the Ω-algebra projections from $\Omega \otimes_R S$ to Ω given by this decomposition. Because S is flat over R, we can identify S with the subring $1 \otimes S \subseteq \Omega \otimes_R S$. Let σ_i be the restriction of π_i to $1 \otimes S$. Then $\sigma_1, \dots, \sigma_n$ are distinct R-algebra homomorphisms from S to Ω. If S is connected, then by Corollary 4.4.7 each σ_i is a monomorphism.

(3): Let Ω be a separable closure of R and let Ω' be any locally separable extension of R. Let S be a separable R-algebra in Ω' which is an R-progenerator, and $G(S) = \operatorname{Hom}_{R-\mathrm{Alg}}(S, \Omega)$ the set of all R-algebra homomorphisms from S to Ω. By (2), $G(S)$ has $\operatorname{Rank}_R(S)$ elements. We make $G(S)$ into a topological space by giving $G(S)$ the discrete topology (point sets are open). In this topology $G(S)$ is a compact Hausdorff space. If T is another separable R-algebra in Ω' which is an R-progenerator and $T \subseteq S$, then restriction gives a continuous function from $G(S)$ to $G(T)$. Let $\mathfrak{S}$ be the collection of all separable R-algebras in Ω' which are R-progenerator modules. If $\mathfrak{S}$ is ordered by set inclusion, then $\mathfrak{S}$ is a directed set and $\{G(S) \mid S \in \mathfrak{S}\}$ is an inverse system of topological spaces. By Lemma 12.8.3, $\varprojlim G(S) \neq \emptyset$.

Now we show that there is a one-to-one correspondence

$$\varprojlim_{S \in \mathfrak{S}} G(S) \to \operatorname{Hom}_{R-\mathrm{Alg}}(\Omega', \Omega). \tag{12.8}$$

Given $\bar{\sigma} \in \varprojlim G(S)$, the associated R-algebra homomorphism $h_{\bar{\sigma}} : \Omega' \to \Omega$ is defined in the following way. If $x \in \Omega'$, then there is a separable R-algebra S in Ω' which is an R-progenerator such that $x \in S$. Under the natural map $\alpha_S : \varprojlim G(S) \to G(S)$, let σ be the image of $\bar{\sigma}$. Define $h_{\bar{\sigma}}(x) = \sigma(x)$. The reader should verify that $h_{\bar{\sigma}}$ is a well defined R-algebra homomorphism and the assignment $\bar{\sigma} \mapsto h_{\bar{\sigma}}$ is a one-to-one correspondence between the sets in (12.8).

By the argument above, we know there exists at least one R-algebra homomorphism $h : \Omega' \to \Omega$. We show that h is one-to-one. Suppose x is in the kernel of h. Then $x \in S$ for some $S \in \mathfrak{S}$. By (2), the restriction of h to S must be a monomorphism, hence $x = 0$. Thus every algebra homomorphism from Ω' to Ω is a monomorphism. If Ω' is also a separable closure of R, then there are R-algebra monomorphisms $h : \Omega' \to \Omega$ and $g : \Omega \to \Omega'$.

Let $\Omega'' = hg(\Omega)$. Then Ω'' is a locally separable extension of R in Ω. If $x \in \Omega$, then x is contained in a separable R-subalgebra S of Ω which is an R-progenerator. There are $\operatorname{Rank}_R(S)$ monomorphisms of S into Ω and $\operatorname{Rank}_R(S)$ monomorphisms of S into Ω'', since hg is one-to-one. Therefore $S \subseteq \Omega''$, and $x \in \Omega''$. Thus h is an isomorphism and Ω is unique. □

Corollary 12.8.5 (T. McKenzie). *Let R be a local ring and Ω the separable closure of R. Then Ω is a splitting ring (Definition* 12.7.10*) for the set of separable polynomials in $R[x]$.*

Proof. By Theorem 12.7.8, every separable polynomial f in $R[x]$ factors into a product of linear factors in $\Omega[x]$. Let $\alpha \in \Omega$. There is a separable R-subalgebra S in Ω which is an R-progenerator and $\alpha \in S$. By Corollaries 12.7.3 and 12.7.4, for some separable polynomial f, $R[x]/(f)$ is connected and contains an R-subalgebra isomorphic to S. Theorem 12.8.4 implies there is an R-algebra monomorphism $R[x]/(f) \to \Omega$ which contains S. □

8.2. The Fundamental Theorem of Infinite Galois Theory.

Corollary 12.8.6. *Let R be a connected commutative ring and Ω the separable closure of R.*

(1) *If $\sigma : \Omega \to \Omega$ is an R-algebra homomorphism, then σ is an automorphism.*
(2) *Let S be a separable R-subalgebra of Ω which is an R-progenerator module. Any R-algebra homomorphism $\sigma : S \to \Omega$ extends to an R-algebra automorphism of Ω.*

Proof. (1): This was proved in Part (3) of Theorem 12.8.4.

(2): Since $S \subseteq \Omega$ is a subring, Ω is an S-algebra. If τ is the set inclusion homomorphism, then $\tau\sigma$

$$S \xrightarrow{\sigma} \sigma(S) \xrightarrow{\tau} \Omega$$

makes Ω into another S-algebra, which is denoted ${}_\sigma\Omega$. Because Ω is a separable closure of R, Ω is also a separable closure of S and $\sigma(S)$. Consequently, ${}_\sigma\Omega$ is a separable closure of S. By Theorem 12.8.4, there is an S-algebra isomorphism $f : \Omega \to {}_\sigma\Omega$. The reader should verify that f is an extension of σ. □

Corollary 12.8.7. *Let R be a connected commutative ring and Ω the separable closure of R. Let $G = \operatorname{Aut}_R \Omega$. Then $\Omega^G = R$.*

Proof. Let $x \in \Omega - R$. Let S be an R-subalgebra of Ω containing x such that S is a separable R-algebra and an R-progenerator module. By Theorem 12.6.3, S can be embedded in a connected Galois extension N/R

with finite group $\operatorname{Aut}_R N$. By Theorem 12.8.4, N may be viewed as an R-subalgebra of Ω containing S. There is $\sigma \in \operatorname{Aut}_R N$ such that $\sigma(x) \neq x$. By Corollary 12.8.6, σ can be extended to an R-automorphism of Ω. □

Let R be a connected commutative ring and Ω the separable closure of R. Let $G = \operatorname{Aut}_R \Omega$ be the full group of R-algebra automorphisms of Ω. We have seen in the proof of Theorem 12.8.4 that $G = \varprojlim G(S)$ where S ranges over the separable R-subalgebras in Ω that are progenerator modules. Thus G is a compact Hausdorff topological space. One can check that the operation of composition in G is continuous, so G is a topological group. Let S be an extension of R in Ω and

$$G_S = \{\sigma \in G \mid \sigma(x) = x, \text{ for all } x \in S\}$$

the subgroup of G fixing S. Then one can check that G_S is closed if and only if S is locally separable and that G_S is a normal subgroup of G if and only if S is Galois over R.

Theorem 12.8.8 (Fundamental Theorem of Infinite Galois Theory). *Let R be a connected commutative ring and Ω the separable closure of R. Let $G = \operatorname{Aut}_R \Omega$ be the group of all R-algebra automorphisms of Ω. Then G is a compact Hausdorff topological group and there is a one-to-one correspondence*

$$H \mapsto \Omega^H$$
$$S \mapsto G_S$$

between the closed subgroups H of G and the locally separable extensions S of R in Ω. Moreover H is a closed normal subgroup of G if and only if S is a Galois extension of R. In this case G/H is the full group of automorphisms of S. Also S is a finitely generated R-module if and only if H is a subgroup of finite index in G and in this case $\operatorname{Rank}_R(S) = [G : H]$.

Proof. The proof is left to the reader. □

8.3. Exercises.

Exercise 12.8.9. Let k be a field and Ω the separable closure of k. Show that Ω is isomorphic to the separable closure of k in an algebraically closed extension field of k.

Exercise 12.8.10. Let R and T be connected commutative rings and $\theta : R \to T$ a homomorphism of rings. Let Ω_R and Ω_T be the separable closures of R and T respectively. Using θ, make T and Ω_T into R-algebras. Let $\mathfrak{S}$ be the collection of all separable R-algebras in Ω_R which are R-progenerator modules. For $S \in \mathfrak{S}$, let $G(S) = \operatorname{Hom}_{R\text{-Alg}}(S, \Omega_T)$ be the set of all R-algebra homomorphisms from S to Ω_T. Prove the following.

(1) $G(S)$ has $\operatorname{Rank}_R(S)$ elements.

(2) If $\mathfrak{S}$ is ordered by set inclusion, then $\mathfrak{S}$ is a directed set and $\{G(S) \mid S \in \mathfrak{S}\}$ is an inverse system of topological spaces.

(3) $\varprojlim G(S) \neq \emptyset$.

(4) There is a one-to-one correspondence

$$\varprojlim_{S \in \mathfrak{S}} G(S) \to \operatorname{Hom}_{R-\mathrm{Alg}}(\Omega_R, \Omega_T).$$

(5) There exists a homomorphism ψ for which the diagram

$$\begin{array}{ccc} \Omega_R & \xrightarrow{\psi} & \Omega_T \\ \uparrow & & \uparrow \\ R & \xrightarrow{\theta} & T \end{array}$$

commutes.

Exercise 12.8.11. Let $\mathfrak{R}$ be the category with objects being connected commutative rings, and morphisms being homomorphisms of rings. Let $\mathfrak{T}$ be the category with objects being compact Hausdorff topological groups and with morphisms being described as follows. A morphism from a group G to a group H is an equivalence class of topological homomorphisms from G to H where two homomorphisms g, h are equivalent if there is an element $y \in H$ such that $g(x) = y^{-1}h(x)y$ for all $x \in G$.

(1) As in Exercise 12.8.10, let $\theta : R \to T$ be in $\mathfrak{R}$ and $\psi : \Omega_R \to \Omega_T$ an extension of θ.
   (a) Show that to each $\tau \in \operatorname{Aut}_T(\Omega_T)$ there is a unique $\sigma \in \operatorname{Aut}_R(\Omega_R)$ such that $\tau\psi = \psi\sigma$ and that $\tau \mapsto \sigma$ defines a homomorphism of groups $h_\psi : \operatorname{Aut}_T(\Omega_T) \to \operatorname{Aut}_R(\Omega_R)$.
   (b) Show that different extensions of θ in $\operatorname{Hom}_{R-\mathrm{Alg}}(\Omega_R, \Omega_T)$ give rise to equivalent homomorphisms of groups $\operatorname{Aut}_T(\Omega_T) \to \operatorname{Aut}_R(\Omega_R)$.

(2) Show that $R \mapsto \operatorname{Aut}_R(\Omega_R)$ defines a contravariant functor $\mathfrak{F} : \mathfrak{R} \to \mathfrak{T}$.

Exercise 12.8.12. Let X be a topological space. We say that a family $\{Z_i \subseteq X \mid i \in D\}$ of closed subsets of X has the *finite intersection property* if for every finite subset $J \subseteq D$, $\bigcap_{j \in J} Z_j \neq \emptyset$. Show that the following are equivalent:

(1) X is compact.

(2) For any family $\{Z_i \subseteq X \mid i \in D\}$ of closed subsets of X with the finite intersection property, $\bigcap_{i \in D} Z_i \neq \emptyset$.

9. Cyclic Extensions

This section focuses on Galois extensions S/R with a finite cyclic Galois group. For the most part, our attention is limited to two special cases: Kummer extensions and Artin-Schreier extensions. For more on the topic of cyclic extensions, the interested reader is referred to the monograph of Greither [**Gre92**]. In Section 9.4 we derived some important results on cyclic extensions which are ramified at a finite number of height one primes. These results will now be applied in this section.

9.1. Kummer Theory. Let $n > 1$ be an integer and ζ a primitive nth root of 1 in $\mathbb{C}$. Consider $\mathbb{Z}[n^{-1}]$ and $A = \mathbb{Z}[n^{-1}, \zeta]$, which we view as $\mathbb{Z}$-subalgebras of $\mathbb{C}$. In this section we study cyclic Galois extensions of degree n of a commutative A-algebra R. Such Galois extensions are also called Kummer extensions. The set of all Kummer extensions of R of degree n form an abelian group, denoted $\operatorname{Gal}(R, \mathbb{Z}/n)$. This group fits into a short exact sequence, called the Kummer sequence, which turns out to be a very useful tool for computing $\operatorname{Gal}(R, \mathbb{Z}/n)$. The interested reader is referred to [**Mil80**] for Kummer sequences of higher étale cohomology groups.

Example 12.9.1. Let R be a commutative ring, $a \in R$, and $n > 1$. By Exercise 4.6.20, the polynomial $\phi(x) = x^n - a \in R[x]$ is separable if and only if na is invertible in R. Thus by Theorem 12.7.9, if R is connected and n is invertible in R, then there are at most n distinct nth roots of 1 in R. This need not be the case in general. For example, the group ring $\mathbb{Z}(\mathbb{Z}/2)$ of the abelian group of order two over the integers is a commutative ring in which there are more than 2 distinct square roots of 1. This commutative ring also has no idempotents other than 0 and 1.

Example 12.9.2. In the notation established above, let $n > 1$ be an integer, ζ a primitive nth root of 1 in $\mathbb{C}$, and $A = \mathbb{Z}[n^{-1}, \zeta]$. By Exercise 4.6.20, $\phi(x) = x^n - 1$ is a separable polynomial in $\mathbb{Z}[n^{-1}][x]$. The elements $1, \zeta, \dots, \zeta^{n-1}$ are distinct roots of ϕ in A. By Lemma 12.7.7, whenever $i \not\equiv j \pmod{n}$, the element $\zeta^i - \zeta^j$ is invertible in A and the ideals $(x - \zeta^i)$ and $(x - \zeta^j)$ are comaximal in $A[x]$. Let $B = A[x]/(x^n - 1)$. By the Chinese Remainder Theorem

$$\begin{aligned} B = \frac{A[x]}{(x^n - 1)} &\cong \frac{A[x]}{(x-1)} \oplus \frac{A[x]}{(x-\zeta)} \oplus \cdots \oplus \frac{A[x]}{(x-\zeta^{n-1})} \\ &\cong \bigoplus_{i=0}^{n-1} A \end{aligned} \tag{12.9}$$

where the last isomorphism sends x to the n-tuple $(1, \zeta, \dots, \zeta^{n-1})$. Let σ be the A-algebra automorphism of B which maps x to ζx. In the isomorphism of (12.9), ζx is mapped to $(\zeta, \zeta^2, \dots, \zeta^{n-1}, 1)$. Thus σ acts on A^n as a cyclic permutation of order n. If $e_0, \dots, e_{n-1}$ are the idempotents corresponding to the direct sum in (12.9), then $\sigma(e_i) = e_{i-1}$, where the subscripts are treated as elements of $\mathbb{Z}/n$. By Example 12.2.5, we see that B is isomorphic to $E_G(A)$, the trivial Galois extension of A with group $G = \langle \sigma \rangle$.

Theorem 12.9.3. *In the notation established above, let $A = \mathbb{Z}[n^{-1}, \zeta]$, R a commutative A-algebra, $G = \langle \sigma \rangle$ a cyclic group of order n, and $E = E_G(R)$ the trivial G-Galois extension of R (see Example 12.2.5). Then the following are true.*

(1) *E is generated as an R-algebra by a primitive element t and $\sigma(t) = \zeta t$.*
(2) *If τ is an R-algebra automorphism of E which is also an $R(G)$-module homomorphism, then $\tau \in G$.*
(3) *The group $\check{\mathrm{H}}^1_{\mathrm{et}}(R, \mathbb{Z}/n)$ classifies up to isomorphism the Galois extensions of R with finite cyclic Galois group $G \cong \mathbb{Z}/n$.*

Proof. (1): As shown in Example 12.9.2, $B = A[x]/(x^n - 1)$ is isomorphic to $E_G(A)$, the trivial G-Galois extension of A, both as an A-algebra, and an $A(G)$-module. Tensoring both sides of (12.9) with $(\,) \otimes_A R$ shows that E is isomorphic to $A[x]/(x^n - 1) \otimes_A R = R[x]/(x^n - 1)$ both as an R-algebra, and an $R(G)$-module. For t, take the image of x.

(2): Let $e_0, \dots, e_{n-1}$ be the idempotents of E corresponding to the direct summands in

$$E \cong \frac{R[x]}{(x^n - 1)} \cong \frac{R[x]}{(x-1)} \oplus \frac{R[x]}{(x-\zeta)} \oplus \cdots \oplus \frac{R[x]}{(x - \zeta^{n-1})}.$$

Given any maximal ideal $\mathfrak{m}$ in R, $E/\mathfrak{m}E$ has n unique primitive idempotents, namely the images of $e_0, \dots, e_{n-1}$. Hence, if $\tau \in \mathrm{Aut}_R(E)$, then τ permutes the idempotents $e_0, \dots, e_{n-1}$. By Example 12.9.2, $\sigma(e_i) = e_{i-1}$. If τ is also a G-module homomorphism, then $\tau(\sigma e_0) = \sigma(\tau(e_0))$ implies τ is a cyclic permutation of $e_0, \dots, e_{n-1}$, and is completely determined by $\tau(e_0)$. Suppose $\tau(e_0) = e_{\tau(0)}$. Then $\tau = \sigma^{n-\tau(0)}$ is in G.

(3): If S/R is a G-Galois extension, then by Theorem 12.2.10, S is a twisted form of E, for some faithfully flat étale R-algebra U. By (2), the automorphism group of $U \otimes_R E$ as a G-Galois extension of U is the cyclic group G. The group $\mathrm{H}^1(U/R, G)$ classifies the twisted forms of the G-Galois extension E/R for U/R (Proposition 5.5.7). Taking the limit over all étale covers U/R gives (3). □

Remark 12.9.4. For any commutative $\mathbb{Z}[n^{-1}, \zeta]$-algebra R, Part (3) of Theorem 12.9.3 shows that the cyclic Kummer extensions of R of degree n are classified by the abelian group $\check{\mathrm{H}}^1(R, \mathbb{Z}/n)$. Let R be any commutative ring and G any finite abelian group. Denote by $\mathrm{Gal}(R, G)$ the set of all commutative Galois extensions of R with group G. Then Harrison has shown that $\text{Gal}(R, G)$ is an abelian group (see Exercise 12.9.16). The Harrison product of two elements S and T in $\text{Gal}(R, G)$ is $(S \otimes_R T)^{\ker \mu}$, where $\mu : G \times G \to G$ is the multiplication homomorphism defined by $(x, y) \mapsto xy$. In Exercise 12.9.17 the reader is asked to verify that if R is a $\mathbb{Z}[n^{-1}, \zeta]$-algebra, and G a cyclic group of order n, then the group operation on $\check{\text{H}}^1(R, G)$ agrees with the Harrison product.

Example 12.9.5. In the notation established above, let $A = \mathbb{Z}[n^{-1}, \zeta]$, and R any commutative A-algebra. Let a be an invertible element in R. By Exercise 4.6.20,

$$S = \frac{R[x]}{(x^n - a)}$$

is a separable extension of R, S is a free R-module with basis $1, x, \dots, x^{n-1}$, and x is a unit in S. If σ is the R-algebra automorphism of S defined by $\sigma(x) = \zeta x$, then $S^{\langle \sigma \rangle} = R$. We now show that S/R is a Galois extension with cyclic group $G = \langle \sigma \rangle$. To do this, we show that $\ell : S \otimes_R S \to \nabla(S, G)$ is an isomorphism and apply Theorem 12.2.9 (5). Over S, the polynomial $y^n - a$ splits into $(y - x)(y - \zeta x) \dots (y - \zeta^{n-1} x)$. If $i \not\equiv j \pmod{n}$, then $\zeta^i - \zeta^j$ is invertible in A (Lemma 12.7.7). Hence, $\zeta^i x - \zeta^j x = (\zeta^i - \zeta^j)x$ is invertible in S and

$$\begin{aligned} S \otimes_R S &\cong S \otimes_R \frac{R[y]}{(y^n - a)} \\ &\cong \frac{S[y]}{((y - x)(y - \zeta x) \cdots (y - \zeta^{n-1} x))} \\ &\cong \bigoplus_{i=0}^{n-1} S \end{aligned}$$

is an isomorphism of S-algebras. The last isomorphism is by the Chinese Remainder Theorem and sends y to the n-tuple $(x, \zeta x, \dots, \zeta^{n-1} x)$. Notice that this agrees with the definition of the map ℓ.

Example 12.9.6. In the notation established above, let $A = \mathbb{Z}[n^{-1}, \zeta]$, and R any commutative A-algebra. In this example, we show that an element of order n in $\text{Pic}(R)$, the Picard group of R, gives rise to a cyclic Galois extension of R of degree n. Let M be an R-progenerator of constant rank one and $T_R(M)$ the tensor algebra. By Exercise 5.2.4, $T_R(M)$ is a commutative R-algebra. Assuming that M defines an element annihilated by n in $\text{Pic}(R)$, there is an R-module isomorphism $\phi : T_R^n(M) \to R$. If $x \in T_R^n(M)$, we view $\phi(x)$ as an element of $R = T_R^0(M)$, so that $x - \phi(x)$ makes sense in $T_R(M)$.

Given R, M and ϕ, we define an R-algebra

$$\Gamma = \Gamma(R, M, \phi) = \frac{T_R(M)}{(\{x - \phi(x) \mid x \in T_R^n(M)\})}.$$

Let Γ_i denote the submodule of Γ generated by the image of $T_R^i(M)$. For $i = 0, \dots, n-1$, the reader should verify that as an R-module, Γ_i is isomorphic to $T_R^i(M)$. Then as R-modules, Γ is isomorphic to $\Gamma_0 \oplus \cdots \oplus \Gamma_{n-1}$. In particular, Γ is an R-progenerator of rank n. If $x_i \in \Gamma_i$ and $y_j \in \Gamma_j$, then $x_i y_j$ is in Γ_{i+j}, where the subscripts are viewed as elements of $\mathbb{Z}/n$. We say that Γ is a *$\mathbb{Z}/n$-graded ring*. By Proposition 5.2.1, the R-algebra Γ is generated by elements from $M = \Gamma_1$.

Define $\sigma : \Gamma \to \Gamma$ by sending a homogeneous element $x_i \in \Gamma_i$ to $\zeta^i x_i$. The reader should verify that σ is an R-algebra automorphism of Γ and $\Gamma^{\langle \sigma \rangle} = R$. To show that Γ/R is Galois with cyclic group $G = \langle \sigma \rangle$, we show that

$$\ell : \Gamma \otimes_R \Gamma \to \nabla(\Gamma, G)$$

is an isomorphism and apply Theorem 12.2.9 (5). By Exercise 2.4.12, it suffices to show that ℓ is locally an isomorphism. Because M is locally free of rank one, it suffices to prove that ℓ is an isomorphism when M is a free module of rank one. If M is principal, say $M = Ra$, then by Exercise 5.2.3, the R-algebra $T_R(M)$ is isomorphic to the polynomial ring $R[x]$. The R-module $T_R^n(M)$ is generated by $a \otimes \cdots \otimes a$. Therefore, $\phi(a \otimes \cdots \otimes a) = \alpha$ is a unit of R. Let $(a)_1$ denote the element of degree 1 in Γ represented by a. Then in Γ we have $(a)_1^n = \phi(a \otimes \cdots \otimes a) = \alpha$, and $\Gamma \cong R[x]/(x^n - \alpha)$, which is a Galois extension of R by Example 12.9.5. Thus ℓ is locally an isomorphism.

Lemma 12.9.7. *Let $f : R \to S$ be a homomorphism of arbitrary commutative rings and A/R a Galois extension with finite abelian group G. Let $E_G(S)$ be the trivial G-Galois extension of S (Example 12.2.5). There exists an isomorphism $g : A \otimes_R S \to E_G(S)$ of S-algebras and $S(G)$-modules, if and only if there exists an R-algebra homomorphism $\bar{f}$ such that the diagram*

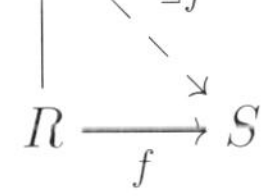

commutes. In particular, A is isomorphic to $E_G(R)$, the trivial extension of R, if and only if there is an R-algebra homomorphism $\bar{f} : A \to R$.

Proof. If $A \otimes_R S \cong S^n$, then take $\bar{f}$ to be the composition

$$A \xrightarrow{1 \otimes f} A \otimes_R S \xrightarrow{\cong} S \oplus \cdots \oplus S \xrightarrow{\pi_1} S$$

where π_1 is the projection onto the first summand. Conversely, assume $\bar{f} : A \to S$ is an extension of f. Define $g : A \otimes_R S \to \sum_{\sigma \in G} Se_\sigma$ by the rule

$$g(a \otimes s) = \sum_{\sigma \in G} \bar{f}(\sigma^{-1}(a))se_\sigma.$$

Using the fact that G is abelian, the reader should verify that g defines an S-algebra homomorphism, and an $S(G)$-module homomorphism. By Exercise 12.5.12, g is an isomorphism. □

Proposition 12.9.8. *In the notation of Section* 12.9.1, *let* $A = \mathbb{Z}[n^{-1}, \zeta]$, *$R$ any commutative A-algebra, $G = \langle \sigma \rangle$ a cyclic group of order n, and $S \in \operatorname{Gal}(R, G)$ a G-Galois extension of R.*

(1) *There exists an R-progenerator M of constant rank one whose class in $\operatorname{Pic}(R)$ is annihilated by n and S is isomorphic to the G-Galois extension $\Gamma(R, M, \phi)$ of Example* 12.9.6.
(2) *If the R-module M of (1) is free, then S is isomorphic to $R[x]/(x^n - a)$ for some invertible element a in R.*
(3) *There exists an isomorphism $g : S \to E_G(R)$ of R-algebras and $R(G)$-modules, if and only if the R-module M of (1) is free, and the polynomial $x^n - a$ of (2) has a root in R.*

Proof. (1): Because S is an R-progenerator module (Lemma 12.2.2), S is faithfully flat over R. By Theorem 12.2.10, $S \otimes_R S$ is isomorphic to the trivial G-Galois extension $E_G(S)$. In $\Delta(S/R, G, 1) \cong \operatorname{Hom}_R(S, S)$ there are the R-module homomorphisms $\sigma - \zeta^i$. Let S_i denote the kernel of $\sigma - \zeta^i$. Tensor the exact sequence of R-modules

$$0 \to S_i \to S \xrightarrow{\sigma - \zeta^i} S$$

with $S \otimes_R ()$ to get the exact sequence

$$0 \to S \otimes_R S_i \to S \otimes_R S \xrightarrow{1 \otimes (\sigma - \zeta^i)} S \otimes_R S$$

of S-modules. By Theorem 12.9.3, $S \otimes_R S_i$ is a free S-module of rank one. Hence, by Theorem 2.6.1, S_i is an R-progenerator module of rank one. Upon tensoring with $S \otimes_R ()$, the sum map $\bigoplus_{i=0}^{n-1} S_i \to S$ becomes onto. Therefore, $S_0 + S_1 + \cdots + S_{n-1} = S$. Using similar arguments, the reader should verify that

(a) S is the internal direct sum $S_0 \oplus S_1 \oplus \cdots \oplus S_{n-1}$,
(b) $S_i S_j = S_{i+j}$, where the subscripts are treated as elements of $\mathbb{Z}/n$,
(c) S is generated as an $S_0 = R$-algebra by S_1, and
(d) the class of S_1 in $\operatorname{Pic}(R)$ is annihilated by n.

Let $M = S_1$ and ϕ the isomorphism $T^n_R(M) \cong R$ of (d). Then $S \cong \Gamma(R, M, \phi)$.

(2): If $M = R\alpha$ is free, then α^n generates the unit ideal of R, and we are in the context of Example 12.9.5.

(3): By (1), if S is isomorphic to the trivial extension of R, then M is free. The other cases can be proved using (2) and Lemma 12.9.7. $\square$

In Theorem 12.9.9 we denote by ${}_n\mathrm{Pic}(R)$ the kernel of the nth power map $\mathrm{Pic}\, R \to \mathrm{Pic}\, R$.

Theorem 12.9.9 (Kummer Sequence)**.** *In the notation of Section* 12.9.1*, let $A = \mathbb{Z}[n^{-1}, \zeta]$, R any commutative A-algebra, and $G = \langle \sigma \rangle$ a cyclic group of order n. There is an exact sequence*

$$1 \to \frac{R^*}{(R^*)^n} \xrightarrow{\delta_0} \mathrm{Gal}(R, G) \xrightarrow{\delta_1} {}_n\mathrm{Pic}(R) \to 0$$

of abelian groups. The map δ_0 is defined as in Example 12.9.5*. That is, $a \in R^*$ is mapped to $R[x]/(x^n - a)$. The group operation on* $\mathrm{Gal}(R, G)$ *is the Harrison product which is defined in Remark* 12.9.4*. The map δ_1 is defined as in Proposition* 12.9.8*. That is, $S \in \mathrm{Gal}(R, G)$ is mapped to $M = \{x \in S \mid \sigma(x) = \zeta x\}$.*

Proof. Let S and T be in $\mathrm{Gal}(R, G)$. Because $G = \langle \sigma \rangle$ is cyclic, the Harrison product of S and T is $(S \otimes_R T)^{(\sigma, \sigma^{-1})}$. Write $S = R \oplus S_1 \oplus \cdots \oplus S_{n-1}$ and $T = R \oplus T_1 \oplus \cdots \oplus T_{n-1}$. Then the degree 1 component of $(S \otimes_R T)^{(\sigma, \sigma^{-1})}$ is $S_1 \otimes_R T_1$. The rest is left to the reader. $\square$

Remark 12.9.10. Let R, S, G, and M be as in Proposition 12.9.8. Since $S \otimes_R M$ is free of rank one, M represents a class in the image of α_1:

$$1 \to \mathrm{H}^1(G, S^*) \xrightarrow{\alpha_1} \mathrm{Pic}\, R \xrightarrow{\alpha_2} (\mathrm{Pic}\, S)^G$$

(see Theorem 13.3.1).

9.2. Artin-Schreier Extensions.

Example 12.9.11. Let k be a field of positive characteristic p. For any $a \in k$, the polynomial $f = x^p - x - a \in k[x]$ is separable over k. To see this, assume u is a root of f in any extension field F/k. Let $i \in \mathbb{Z}/p$ be any element of the prime field of k. Then $f(u + i) = (u + i)^p - (u + i) - a = u^p + i - u - i - a = f(u) = 0$. Therefore, f has p distinct roots in F, namely $u, u + 1, \dots, u + p - 1$.

Example 12.9.12. Let R be a commutative ring of characteristic p, where p is prime. Let $a \in R$ and $f(x) = x^p - x - a$. Then $f'(x) = -1$, and by Proposition 4.6.1, $A = R[x]/(f(x))$ is separable over R. As in Example 12.9.11, over

A the polynomial $y^p - y - a$ splits into $(y-x)(y-(x+1))\cdots(y-(x+p-1))$. Define $\sigma : S \to S$ by $x \mapsto x+1$. The reader should verify that σ is an R-algebra automorphism of A and $A^{\langle\sigma\rangle} = R$. To show that A/R is Galois with cyclic group $G = \langle\sigma\rangle$ we show that $\ell : A \otimes_R A \to \nabla(A, G)$ is an isomorphism and apply Theorem 12.2.9 (5). If $i \not\equiv j \pmod{p}$, then $(x+i)-(x+j) = i-j$ is a unit in A. Then

$$\begin{aligned} A \otimes_R A &\cong A \otimes_R \frac{R[y]}{(y^p - y - a)} \\ &\cong \frac{A[y]}{((y-x)(y-(x+1))\cdots(y-(x+p-1)))} \\ &\cong \bigoplus_{i=0}^{p-1} A \end{aligned}$$

where the last isomorphism is by the Chinese Remainder Theorem, and sends y to the p-tuple $(x, x+1, \dots, x+p-1)$. Notice that this agrees with the definition of the map ℓ.

9.3. Exercises.

Exercise 12.9.13. Let A be an R-algebra and $b \in A$. Assume $R[b]$ is a separable R-algebra and an R-progenerator module of constant rank n. Show that there is a separable polynomial $f \in R[x]$ satisfying the following.

(1) $R[b] \cong R[x]/(f)$.
(2) $\deg(f) = n$.
(3) $f(b) = 0$.
(4) $R[b]$ is a free R-module of rank n.

Exercise 12.9.14. In the notation of Section 12.9.1, let R be a noetherian normal integral domain with field of fractions K which is a $\mathbb{Z}[n^{-1}, \zeta]$-algebra. In Example 12.9.6 we saw that a rank one R-progenerator M whose class in $\operatorname{Pic}(R)$ is annihilated by n gives rise to a $\mathbb{Z}/n$-graded commutative separable R-algebra $A = A_0 \oplus \cdots \oplus A_{n-1}$. Moreover, A has an R-algebra automorphism σ satisfying $A^{\langle\sigma\rangle} = R$. Assuming M is a rank one reflexive R module which defines a class annihilated by n in $\operatorname{Cl}(R)$, show how to construct from the tensor algebra $T(M)$ a commutative $\mathbb{Z}/n$-graded R-algebra $A = A_0 \oplus \cdots \oplus A_{n-1}$ which is a reflexive R-module, and which agrees with Example 12.9.6 in case M is projective. For some specific examples of "reflexive Kummer extensions", see Exercises 6.3.22 and 6.3.23.

Exercise 12.9.15. Let R be a commutative connected ring. Let $n > 1$, $\alpha \in R$, and $f(x) = x^n - \alpha$. As in Theorem 12.7.8 (4), let $S = R[x]/(f(x))$

and $T_R^S : S \to R$ the trace map. Show that the determinant of the matrix $(T_R^S(x^{i-1}x^{j-1}))$ is $\pm n(n\alpha)^{n-1}$.

Exercise 12.9.16. Let R be a commutative ring and G a finite abelian group. Two G-Galois extensions of R, say S_1 and S_2, are isomorphic if there is an R-algebra isomorphism $f : S_1 \to S_2$ which is also an $R(G)$-module isomorphism. Let $\operatorname{Gal}(R, G)$ denote the set of all isomorphism classes of G-Galois extensions of R. In this exercise we outline Harrison's construction of an abelian group structure on $\operatorname{Gal}(R, G)$. For more hints, the interested reader is referred to [**Har65**] and [**Gre92**]. Denote by $E_G(R)$ the trivial G-Galois extension of R (Example 12.2.5). Prove the following.

(1) If $A \in \operatorname{Gal}(R, G)$ and $R \to S$ a homomorphism of commutative rings, then $A \otimes_R S \in \operatorname{Gal}(S, G)$. Then $R \mapsto \operatorname{Gal}(R, G)$ defines a covariant functor from commutative rings to sets.

(2) Associated to a homomorphism of abelian groups $\theta : G \to H$, is an exact sequence

$$1 \to \ker(\theta \times 1) \to G \times H \xrightarrow{\theta \times 1} H \to 1.$$

The multiplication homomorphism $\theta \times 1$ is defined by $(x, y) \mapsto \theta(x)y$. If $A \in \operatorname{Gal}(R, G)$, then $A \otimes_R E_H(R) \in \operatorname{Gal}(R, G \times H)$ (Exercise 12.5.9). By Theorem 12.5.1, $(A \otimes_R E_H(R))^{\ker(\theta\times 1)}$ is a G-Galois extension of R. In this way, $G \mapsto \operatorname{Gal}(R, G)$ defines a covariant functor from abelian groups to sets.

(3) There is a one-to-one correspondence $\operatorname{Gal}(R, G\times H) \to \operatorname{Gal}(R, G) \times \operatorname{Gal}(R, H)$.

(4) Let $\mu : G \times G \to G$ be the multiplication homomorphism which is defined by $(\sigma, \tau) \mapsto \sigma\tau$. If S and T are in $\operatorname{Gal}(R, G)$, then the *Harrison product* of S and T is defined to be $(S \otimes_R T)^{\ker \mu}$. Prove that this product makes $\operatorname{Gal}(R, G)$ into an abelian group, with identity element $E_G(R)$. Hints: Apply the functor $\operatorname{Gal}(R, \cdot)$ to the three commutative diagrams

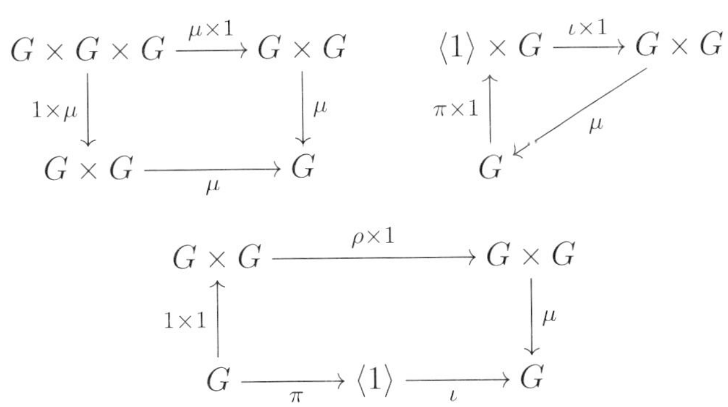

of groups, where $\iota : \langle 1 \rangle \to G$, $\pi : G \to \langle 1 \rangle$, and $\rho : G \to G$, is defined by $x \mapsto x^{-1}$.

(5) $R \mapsto \operatorname{Gal}(R, G)$ is a functor from commutative rings to abelian groups, and $G \mapsto \operatorname{Gal}(R, G)$ is a functor from abelian groups to abelian groups.

Exercise 12.9.17. In the notation of Section 12.9.1, let R be a commutative ring which is a $\mathbb{Z}[n^{-1}, \zeta]$-algebra. Let $G = \langle \sigma \rangle$ be a cyclic group of order n. Show that the abelian group $\check{\mathrm{H}}^1(R, G)$ is isomorphic to $\operatorname{Gal}(R, G)$.

Exercise 12.9.18. In the notation of Section 12.9.1, let k be a field which is a $\mathbb{Z}[n^{-1}, \zeta]$-algebra.

(1) Let $R = k[x_1, \dots, x_m]$ be a polynomial ring over k, and $G = \langle \sigma \rangle$ a cyclic group of order n. Show that $\operatorname{Gal}(R, G) = k^*/(k^*)^n$.

(2) Let f be a homogeneous irreducible polynomial in $S = k[x, y, z]$, and R the subring of $S[f^{-1}]$ consisting of all homogeneous elements of degree 0. As shown in Example 6.5.12, $\operatorname{Pic} R = \langle \mathfrak{p} \rangle$ is a cyclic group of order n. Show that there is a split exact sequence $1 \to k^*/(k^*)^n \to \operatorname{Gal}(R, G) \to \langle \mathfrak{p} \rangle \to 1$ of abelian groups.

Exercise 12.9.19. In the notation of Section 12.9.1, let R be a noetherian normal integral domain which is a $\mathbb{Z}[n^{-1}, \zeta]$-algebra. Let $G = \langle \sigma \rangle$ be a cyclic group of order n and let $T = R[t_1, t_1^{-1}, \dots, t_v, t_v^{-1}]$ be the Laurent polynomial ring in v indeterminates over R. Using the Kummer Sequence (Theorem 12.9.9) and Exercise 6.5.14, prove that there is a split exact sequence

$$1 \to \bigoplus_{i=1}^{v} \frac{\langle t_i \rangle}{\langle t_i^n \rangle} \to \operatorname{Gal}(T, G) \to \operatorname{Gal}(R, G) \to 1$$

of abelian groups.

Exercise 12.9.20. Let $n \geq 2$ be an integer, k a field in which n is invertible, and assume k contains a primitive nth root of unity, ζ. Let f be an irreducible polynomial in $A = k[x_1, \dots, x_m]$ and set $T = A[z]/(z^n - f)$, $S = T[z^{-1}]$, $R = A[f^{-1}]$. The assignment $z \mapsto \zeta z$ defines an A-algebra automorphism $\sigma : T \to T$ as well as an R-algebra automorphism $\sigma : S \to S$, and if $G = \langle \sigma \rangle$, then G has order n. Prove the following.

(1) $R^* = k^* \times \langle f \rangle$.

(2) $S^* = k^* \times \langle z \rangle$.

(3) $T^* = k^*$.

(4) $\operatorname{Cl}(T) = \operatorname{Cl}(S)$.

$$(5)\ \mathrm{H}^r(\langle\sigma\rangle, T^*) = \begin{cases} k^* & \text{if } r = 0, \\ k^*/(k^*)^n & \text{if } r > 0 \text{ is even}, \\ \langle\zeta\rangle \cong \mathbb{Z}/n & \text{if } r \text{ is odd}. \end{cases}$$

$$(6)\ \mathrm{H}^r(\langle\sigma\rangle, S^*) = \begin{cases} k^* \times \langle f\rangle & \text{if } r = 0, \\ k^*/(k^*)^n & \text{if } r > 0 \text{ is even}, \\ \langle 1\rangle & \text{if } r \text{ is odd}. \end{cases}$$

The ring R is also the subject of Exercise 13.3.17. (Hints: Use Theorem 6.2.4 for (1) – (4). Use Corollary 5.1.18 for (5). Use Theorem 5.1.17 for (6).)

Exercise 12.9.21 ([**Har77**, Exercise III.10.6]). Let k be a field in which 2 is invertible. Let $R = k[x, y]/(y^2 - x^2(x + 1))$ be the affine coordinate ring of a nodal cubic curve. Let $S = k[v, w]/(v^2 - (w^2 - 1)^2)$ be the affine coordinate ring of the union of two irreducible conics that intersect in two points. Use the following outline to show that S can be made into a Galois extension of R of degree two.

(1) Show that there is a k-algebra homomorphism $\beta : R \to S$ defined by $x \mapsto w^2 - 1$, $y \mapsto vw$.
(2) Prove that β makes S into a faithfully flat étale R-algebra. (Hints: For separable, use Theorem 8.1.24, for flat use Corollary 2.6.6.)
(3) Show that there is a k-algebra automorphism $\sigma \in \mathrm{Aut}_k(S)$ defined by $v \mapsto -v$, $w \mapsto -w$.
(4) Show that $R = S^{\langle\sigma\rangle}$.

For a continuation of this exercise, see Exercise 14.3.14.

Exercise 12.9.22. Generalize Exercise 12.9.21. Let k be an algebraically closed field of characteristic zero. In $k[x]$, let $q(x) = (x-1)(x-2)\cdots(x-n)$. Let $R = k[x, y]/(y^2 - x(q(x))^2)$ be the affine coordinate ring of an irreducible hyperelliptic curve with n nodes. Show how to construct a Galois extension of R of degree two by gluing together two copies of the affine line $\mathbb{A}^1$.

Chapter 13

Crossed Products and Galois Cohomology

If S/R is a Galois extension of commutative rings with finite group G and ξ is a 2-cocycle in $\mathrm{Z}^2(G, S^*)$, then the crossed product algebra $\Delta(S/R, G, \xi)$ is an Azumaya R-algebra and S is a splitting ring. The assignment $\xi \mapsto \Delta(S/R, G, \xi)$ induces a homomorphism of groups $\alpha_4 : \mathrm{H}^2(G, S^*) \to \mathrm{B}(S/R)$ which is called the crossed product homomorphism. In general α_4 is neither one-to-one nor onto. The kernel and cokernel of α_4 are described by the seven term exact sequence of Chase, Harrison and Rosenberg. This exact sequence is actually a special case of the Hochschild-Serre spectral sequence (see, for example, [**Mil80**, Theorem III.2.20]). The construction of the sequence presented here is based on a proof by Kanzaki [**Kan68**] and involves explicit descriptions of the homomorphisms. Kanzaki's approach is based on generalized crossed product algebras. If A is an Azumaya R-algebra which contains S as a maximal commutative subring, then for each $\sigma \in G$ there is an $S \otimes_R S$-submodule $\Phi_\sigma \subseteq A$ such that A decomposes into the internal direct sum $\bigoplus_{\sigma \in G} \Phi_\sigma$ and $\sigma \mapsto \Phi_\sigma$ defines a homomorphism of groups $\Phi : G \to \mathrm{Pic}_R(S)$. In addition, multiplication in A induces $S \otimes_R S$-module isomorphisms $f_{\sigma,\tau} : \Phi_\sigma \otimes_S \Phi_\tau \to \Phi_{\sigma\tau}$. Conversely, if there is a homomorphism of groups $\Phi : G \to \mathrm{Pic}_R(S)$ and a family of isomorphisms $f_{\sigma,\tau}$ such that $\bigoplus_{\sigma \in G} \Phi_\sigma$ becomes an R-algebra, then we call the resulting R-algebra a generalized crossed product, and denote it by $\Delta(S/R, \Phi, f)$.

Section 13.4 is devoted to cyclic crossed product algebras associated to a Galois extension S/R with finite cyclic group $G = \langle \sigma \rangle$. As an application, we give Azumaya's proof that the Brauer group of a henselian local ring is isomorphic to the Brauer group of the residue field. In Section 13.5 we study generalized cyclic crossed product algebras.

The topic of Section 13.5 is the Brauer group of a graded ring. This includes the Brauer group of a polynomial ring and the Brauer group of a Laurent polynomial ring. This section concludes with some interesting computations. There is an example of a ring R with $\operatorname{Pic}(R) \cong \mathbb{Z}/n$ and $\mathrm{B}(R) = (0)$. For this ring, we construct the Kummer extension T/R corresponding to the generator of $\operatorname{Pic}(R)$. We also prove that $\mathrm{B}(R[t, t^{-1}])$ is cyclic of order n and describe the Azumaya algebra which represents the generator.

1. Crossed Product Algebras

Given a Galois extension S/R of commutative rings with finite group G, together with a factor set ξ, the crossed product algebra $\Delta(S/R, G, \xi)$ was defined in Definition 12.1.2. Our convention is to assume the factor set ξ is a normalized 2-cocycle in $\mathrm{Z}^2(G, S^*)$.

Lemma 13.1.1. *Let S/R be a Galois extension of commutative rings with finite group G. Let $\xi \in \mathrm{Z}^2(G, R^*)$ be a normalized 2-cocycle.*

(1) *S is a maximal commutative R-subalgebra of $\Delta(S/R, G, \xi)$.*
(2) *The center of $\Delta(S/R, G, \xi)$ is R.*
(3) *If R is a field, then the crossed product algebra $\Delta(S/R, G, \xi)$ is an Azumaya R-algebra.*

Proof. (1): We identify S with $S \cdot 1$. Let $u = \sum_{\sigma \in G} s_\sigma u_\sigma$ represent an element of the center of $\Delta(S/R, G, \xi)$. Let a be an arbitrary element of S. Then

$$0 = au - ua = \sum_\sigma (a - \sigma(a)) s_\sigma u_\sigma$$

implies $(a - \sigma(a)) s_\sigma = 0$ for all $\sigma \in G$. If τ is an element of $G - \{1\}$, then $\{a - \tau(a) \mid a \in S\}$ generates the unit ideal in S, by Theorem 12.2.9 (6). We conclude that $s_\tau = 0$. It follows that $\Delta(S/R, G, \xi)^S = S$.

(2): It follows from (1) and $S^G = R$ that the center of $\Delta(S/R, G, \xi)$ is equal to R.

(3): If R is a field, we show that $\Delta(S/R, G, \xi)$ is simple, and apply Corollary 4.5.4. Let U be a nonzero two-sided ideal in $\Delta(S/R, G, \xi)$. Let $u = \sum_{\sigma \in G} s_\sigma u_\sigma$ represent an element of U, chosen so that the number of nonzero coefficients is minimal. If necessary, multiply u by one of the units u_σ and assume $s_1 \neq 0$. For any $a \in S$, $au - ua$ is a shorter element of U. By the choice of u, this implies $(a - \sigma(a)) s_\sigma = 0$ for all $\sigma \in G$. By the argument in Part (1) above, we conclude that $s_\tau = 0$ for all $\tau \in G - \{1\}$. This proves

U contains a nonzero element of S. Let $I = U \cap S$. For each $\sigma \in G$, we have $u_\sigma I u_\sigma^{-1} \subseteq I$. By Lemma 12.2.2 (7), I contains a nonzero element of R. Since R is a field, this implies I is the unit ideal. Hence U is the unit ideal. □

Theorem 13.1.2. *Let S/R be a Galois extension of commutative rings with finite group G. Let $\xi \in \mathrm{Z}^2(G, S^*)$ be a normalized 2-cocycle. Then the crossed product algebra $\Delta(S/R, G, \xi)$ is an Azumaya R-algebra which is split by S.*

Proof. By Lemma 12.2.2, S is an R-progenerator. By Definition 12.1.2, the crossed product $\Delta(S/R, G, \xi)$ is a free S-module. Therefore, $\Delta(S/R, G, \xi)$ is an R-module progenerator. Let $\mathfrak{m}$ be a maximal ideal of R. The reader should verify that

$$\Delta(S/R, G, \xi) \otimes_R R/\mathfrak{m} \cong \Delta((S/\mathfrak{m}S)/(R/\mathfrak{m}), G, g)$$

where $g : G \times G \to (S/\mathfrak{m}S)^*$ is the 2-cocycle defined by reducing ξ modulo $\mathfrak{m}S$. By Lemma 13.1.1, $\Delta((S/\mathfrak{m}S)/(R/\mathfrak{m}), G, g)$ is an Azumaya $R/\mathfrak{m}$-algebra. The proof follows from Corollary 8.1.25 and Theorem 7.4.2. □

Lemma 13.1.3. *Let S/R be a Galois extension of commutative rings with finite group G. Let $\xi \in \mathrm{Z}^2(G, S^*)$ be a normalized 2-cocycle and $\Delta = \Delta(S/R, G, \xi)$ the crossed product algebra. Then the opposite algebra Δ^o is isomorphic to the crossed product algebra $\Delta(S/R, G, \xi^{-1})$ with factor set ξ^{-1}, both as an R-algebra and an $S \otimes_R S$-module.*

Proof. Write $\Delta = \Delta(S/R, G, \xi) = \bigoplus_{\sigma \in G} S u_\sigma$, and $\Delta(S/R, G, \xi^{-1}) = \bigoplus_{\sigma \in G} S v_\sigma$. Define $\psi : \Delta^o \to \Delta(S/R, G, \xi^{-1})$ by $\psi(s u_\sigma) = \sigma^{-1}(s)(v_\sigma)^{-1}$. Let $a \in S$, $b \in S^*$. The reader should verify the formulas:

$$\begin{aligned}(v_\sigma)^{-1} &= \sigma^{-1}(\xi(\sigma, \sigma^{-1}))v_{\sigma^{-1}} \\ (v_\sigma)^{-1}a &= \sigma^{-1}(a)(v_\sigma)^{-1} \\ (bv_\sigma)^{-1} &= \sigma^{-1}(b^{-1})(v_\sigma)^{-1}.\end{aligned}$$

If $a, b \in S$, then from

$$\begin{aligned}\psi(a \cdot bu_\sigma) &= \psi(\sigma(a)(bu_\sigma)) \\ &= a\sigma^{-1}(b)(v_\sigma)^{-1} \\ &= a\psi(bu_\sigma)\end{aligned}$$

and

$$\begin{aligned}\psi(bu_\sigma) \cdot a &= \psi(abu_\sigma) \\ &= \sigma^{-1}(a)\sigma^{-1}(b)(v_\sigma)^{-1} \\ &= \sigma^{-1}(b)(v_\sigma)^{-1}a \\ &= \psi(bu_\sigma)a\end{aligned}$$

one verifies that ψ is an isomorphism of left $S \otimes_R S$-modules. From the chain of equalities

$$\begin{aligned}\psi((au_\sigma)\cdot(bu_\tau)) &= \psi(b\tau(a)\xi(\tau,\sigma)u_{\tau\sigma})\\ &= \sigma^{-1}\tau^{-1}(b\tau(a)\xi(\tau,\sigma))(v_{\tau\sigma})^{-1}\\ &= \sigma^{-1}\tau^{-1}(b)\sigma^{-1}(a)(v_\tau v_\sigma)^{-1}\\ &= \sigma^{-1}(a)(v_\sigma)^{-1}\tau^{-1}(b)(v_\tau)^{-1}\\ &= \psi(au_\sigma)\psi(bu_\tau)\end{aligned}$$

we see that ψ is an R-algebra isomorphism. □

Theorem 13.1.4. *Let S/R be a Galois extension of commutative rings with finite group G.*

(1) *If ξ and ζ are 2-cocycles in $\mathrm{Z}^2(G,S^*)$, then the Azumaya R-algebras $\Delta(S/R,G,\xi)\otimes_R\Delta(S/R,G,\zeta)$ and $\Delta(S/R,G,\xi\zeta)$ are Brauer equivalent.*

(2) *If ξ is a normalized 2-coboundary in $\mathrm{B}^2(G,S^*)$, then the R-algebras $\Delta(S/R,G,\xi)$ and $\Delta(S/R,G,1)$ are isomorphic.*

(3) *The assignment $\xi \mapsto \Delta(S/R,G,\xi)$ induces a homomorphism of abelian groups $\alpha_4 : \mathrm{H}^2(G,S^*) \to \mathrm{B}(S/R)$ which is called the crossed product homomorphism.*

Proof. Part (2) follows from Lemma 12.1.3.

(3): By Theorem 13.1.2, if ξ is a normalized 2-cocycle in $\mathrm{Z}^2(G,S^*)$, then the crossed product $\Delta(S/R,G,\xi)$ is an Azumaya R-algebra that is split by S. The rest of (3) follows from Parts (1) and (2).

(1): For notational simplicity, for any 2-cocycle f in $\mathrm{Z}^2(G,S^*)$, write Δ_f instead of $\Delta(S/R,G,f)$. Let c and d be normalized 2-cocycles in $\mathrm{Z}^2(G,S^*)$. Let $\Delta_c = \bigoplus_{\sigma\in G} Su_\sigma$, and $\Delta_d = \bigoplus_{\sigma\in G} Sv_\sigma$. We identify $S\otimes_R S$ with the subring $Su_1\otimes_R Sv_1 \subseteq \Delta_c\otimes_R\Delta_d$. Let $e \in S\otimes_R S$ be the (unique) separability idempotent (Definition 4.1.3). Let $A = e(\Delta_c\otimes_R\Delta_d)e$. By Exercise 7.6.4, it suffices to prove $\Delta(S/R,G,cd) \cong A$ as R-algebras.

The map $\mu_{S/R} : S\otimes_R S \to S$ is an R-algebra homomorphism. The restriction of $\mu_{S/R}$ is an R-algebra isomorphism: $e(S\otimes_R S)e \to S$. So we identify S with the subalgebra $e(S\otimes_R 1)e = e(S\otimes_R S)e \subseteq A$. Let $\sigma,\tau\in G$. The reader should verify that A is generated as a left S-module by the elements $e(u_\sigma\otimes v_\tau)e$. If $s\in S$, then

$$e(u_\sigma\otimes u_\tau)e(s\otimes 1) = (\sigma(s)\otimes 1)e(u_\sigma\otimes u_\tau)e.$$

Since $(s \otimes 1 - 1 \otimes s)e = 0$,

$$\begin{aligned} e(u_\sigma \otimes u_\tau)e(s \otimes 1) &= e(u_\sigma \otimes u_\tau)e(1 \otimes s) \\ &= (1 \otimes \tau(s))e(u_\sigma \otimes u_\tau)e \\ &- (\tau(s) \otimes 1)e(u_\sigma \otimes u_\tau)e. \end{aligned}$$

Combining the above,

$$((\sigma(s) - \tau(s)) \otimes 1)e(u_\sigma \otimes u_\tau)e = 0.$$

By Theorem 12.2.9 (6), the set $\{\sigma(s) - \tau(s) \mid s \in S\}$ generates the unit ideal in S, if $\sigma \neq \tau$, and in this case,

$$e(u_\sigma \otimes u_\tau)e = 0. \tag{13.1}$$

This implies A is generated as a left S-module by the elements $w_\sigma = e(u_\sigma \otimes v_\sigma)e$.

By Exercise 4.1.8, the separability idempotent e is unique. Using this fact, we see that $(\sigma \otimes \sigma)(e) = e$. Therefore,

$$(u_\sigma \otimes v_\sigma)e = ((\sigma \otimes \sigma)(e))(u_\sigma \otimes v_\sigma) = e(u_\sigma \otimes v_\sigma).$$

From this it follows that

$$\begin{aligned} w_\sigma w_\tau &= e(u_\sigma \otimes v_\sigma)e(u_\tau \otimes v_\tau)e \\ &= e(u_\sigma u_\tau \otimes v_\sigma v_\tau)e \\ &= e(c(\sigma,\tau)u_{\sigma\tau} \otimes d(\sigma,\tau)v_{\sigma\tau})e \\ &= (c(\sigma,\tau) \otimes d(\sigma,\tau))e(u_{\sigma\tau} \otimes v_{\sigma\tau})e \\ &= c(\sigma,\tau)d(\sigma,\tau))w_{\sigma\tau}. \end{aligned} \tag{13.2}$$

If we write $\Delta(S/R, G, cd) = \bigoplus_{\sigma \in G} Sz_\sigma$, then the reader should verify that the assignment $z_\sigma \mapsto w_\sigma$ defines an R-algebra isomorphism $\Delta(S/R, G, cd) \cong A$. □

2. Generalized Crossed Product Algebras

Let S/R be a Galois extension of commutative rings with finite group G (Definition 12.2.1). Using 1-cocycles in $\mathrm{Z}^1(G, \operatorname{Pic} S)$, Kanzaki ([**Kan68**]) showed how to construct Azumaya R-algebras that represent Brauer classes that are split by S. In this section we introduce this so-called generalized crossed product construction. In Proposition 13.2.3 (3) we show that this construction is a generalization of the crossed product algebra of Definition 12.1.2. Most of the results in this section are from [**Kan68**].

We view G as a subgroup of $\operatorname{Aut}_R(S)$. Let $\Phi_0 : \operatorname{Aut}_R S \to \operatorname{Pic}_R S$ be the homomorphism defined by $\sigma \mapsto [{}_1A_\sigma]$. By Lemma 7.8.7, we know that Φ_0 has a left inverse denoted Ψ and $\operatorname{Aut}_R S$ acts as a group on $\operatorname{Pic}_R S$ by $\tau[M] = [{}_{\tau^{-1}}M_{\tau^{-1}}]$.

Lemma 13.2.1. *Let S/R be a Galois extension of commutative rings with finite group G. There is a one-to-one correspondence between $\mathrm{Z}^1(G, \mathrm{Pic}(S))$, the set of all 1-cocycles of G with coefficients in $\mathrm{Pic}(S)$, and the set of all group homomorphisms $\Phi : G \to \mathrm{Pic}_R(S)$ such that the diagram*

$$\begin{array}{ccccc} & & & & G \\ & & & \overset{\Phi}{\swarrow} & \downarrow \subseteq \\ \operatorname{Pic} S & \longrightarrow & \operatorname{Pic}_R S & \xrightarrow{\Psi} & \operatorname{Aut}_R S \end{array}$$

commutes.

Proof. By Lemma 7.8.7 (4), associated to Φ is the function $\varphi : G \to \operatorname{Pic} S$ defined by $\varphi(\tau) = \Phi(\tau)\Phi_0(\tau^{-1})$. Since Φ and Φ_0 are homomorphisms, we have $\varphi(\sigma\tau) = \varphi(\sigma)\sigma(\varphi(\tau))$, so φ is a 1-cocycle in $\mathrm{Z}^1(G, \operatorname{Pic} S)$. Conversely, given a 1-cocycle $\phi : G \to \operatorname{Pic} S$ in $\mathrm{Z}^1(G, \operatorname{Pic} S)$, the formula $\tau \mapsto \phi(\tau)\Phi_0(\tau)$ defines a function $f_\phi : G \to \operatorname{Pic}_R S$. Using Lemma 7.8.7 we have

$$\begin{aligned} f_\phi(\sigma\tau) &= \phi(\sigma\tau)\Phi_0(\sigma\tau) \\ &= \phi(\sigma)\sigma\phi(\tau)\Phi_0(\sigma)\Phi_0(\tau) \\ &= \phi(\sigma)\Phi_0(\sigma)\phi(\tau)\Phi_0(\sigma^{-1})\Phi_0(\sigma)\Phi_0(\tau) \\ &= f_\phi(\sigma)f_\phi(\tau). \end{aligned}$$

Hence f_ϕ is a group homomorphism. □

Definition 13.2.2. Let Φ be a homomorphism of groups as in Lemma 13.2.1 and for each $\tau \in G$ write $\Phi(\tau) = [\Phi_\tau]$. A *factor set* related to Φ is a family of $S \otimes_R S$-isomorphisms $f = \{f_{\sigma,\tau} : \Phi_\sigma \otimes_S \Phi_\tau \to \Phi_{\sigma\tau}\}$ indexed by $G \times G$ such that the diagram

$$\begin{array}{ccc} \Phi_\sigma \otimes_S \Phi_\tau \otimes_S \Phi_\gamma & \xrightarrow{1 \otimes f_{\tau,\gamma}} & \Phi_\sigma \otimes_S \Phi_{\tau\gamma} \\ {\scriptstyle f_{\sigma,\tau} \otimes 1}\downarrow & & \downarrow{\scriptstyle f_{\sigma,\tau\gamma}} \\ \Phi_{\sigma\tau} \otimes_S \Phi_\gamma & \xrightarrow[f_{\sigma\tau,\gamma}]{} & \Phi_{\sigma\tau\gamma} \end{array} \tag{13.3}$$

commutes for every triple $(\sigma, \tau, \gamma) \in G \times G \times G$. Given a Galois extension S/R, a homomorphism of groups Φ, and a factor set f, we define the *generalized crossed product algebra* $\Delta(S/R, \Phi, f)$. As an $S \otimes_R S$-module, $\Delta(S/R, \Phi, f)$ is the direct sum $\bigoplus_{\tau \in G} \Phi_\tau$. Using the factor set f, the multiplication rule is defined on homogeneous elements $x_\sigma \in \Phi_\sigma$, $x_\tau \in \Phi_\tau$ by the rule $x_\sigma x_\tau = f_{\sigma,\tau}(x_\sigma \otimes x_\tau)$.

Proposition 13.2.3. *In the context of Definition 13.2.2, the generalized crossed product algebra $\Delta(S/R, \Phi, f)$ satisfies the following:*

(1) *$\Delta(S/R, \Phi, f)$ is an R-algebra.*

(2) *(Crossed Products Commute with Change of Base Ring) If* $\theta : R \to T$ *is a homomorphism of commutative rings, then* $\Delta(S/R, \Phi, f) \otimes_R T$ *is isomorphic (as a* T*-algebra) to the generalized crossed product* $\Delta((S \otimes_R T)/T, \Phi_T, f_T)$ *where* $\Phi_T(\tau)$ *is equal to* $\Phi_\tau \otimes_R T$ *for each* $\tau \in G$*, and* $(f_T)_{\sigma,\tau} : \Phi_\sigma \otimes_S \Phi_\tau \otimes_R T \to \Phi_{\sigma\tau} \otimes_R T$ *is the factor set obtained by tensoring* f *with* T*.*

(3) *If the image of* Φ *is contained in the image of* Φ_0*, then for some 2-cocycle* $g \in Z^2(G, S^*)$*, the generalized crossed product* $\Delta(S/R, \Phi, f)$ *is a crossed product* $\Delta(S/R, G, g)$ *as defined in Definition 12.1.2. Conversely, a crossed product* $\Delta(S/R, G, g)$ *is a generalized crossed product* $\Delta(S/R, \Phi, f)$ *where* $\Phi(\tau) = \Phi_0(\tau)$ *and the factor set* f *is defined as in Lemma 7.8.5.*

(4) $\Delta(S/R, \Phi, f)$ *is an Azumaya* R*-algebra.*

(5) *The ring* S *can be identified with the component* Φ_1*, and is a maximal commutative subalgebra.*

Proof. Part (2) is left to the reader.

(1): The associative law for multiplication holds as a result of the commutative diagram (13.3). Because Φ is a homomorphism, Φ_1 is isomorphic to the free S-module ${}_1S_1$. If u_1 is a generator, then $\Phi_1 = Su_1$. For some $f_{11} \in S$, $f_{1,1}(u_1 \otimes u_1) = f_{11}u_1$. Since $f_{1,1}$ is an isomorphism, f_{11} is a unit in S. As in Exercise 12.5.5, the reader should verify that the multiplicative identity of $\Delta(S/R, \Phi, f)$ is $f_{11}^{-1}u_1$.

(3): Lemma 13.2.1 implies that $\Phi_\tau \cong {}_1S_\tau$, for each $\tau \in G$. Let u_τ be a generator for the free left S-module ${}_1S_\tau$. By Lemma 7.8.5, ${}_1S_\sigma \otimes_S {}_1S_\tau \cong {}_1S_{\sigma\tau}$. Therefore, $f(\sigma, \tau)$ corresponds to an invertible element of $\operatorname{Hom}_S({}_1S_{\sigma\tau}, {}_1S_{\sigma\tau})$, hence is a unit in S. The reader should verify that the commutative diagram (13.3) translates into the 2-cocycle identity

$$f(\sigma, \tau)f(\sigma\tau, \gamma) = \sigma f(\tau, \gamma)f(\sigma, \tau\gamma)$$

in the group of units S^*. For any $s, t \in S$, and $\sigma, \tau \in G$, we have

$$(su_\sigma)(tu_\tau) = s(u_\sigma \cdot t)u_\tau = s(u_\sigma\sigma(t))u_\tau = s\sigma(t)(u_\sigma u_\tau) = s\sigma(t)f(\sigma, \tau)u_{\sigma\tau}.$$

This shows that $\Delta(S/R, \Phi, f)$ is a crossed product algebra, as defined in Definition 12.1.2. The converse is left to the reader.

(4): Since S is an R-progenerator (Lemma 12.2.2), $\Delta(S/R, \Phi, f)$ is as well. Let $\mathfrak{m}$ be a maximal ideal of R. Let $R_\mathfrak{m}$ be the local ring, and write $S_\mathfrak{m}$ for $S \otimes_R R_\mathfrak{m}$. If we show $\Delta(S_\mathfrak{m}/R_\mathfrak{m}, \Phi_\mathfrak{m}, f_\mathfrak{m})$ is an Azumaya $R_\mathfrak{m}$-algebra, then (4) follows from (2) and Corollary 8.1.23. We have reduced to the case where R is a local ring, with maximal ideal $\mathfrak{m}$. By Theorem 8.3.7, there are only a finite number of primes of S that contain $\mathfrak{m}S$, and each is a maximal ideal. Therefore, S is a semilocal ring, and by Exercise 3.1.8 it follows that

$\operatorname{Pic}(S) = 0$. Lemma 13.2.1 implies that $\Phi_\tau \cong {}_1S_\tau$, for each $\tau \in G$. By (3), $\Delta(S/R, \Phi, f)$ is a crossed product algebra, as defined in Definition 12.1.2. By Theorem 13.1.2, we are done.

(5): By (1), S can be identified with Φ_1. By (3) we know $S_\mathfrak{m}$ is a maximal commutative subalgebra of $\Delta(S/R, \Phi, f) \otimes_R R_\mathfrak{m}$, for each $\mathfrak{m} \in \operatorname{Max}(R)$. Let T be a maximal commutative subalgebra of $\Delta(S/R, \Phi, f)$ which contains S. Using Exercise 2.4.12, we see that $S = T$. $\square$

By Theorem 13.2.4, any class $[\Lambda]$ in $\mathrm{B}(S/R)$ contains a generalized crossed product $\Delta(S/R, \Phi, f)$.

Theorem 13.2.4. *Let S/R be a Galois extension of commutative rings with group G and let Λ be an Azumaya R-algebra that is split by S. Then Λ is Brauer equivalent to a generalized crossed product $\Delta(S/R, \Phi, f)$ for some $\Phi : G \to \operatorname{Pic}_R S$ and some factor set f.*

Proof. By Theorem 7.4.3, if necessary, we replace Λ with an algebra in the same Brauer class and assume Λ contains S as a maximal commutative subalgebra. It follows from Theorem 7.4.2 that Λ is a left S-progenerator and $S \otimes_R \Lambda^o \cong \operatorname{Hom}_S(\Lambda, \Lambda)$. Note that the left $S \otimes_R \Lambda^o$-module structure on Λ is given by the rule $(a \otimes b)x = axb$. There is a Morita equivalence between the category of right S-modules and the category of left $S \otimes_R \Lambda^o$-modules given by the assignments

$$M \mapsto M \otimes_S \Lambda \tag{13.4}$$

$$N \mapsto \Lambda^* \otimes_{S \otimes_R \Lambda^o} N \cong \operatorname{Hom}_{S \otimes_R \Lambda^o}(\Lambda, N) \cong (N)^S \tag{13.5}$$

(Exercise 7.4.7). Recall that $\Lambda^* = \operatorname{Hom}_S(\Lambda, S)$ and N^S is the set of all $x \in N$ such that $s \otimes 1 \cdot x = 1 \otimes s \cdot x$ for all $s \in S$. Combining the two functors we have

$$N^S \otimes_S \Lambda \cong N \tag{13.6}$$

which is an isomorphism of left $S \otimes_R \Lambda^o$-modules. It follows from Corollary 7.1.6 that on generators the isomorphism (13.6) is defined by $n \otimes x \mapsto nx$. Notice that on the left hand side of (13.6) the ring $S \otimes_R \Lambda^o$ acts on the Λ factor: $(a \otimes b)(x \otimes y) = x \otimes ayb$.

For any $\tau \in G$, $\tau \otimes 1$ is an R-algebra automorphism of $S \otimes_R \Lambda^o$. Therefore, ${}_\tau(S \otimes_R \Lambda^o)_1$ is an invertible two-sided $S \otimes_R \Lambda^o$-bimodule over R (Lemma 7.8.6), and

$${}_\tau\Lambda_1 = {}_\tau(S \otimes_R \Lambda^o)_1 \otimes_{S \otimes_R \Lambda^o} \Lambda \tag{13.7}$$

is a left $S \otimes_R \Lambda^o$-module. In (13.6) take N to be ${}_\tau\Lambda_1$, and denote N^S by Φ_τ. We have $\Phi_\tau = \{\lambda \in \Lambda \mid \tau(a)\lambda = \lambda a, \text{ for all } a \in S\}$. Combining (13.5) and (13.7), we have

$$\Phi_\tau = \Lambda^* \otimes_{S\otimes_R\Lambda^o} {}_\tau(S \otimes_R \Lambda^o)_1 \otimes_{S\otimes_R\Lambda^o} \Lambda. \tag{13.8}$$

Under the isomorphism of groups $\operatorname{Pic}_R(S) \cong \operatorname{Pic}_R(S\otimes_R\Lambda^o)$ of Theorem 7.8.8, Φ_τ is the invertible $S \otimes_R S$-module corresponding to ${}_\tau(S \otimes_R \Lambda^o)_1$. Using Lemma 7.8.5 and Theorem 7.8.8, it follows that there is an $S \otimes_R S$-module isomorphism

$$f_{\sigma,\tau} : \Phi_\sigma \otimes_S \Phi_\tau \cong \Phi_{\sigma\tau} \tag{13.9}$$

induced by $x \otimes y \mapsto xy$. Hence, the map $\Phi : G \to \operatorname{Pic}_R S$ defined by $\Phi(\tau) = \Phi_\tau$ is a homomorphism of groups. It follows from the definition of Ψ in Lemma 7.8.3 that $\Psi([\Phi_\tau]) = \tau$, and Φ factors as in Lemma 13.2.1. The reader should verify that the set of isomorphisms $\{f_{\sigma,\tau}\}$ is a factor set related to Φ. Let

$$\theta : \Delta(S/R, \Phi, f) \to \Lambda$$

be the map which on homogeneous components is defined by $\Phi_\tau \subseteq \Lambda$. The reader should verify that θ is a well defined R-algebra homomorphism. By Theorem 7.4.3, for each prime $\mathfrak{p} \in \operatorname{Spec} R$, $\operatorname{Rank}_{R_\mathfrak{p}}(\Lambda_\mathfrak{p}) = (\operatorname{Rank}_{R_\mathfrak{p}}(S_\mathfrak{p}))^2$, which is equal to the rank of $\Delta(S/R, \Phi, f)_\mathfrak{p}$ over $R_\mathfrak{p}$. It follows from Theorem 7.6.1 that θ is an isomorphism. □

As in Definition 13.2.2, let $\Delta(S/R, \Phi, f)$ be a generalized crossed product. Proposition 13.2.5 shows that the factor set f is unique up to a cohomology class in $\mathrm{H}^2(G, S^*)$.

Proposition 13.2.5. *Let S/R be a Galois extension of commutative rings with finite group G. Let $\Delta(S/R, \Phi, f)$ and $\Delta(S/R, \Phi, g)$ be two generalized crossed products with the same $\Phi : G \to \operatorname{Pic}_R(S)$, but possibly different factor sets.*

1. *There is a 2-cocycle $\xi \in \mathrm{Z}^2(G, S^*)$ such that $g = \xi f$.*
2. *In the context of (1), ξ is a 2-coboundary in $\mathrm{B}^2(G, S^*)$ if and only if $\Delta(S/R, \Phi, f)$ and $\Delta(S/R, \Phi, \xi f)$ are isomorphic as R-algebras and $S \otimes_R S$-modules.*

Proof. (1): We will utilize the notation of Definition 13.2.2. For each pair $\sigma, \tau \in \operatorname{Aut}_R(S)$, $f_{\sigma,\tau}$ and $g_{\sigma,\tau}$ are isomorphisms from $\Phi_\sigma \otimes_S \Phi_\tau$ to $\Phi_{\sigma\tau}$. Then $g_{\sigma,\tau} f_{\sigma,\tau}^{-1}$ is a unit in $\operatorname{Hom}_S(\Phi_{\sigma\tau}, \Phi_{\sigma\tau})$. The left S-module $\Phi(\sigma) = \Phi_\sigma$ is invertible, and by Lemma 2.6.7, there exists a unit $\xi(\sigma, \tau) \in S^*$ such that $g_{\sigma,\tau} f_{\sigma,\tau}^{-1}$ is left multiplication by $\xi(\sigma, \tau)$. The fact that the diagram (13.3) commutes translates into the 2-cocycle identity for ξ. Therefore, $g = \xi f$.

(2): Assume $\theta : \Delta(S/R, \Phi, f) \to \Delta(S/R, \Phi, \xi f)$ is an isomorphism of both R-algebras and $S \otimes_R S$-modules. Let $\sigma \in G$. First we show that

$\theta(\Phi_\sigma) = \Phi_\sigma$. Let $x \in \Phi_\sigma$ and write $\theta(x) = \sum_{\tau \in G} z_\tau$. We show that $z_\tau = 0$ for all $\tau \neq \sigma$. For every $a \in S$, we have

$$\theta(x)a = \sum_{\tau \in G} z_\tau a = \sum_{\tau \in G} \tau(a) z_\tau$$

which is equal to

$$\theta(xa) = \theta(\sigma(a)x) = \sigma(a)\theta(x) = \sum_{\tau \in G} \sigma(a) z_\tau.$$

Therefore, $\sigma(a)z_\tau = \tau(a)z_\tau$, for all $a \in S$ and $\tau \in G$. Pick elements $x_1, \dots, x_n, y_1, \dots, y_n$ in S such that

$$\sum_{j=1}^{n} x_j \rho(y_j) = \begin{cases} 1 & \text{if } \rho = 1, \\ 0 & \text{otherwise,} \end{cases}$$

for all $\rho \in G$ (Theorem 12.2.9). If $\tau \neq \sigma$, then

$$z_\tau = \sum_{j=1}^{n} z_\tau x_j y_j = \sum_{j=1}^{n} \tau(x_j)\tau(y_j) z_\tau =$$

$$\sum_{j=1}^{n} \tau(x_j)\sigma(y_j) z_\tau = \tau\left(\sum_{j=1}^{n} x_j \tau^{-1}\sigma(y_j)\right) z_\tau = 0.$$

This proves that $\theta(\Phi_\sigma) = \Phi_\sigma$. Hence θ determines a unit in the ring $\operatorname{Hom}_S(\Phi_\sigma, \Phi_\sigma)$. By Lemma 2.6.7, there exists a unit $u_\sigma \in S^*$ such that $\theta : \Phi_\sigma \to \Phi_\sigma$ is left multiplication by u_σ. Since $\Phi_1 = {}_1S_1$, we know $u_1 = 1$. Because θ is an R-algebra isomorphism, for all $x \otimes y \in \Phi_\sigma \otimes_S \Phi_\tau$, we have

$$\begin{aligned} \theta(x \cdot y) &= \theta(f_{\sigma,\tau}(x \otimes y)) \\ &= u_{\sigma\tau} f_{\sigma,\tau}(x \otimes y) \end{aligned}$$

and

$$\begin{aligned} \theta(x) \cdot \theta(y) &= (\xi f)_{\sigma,\tau}(h(x) \otimes h(y)) \\ &= \xi(\sigma,\tau) f_{\sigma,\tau}(u_\sigma x \otimes u_\tau y) \\ &= \xi(\sigma,\tau) u_\sigma \sigma(u_\tau) f_{\sigma,\tau}(x \otimes y) \end{aligned}$$

are equal. Consequently, $\xi(\sigma,\tau) = u_{\sigma\tau} u_\sigma^{-1} (\sigma u_\tau)^{-1}$. This proves that ξ is a 2-coboundary (Proposition 5.1.6).

Conversely, assume ξ is a normalized 2-coboundary. By Corollary 5.1.7, there is a set of units $\{u_\sigma \in S^* \mid \sigma \in G\}$ such that $u_1 = 1$ and $\xi(\sigma,\tau) = u_{\sigma\tau} u_\sigma^{-1} (\sigma u_\tau)^{-1}$ for all $(\sigma,\tau) \in G \times G$. Define

$$\theta : \Delta(S/R, \Phi, f) \to \Delta(S/R, \Phi, \xi f)$$

by mapping a homogeneous element $x \in \Phi_\sigma$ to $\theta(x) = u_\sigma x$. The reader should verify that this extends to an R-algebra and $S \otimes_R S$-module isomorphism. (Notice that this agrees with the proof of Lemma 12.1.3.) □

Proposition 13.2.6. *Let S/R be a Galois extension of commutative rings with finite group G. Let $\Delta(S/R, \Phi, f)$ be a generalized crossed product algebra which is split. That is, $\Delta(S/R, \Phi, f) \cong \operatorname{Hom}_R(P, P)$ for some R-progenerator module P. Then*

(1) *P is a rank one S-progenerator module, hence represents a class in* $\operatorname{Pic} S$.
(2) *$\Phi(\sigma)[P] = [P]\Phi_0(\sigma)$ for each $\sigma \in G$.*
(3) *If $\varphi(\sigma) = \sigma[P^*][P]$, then $\Phi(\sigma) = \varphi(\sigma)\Phi_0(\sigma)$. In the context of Lemma 13.2.1, Φ corresponds to a 1-coboundary in $\mathrm{B}^1(G, \operatorname{Pic} S)$.*

Proof. For notational simplicity, write Δ instead of $\Delta(S/R, \Phi, f)$. The proof of Part (3) is left to the reader.

(1): Since S is an R-subalgebra of $\Delta \cong \operatorname{Hom}_R(P, P)$, we view P as a left S-module. Since S is a maximal commutative subalgebra, $S = (\operatorname{Hom}_R(P, P))^S = \operatorname{Hom}_S(P, P)$. In particular, P is a faithful S-module. By Theorem 4.4.1, P is a finitely generated projective S-module. The rest follows from Lemma 2.6.7.

(2): By the proof of Theorem 13.2.4, $\Phi(\tau) = \Phi_\tau = ({}_\tau\Delta_1)^S$, which we identify as $\{f \in \operatorname{Hom}_R(P, P) \mid \tau(s)f(x) = f(sx) \text{ for all } s \in S \text{ and } x \in P\}$. Let u_σ denote a generator for the free left S-module ${}_1S_\sigma$. Define a function

$$\theta : \Phi_\sigma \otimes_S P \to P \otimes_S {}_1S_\sigma$$

by $\theta(f \otimes x) = f(x) \otimes u_\sigma$. The reader should verify that θ is a $\mathbb{Z}$-module homomorphism. For any $a \in S$,

$$\begin{aligned} \theta(f \otimes xa) &= f(ax) \otimes u_\sigma \\ &= \sigma(a)f(x) \otimes u_\sigma \\ &= f(x)\sigma(a) \otimes u_\sigma \\ &= f(x) \otimes \sigma(a)u_\sigma \\ &= \theta(f \otimes x)a. \end{aligned}$$

Since $\theta(af \otimes x) = af(x) \otimes u_\sigma$, this shows that θ is a left $S \otimes_R S$-module homomorphism. To show θ is an isomorphism, it suffices to localize at a maximal ideal (Exercise 2.4.12). That is, we assume R is a local ring. As in Proposition 13.2.3 (4), Δ is a crossed product and $\Phi_\tau = Sf_\tau$ for some unit $f_\tau \in \Delta$. From this, the reader should verify that θ is an isomorphism. □

Corollary 13.2.7. *Let S/R be a Galois extension of commutative rings with finite group G. Let P be a rank one S-progenerator module. The following are equivalent:*

(1) $\operatorname{Hom}_R(P,P)$ *is isomorphic as both an R-algebra and left $S \otimes_R S$-module to a generalized crossed product $\Delta(S/R, \Phi_0, f)$ for some factor set related to Φ_0.*

(2) $[P] \in (\operatorname{Pic} S)^G$.

If either (1) or (2) is satisfied, then there is a 2-cocycle $\xi \in \mathrm{Z}^2(G, S^*)$ such that $\operatorname{Hom}_R(P,P)$ is isomorphic as both an R-algebra and left $S \otimes_R S$-module to the crossed product $\Delta(S/R, G, \xi)$.

Proof. (1) implies (2): By Proposition 13.2.6, $\Phi_0(\sigma)[P] = [P]\Phi_0(\sigma)$ for all $\sigma \in G$. By Lemma 7.8.7 (4), $[P] \in (\operatorname{Pic} S)^G$.

(2) implies (1): Assume $\Phi_0(\sigma)[P] = [P]\Phi_0(\sigma)$ for all $\sigma \in G$. By Lemma 2.6.7, $\operatorname{Hom}_R(P,P)^S = \operatorname{Hom}_S(P,P) = S$. Hence, S is a maximal commutative subalgebra of $\operatorname{Hom}_R(P,P)$. By Theorem 13.2.4 it follows that $\operatorname{Hom}_R(P,P) = \Delta(S/R, \Phi, f)$ is a generalized crossed product. Proposition 13.2.6 implies that $\Phi(\sigma)[P] = [P]\Phi_0(\sigma) = \Phi_0(\sigma)[P]$. Then $\Phi(\sigma) = \Phi_0(\sigma)$ for all $\sigma \in G$.

The last claim follows from Proposition 13.2.3 (3). By Proposition 13.2.5, ξ determines a unique cohomology class $[\xi] \in \mathrm{H}^2(G, S^*)$. □

Lemma 13.2.8. *Let S/R be a Galois extension of commutative rings with finite group G. Let $\Delta(S/R, \Phi_1, f_1)$ and $\Delta(S/R, \Phi_2, f_2)$ be two generalized crossed product algebras.*

(1) $\Delta(S/R, \Phi_1, f_1) \otimes_R \Delta(S/R, \Phi_2, f_2)$ *is Brauer equivalent to the generalized crossed product $\Delta = \Delta(S/R, \Phi, f)$, where*
   (a) *the homomorphism $\Phi : G \to \operatorname{Pic}_R(S)$ is defined by*

$$\Phi(\sigma) = [\Phi_1(\sigma)][\Phi_0(\sigma^{-1})][\Phi_2(\sigma)]$$

   (b) *and the factor set is $f = f_1 \otimes \eta \otimes f_2$, where*

$$\eta : {}_1S_{\sigma^{-1}} \otimes_S {}_1S_{\tau^{-1}} \to {}_1S_{(\tau\sigma)^{-1}}$$

   is the isomorphism of Lemma 7.8.5.

(2) *For $i = 1, 2$, let φ_i be the 1-cocycle in $\mathrm{Z}^1(G, \operatorname{Pic} S)$ that is associated to Φ_i (see Lemma 13.2.1). Then the 1-cocycle associated to Φ is the product $\varphi_1\varphi_2$.*

Proof. Let $e \in S \otimes_R S$ be the separability idempotent. One shows that Δ is isomorphic to $e(\Delta(S/R, \Phi_1, f_1) \otimes_R \Delta(S/R, \Phi_2, f_2))e$ as in the proof of Theorem 13.1.4 (1). The details are left to the reader. □

Proposition 13.2.9. *Let S/R be a Galois extension of commutative rings with finite group G and $\Delta = \Delta(S/R, \Phi, f)$ a generalized crossed product algebra.*

(1) *Δ^o, the opposite R-algebra, is isomorphic to a generalized crossed product $\Delta(S/R, \Phi^o, f^o)$.*

(2) *If $\varphi \in \mathrm{Z}^1(G, \operatorname{Pic} S)$ is the 1-cocycle associated to Φ (Lemma 13.2.1), then the 1-cocycle associated to Φ^o is φ^{-1}.*

Proof. We know from Lemma 13.2.1 that $\varphi(\sigma) = \Phi(\sigma)\Phi_0(\sigma^{-1})$. Now define $\Phi^o(\sigma)$ to be $\varphi(\sigma)^{-1}\Phi_0(\sigma)$. Let $\Phi(\sigma) = [\Phi_\sigma]$, and $\Phi^o(\sigma) = [\Phi^o_\sigma]$. Then $[\Phi^o_\sigma] = \Phi_0(\sigma)[\Phi_\sigma]^{-1}\Phi_0(\sigma)$. Because Φ and Φ_0 are homomorphisms,

$$[\Phi^o_{\sigma^{-1}}] = \Phi_0(\sigma^{-1})[\Phi_\sigma]\Phi_0(\sigma^{-1}) = [{}_\sigma S_1 \otimes_S \Phi_\sigma \otimes_S {}_\sigma S_1].$$

Define

$$\Phi_\sigma \xrightarrow{g_\sigma} {}_\sigma S_1 \otimes_S \Phi_\sigma \otimes_S {}_\sigma S_1$$

by $x \mapsto 1 \otimes x \otimes 1$. Then $g_\sigma(xa) = 1 \otimes xa \otimes 1 = 1 \otimes \sigma(a)x \otimes 1 = \sigma(a) \otimes x \otimes 1 = a g_\sigma(x)$, and $g_\sigma(ax) = 1 \otimes ax \otimes 1 = 1 \otimes x\sigma^{-1}(a) \otimes 1 = 1 \otimes x \otimes a = g_\sigma(x)a$. The maps g_σ induce an isomorphism of R-modules

$$g : \bigoplus_\sigma \Phi_\sigma \to \bigoplus_\sigma \Phi^o_{\sigma^{-1}}.$$

If ϕ is the switch map, then $f^o_{\sigma,\tau}$ is defined to be the composite

$$\Phi^o_\sigma \otimes_S \Phi^o_\tau \xrightarrow{\phi} \Phi^o_\tau \otimes_S \Phi^o_\sigma \xrightarrow{g^{-1}\otimes g^{-1}} \Phi_{\tau^{-1}} \otimes_S \Phi_{\sigma^{-1}} \xrightarrow{f_{\tau^{-1},\sigma^{-1}}} \Phi_{(\sigma\tau)^{-1}} \xrightarrow{g} \Phi^o_{\sigma\tau}$$

which is an isomorphism of R-modules. Since

$$\begin{aligned}
f^o_{\sigma,\tau}(ax \otimes y) &= g(f_{\tau^{-1},\sigma^{-1}}(g^{-1}(y) \otimes g^{-1}(ax))) \\
&= g(f_{\tau^{-1},\sigma^{-1}}(g^{-1}(y) \otimes g^{-1}(x)a)) \\
&= g(f_{\tau^{-1},\sigma^{-1}}(g^{-1}(y) \otimes g^{-1}(x))a) \\
&= ag(f_{\tau^{-1},\sigma^{-1}}(g^{-1}(y) \otimes g^{-1}(x))) \\
&= af^o_{\sigma,\tau}(ax \otimes y)
\end{aligned}$$

and similarly, $f^o_{\sigma,\tau}(x \otimes ya) = f^o_{\sigma,\tau}(x \otimes y)a$, we see that $f^o_{\sigma,\tau}$ is an $S \otimes_R S$-module isomorphism. Since f is a factor set,

$$\begin{aligned}
f^o_{\sigma\tau,\gamma}(f^o_{\sigma,\tau}(x \otimes y) \otimes z) &= g(f_{\gamma^{-1},(\sigma\tau)^{-1}}(g^{-1}(z) \otimes g^{-1}(f^o_{\sigma,\tau}(x \otimes y)))) \\
&= g(f_{\gamma^{-1},(\sigma\tau)^{-1}}(g^{-1}(z) \otimes f_{\tau^{-1},\sigma^{-1}}(g^{-1}(y) \otimes g^{-1}(x))))
\end{aligned}$$

is equal to

$$\begin{aligned}
f^o_{\sigma,\tau\gamma}(x \otimes f^o_{\tau,\gamma}(y \otimes z)) &= g(f_{(\tau\gamma)^{-1},\sigma^{-1}}(g^{-1}(f^o_{\tau,\gamma}(y \otimes z)) \otimes g^{-1}(x))) \\
&= g(f_{(\tau\gamma)^{-1},\sigma^{-1}}(f_{\gamma^{-1},\tau^{-1}}(g^{-1}(z) \otimes g^{-1}(y)) \otimes g^{-1}(x)))
\end{aligned}$$

which proves that f^o is a factor set. Since $g(f_{\sigma,\tau}(x \otimes y)) = f^o_{\tau^{-1},\sigma^{-1}}(g(y) \otimes g(x))$, we see that g defines an isomorphism between the opposite algebra Δ^o and $\Delta(S/R, \Phi^o, f^o)$. □

Proposition 13.2.10. *Let S/R be a Galois extension of commutative rings with finite group G. Let $\xi \in \mathrm{Z}^2(G, S^*)$ be a normalized 2-cocycle. In the notation of Exercise 13.2.12,*

(1) $(\xi, 1) \in \mathrm{B}^2(G \times 1, (S \otimes_R S)^*)$,
(2) $(1, \zeta) \in \mathrm{B}^2(1 \times G, (S \otimes_R S)^*)$, *and*
(3) $(\xi, \xi^{-1}) \in \mathrm{B}^2(G \times G, (S \otimes_R S)^*)$.

Proof. The proof is based on [**AG60a**, Proposition A.11].

(1): By Theorem 12.2.9 (5), $\ell : S \otimes_R S \to \nabla(S, G)$ is an S-algebra isomorphism (see Example 12.2.6). So there are orthogonal idempotents $\{e_\sigma \mid \sigma \in G\}$ in $S \otimes_R S$ defined by $\ell(e_\sigma) = e_\sigma$. Using the isomorphism ℓ, the reader should verify that $(1 \otimes \sigma(x))e_\sigma = (x \otimes 1)e_\sigma$ and $(\sigma, \tau)e_\rho = e_{\tau\rho\sigma^{-1}}$ for all $x \in S$, $\sigma, \tau, \rho \in G$. Let $h : G \times 1 \to (S \otimes_R S)^*$ be defined by

$$h(\sigma, 1) = \sum_{\tau \in G} (1 \otimes \xi(\tau, \sigma))e_\tau.$$

Identify $G \times 1$ with G. If we define a 2-coboundary by the formula

$$\phi(\sigma, \tau) = h(\sigma\tau)h(\sigma)^{-1}\sigma h(\tau)^{-1},$$

then we see that $\phi(\sigma, \tau)$ is equal to

$$\begin{aligned}
&\left[\sum_{\rho \in G} (1 \otimes \xi(\rho, \sigma\tau))\, e_\rho\right]\left[\sum_{\beta \in G} \left(1 \otimes \xi(\beta, \sigma)^{-1}\right) e_\beta\right]\sigma\left[\sum_{\gamma \in G} \left(1 \otimes \xi(\gamma, \tau)^{-1}\right) e_\gamma\right] \\
&\quad = \left[\sum_{\rho \in G} \left(1 \otimes \xi(\rho, \sigma\tau)\xi(\rho, \sigma)^{-1}\right) e_\rho\right]\left[\sum_{\gamma \in G} \left(1 \otimes \xi(\gamma, \tau)^{-1}\right) e_{\gamma\sigma^{-1}}\right] \\
&\quad = \sum_{\rho \in G} \left(1 \otimes \xi(\rho, \sigma\tau)\xi(\rho, \sigma)^{-1}\xi(\rho\sigma, \tau)^{-1}\right) e_\rho \\
&\quad = \sum_{\rho \in G} \left(1 \otimes \xi(\rho, \sigma\tau)\xi(\rho, \sigma\tau)^{-1}\rho\xi(\sigma, \tau)^{-1}\right) e_\rho \\
&\quad = \sum_{\rho \in G} (1 \otimes \rho)\xi(\sigma, \tau)^{-1} e_\rho \\
&\quad = \sum_{\rho \in G} \left(\xi(\sigma, \tau)^{-1} \otimes 1\right) e_\rho \\
&\quad = \xi(\sigma, \tau)^{-1} \otimes 1 \\
&\quad = (\xi, 1)^{-1}\left((\sigma, 1), (\tau, 1)\right).
\end{aligned}$$

This proves $(\xi, 1)^{-1} \in \mathrm{B}^2(G \times 1, (S \otimes_R S)^*)$, from which (1) follows.

(2): This is left to the reader.

(3): Let

$$\eta : \Delta(S/R, G, \xi) \to S \otimes_R S$$

be defined by

$$\sum_{\tau \in G} s_\tau u_\tau \mapsto \sum_{\tau \in G} (s_\tau \otimes 1) e_{\tau^{-1}}$$

where the e_τ are the orthogonal idempotents in $S \otimes_R S$ defined in the proof of Part (1). It is clear that η is an R-module isomorphism. Consider the chain of R-algebra isomorphisms

$$\begin{aligned}
\Delta\left((S \otimes_R S)/R, G \times G, (\xi, \xi^{-1})\right) &\xrightarrow{\theta} \Delta(S/R, G, \xi) \otimes_R \Delta(S/R, G, \xi^{-1}) \\
&\xrightarrow{1 \otimes \psi^{-1}} \Delta(S/R, G, \xi) \otimes_R \Delta(S/R, G, \xi)^o \\
&\xrightarrow{\varphi} \operatorname{Hom}_R\left(\Delta(S/R, G, \xi), \Delta(S/R, G, \xi)\right) \\
&\xrightarrow{\mathrm{H}(\eta)} \operatorname{Hom}_R(S \otimes_R S, S \otimes_R S) \\
&\xrightarrow{\phi^{-1}_{S \otimes S/R}} \Delta((S \otimes_R S)/R, G \times G, 1).
\end{aligned}$$

The isomorphism θ is defined in Exercise (13.2.12), ψ is from Lemma 13.1.3, φ is from Theorem 7.1.4, and $\phi_{S \otimes S/R}$ is given in Theorem 12.2.9 (4). Explicitly, the composition is given by

$$\begin{aligned}
(s_\sigma \otimes s_\tau) u_{(\sigma,\tau)} &\mapsto s_\sigma u_\sigma \otimes s_\tau u_\tau \\
&\mapsto \delta = s_\sigma u_\sigma \otimes \tau^{-1}(s_\tau) \xi(\tau^{-1}, \tau)^{-1} u_{\tau^{-1}} \\
&\mapsto \varphi(\delta) \\
&\mapsto \eta \varphi(\delta) \eta^{-1} \\
&\mapsto \phi^{-1}_{S \otimes S/R}(\eta \varphi(\delta) \eta^{-1}).
\end{aligned}$$

Take $z \otimes w \in S \otimes S$. By a straightforward computation (prove it first for $z \otimes w = e_\gamma$) we have

$$\eta \varphi(\delta) \eta^{-1}(z \otimes w) = \phi_{S/R}\big(h(\sigma, \tau)(s_\sigma \otimes s_\tau) u_{(\sigma,\tau)}\big)(z \otimes w)$$

where for all $(\sigma, \tau) \in G \times G$ we define

$$h(\sigma, \tau) = \sum_{\rho \in G} \left(\xi(\sigma, \rho) \xi(\sigma\rho, \tau^{-1}) \otimes \tau \xi(\tau^{-1}, \tau)^{-1}\right) e_{\tau \rho^{-1} \sigma^{-1}}.$$

This shows that the correspondence

$$(s_\sigma \otimes s_\tau) u_{(\sigma,\tau)} \mapsto (s_\sigma \otimes s_\tau) h(\sigma, \tau) u_{(\sigma,\tau)}$$

is an R-algebra isomorphism

$$\Delta((S \otimes_R S)/R, G \times G, (\xi, \xi^{-1})) \cong \Delta((S \otimes_R S)/R, G \times G, 1).$$

One can check that this is a two-sided $S \otimes_R S$-module homomorphism as well. It follows from Proposition 13.2.5 that $(\xi, \xi^{-1}) \in \mathrm{B}^2(G \times G, (S \otimes_R S)^*)$. $\square$

2.1. Exercises.

Exercise 13.2.11. Let S/R be a Galois extension of commutative rings with finite group G. Let $\Delta(S/R, \Phi_1, f_1)$ and $\Delta(S/R, \Phi_2, f_2)$ be generalized crossed products. Prove:

(1) $S \otimes_R S$ is a Galois extension of R with group $G \times G$. (See Exercise 12.5.9.)

(2) There is a homomorphism $\Phi_1 \otimes \Phi_2 : G \times G \to \operatorname{Pic}_R(S \otimes_R S)$ defined by $(\sigma, \tau) \mapsto \Phi_1(\sigma) \otimes_R \Phi_2(\tau)$.

(3) There is a factor set for $\Phi_1 \otimes \Phi_2$ defined by $f_1 \otimes f_2((\sigma, \mu), (\tau, \nu)) = f_1(\sigma, \tau) \otimes f_2(\mu, \nu)$.

(4) There is an isomorphism

$$\Delta(S/R, \Phi_1, f_1) \otimes_R \Delta(S/R, \Phi_2, f_2) \cong \Delta((S \otimes_R S)/R, \Phi_1 \otimes \Phi_2, f_1 \otimes f_2)$$

of R-algebras.

Exercise 13.2.12. Let S/R be a Galois extension of commutative rings with finite group G. Let $\xi, \zeta \in \mathrm{Z}^2(G, S^*)$ be normalized 2-cocycles. Define $(\xi, 1) : G \times 1 \to (S \otimes_R S)^*$ by

$$(\xi, 1)\,((\sigma, 1), (\tau, 1)) = \xi(\sigma, \tau) \otimes 1,$$

$(1, \zeta) : 1 \times G \to (S \otimes_R S)^*$ by

$$(1, \zeta)\,((1, \alpha), (1, \beta)) = 1 \otimes \zeta(\alpha, \beta),$$

and $(\xi, \zeta) : G \times G \to (S \otimes_R S)^*$ by

$$(\xi, \zeta)\,((\sigma, \alpha), (\tau, \beta)) = \xi(\sigma, \tau) \otimes \zeta(\alpha, \beta).$$

Prove the following.

(1) $(\xi, 1)$ and $(1, \zeta)$ agree with the restriction homomorphisms of Definition 5.1.12, hence $(\xi, 1) \in \mathrm{Z}^2(G \times 1, (S \otimes_R S)^*)$, $(1, \zeta) \in \mathrm{Z}^2(1 \times G, (S \otimes_R S)^*)$.

(2) $(\xi, \zeta) \in \mathrm{Z}^2(G \times G, (S \otimes_R S)^*)$.

(3) There is an R-algebra isomorphism

$$\theta : \Delta(S/R, G, \xi) \otimes_R \Delta(S/R, G, \zeta) \cong \Delta\left((S \otimes_R S)/R, G \times G, (\xi, \zeta)\right)$$

defined by the correspondence $(s_\sigma u_\sigma \otimes s_\tau u_\tau) \mapsto (s_\sigma \otimes s_\tau) u_{(\sigma,\tau)}$.

Exercise 13.2.13. This exercise is based on [**DeM81**]. Let R be a commutative ring and $\operatorname{Aut}(R)$ the group of ring automorphisms of R. In this exercise we outline a proof that $\operatorname{Aut}(R)$ acts as a group of automorphisms of $\mathrm{B}(R)$. In the following, α is always an automorphism of R.

(1) Let A be an R-algebra. Show that A is a separable R-algebra if and only if ${}_\alpha A_\alpha$ is a separable R-algebra. (Hint: The same separability idempotent works in both rings.)
(2) Let M be an R-module (assume the left and right actions are equal). Show that M is an R-progenerator module if and only if ${}_\alpha M_\alpha$ is an R-progenerator module.
(3) Show that $\operatorname{Aut}(R)$ acts on $\mathrm{B}(R)$ as a group of automorphisms, where $\alpha[A] = [{}_{\alpha^{-1}}A_{\alpha^{-1}}]$ for any Azumaya R-algebra A.

The group action by $\operatorname{Aut}(R)$ on $\mathrm{B}(R)$ can be generalized to a Hecke action on $\mathrm{B}(R)$. For more on this subject, the interested reader is referred to [**For84**]. For Hecke actions on Picard groups, see [**RS82**].

3. The Seven Term Exact Sequence of Galois Cohomology

3.1. The Theorem and Its Corollaries. The seven term exact sequence of Chase, Harrison, and Rosenberg [**CHR65**, Corollary 5.5, p. 17] appears in Theorem 13.3.1 below. It is a sequence of abelian groups. In this section we define the homomorphisms $\alpha_1, \dots, \alpha_6$ explicitly and prove exactness at each term. We make $\operatorname{Pic} S$ into a $\mathbb{Z}G$-module as in Lemma 7.8.7 (4).

Theorem 13.3.1 (The Seven Term Exact Sequence of Galois Cohomology)**.** *If S/R is a Galois extension of commutative rings with finite group G, then there exists an exact sequence*

$$1 \to \mathrm{H}^1(G, S^*) \xrightarrow{\alpha_1} \operatorname{Pic}(R) \xrightarrow{\alpha_2} (\operatorname{Pic} S)^G \xrightarrow{\alpha_3}$$
$$\mathrm{H}^2(G, S^*) \xrightarrow{\alpha_4} \mathrm{B}(S/R) \xrightarrow{\alpha_5} \mathrm{H}^1(G, \operatorname{Pic} S) \xrightarrow{\alpha_6} \mathrm{H}^3(G, S^*)$$

of abelian groups.

Corollary 13.3.2 (Hilbert's Theorem 90)**.** *If* $\operatorname{Pic} R = (0)$, *then* $\mathrm{H}^1(G, S^*) = \langle 1 \rangle$.

Corollary 13.3.3 (The Crossed Product Theorem)**.** *If* $\operatorname{Pic} S = (0)$, *then* $\mathrm{H}^2(G, S^*) \cong \mathrm{B}(S/R)$.

The proof of Theorem 13.3.1 consists of a series of lemmas. We systematically define each of the homomorphisms $\alpha_1, \dots, \alpha_6$ and prove exactness at each term of the sequence. The homomorphisms α_1 and α_2 will be defined with the assistance of the next lemma.

Lemma 13.3.4. *In the context of Theorem* 13.3.1, *let* $\Delta = \Delta(S/R, G, 1)$ *be the trivial crossed product (see Definition* 12.1.4*), and* ξ *a normalized* 1*-cocycle in* $\mathrm{Z}^1(G, S^*)$ *(see Corollary* 5.1.7*). Then the following are true.*

(1) *There is an* R*-algebra automorphism* $\theta_\xi : \Delta \to \Delta$ *which is defined on the* S*-basis element* u_σ *by the assignment* $\theta_\xi(u_\sigma) = \xi(\sigma)u_\sigma$.

(2) *The mapping* $\xi \mapsto \theta_\xi$ *induces a monomorphism of abelian groups*

$$\mathrm{H}^1(G, S^*) \xrightarrow{\theta} \mathrm{Aut}_R(\Delta)/\operatorname{Inn}(\Delta).$$

The image of θ *consists of those cosets which are represented by elements of* $\mathrm{Aut}_R(\Delta)$ *that are both* R*-algebra and* S*-module homomorphisms.*

Proof. (1): By Corollary 5.1.7, $\xi(1) = 1$ and $\xi(\sigma\tau) = \xi(\sigma)\sigma\xi(\tau)$. Using this and Theorem 7.6.1, the reader should verify that θ_ξ induces an R-algebra automorphism of Δ.

(2): The groups are abelian, by Corollary 7.8.14. The reader should verify that $\theta_\xi\theta_\zeta = \theta_{\xi\zeta}$ for two 1-cocycles $\xi, \zeta \in \mathrm{Z}^1(G, S^*)$. If ξ is a 1-coboundary in $\mathrm{B}^1(G, S^*)$, then there is a unit $a \in S^*$ such that $\xi(\sigma) = \sigma(a)a^{-1}$ for all $\sigma \in G$. Thus $\theta_\xi(su_\sigma) = a^{-1}(su_\sigma)a$ is an inner automorphism of Δ.

Now we assume θ_ξ is an inner automorphism and prove that ξ is a 1-coboundary. There exists an invertible element $w = \sum_{\tau \in G} a_\tau u_\tau$ in Δ such that $\theta_\xi(v) = w^{-1}vw$ for all $v \in \Delta$. By assumption $\xi(1) = 1$, and by Definition 12.1.2, the multiplicative identity of Δ is u_1. If we take $v = su_1$ for some $s \in S$, then $\theta_\xi(su_1) = su_1 = w^{-1}su_1w$. Thus

$$\sum_{\tau \in G} a_\tau u_\tau s = \sum_{\tau \in G} a_\tau \tau(s) u_\tau = \sum_{\tau \in G} a_\tau s u_\tau.$$

Therefore, $a_\tau\tau(s) = a_\tau s$ for all $s \in S$ and all $\tau \in G$. By Theorem 12.2.9 (6), the set $\{\tau(s) - s \mid s \in S\}$ generates the unit ideal, if $\tau \neq 1$. So $w = a_1u_1$, and a_1 is a unit in S. Lastly,

$$\theta_\xi(u_\sigma) = \xi(\sigma)u_\sigma = a_1^{-1}u_\sigma a = a_1^{-1}\sigma(a)u_\sigma$$

implies $\xi(\sigma) = \sigma(a_1)a_1^{-1}$ for all $\sigma \in G$. This proves that ξ is a 1-coboundary in $\mathrm{B}^1(G, S^*)$.

Let $\phi \in \mathrm{Aut}_R(\Delta)$ be both an R-algebra and S-module automorphism. For any $\sigma \in G$ and $s \in S$,

$$\phi(u_\sigma)su_1 = \phi(u_\sigma)\phi(su_1) = \phi(u_\sigma s) = \phi(\sigma(s)u_\sigma) = \sigma(s)\phi(u_\sigma).$$

By Theorem 12.2.9 (4), we view $\Delta \cong \mathrm{Hom}_R(S, S)$. Let $\xi(\sigma) = \phi(u_\sigma)(1)$. Then $\phi(u_\sigma)(s) = \sigma(s)\xi(\sigma)$ for all $s \in S$. Therefore, in Δ we have $\phi(u_\sigma) = \xi(\sigma)u_\sigma$. Since u_σ is a unit and ϕ is an automorphism, this implies $\xi(\sigma)S = S$, or $\xi(\sigma) \in S^*$. It follows from $\phi(u_{\sigma\tau}) = \phi(u_\sigma)\phi(u_\tau)$ that $\xi(\sigma\tau) = \xi(\sigma)\sigma\xi(\tau)$, hence $\xi \in \mathrm{Z}^1(G, S^*)$. Since $\phi = \theta_\xi$, we are done. □

Lemma 13.3.5. *In the context of Theorem* 13.3.1, *the sequence*

$$0 \to \mathrm{H}^1(G, S^*) \xrightarrow{\alpha_1} \operatorname{Pic} R \xrightarrow{\alpha_2} (\operatorname{Pic} S)^G$$

of abelian groups is exact. The map α_1 is defined to be the homomorphism θ of Lemma 13.3.4 *(2) followed by the homomorphism α of Corollary* 7.8.12. *The Picard group functor of Proposition* 2.6.9 *applied to the natural map $R \to S$ defines $\alpha_2 : \operatorname{Pic} R \to \operatorname{Pic} S^G$.*

Proof. First we show that $\operatorname{im} \alpha_1 \subseteq \ker \alpha_2$. Let $\xi \in \mathrm{Z}^1(G, S^*)$. Write σ for the R-automorphism θ_ξ defined in Lemma 13.3.4. By Corollary 7.8.12 and Exercise 13.3.15, $\alpha_2\alpha_1(\xi) = \alpha_2[({}_1\Delta_\sigma \otimes_\Delta S)^G]$. By Proposition 12.4.2, $\alpha_2[({}_1\Delta_\sigma \otimes_\Delta S)^G] = [{}_1\Delta_\sigma \otimes_\Delta S]$. From the definition of $\sigma = \theta_\xi$, we see that ${}_1\Delta_\sigma \otimes_\Delta S$ is a free left S-module of rank one. This proves $\operatorname{im} \alpha_1 \subseteq \ker \alpha_2$.

Now we show $\operatorname{im} \alpha_1 \supseteq \ker \alpha_2$. Let E be a rank one R-progenerator and assume there is an S-module isomorphism $\phi : E \otimes_R S \to S$. Define $\psi : \Delta \to \Delta$ by $w \mapsto \phi(1 \otimes w)\phi^{-1}$. The reader should verify that ψ is both an R-algebra and S-module homomorphism. By Theorem 7.6.1, ψ is an automorphism. By Lemma 13.3.4 (2), $\psi = \theta_\xi$, for some 1-cocycle ξ. To finish this step, we show $E \otimes_R S$ is isomorphic as a left Δ-module to ${}_{\theta_\xi}S_1$. Recall that $E \otimes_R S$ is a Δ module by virtue of the S factor. For any $z \in {}_{\theta_\xi}S_1$, $s \in S$, and $\sigma \in G$, we have

$$su_\sigma \cdot z = s\xi(\sigma)z = \phi(1 \otimes su_\sigma)\phi^{-1}(z).$$

Applying ϕ^{-1} to both sides yields

$$\phi^{-1}(su_\sigma \cdot z) = su_\sigma \cdot \phi^{-1}(z).$$

This shows $\phi^{-1} : {}_{\theta_\xi}S_1 \to E \otimes_R S$ is an isomorphism of left Δ modules. By Exercise 13.3.15, and the definition of α_1, this shows E is in the image of α_1.

Lastly we show that $\operatorname{im} \alpha_2 \subseteq (\operatorname{Pic} S)^G$. We make $S \otimes_R P$ into a two sided S/R-module by multiplication on the S factor: $a \otimes b \cdot x \otimes y = axb \otimes y$. Therefore, if $\sigma \in G$, $\sigma[S \otimes_R P] = [{}_1S_\sigma \otimes_S (S \otimes_R P) \otimes_S {}_\sigma S_1] = [{}_1S_\sigma \otimes_S S \otimes_S {}_\sigma S_1 \otimes_R P] = [S \otimes_R P]$. So the image of α_2 is a subgroup of $(\operatorname{Pic} S)^G$. □

The homomorphism α_3 will be defined in the next lemma. In the context of Theorem 13.3.1, let $[P] \in (\operatorname{Pic} S)^G$. By Corollary 13.2.7, there is a 2-cocycle $\xi \in \mathrm{Z}^2(G, S^*)$ such that $\operatorname{Hom}_R(P, P)$ is isomorphic as both an R-algebra and left $S \otimes_R S$-module to the crossed product $\Delta(S/R, G, \xi)$. By Proposition 13.2.5, ξ determines a unique cohomology class $[\xi] \in \mathrm{H}^2(G, S^*)$.

Lemma 13.3.6. *In the above context, define* $\alpha_3 : (\operatorname{Pic} S)^G \to \mathrm{H}^2(G, S^*)$ *by* $[P] \mapsto [\xi]$. *Then the following are true.*

(1) α_3 *is a homomorphism of abelian groups.*

(2) *There is an exact sequence*

$$\operatorname{Pic} R \xrightarrow{\alpha_2} (\operatorname{Pic} S)^G \xrightarrow{\alpha_3} \mathrm{H}^2(G, S^*)$$

of abelian groups.

Proof. (1): By the remarks in the preceding paragraph, the function α_3 is well defined. We show α_3 is a homomorphism. Let $[P], [Q] \in (\operatorname{Pic} S)^G$. By Corollary 13.2.7, there are 2-cocycles $\xi, \zeta \in \mathrm{Z}^2(G, S^*)$ such that $\operatorname{Hom}_R(P, P)$ is isomorphic as both an R-algebra and left $S \otimes_R S$-module to the crossed product algebra $\Delta(S/R, G, \xi)$. Then there is a free basis $\{f_\sigma \mid \sigma \in G\}$ for the left S-module $\operatorname{Hom}_R(P, P)$ and $f_\sigma f_\tau = \xi(\sigma, \tau) f_{\sigma\tau}$ for all $(\sigma, \tau) \in G \times G$. Likewise, there are isomorphisms $\operatorname{Hom}_R(Q, Q) \cong \Delta(S/R, G, \zeta) \cong \bigoplus_{\sigma \in G} S g_\sigma$, and $g_\sigma g_\tau = \zeta(\sigma, \tau) g_{\sigma\tau}$ for all $(\sigma, \tau) \in G \times G$. Then $\{f_\sigma \otimes g_\sigma \mid \sigma \in G\}$ is a set of invertible elements in $\operatorname{Hom}_R(P \otimes_S Q, P \otimes_S Q)$ which makes up a free S-basis. Also $\xi\zeta \in \mathrm{Z}^2(G, S^*)$ and $(f_\sigma \otimes g_\sigma)(f_\tau \otimes g_\tau) = \xi(\sigma, \tau)\zeta(\sigma, \tau) f_{\sigma\tau} \otimes g_{\sigma\tau}$. Therefore,

$$\operatorname{Hom}_R(P \otimes_S Q, P \otimes_S Q) \cong \Delta(S/R, G, \xi\zeta) \cong \bigoplus_{\sigma \in G} S f_\sigma \otimes g_\sigma$$

which shows $\alpha_3([P \otimes_S Q]) = [\xi][\zeta]$.

(2): It follows from Corollary 12.4.5 that $\operatorname{im} \alpha_2 = \ker \alpha_3$. □

Now we prove that the image of α_3 is equal to the kernel of α_4.

Lemma 13.3.7. *In the context of Theorem* 13.3.1, *there is an exact sequence*

$$(\operatorname{Pic} S)^G \xrightarrow{\alpha_3} \mathrm{H}^2(G, S^*) \xrightarrow{\alpha_4} \mathrm{B}(S/R)$$

of abelian groups, where α_3 *is defined in Lemma* 13.3.6 *and* α_4 *is defined in Theorem* 13.1.4.

Proof. First we show that $\operatorname{im} \alpha_3 \subseteq \ker \alpha_4$. Let $[V] \in (\operatorname{Pic} S)^G$. By the definition of α_3, $\operatorname{Hom}_R(V, V) \cong \Delta(S/R, G, \xi)$, $\alpha_3[V] = [\xi]$, and $\alpha_4[\xi]$ is equal to the Brauer class containing the split Azumaya R-algebra $\Delta(S/R, G, \xi) \cong \operatorname{Hom}_R(V, V)$.

Now we show that $\operatorname{im} \alpha_3 \supseteq \ker \alpha_4$. Let $[\xi] \in \ker \alpha_4$. Then there is an R-progenerator module V such that $\Delta(S/R, G, \xi) \cong \operatorname{Hom}_R(V, V)$. We know from Proposition 13.2.6 that V is an invertible S-module. By Corollary 13.2.7, $[V] \in (\operatorname{Pic} S)^G$. Therefore, $[\xi] = \alpha_3([V])$. □

We are now ready to define the homomorphism α_5. By Theorem 13.2.4, if S/R is a Galois extension of commutative rings with group G and Λ is an Azumaya R-algebra that is split by S, then Λ is Brauer equivalent to a generalized crossed product $\Delta(S/R, \Phi, f)$ for some homomorphism of groups $\Phi : G \to \operatorname{Pic}_R S$ and some factor set f. By Lemma 13.2.1, there is associated to Φ a 1-cocycle φ in $\mathrm{Z}^1(G, \operatorname{Pic} S)$. Define the homomorphism $\alpha_5 : \mathrm{B}(S/R) \to \mathrm{H}^1(G, \operatorname{Pic} S)$ by the assignment $[\Lambda] \mapsto [\varphi]$.

Lemma 13.3.8. *The assignment $[\Lambda] \mapsto [\varphi]$ is a well defined homomorphism of abelian groups $\alpha_5 : \mathrm{B}(S/R) \to \mathrm{H}^1(G, \operatorname{Pic} S)$.*

Proof. By Lemma 13.2.8, if the function is well defined, it is a homomorphism of groups. Let $\Delta(S/R, \Phi_1, f_1)$ and $\Delta(S/R, \Phi_2, f_2)$ be two generalized crossed products and assume they are Brauer equivalent. Let φ_i in $\mathrm{Z}^1(G, \operatorname{Pic} S)$ be the 1-cocycle corresponding to Φ_i. We show that $\varphi_1\varphi_2^{-1}$ is a 1-coboundary in $\mathrm{B}^1(G, \operatorname{Pic} S)$. By Proposition 13.2.9, the opposite algebra $\Delta(S/R, \Phi_2, f_2)^o$ is a generalized crossed product $\Delta(S/R, \Phi^o, f^o)$. By Proposition 7.3.4, there exists an R-progenerator P such that

$$\begin{aligned}\operatorname{Hom}_R(P, P) &\cong \Delta(S/R, \Phi_1, f_1) \otimes_R \Delta(S/R, \Phi_2, f_2)^o \\ &= \Delta(S/R, \Phi_1, f_1) \otimes_R \Delta(S/R, \Phi^o, f^o) \\ &\cong \Delta((S \otimes_R S)/R, \Phi_1 \otimes_R \Phi^o, f_1 \otimes f^o)\end{aligned}$$

where the last isomorphism is from Exercise 13.2.11. By Proposition 13.2.6, P is a rank one $S \otimes_R S$-progenerator module. For $(\sigma, \tau) \in G \times G$, we have

$$(\Phi_1(\sigma) \otimes_R \Phi^o(\tau))[P] = [P](\Phi_0(\sigma) \otimes_R \Phi_0(\tau))$$

in $\operatorname{Pic}_R(S \otimes_R S)$ and

$$(\varphi_1 \otimes \varphi_2^{-1})(\sigma, \tau) = (\sigma, \tau)[P^*][P]$$

in $\operatorname{Pic}(S \otimes_R S)$. In particular, taking $\Phi_1 = \Phi_2$,

$$(\varphi_1^{-1} \otimes \varphi_1)(\sigma, \tau) = (\sigma, \tau)[Q^*][Q]$$

for some $[Q]$ in $\operatorname{Pic}(S \otimes_R S)$. Let $[P_\tau] = \varphi_1(\tau)\varphi_2^{-1}(\tau)$ in $\operatorname{Pic}(S)$. Let $[R] = [P \otimes_{S \otimes_R S} Q]$ in $\operatorname{Pic}(S \otimes_R S)$. Then

$$[S \otimes_R P_\tau] = (\sigma, \tau)[R^*][R]$$

in $\operatorname{Pic}(S \otimes_R S)$. Taking $\tau = 1$, we have $[P_1] = [S]$, and

$$[S \otimes_R S] = (\sigma, 1)[R^*][R]$$

shows $(\sigma, 1)[R] = [R]$ for each $\sigma \in G$. Since $S \otimes_R S$ is a trivial Galois extension with group $G \times \langle 1 \rangle$, there is $[R_0] \in \operatorname{Pic} S$ such that $[R] = [(S \otimes_R S) \otimes_S R_0] = [S \otimes_R R_0]$ in $\operatorname{Pic}(S \otimes_R S)$ (Exercise 13.3.12). Then

$$[S \otimes_R P_\tau] = [S \otimes_R R_0](\sigma, \tau)[S \otimes_R R_0]^{-1}$$

implies there are isomorphisms

$$\begin{aligned} S \otimes_R P_\tau &\cong (S \otimes_R R_0) \otimes_{S \otimes_R S} \\ &\qquad ((_1S_\sigma \otimes_R {}_1S_\tau) \otimes_{S\otimes_R S} (S \otimes_R R_0^*) \otimes_{S \otimes_R S} (_1S_{\sigma^{-1}} \otimes_R {}_1S_{\tau^{-1}})) \\ &\cong (S \otimes_R R_0) \otimes_{S\otimes_R S} (S \otimes_R (_1S_\tau \otimes_S R_0^* \otimes_S {}_1S_{\tau^{-1}})) \\ &\cong S \otimes_R (R_0 \otimes_S {}_{\tau^{-1}}(R_0^*)_{\tau^{-1}}) \end{aligned}$$

of $S \otimes_R S$-modules. Hence there is an isomorphism

$$S \otimes_R (S \otimes_S P_\tau) \cong S \otimes_R (S \otimes_S R_0 \otimes_S {}_{\tau^{-1}}(R_0^*)_{\tau^{-1}})$$

of $S\otimes_R S$-modules. Because $S\otimes_R S = \nabla(S, G) = \bigoplus_{\sigma \in G} Se_\sigma$ is a split Galois extension, for each $\sigma \in G$ there is an isomorphism

$$Se_\sigma \otimes_S P_\tau \cong Se_\sigma \otimes_S R_0 \otimes_S {}_{\tau^{-1}}(R_0^*)_{\tau^{-1}}$$

of $S \otimes_R S$-modules. As S-modules

$$Se_\sigma \otimes_S P_\tau \cong P_\tau$$

and

$$Se_\sigma \otimes_S R_0 \otimes_S {}_{\tau^{-1}}(R_0^*)_{\tau^{-1}} \cong R_0 \otimes_S {}_{\tau^{-1}}(R_0^*)_{\tau^{-1}}$$

are isomorphisms. Therefore, in $\operatorname{Pic} S$, $[P_\tau] = [R_0]\tau[R_0]^{-1} = (\varphi_1\varphi_2^{-1})(\tau)$. This proves $\varphi_1\varphi_2^{-1} \in \mathrm{B}^1(G, \operatorname{Pic} S)$. □

Lemma 13.3.9. *In the context of Theorem* 13.3.1, *there is an exact sequence*

$$\mathrm{H}^2(G, S^*) \xrightarrow{\alpha_4} \mathrm{B}(S/R) \xrightarrow{\alpha_5} \mathrm{H}^1(G, \operatorname{Pic} S)$$

of abelian groups. The homomorphism α_4 is defined in Theorem 13.1.4 *and α_5 in Lemma* 13.3.8.

Proof. First we show that $\operatorname{im} \alpha_4 \subseteq \ker \alpha_5$. If $f \in \mathrm{Z}^2(G, S^*)$, then $\alpha_4[f] = \Delta(S/R, \Phi_0, f)$, by Proposition 13.2.3 (3). By Lemma 13.2.1, the 1-cocycle associated to Φ_0 is the constant function $\varphi(\sigma) = [S]$.

Now we show that $\operatorname{im} \alpha_4 \supseteq \ker \alpha_5$. Suppose Λ is an Azumaya R-algebra that is split by S. By Theorem 13.2.4, Λ is Brauer equivalent to a generalized crossed product $\Delta(S/R, \Phi, f)$ for some homomorphism of groups $\Phi : G \to \operatorname{Pic}_R S$ and some factor set f. Assume $[\Lambda] \in \ker \alpha_5$. That is, $\Phi = \varphi\Phi_0$ and φ in $\mathrm{B}^1(G, \operatorname{Pic} S)$. Then there is $[P] \in \operatorname{Pic} S$ such that $\varphi(\sigma) = [P]\sigma[P]^{-1}$, for all $\sigma \in G$. Then $\operatorname{Hom}_R(P, P)$ is an Azumaya R-algebra with maximal commutative subalgebra $S = \operatorname{Hom}_S(P, P)$. By Theorem 13.2.4, $\operatorname{Hom}_R(P, P)$ is isomorphic as an S-module and as an R-algebra to a generalized crossed product $\Delta(S/R, \Phi_1, f_1)$. By Proposition 13.2.6, $\Phi_1(\sigma)[P] = [P]\Phi_0(\sigma)$ for all $\sigma \in G$. Therefore,

$$\varphi_1(\sigma) = \Phi_1(\sigma)\Phi_0(\sigma^{-1}) = [P]\Phi_0(\sigma)[P]^{-1}\Phi_0(\sigma^{-1}) = [P]\sigma[P]^{-1} = \varphi(\sigma).$$

By Lemma 13.2.1, $\Phi = \Phi_1$. By Proposition 13.2.5, there is a 2-cocycle $\xi \in \mathrm{Z}^2(G, S^*)$ such that $f = \xi f_1$. It follows from Proposition 13.2.10 that $\xi \otimes \xi^{-1}$ is in $\mathrm{B}^2(G \times G, (S \otimes_R S)^*)$. It follows from Proposition 13.2.5 that there is an isomorphism

$$\begin{aligned}\Delta\left((S \otimes_R S)/R, \Phi \otimes_R \Phi_0, (\xi \otimes \xi^{-1})(1 \otimes \xi)(f_1 \otimes 1)\right)& \\ \cong \Delta\left((S \otimes_R S)/R, \Phi \otimes_R \Phi_0, (1 \otimes \xi)(f_1 \otimes 1)\right)&\end{aligned}$$

of both R-algebras and $S \otimes_R S$-modules. On the other hand,

$$\begin{aligned}\Delta\left((S \otimes_R S)/R, \Phi \otimes_R \Phi_0, (\xi \otimes \xi^{-1})(1 \otimes \xi)(f_1 \otimes 1)\right)& \\ \cong \Delta(S/R, \Phi, \xi f_1) \otimes_R \Delta(S/R, \Phi_0, 1)& \\ = \Delta(S/R, \Phi, f) \otimes_R \Delta(S/R, G, 1)&\end{aligned}$$

and

$$\begin{aligned}\Delta\left((S \otimes_R S)/R, \Phi \otimes_R \Phi_0, (1 \otimes \xi)(f_1 \otimes 1)\right)& \\ \cong \Delta(S/R, \Phi, f_1) \otimes_R \Delta(S/R, \Phi_0, \xi)& \\ = \operatorname{Hom}_R(P, P) \otimes_R \Delta(S/R, G, \xi).&\end{aligned}$$

Hence, $[\Lambda] = [\Delta(S/R, \Phi, f)] = [\Delta(S/R, G, \xi)]$ is the image of ξ under α_4. □

In the context of Theorem 13.3.1, let S/R be a Galois extension of commutative rings. As in Lemma 13.2.1, let $\Phi : G \to \operatorname{Pic}_R(S)$ be a homomorphism of groups such that $\Psi\Phi(\sigma) = \Phi_0(\sigma)$ for each $\sigma \in G$. Let $\varphi(\tau) = \Phi(\tau)\Phi_0(\tau^{-1})$ be the corresponding 1-cocycle in $\mathrm{Z}^1(G, \operatorname{Pic} S)$. Write $\Phi(\tau) = [\Phi_\tau]$. Since $\Phi(\sigma\tau) = \Phi(\sigma)\Phi(\tau)$, there is a family of $S \otimes_R S$-module isomorphisms $\{f_{\sigma,\tau} : \Phi_\sigma \otimes_S \Phi_\tau \to \Phi_{\sigma\tau}\}$ indexed by $G \times G$. For each triple $(\sigma, \tau, \gamma) \in G \times G \times G$, define

$$\omega(\sigma, \tau, \gamma) = f_{\sigma\tau,\gamma} \circ f_{\sigma,\tau} \otimes 1 \circ (1 \otimes f_{\tau,\gamma})^{-1} \circ f_{\sigma,\tau\gamma}^{-1}$$

which is an isomorphism in $\operatorname{Hom}_S(\Phi_{\sigma\tau\gamma}, \Phi_{\sigma\tau\gamma}) \cong S$, hence corresponds to left multiplication by a unit in S. That is, $\omega : G \times G \times G \to S^*$.

Lemma 13.3.10. *In the above context, the assignment $[\varphi] \mapsto [\omega]$ defines a homomorphism of abelian groups $\alpha_6 : \mathrm{H}^1(G, \operatorname{Pic} S) \to \mathrm{H}^3(G, S^*)$.*

Proof. The proof is left to the reader. □

Lemma 13.3.11. *In the context of Theorem 13.3.1, there is an exact sequence*

$$\mathrm{B}(S/R) \xrightarrow{\alpha_5} \mathrm{H}^1(G, \operatorname{Pic} S) \xrightarrow{\alpha_6} \mathrm{H}^3(G, S^*)$$

of abelian groups. The homomorphism α_5 is defined in Lemma 13.3.8 and α_6 in Lemma 13.3.10.

Proof. The proof is left to the reader. □

3.2. Exercises.

Exercise 13.3.12. Let R be a commutative ring, G a finite group, and $S = \bigoplus_{\sigma \in G} Re_\sigma$ the trivial Galois extension of R with group G (Example 12.2.5).

(1) Using Exercise 2.6.14, show that the map $\alpha_2 : \mathrm{Pic}(R) \to (\mathrm{Pic}\, S)^G$ is an isomorphism.

(2) Show that $\mathrm{B}(S/R) = (0)$. (Hint: Exercise 7.4.8.)

(3) Show that $\mathrm{H}^1(G, S^*) = \mathrm{H}^2(G, S^*) = \langle 1 \rangle$.

Exercise 13.3.13. Prove that the Brauer group of $\mathbb{R}$ is a cyclic group of order two. Show that $\mathrm{B}(\mathbb{R})$ is generated by the class represented by the ring of real quaternions. (Hints: Example 1.1.1, Exercise 4.1.13, Theorem 7.5.1, Exercise 5.1.23.)

Exercise 13.3.14. Let S/R be a Galois extension of commutative rings and $\Delta = \Delta(S/R, G, 1)$ the trivial crossed product algebra. Show that there is an isomorphism of groups $\mathrm{Pic}\, R \to \mathrm{Pic}_R(\Delta)$ defined by the assignment $[P] \mapsto [S^* \otimes_R P \otimes_R S]$, where $S^* = \mathrm{Hom}_R(S, R)$. The inverse map is defined by $[P] \mapsto [S^* \otimes_R P \otimes_R S]$.

Exercise 13.3.15. Let S/R be a Galois extension of commutative rings and $\Delta = \Delta(S/R, G, 1)$ the trivial crossed product algebra. Let $\sigma \in \mathrm{Aut}_R(\Delta)$. Show that there is an isomorphism of R-modules $({}_1\Delta_\sigma)^\Delta \cong ({}_1\Delta_\sigma \otimes_\Delta S)^G$. (Hint: The category of right R-modules is equivalent to the category of left Δ-modules (Remark 12.2.3) as well as to the category of left Δ^e-modules (Corollary 7.1.6).)

Exercise 13.3.16. Let R be a noetherian regular integral domain with field of fractions K. Let L/K be a Galois extension of fields with finite group G. Let S be any R-subalgebra of L. Show that the natural homomorphism $\mathrm{B}(S/R) \to \mathrm{B}(L/K)$ on relative Brauer groups is one-to-one. If G has order n, show that $\mathrm{B}(S/R)$ is a $\mathbb{Z}/n$-module.

Exercise 13.3.17. This exercise is a continuation of Exercise 12.9.20. Let $n \geq 2$ be an integer, k a field in which n is invertible, and assume k contains a primitive nth root of unity, ζ. Let f be an irreducible polynomial in $A = k[x_1, \dots, x_m]$ and set $T = A[z]/(z^n - f)$, $S = T[z^{-1}]$, $R = A[f^{-1}]$. By Kummer Theory (Example 12.9.5), S/R is a Galois extension with group $G = \langle \sigma \rangle$, where $\sigma(z) = \zeta z$, and G is a cyclic group of order n. Assume f is geometrically irreducible. In other words, assume f is irreducible in $\bar{k}[x_1, \dots, x_m]$, if $\bar{k}$ is an algebraic closure of k. Prove that

$$0 \to \mathrm{H}^2(G, S^*) \xrightarrow{\alpha_4} \mathrm{B}(S/R) \xrightarrow{\alpha_5} \mathrm{H}^1(G, \mathrm{Pic}\, S) \to 0$$

is an exact sequence of $\mathbb{Z}/n$-modules. In fact, it is not necessary to assume f is geometrically irreducible. Let F be an algebraic extension field of k

which contains a splitting field for the set of polynomials $\{x^n - a \mid a \in k^*\}$. Show that if f is irreducible in $F[x_1, \ldots, x_m]$, then the above sequence is exact.

3.3. Galois Cohomology Agrees with Amitsur Cohomology. The following result, which is from [**CHR65**], shows that Galois cohomology agrees with Amitsur cohomology.

Theorem 13.3.18. *Let S be a Galois extension of R with finite group G. Let $\mathfrak{F}$ be a functor from the category of commutative R-algebras to the category of abelian groups such that $\mathfrak{F}$ distributes across finite direct sums:*

$$\mathfrak{F}(S_1 \oplus S_2) = \mathfrak{F}(S_1) \oplus \mathfrak{F}(S_2).$$

(1) *The nth Amitsur cohomology group agrees with the corresponding Galois cohomology group. That is,*

$$\mathrm{H}^n(S/R, \mathfrak{F}) \cong \mathrm{H}^n(G, \mathfrak{F}(S))$$

for all $n \geq 0$.

(2) *If H is a normal subgroup of G, then for each $n \geq 0$ the diagram of abelian groups*

$$\begin{array}{ccc} \mathrm{H}^n(S^H/R, \mathfrak{F}) & \xrightarrow{\cong} & \mathrm{H}^n(G/H, \mathfrak{F}(S^H)) \\ \downarrow{\scriptstyle i^*} & & \downarrow{\scriptstyle \mathrm{Inf}} \\ \mathrm{H}^n(S/R, \mathfrak{F}) & \xrightarrow{\cong} & \mathrm{H}^n(G, \mathfrak{F}(S)) \end{array}$$

is commutative, where i^* is induced by $i : S^H \to S$ (Theorem 5.5.5), and Inf *is the inflation homomorphism (Definition 5.1.12).*

Proof. (1): Let $Q_\bullet \to \mathbb{Z}$ be the standard (homogeneous) resolution of $\mathbb{Z}$ as a $\mathbb{Z}G$-module from Definition 5.1.2. As in Section 5.1, if A is a $\mathbb{Z}G$-module, the cohomology groups $\mathrm{H}^n(G, A)$ are equal to the homology of the cochain complex $C^\bullet(G, A) = \mathrm{Hom}_{\mathbb{Z}G}(Q_\bullet, A)$. To prove the theorem, we construct an isomorphism of cochain complexes $\theta : C^\bullet(S/R, \mathfrak{F}) \to C^\bullet(G, \mathfrak{F}(S))$. For $n \geq 0$, Q_n is the free $\mathbb{Z}G$-module on G^n (recall that G^0 is the singleton set). We therefore identify $\mathrm{Hom}_{\mathbb{Z}G}(Q_n, A)$ with $\mathrm{Map}(G^n, A)$, the set of all functions mapping G^n to A. In particular, S is a left G module, so we can take A to be S and we can make $\mathrm{Map}(G^n, S)$ into a ring by pointwise addition and multiplication of functions. The constant functions form a subring isomorphic to S, so $\mathrm{Map}(G^n, S)$ is an S-algebra. Recall that in Example 12.2.6, $\mathrm{Map}(G, S)$ was denoted $\nabla(S, G)$. For $n \geq 1$, define $\theta^n : S^{\otimes(n+1)} \to \mathrm{Map}(G^n, S)$ by

$$\theta^n(s_0 \otimes \cdots \otimes s_n)(\sigma_1, \ldots, \sigma_n) = s_0\sigma_1(s_1)\sigma_1\sigma_2(s_2) \cdots (\sigma_1 \cdots \sigma_n)(s_n).$$

We show that θ^n is an S-algebra isomorphism for all $n \geq 1$. The reader should verify that θ^n is an S-algebra homomorphism. For $n = 1$, θ^1 is the isomorphism ℓ of Theorem 12.2.9 (5). Inductively, assume $n > 1$ and that θ^{n-1} is an S-algebra isomorphism. For $v : G^n \to S$ and $\sigma \in G$, define $\sigma v : G^n \to S$ by

$$(\sigma v)(\tau_1, \dots, \tau_n) = \sigma(v(\sigma^{-1}\tau_1, \tau_2, \dots, \tau_n)).$$

Using the method of Example 12.4.1, the reader should verify that this makes $\operatorname{Map}(G^n, S)$ into a left $\Delta(S/R, G, 1)$-module. Notice that $\operatorname{Map}(G^n, S)^G$ is equal to the set of all functions $v : G^n \to S$ such that $\sigma(v(\tau_1, \dots, \tau_n) = v(\sigma\tau_1, \tau_2, \dots, \tau_n)$ for all $\sigma \in G$. Define $\psi : \operatorname{Map}(G^{n-1}, S) \to \operatorname{Map}(G^n, S)^G$ by $(\psi g)(\tau_1, \dots, \tau_n) = \tau_1(g(\tau_2, \dots, \tau_n))$. Notice that the inverse of ψ is defined by $(\psi^{-1}v)(\tau_2, \dots, \tau_n) = v(1, \tau_2, \dots, \tau_n)$. The reader should verify that ψ is an R-algebra isomorphism. We have a sequence of S-algebra isomorphisms

$$S^{\otimes(n+1)} \to S \otimes_R S^{\otimes n} \xrightarrow{1\otimes\theta^{n-1}} S \otimes_R \operatorname{Map}(G^{n-1}, S) \xrightarrow{1\otimes\psi} S \otimes_R \operatorname{Map}(G^n, S)^G \xrightarrow{\mu} \operatorname{Map}(G^n, S)$$

and the reader should verify that the composite map is equal to θ^n. The map μ is from Theorem 12.4.3. We view $\operatorname{Map}(G^n, S)$ as a product of copies of S. Because G is finite and $\mathfrak{F}$ commutes with finite direct sums, we have $\mathfrak{F}(\operatorname{Map}(G^n, S)) = \operatorname{Map}(G^n, \mathfrak{F}(S))$. For $n \geq 1$ we have an isomorphism of abelian groups

$$\theta^n : \mathfrak{F}(S^{\otimes(n+1)}) \to \operatorname{Map}(G^n, \mathfrak{F}(S)).$$

Identify $\operatorname{Map}(G^0, S)$ with the constant functions, S, and define $\theta^0 : \mathfrak{F}(S) \to \mathfrak{F}(S)$ to be the identity map on $\mathfrak{F}(S)$. Using the descriptions of the coboundary maps given in Sections 5.1.1, and 5.5, the reader should verify that the diagram

$$\begin{array}{ccc} \mathfrak{F}(S^{\otimes(n+1)}) & \xrightarrow{d^n} & \mathfrak{F}(S^{\otimes(n+2)}) \\ \downarrow{\scriptstyle\theta^n} & & \downarrow{\scriptstyle\theta^{n+1}} \\ \operatorname{Map}(G^n, \mathfrak{F}(S)) & \xrightarrow{d^n} & \operatorname{Map}(G^{n+1}, \mathfrak{F}(S)) \end{array}$$

commutes for all $n \geq 0$. Therefore $\theta : C^\bullet(S/R, \mathfrak{F}) \to C^\bullet(G, \mathfrak{F}(S))$ is an isomorphism of cochain complexes. Consequently the cohomology groups are isomorphic.

(2): By Theorem 12.5.1, S^H/R is Galois with group G/H. The description of the inflation homomorphism given in Definition 5.1.12 shows that it agrees with i^*. The rest is left to the reader. □

3.4. Galois Cohomology and the Brauer Group. When S/R is a Galois extension of fields, a proof of Theorem 13.3.19 (1) involving an explicit isomorphism is given in [**Rei75**, Theorem (29.16)].

Theorem 13.3.19. *Let S/R be a Galois extension of commutative rings with finite group G and H a normal subgroup of G.*

(1) *Let $f \in \mathrm{Z}^2(G/H, (S^H)^*)$ be a normalized 2-cocycle. If $g \in \mathrm{Z}^2(G, S^*)$ is the 2-cocycle defined by the inflation of f (Definition 5.1.12), then the crossed product $\Delta(S^H/R, G/H, f)$ is Brauer equivalent to the crossed product $\Delta(S/R, G, g)$.*

(2) *The diagram of abelian groups*

$$\begin{array}{ccc} \mathrm{H}^2(G/H, (S^H)^*) & \xrightarrow{\alpha_4} & \mathrm{B}(S^H/R) \\ \Big\downarrow \mathrm{Inf} & & \Big\downarrow \subseteq \\ \mathrm{H}^2(G, S^*) & \xrightarrow{\alpha_4} & \mathrm{B}(S/R) \end{array}$$

commutes.

Proof. Part (2) follows immediately from Part (1).

(1): Let $\{\tau_1, \dots, \tau_m\}$ be a full set of left coset representatives for H in G. Write Δ_f instead of $\Delta(S^H/R, G/H, f)$ and Δ_g instead of $\Delta(S/R, G, g)$. Let $A = \Delta_g \otimes_R \Delta_f^o$. By Lemma 13.1.3, $\Delta_f^o \cong \Delta(S^H/R, G/H, f^{-1})$. Let $\Delta_g = \bigoplus_{\sigma \in G} Su_\sigma$, and $\Delta_f^o = \bigoplus_{i=1}^n S^H v_{\bar{\tau}_i}$. We identify $S \otimes_R S^H$ with the subring $Su_1 \otimes_R S^H \bar{u_1} \subseteq \Delta_g \otimes_R \Delta_f^o$. Let $e \in S^H \otimes_R S^H$ be the (unique) separability idempotent (Definition 4.1.3). By Exercise 7.6.4, it suffices to prove $\Delta(S/R, G, 1) \cong eAe$ as R-algebras. The proof is very similar to the second proof of Theorem 13.1.4 (1) and some of the details are left to the reader. Since $(S^H \otimes_R S^H)e = (S^H \otimes_R 1)e$, it follows that $(S \otimes_R S^H)e = (S \otimes_R 1)e$. As in Eq. (13.1),

$$e(u_\sigma \otimes v_{\bar{\tau}})e = 0$$

if $\bar{\sigma} \neq \bar{\tau}$. For each $\sigma \in G$, set $w_\sigma = e(u_\sigma \otimes v_{\bar{\sigma}})e$. The reader should verify that eAe is generated as a left $S = e(S \otimes_R 1)e$-module by $\{w_\sigma \mid \sigma \in G\}$. As in Eq. (13.2), $w_\sigma w_\tau = w_{\sigma\tau}$. Therefore, $\Delta(S/R, G, 1) \cong eAe$. □

Lemma 13.3.20. *Let k be a field and A a central simple k algebra. Assume n, the order of $[A]$ in $\mathrm{B}(k)$, is invertible in k. Then there is a finite Galois field extension K/k such that K is a splitting field for A, A is Brauer equivalent to a crossed product $\Delta(K/k, G, \xi)$, and ξ takes its values in μ_n, the group of nth roots of unity in K.*

Proof. Let F/k be a Galois extension of fields with finite group G such that F is a splitting field for A (Corollary 12.6.4). By Corollary 13.3.3,

$\mathrm{H}^2(G, F^*) \cong \mathrm{B}(F/k)$. Let $\xi \in \mathrm{Z}^2(G, F^*)$ be a 2-cocycle that represents the cohomology class corresponding to $[A]$. Let $\mu_n(F)$ be the group of nth roots of unity in F and $\varphi : F^* \to F^*$ the nth power map. Factoring φ into a projection followed by an injection gives rise to two short exact sequences

$$1 \to \mu_n(F) \to F^* \xrightarrow{\pi} (F^*)^n \to 1 \tag{13.10}$$

$$1 \to (F^*)^n \xrightarrow{\iota} F^* \to F^*/(F^*)^n) \to 1 \tag{13.11}$$

of G-modules. The long exact sequences of cohomology associated to (13.10) and (13.11) give rise to the commutative diagram

$$\begin{array}{ccccc} & & \mathrm{H}^2(G, \mu_n(F)) & & \\ & & \downarrow & & \\ & & \mathrm{H}^2(G, F^*) & & \\ & & \pi \downarrow & \searrow^{\varphi} & \\ \mathrm{H}^1(G, F^*/(F^*)^n) & \xrightarrow{\partial} & \mathrm{H}^2(G, (F^*)^n) & \xrightarrow{\iota} & \mathrm{H}^2(G, F^*) \end{array} \tag{13.12}$$

of cohomology groups where the horizontal row and the vertical column in (13.12) are exact. The order of $[\xi]$ in $\mathrm{H}^2(G, F^*)$ is n. Since φ is the nth power map, $[\xi]$ is in the kernel of $\varphi = \iota\pi$. If $[\xi]$ is in the kernel of π, then we are done. Assume otherwise. Since $\pi([\xi])$ is in the image of ∂, let $\nu \in \mathrm{Z}^1(G, F^*/(F^*)^n)$ be a 1-cocycle that maps onto $\pi([\xi])$. By adjoining to F a finite number of nth roots of elements, we can find a Galois extension K/k containing F/k with finite group G' such that the values of ν are in $(K^*)^n$. The natural map $F^* \to K^*$ together with the inflation maps give rise to a commutative square

$$\begin{array}{ccc} \mathrm{H}^1(G, F^*/(F^*)^n) & \xrightarrow{\partial} & \mathrm{H}^2(G, (F^*)^n) \\ \downarrow \mathrm{Inf} & & \downarrow \mathrm{Inf} \\ \mathrm{H}^1(G', K^*/(K^*)^n) & \xrightarrow{\partial} & \mathrm{H}^2(G', (K^*)^n) \end{array} \tag{13.13}$$

and the left hand vertical map in (13.13) sends ν to the 1-cocycle with constant value 1. Therefore, $\mathrm{Inf}(\pi([\xi]))$ is equal to 1 in $\mathrm{H}^2(G', (K^*)^n)$. It follows from the commutative square

$$\begin{array}{ccc} \mathrm{H}^2(G, F^*) & \xrightarrow{\pi} & \mathrm{H}^2(G, (F^*)^n) \\ \downarrow \mathrm{Inf} & & \downarrow \mathrm{Inf} \\ \mathrm{H}^2(G', K^*) & \xrightarrow{\pi} & \mathrm{H}^2(G', (K^*)^n) \end{array} \tag{13.14}$$

that $\pi(\mathrm{Inf}([\xi]))$ is equal to 1 in $\mathrm{H}^2(G', (K^*)^n)$. In (13.12), replace F with K, and we are done. □

3.5. Exercise.

Exercise 13.3.21. Let S/R be a Galois extension of commutative rings with finite group G and H a normal subgroup of G. Show that there are homomorphisms connecting the seven term exact sequence for S^H/R with the sequence for S/R and that the diagram

$$\begin{array}{ccccccc}
1 & \longrightarrow & \mathrm{H}^1(G/H,(S^H)^*) & \xrightarrow{\alpha_1} & \mathrm{Pic}(R) & \xrightarrow{\alpha_2} & \mathrm{Pic}(S^H)^{G/H} \\
 & & \downarrow{\scriptstyle\theta_1} & & \downarrow{\scriptstyle\theta_2} & & \downarrow{\scriptstyle\theta_3} \\
1 & \longrightarrow & \mathrm{H}^1(G,S^*) & \xrightarrow{\alpha_1} & \mathrm{Pic}(R) & \xrightarrow{\alpha_2} & \mathrm{Pic}(S)^G
\end{array}$$

$$\begin{array}{ccccc}
\mathrm{Pic}(S^H)^{G/H} & \xrightarrow{\alpha_3} & \mathrm{H}^2(G/H,(S^H)^*) & \xrightarrow{\alpha_4} & \mathrm{B}(S^H/R) \\
\downarrow{\scriptstyle\theta_3} & & \downarrow{\scriptstyle\theta_4} & & \downarrow{\scriptstyle\theta_5} \\
\mathrm{Pic}(S)^G & \xrightarrow{\alpha_3} & \mathrm{H}^2(G,S^*) & \xrightarrow{\alpha_4} & \mathrm{B}(S/R)
\end{array}$$

$$\begin{array}{ccccc}
\mathrm{B}(S^H/R) & \xrightarrow{\alpha_5} & \mathrm{H}^1(G/H,\mathrm{Pic}\, S^H) & \xrightarrow{\alpha_6} & \mathrm{H}^3(G/H,(S^H)^*) \\
\downarrow{\scriptstyle\theta_5} & & \downarrow{\scriptstyle\theta_6} & & \downarrow{\scriptstyle\theta_7} \\
\mathrm{B}(S/R) & \xrightarrow{\alpha_5} & \mathrm{H}^1(G,\mathrm{Pic}\, S) & \xrightarrow{\alpha_6} & \mathrm{H}^3(G,S^*)
\end{array}$$

commutes. The vertical maps on cohomology (θ_1, θ_4, θ_6, θ_7) are the inflation homomorphisms (Definition 5.1.12). The map θ_2 is the identity map and θ_5 is the set inclusion map. The map θ_3 is induced by the Picard group functor. (Hints: For the first square use Theorem 13.3.18 and Theorem 10.4.16. For the third square, use the definition of α_3. The fourth square is Theorem 13.3.19. For squares five and six, use the definitions of α_5 and α_6.)

4. Cyclic Crossed Product Algebras

Definition 13.4.1. Let S be a commutative R-algebra and $\sigma \in \mathrm{Aut}_R(S)$. If v is an indeterminate, then by $S[v,\sigma]$ we denote the *twisted polynomial ring* which is equal to the ordinary polynomial ring $S[v]$ as a left S-module, and on monomials the multiplication rule satisfies

$$(sv^i)(tv^j) = s\sigma^i(t)v^{i+j} \tag{13.15}$$

for all $s,t \in S$. The reader should verify that the following are true for $S[v,\sigma]$.

(1) The multiplication formula (13.15) makes the left S-module $S[u,\sigma]$ into a ring containing $S = S \cdot 1$ as a subring.

(2) $R = R \cdot 1$ is a subring of the center, hence $S[v,\sigma]$ is an R-algebra.

(3) $S[u,\sigma]$ is noncommutative if and only if $\sigma \neq 1$.

We make $S[v,\sigma]$ into a left $S \otimes_R S$-module by virtue of S being a subring of $S[v,\sigma]$ which contains R (see Definition 4.1.1).

Definition 13.4.2. Retain the notation from Definition 13.4.1. Assume $\sigma^n = 1$ and let a be a nonzero element in R. Using (13.15) one verifies that $v^n - a$ is a central element in $S[u,\sigma]$. By $\Delta(S/R,\sigma,a)$ we denote the quotient $S[v,\sigma]/(v^n - a)$ of the twisted polynomial ring modulo the principal ideal generated by the central element $v^n - a$. The R-algebra $\Delta(S/R,\sigma,a)$ is called a *cyclic crossed product.*

Let S/R be a Galois extension of commutative rings with finite group G. We say S/R is a *cyclic* extension if $G = \langle\sigma\rangle$ is a cyclic group. Lemma 13.4.3 shows that a cyclic crossed product is a crossed product. Theorem 13.4.4 proves the converse: for a cyclic extension every crossed product is a cyclic crossed product.

Lemma 13.4.3. *Let S/R be a cyclic Galois extension of commutative rings with finite group $G = \langle\sigma\rangle$. Let $a \in R^*$, and let $\Delta = \Delta(S/R,\sigma,a)$ be the cyclic crossed product. Assume G has order n.*

(1) *As a left S-module, Δ is free of rank n, and $1, v, \dots, v^{n-1}$ is a basis.*
(2) *As an R-algebra, Δ is generated by S and v.*
(3) *Δ is isomorphic to the crossed product $\Delta(S/R,G,\xi)$, where $\xi : G \times G \to S^*$ is the factor set defined by*

$$\xi(\sigma^i,\sigma^j) = \begin{cases} 1 & \text{if } i+j < n, \\ a & \text{if } i+j \geq n, \end{cases}$$

for all i and j in $\{0,1,\dots,n-1\}$.
(4) *Δ is an Azumaya R-algebra, is split by S, and S is a maximal commutative subring.*

Proof. Parts (1) and (2) are left to the reader.

(3): By Exercise 5.1.28, ξ is a factor set. Let $u_1, u_\sigma, \dots, u_{\sigma^{n-1}}$ be the basis elements for $\Delta(S/R,G,\xi)$. Then the R-algebra isomorphism is induced by assignment $v^i \mapsto u_{\sigma^i}$, for $0 \leq i < n$.

Part (4) follows from Part (3), Lemma 13.1.1, and Theorem 13.1.2. □

Theorem 13.4.4. *Let S/R be a cyclic Galois extension of commutative rings with group $G = \langle\sigma\rangle$ of order n. Let $\xi \in \mathrm{Z}^2(G,S^*)$ be a normalized 2-cocycle and*

$$\Delta = \Delta(S/R,G,\xi) = \bigoplus_{i=0}^{n-1} S u_{\sigma^i}$$

the crossed product algebra. Then

(1) $1, u_\sigma, u_\sigma^2, \dots, u_\sigma^{n-1}$ *is a free basis for the* S*-module* Δ.

(2) *Let* $a = \prod_{j=0}^{n-1} \xi(\sigma^j, \sigma)$. *Then* $a \in R^*$ *and* $u_\sigma^n = a$.

(3) Δ *is isomorphic to the cyclic crossed product* $\Delta(S/R, \sigma, a)$.

Proof. (1) and (2): By Exercise 5.1.28, $a \in R^*$. For $0 \le i, j < n$, we have $u_{\sigma^i} u_{\sigma^j} = \xi(\sigma^i, \sigma^j) u_{\sigma^{i+j}}$. Computing the powers of u_σ, we find

$$u_\sigma^2 = \xi(\sigma, \sigma) u_{\sigma^2}$$
$$u_\sigma^3 = (\xi(\sigma, \sigma) u_{\sigma^2}) u_\sigma = \xi(\sigma, \sigma)\xi(\sigma^2, \sigma) u_{\sigma^3}.$$

Iterating, we get $u_\sigma^j = \xi(\sigma, \sigma) \cdots \xi(\sigma^{j-1}, \sigma) u_{\sigma^j}$. This proves (1). Taking $j = n$, it follows that

$$u_\sigma^n = \xi(\sigma, \sigma) \cdots \xi(\sigma^{n-1}, \sigma) u_{\sigma^n} = a u_1 = a$$

which proves (2).

(3): Mapping u_σ to the generator of the cyclic crossed product defines the R-algebra isomorphism. □

Theorem 13.4.5. *S/R be a cyclic Galois extension of commutative rings with group $G = \langle \sigma \rangle$ of order n. Let $N_R^S : S^* \to R^*$ be the norm map, where $N_R^S(c) = c\sigma(c) \cdots \sigma^{n-1}(c)$. Let $a, b \in S^*$. Then*

(1) *If* $\gcd(k, n) = 1$, *then* $\Delta(S/R, \sigma, a) \cong \Delta(S/R, \sigma^k, a^k)$.

(2) *If* $c \in S^*$, *then* $\Delta(S/R, \sigma, a) \cong \Delta(S/R, \sigma, a N_R^S(c))$.

(3) $\Delta(S/R, \sigma, a) \otimes_R \Delta(S/R, \sigma, b)$ *is Brauer equivalent to* $\Delta(S/R, \sigma, ab)$.

(4) *The cyclic crossed product map* $a \mapsto \Delta(S/R, \sigma, a)$ *induces a homomorphism of abelian groups*

$$(S^*)^G / N_R^S(S^*) \to \mathrm{B}(S/R)$$

which is called the norm residue *map.*

Proof. (1): Let

$$\Delta = \Delta(S/R, \sigma, a) = \bigoplus_{i=0}^{n-1} S u^i$$

where for all $x \in S$, $ux = \sigma(x)u$, and $u^n = a$. If we set $w = u^k$, then $wx = \sigma^k(x)w$, for all $x \in S$, and $w^n = u^{kn} = a^k$. Let

$$\Delta(S/R, \sigma^k, a^k) = \bigoplus_{i=0}^{n-1} S v^i$$

and define $\theta : \Delta(S/R, \sigma^k, a^k) \to \Delta(S/R, \sigma, a)$ by $v^i \mapsto w^i$. The reader should verify that θ is an R-algebra isomorphism.

(2): This is a special case of Lemma 12.1.3. Let $\Delta(S/R, \sigma, aN_R^S(c)) = \bigoplus_{i=1}^n Sv^i$. Define

$$\theta : \Delta(S/R, \sigma, a) \to \Delta(S/R, \sigma, aN_R^S(c))$$

by sending the basis element u to $c^{-1}v$. The reader should verify that this extends to an R-algebra isomorphism.

(3): This follows from Lemma 13.4.3 and Theorem 13.1.4.

(4): This follows from Parts (2) and (3), and Lemma 13.4.3 (4). □

4.1. Symbol Algebras. Suppose R is a connected commutative ring such that $n > 1$ is invertible in R. Moreover assume R contains a primitive nth root of unity, ζ. Let a and b be units in R. As shown in Example 12.9.2, the Kummer extension $R[a^{1/n}] = R[x]/(x^n - a)$ is a cyclic Galois extension of R with group $\langle\sigma\rangle$, where $\sigma(x) = \zeta x$. Given R, a, b, n, and ζ, we define the *symbol algebra*, denoted $(a, b)_{n,R,\zeta}$, to be the cyclic crossed product algebra $\Delta(R[a^{1/n}]/R, \sigma, b)$. By Lemma 13.4.3, the symbol algebra $(a, b)_{n,R,\zeta}$ is an Azumaya R-algebra which is split by $R[a^{1/n}]$.

Lemma 13.4.6. *Let R, n, ζ be as above. For units a, b in R, write (a, b) for the symbol algebra $(a, b)_{n,R,\zeta}$.*

1. *As an R-algebra, (a, b) is generated by two invertible elements α, β which satisfy the relations: $\alpha^n = a$, $\beta^n = b$, $\beta\alpha = \zeta\alpha\beta$. As a left R-module, (a, b) is free with basis $\{\alpha^i\beta^j \mid 0 \le i < n, 0 \le j < n\}$.*
2. *The R-algebra (a, b) is isomorphic to the opposite ring of (b, a).*
3. *$(a, b) \otimes_R (a, c)$ is Brauer equivalent to (a, bc).*
4. *Let $N : R[a^{1/n}] \to R$ be the norm map. If $c \in R[a^{1/n}]^*$, then the R-algebra (a, b) is isomorphic to $(a, bN(c))$.*
5. *If $\gcd(k, n) = 1$ and $k\ell \equiv 1 \pmod{n}$, then the R-algebra $(a, b)_{n,R,\zeta^k}$ is isomorphic to $(a, b^\ell)_{n,R,\zeta}$.*

Proof. (1): Let α be the image of x in $R[x]/(x^n - a)$. Then $R[\alpha]$ is a cyclic Galois extension of R with group $\langle\sigma\rangle$, where $\sigma(\alpha) = \zeta\alpha$. The R-module $R[\alpha]$ is free with basis $1, \alpha, \dots, \alpha^{n-1}$. Let $1, \beta, \dots, \beta^{n-1}$ be a free left $R[\alpha]$-module basis for the cyclic crossed product $\Delta(R[\alpha]/R, \sigma, b)$. The rest is left to the reader.

(2): In (b, a) the multiplication is $\alpha\beta = \zeta\beta\alpha$. Therefore in the opposite ring $(b, a)^o$ we have $\beta * \alpha = \zeta\alpha * \beta$.

(3), (4) and (5): Follow from Theorem 13.4.5. □

4.2. Cyclic Algebras in Characteristic p. In this section, p is a prime number. Most of the material in this section comes from [**Sal76**].

Proposition 13.4.7. *Let R be a commutative ring of characteristic p. Let a and b be elements of R, with b invertible. Let $(a,b)_R$ be the elementary p-algebra of degree p defined in Section 8.2.1. Then $(a,b)_R$ is a cyclic crossed product algebra.*

Proof. Let $S = R[x]/(x^p - x - ab)$. By Example 12.9.12, S is a cyclic Galois extension of R with group $G = \langle\sigma\rangle$, and $\sigma(v) = v+1$, where $v \in S$ is the coset represented by x. Over S the polynomial $z^p - z - ab$ splits into $(z-v)(z-(v+1))\cdots(z-(v+p-1))$, hence $(v+p-1)\cdots(v+1)v = ab$. The cyclic crossed product algebra $\Delta(S/R,\sigma,b)$ is an Azumaya R-algebra and as a left S-module is free with basis $1, w, \dots, w^{p-1}$. By Eq. (13.15), $wv = \sigma(v)w$, or equivalently, $vw^{-1} = w^{-1}\sigma(v)$. Recursively, we get

$$\begin{aligned}(vw^{-1})^p &= (vw^{-1})(vw^{-1})\cdots(vw^{-1})\\ &= w^{-p}\sigma^p(v)\cdots\sigma^2(v)\sigma(v)\\ &= b^{-1}v(v+p-1)\cdots(v+1)\\ &= b^{-1}ab\\ &= a.\end{aligned}$$

Let $(a,b)_R$ be the elementary p-algebra with generators α and β, where $\alpha^p = a$, $\beta^p = b$, and $\beta\alpha = \alpha\beta + 1$. Define

$$(a,b)_R \xrightarrow{\theta} \Delta(S/R,\sigma,b)$$

by $\theta(\alpha) = vw^{-1}$ and $\theta(\beta) = w$. Then

$$\begin{aligned}&\theta(\beta\alpha) = wvw^{-1} = \sigma(v) = v+1 = (vw^{-1})w + 1 = \theta(\alpha\beta+1),\\ &\theta(\alpha)^p = (vw^{-1})^p = a, \text{ and}\\ &\theta(\beta)^p = w^p = b.\end{aligned}$$

Therefore, θ is an R-algebra homomorphism. Because both algebras are Azumaya R-algebras of rank p^2, θ is an isomorphism (Theorem 7.6.1). □

Now consider p-torsion in the Brauer group of a field. Let k be a field and K/k a Galois extension with finite group G. By the Crossed Product Theorem (Corollary 13.3.3), $\mathrm{H}^2(G,K^*) \cong \mathrm{B}(K/k)$. If $G = \langle\sigma\rangle$ is a cyclic group, then every k-Azumaya algebra that is split by K is Brauer equivalent to a cyclic crossed product $\Delta(K/k,\sigma,a)$ for some $a \in k^*$ (Theorem 13.4.4). A proof of Theorem 13.4.8 appears in [**Alb39**]. In [**Yua67**], S. Yuan claims to prove it using Amitsur cohomology.

Theorem 13.4.8. *Let k be a field of characteristic p. The subgroup of $\mathrm{B}(k)$ consisting of those elements which are annihilated by some power of p is generated by the subgroups $\mathrm{B}(K/k)$, where K runs through all of the finite cyclic extension fields K/k of degree p^e for some $e \geq 1$. If A is a k-central* *simple algebra of order p^e for some $e \geq 1$, then A is Brauer equivalent to a product of cyclic crossed product algebras.*

Proof. Let A be a k-central simple algebra of order p^e for some $e \geq 1$. The proof is by induction on the order of A. Write $A^{\otimes i}$ for the tensor product of i copies of A. By $A \sim B$ we mean B is an Azumaya k-algebra which is Brauer equivalent to A. The Brauer class of A is denoted $[A]$.

Basis Step: If A has order p, then the result is true by Corollary 11.4.10 and Proposition 13.4.7.

Induction Step: Assume $n \geq 1$, the order of A is p^{n+1}, and that the result is true for all Azumaya k-algebras of order less than or equal to p^n. Then $A^{\otimes p} \sim A_1 \otimes_k \cdots \otimes_k A_m$ where each A_i is a cyclic crossed product over k. Write $A_i = \Delta(C_i/k, \sigma_i, a_i)$, where C_i/k is a cyclic extension of fields of degree p^{e_i} with group $\langle \sigma_i \rangle$, $1 \leq e_i \leq n$, and $a_i \in k$. By Exercise 13.4.12, for each i there is a cyclic extension field Z_i/k which contains C_i as a proper intermediate field. By Corollary 5.1.19 and Theorem 13.3.19, under the natural map $\mathrm{B}(C_i/k) \to \mathrm{B}(Z_i/k)$, the image of $[A]$ is p-divisible. That is, there exists an Azumaya k-algebra B_i which is split by Z_i and $A_i \sim B_i^{\otimes p}$. Then $B = B_1 \otimes_k \cdots \otimes_k B_m$ is a product of cyclic crossed product algebras. In $\mathrm{B}(k)$ we have $([B]^{-1}[A])^p = 1$. By the Basis Step, $B^o \otimes_k A$ is similar to a product of cyclic crossed product algebras. Since $A \sim B \otimes_k (B^o \otimes_k A)$, we are done. □

4.3. The Brauer Group of a Henselian Local Ring. Let R be a henselian local ring with maximal ideal $\mathfrak{m}$ and residue field k. In this section we prove that the natural map $\mathrm{B}(R) \to \mathrm{B}(k)$ is an isomorphism. The theorem and its proof are due to Azumaya ([**Azu51**, Theorem 31]), although Nakayama is credited with the idea. Because Theorem 7.9.1 can be proved for a henselian local ring, Theorem 13.4.9 is in fact a special case of Theorem 7.9.2.

Theorem 13.4.9. *Let R be a henselian local ring with maximal ideal $\mathfrak{m}$ and residue field k. Then the natural map $\mathrm{B}(R) \to \mathrm{B}(k)$ is an isomorphism.*

Proof. By Theorem 10.3.2, the natural map $\eta : \mathrm{B}(R) \to \mathrm{B}(k)$ is one-to-one. Assume A is an Azumaya k-algebra and the order of $[A]$ in $\mathrm{B}(k)$ is n. First assume k has positive characteristic p. Let ${}_p\mathrm{B}(k)$ denote the subgroup of $\mathrm{B}(k)$ consisting of those elements of order some power of p. By Theorem 13.4.8, ${}_p\mathrm{B}(k)$ is generated by the classes of cyclic crossed product algebras. By Exercise 13.4.11, ${}_p\mathrm{B}(k)$ is contained in the image of $\eta : \mathrm{B}(R) \to \mathrm{B}(k)$. In the Brauer group of k, we have a factorization $[A] = [A_1][A_2]$ where $[A_1]$ has order p^e for some e, and the order of $[A_2]$ is relatively prime to p.

Since $[A_1]$ is in the image of η, to finish the proof, it suffices to replace A with A_2 and assume n is invertible in k.

Let F/k be a Galois extension of fields with finite group G such that F splits A (Corollary 12.6.4). By Corollary 13.3.3, $\mathrm{H}^2(G, F^*) \cong \mathrm{B}(F/k)$. Let $\xi \in \mathrm{Z}^2(G, F^*)$ be a 2-cocycle that represents the cohomology class corresponding to $[A]$. By Lemma 13.3.20, we can assume the cohomology class $[\xi]$ is in the image of the natural map $\mathrm{H}^2(G, \mu_n) \to \mathrm{H}^2(G, F^*)$, where μ_n denotes the group of nth roots of unity in F. By Exercise 12.5.8, there is a Galois extension S/R where S is a henselian local ring, $F = S \otimes_R k$, and $\mathrm{Aut}_R(S) = G$. By Corollary 13.3.3, $\mathrm{H}^2(G, S^*) \cong \mathrm{B}(S/R)$. Since n is invertible in k, the polynomial $x^n - 1$ has only simple roots in F. As shown in Lemma 10.1.9, each element of μ_n lifts to a unique element of S. The commutative diagram

$$\begin{array}{ccc} \mathrm{H}^2(G, \mu_n) & \xrightarrow{\cong} & \mathrm{H}^2(G, \mu_n) \\ \downarrow & & \downarrow \\ \mathrm{H}^2(G, S^*) & \longrightarrow & \mathrm{H}^2(G, F^*) \end{array}$$

shows that the class of ξ in $\mathrm{H}^2(G, F^*)$ comes from $\mathrm{H}^2(G, S^*)$. □

4.4. Exercises.

Exercise 13.4.10. Let R be a commutative local ring which contains a field k as a subring. Assume $n > 1$ is invertible in k and that k contains a primitive nth root of unity, ζ. Let $\mathfrak{m}$ be the maximal ideal in R and $F = R/\mathfrak{m}$ the residue field. If $\bar{a}$, $\bar{b}$ are units in F, show that the F-symbol algebra $(\bar{a}, \bar{b})_{n,F,\zeta}$ lifts to an R-symbol algebra. That is, there exist units $a, b \in R^*$ such that $(\bar{a}, \bar{b})_{n,F,\zeta} \cong (a, b)_{n,R,\zeta} \otimes_R F$.

Exercise 13.4.11. Let R be a henselian local ring with maximal ideal $\mathfrak{m}$ and residue field k. Let F/k be a cyclic Galois extension of fields with finite group $G = \langle \sigma \rangle$ of order n. Let $\Delta = \Delta(F/k, \sigma, \bar{\beta})$ be a cyclic crossed product algebra, for some $\bar{\beta} \in k^*$. Use Exercise 12.5.8 to show that Δ lifts to a cyclic crossed product over R. That is, there exist a cyclic Galois extension S/R with group $\langle \tau \rangle$ and $\beta \in R^*$ such that $\Delta \cong \Delta(S/R, \tau, \beta) \otimes_R k$.

Exercise 13.4.12. Let k be a field of positive characteristic p.

(1) The map $a \mapsto a^p - a$ defines a homomorphism of additive groups $\varphi : k \to k$. Prove that a cyclic extension field E/k exists if and only if the map φ is not onto.

(2) Below is an outline of a proof that a cyclic extension field E/k can be embedded in a larger cyclic extension field F/k. For the

complete classification of cyclic extensions F/k of degree p^n, the interested reader is referred to [**Alb34**]. Assume $n > 1$, E/k is cyclic of degree p^{n-1}, and $\mathrm{Aut}_k(E) = \langle \sigma \rangle$.

(a) Show that there exists $a, b \in E$ satisfying: $T_k^E(a) = 1$ and $\sigma(b) - b = a^p - a$.
(b) Show that $x^p - x - a$ is irreducible in $E[x]$.
(c) Let $F = E[x]/(x^p - x - a)$. Show that F/E is cyclic of degree p and F/k is cyclic of degree p^n.

5. Generalized Cyclic Crossed Product Algebras

In this section we define a generalized cyclic crossed product algebra. The definition is based on a construction involving the tensor algebra that is a generalization of the cyclic crossed product defined in Definition 13.4.2. Most of the results in this section were first published in [**For16**].

Let A be an R-algebra, M a left $A \otimes_R A$-module, and $T(M)$ the tensor algebra (Section 5.2). Suppose $n \geq 1$ and $\theta : T^n(M) \to A$ is an $A \otimes_R A$-module homomorphism. Denote by $\Delta(A/R, M, \theta)$ the R-algebra quotient $T(M)/\Gamma(M, \theta)$, where $\Gamma(M, \theta)$ is the two-sided ideal of $T(M)$ generated by the set

$$\{y - \theta(y) \mid y \in T^n(M)\}.$$

Lemma 13.5.1. *As an $A \otimes_R A$-module, $\Delta(A/R, M, \theta)$ is equal to the homomorphic image of $T^0(M) \oplus T^1(M) \oplus T^2(M) \oplus \cdots \oplus T^{n-1}(M)$ under the natural map $T(M) \to \Delta(A/R, M, \theta)$.*

Proof. Let $i \geq 0$. If y is an element of $T^{n+i}(M) = T^i(M) \otimes_A T^n(M)$, then y can be written as a sum of elements of the form $z = x_i x_n$. Hence $z - x_i(x_n - \theta(x_n)) \in T^i(M)$. By induction, we find that $\Delta(A/R, M, \theta)$ is a homomorphic image of $T^0(M) \oplus T^1(M) \oplus T^2(M) \oplus \cdots \oplus T^{n-1}(M)$. $\square$

First we establish some notation that will be in effect throughout this section. Let A be a noetherian integrally closed integral domain with quotient field K. Let L be a cyclic Galois extension of K with group $G = \langle \sigma \rangle$ and $\dim_K(L) = n$. Let T be the integral closure of A in L.

Lemma 13.5.2. *In the above context, the following are true.*

(1) *T is finitely generated as an A-module and L is the quotient field of T.*
(2) *G acts on T as a group of A-algebra automorphisms and $T^G = A$.*
(3) *If T is separable over A, then T is Galois over A with group G.*
(4) *There exists $\alpha \in A$ such that $T[\alpha^{-1}]$ is Galois over $A[\alpha^{-1}]$.*

Proof. Part (1) follows from Theorem 4.6.10. Part (2) is left to the reader. Part (3) follows from Part (1) and Theorem 12.2.9 (2). Using the fact that L/K is Galois, Part (4) follows from Definition 12.2.1 and Lemma 10.3.7. □

Since T is a noetherian normal integral domain, we are in the context of Chapter 6. By $\operatorname{Reflex}(T)$ we denote the group of reflexive fractional ideals of T in L. The binary operation on $\operatorname{Reflex}(T)$ is defined by multiplying the ideals and taking double duals: $I * J = T : (T : IJ)$. The class group of rank one reflexive T-modules, denoted $\operatorname{Cl}(T)$, is defined to be the quotient of $\operatorname{Reflex}(T)$ modulo $\operatorname{Prin}(T)$, the subgroup of principal fractional ideals (Exercise 6.4.16). Contained in $\operatorname{Cl}(T)$ as a subgroup is the Picard group $\operatorname{Pic}(T)$ which is the subgroup generated by the invertible, or projective fractional ideals of T in L.

Lemma 13.5.3. *In the above context, let I be a reflexive fractional ideal of T in L.*

(1) *$\sigma(I)$ is a reflexive fractional ideal of T in L.*
(2) *As $T \otimes_A T$-modules, $\sigma^{-1}(I) \cong {}_\sigma I_\sigma$ (see Definition 7.8.4 for the definition of twisted modules).*
(3) *G acts as a group of automorphisms on $\operatorname{Cl}(T)$. On a divisor class $[I]$, the action is defined by $\sigma[I] = [\sigma^{-1}(I)]$.*

Proof. The proof is left to the reader. □

Lemma 13.5.4. *Let I be a fractional ideal of T in L. For any $i \geq 1$, the map*

$$T^i({}_1I_\sigma) \xrightarrow{\theta_i} {}_1\left(I\sigma(I)\cdots\sigma^{i-1}(I)\right)_{\sigma^i}$$
$$x_0 \otimes x_1 \otimes \cdots \otimes x_{i-1} \mapsto x_0\sigma(x_1)\cdots\sigma^{i-1}(x_{i-1})$$

is a $T \otimes_A T$-module homomorphism. The map θ_i is an isomorphism if I is an invertible fractional ideal (equivalently, I is projective as a T-module).

Proof. First we show that the map

$$\begin{aligned} {}_1\left(I\sigma(I)\cdots\sigma^{i-1}(I)\right)_{\sigma^i} \otimes_T {}_1I_\sigma &\xrightarrow{\mu_i} {}_1\left(I\sigma(I)\cdots\sigma^{i}(I)\right)_{\sigma^{i+1}} \\ x \otimes y &\mapsto x\sigma^i(y) \end{aligned} \tag{13.16}$$

is a $T \otimes_A T$-module homomorphism. Let a and b denote elements of T. Let x and y be elements of ${}_1\left(I\sigma(I)\cdots\sigma^{i-1}(I)\right)_{\sigma^i}$ and ${}_1I_\sigma$ respectively. We have

$$\begin{aligned} \mu_i(x \cdot a \otimes y) &= \mu_i(x\sigma^i(a) \otimes y) \\ &= x\sigma^i(a)\sigma^i(y) \\ &= x\sigma^i(ay) \\ &= \mu_i(x \otimes a \cdot y) \end{aligned} \tag{13.17}$$

and

$$(13.18)\quad \begin{aligned} \mu_i(a \otimes b \cdot x \otimes y) &= \mu_i(a \cdot x \otimes y \cdot b) \\ &= \mu_i\left(ax \otimes y\sigma(b)\right) \\ &= ax\sigma^i\left(y\sigma(b)\right) \\ &= ax\sigma^i(y)\sigma^{i+1}(b) \\ &= a \otimes b \cdot x\sigma^i(y) \\ &= a \otimes b \cdot \mu_i(x \otimes y). \end{aligned}$$

From (13.17) and (13.18) we see that μ_i is a well-defined $T \otimes_A T$-module homomorphism. Lastly, applying (13.16) inductively we see that θ_i is a well-defined $T \otimes_A T$-module homomorphism. □

Definition 13.5.5. Let I be a reflexive fractional ideal of T in L such that the norm of I, $T : (T : I\sigma(I) \cdots \sigma^{n-1}(I))$, is a principal ideal Tg for some g in K^*. In the notation of Lemma 13.5.4 we have

$$\theta_n\left(T^n({}_1I_\sigma)\right) \subseteq I\sigma(I)\sigma^2(I) \cdots \sigma^{n-1}(I) \subseteq Tg.$$

Following θ_n by "multiplication by g^{-1}" defines a $T \otimes_A T$-module homomorphism

$$(13.19)\quad \begin{aligned} T^n({}_1I_\sigma) &\xrightarrow{\lambda_0} T^0({}_1I_\sigma) = T \\ x &\mapsto \theta_n(x)g^{-1} \end{aligned}$$

which is an isomorphism if and only if I is projective (Lemma 6.4.3). As in Lemma 13.5.1, the *generalized cyclic crossed product algebra* $\Delta(T/A, I, g)$ is defined to be the A-algebra quotient $T({}_1I_\sigma)/\Gamma(I, g)$, where $\Gamma(I, g)$ is the two-sided ideal of the tensor algebra $T({}_1I_\sigma)$ generated by the set

$$(13.20)\quad \left\{x - \lambda_0(x) \mid x \in T^n({}_1I_\sigma)\right\}.$$

Remark 13.5.6. In Definition 13.5.5, the element g plays an important role. We remark that it is tempting to define the algebra $\Delta(T/A, I, g)$ when I satisfies the weaker hypothesis $N(I) = Tg$ for some $g \in L$. However, our goal is to construct an Azumaya algebra from $\Delta(T/A, I, g)$, and as we will see in Theorem 13.5.10, g^{-1} is the factor set for the cyclic crossed product $(L/K, \sigma, g^{-1})$ which is a K-central simple algebra if and only if g is fixed by the group G. For this reason we assume in Definition 13.5.5 that $g \in K = L^G$. In Example 13.6.14 we see that there exist ideals I with trivial norm but for which there is no $g \in K^*$ such that $N(I) = Tg$.

Lemma 13.5.7. *Let $\Delta(T/A, I, g)$ be a generalized cyclic crossed product as in Definition 13.5.5. Then $\Delta(T/A, I, g)$ is a finitely generated A-module.*

Proof. The proof is left to the reader. □

In Lemma 13.5.8 we see how the generalized cyclic crossed product algebra $\Delta(T/A, I, g)$ depends on the isomorphism class of I as a T-module.

Lemma 13.5.8. *Let $\Delta(T/A, I, g)$ be a generalized cyclic crossed product as in Definition 13.5.5. Let α be a nonzero element in L. Let $\ell_\alpha : I \to \alpha I$ be the "left multiplication by α" map. Then ℓ_α induces*

$$\Delta(T/A, I, g) \to \Delta(T/A, \alpha I, N(\alpha)g)$$

which is an isomorphism of both A-algebras and $T \otimes_A T$-modules.

Proof. The induced map $\ell_\alpha : {}_1I_\sigma \to {}_1(\alpha I)_\sigma$ is an isomorphism of $T \otimes_A T$-modules. This induces an isomorphism on the tensor algebras $\ell_\alpha : T({}_1I_\sigma) \to T({}_1(\alpha I)_\sigma)$ (Proposition 5.2.1). Let $x = x_0 \otimes \cdots \otimes x_{n-1} \in T^n({}_1I_\sigma)$. Since

$$\begin{aligned}\theta_n(\ell_\alpha(x)) &= \theta_n(\alpha x_0 \otimes \cdots \otimes \alpha x_{n-1}) \\ &= \alpha x_0 \sigma(\alpha)\sigma(x_1)\cdots\sigma^{n-1}(\alpha)\sigma^{n-1}(x_{i-1}) \\ &= N(\alpha)\theta_n(x)\end{aligned}$$

it follows that the diagram

$$\begin{array}{ccc} T^n({}_1I_\sigma) & \xrightarrow{\lambda_0} & T \\ {\scriptstyle \ell_\alpha}\downarrow & & \downarrow{\scriptstyle =} \\ T^n({}_1(\alpha I)_\sigma) & \xrightarrow{\lambda_0} & T \end{array} \tag{13.21}$$

commutes. The rest follows from this and Definition 13.5.5. □

Lemma 13.5.9. *Let $\Delta(T/A, I, g)$ be a generalized cyclic crossed product as in Definition 13.5.5. Let R be a localization of A in K, $S = T \otimes_A R$, and assume $SI = Sa$ is principal, for some $a \in L$. The assignment*

$$x_0 a \otimes \cdots \otimes x_{i-1} a \mapsto x_0 \sigma(x_1)\cdots\sigma^{i-1}(x_{i-1})v^i$$

defines an $S \otimes_R S$-module isomorphism

$$T^i({}_1(SI)_\sigma) = {}_1(SI)_\sigma \otimes_S \cdots \otimes_S {}_1(SI)_\sigma \xrightarrow{\phi_i} Sv^i$$

for all $i \geq 1$. Take ϕ_0 to be the identity map $S \to S$. If $S[v, \sigma]$ is the twisted polynomial ring (Definition 13.4.1), then the sum $\phi = \phi_0 + \phi_1 + \ldots$ is an isomorphism

$$T({}_1(SI)_\sigma) \xrightarrow{\phi} S[v, \sigma]$$

of graded R-algebras.

Proof. For $i = 0$ there is nothing to prove. Suppose $0 \leq i$. We show that

$$\begin{aligned} Sv^i \otimes_S {}_1(SI)_\sigma &\xrightarrow{\mu_i} Sv^{i+1} \\ xv^i \otimes ya &\mapsto x\sigma^i(y)v^{i+1}. \end{aligned}$$

is an $S \otimes_R S$-module isomorphism. In the following, $x, y, x_0, y_0, \dots$ represent elements of S. For all $s \in S$,

$$\begin{aligned}\mu_i(xv^i \cdot s \otimes ya) &= \mu_i(x\sigma^i(s)v^i \otimes ya)\\ &= x\sigma^i(s)\sigma^i(y)v^{i+1}\\ &= x\sigma^i(sy)v^{i+1}\\ &= \mu_i(xv^i \otimes s \cdot ya).\end{aligned}$$

For any $s, t \in S$,

$$\begin{aligned}\mu_i(s \otimes t \cdot xv^i \otimes ya) &= \mu_i(s \cdot xv^i \otimes ya \cdot t)\\ &= \mu_i((sx)v^i \otimes y\sigma(t)a)\\ &= (sx)\sigma^i(y)\sigma^{i+1}(t)v^{i+1}\\ &= (sx)\sigma^i(y)v^{i+1} \cdot t\\ &= s \otimes t \cdot x\sigma^i(y)v^{i+1}\\ &= s \otimes t \cdot \mu_i(xv^i \otimes ya).\end{aligned}$$

By induction on i, we see that ϕ_i is an $S \otimes_R S$-module isomorphism. The diagram

$$\begin{array}{ccc} T^i({}_1(SI)_\sigma) \otimes_S T^j({}_1(SI)_\sigma) & \xrightarrow{\eta_{i,j}} & T^{i+j}({}_1(SI)_\sigma)\\ \downarrow{\scriptstyle \phi_i \otimes \phi_j} & & \downarrow{\scriptstyle \phi_{i+j}}\\ Sv^i \otimes_S Sv^j & \longrightarrow & Sv^{i+j}\end{array}$$

commutes since

$$\begin{aligned}&\phi_i(x_0a \otimes \cdots \otimes x_{i-1}a)\phi_j(y_0a \otimes \cdots \otimes y_{j-1}a)\\ &\quad = x_0\sigma(x_1)\cdots\sigma^{i-1}(x_{i-1})v^iy_0\sigma(y_1)\cdots\sigma^{j-1}(y_{j-1})v^j\\ &\quad = x_0\sigma(x_1)\cdots\sigma^{i-1}(x_{i-1})\sigma^i(y_0)\sigma^{i+1}(y_1)\cdots\sigma^{i+j-1}(y_{j-1})v^{i+j}\\ &\quad = \phi_{i+j}(x_0a \otimes \cdots \otimes x_{i-1}a \otimes y_0a \otimes \cdots \otimes y_{j-1}a).\end{aligned}$$

It follows that the induced map ϕ is multiplicative. □

Theorem 13.5.10. *Let $\Delta(T/A, I, g)$ be a generalized cyclic crossed product as in Definition 13.5.5. Let R be a localization of A in K, $S = T \otimes_A R$, and assume $SI = Sa$ is principal, for some $a \in L$.*

(1) *$N(a)g^{-1}$ is an invertible element in R.*

(2) *There is a monomorphism of both A-algebras and $T \otimes_A T$-modules*

$$\psi : \Delta(T/A, I, g) \to (S/R, \sigma, N(a)g^{-1})$$

which is induced by the inclusion $I \subseteq SI = Sa$ and the isomorphism ϕ from Lemma 13.5.9.

(3) *Let a be any nonzero element in I. The image of*
$$\psi : \Delta(T/A, I, g) \to (L/K, \sigma, N(a)g^{-1})$$
is an A-order in the central simple K-algebra $(L/K, \sigma, N(a)g^{-1})$. That is, the image of ψ is a finitely generated A-module and contains a basis for the K-vector space $(L/K, \sigma, N(a)g^{-1})$.

Proof. (1): We are given that $N(I) = Tg$, where $g \in K^*$, and $SI = Sa$, where $a \in L^*$. Using Proposition 6.3.5, we have

$$\begin{aligned} Sg &= S\left(T : (T : I\sigma(I)\cdots\sigma^{n-1}(I))\right) \\ &= S : (S : (SI)\sigma(SI)\cdots\sigma^{n-1}(SI)) \\ &= S : (S : Sa\sigma(Sa)\cdots\sigma^{n-1}(Sa)) \\ &= S : (S : SN(a)) \\ &= SN(a). \end{aligned} \tag{13.22}$$

It follows that $N(a)g^{-1}$ is an invertible element of S. But $N(a)g^{-1} \in K^* \cap S^* = R^*$.

(2): Viewing I as a T-submodule of SI, there is an A-algebra monomorphism $T({}_1I_\sigma) \to T({}_1(SI)_\sigma)$. Let $y = y_0 \otimes \cdots \otimes y_{n-1}$ be a typical generator of $T^n({}_1I_\sigma)$. Since $SI = Sa$, there are elements $x_i \in S$ such that $y = x_0 a \otimes \cdots \otimes x_{n-1}a$. Under the composite map

$$T({}_1I_\sigma) \subseteq T({}_1(SI)_\sigma) \xrightarrow{\phi} S[v, \sigma] \xrightarrow{\eta} (S/R, \sigma, N(a)g^{-1})$$

we have

$$\begin{aligned} \eta\phi(y) &= \eta\left(x_0\sigma(x_1)\cdots\sigma^{n-1}(x_{n-1})v^n\right) \\ &= x_0\sigma(x_1)\cdots\sigma^{n-1}(x_{n-1})N(a)g^{-1}. \end{aligned} \tag{13.23}$$

Also,

$$\begin{aligned} \lambda_0(y) &= y_0\sigma(y_1)\cdots\sigma^{n-1}(y_{n-1})g^{-1} \\ &= x_0 a\sigma(x_1)\sigma(a)\cdots\sigma^{n-1}(x_{n-1})\sigma^{n-1}(a)g^{-1} \\ &= x_0\sigma(x_1)\cdots\sigma^{n-1}(x_{n-1})a\sigma(a)\cdots\sigma^{n-1}(a)g^{-1} \\ &= x_0\sigma(x_1)\cdots\sigma^{n-1}(x_{n-1})N(a)g^{-1}. \end{aligned} \tag{13.24}$$

Equations (13.23) and (13.24) show that the ideal $\Gamma(I, g)$ defined by (13.20) is contained in the kernel of $\eta\phi$. So there is an A-algebra homomorphism ψ such that the diagram

$$\begin{array}{ccc} T({}_1I_\sigma) & \xrightarrow{\ \rho\ } & \Delta(T/A, I, g) \\ {\scriptstyle\subseteq}\downarrow & & \downarrow{\scriptstyle\psi} \\ T({}_1(SI)_\sigma) & \xrightarrow{\ \phi\eta\ } & (S/R, \sigma, N(a)g^{-1}) \end{array} \tag{13.25}$$

commutes. The diagram

(13.26)

$$\begin{array}{ccc} T \oplus T^1({}_1I_\sigma) \oplus T^2({}_1I_\sigma) \oplus \cdots \oplus T^{n-1}({}_1I_\sigma) & \longrightarrow & S \oplus Sv \oplus \cdots \oplus Sv^{n-1} \\ \rho \downarrow & & \downarrow \cong \\ \Delta(T/A, I, g) & \xrightarrow{\psi} & (S/R, \sigma, N(a)g^{-1}) \end{array}$$

commutes, where the top row is the $T \otimes_A T$-module monomorphism induced by restricting the composite map $T({}_1I_\sigma) \subseteq T({}_1(SI)_\sigma) \cong S[v, \sigma]$. The right vertical arrow is an isomorphism of S-modules. Therefore, the map ρ is one-to-one. By Lemma 13.5.1 the map ρ is a $T \otimes_A T$-module epimorphism. It follows that the A-algebra homomorphism ψ is a monomorphism.

(3): By Lemma 13.5.7, $\Delta(T/A, I, g)$ is a finitely generated A-module. The map ψ identifies $\Delta(T/A, I, g)$ with an A-subalgebra (and $T \otimes_A T$-submodule) of $(L/K, \sigma, N(a)g^{-1})$. By Lemma 13.4.3, $(L/K, \sigma, N(a)g^{-1})$ is a central simple K-algebra which is generated by $L = T \otimes_A K$ and $v = \psi(a)$. Therefore, the image of ψ is an A-order. □

Corollary 13.5.11. *Let $\Delta(T/A, I, g)$ be a generalized cyclic crossed product as in Definition 13.5.5. Let R be any localization of A in K and let $S = T \otimes_A R$. There is an R-algebra isomorphism $\Delta(T/A, I, g) \otimes_A R \cong \Delta(S/R, SI, g)$.*

Proof. Let a be any nonzero element of I. Let

$$\psi_A : \Delta(T/A, I, g) \to (L/K, \sigma, N(a)g^{-1}), \text{ and}$$
$$\psi_R : \Delta(S/R, SI, g) \to (L/K, \sigma, N(a)g^{-1})$$

be the A-algebra and R-algebra homomorphisms (respectively) given by Theorem 13.5.10 (3). From the proof of Theorem 13.5.10, the image of ψ_A is a subalgebra of the image of ψ_R. Since the natural map $T({}_1I_\sigma) \otimes_A R \to T({}_1(SI)_\sigma)$ is an R-algebra isomorphism, the natural map $\Delta(T/A, I, g) \otimes_A R \to \Delta(S/R, SI, g)$ is an R-algebra isomorphism. □

Corollary 13.5.12. *Let $\Lambda = \Delta(T/A, I, g)$ be a generalized cyclic crossed product as in Definition 13.5.5. Let R be a localization of A in K such that $S = T \otimes_A R$ is Galois over R and $SI = Sa$ is principal for some $a \in L$. Then $u = N(a)g^{-1}$ is a unit in R. There is an isomorphism of both Azumaya R-algebras and $S \otimes_R S$-modules $\Lambda \otimes_A R \cong (S/R, \sigma, u)$.*

Proof. By Corollary 13.5.11, $\Delta(T/A, I, g) \otimes_A R \cong \Delta(S/R, SI, g)$. We are given that $SI = Sa$. By Theorem 13.5.10 (1), $u = N(a)g^{-1}$ is a unit in R. In diagram (13.26) substitute S for T and R for A. Then the top row is an isomorphism, hence the bottom row is an isomorphism. By Lemma 13.4.3, the cyclic crossed product $(S/R, \sigma, u)$ is an Azumaya R-algebra which is split by S. □

Corollary 13.5.13. *Let $\Lambda = \Delta(T/A, I, g)$ be a generalized cyclic crossed product as in Definition 13.5.5. Then $\Lambda \otimes_A K$ is a central simple K-algebra, is split by L, and is isomorphic to the cyclic crossed product $(L/K, \sigma, g^{-1})$.*

Proof. For any nonzero $a \in I$, $\Lambda \otimes_A K$ is isomorphic to $(L/K, \sigma, N(a)g^{-1})$, which is isomorphic to $(L/K, \sigma, g^{-1})$, by Theorem 13.4.5. □

Corollary 13.5.14. *Let R be a localization of A in K such that $S = T \otimes_A R$ is Galois over R. If $\alpha \in A^*$, then $\Delta(T/A, T, \alpha) \otimes_A R$ is isomorphic to the cyclic crossed product $(S/R, \sigma, \alpha^{-1})$, hence is an Azumaya R-algebra which is split by S.*

Proof. In Corollary 13.5.12 take $I = T$, $g = \alpha$, and $a = 1$. □

In Proposition 13.5.15 we show that the generalized crossed product of Definition 13.5.5 agrees with the generalized crossed product defined in Definition 13.2.2.

Proposition 13.5.15. *Let S/R be a Galois extension of noetherian integrally closed integral domains with cyclic Galois group $G = \langle \sigma \rangle$, $[G : 1] = n$. Let L/K be the quotient field extension. If Λ is an Azumaya R-algebra that is split by S, then there exists a projective fractional ideal I of S in L such that $N(I) = I\sigma(I) \cdots \sigma^{n-1}(I) = Sg$ for some $g \in K^*$ and Λ is Brauer equivalent to $\Delta(S/R, I, g)$.*

Proof. We are in the context of Theorem 13.2.4. Continuing from where that proof ended, now we use the fact that $G = \langle \sigma \rangle$ is cyclic. From Theorem 13.2.4, Λ is generated as an R-algebra by S and $\Phi_\sigma = \{\lambda \in \Lambda \mid \sigma(x)\lambda = \lambda x, \text{ for all } x \in S\}$. Pick a nonzero $v \in \Phi_\sigma$. Then $\Lambda \otimes_R K = L \oplus Lv \oplus \cdots \oplus Lv^{n-1}$ is a cyclic crossed product. Let $v^n = a$. Then $\sigma(a)v = va = vv^n = v^n v = av$ shows that $\sigma(a) = a$, hence $a \in K^*$. Using v, we can map the left S-module Φ_σ isomorphically onto an invertible fractional ideal of S in L by sending $y \in \Phi_\sigma$ to $yv^{-1} \in L$. Let $I = \{yv^{-1} \mid y \in \Phi_\sigma\}$, a fractional ideal of S in L. The reader should verify that $\Phi_\sigma \cong {}_1I_\sigma$ as left $S \otimes_R S$-modules. By Lemma 13.5.4, for any $i > 0$ there are isomorphisms

$$T^i(\Phi_\sigma) \cong T^i({}_1I_\sigma) \cong {}_1\left(I\sigma(I) \cdots \sigma^{i-1}(I)\right)_{\sigma^i}$$

induced by the assignments

$$\begin{aligned} x_0 \otimes \cdots \otimes x_{i-1} &\mapsto x_0 v^{-1} \otimes \cdots \otimes x_{i-1} v^{-1} \\ &\mapsto (x_0 v^{-1})\sigma(x_1 v^{-1}) \cdots \sigma^{i-1}(x_{i-1} v^{-1}). \end{aligned}$$

For any y_1, y_2 in Φ_σ, $y_2 v^{-1}$ is in L. It follows that in $\Lambda \otimes_A K$ we have

$$(y_1 y_2)v^{-2} = (y_1 v^{-1})\sigma(y_2 v^{-1}). \tag{13.27}$$

Applying (13.27) recursively,

$$(x_0x_1\cdots x_{i-1})v^{-i} = (x_0v^{-1})\sigma(x_1v^{-1})\cdots\sigma^{i-1}(x_{i-1}v^{-1}).$$

From this it follows that $\Phi_\sigma^i v^{-i} = I\sigma(I)\cdots\sigma^{i-1}(I)$. Since $\Phi_\sigma^n = S$, we can compute the norm of I to be $N(I) = I\sigma(I)\cdots\sigma^{n-1}(I) = \Phi_\sigma^n v^{-n} = Sv^{-n} = Sa^{-1}$. If we set $g = a^{-1}$, then $N(I) = Sg$. By Definition 13.5.5, it follows that mapping $S \to S$ and $I \to \Phi_\sigma$ induces an isomorphism between the generalized crossed product R-algebra $\Delta(S/R, I, g)$ and Λ. □

Corollary 13.5.16. *Let $\Lambda = \Delta(T/A, I, g)$ be a generalized cyclic crossed product as in Definition 13.5.5. Let R be a localization of A in K such that $S = T \otimes_A R$ is Galois over R and SI is projective. Then $\Lambda \otimes_A R$ is an Azumaya R-algebra, is split by S, and contains S as a maximal commutative subalgebra, and under the homomorphism*

$$\alpha_5 : \mathrm{B}(S/R) \to \mathrm{H}^1(G, \mathrm{Pic}\, S)$$

of Theorem 13.3.1, the Brauer class of $\Lambda \otimes_A R$ is mapped to the cohomology class represented by I.

Proof. Let $\mathfrak{m} \in \mathrm{Max}\, R$. Then $S_\mathfrak{m} = S \otimes_R R_\mathfrak{m}$ is a semilocal normal domain (Corollary 3.3.6). Since SI is a rank one projective S-module, $I_\mathfrak{m} = I \otimes_R R_\mathfrak{m}$ is a free $S_\mathfrak{m}$-module of rank one (Exercise 3.1.8). By Corollary 13.5.12, $\Lambda \otimes_A R_\mathfrak{m}$ is an Azumaya $R_\mathfrak{m}$-algebra. It follows from Corollary 8.1.23 that $\Lambda \otimes_A R$ is an Azumaya R-algebra. Since S is integrally closed in L, and L is a maximal commutative subring of $\Lambda \otimes_A K$, S is a maximal commutative subalgebra of $\Lambda \otimes_A R$. By Theorem 7.4.2 S is a splitting ring for $\Lambda \otimes_A R$. □

Remark 13.5.17. For $0 \le i < n$, the image of $T^i({}_1I_\sigma)$ in $\Delta(T/A, I, g)$ is denoted $\Delta_{\sigma^i}(T/A, I, g)$. By the proof of Theorem 13.5.10, $\Delta_{\sigma^i}(T/A, I, g)$ and $T^i({}_1I_\sigma)$ are naturally isomorphic as $T \otimes_A T$-modules. The A-algebra $\Delta(T/A, I, g)$ is G-graded:

$$\Delta(T/A, I, g) = \bigoplus_{i=0}^{n-1} \Delta_{\sigma^i}(T/A, I, g). \tag{13.28}$$

Multiplication in the tensor algebra is based on the natural $T \otimes_A T$-module isomorphisms

$$T^i({}_1I_\sigma) \otimes_T T^j({}_1I_\sigma) \xrightarrow{\eta_{i,j}} T^{i+j}({}_1I_\sigma). \tag{13.29}$$

The $T \otimes_A T$-module homomorphism λ_i is defined by the diagram

$$\begin{array}{ccc} T^{n+i}({}_1I_\sigma) & \xrightarrow{\lambda_i} & T^i({}_1I_\sigma) \\ \uparrow{\scriptstyle \eta_{n,i}} & & \uparrow{\scriptstyle \eta_{0,i}} \\ T^n({}_1I_\sigma) \otimes_T T^i({}_1I_\sigma) & \xrightarrow{\lambda_0 \otimes 1} & T^0({}_1I_\sigma) \otimes_T T^i({}_1I_\sigma) \end{array} \tag{13.30}$$

Then λ_i is an isomorphism if and only if I is projective (Lemma 6.4.3). If $0 \leq i + j < n$, then we define $f_{i,j}$ to be the natural map $\eta_{i,j}$ of (13.29). If $i + j \geq n$, define $f_{i,j}$ to be $\lambda_{i+j-n} \circ \eta_{i,j}$. Therefore, the multiplication rule on $\Delta(T/A, I, g)$ is defined on homogeneous components by

$$\Delta_{\sigma^i}(T/A, I, g) \otimes_T \Delta_{\sigma^j}(T/A, I, g) \xrightarrow{f_{i,j}} \Delta_{\sigma^{i+j}}(T/A, I, g). \tag{13.31}$$

So if I is a projective fractional ideal, in the terminology of Definition 13.2.2, this shows that $\{f_{i,j}\}$ is a factor set and the A-algebra $\Delta(T/A, I, g)$ of Definition 13.5.5 is a generalized crossed product.

6. The Brauer Group of a Polynomial Ring

Let R be a commutative ring and x an indeterminate. Let $\eta : R \to R[x]$ be the natural map and $\pi_0 : R[x] \to R$ the homomorphism which maps x to 0. Then $\pi_0\eta$ is the identity homomorphism on R. Thus $\pi_0 : \mathrm{B}(R[x]) \to \mathrm{B}(R)$ is a left inverse to

$$0 \to \mathrm{B}(R) \xrightarrow{\mathrm{B}(\eta)} \mathrm{B}(R[x])$$

so the sequence splits. If $\mathrm{B}(R/R[x])$ denotes the kernel of $\mathrm{B}(\pi_0)$, then there is an isomorphism of abelian groups $\mathrm{B}(R[x]) \cong \mathrm{B}(R) \oplus \mathrm{B}(R/R[x])$.

Lemma 13.6.1. *Let k be a field and A an Azumaya $k[x]$-algebra. The following are equivalent.*

(1) *The Brauer class $[A]$ is in the image of $\eta : \mathrm{B}(k) \to \mathrm{B}(k[x])$.*
(2) *There is a finite separable extension of fields K/k such that $K[x]$ splits A.*
(3) *There is a Galois extension of fields K/k with finite group G such that $K[x]$ splits A.*

Proof. We prove (3) implies (1). The rest is left to the reader. Assume K/k is a Galois extension of fields with finite group G. Then $K[x]/k[x]$ is Galois with group G (Lemma 12.2.7) and the diagram

$$\begin{array}{ccc} \mathrm{H}^2(G, K^*) & \xrightarrow{\alpha_4} & \mathrm{B}(K/k) \\ \downarrow & & \downarrow \gamma \\ \mathrm{H}^2(G, K[x]^*) & \xrightarrow{\alpha_4} & \mathrm{B}(K[x]/k[x]) \end{array} \tag{13.32}$$

commutes. The map γ is the restriction of η to the subgroup $\mathrm{B}(K/k)$, hence is one-to-one. The horizontal arrows in (13.32) are the crossed product homomorphism. Notice that $K[x]$ is a principal ideal domain, hence $\mathrm{Pic}(K[x]) = (0)$ (Example 2.6.11). By Corollary 13.3.3, both horizontal arrows in (13.32) are isomorphisms. Since $K^* = K[x]^*$, the left vertical map in (13.32) is an isomorphism. This proves that γ is an isomorphism. If A is split by $K[x]$, then $[A]$ is in the image of γ. $\square$

Theorem 13.6.2. *Let k be a field. In the notation established above, $\eta : \mathrm{B}(k) \to \mathrm{B}(k[x])$ is the natural homomorphism and $\mathrm{B}(\pi_0) : \mathrm{B}(k[x]) \to \mathrm{B}(k)$ is the homomorphism induced by $x \mapsto 0$.*

(1) *If k is a perfect field, then η is an isomorphism.*

(2) *Assume $\mathrm{char}(k) = p > 0$ and A is an Azumaya $k[x]$-algebra. If the order of A is relatively prime to p, then $[A]$ is in the image of η.*

(3) *Assume $\mathrm{char}(k) = p > 0$. Then $\mathrm{B}(k/k[x])$, the kernel of $\mathrm{B}(\pi_0)$, is a p-group.*

Proof. (1): As mentioned above, η is always one-to-one. Let A be an Azumaya $k[x]$-algebra. Let $\bar{k}$ be an algebraic closure of k. By Tsen's Theorem (Corollary 11.1.9), A is split by $\bar{k}(x)$. Since $\bar{k}[x]$ is a PID, it is a regular domain, and $\mathrm{B}(\bar{k}[x] \to \mathrm{B}(\bar{k}(x))$ is one-to-one (Theorem 11.3.18). Therefore, A is split by $\bar{k}[x]$. By Exercise 7.7.14, there is an intermediate field $k \subseteq K \subseteq \bar{k}$ such that K/k is finitely generated and A is split by $K[x]$. Since k is perfect, K/k is separable. By Lemma 13.6.1, η is onto.

(2): As in the proof of Part (1), let K/k be a finitely generated algebraic extension of fields such that A is split by $K[x]$. Let S be the separable closure of k in K. Since K/S is finite dimensional and purely inseparable, $\dim_S(K) = p^e$ for some $e \geq 0$ (see, for example, [**Hun80**, Corollary V.6.5]). The reader should verify that $\mathrm{Rank}_{S[x]}(K[x]) = p^e$ and $\dim_{S(x)}(K(x)) = p^e$. The diagram

$$\begin{array}{ccc} \mathrm{B}(k(x)) & \xrightarrow{\gamma} & \mathrm{B}(K(x)) \\ & {\scriptstyle\alpha}\searrow \quad \nearrow{\scriptstyle\beta} & \\ & \mathrm{B}(S(x)) & \end{array} \tag{13.33}$$

commutes and $[A \otimes_{k[x]} k(x)]$ is in the kernel of γ. By Theorem 11.2.7 and Corollary 7.5.2, the kernel of β is a p-group. The order of A is relatively prime to p, hence $[A \otimes_{k[x]} k(x)]$ is in the kernel of α. That is, A is split by $S(x)$. Since $\mathrm{B}(S[x]) \to \mathrm{B}(S(x))$ is one-to-one (Theorem 11.3.18), A is split by $S[x]$. By Lemma 13.6.1, A is in the image of η.

(3): Let A be an Azumaya $k[x]$-algebra and assume $[A]$ is in the kernel of π_0. As in the proof of Part (2), let K/k be a finitely generated algebraic extension of fields such that A is split by $K[x]$ and let S be the separable closure of k in K. The diagram of abelian groups

$$
\begin{array}{ccccccccc}
0 & \longrightarrow & \mathrm{B}(k/k[x]) & \longrightarrow & \mathrm{B}(k[x]) & \xrightarrow{\pi_0} & \mathrm{B}(k) & \longrightarrow & 0 \\
 & & \downarrow{\scriptstyle \zeta} & & \downarrow{\scriptstyle \rho} & & \downarrow & & \\
0 & \longrightarrow & \mathrm{B}(S/S[x]) & \longrightarrow & \mathrm{B}(S[x]) & \xrightarrow{\pi_0} & \mathrm{B}(S) & \longrightarrow & 0
\end{array}
\tag{13.34}
$$

commutes and the rows are exact sequences. Since S/k is a finite separable extension of fields, by Lemma 13.6.1, the kernel of ρ is a subgroup of the image of η. This implies ζ is one-to-one. The diagram of abelian groups

$$
\begin{array}{ccccc}
0 & \longrightarrow & \mathrm{B}(S[x]) & \longrightarrow & \mathrm{B}(S(x)) \\
 & & \downarrow{\scriptstyle \delta} & & \downarrow{\scriptstyle \beta} \\
0 & \longrightarrow & \mathrm{B}(K[x]) & \longrightarrow & \mathrm{B}(K(x))
\end{array}
\tag{13.35}
$$

commutes and the rows are exact sequences. The map β is from (13.33). Since A is split by $K[x]$, $\zeta([A])$ is in the kernel of δ. As mentioned already, the kernel of β is a p-group. By the Snake Lemma (Theorem 1.4.1) the kernel of δ is a p-group. Since ζ is one-to-one, this implies the order of A is a power of p. □

The next theorem is due to Auslander and Goldman [**AG60a**].

Theorem 13.6.3. *Let R be a noetherian regular integral domain of characteristic zero. Then $\eta : \mathrm{B}(R) \to \mathrm{B}(R[x])$ is an isomorphism.*

Proof. Let K be the field of fractions of R. Then K has characteristic zero. By Theorem 6.1.9, $R[x]$ is a noetherian regular integral domain. The diagram of abelian groups

$$
\begin{array}{ccc}
\mathrm{B}(R[x]) & \xrightarrow{\pi_0} & \mathrm{B}(R) \\
{\scriptstyle \rho}\downarrow & & \downarrow \\
\mathrm{B}(K[x]) & \xrightarrow{\pi_0} & \mathrm{B}(K)
\end{array}
\tag{13.36}
$$

commutes. By Theorem 13.6.2 (1), the map π_0 in the bottom row of (13.36) is an isomorphism. The diagram

$$
\begin{array}{ccccc}
\mathrm{B}(R[x]) & & \xrightarrow{\quad \phi \quad} & & \mathrm{B}(K(x)) \\
 & {\scriptstyle \rho}\searrow & & \nearrow{\scriptstyle \psi} & \\
 & & \mathrm{B}(K[x]) & &
\end{array}
\tag{13.37}
$$

commutes. Since $K(x)$ is the field of fractions of both $R[x]$ and $K[x]$, by Theorem 11.3.18 the arrows ϕ and ψ are one-to-one. Then in (13.37) and

(13.36) ρ is one-to-one. In the top row of (13.36), π_0 is one-to-one. Since π_0 is a left inverse to η, this proves η is an isomorphism. □

Concerning the Brauer group of a polynomial ring, the most general result is the following theorem of R. Hoobler. For any $\mathbb{Z}$-module M, by ${}_nM$ we denote the subgroup annihilated by n. In the terminology of Remark 6.5.4, Theorem 13.6.4 shows that the functor ${}_n\mathrm{B}(\ )$ is a homotopy functor on the category of noetherian commutative rings. The reader is referred to [**Hoo84**] for the proof, which utilizes cohomological methods. In particular, Gabber's Theorem (Theorem 10.4.13) and the Smooth Base Change Theorem for étale cohomology ([**Mil80**, Corollary VI.4.15]) play critical roles.

Theorem 13.6.4. *Let R be a commutative noetherian ring and n an integer that is invertible in R. Then the natural map ${}_n\mathrm{B}(R) \to {}_n\mathrm{B}(R[x])$ is an isomorphism.*

6.1. The Brauer Group of a Graded Ring.

Proofs of the following result have appeared in various places, including [**Wei77**] and [**Wei81**, Proposition 3.7, p. 477].

Theorem 13.6.5 (C. Weibel). *Let $\mathfrak{F}$ be a covariant functor from the category of commutative rings to abelian groups. Let $A = A_0 \oplus A_1 \oplus A_2 \oplus \cdots$ be a commutative graded ring and suppose that the natural map $\eta : A \to A[x]$ gives rise to an isomorphism $\mathfrak{F}(\eta) : \mathfrak{F}(A) \cong \mathfrak{F}(A[x])$. Then $\mathfrak{F}(A_0) \cong \mathfrak{F}(A)$.*

Proof. Using the grading on A, define $\pi : A \to A_0$ by the assignment $\sum_i a_i \mapsto a_0$. Let $\iota : A_0 \to A$ be the natural map. Then $\pi\iota = 1_{A_0}$, so $\mathfrak{F}(\iota) : \mathfrak{F}(A_0) \to \mathfrak{F}(A)$ is one-to-one. Therefore, it is enough to show $\mathfrak{F}(\iota)$ is onto. Using the grading on A, define $\phi : A \to A[x]$ by the assignment $\sum_i a_i \mapsto \sum_i a_i x^i$. Define $\pi_0 : A[x] \to A$ by $x \mapsto 0$, and $\pi_1 : A[x] \to A$ by $x \mapsto 1$. Then $\pi_0\eta = 1_A$ and $\pi_1\eta = 1_A$. Hence, $\mathfrak{F}(\pi_0)$ and $\mathfrak{F}(\pi_1)$ are both inverses to the isomorphism $\mathfrak{F}(\eta)$. Therefore, the maps $\mathfrak{F}(\pi_0)$ and $\mathfrak{F}(\pi_1)$ are equal. Notice that $\pi_0\phi = \iota\pi$ and $\pi_1\phi = 1_A$ is the identity map on A. Therefore,

$$\mathfrak{F}(1_A) = \mathfrak{F}(\pi_1)\mathfrak{F}(\phi) = \mathfrak{F}(\pi_0)\mathfrak{F}(\phi) = \mathfrak{F}(\iota)\mathfrak{F}(\pi)$$

which shows $\mathfrak{F}(\iota)$ is onto. □

Corollary 13.6.6. *Let $A = A_0 \oplus A_1 \oplus A_2 \oplus \cdots$ be a commutative noetherian graded ring.*

(1) *If A is a normal ring, then $\mathrm{Pic}(A_0) \to \mathrm{Pic}(A)$ is an isomorphism.*
(2) *If A is a noetherian regular integral domain of characteristic zero, then $\mathrm{B}(A_0) \to \mathrm{B}(A)$ is an isomorphism.*

(3) *If* n *is an integer that is invertible in* A, *then* ${}_n\mathrm{B}(A_0) \to {}_n\mathrm{B}(A)$ *is an isomorphism.*

(4) *If* $\mathrm{Rad}_A(0) = (0)$, *then* $A_0^* = A^*$. *For this part,* A *is not required to be noetherian.*

Proof. Parts (1), (2) and (3) follow from Theorem 13.6.5 using Theorem 6.5.3, Theorem 13.6.3, and Theorem 13.6.4, respectively.

(4): Since $\mathrm{Rad}_A(0) = (0)$, the groups of units A^* and $A[x]^*$ are equal. Therefore, Theorem 13.6.5 applies. □

Example 13.6.7. Let k be a field and $f \in k[x_1, \dots, x_m]$ a homogeneous polynomial of degree $n \geq 1$. Let $d \geq 1$ and set $T = k[x_1, \dots, , x_m, z]/(z^d - f)$. The reader should verify that if we set $\deg(x_1) = \dots = \deg(x_m) = d$ and $\deg(z) = n$, then this makes T into a graded ring, with $T_0 = k$. The exceptional ideal is the maximal ideal $\mathfrak{m} = (x_1, \dots, x_m, z)$. By Corollary 13.6.6 (4), $T^* = k^*$. If r is an integer that is invertible in k, then ${}_r\mathrm{B}(k) \to {}_r\mathrm{B}(T)$ is an isomorphism (Corollary 13.6.6 (3)). Assume from now on that f is square-free and that d is invertible in k. Then by Theorem 9.4.4, T is an integrally closed integral domain. By Corollary 13.6.6 (1), $\mathrm{Pic}(T) = \mathrm{Pic}(k) = (0)$.

Example 13.6.8. Let k be a field.

(1) Let $n \geq 2$ be an integer that is invertible in k, and set R equal to $k[x, y, z]/(z^n - xy)$. By Exercise 6.2.9, $\mathrm{Cl}(R) \cong \mathbb{Z}/n$. By Example 13.6.7, $\mathrm{Pic}(R) = (0)$.

(2) Let $n \geq 2$ be an integer such that $2n$ is invertible in k. Also assume k contains a primitive $2n$th root of unity. By Exercise 9.4.10, if $T = k[x, y, z]/(z^n - x^{n-1} + y^n)$, then $\mathrm{Cl}(T)$ is a free $\mathbb{Z}/(n-1)$ module of rank $n - 1$. By Example 13.6.7, $\mathrm{Pic}(T) = (0)$.

6.2. The Brauer Group of a Laurent Polynomial Ring. Let R be a commutative noetherian ring and t an indeterminate. In this section we state some known results concerning the Brauer group of the Laurent polynomial ring, $R[t, t^{-1}]$. For similar results concerning the Picard group functor, the interested reader is referred to [**Wei91**]. Under various hypotheses on R, and utilizing various methods and techniques, computations of $\mathrm{B}(R[t, t^{-1}])$ appear in [**DeM76**], [**For81**], [**Mag78**], and [**Hoo84**]. The most general result is in [**For88**]. Assuming a primitive nth root of unity is present, it takes the following convenient form.

Theorem 13.6.9. *Let* R *be a connected commutative noetherian ring and* t *an indeterminate. Let* $n > 1$ *and assume* R *is a* $\mathbb{Z}[n^{-1}, \zeta]$*-algebra, where*

ζ is a primitive nth root of unity in $\mathbb{C}$. Then there is a split exact sequence

$$0 \to {}_n\mathrm{B}(R) \to {}_n\mathrm{B}(R[t,t^{-1}]) \to \frac{\mathrm{Gal}(R,\mathbb{Z}/n)}{C/nC} \to 1$$

where C denotes the cokernel of the natural map $\operatorname{Pic} R[t] \to \operatorname{Pic} R[t,t^{-1}]$.

Proof. See [**For88**]. The proof utilizes étale cohomology, in particular Gabber's Theorem (Theorem 10.4.13), and Cohomological Purity ([**Mil80**, Chapter VI, § 5]). If S/R is a cyclic Galois extension of degree n, with group $\langle\sigma\rangle$, then the splitting map is induced by sending $S \in \mathrm{Gal}(R,\mathbb{Z}/n)$ to the cyclic crossed product algebra

$$\Delta(S[t,t^{-1}]/R[t,t^{-1}],\sigma,t)$$

which is an Azumaya $R[t,t^{-1}]$-algebra. □

When R is a noetherian normal integral domain, the natural maps $\operatorname{Pic} R \to \operatorname{Pic} R[t] \to \operatorname{Pic} R[t,t^{-1}]$ are isomorphisms (Exercise 6.5.14). Hence we have the following important special case of Theorem 13.6.9.

Corollary 13.6.10. *Let R be a noetherian integrally closed integral domain and t an indeterminate. Let $n > 1$ and assume R is a $\mathbb{Z}[n^{-1},\zeta]$-algebra, where ζ is a primitive nth root of unity in $\mathbb{C}$. Then there is a split exact sequence*

$$0 \to {}_n\mathrm{B}(R) \to {}_n\mathrm{B}(R[t,t^{-1}]) \to \mathrm{Gal}(R,\mathbb{Z}/n) \to 1.$$

Corollary 13.6.11 is a special case of [**Hoo84**, Corollary 2.6]. It can be proved by a recursive application of Corollary 13.6.10 and Exercise 12.9.19.

Corollary 13.6.11. *Let R be a noetherian integrally closed integral domain. Let $n > 1$ and assume R is a $\mathbb{Z}[n^{-1},\zeta]$-algebra, where ζ is a primitive nth root of unity in $\mathbb{C}$. If $T = R[t_1,t_1^{-1},\dots,t_v,t_v^{-1}]$ is the Laurent polynomial ring in v indeterminates over R, then ${}_n\mathrm{B}(T)$ is isomorphic to*

$${}_n\mathrm{B}(R) \oplus \left(\mathrm{Gal}(R,\mathbb{Z}/n)\right)^{(v)} \oplus \left(\frac{\langle t_1\rangle}{\langle t_1^n\rangle}\right)^{(v-1)} \oplus \left(\frac{\langle t_2\rangle}{\langle t_2^n\rangle}\right)^{(v-2)} \oplus \cdots \oplus \frac{\langle t_{v-1}\rangle}{\langle t_{v-1}^n\rangle}.$$

If k is an algebraically closed field, then a Laurent polynomial ring over k in v indeterminates is a toric variety over k. For the computation of the Brauer group of a nonsingular toric variety, the interested reader is referred to [**DF93**] and [**DFM93a**].

6.3. Examples of Brauer Groups. The purpose of this section is to compute the terms in the exact sequence of Galois cohomology (Theorem 13.3.1) and the Kummer sequence (Theorem 12.9.9) for some nontrivial examples. Along the way, many preliminary computations are made. The examples we consider are based on the affine coordinate ring R of the open affine complement of the cuspidal curve $x^{n-1}z = y^n$ in the projective plane over an algebraically closed field. First we compute the Brauer group of R and classify the cyclic Galois extensions of R (Example 13.6.12). Secondly we compute all of the terms in the exact sequence of Galois cohomology for a cyclic Galois extension of R of degree n (Example 13.6.13). Lastly, we show that the Brauer group of the ring of Laurent polynomials $R[t, t^{-1}]$ is a cyclic group of order n (Example 13.6.15). By taking direct sums, this proves that any finite abelian group appears as the Brauer group of the affine coordinate ring of an algebraic 3-fold. The ring $R[t, t^{-1}]$ can be viewed as the affine coordinate ring of a dense open subset of $\mathbb{P}^2 \times \mathbb{A}^1$, hence is a regular integral domain, with quotient field isomorphic to $k(x, y, z)$. Therefore, the natural map $\mathrm{B}(R[t, t^{-1}]) \to \mathrm{B}(k(x, y, z))$ is one-to-one (Theorem 11.3.18). Another example of an integral domain A with Brauer group isomorphic to $\mathbb{Z}/n$ is presented in Section 14.3.2. In the latter example, A is a nonnormal domain, and the Azumaya algebra which generates the Brauer group of A turns out to be generically trivial.

Example 13.6.12. In this example we exhibit a regular integral domain R such that $\mathrm{Pic}(R) \cong \mathbb{Z}/n$ and $\mathrm{B}(R) = (0)$. Let k be an algebraically closed field of characteristic zero and $n \geq 2$. The basis for this example is the general construction of Example 6.5.12. In the polynomial ring in three variables $S = k[x, y, z]$, let $f = x^{n-1}z - y^n$. Let $R = S[f^{-1}]_0$ be the subring of the localized ring $S[f^{-1}]$ consisting of those homogeneous elements of degree zero. Then the following are true:

(1) R is a regular integral domain, a finitely generated k-algebra, and the Krull dimension of R is two.

(2) $\mathrm{Pic}(R) = \mathrm{Cl}(R) \cong \mathbb{Z}/n$.

(3) $R^* = k^*$.

(4) $\mathrm{Gal}(R, \mathbb{Z}/n) \cong \mathbb{Z}/n$.

(5) $\mathrm{B}(R) = (0)$.

For (1), (2) and (3), see Example 6.5.12. Part (4) follows from (2), (3), and the Kummer Sequence (Theorem 12.9.9). We now prove (5). Consider the localization of R obtained by inverting $x^n f^{-1}$. By Equation (6.25), there is an isomorphism of k-algebras

$$R[fx^{-n}] \to k[u, v][(v - u^n)^{-1}] \tag{13.38}$$

which is defined by $x \mapsto 1$, $y \mapsto u$, $z \mapsto v$. The assignments $u \mapsto s$, $v \mapsto t+s^n$ induce another isomorphism of k-algebras

$$k[u, v][(v - u^n)^{-1}] \to k[s, t][t^{-1}]. \tag{13.39}$$

By Tsen's Theorem (Corollary 11.1.9), $\mathrm{B}(k(t)) = (0)$. Since $\mathrm{B}(k[t,t^{-1}]) \to \mathrm{B}(k(t))$ is one-to-one (Theorem 11.3.18), this proves $\mathrm{B}(k[t,t^{-1}]) = (0)$. Theorem 13.6.3 implies that $\mathrm{B}(k[t,t^{-1}][s] = (0)$. By the isomorphisms (13.38) and (13.39), this implies $\mathrm{B}(R[fx^{-n}]) = (0)$. If K is the quotient field of R, then the diagram

$$\begin{array}{ccc} \mathrm{B}(R) & \longrightarrow & \mathrm{B}(K) \\ & \searrow \quad \nearrow & \\ & \mathrm{B}(R[fx^{-n}]) & \end{array} \tag{13.40}$$

commutes. By Theorem 11.3.18, all of the maps in (13.40) are one-to-one. Therefore, $\mathrm{B}(R) = (0)$.

Example 13.6.13. Let n, k, f and R be as in Example 13.6.12. Let $\zeta \in k$ be a fixed primitive nth root of unity. In this example we describe a cyclic Galois extension T/R of degree n which is a generator for the group $\mathrm{Gal}(R, \mathbb{Z}/n)$. As in Example 6.5.12, let $\mathfrak{p}$ denote the unique minimal prime ideal of R containing $x^n f^{-1}$. Then $\mathfrak{p}$ is the element of order n which generates $\operatorname{Pic} R$. Since $\mathrm{Div}(x^n f^{-1}) = n\mathfrak{p}$, it follows that $\mathfrak{p}^n$ is equal to the principal ideal generated by $x^n f^{-1}$. In the notation of Example 12.9.6, let $T = \Gamma(R, \mathfrak{p}, \phi)$ be the cyclic Galois extension of R corresponding to the isomorphism $\phi : \mathfrak{p}^n \to R$. Let $G = \langle \sigma \rangle$ be the Galois group of T/R. We will prove the following.

(1) T is a finitely generated k-algebra, a regular noetherian integral domain, and $\dim(T) = 2$.
(2) $T^* = k^*$.
(3) $\operatorname{Pic} T = \mathrm{Cl}(T) \cong \mathbb{Z}^{(n-1)}$.
(4) For every $m \geq 2$, $\mathrm{Gal}(T, \mathbb{Z}/m) = \langle 1 \rangle$.
(5) $\mathrm{B}(T) = (0)$.
(6) $\mathrm{H}^j(G, \mathrm{Pic}(T)) \cong \begin{cases} (0) & \text{if } j \text{ is even,} \\ \mathbb{Z}/n & \text{if } j \text{ is odd.} \end{cases}$
(7) $\mathrm{H}^j(G, T^*) = \begin{cases} k^* & \text{if } j = 0, \\ \langle 1 \rangle & \text{if } j \geq 2 \text{ is even,} \\ \langle \zeta \rangle \cong \mathbb{Z}/n & \text{if } j \text{ is odd.} \end{cases}$
(8) In the seven term exact sequence for T/R, (Theorem 13.3.1), α_1 and α_6 are isomorphisms and $\alpha_2 = \alpha_3 = \alpha_4 = \alpha_5 = 0$.

Since T/R is Galois, T is faithfully flat étale over R (Theorem 12.2.10). By Example 13.6.12, R is a finitely generated k-algebra, regular, and $\dim(R) = 2$. By Proposition 9.3.12, T has the same attributes. To finish (1) we show T

is an integral domain. A typical element of R is pf^{-d}, where $p \in S_d$. Write R_x for the ring $R[fx^{-n}]$, and R_y for $R[fy^{-n}]$. Using the computations in (6.24), we find that

$$(13.41)\qquad \begin{aligned} R_x &= R\left[fx^{-n}\right] = k\left[yx^{-1}, zx^{-1}, x^n f^{-1}\right] \\ R_y &= R\left[fy^{-1}\right] = k\left[xy^{-1}, zy^{-1}, y^n f^{-1}\right] \end{aligned}$$

where all of the rings are viewed as subrings of $S[f^{-1}]$. It follows from

$$\begin{aligned} 1 = \frac{f^2}{f^2} &= \frac{(x^{n-1}z - y^n)^2}{f^2} \\ &= \frac{x^{2n-2}z^2 - 2x^{n-1}zy^n + y^{2n}}{f^2} \\ &= \frac{x^n}{f}\frac{x^{n-2}z^2}{f} - \frac{2x^{n-1}z}{f}\frac{y^n}{f} + \left(\frac{y^n}{f}\right)^2 \end{aligned}$$

that $x^n f^{-1}$ and $y^n f^{-1}$ generate the unit ideal in R. Hence $R_x \oplus R_y$ is a faithfully flat extension of R (Exercise 2.6.15). Since $\mathfrak{p}R_x$ is the unit ideal, and $\mathfrak{p}R_y$ is the principal ideal (xy^{-1}), as shown in Example 12.9.6, the extension T/R is split by the faithfully flat extension $R_x \oplus R_y$ and upon restriction to $R_x \oplus R_y$, T corresponds to adjoining the nth root of $x^n f^{-1}$. Let $f^{1/n}$ be a fixed root of the irreducible polynomial $w^n - f$ in an extension field of $k(x, y, z)$. We see that

$$(13.42)\qquad \begin{aligned} T_x &= T[fx^{-n}] = R_x[xf^{-1/n}] = k\left[yx^{-1}, zx^{-1}, xf^{-1/n}, f^{1/n}x^{-1}\right] \\ T_y &= T[fy^{-n}] = R_y[yf^{-1/n}] = k\left[xy^{-1}, zy^{-1}, yf^{-1/n}, f^{1/n}y^{-1}\right]. \end{aligned}$$

The proof of (1) is finished because T is a subring of the field $k(x, y, z)[f^{1/n}]$. The action of σ on T_x is defined by $xf^{-1/n} \mapsto \zeta xf^{-1/n}$ and on T_y by $yf^{-1/n} \mapsto \zeta yf^{-1/n}$. Using the isomorphisms (13.38) and (13.39), the reader should verify that T_x is isomorphic as a k-algebra to R_x. Thus

$$(13.43)\qquad \begin{aligned} T_x^* &= k^* \times \left\langle xf^{-1/n}\right\rangle \\ \mathrm{Cl}(T_x) &= (0) \\ \mathrm{B}(T_x) &= (0). \end{aligned}$$

The assignments $xy^{-1} \mapsto a$, $zy^{-1} \mapsto b$, induce isomorphisms

$$(13.44)\qquad \begin{aligned} R_y &\cong k\left[a, b, \frac{1}{a^{n-1}b - 1}\right] \\ T_y &\cong \frac{k[a, b, c][c^{-1}]}{(c^n - a^{n-1}b + 1)}. \end{aligned}$$

Using (13.44) and Exercise 9.4.9, the reader should verify that

$$\begin{aligned} T_y^* &= k^* \times \left\langle yf^{-1/n} \right\rangle \\ \mathrm{Cl}(T_y) &= \mathbb{Z}^{(n-1)}. \end{aligned} \tag{13.45}$$

Moreover, $\mathrm{Cl}(T_y)$ is generated by $Q_1, \dots, Q_n$, the minimal prime ideals containing xy^{-1}, subject to the relation $\mathrm{Div}\,(xy^{-1}) = Q_1 + \cdots + Q_n$. Notice $Q_1, \dots, Q_n$ are also the minimal primes containing $xf^{-1/n}$. From (13.42) we see that T_x is the localization of T obtained by inverting $xf^{-1/n}$. Therefore, $\mathrm{Div}\,(xf^{-1/n})$ can be computed in $\mathrm{Div}\, T_y$. Then, in the notation of Exercise 9.4.9, we have $\mathrm{Div}\,(xf^{-1/n}) = \mathrm{Div}\,(xy^{-1}) = Q_1 + \cdots + Q_n$. Since $\mathrm{Cl}(T_x) = (0)$, Nagata's sequence becomes

$$1 \to T^* \to T_x^* \xrightarrow{\mathrm{Div}} \bigoplus_{i=1}^{n} \mathbb{Z}Q_i \to \mathrm{Cl}(T) \to 0 \tag{13.46}$$

(Theorem 6.2.4). From (13.46) we conclude that

$$\begin{aligned} T^* &= k^* \\ \mathrm{Cl}(T) &= \mathbb{Z}^{(n-1)} \end{aligned} \tag{13.47}$$

and $\mathrm{Cl}(T)$ is generated by $Q_1, \dots, Q_n$, subject to the relation $\mathrm{Div}\,(xf^{-1/n}) = Q_1 + \cdots + Q_n$. By Theorem 6.5.11, $\mathrm{Pic}(T) = \mathrm{Cl}(T)$, so we have proved (2) and (3). Part (4) follows from the Kummer Sequence (Theorem 12.9.9). Part (5) is proved by the same argument that was used to prove Example 13.6.12 (5). The proof of (6) is similar to Exercise 5.1.30 and is left to the reader. Part (7) follows from Corollary 5.1.18.

Example 13.6.14. This example is based on [**For16**, Remark 4.12]. Let T/R be the cyclic Galois extension of degree n of Example 13.6.13, with group $G = \langle \sigma \rangle$. Let L be the quotient field of T, and K the quotient field of R. By the computations in Examples 13.6.12 and 13.6.13, we see that in the exact sequence of Galois cohomology (Theorem 13.3.1, α_1 and α_6 are isomorphisms and $\alpha_2 = \alpha_3 = \alpha_4 = \alpha_5 = 0$. Since T is regular, a reflexive fractional ideal of T in L is invertible. By Exercise 5.1.30, the norm $N = 1 + \sigma + \cdots + \sigma^{n-1}$ is the zero map on $\mathrm{Pic}(T)$. Thus, for every reflexive fractional ideal I of T in L, the ideal $I\sigma(I) \cdots \sigma^{n-1}(I)$ is a principal ideal Tg for some g in L^*. If g is in K^*, then the generalized cyclic crossed product $\Delta(T/R, I, g)$ represents a class in $\mathrm{B}(T/R)$, which is split since $\mathrm{B}(R) = (0)$. Since the map $\alpha_5 : \mathrm{B}(T/R) \to \mathrm{H}^1(G, \mathrm{Pic}(T))$ is the zero map, Corollary 13.5.16 implies I represents a coboundary in $\mathrm{B}^1(G, \mathrm{Pic}\,(T))$. Since $\mathrm{H}^1(G, \mathrm{Pic}(T))$ has order n, this proves that there exist reflexive fractional ideals I for which the norm $N(I) = Tg$ is principal, but g is not in K^*. This also elucidates a comment made in Remark 13.5.6.

Example 13.6.15. This example is from [**For89**, Example 3]. Let n, k, R, and T be as in Example 13.6.13, and t an indeterminate. Then T is a cyclic Galois extension of R with group $G = \langle\sigma\rangle$ of order n. By Theorem 6.1.9, $T[t]$ is a regular integral domain. Consequently, $T[t, t^{-1}]$ is a regular integral domain. Upon tensoring T and R with $R[t, t^{-1}]$, we view $T[t, t^{-1}]$ as a cyclic Galois extension of $R[t, t^{-1}]$ with group G. We will now show that the following are true:

(1) $R[t, t^{-1}]^* = T[t, t^{-1}]^* = k^* \times \langle t\rangle$.

(2) $\operatorname{Pic}(R[t, t^{-1}]) = \operatorname{Pic}(R) \cong \mathbb{Z}/n$.

(3) $\operatorname{Pic}(T[t, t^{-1}]) = \operatorname{Pic}(T) = \mathbb{Z}^{(n-1)}$.

(4) $\mathrm{B}(T[t, t^{-1}]) = (0)$.

(5) $\mathrm{B}(R[t, t^{-1}]) \cong \mathbb{Z}/n$, a generator for the group is the Brauer class of the cyclic crossed product algebra $\Delta(T[t, t^{-1}]/R[t, t^{-1}], \sigma, t)$.

By Theorem 6.5.11, $\operatorname{Pic}(\cdot) = \operatorname{Cl}(\cdot)$ on all of the rings involved here. Parts (1), (2) and (3) all follow from Exercise 6.5.14. Part (4) follows from Example 13.6.13 and Corollary 13.6.10. Part (5) could be proved directly from Corollary 13.6.10, but instead we use the seven term exact sequence of Galois cohomology. Following Example 13.6.13, the reader should verify the following.

$$\mathrm{H}^j(G, T[t, t^{-1}]^*) = \begin{cases} k^* \times \langle t\rangle & \text{if } j = 0, \\ \frac{\langle t\rangle}{\langle t^n\rangle} \cong \mathbb{Z}/n & \text{if } j \geq 2 \text{ is even}, \\ \langle\zeta\rangle \cong \mathbb{Z}/n & \text{if } j \text{ is odd}. \end{cases} \tag{13.48}$$

$$\mathrm{H}^j(G, \operatorname{Pic}(T[t, t^{-1}])) = \begin{cases} (0) & \text{if } j \text{ is even}, \\ \operatorname{Pic}(T)/D\operatorname{Pic}(T) \cong \mathbb{Z}/n & \text{if } j \text{ is odd}. \end{cases} \tag{13.49}$$

By Part (4), every Azumaya $R[t, t^{-1}]$-algebra is split by $T[t, t^{-1}]$. We have computed six of the seven terms in the exact sequence of Theorem 13.3.1. By the computations in Example 13.6.13, we conclude that α_3 is the zero map and α_6 is an isomorphism. Therefore, α_5 is the zero map and α_4 is an isomorphism.

$$\begin{array}{ccccc} \mathrm{H}^1(G, T[t, t^{-1}]^*) & \xrightarrow{\alpha_1} & \operatorname{Pic}(R[t, t^{-1}]) & \xrightarrow{\alpha_2} & \operatorname{Pic}(T[t, t^{-1}])^G \\ \downarrow \cong & & \downarrow \cong & & \downarrow = \\ \mathbb{Z}/n & \xrightarrow{\cong} & \mathbb{Z}/n & \longrightarrow & 0 \end{array}$$

$$\begin{array}{ccccc}
\operatorname{Pic}(T[t,t^{-1}])^G & \xrightarrow{\alpha_3} & \mathrm{H}^2(G,T[t,t^{-1}]^*) & \xrightarrow{\alpha_4} & \mathrm{B}(R[t,t^{-1}]) \\
\downarrow = & & \downarrow \cong & & \downarrow \cong \\
0 & \longrightarrow & \mathbb{Z}/n & \xrightarrow{\cong} & \mathbb{Z}/n
\end{array}$$

$$\begin{array}{ccccc}
\mathrm{B}(R[t,t^{-1}]) & \xrightarrow{\alpha_5} & \mathrm{H}^1(G,\operatorname{Pic} T[t,t^{-1}]) & \xrightarrow{\alpha_6} & \mathrm{H}^3(G,T[t,t^{-1}]^*) \\
\downarrow \cong & & \downarrow \cong & & \downarrow \cong \\
\mathbb{Z}/n & \longrightarrow & \mathbb{Z}/n & \xrightarrow{\cong} & \mathbb{Z}/n.
\end{array}$$

Since α_4 is the crossed product homomorphism, we see from (13.48) that the cyclic crossed product $\Delta(T[t,t^{-1}]/R[t,t^{-1}],\sigma,t)$ is an Azumaya $R[t,t^{-1}]$-algebra whose Brauer class has order n. We have shown that $R[t,t^{-1}]$ is a regular noetherian integral domain whose Brauer group is cyclic of order n.

6.4. Exercises.

Exercise 13.6.16. This exercise is based on [**For17**, §3]. Let k be a field, and $\bar{k}$ an algebraic closure of k. Let d and n be integers such that $d \geq 2$ and d divides $n-1$. Assume d is invertible in k and that k contains ζ, a primitive dth root of unity. Fix $f \in k[x,y]$, a homogeneous polynomial of degree n such that the image of f in the ring $\bar{k}[x,y]$ is square-free. If $R = k[x,y][f^{-1}]$, $T = k[x,y,z]/(z^d - f(x,y))$, and $S = T[z^{-1}]$, then T is an integrally closed integral domain (see Example 13.6.7) and S is a cyclic Galois extension of R of degree d, with group $G = \langle\sigma\rangle$, where $\sigma(z) = \zeta z$ (see Example 12.9.5).

(1) Consider the two k-algebra homomorphisms

$$S[x^{-1}] = \frac{k[x,y,z][x^{-1},z^{-1}]}{(z^d - f(x,y))} \xrightarrow{\alpha} \frac{k[x,v,s][x^{-1},s^{-1}]}{(s^d - xf(1,v))} \xrightarrow{\beta} k[v,s][s^{-1},f(1,v)^{-1}]$$

where α is defined by $x \mapsto x$, $y \mapsto xv$, $z \mapsto x^{(n-1)/d}s$, and β is defined by $x \mapsto s^d f(1,v)^{-1}$, $v \mapsto v$, $s \mapsto s$. Prove that α and β are isomorphisms.

(2) Prove: $\operatorname{Cl}(S) = \operatorname{Pic}(S) = (0)$. Conclude that the crossed product homomorphism $\alpha_4 : \mathrm{H}^2(G,S^*) \to \mathrm{B}(S/R)$ is an isomorphism.

(3) In $k[x,y]$, let $f = f_1 \cdots f_v$ be the unique factorization of f into irreducibles. Prove:

$$R^* = k^* \times \langle f_1\rangle \times \langle f_2\rangle \cdots \times \langle f_v\rangle$$
$$S^* = k^* \times \langle z\rangle \times \langle f_2\rangle \cdots \times \langle f_v\rangle.$$

(4) Prove:

$$\mathrm{H}^i(G, S^*) \cong \begin{cases} R^* = k^* \times \langle f_1 \rangle \cdots \times \langle f_v \rangle & \text{if } i = 0, \\ \frac{k^*}{(k^*)^d} \times \frac{\langle f_2 \rangle}{\langle f_2^d \rangle} \times \cdots \times \frac{\langle f_v \rangle}{\langle f_v^d \rangle} & \text{if } i = 2, 4, 6, \dots, \\ \langle 1 \rangle & \text{if } i = 1, 3, 5, \dots. \end{cases}$$

(5) Let $I_i = (z, f_i)$ be the ideal in T generated by z and f_i. Show that $I_1, \dots, I_v$ are height one prime ideals, and $\mathrm{Cl}(T)$ is generated by $I_1, \dots, I_v$, subject to the relations $\mathrm{Div}(z) = I_1 + \cdots + I_v$, $\mathrm{Div}(f_1) = dI_1, \dots, \mathrm{Div}(f_v) = dI_v$. Conclude that $\mathrm{Cl}(T) \cong (\mathbb{Z}/d)^{(v-1)}$.

(6) Show that the group $G = \langle \sigma \rangle$ acts trivially on $\mathrm{Cl}(T)$ and

$$\mathrm{H}^i(G, \mathrm{Cl}(T)) \cong \begin{cases} \mathrm{Cl}(T) \otimes \mathbb{Z}/d \cong (\mathbb{Z}/d)^{(v-1)} & \text{if } i \text{ is odd}, \\ \mathrm{Cl}(T)^G = \mathrm{Cl}(T) \cong (\mathbb{Z}/d)^{(v-1)} & \text{if } i \text{ is even}. \end{cases}$$

Exercise 13.6.17. This example is from [**For16**, §5.2]. Let k be a field with a fixed algebraic closure $\bar{k}$. Let $n \geq 2$ be an integer that is invertible in k and assume k contains ζ, a primitive nth root of unity in $\bar{k}$. Let c be an element of k such that $z^n - c$ is irreducible in $k[z]$. Let $A = k[x_1, \dots, x_m]$, $f = x_1 \cdots x_m + c$, and $T = A[z]/(z^n - f)$. Let σ be the A-automorphism of T defined by $\sigma(z) = \zeta z$, $R = A[f^{-1}]$, $S = T[z^{-1}]$. Prove:

(1) T is a regular integral domain of dimension m.

(2) $$\mathrm{H}^r(\langle \sigma \rangle, T^*) = \begin{cases} k^* & \text{if } r = 0, \\ k^*/(k^*)^n & \text{if } r > 0 \text{ is even}, \\ \langle \zeta \rangle \cong \mathbb{Z}/n & \text{if } r \text{ is odd}. \end{cases}$$

(3) $$\mathrm{H}^r(\langle \sigma \rangle, S^*) = \begin{cases} k^* \times \langle f \rangle & \text{if } r = 0, \\ k^*/(k^*)^n & \text{if } r > 0 \text{ is even}, \\ \langle 1 \rangle & \text{if } r \text{ is odd}. \end{cases}$$

(4) $\mathrm{Pic}\,(T) = \mathrm{Cl}(T) = (0)$.

(5) $\mathrm{Pic}\,(S) = \mathrm{Cl}(S) = (0)$.

(6) The crossed product map $\alpha_4 : \mathrm{H}^2(\langle \sigma \rangle, S^*) \to \mathrm{B}(S/R)$ is an isomorphism.

Exercise 13.6.18. This exercise is a continuation of Exercise 13.6.17 and we continue to use the same notation. Let θ be a root of $z^n - c$ in $\bar{k}$. By Kummer Theory, $k(\theta)/k$ is a cyclic Galois extension of fields with group $\mathrm{Aut}_k\, k(\theta) = \langle \tau \rangle$ where $\tau(\theta) = \zeta^{-1}\theta$. Set $A(\theta) = A \otimes_k k(\theta)$, $R(\theta) = R \otimes_k k(\theta)$, $T(\theta) = T \otimes_k k(\theta)$, $S(\theta) = S \otimes_k k(\theta)$. Prove:

(1) $T(\theta)$ is a regular integral domain of dimension m.

(2) $T(\theta)^* = k(\theta)^*$.

(3) $$\mathrm{H}^r(\langle\sigma\rangle, T(\theta)^*) = \begin{cases} k(\theta)^* & \text{if } r = 0, \\ k(\theta)^*/(k(\theta)^*)^n & \text{if } r > 0 \text{ is even}, \\ \langle\zeta\rangle \cong \mathbb{Z}/n & \text{if } r \text{ is odd}. \end{cases}$$

(4) $$\mathrm{H}^r(\langle\tau\rangle, T(\theta)^*) = \begin{cases} k^* & \text{if } r = 0, \\ \mathrm{B}(k(\theta)/k) & \text{if } r > 0 \text{ is even}, \\ \langle 1\rangle & \text{if } r \text{ is odd}. \end{cases}$$

(5) For $1 \le i < m$ and $0 \le j < n$, define the ideal $I_{i,j} = (x_i, \zeta^j z - \theta)$ in $T(\theta)$. Then $\mathrm{Pic}\,(T(\theta)) = \mathrm{Cl}(T(\theta))$ is a free $\mathbb{Z}$-module of rank $(n-1)(m-1)$ generated by the classes of the ideals $I_{i,j}$, $1 \le i < m$, $0 \le j < n-1$.

(6) $$\mathrm{H}^r(\langle\sigma\rangle, \mathrm{Cl}(T(\theta))) = \begin{cases} (0) & \text{if } r \text{ is even}, \\ \bigoplus_{i=1}^{m-1} \mathbb{Z}/n[I_{i,0}] \cong (\mathbb{Z}/n)^{m-1} & \text{if } r \text{ is odd}. \end{cases}$$

(7) $$\mathrm{H}^r(\langle\tau\rangle, \mathrm{Cl}(T(\theta))) = \begin{cases} (0) & \text{if } r \text{ is even}, \\ \bigoplus_{i=1}^{m-1} \mathbb{Z}/n[I_{i,0}] \cong (\mathbb{Z}/n)^{m-1} & \text{if } r \text{ is odd}. \end{cases}$$

(8) There is an exact sequence

$$(0) \to \mathrm{B}(k(\theta)/k) \to \mathrm{B}(T(\theta)/T) \xrightarrow{\alpha_5} \mathrm{H}^1(\langle\tau\rangle, \mathrm{Cl}(T(\theta))) \to (0)$$

of abelian groups. In particular, if k is algebraically closed, α_5 is an isomorphism. Conclude that in general there is a nontrivial subgroup of $\mathrm{B}(T)$ that is split by the cyclic extension $k(\theta)/k$ of the ground field.

(9) There is an exact sequence

$$\langle 1\rangle \to \mathrm{H}^2(\langle\sigma\rangle, T(\theta)^*) \xrightarrow{\alpha_4} \mathrm{B}(S(\theta)/R(\theta)) \xrightarrow{\alpha_5} \mathrm{H}^1(\langle\sigma\rangle, \mathrm{Cl}(T(\theta))) \to (0)$$

of abelian groups. In particular, if k is algebraically closed, α_5 is an isomorphism.

Exercise 13.6.19. This example is from [**Lor95**] and [**For16**, §5.4]. Let k be a field in which 2 is invertible. Let

$$A = k[x, y] \text{ and } T = A[z]/(z^2 - (x^2 - 4)(y^2 - 4)).$$

Let $T_2 = A[u, v]/(u^2 - (x^2 - 4), v^2 - (y^2 - 4))$.

(1) Show that T_2 is isomorphic to a Laurent polynomial ring over k in two indeterminates. That is, $T_2 \cong k[s, s^{-1}, t, t^{-1}]$.

(2) Show that the assignment $z \mapsto uv$ maps T isomorphically onto the A-subalgebra of T_2 generated by uv. Let ρ be the A-algebra automorphism of T_2 induced by $u \mapsto -u$ and $v \mapsto -v$. Show that $T = T_2^{\langle\rho\rangle}$.

(3) Show that $\mathrm{Cl}(T)$ is generated by the four prime divisors

$$I_1 = (z, x-2), \quad I_2 = (z, x+2), \quad I_3 = (z, y-2), \quad I_4 = (z, y+2),$$

and $\mathrm{Cl}(T) \cong \mathbb{Z}/2 \oplus \mathbb{Z}/2 \oplus \mathbb{Z}/2$.

(4) Show that $\mathrm{Sing}(T) = \{\mathfrak{m}_1, \mathfrak{m}_2, \mathfrak{m}_3, \mathfrak{m}_4\}$, where $\mathfrak{m}_1 = I_1 + I_3$, $\mathfrak{m}_2 = I_1 + I_4$, $\mathfrak{m}_3 = I_2 + I_3$, $\mathfrak{m}_4 = I_2 + I_4$. Show that the singularity of T at each $\mathfrak{m}_i$ is a rational double point of type A_1 and $\mathrm{Cl}(T_{\mathfrak{m}_i}) \cong \mathbb{Z}/2$. (Hint: Exercise 10.1.18.)

(5) Show that $\mathrm{Pic}(T) = (0)$.

(6) As in Section 11.3.7, show that $\mathrm{B}(T)$ contains an element of order two which is locally trivial. As in Proposition 11.3.26, show that T has a reflexive lattice M which has rank 2, and $\Lambda = \mathrm{Hom}_T(M, M)$ is an Azumaya T-algebra which has order 2 in $\mathrm{B}(T)$.

Chapter 14

Further Topics

Let R be a commutative ring and S a commutative separable R-algebra which as a module is an R-progenerator of constant rank. A corestriction functor from S-modules and S-algebras to R-modules and R-algebras is defined. We show that this functor defines corestriction homomorphisms on the Picard groups $\operatorname{Cor}_R^S : \operatorname{Pic}(S) \to \operatorname{Pic}(R)$, the Brauer groups $\operatorname{Cor}_R^S : \mathrm{B}(S) \to \mathrm{B}(R)$ and on the Čech étale cohomology groups $\operatorname{Cor}_R^S : \check{\mathrm{H}}^r_{\mathrm{et}}(S, \mathbb{G}_m) \to \check{\mathrm{H}}^r_{\mathrm{et}}(R, \mathbb{G}_m)$. These corestriction maps agree with their counterparts for Galois cohomology.

In Section 14.2, Milnor's six term Mayer-Vietoris sequence for the groups of units and Picard groups is constructed. Then we give sufficient conditions such that this sequence can be extended to a Mayer-Vietoris sequence for Brauer groups.

Section 14.3 consists of applications of the Mayer-Vietoris sequence. An example is given of a ring R whose Brauer group is cyclic of order n, for any n. The Brauer group of a ring of integers R in a global field K is described in terms of the number of primes at infinity. As a special case, we find that the Brauer group of the ring $\mathbb{Z}$ is the trivial group.

1. Corestriction

Let R be a commutative ring and S a commutative separable R-algebra which is a projective R-module of constant rank. First a general corestriction functor from S-modules to R-modules is defined. Then we derive a set of axioms that uniquely determine the corestriction functor. The corestriction functor induces group homomorphisms $\operatorname{Pic}(S) \to \operatorname{Pic}(R)$ and $\mathrm{B}(S) \to \mathrm{B}(R)$ as well as for Čech cohomology $\check{\mathrm{H}}^r_{\mathrm{et}}(S, \mathbb{G}_m) \to \check{\mathrm{H}}^r_{\mathrm{et}}(R, \mathbb{G}_m)$. In degrees one and two, the corestriction homomorphisms defined in this chapter commute with the natural maps $\operatorname{Pic}(R) \to \check{\mathrm{H}}^1_{\mathrm{et}}(R, \mathbb{G}_m)$ and $\mathrm{B}(R) \to \check{\mathrm{H}}^2_{\mathrm{et}}(R, \mathbb{G}_m)$ defined in Section 10.4.2. In a suitably defined context, the corestriction homomorphisms defined in this chapter commute with the usual corestriction homomorphism for Galois cohomology. The presentation given below is based on [**KO75**] and [**Sal99**, Chapter 8]. Historically, the corestriction homomorphism $\mathrm{B}(K) \to \mathrm{B}(F)$ was defined on the Brauer groups of a separable field extension K/F using the crossed product homomorphism and the corestriction homomorphism on the Galois cohomology groups. In this sense, our approach is the reverse of the classical one.

Definition 14.1.1. Let R be a commutative ring and S a commutative separable R-algebra which as a module is an R-progenerator of constant rank n. For any S-module M, the tensor module $T^n_R(M)$ is a module over the R-algebra $T^n_R(S)$ (Exercise 5.2.5). Let $G = \Sigma_n$ be the symmetric group on n-letters. Then G acts on both $T^n_R(S)$ and $T^n_R(M)$ by permuting the factors. If $\sigma \in G$, then on a basic tensor element, we have

$$\sigma(x_1 \otimes \cdots \otimes x_n) = x_{\sigma^{-1}(1)} \otimes \cdots \otimes x_{\sigma^{-1}(n)}.$$

In the proof of Theorem 12.6.1 we saw that there is a canonically defined idempotent $f \in T^n_R(S)$ which is fixed by every $\sigma \in G$, and the ring $T^n_R(S)f$ is Galois over R with group G. The reader should verify that $T^n_R(M)f$ is a left G-module, and that this action makes $T^n_R(M)f$ into a left $\Delta(T^n_R(S)f/R, G, 1)$-module. By Galois descent (Proposition 12.4.2), $(T^n_R(M)f)^G$ is an R-module, which is denoted $\operatorname{Cor}^S_R(M)$ and is called the *corestriction of M from S to R*. In [**KO75**] the corestriction of M is called the *norm of M*.

Example 14.1.2. Let R be a commutative ring and $S = R^n = R \oplus \cdots \oplus R$ the trivial separable extension of degree n (Section 4.6.3). Following the construction in the proof of Theorem 12.6.1, let $S_i = R \otimes_R \cdots \otimes_R S \otimes_R \cdots \otimes_R R$ be the R-subalgebra of $T^n_R(S)$ corresponding to the ith tensor factor. For every pair $i < j$, let S_{ij} denote the R-subalgebra of T generated by S_i and S_j. Then S_{ij} is isomorphic to $S \otimes_R S$. Let $e_1, \dots, e_n$ be the canonical orthogonal idempotents in S. Using Exercise 4.1.11, we see that the separability idempotent in $S \otimes_R S$ is $e = \sum_{i=1}^n e_i \otimes e_i$. Denote by e_{ij} the corresponding separability idempotent in S_{ij}, and set

$$f = \prod_{i=1}^{n-1} \prod_{j=i+1}^{n} (1 - e_{ij}).$$

As above, let $G = \Sigma_n$ be the symmetric group on n letters. The set $\{e_{\sigma(1)} \otimes e_{\sigma(2)} \otimes \cdots \otimes e_{\sigma(n)} \mid \sigma \in G\}$ is a free R-basis for $T^n_R(S)f$, and $f = \sum_{\sigma \in G} e_{\sigma(1)} \otimes e_{\sigma(2)} \otimes \cdots \otimes e_{\sigma(n)}$. The subring fixed by G is completely determined by the summand corresponding to $\sigma = 1$. That is, the image of the R-module homomorphism

$$Se_1 \otimes_R Se_2 \otimes_R \cdots \otimes_R Se_n \to T^n_R(S)f$$
$$x_1e_1 \otimes \cdots \otimes x_ne_n \mapsto \sum_{\sigma \in G} x_{\sigma(1)}e_{\sigma(1)} \otimes x_{\sigma(2)}e_{\sigma(2)} \otimes \cdots \otimes x_{\sigma(n)}e_{\sigma(n)}$$

is $(T^n_R(S)f)^G$. Thus $\operatorname{Cor}^S_R(S)$ is isomorphic to $Se_1 \otimes_R Se_2 \otimes_R \cdots \otimes_R Se_n \cong R$ as an R-module.

Let M be an S-module. As in Exercise 2.3.13, the S-module M decomposes into the internal direct sum $Me_1 \oplus Me_2 \oplus \cdots \oplus Me_n$. Since $Se_i = R$, each Me_i is an R-module. Then

$$T^n_R(M)f = \bigoplus_{\sigma \in G} \left(Me_{\sigma(1)} \otimes_R Me_{\sigma(2)} \otimes_R \cdots \otimes_R Me_{\sigma(n)}\right).$$

The subset fixed by G is completely determined by the summand corresponding to $\sigma = 1$. That is, the image of the R-module homomorphism

$$Me_1 \otimes_R Me_2 \otimes_R \cdots \otimes_R Me_n \to T^n_R(M)f$$
$$x_1e_1 \otimes \cdots \otimes x_ne_n \mapsto \sum_{\sigma \in G} x_{\sigma(1)}e_{\sigma(1)} \otimes \cdots \otimes x_{\sigma(n)}e_{\sigma(n)}.$$

is equal to $(T^n_R(M)f)^G$. Thus, $\operatorname{Cor}^S_R(M)$ is isomorphic to $Me_1 \otimes_R Me_2 \otimes_R \cdots \otimes_R Me_n$ as an R-module.

Theorem 14.1.3. *Let R be a commutative ring and S a commutative separable R-algebra which as a module is an R-progenerator of constant rank n. The assignment $M \mapsto \operatorname{Cor}^S_R(M)$ induces a covariant functor ${}_S\mathfrak{M} \to {}_R\mathfrak{M}$.*

(1) $\operatorname{Cor}^S_R(S) = R$.

(2) *Suppose Q and T are commutative rings such that T and S are algebras over Q. Then $T \otimes_Q \operatorname{Cor}^S_R(M) \cong \operatorname{Cor}^{T\otimes_Q S}_{T\otimes_Q R}(T \otimes_Q M)$.*

(3) *If $S = R^n$ is the trivial extension of R and $e_1, \ldots, e_n$ are the canonical orthogonal idempotents of S, then $\operatorname{Cor}^S_R(M) \cong Me_1 \otimes_R Me_2 \otimes_R \cdots \otimes_R Me_n$.*

(4) *If M is finitely generated, then $\operatorname{Cor}^S_R(M)$ is finitely generated.*

(5) *If M is projective, then $\operatorname{Cor}^S_R(M)$ is projective.*

(6) *If M is a generator module, then $\operatorname{Cor}^S_R(M)$ is a generator module.*

(7) *If A is an S-algebra, then $\operatorname{Cor}^S_R(A)$ is an R-algebra.*

Proof. Let $\theta : M \to N$ be an S-module homomorphism. The reader should verify that θ induces a left $\Delta(T^n_R(S)f/R, G, 1)$-module homomorphism $\theta : T^n_R(M)f \to T^n_R(N)f$. By the categorical equivalence of Proposition 12.4.2, there is a corresponding R-module homomorphism $\mathrm{Cor}_R^S(M) \to \mathrm{Cor}_R^S(N)$. The reader should verify that Cor_R^S defines a functor.

Part (1) follows directly from Theorem 12.6.1.

(2): The diagram of commutative rings

$$\begin{array}{ccccc} T & \longrightarrow & T \otimes_Q R & \longrightarrow & T \otimes_Q S \\ \uparrow & & \uparrow & & \uparrow \\ Q & \longrightarrow & R & \longrightarrow & S \end{array}$$

commutes, where an arrow is a homomorphism of rings. We are given that S is a separable algebra and a progenerator module over R. By change of base, $T \otimes_Q S$ is a separable algebra and progenerator module over $T \otimes_Q R$. Tensoring the split exact sequence

$$(14.1) \qquad 0 \to J_{S/R} \to S \otimes_R S \xrightarrow{\mu} S \to 0$$

of $S \otimes_R S$-modules with $T \otimes_Q ()$ gives rise to the split exact sequence

$$(14.2) \qquad 0 \to J_{T\otimes_Q S/T\otimes_Q R} \to T \otimes_Q S \otimes_R S \xrightarrow{\mu} T \otimes_Q S \to 0.$$

The ideal $J_{S/R}$ is generated by the idempotent $1 - e$, where e is the separability idempotent for S/R. Then $J_{T\otimes_Q S/T\otimes_Q R}$ is also generated by $1 - e$. Let $f_{S/R} \in T_R^n(S)$ be the idempotent of Definition 14.1.1. Under the isomorphism of rings

$$T \otimes_Q T_R^n(S) \xrightarrow{\psi} T_R^n(T \otimes_Q S)$$

we have $\psi(1 \otimes f_{S/R}) = f = f_{T\otimes_Q S/T\otimes_Q R}$. The isomorphism of $T \otimes_Q T_R^n(S)$-modules

$$T \otimes_Q T_R^n(M) \xrightarrow{\phi} T_{T\otimes_Q R}^n(T \otimes_Q M)$$

is defined on basic tensors by the rule $t \otimes x_1 \otimes \cdots \otimes x_n \mapsto (t \otimes x_1) \otimes (1 \otimes x_2) \otimes \cdots \otimes (1 \otimes x_n)$. If σ is a permutation in the symmetric group $G = S_n$, then

$$\begin{aligned} \phi\left(\sigma(t \otimes x_1 \otimes \cdots \otimes x_n)\right) &= \phi(t \otimes x_{\sigma^{-1}(1)} \otimes \cdots \otimes x_{\sigma^{-1}(n)}) \\ &= (t \otimes x_{\sigma^{-1}(1)}) \otimes (1 \otimes x_{\sigma^{-1}(2)}) \otimes \cdots \otimes (1 \otimes x_{\sigma^{-1}(n)}) \\ &= (1 \otimes x_{\sigma^{-1}(1)}) \otimes \cdots \otimes (t \otimes x_1) \otimes \cdots \otimes (1 \otimes x_{\sigma^{-1}(n)}) \\ &= \sigma\left((t \otimes x_1) \otimes (1 \otimes x_2) \otimes \cdots \otimes (1 \otimes x_n)\right) \\ &= \sigma\left(\phi(t \otimes x_1 \otimes \cdots \otimes x_n)\right) \end{aligned}$$

which shows ϕ is an isomorphism of G-modules. Therefore ϕ induces isomorphisms

$$\begin{aligned} T \otimes_Q T_R^n(M)f &\cong T_{T\otimes_Q R}^n(T \otimes_Q M)f, \text{ and} \\ T \otimes_Q (T_R^n(M)f)^G &\cong (T_R^n(T \otimes_Q M)f)^G. \end{aligned}$$

Part (3) is proved in Example 14.1.2. Parts (4), (5), (6) and (7) are left to the reader. (See Exercise 5.2.5.) □

1.1. Norms of Modules and Algebras. Let S be a commutative R-algebra and $\operatorname{Aut}_R(S)$ the group of all R-algebra automorphisms of S. Let M and N be two S-modules, and $\alpha \in \operatorname{Aut}_R(S)$. By ${}_\alpha N$ we denote the left S-module with action twisted by α (Definition 7.8.4). In the literature, a homomorphism h in $\operatorname{Hom}_S(M, {}_\alpha N)$ is also said to be *semi-linear relative to* α ([**Bou89a**, Chap. II, § 1, no. 13]), or α*-semilinear* ([**KO75**]). Then $h(x+y) = h(x) + h(y)$, and $h(sx) = \alpha(s)h(x)$, for all $x, y \in M$ and $s \in S$.

Example 14.1.4. Let R be a commutative ring and $S = R^n = R \oplus \cdots \oplus R$ the trivial separable extension of degree n (Section 4.6.3). Let $e_1, \ldots, e_n$ be the canonical orthogonal idempotents in S. As shown in Theorem 14.1.3, the corestriction of an S-module M is the R-module

$$\operatorname{Cor}_R^S(M) = Me_1 \otimes_R \cdots \otimes_R Me_n. \tag{14.3}$$

If $f : M \to N$ is a homomorphism of S-modules, then

$$Me_1 \otimes_R \cdots \otimes_R Me_n \xrightarrow{\operatorname{Cor}_R^S(f)} Ne_1 \otimes_R \cdots \otimes_R Ne_n \tag{14.4}$$

is induced by the assignments $xe_i \mapsto f(x)e_i$. Suppose Q and T are commutative rings such that T and S are Q-algebras. Then $T \otimes_Q S = T \otimes_Q R^n = (T \otimes_Q R)^n = \bigoplus_{i=1}^n (T \otimes_Q S)(1 \otimes e_i)$. Hence $T \otimes_Q M = \bigoplus_{i=1}^n T \otimes_Q Me_i$ and it follows that

$$T \otimes_Q \operatorname{Cor}_R^S(M) \cong \operatorname{Cor}_{T \otimes_Q R}^{T \otimes_Q S}(T \otimes_Q M). \tag{14.5}$$

If L is an R-module, then

$$\operatorname{Cor}_R^S(S \otimes_R L) = \operatorname{Cor}_R^S(L \oplus \cdots \oplus L) = L \otimes_R \cdots \otimes_R L = T_R^n(L). \tag{14.6}$$

In particular, $\operatorname{Cor}_R^S(S) = T_R^n(R) = \bigwedge^n(R^n) = R$ (see [**DF04**, Section 11.5], for example). If $s \in S$ and $\lambda_s : S \to S$ is "left multiplication by s", then

$$\operatorname{Cor}_R^S(\lambda_s) : R \to R \tag{14.7}$$

is "left multiplication by $\det(s)$" (Exercise 5.2.5). If M and N are S-modules, the reader should verify that

$$\operatorname{Cor}_R^S(M \otimes_S N) = \operatorname{Cor}_R^S(M) \otimes_R \operatorname{Cor}_R^S(N). \tag{14.8}$$

If T is a separable commutative S-algebra which is an S-module progenerator of constant rank, and M is a T-module the reader should verify that

$$\operatorname{Cor}_R^T(M) = \operatorname{Cor}_R^S(\operatorname{Cor}_S^T(N)). \tag{14.9}$$

By Proposition 4.6.11 we know that every commutative separable R-algebra which is a progenerator module of constant rank is a twisted form of the trivial extension of R by a faithfully flat R-algebra. In Theorem 14.1.5

below, we use faithfully flat descent to show that the corestriction functor of Theorem 14.1.3 is completely determined by the three properties (14.3), (14.4), (14.5). Faithfully flat descent can then be used to show that the corestriction functor also satisfies the properties (14.6), (14.7), (14.8), (14.9) for S in general.

Theorem 14.1.5. *Let R be a commutative ring. To every commutative separable R-algebra S which as a module is an R-progenerator of constant rank n, there is associated a covariant functor $\mathcal{N}_R^S : {}_S\mathfrak{M} \to {}_R\mathfrak{M}$ which satisfies up to natural isomorphisms the following properties. Up to natural isomorphism, $\mathcal{N}_R^S$ is determined by properties (1) and (2).*

(1) *If $S = R^n$ is the trivial extension, then $\mathcal{N}_R^S$ agrees with the corestriction functor Cor_R^S of Theorem 14.1.3.*
(2) *Suppose Q and T are commutative rings such that T and S are algebras over Q. Then $T \otimes_Q \mathcal{N}_R^S(M) \cong \mathcal{N}_{T\otimes_Q R}^{T\otimes_Q S}(T \otimes_Q M)$.*
(3) *If $L \in {}_R\mathfrak{M}$, then $\mathcal{N}_R^S(S \otimes_R L) \cong T_R^n(L) = L \otimes_R \cdots \otimes_R L$.*
(4) *If $M, N \in {}_S\mathfrak{M}$, then $\mathcal{N}_R^S(M \otimes_S N) \cong \mathcal{N}_R^S(M) \otimes_R \mathcal{N}_R^S(N)$.*
(5) *If T is another commutative separable R-algebra which as a module is an R-progenerator of constant rank, $M \in {}_S\mathfrak{M}$, and $N \in {}_T\mathfrak{M}$, then $\mathcal{N}_R^{S\oplus T}(M \oplus N) \cong \mathcal{N}_R^S(M) \otimes_R \mathcal{N}_R^T(N)$.*
(6) *If T is a commutative separable S-algebra which as a module is an S-progenerator of constant rank, and $M \in {}_T\mathfrak{M}$, then*
$$\mathcal{N}_R^T(M) \cong \mathcal{N}_R^S\left(\mathcal{N}_S^T(M)\right).$$
(7) *If $s \in S$ and $\ell_s : S \to S$ is the "left multiplication by s" homomorphism, then $\mathcal{N}_R^S(\ell_s) : R \to R$ is the "left multiplication by $\det(s)$" homomorphism.*

Proof. Step 1: Define $\mathcal{N}_R^S(M)$, if M is an S-module. By Proposition 4.6.11, S is a twisted form of the trivial extension R^n. In other words, there is a faithfully flat R-algebra T and an isomorphism $\beta : S \otimes_R T \to R^n \otimes_R T = T^n$ of T-algebras. Let M be an S-module. Then $M \otimes_R T$ is an $S \otimes_R T$-module, and by the homomorphism β^{-1}, we make $M \otimes_R T$ into a T^n-module. There is a T^n-module decomposition into an internal direct sum $M \otimes_R T = M_1 \oplus \cdots \oplus M_n$. By (1), $\mathcal{N}_T^{T^n}(M \otimes_R T) = M_1 \otimes_T \cdots \otimes_T M_n$. We now define a descent datum from T to R on $M \otimes_R T = M_1 \otimes_T \cdots \otimes_T M_n$. In a departure from the notation of Proposition 5.3.4, we write β_i instead of $\mathfrak{F}_i(\beta)$. Define θ to be $\beta_1 \beta_0^{-1}$. So θ is the $T \otimes_R T$-algebra automorphism of $R^n \otimes_R T \otimes_R T$

which makes the diagram

(14.10)
$$\begin{array}{ccc} & & R^n \otimes_R T \otimes_R T \\ & \overset{\beta_0}{\nearrow} & \\ S \otimes_R T \otimes_R T & & \downarrow \theta \\ & \underset{\beta_1}{\searrow} & \\ & & R^n \otimes_R T \otimes_R T \end{array}$$

commute. As in the proof of Proposition 5.5.7, $\theta_2\theta_0 = \theta_1$. The corresponding diagram for $\gamma : M \otimes_R T \cong M_1 \oplus \cdots \oplus M_n$ is

(14.11)
$$\begin{array}{ccc} & & M \otimes_R T \otimes_R T \cong (M_1 \otimes_R T) \oplus \cdots \oplus (M_n \otimes_R T) \\ & \overset{\gamma_0}{\nearrow} & \\ M \otimes_R T \otimes_R T & & \downarrow \alpha \\ & \underset{\gamma_1}{\searrow} & \\ & & M \otimes_R T \otimes_R T \cong (T \otimes_R M_1) \oplus \cdots \oplus (T \otimes_R M_n). \end{array}$$

As above, $\alpha_2\alpha_0 = \alpha_1$. We have come to what is probably the only part of this proof that is not straightforward. In (14.11), α is not a $(T \otimes_R T)^n$-module homomorphism. However, α is a θ-semilinear automorphism of $M \otimes_R T \otimes_R T$. That is,

$$M \otimes_R T \otimes_R T \xrightarrow{\alpha} {}_\theta(M \otimes_R T \otimes_R T)$$

is an isomorphism of $(T \otimes_R T)^n$-modules, where the module on the right hand side is twisted by θ. By (1), the functor $\mathcal{N}_{T\otimes_R T}^{(T\otimes_R T)^n}(\,)$ agrees with the corestriction $\operatorname{Cor}_{T\otimes_R T}^{(T\otimes_R T)^n}(\,)$. By Theorem 14.1.3 (3),

$$\begin{aligned} \operatorname{Cor}_{T\otimes_R T}^{(T\otimes_R T)^n} & \big((M_1 \otimes_R T) \oplus \cdots \oplus (M_n \otimes_R T)\big) \\ &= (M_1 \otimes_R T) \otimes_{T\otimes_R T} \cdots \otimes_{T\otimes_R T} (M_n \otimes_R T). \end{aligned}$$

Applying this to α, we have the $T \otimes_R T$-module isomorphism

(14.12)
$$\begin{aligned} (M_1 \otimes_T \cdots \otimes_T M_n) \otimes_R T &= (M_1 \otimes_R T) \otimes_{T\otimes_R T} \cdots \otimes_{T\otimes_R T} (M_n \otimes_R T) \\ &\xrightarrow{\mathcal{N}(\alpha)} (T \otimes_R M_1) \otimes_{T\otimes_R T} \cdots \otimes_R (T \otimes_{T\otimes_R T} M_n) \\ &= T \otimes_R (M_1 \otimes_T \cdots \otimes_T M_n). \end{aligned}$$

By (1) and Theorem 14.1.3, $\mathcal{N}(\alpha)_i = \mathcal{N}(\alpha_i)$, for $i = 0, 1, 2$. From the commutative square

$$\begin{array}{ccc} M \otimes_R T^3_R(T) & \xrightarrow{\alpha_0} & \theta_0 \left(M \otimes_R T^3_R(T)\right) \\ {\scriptstyle \alpha_1}\downarrow & & \downarrow{\scriptstyle \alpha_2} \\ \theta_1 \left(M \otimes_R T^3_R(T)\right) & \xrightarrow{=} & \theta_2\theta_0 \left(M \otimes_R T^3_R(T)\right) \end{array}$$

we see that $\mathcal{N}(\alpha)_1 = \mathcal{N}(\alpha)_2\mathcal{N}(\alpha)_0$, hence $\mathcal{N}(\alpha)$ is a descent datum. By Faithfully Flat Descent (Theorem 5.3.7), $M_1 \otimes_T \cdots \otimes_T M_n$ comes from an R-module, which we denote $\mathcal{N}^S_R(M)$. There is an isomorphism of T-modules

$$\nu : \mathcal{N}^S_R(M) \otimes_R T \xrightarrow{\cong} M_1 \otimes_T \cdots \otimes_T M_n. \tag{14.13}$$

Step 2: If $\phi : M \to N$ is an S-module homomorphism, we show that there is a uniquely defined R-module homomorphism $\mathcal{N}^S_R(\phi) : \mathcal{N}^S_R(M) \to \mathcal{N}^S_R(N)$. Retain the notation derived above for T and M. Substituting N into (14.11), let $\beta : N \otimes_R T \otimes_R T \to N \otimes_R T \otimes_R T$ be the counterpart of α. The diagram

$$\begin{array}{ccc} M \otimes_R T \otimes_R T & \xrightarrow{\alpha} & \theta\,(M \otimes_R T \otimes_R T) \\ {\scriptstyle \phi\otimes 1\otimes 1}\downarrow & & \downarrow{\scriptstyle \phi\otimes 1\otimes 1} \\ N \otimes_R T \otimes_R T & \xrightarrow{\beta} & \theta\,(N \otimes_R T \otimes_R T) \end{array}$$

commutes. The norm map (14.12) and its counterpart for N yield the commutative diagram

$$\begin{array}{ccc} \mathcal{N}(M \otimes_R T) \otimes_R T & \xrightarrow{\mathcal{N}(\alpha)} & T \otimes_R \mathcal{N}(M \otimes_R T) \\ {\scriptstyle \mathcal{N}(\phi\otimes 1)\otimes 1}\downarrow & & \downarrow{\scriptstyle 1\otimes\mathcal{N}(\phi\otimes 1)} \\ \mathcal{N}(N \otimes_R T) \otimes_R T & \xrightarrow{\mathcal{N}(\beta)} & T \otimes_R \mathcal{N}(N \otimes_R T). \end{array}$$

The isomorphism (14.13) and its counterpart for N yield the commutative diagram

$$\begin{array}{ccc} \mathcal{N}(M) \otimes_R T \otimes_R T & \longrightarrow & T \otimes_R \mathcal{N}(M) \otimes_R T \\ {\scriptstyle \mathcal{N}(\phi\otimes 1)\otimes 1}\downarrow & & \downarrow{\scriptstyle 1\otimes\mathcal{N}(\phi\otimes 1)} \\ \mathcal{N}(N) \otimes_R T \otimes_R T & \longrightarrow & T \otimes_R \mathcal{N}(N) \otimes_R T \end{array}$$

where the horizontal maps are the switches $a \otimes b \otimes c \mapsto b \otimes a \otimes c$. In the notation of Proposition 5.3.4, $\mathfrak{F}_0(\mathcal{N}(\phi \otimes 1)) = \mathfrak{F}_1(\mathcal{N}(\phi \otimes 1))$. By Proposition 5.3.4, $\mathcal{N}(\phi \otimes 1)$ comes from a unique R-module homomorphism $\mathcal{N}^S_R(\phi) : \mathcal{N}^S_R(M) \to \mathcal{N}^S_R(N)$.

Step 3: Show that $\mathcal{N}^S_R(M)$ does not depend on the isomorphism $\beta : S \otimes_R T \to R^n \otimes_R T$. As in Step 1, using β, the T^n-module $M \otimes_R T$

decomposes into $M_1 \oplus \cdots \oplus M_n$. Assume $\beta' : S \otimes_R T \to R^n \otimes_R T$ is another T-algebra isomorphism. Using β', the T^n-module $M \otimes_R T$ decomposes into $M_1' \oplus \cdots \oplus M_n'$. Let $\gamma = \beta'\beta^{-1}$. The identity map induces a T^n-module isomorphism $\sigma : M_1 \oplus \cdots \oplus M_n \to {}_\gamma(M_1' \oplus \cdots \oplus M_n')$. Let α be as in (14.11) and let α' be the counterpart for β'. The diagram

$$\begin{array}{ccc}
(M_1 \otimes_T \cdots \otimes_T M_n) \otimes_R T & \xrightarrow{\mathcal{N}(\sigma)\otimes 1} & (M_1' \otimes_T \cdots \otimes_T M_n') \otimes_R T \\
\downarrow{\scriptstyle \mathcal{N}(\alpha)} & & \downarrow{\scriptstyle \mathcal{N}(\alpha')} \\
T \otimes_R (M_1 \otimes_T \cdots \otimes_T M_n) & \xrightarrow{1\otimes\mathcal{N}(\sigma)} & T \otimes_R (M_1' \otimes_T \cdots \otimes_R M_n')
\end{array}$$

commutes. From Step 1 and Theorem 5.3.7,

$$\mathcal{N}_R^S(M) = \{x \in M_1 \otimes_T \cdots \otimes_T M_n \mid \mathcal{N}(\alpha)(x \otimes 1) = 1 \otimes x\}.$$

If $\mathcal{N}'(M)$ denotes the R-module defined by the descent datum $\mathcal{N}(\alpha')$, then

$$\mathcal{N}'(M) = \{x \in M_1' \otimes_T \cdots \otimes_T M_n' \mid \mathcal{N}(\alpha')(x \otimes 1) = 1 \otimes x\}$$

and $\mathcal{N}(\sigma) : \mathcal{N}_R^S(M) \to \mathcal{N}'(M)$ is the desired isomorphism.

Step 4: Show that $\mathcal{N}_R^S(M)$ does not depend on the choice of T. Let T and T' be two faithfully flat R-algebras. As in Step 1, let $\beta : S \otimes_R T \to R^n \otimes_R T$ be a T-algebra isomorphism and $\beta' : S \otimes_R T' \to R^n \otimes_R T'$ a T'-algebra isomorphism. Let $M_1, \ldots, M_n$ be the T-modules derived in Step 1 using β, and

$$\nu : \mathcal{N}_R^S(M) \otimes_R T \to M_1 \otimes_T \cdots \otimes_T M_n$$

the isomorphism of (14.13).

Step 4.1: Assume T is a subalgebra of T', and $\beta' = \beta \otimes 1$. Then $M \otimes_R T' = M \otimes_R T \otimes_R T'$ decomposes into $M_1 \otimes_T T' \oplus \cdots M_n \otimes_T T'$. Let $\mathcal{N}'(M)$ denote the R-module derived by applying Step 1 using T' and β'. Since

$$\begin{aligned}
(M_1 \otimes_T T') \otimes_{T'} \cdots \otimes_{T'} (M_n \otimes_T T') &= M_1 \otimes_T \cdots \otimes_T M_n \otimes_T T' \\
&= \mathcal{N}_R^S(M) \otimes_R T \otimes_T T' \\
&= \mathcal{N}_R^S(M) \otimes_R T'
\end{aligned}$$

it follows from Step 1 and Theorem 5.3.7 that $\mathcal{N}'(M) = \mathcal{N}_R^S(M)$.

Step 4.2: The general case. Applying Step 4.1 twice (first with T and $T \otimes_R T'$, then with T' and $T \otimes_R T'$) shows that $\mathcal{N}_R^S(M)$ does not depend on the choice of the splitting ring T.

Step 5: Conclusion. Steps 1 through 4 show that $\mathcal{N}_R^S$ defines a functor which satisfies (1). By Step 1, the descent datum defining $\mathcal{N}_R^S(M)$ is completely determined by (1) and (2). Because $\mathcal{N}_R^S$ is independent of the choice of the splitting for S, we see that $\mathcal{N}_R^S$ commutes with change of base, so (2) holds in general. The same argument can be used to prove (5). The reader should verify (3), (4), (6) and (7) when S is the trivial extension (Example 14.1.4). Using this and faithfully flat descent, we are done. $\square$

Corollary 14.1.6. *Let R be a commutative ring and S a commutative separable R-algebra which as a module is an R-progenerator of constant rank n. The functor $\mathrm{Cor}_R^S : {}_S\mathfrak{M} \to {}_R\mathfrak{M}$ defined in Definition 14.1.1 is naturally isomorphic to the functor $\mathcal{N}_R^S$ of Theorem 14.1.5.*

1.2. Applications of Corestriction.

Lemma 14.1.7. *Let R be a commutative ring and S a commutative separable R-algebra which as a module is an R-progenerator of constant rank n.*

(1) *If A is a separable S-algebra, then $\mathrm{Cor}_R^S(A)$ is a separable R-algebra.*
(2) *If M is a faithfully flat S-module, then $\mathrm{Cor}_R^S(M)$ is a faithfully flat R-module.*
(3) *If A is a commutative finitely generated S-algebra, then $\mathrm{Cor}_R^S(A)$ is a finitely generated R-algebra.*
(4) *If R is noetherian and A is a faithfully flat étale S-algebra, then $\mathrm{Cor}_R^S(A)$ is a faithfully flat étale R-algebra.*

Proof. (1): By Corollary 4.3.2, $T_R^n(A)f$ is a separable algebra over $T_R^n(S)f$. Since $T_R^n(S)f$ is Galois over R, it is faithfully flat (Lemma 12.2.2). Since

$$\mathrm{Cor}_R^S(A) \otimes_R T_R^n(S)f = T_R^n(A)f$$

by Theorem 8.1.20, $\mathrm{Cor}_R^S(A)$ is separable over R.

(2): Using Exercise 2.4.14, the reader should verify that $T_R^n(M)f$ is faithfully flat over $T_R^n(S)f$. Since $\mathrm{Cor}_R^S(M) \otimes_R T_R^n(S)f = T_R^n(M)f$, $\mathrm{Cor}_S^R(M)$ is faithfully flat over R (Theorem 2.6.1 (5)).

Part (3) is left to the reader.

(4): Since S is a finitely generated R-module, S is noetherian, by the Hilbert Basis Theorem. The rest follows from (1), (2), and (3). $\square$

Theorem 14.1.8. *Let R be a commutative ring and S a commutative separable R-algebra which as a module is an R-progenerator of constant rank n.*

(1) *If P is a rank one projective S-module, then $\mathrm{Cor}_R^S(P)$ is a rank one projective R-module.*
(2) *The assignment $P \mapsto \mathrm{Cor}_R^S(P)$ induces a homomorphism of abelian groups $\mathrm{Cor}_R^S : \mathrm{Pic}(S) \to \mathrm{Pic}(R)$.*

(3) *The composite map*

$$\mathrm{Pic}(R) \xrightarrow{\mathrm{Res}_R^S} \mathrm{Pic}(S) \xrightarrow{\mathrm{Cor}_R^S} \mathrm{Pic}(R)$$

is the n*th power map, where* Res_R^S *is the natural map induced by* $R \to S$.

(4) *If* $\theta : R \to R'$ *is a homomorphism of commutative rings, then the diagram of abelian groups*

$$\begin{array}{ccc} \mathrm{Pic}(S) & \xrightarrow{\mathrm{Cor}_R^S} & \mathrm{Pic}(R) \\ {\scriptstyle \mathrm{Pic}(1\otimes\theta)}\downarrow & & \downarrow{\scriptstyle \mathrm{Pic}(\theta)} \\ \mathrm{Pic}(S \otimes_R R') & \xrightarrow{\mathrm{Cor}_{R'}^{S\otimes_R R'}} & \mathrm{Pic}(R') \end{array}$$

commutes.

Proof. Let M and N be invertible S modules. By Lemma 2.6.7, $S \cong M \otimes_S M^*$. By Theorem 14.1.5, $R = \mathrm{Cor}_R^S(M \otimes_S M^*) = \mathrm{Cor}_R^S(M) \otimes_R \mathrm{Cor}_R^S(M^*)$. Again by Lemma 2.6.7, $\mathrm{Cor}_R^S(M)$ is an invertible R-module. The rest follows from Theorem 14.1.5. □

Theorem 14.1.9. *Let* R *be a commutative ring and* S *a commutative separable* R*-algebra which as a module is an* R*-progenerator of constant rank* n.

(1) *If* A *is an Azumaya* S*-algebra, then* $\mathrm{Cor}_R^S(A)$ *is an Azumaya* R*-algebra.*

(2) *The assignment* $A \mapsto \mathrm{Cor}_R^S(A)$ *induces a homomorphism of abelian groups*

$$\mathrm{Cor}_R^S : \mathrm{B}(S) \to \mathrm{B}(R).$$

(3) *The composite map*

$$\mathrm{B}(R) \xrightarrow{\mathrm{Res}_R^S} \mathrm{B}(S) \xrightarrow{\mathrm{Cor}_R^S} \mathrm{B}(R)$$

is the n*th power map, where* Res_R^S *is the natural map induced by* $R \to S$.

(4) *If* $\theta : R \to R'$ *is a homomorphism of commutative rings, then the diagram of abelian groups*

$$\begin{array}{ccc} \mathrm{B}(S) & \xrightarrow{\mathrm{Cor}_R^S} & \mathrm{B}(R) \\ {\scriptstyle \mathrm{B}(1\otimes\theta)}\downarrow & & \downarrow{\scriptstyle \mathrm{B}(\theta)} \\ \mathrm{B}(S \otimes_R R') & \xrightarrow{\mathrm{Cor}_{R'}^{S\otimes_R R'}} & \mathrm{B}(R') \end{array}$$

commutes.

Proof. (1): Let $A_i = S \otimes_R \cdots \otimes_R A \otimes_R \cdots \otimes_R S$ be the $T^n_R(S)$-subalgebra of $T^n_R(A)$ where the ith tensor factor is A and the other factors are copies of S. By Corollary 7.1.9, $A \otimes_R S = A \otimes_S (S \otimes_R S)$ is an Azumaya $S \otimes_R S$-algebra and by a recursive argument we see that A_i is an Azumaya $T^n_R(S)$-algebra. If $i \neq j$, then elements of A_i commute with elements of A_j. By Lemma 7.1.1, $T^n_R(S) \cdot 1$ is a $T^n_R(S)$-module direct summand of $T^n_R(A)$. Using the proof of Theorem 7.6.1, it follows that the $T^n_R(S)$-algebra homomorphism

$$A_1 \otimes_{T^n_R(S)} \cdots \otimes_{T^n_R(S)} A_n \to T^n_R(A)$$

is an isomorphism. In particular, $T^n_R(A)$ is an Azumaya $T^n_R(S)$-algebra and by Lemma 7.1.1, $T^n_R(A)f$ is $T^n_R(S)f$-Azumaya. We have

$$\operatorname{Cor}^S_R(A) \otimes_R T^n_R(S)f = T^n_R(A)f.$$

Since $T^n_R(S)f$ is Galois over R, it is faithfully flat (Lemma 12.2.2). By Corollary 8.1.21, $\operatorname{Cor}^S_R(A)$ is R-Azumaya.

(2): Let P be an S-progenerator module. By Theorem 1.3.26,

$$T^n_R\left(\operatorname{Hom}_S(P, P)\right) = \operatorname{Hom}_{T^n_R(S)}(T^n_R(P), T^n_R(P)).$$

Using this and Corollary 1.3.27, there are natural isomorphisms

$$\begin{aligned}\operatorname{Hom}_R(\operatorname{Cor}^S_R(P), \operatorname{Cor}^S_R(P)) \otimes_R T^n_R(S)f &= \operatorname{Hom}_{T^n_R(S)f}(T^n_R(P)f, T^n_R(P)f)\\ &= \operatorname{Hom}_{T^n_R(S)}(T^n_R(P), T^n_R(P)) \otimes_S Sf\\ &= T^n_R\left(\operatorname{Hom}_S(P, P)\right) \otimes_S Sf.\end{aligned}$$

Proposition 12.4.2 implies $\operatorname{Cor}^S_R(\operatorname{Hom}_S(P, P)) = \operatorname{Hom}_R(\operatorname{Cor}^S_R(P), \operatorname{Cor}^S_R(P))$. Part (2) follows from this, (1) and Theorem 14.1.5 (4).

Parts (3) and (4) follow from (2) and Theorem 14.1.5. □

Remark 14.1.10. For any abelian group M, let ${}_2M$ denote the kernel of the map $M \to M$ defined by $x \mapsto x^2$. By Theorem 14.1.9 (3), there is a complex

$${}_2\mathrm{B}(R) \xrightarrow{\operatorname{Res}^S_R} {}_2\mathrm{B}(S) \xrightarrow{\operatorname{Cor}^S_R} {}_2\mathrm{B}(R)$$

of abelian groups. Knus, Parimala and Srinivas (see [**KPS90**]) show that in general the homology of this complex is nontrivial. If R and S are fields and the characteristic of R is not equal to 2, then the exactness of this complex is proved in [**Ara75**, Corollary 4.6].

1.3. Corestriction and Galois Descent. Let S be a commutative ring and G finite group of automorphisms of S. When convenient, we will assume $\{\sigma_1, \ldots, \sigma_n\}$ is an enumeration of G. Let R be a subring of S^G. Recall from Example 12.2.6 that $\nabla(S, G)$, the set of all functions $v : G \to S$, is isomorphic to the trivial extension of S of rank n. In fact,

$$\nabla(S, G) = \bigoplus_{\sigma \in G} S e_\sigma \tag{14.14}$$

where $e_{\sigma_i}(\sigma_j) = \delta_{i,j}$ is the Kronecker delta function. By Example 12.4.1, G acts on $\nabla(S, G)$ by the conjugation formula $\sigma \cdot v = \sigma v \sigma^{-1}$. For $\sigma, \tau \in G$, we have $\tau e_\sigma = e_{\tau\sigma}$. There is a commutative diagram of R-algebras

$$\begin{array}{ccccc} & & \nabla(G,S) & & \\ & \overset{\theta_1}{\nearrow} & \uparrow \ell & \overset{\theta_2}{\nwarrow} & \\ S & \xrightarrow[\rho_1]{} & S \otimes_R S & \xleftarrow[\rho_2]{} & S \end{array} \tag{14.15}$$

where ρ_1 and ρ_2 are the R-algebra homomorphisms of Exercise 1.3.8. The other three homomorphisms are given by

$$\begin{aligned} \theta_1(a) &= \sum_{\sigma \in G} a e_\sigma \\ \theta_2(b) &= \sum_{\sigma \in G} \sigma(b) e_\sigma \\ \ell(a \otimes b) &= \sum_{\sigma \in G} a\sigma(b) e_\sigma \end{aligned}$$

and the image of θ_2 is equal to the fixed subring $\nabla(S, G)^G$. For $s \in S$, we have

$$\begin{aligned} \theta_1(s) e_\sigma &= s e_\sigma \\ \theta_2(s) e_\sigma &= \sigma(s) e_\sigma. \end{aligned} \tag{14.16}$$

In Definition 7.8.4 the free left S-module of rank one with action twisted by σ was denoted by ${}_\sigma S_\sigma$. Throughout this section, all left and right module structures agree, hence for notational simplicity we will denote the twisted module by ${}_\sigma S$. From (14.16), we see that using the θ_2-structure, the S-module $S e_\sigma$ is isomorphic to ${}_\sigma S$.

Assume from now on that S is Galois over R with group G. Then ℓ is an isomorphism (Theorem 12.2.9). Viewing $S \otimes_R S$ and $\nabla(S, G)$ as isomorphic S-algebras using the right hand side of (14.15), we have

$$S \otimes_R S \cong \bigoplus_{\sigma \in G} {}_\sigma S. \tag{14.17}$$

If we tensor an S module M with $\rho_2 : S \to S \otimes_R S$, then

$$M \otimes_S (S \otimes_R S) = \bigoplus_{\sigma \in G} M e_\sigma \cong \bigoplus_{\sigma \in G} {}_\sigma M.$$

By Example 14.1.4,

$$\operatorname{Cor}_S^{S\otimes_R S}(M\otimes_S (S\otimes_R S)) = {}_{\sigma_1}M \otimes_S \cdots \otimes_S {}_{\sigma_n}M.$$

Now we show how to make ${}_{\sigma_1}M \otimes_S \cdots \otimes_S {}_{\sigma_n}M$ into a left $\Delta(S/R, G, 1)$-module. If $s \in S$, then left multiplication by s on a basic element of the tensor product satisfies the rules

$$\begin{aligned} s(x_{\sigma_1} \otimes \cdots \otimes x_{\sigma_n}) &= \sigma_1(s)x_{\sigma_1} \otimes \cdots \otimes x_{\sigma_n} \\ &= x_{\sigma_1} \otimes \cdots \otimes \sigma_i(s)x_{\sigma_i} \otimes \cdots \otimes x_{\sigma_n} \\ &= x_{\sigma_1} \otimes \cdots \otimes \sigma_n(s)x_{\sigma_n} \end{aligned}$$

for all $1 \le i \le n$. Let ρ, σ, τ denote elements of G. Let

$$\bar{\sigma} : {}_\rho M \to {}_{\rho\sigma^{-1}}M$$

be the identity mapping $x \mapsto x$. Then $\bar{\sigma}(s\cdot x) = \rho(s)x = \sigma(s)\cdot x = \sigma(s)\cdot\bar{\sigma}(x)$. Hence $\bar{\sigma}$ is a σ-semilinear homomorphism of S-modules. The reader should verify that

$$\bar{\tau}\bar{\sigma} = \overline{\tau\sigma}. \tag{14.18}$$

The map

$$\bar{\sigma} : {}_{\sigma_1}M \otimes_S \cdots \otimes_S {}_{\sigma_n}M \to {}_{\sigma_1}M \otimes_S \cdots \otimes_S {}_{\sigma_n}M \tag{14.19}$$

is defined by tensoring the individual maps $\bar{\sigma} : {}_{\sigma_i}M \to {}_{\sigma_i\sigma^{-1}}M$, and sorting the right hand side into the original order. On a basic tensor element, (14.19) is defined by

$$\bar{\sigma}(x_{\sigma_1} \otimes \cdots \otimes x_{\sigma_n}) = x_{\sigma_1\sigma} \otimes \cdots \otimes x_{\sigma_n\sigma}. \tag{14.20}$$

Thus G acts as a group on the set ${}_{\sigma_1}M \otimes_S \cdots \otimes_S {}_{\sigma_n}M$. Because each $\bar{\sigma}$ is semilinear, the action by G is semilinear, nevertheless we check this directly. Without loss of generality assume $\sigma_1 = 1$. If $s \in S$, then

$$\begin{aligned} \bar{\sigma}(s(x_{\sigma_1} \otimes \cdots \otimes x_{\sigma_n})) &= \bar{\sigma}(sx_{\sigma_1} \otimes \cdots \otimes x_{\sigma_n}) \\ &= \bar{\sigma}(x_{\sigma_1} \otimes \cdots \otimes \sigma(s)x_{\sigma} \otimes \cdots \otimes x_{\sigma_n}) \\ &= \sigma(s)x_{\sigma_1\sigma} \otimes x_{\sigma_2\sigma} \otimes \cdots \otimes x_{\sigma_n\sigma} \\ &= \sigma(s)\bar{\sigma}(x_{\sigma_1} \otimes \cdots \otimes x_{\sigma_n}). \end{aligned}$$

By Section 12.4, this shows ${}_{\sigma_1}M \otimes_S \cdots \otimes_S {}_{\sigma_n}M$ is a left $\Delta(S/R, G, 1)$-module. Let $\mathcal{N}(M) = ({}_{\sigma_1}M \otimes_S \cdots \otimes_S {}_{\sigma_n}M)^G$. By Proposition 12.4.2, $\mathcal{N}(M)$ is the unique R-module such that there is an isomorphism

$$\mathcal{N}(M) \otimes_R S \cong {}_{\sigma_1}M \otimes_S \cdots \otimes_S {}_{\sigma_n}M \tag{14.21}$$

of left $\Delta(S/R, G, 1)$-modules.

Lemma 14.1.11. *Let S/R be a Galois extension of commutative rings with finite group $G = \{\sigma_1, \dots, \sigma_n\}$. The assignment*

$$M \mapsto \mathcal{N}(M) = ({}_{\sigma_1}M \otimes_S \cdots \otimes_S {}_{\sigma_n}M)^G$$

induces a covariant functor ${}_S\mathfrak{M} \to {}_R\mathfrak{M}$. If $F = S^I$ is the free S-module on the index set I, then $\mathcal{N}(F)$ is the free R-module $T^n_R(R^I)$. In particular, $\mathcal{N}(S) = R$.

Proof. The proof is left to the reader. □

Theorem 14.1.12. *Let S/R be a Galois extension of commutative rings with finite group $G = \{\sigma_1, \dots, \sigma_n\}$. For any S-module M, the R-modules $\operatorname{Cor}^S_R(M)$ and $\mathcal{N}(M) = ({}_{\sigma_1}M \otimes_S \cdots \otimes_S {}_{\sigma_n}M)^G$ are isomorphic.*

Proof. For this proof we will denote the symmetric group on n letters by Σ_n. By Theorem 12.6.1, the ring $T^n_R(S)f$ is Galois over R with group Σ_n. The group Σ_n acts on $T^n_R(M)f$ and this makes $T^n_R(M)f$ into a left $\Delta(T^n_R(S)f/R, \Sigma_n, 1)$-module. By Definition 14.1.1, we have $\operatorname{Cor}^S_R(M) = (T^n_R(M)f)^{\Sigma_n}$. Let $H_i \cong \Sigma_{n-1}$ be the subgroup of Σ_n which fixes i. Let $S_i = (T^n_R(S)f)^{H_i}$. From (14.13) and (14.21),

$$\operatorname{Cor}^S_R(M) \otimes_R S_i \cong {}_{\sigma_1}M \otimes_{S_i} \cdots \otimes_{S_i} {}_{\sigma_n}M \cong \mathcal{N}(M) \otimes_R S_i \tag{14.22}$$

as S_i-modules. By Proposition 12.4.2,

$$(T^n_R(M)f)^{H_i} = \operatorname{Cor}^S_R(M) \otimes_R S_i$$

and

$$\operatorname{Cor}^S_R(M) \otimes_R T^n_R(S)f = T^n_R(M)f.$$

Using this and (14.22), we identify $\mathcal{N}(M) \otimes_R 1$ with a subset of

$$\bigcap_{i=1}^{n} (T^n_R(M)f)^{H_i} = (T^n_R(M)f)^{\Sigma_n} = \operatorname{Cor}^S_R(M).$$

But S is faithfully flat over R, and by (14.22), $\mathcal{N}(M) = \operatorname{Cor}^S_R(M)$. □

1.4. Corestriction and Amitsur Cohomology. Let R be a commutative ring and S a commutative separable R-algebra which as a module is an R-progenerator of constant rank n. Let $G = \Sigma_n$ be the symmetric group on n-letters. Let A be an S-algebra and $\lambda : A \to \operatorname{Hom}_S(A, A)$ the left regular representation. If $a \in A$, then from the proof of Theorem 14.1.3 we see that the diagram

$$\begin{array}{ccc} \operatorname{Cor}_R^S(A) & \xrightarrow{\operatorname{Cor}_R^S(\lambda_a)} & \operatorname{Cor}_R^S(A) \\ \downarrow \subseteq & & \downarrow \subseteq \\ T_R^n(A)f & \xrightarrow{\lambda_{(a\otimes\cdots\otimes a)f}} & T_R^n(A)f \end{array}$$

commutes. Because $(a \otimes \cdots \otimes a)f$ is fixed by G, $\lambda_{(a\otimes\cdots\otimes a)f}$ represents an element in the image of the left regular representation map $\operatorname{Cor}_R^S(A) \to \operatorname{Hom}_R(\operatorname{Cor}_R^S(A), \operatorname{Cor}_R^S(A))$. Since Cor_R^S is a functor and $\lambda_{ab} = \lambda_a\lambda_b$, we have

$$\operatorname{Cor}_R^S(\lambda_{ab}) = \operatorname{Cor}_R^S(\lambda_a)\operatorname{Cor}_R^S(\lambda_b).$$

If we identify $\operatorname{Cor}_R^S(A)$ with its image in $\operatorname{Hom}_R(\operatorname{Cor}_R^S(A), \operatorname{Cor}_R^S(A))$, then the assignment $u \mapsto \operatorname{Cor}_R^S(\lambda_u)$ induces a homomorphism on the groups of units

$$\operatorname{Cor}_R^S : A^* \to \operatorname{Cor}_R^S(A)^*.$$

Now assume A is a commutative S-algebra. We denote by $\mathbb{G}_m$ the group of units functor. Then $\mathbb{G}_m(A) = A^*$. As in Section 5.3.1, for $0 \le j \le n+1$, there is an S-algebra homomorphism

$$A^{\otimes(n+1)} \xrightarrow{e_j} A^{\otimes(n+2)}$$

$$(x_0 \otimes \cdots \otimes x_n) \mapsto x_0 \otimes \cdots \otimes x_{j-1} \otimes 1 \otimes x_j \otimes \cdots \otimes x_n.$$

The Amitsur complex for A/S with coefficients in $\mathbb{G}_m$ is

$$1 \to A^* \xrightarrow{d^0} (A^{\otimes 2})^* \xrightarrow{d^1} (A^{\otimes 3})^* \xrightarrow{d^2} \cdots$$

where the coboundary map $d^r : (A^{\otimes(r+1)})^* \to (A^{\otimes(r+2)})^*$ is defined to be

$$d^r = \prod_{j=0}^{r+1} e_j^{(-1)^j}.$$

Denote this complex of abelian groups by $C^\bullet(A/S, \mathbb{G}_m)$. By the proof of Theorem 14.1.3, the maps e_j commute with the corestriction functor. It follows from Theorem 14.1.5 (4) that

$$\operatorname{Cor}_R^S(T_S^n(A)) = T_R^n(\operatorname{Cor}_R^S(A)).$$

Therefore, the coboundary maps commute with the corestriction on groups of units, and there is a morphism of complexes

$$\begin{array}{ccc} (A^{\otimes(r+1)})^* & \xrightarrow{d^r} & (A^{\otimes(r+2)})^* \\ \downarrow \operatorname{Cor}_R^S & & \downarrow \operatorname{Cor}_R^S \\ \left(\operatorname{Cor}_R^S(A)^{\otimes(r+1)}\right)^* & \xrightarrow{d^r} & \left(\operatorname{Cor}_R^S(A)^{\otimes(r+2)}\right)^* \end{array} \tag{14.23}$$

and a group homomorphism

$$\operatorname{Cor}_R^S : \mathrm{H}^r(A/S, \mathbb{G}_m) \to \mathrm{H}^r(\operatorname{Cor}_R^S(A)/R, \mathbb{G}_m)$$

for every $r \geq 0$. If $\theta : A \to A'$ is a homomorphism of commutative S-algebras, then

$$\theta^* : \mathrm{H}^r(A/S, \mathbb{G}_m) \to \mathrm{H}^r(A'/S, \mathbb{G}_m) \text{ and}$$
$$\theta^* : \mathrm{H}^r(\operatorname{Cor}_R^S(A)/R, \mathbb{G}_m) \to \mathrm{H}^r(\operatorname{Cor}_R^S(A')/R, \mathbb{G}_m)$$

are homomorphisms of abelian groups (Theorem 5.5.5). The reader should verify that the diagram (14.23) and its counterpart for A'/S induce a commutative diagram

$$\begin{array}{ccc} \mathrm{H}^r(A/S, \mathbb{G}_m) & \xrightarrow{\operatorname{Cor}_R^S} & \mathrm{H}^r(\operatorname{Cor}_R^S(A)/R, \mathbb{G}_m) \\ \downarrow{\scriptstyle \theta^*} & & \downarrow{\scriptstyle \theta^*} \\ \mathrm{H}^r(A'/S, \mathbb{G}_m) & \xrightarrow{\operatorname{Cor}_R^S} & \mathrm{H}^r(\operatorname{Cor}_R^S(A')/R, \mathbb{G}_m) \end{array} \tag{14.24}$$

for every $r \geq 0$.

Theorem 14.1.13. *Let R be a commutative noetherian ring and S a commutative separable R-algebra which is an R-module progenerator of constant rank n.*

(1) *The corestriction functor induces a homomorphism on the Čech cohomology groups*

$$\operatorname{Cor}_R^S : \check{\mathrm{H}}^r_{\mathrm{et}}(S, \mathbb{G}_m) \to \check{\mathrm{H}}^r_{\mathrm{et}}(R, \mathbb{G}_m)$$

for every $r \geq 0$.

(2) *If Res_R^S is the homomorphism induced by $R \to S$, then the composite map*

$$\check{\mathrm{H}}^r_{\mathrm{et}}(R, \mathbb{G}_m) \xrightarrow{\operatorname{Res}_R^S} \check{\mathrm{H}}^r_{\mathrm{et}}(S, \mathbb{G}_m) \xrightarrow{\operatorname{Cor}_R^S} \check{\mathrm{H}}^r_{\mathrm{et}}(R, \mathbb{G}_m)$$

is the nth power map.

(3) *The diagram of abelian groups*

$$\begin{array}{ccc} \operatorname{Pic}(S) & \longrightarrow & \check{\mathrm{H}}^1_{\mathrm{et}}(S, \mathbb{G}_m) \\ \downarrow{\scriptstyle \operatorname{Cor}_R^S} & & \downarrow{\scriptstyle \operatorname{Cor}_R^S} \\ \operatorname{Pic}(R) & \longrightarrow & \check{\mathrm{H}}^1_{\mathrm{et}}(R, \mathbb{G}_m) \end{array}$$

commutes. The horizontal arrows are the natural isomorphisms of Proposition 10.4.8. *The left vertical arrow is from Theorem* 14.1.8 *and the right vertical arrow is from Part (1).*

(4) *The diagram of abelian groups*

$$\begin{array}{ccc} \mathrm{B}(S) & \longrightarrow & \check{\mathrm{H}}^2_{\mathrm{et}}(S, \mathbb{G}_m) \\ {\scriptstyle \mathrm{Cor}^S_R}\downarrow & & \downarrow{\scriptstyle \mathrm{Cor}^S_R} \\ \mathrm{B}(R) & \longrightarrow & \check{\mathrm{H}}^2_{\mathrm{et}}(R, \mathbb{G}_m) \end{array}$$

commutes. The horizontal arrows are the natural monomorphisms of Theorem 10.4.11. *The left vertical arrow is from Theorem* 14.1.9 *and the right vertical arrow is from Part (1).*

Proof. (1): By Definition 10.4.1, the rth Čech cohomology group of S with coefficients in $\mathbb{G}_m$ for the étale topology is the direct limit

$$\check{\mathrm{H}}^r_{\mathrm{et}}(S, \mathbb{G}_m) = \varinjlim_{A \in \mathfrak{C}_{\mathrm{et}}(S)} \mathrm{H}^r(A/S, \mathbb{G}_m),$$

where the limit is over all faithfully flat étale extensions of S. We see from Lemma 14.1.7 that if $A \in \mathfrak{C}_{\mathrm{et}}(S)$, then $\mathrm{Cor}^S_R(A) \in \mathfrak{C}_{\mathrm{et}}(R)$. From diagram (14.24), there are homomorphisms

$$\mathrm{H}^r(A/S, \mathbb{G}_m) \xrightarrow{\mathrm{Cor}^S_R} \mathrm{H}^r(\mathrm{Cor}^S_R(A)/R, \mathbb{G}_m) \to \check{\mathrm{H}}^r_{\mathrm{et}}(R, \mathbb{G}_m)$$

which commute with the transition homomorphisms θ^* for each $\theta : A \to A'$. Hence the desired homomorphism exists.

(2): Let A be a faithfully flat étale R-algebra. The multiplication map $\mu : A^{\otimes n} \to A$ defined by $x_1 \otimes \cdots \otimes x_n \mapsto x_1 \cdots x_n$ is an R-algebra homomorphism. From Theorem 14.1.5 we know $\mathrm{Cor}^S_R(A \otimes_R S) = A^{\otimes n}$. Consider the diagram

$$\begin{array}{ccccc} \mathrm{H}^r(A \otimes_R S/S, \mathbb{G}_m) & \xrightarrow{\mathrm{Cor}^S_R} & \mathrm{H}^r(A^{\otimes n}/R, \mathbb{G}_m) & \xrightarrow{\alpha} & \check{\mathrm{H}}^r_{\mathrm{et}}(R, \mathbb{G}_m) \\ {\scriptstyle \mathrm{Res}^S_R}\uparrow & & \downarrow{\scriptstyle \mu^*} & \nearrow{\scriptstyle \beta} & \\ \mathrm{H}^r(A/R, \mathbb{G}_m) & \xrightarrow[\gamma]{} & \mathrm{H}^r(A/R, \mathbb{G}_m) & & \end{array}$$

of abelian groups. The maps α and β are the transition maps in the direct limit, hence $\alpha = \beta\mu^*$. The map γ is defined to be the composite $\mu^* \circ \mathrm{Cor}^S_R \circ \mathrm{Res}^S_R$, so the diagram commutes. On cochain complexes, γ is induced by the composite map

$$\left(A^{\otimes(r+1)}\right)^* \xrightarrow{\mathrm{Res}^S_R} \left((A \otimes_R S)^{\otimes(r+1)}\right)^*$$
$$\xrightarrow{\mathrm{Cor}^S_R} \left((A^{\otimes n})^{\otimes(r+1)}\right)^* \xrightarrow{\mu^{\otimes(r+1)}} \left(A^{\otimes(r+1)}\right)^*$$

which is the nth power map. This shows γ is the nth power map. The map in Part (2) is the limit over all A of the corresponding maps γ.

(3): Let P be a rank one projective S-module. By Exercise 2.6.16 there is a faithfully flat étale S-algebra T and an T-module isomorphism $\beta : P \otimes_S T \to T$. Let $\theta = \beta_1\beta_0^{-1}$. Then the diagram

$$
\begin{array}{ccc}
 & & T \otimes_S T \\
 & \overset{\beta_0}{\nearrow} & \\
P \otimes_S T \otimes_S T & & \Big\downarrow \theta \\
 & \underset{\beta_1}{\searrow} & \\
 & & T \otimes_S T
\end{array}
\tag{14.25}
$$

commutes and θ is a 1-cocycle in $\mathrm{Z}^1(T/S, \mathbb{G}_m)$. The assignment $[P] \mapsto [\theta]$ defines the homomorphism $\mathrm{Pic}(T/S) \to \mathrm{H}^1(T/S, \mathbb{G}_m)$ (Proposition 5.5.7 and Corollary 5.5.9). By Lemma 14.1.7, $\mathrm{Cor}_R^S(T)$ is a faithfully flat étale R-algebra. By Theorem 14.1.5,

$$\mathrm{Cor}_R^S(\beta) : \mathrm{Cor}_R^S(P) \otimes_R \mathrm{Cor}_R^S(T) \to \mathrm{Cor}_R^S(T)$$

is an isomorphism, so $\mathrm{Cor}_R^S(P)$ is a rank one projective R-module which is split by $\mathrm{Cor}_R^S(T)$. The commutative diagram

$$
\begin{array}{ccc}
 & & \mathrm{Cor}_R^S(T) \otimes_R \mathrm{Cor}_R^S(T) \\
 & \overset{\mathrm{Cor}_R^S(\beta)_0}{\nearrow} & \\
\mathrm{Cor}_R^S(P) \otimes_R \mathrm{Cor}_R^S(T) \otimes_R \mathrm{Cor}_R^S(T) & & \Big\downarrow \mathrm{Cor}_R^S(\theta) \\
 & \underset{\mathrm{Cor}_R^S(\beta)_1}{\searrow} & \\
 & & \mathrm{Cor}_R^S(T) \otimes_R \mathrm{Cor}_R^S(T)
\end{array}
$$

is obtained by applying Cor_R^S to diagram (14.25). As in Proposition 5.5.7, $\mathrm{Cor}_R^S(\theta)$ is a 1-cocycle in $\mathrm{Z}^1(\mathrm{Cor}_R^S(T)/R, \mathbb{G}_m)$ and represents the image of $\mathrm{Cor}_R^S(P)$ under the homomorphism

$$\mathrm{Pic}(\mathrm{Cor}_R^S(T)/R) \to \mathrm{H}^1(\mathrm{Cor}_R^S(T)/R, \mathbb{G}_m).$$

Therefore the diagram

$$
\begin{array}{ccc}
\mathrm{Pic}(T/S) & \longrightarrow & \mathrm{H}^1(T/S, \mathbb{G}_m) \\
\Big\downarrow \mathrm{Cor}_R^S & & \Big\downarrow \mathrm{Cor}_R^S \\
\mathrm{Pic}(\mathrm{Cor}_R^S(T)/R) & \longrightarrow & \mathrm{H}^1(\mathrm{Cor}_R^S(T)/R, \mathbb{G}_m) \\
\Big\downarrow \subseteq & & \Big\downarrow \\
\mathrm{Pic}(R) & \longrightarrow & \check{\mathrm{H}}^1_{\mathrm{et}}(R, \mathbb{G}_m)
\end{array}
\tag{14.26}
$$

commutes. The commutative diagram of Part (3) is obtained from (14.26) by taking the direct limit over all $T \in \mathfrak{C}_{\text{et}}(S)$.

(4): First we review the construction of the homomorphism $\mathrm{B}(S) \to \check{\mathrm{H}}^2_{\text{et}}(S, \mathbb{G}_m)$. For the details, the reader is referred to Proposition 5.5.7, Proposition 10.4.7, and Theorem 10.4.10. Let A be an Azumaya S-algebra of constant rank m^2. There is a faithfully flat étale S-algebra T and an isomorphism of T-algebras $\beta : A \otimes_S T \to M_m(T)$. The diagram of $T \otimes_S T$-algebras

$$
\begin{array}{ccc}
 & & M_m(T \otimes_S T) \\
 & \overset{\beta_0}{\nearrow} & \\
A \otimes_S T \otimes_S T & & \downarrow \theta \\
 & \underset{\beta_1}{\searrow} & \\
 & & M_m(T \otimes_S T)
\end{array}
\tag{14.27}
$$

commutes, where $\theta = \beta_1 \beta_0^{-1}$, by definition. If necessary, we can replace T by a larger S-algebra and assume θ is conjugation by $f \in \mathrm{GL}_m(T \otimes_S T)$. Then $f_1^{-1} f_2 f_0$ is a 2-cocycle in $\mathrm{Z}^2(T/S, \mathbb{G}_m)$. The homomorphism $\mathrm{B}(T/S) \to \mathrm{H}^2(T/S, \mathbb{G}_m)$ is induced by $[A] \mapsto [f_1^{-1} f_2 f_0]$. It follows from Theorem 14.1.9 that $\mathrm{Cor}_R^S(A)$ is an Azumaya R-algebra and $\mathrm{Cor}_R^S(M_m(T)) = M_m(\mathrm{Cor}_R^S(T))$. By Theorem 14.1.5

$$
\mathrm{Cor}_R^S(\beta) : \mathrm{Cor}_R^S(A) \otimes_R \mathrm{Cor}_R^S(T) \to M_m(\mathrm{Cor}_R^S(T))
$$

is an isomorphism, so $\mathrm{Cor}_R^S(A)$ is split by $\mathrm{Cor}_R^S(T)$. The commutative diagram

$$
\begin{array}{ccc}
 & & M_m(\mathrm{Cor}_R^S(T) \otimes_R \mathrm{Cor}_R^S(T)) \\
 & \overset{\mathrm{Cor}_R^S(\beta)_0}{\nearrow} & \\
\mathrm{Cor}_R^S(A) \otimes_R \mathrm{Cor}_R^S(T) \otimes_R \mathrm{Cor}_R^S(T) & & \downarrow \mathrm{Cor}_R^S(\theta) \\
 & \underset{\mathrm{Cor}_R^S(\beta)_1}{\searrow} & \\
 & & M_m(\mathrm{Cor}_R^S(T) \otimes_R \mathrm{Cor}_R^S(T))
\end{array}
$$

is obtained by applying Cor_R^S to diagram (14.27). The reader should verify that $\mathrm{Cor}_R^S(f) \in \mathrm{GL}_m(\mathrm{Cor}_R^S(T) \otimes_R \mathrm{Cor}_R^S(T))$ and $\mathrm{Cor}_R^S(\theta)$ is conjugation by $\mathrm{Cor}_R^S(f)$. It follows that

$$
\mathrm{Cor}_R^S(f)_1^{-1} \, \mathrm{Cor}_R^S(f)_2 \, \mathrm{Cor}_R^S(f)_0 = \mathrm{Cor}_R^S(f_1^{-1} f_2 f_0)
$$

is a 2-cocycle in $\mathrm{Z}^2(\mathrm{Cor}_R^S(T)/R, \mathbb{G}_m)$ and represents the image of $\mathrm{Cor}_R^S(A)$ under the homomorphism

$$\mathrm{B}(\mathrm{Cor}_R^S(T)/R) \to \mathrm{H}^2(\mathrm{Cor}_R^S(T)/R, \mathbb{G}_m).$$

The rest of the proof is similar to Part (3) and is left to the reader. □

1.5. Corestriction and Galois Cohomology. First we establish notation that will be in effect throughout this section. Let S be a commutative separable R-algebra which is an R-module progenerator of constant rank n. Let L/R be a Galois extension of commutative rings with finite group $G = \mathrm{Aut}_R(L)$. Finally, let $L' = L \otimes_R S$. Then L' is Galois over S with group G (Lemma 12.2.7). The diagram

$$\begin{array}{ccc} L & \longrightarrow & L' = L \otimes_R S \\ \uparrow & & \uparrow \\ R & \longrightarrow & S \end{array} \tag{14.28}$$

commutes, where each arrow is a homomorphism of commutative rings.

Lemma 14.1.14. *In the above context, the following are true.*

(1) $\mathrm{Cor}_L^{L'} : (L')^* \to L^*$ *is a homomorphism of* $\mathbb{Z}G$*-modules.*

(2) *Let* M *be an* L'*-module and* $\sigma \in G = \mathrm{Aut}_R(L)$*. Then there is an isomorphism*

$${}_\sigma \mathrm{Cor}_L^{L'}(M)_\sigma \cong \mathrm{Cor}_L^{L'}({}_\sigma M_\sigma)$$

L-modules. (See Definition 7.8.4 *for the definition of a module with action twisted by* σ*.)*

(3) *The assignment* $[M] \mapsto [\mathrm{Cor}_L^{L'}(M)]$ *induces homomorphisms*

$$\mathrm{Cor}_L^{L'} : \mathrm{Pic}(L') \to \mathrm{Pic}(L)$$
$$\mathrm{Cor}_L^{L'} : \mathrm{B}(L') \to \mathrm{B}(L)$$

of $\mathbb{Z}G$*-modules.*

(4) *For each* $n \geq 0$ *there are homomorphisms*

$$\mathrm{Cor}_L^{L'} : \mathrm{H}^n(G, (L')^*) \to \mathrm{H}^n(G, L^*)$$
$$\mathrm{Cor}_L^{L'} : \mathrm{H}^n(G, \mathrm{Pic}(L')) \to \mathrm{H}^n(G, \mathrm{Pic}(L))$$
$$\mathrm{Cor}_L^{L'} : \mathrm{H}^n(G, \mathrm{B}(L')) \to \mathrm{H}^n(G, \mathrm{B}(L))$$

of abelian groups.

Proof. (1): Let $x \in (L')^*$. Then $\operatorname{Cor}_L^{L'}(\sigma(x))\operatorname{Cor}_L^{L'}(\ell_{\sigma(x)})$ is left multiplication by the determinant of $\sigma(x)$ (Theorem 14.1.5 (7)). It is enough to show that $\det(\ell_{\sigma(x)}) = \sigma(\det(\ell_x))$ in L. Following the construction in Example 5.3.3, let $\tilde{L}$ be a faithfully flat extension of L such that $\tilde{L} \otimes_L L'$ is free of rank n. Then $\ell_x : \tilde{L} \otimes_L L' \to \tilde{L} \otimes_L L'$ can be represented by a matrix with entries in $\tilde{L}$. The matrix of $\ell_{\sigma(x)}$ is obtained by applying σ element-wise to the matrix of ℓ_x.

(2): Step 1: Assume $L' = Le_1 \oplus \cdots \oplus Le_r$ is the trivial extension. Using Lemma 7.8.5, there is an isomorphism of L-modules

$$\begin{aligned}
{}_\sigma L_1 \otimes_L (Me_1 \otimes_L \cdots \otimes_L Me_r) \otimes_L {}_1L_\sigma \\
\cong ({}_\sigma L_1 \otimes_L Me_1 \otimes_L {}_1L_\sigma) \otimes_L ({}_\sigma L_1 \otimes_L Me_2 \otimes_L {}_1L_\sigma) \\
\otimes_L \cdots \otimes_L ({}_\sigma L_1 \otimes_L Me_r \otimes_L {}_1L_\sigma)\,.
\end{aligned}$$

From this and the formula in Example 14.1.4, (3) follows when L' is the trivial extension.

Step 2: Let T be a faithfully flat L-algebra such that $T \otimes_L L'$ is the trivial extension of T. From Theorem 14.1.3 and Step 1, the identity map on sets induces the isomorphism

$$T \otimes_L {}_\sigma \operatorname{Cor}_L^{L'}(M)_\sigma \cong T \otimes_L \operatorname{Cor}_L^{L'}({}_\sigma M_\sigma)$$

of T-modules.

(3): The multiplication rule $\sigma \cdot [M] = [{}_{\sigma^{-1}}M_{\sigma^{-1}}]$ defines the action by the group G on the Picard groups (Lemma 7.8.7) and the Brauer groups (Exercise 13.2.13). Therefore (3) follows from (2).

Part (4) follows from Parts (1), (2), (3), and Lemma 5.1.11. □

Theorem 14.1.15. *In the above context, the diagram of abelian groups*

$$\begin{array}{ccc}
\mathrm{H}^2(G,(L')^*) & \xrightarrow{\alpha_4} & \mathrm{B}(L'/S) \\
\Big\downarrow{\scriptstyle \operatorname{Cor}_L^{L'}} & & \Big\downarrow{\scriptstyle \operatorname{Cor}_R^{S}} \\
\mathrm{H}^2(G,L^*) & \xrightarrow{\alpha_4} & \mathrm{B}(L/R)
\end{array}$$

commutes. Both horizontal maps are the crossed product homomorphism α_4 of Theorem 13.3.1. The left vertical map $\operatorname{Cor}_L^{L'}$ is from Lemma 14.1.14, and the right vertical map Cor_R^S is from Theorem 14.1.9. If $\xi \in \mathrm{Z}^2(G,(L')^*)$ is a normalized 2-cocycle, and $A = \Delta(L'/S,G,\xi)$, then $\operatorname{Cor}_R^S(A)$ is Brauer equivalent to $\Delta\big(L/R,G,\operatorname{Cor}_L^{L'}(\xi)\big)$.

Proof. Step 1: Verify that the maps of the diagram are well defined. From Theorem 14.1.9 the diagram

$$\begin{array}{ccccccc} 0 & \longrightarrow & \mathrm{B}(L'/S) & \longrightarrow & \mathrm{B}(S) & \longrightarrow & \mathrm{B}(L') \\ & & & & \big\downarrow{\scriptstyle \mathrm{Cor}_R^S} & & \big\downarrow{\scriptstyle \mathrm{Cor}_L^{L'}} \\ 0 & \longrightarrow & \mathrm{B}(L/R) & \longrightarrow & \mathrm{B}(R) & \longrightarrow & \mathrm{B}(L) \end{array}$$

commutes. The rows are exact, so Cor_R^S maps $\mathrm{B}(L'/S)$ into $\mathrm{B}(L/R)$.

Step 2: Construct a candidate for $\mathrm{Cor}_R^S(A)$. If Σ_n denotes the symmetric group on n letters, then the corestriction of A is $\mathrm{Cor}_R^S(A) = (T_R^n(A)f)^{\Sigma_n}$ (Definition 14.1.1). By Theorem 14.1.9, $T_R^n(A)$ is an Azumaya $T_R^n(S)$-algebra and $T_R^n(A)f$ is an Azumaya $T_R^n(S)f$-algebra. We view L' as a maximal commutative S-subalgebra of

$$A = \Delta(L'/S, G, \xi) = \bigoplus_{\sigma \in G} L' u_\sigma$$

(Lemma 13.1.1). Every ring in (14.28) is an R-progenerator, so we can view $T_R^n(L')$ as a subring of $T_R^n(A)$. We have the factorization

$$T_R^n(L') = T_R^n(L \otimes_R S) = T_R^n(L) \otimes_R T_R^n(S)$$

Let e be the idempotent in $T_R^n(L)$ such that $1 - e$ generates the kernel of the product map $\mu : T_R^n(L) \to L$ (Exercise 4.4.18). Then

$$T_R^n(L)e = (L \otimes_R 1 \otimes_R \cdots \otimes_R 1)e. \tag{14.29}$$

Notice that e is fixed by the group Σ_n. By Exercise 7.6.4,

$$B = (e \otimes f) T_R^n(A)(e \otimes f)$$

is an Azumaya $T_R^n(S)f$-algebra and the action by Σ_n induces an action on B. By Galois Descent (Corollary 12.4.4), B^{Σ_n} is Azumaya over $R = (T_R^n(S)f)^{\Sigma_n}$. By Exercise 7.6.4,

$$B^{\Sigma_n} = (e \otimes f)\,(T_R^n(A)f)^{\Sigma_n}\,(e \otimes f)$$

is Brauer equivalent to $\mathrm{Cor}_R^S(A)$. Notice that B contains the commutative subring $T_R^n(L')(e \otimes f) = Le \otimes_R T_R^n(S)f$. Therefore B^{Σ_n} contains $(Le \otimes_R T_R^n(S)f)^{\Sigma_n} \cong L \otimes_R R = L$. In the commutative diagram

$$\begin{array}{ccc} T_R^n(L')(e \otimes f) = Le \otimes_R T_R^n(S)f & \xrightarrow{\ \subseteq\ } & B \\ \big\uparrow{\scriptstyle \subseteq} & & \big\uparrow{\scriptstyle \subseteq} \\ (T_R^n(L')(e \otimes f))^{\Sigma_n} \cong L & \xrightarrow{\ \subseteq\ } & B^{\Sigma_n} \end{array}$$

each arrow corresponds to a subring. To complete the proof, we show that B^{Σ_n} is isomorphic to $\Delta(L/R, G, \operatorname{Cor}_L^{L'}(\xi))$. This is divided into two more steps.

Step 3: Show that if $x \in L'$, then $\operatorname{Cor}_L^{L'}(x)$ is equal to the element $(x \otimes \cdots \otimes x)(e \otimes f)$ in $(T_R^n(L')(e \otimes f))^{\Sigma_n} \cong L$. If suffices to show that the elements are equal after a faithfully flat base change. Therefore, assume S is the trivial extension of R. Write $S = Re_1 \oplus \cdots \oplus Re_n$, where $e_1, \dots, e_n$ are the canonical orthogonal idempotents. We also have $L' = Le_1 \oplus \cdots \oplus Le_n$. Write $x = \sum_{i=1}^n x_i e_i$. By Example 14.1.2, $\operatorname{Cor}_L^{L'}(L')$ is the image of the L-module homomorphism $L'e_1 \otimes_L \cdots \otimes_L L'e_n \to T_L^n(L')f$ defined by $a_1e_1 \otimes \cdots \otimes a_n e_n \mapsto \sum_{\sigma \in \Sigma_n} a_{\sigma(1)} e_{\sigma(1)} \otimes a_{\sigma(2)} e_{\sigma(2)} \otimes \cdots \otimes a_{\sigma(n)} e_{\sigma(n)}$. From this we compute

$$\begin{aligned}
\operatorname{Cor}_L^{L'}(x) &= \sum_{\sigma \in \Sigma_n} x_{\sigma(1)} e_{\sigma(1)} \otimes x_{\sigma(2)} e_{\sigma(2)} \otimes \cdots \otimes x_{\sigma(n)} e_{\sigma(n)} \\
&= x_1 x_2 \cdots x_n \sum_{\sigma \in \Sigma_n} e_{\sigma(1)} \otimes e_{\sigma(2)} \otimes \cdots \otimes e_{\sigma(n)} \\
&= x_1 x_2 \cdots x_n f
\end{aligned}$$

which a direct computation shows is equal to $(x \otimes \cdots \otimes x)(e \otimes f)$.

Step 4: As in the proof of Theorem 13.1.4 (1), the reader should verify that $(e \otimes f)(u_{\sigma_1} \otimes \cdots \otimes u_{\sigma_n}) = 0$ if $\sigma_i = \sigma_j$ for some $i \neq j$. Therefore B is generated by $\{u_\sigma \otimes \cdots \otimes u_\sigma \mid \sigma \in G\}$ as a left $Le \otimes_R T_R^n(S)f$-module. For each $\sigma \in G$, let $v_\sigma = u_\sigma \otimes \cdots \otimes u_\sigma$. Then

$$\begin{aligned}
v_\sigma v_\tau &= u_\sigma u_\tau \otimes \cdots \otimes u_\sigma u_\tau \\
&= \xi(\sigma, \tau) u_{\sigma\tau} \otimes \cdots \otimes \xi(\sigma, \tau) u_{\sigma\tau} \\
&= (\xi(\sigma, \tau) \otimes \cdots \otimes \xi(\sigma, \tau))(e \otimes f) v_{\sigma\tau} \\
&= \operatorname{Cor}_L^{L'}(\xi(\sigma, \tau)) v_{\sigma\tau}.
\end{aligned}$$

Notice that each v_σ is in B^{Σ_n}. Let $\Delta\big(L/R, G, \operatorname{Cor}_L^{L'}(\xi)\big) = \bigoplus_{\sigma \in G} L w_\sigma$ and define

$$\Delta\big(L/R, G, \operatorname{Cor}_L^{L'}(\xi)\big) \xrightarrow{\phi} B^{\Sigma_n}$$

by $w_\sigma \mapsto v_\sigma$. Then ϕ is a well defined homomorphism of Azumaya R-algebras, hence is one-to-one (Theorem 7.6.1). Since

$$1 \otimes \phi : T_R^n(S)f \otimes_R \Delta(L/R, G, \operatorname{Cor}_L^{L'}(\xi)) \to T_R^n(S)f \otimes_R B^{\Sigma_n} = B$$

is onto, and $T_R^n(S)f$ is faithfully flat over R, it follows that ϕ is an isomorphism. $\square$

1.6. Corestriction and Generalized Crossed Products. In this section we retain the notation of Diagram (14.28). Then L/R is Galois with group $G = \operatorname{Aut}_R(L)$, and $L' = L \otimes_R S$ is Galois over S with group $G = \operatorname{Aut}_S(L')$. In the context of Section 13.8.1, there is an exact sequence

$$0 \to \operatorname{Pic}(L') \to \operatorname{Pic}_S(L') \xrightarrow{\Psi} G \to 1$$

of groups and Ψ is split by Φ_0 (Lemma 7.8.7). In other words, $\operatorname{Pic}_S(L')$ is a semidirect product. The action of the group G on $\operatorname{Pic}(L')$ is defined by $\tau[P] = \Phi_0(\tau)[P]\Phi_0(\tau^{-1}) = [{}_{\tau^{-1}}P_{\tau^{-1}}]$. Then we view $\operatorname{Pic}_S(L')$ as the set of ordered pairs $([P], \sigma)$ in $\operatorname{Pic}(L') \times G$, with group action defined by

$$([P], \sigma)([Q], \tau) = ([P]\sigma[Q], \sigma\tau).$$

By Lemma 14.1.14, the assignment $([P], \sigma) \mapsto (\operatorname{Cor}_L^{L'}([P]), \sigma)$ induces a homomorphism

$$\operatorname{Cor}_L^{L'} : \operatorname{Pic}_S(L') \to \operatorname{Pic}_R(L)$$

of groups. This map can also be described on modules. If M is an invertible left $L' \otimes_S L'$-module, then the left and right L'-module actions on M differ by a twist by the automorphism $\Psi([M]) = \alpha_M$. That is, for any $x \in M$ and $a \in L'$, $xa = \alpha_M(a)x$. Hence $M \otimes_{L'} \Phi_0(\alpha_M^{-1})$ is in $\operatorname{Pic}(L')$. For simplicity, let $\tau = \alpha_M$. The corestriction mapping defined above becomes $[M] \mapsto [\operatorname{Cor}_L^{L'}(M_{\tau^{-1}})_\tau]$. So on the left L-module, this definition agrees with that of Theorem 14.1.8. The right L-module structure is inherited from the right L'-structure on M.

Lemma 14.1.16. *In the above context, the following are true.*

(1) *There is a homomorphism of groups* $\operatorname{Cor}_L^{L'} : \operatorname{Pic}_S(L') \to \operatorname{Pic}_R(L)$.
(2) *Let* $\Phi : G \to \operatorname{Pic}_S(L')$ *be a homomorphism of groups such that* $\Psi\Phi = 1$, *and let* f *be a factor set related to* Φ. *Then* $\operatorname{Cor}_L^{L'} \Phi : G \to \operatorname{Pic}_R(L)$ *is a homomorphism of groups such that* $\Psi(\operatorname{Cor}_L^{L'} \Phi) = 1$ *and* $\operatorname{Cor}_L^{L'}(f)$ *defines a factor set for* $\operatorname{Cor}_L^{L'} \Phi$.

Proof. Part (1) is proved already.

(2): It follows straight from the above construction that $\Psi \operatorname{Cor}_L^{L'} \Phi(\sigma) = \sigma$ for any $\sigma \in G$. Assume we are given $L' \otimes_S L'$-isomorphisms $f = \{f_{\sigma,\tau} : \Phi_\sigma \otimes_{L'} \Phi_\tau \to \Phi_{\sigma\tau}\}$ indexed by $G \times G$ such that the diagram

$$\begin{array}{ccc} \Phi_\sigma \otimes_{L'} \Phi_\tau \otimes_{L'} \Phi_\gamma & \xrightarrow{1 \otimes f_{\tau,\gamma}} & \Phi_\sigma \otimes_{L'} \Phi_{\tau\gamma} \\ {\scriptstyle f_{\sigma,\tau} \otimes 1} \downarrow & & \downarrow {\scriptstyle f_{\sigma,\tau\gamma}} \\ \Phi_{\sigma\tau} \otimes_{L'} \Phi_\gamma & \xrightarrow[f_{\sigma\tau,\gamma}]{} & \Phi_{\sigma\tau\gamma} \end{array} \tag{14.30}$$

commutes for every triple $(\sigma, \tau, \gamma) \in G \times G \times G$ (Definition 13.2.2). Viewing all of the modules Φ_σ as left L'-modules, apply the corestriction functor to get a family of left L-module isomorphisms

$$\operatorname{Cor}_L^{L'}(f) = \{\operatorname{Cor}_L^{L'}(f_{\sigma,\tau}) : \operatorname{Cor}_L^{L'}(\Phi_\sigma) \otimes_L \operatorname{Cor}_L^{L'}(\Phi_\tau) \to \operatorname{Cor}_L^{L'}(\Phi_{\sigma\tau})\}$$

for which the diagram (14.30) (corresponding to the L-modules) commutes for every triple (Theorem 14.1.3). Using the construction of the homomorphisms $\operatorname{Cor}_L^{L'}(f_{\sigma,\tau})$, the fact that each $f_{\sigma,\tau}$ is an isomorphism between invertible modules, and Exercise 14.1.18, the reader should verify that $\operatorname{Cor}_L^{L'}(f_{\sigma,\tau})$ is a two-sided L/R-module isomorphism. □

Theorem 14.1.17. *In the above context, the diagram of abelian groups*

$$\begin{array}{ccc} \mathrm{B}(L'/S) & \xrightarrow{\alpha_5} & \mathrm{H}^2(G, \operatorname{Pic}(L')) \\ {\scriptstyle \operatorname{Cor}_R^S}\downarrow & & \downarrow{\scriptstyle \operatorname{Cor}_L^{L'}} \\ \mathrm{B}(L/R) & \xrightarrow{\alpha_5} & \mathrm{H}^2(G, \operatorname{Pic}(L)) \end{array}$$

commutes. The two horizontal maps are the homomorphism α_5 of Theorem 13.3.1, the left vertical map Cor_R^S is from Theorem 14.1.9 and the right vertical map $\operatorname{Cor}_L^{L'}$ is from Lemma 14.1.14. If $A = \Delta(L'/S, \Phi, \xi)$ is a generalized crossed product algebra, then $\operatorname{Cor}_R^S(A)$ is Brauer equivalent to $\Delta\big(L/R, \operatorname{Cor}_L^{L'} \Phi, \operatorname{Cor}_L^{L'}(\xi)\big)$.

Proof. Let $A = \Delta(L'/S, \Phi, \xi)$ be a generalized crossed product. The homomorphism α_5 maps $[A]$ to $[\varphi]$, where $\varphi : G \to \operatorname{Pic} L'$ is the 1-cocycle defined by $\varphi(\tau) = \Phi(\tau)\Phi_0(\tau^{-1})$. By the constructions in Lemma 14.1.16 and Lemma 14.1.14, The homomorphism α_5 maps $\Delta\big(L/R, \operatorname{Cor}_L^{L'} \Phi, \operatorname{Cor}_L^{L'}(\xi)\big)$ to $\operatorname{Cor}_L^{L'}([\varphi])$. To finish the proof, it suffices to show $\operatorname{Cor}_R^S(A)$ is Brauer equivalent to the generalized crossed product $\Delta\big(L/R, \operatorname{Cor}_L^{L'} \Phi, \operatorname{Cor}_L^{L'}(\xi)\big)$.

We have $A = \bigoplus_{\tau \in G} \Phi_\tau$. As in Step 4 of the proof of Theorem 14.1.15,

$$\begin{aligned} B &= (e \otimes f) T_R^n(A) (e \otimes f) \\ &= (e \otimes f) \left(\bigoplus_{\tau \in G} T_R^n(\Phi_\tau) \right) (e \otimes f). \end{aligned}$$

Now

$$\begin{aligned} \operatorname{Cor}_L^{L'}(\Phi_\tau) \otimes_R T_R^n(S) f &= \operatorname{Cor}_L^{L'}(\Phi_\tau) \otimes_L L \otimes_R T_R^n(S) f \\ &= \operatorname{Cor}_L^{L'}(\Phi_\tau) \otimes_L T_R^n(L \otimes_R S) f \\ &= \operatorname{Cor}_L^{L'}(\Phi_\tau) \otimes_L T_R^n(L') f \\ &= T_L^n\big(\Phi(\tau)_{\tau^{-1}}\big)_\tau f \end{aligned}$$

so we define $\psi_\tau : T^n_L\left(\Phi(\tau)_{\tau^{-1}}\right)_\tau f \to (e\otimes f)T^n_R(\Phi_\tau)(e\otimes f)$ by

$$(\phi_1\otimes\cdots\otimes\phi_n)\mapsto(e\otimes f)(\phi_1\otimes\cdots\otimes\phi_n)(e\otimes f).$$

Since the tensor product on the left is over L, and the tensor product on the right is over R, we must verify that ψ_τ is well defined. Let $a\in L$. Using (14.29) we have

$$\begin{aligned}
\psi_\tau(\phi_1\otimes\cdots\otimes a\phi_i\otimes\cdots\otimes\phi_n) &= (e\otimes f)(\phi_1\otimes\cdots\otimes a\phi_i\otimes\cdots\otimes\phi_n)(e\otimes f)\\
&= (e\otimes f)(1\otimes\cdots\otimes a\otimes\cdots\otimes 1)(\phi_1\otimes\cdots\otimes\phi_i\otimes\cdots\otimes\phi_n)(e\otimes f)\\
&= (e\otimes f)(a\otimes 1\otimes\cdots\otimes 1)(\phi_1\otimes\cdots\otimes\phi_i\otimes\cdots\otimes\phi_n)(e\otimes f)\\
&= (e\otimes f)(a\phi_1\otimes\cdots\otimes\phi_i\otimes\cdots\otimes\phi_n)(e\otimes f)\\
&= \psi_\tau(a\phi_1\otimes\cdots\otimes\phi_i\otimes\cdots\otimes\phi_n)
\end{aligned}$$

hence ψ_τ is well defined. Denote by ψ the composite map

$$\begin{aligned}
\Delta\left(L/R,\operatorname{Cor}^{L'}_L\Phi,\operatorname{Cor}^{L'}_L(\xi)\right) &\xrightarrow{\cong}\Delta\left(L/R,\operatorname{Cor}^{L'}_L\Phi,\operatorname{Cor}^{L'}_L(\xi)\right)\otimes_R 1\\
&\to\Delta\left(L/R,\operatorname{Cor}^{L'}_L\Phi,\operatorname{Cor}^{L'}_L(\xi)\right)\otimes_R T^n_R(S)f\\
&\to\bigoplus_{\tau\in G}\operatorname{Cor}^{L'}_L(\Phi_\tau)\otimes_R T^n_R(S)f\\
&\xrightarrow{\oplus_\tau\psi_\tau}(e\otimes f)\left(\bigoplus_{\tau\in G}T^n_R(\Phi_\tau)\right)(e\otimes f)\\
&= B.
\end{aligned}$$

The image of ψ is a subset of B^{Σ_n}. Let $\mathfrak{p}\in\operatorname{Spec}R$. Then $L\otimes_R R_\mathfrak{p}$ is a semilocal ring (Corollary 3.3.6) and the left $L\otimes_R R_\mathfrak{p}$-module $\operatorname{Cor}^{L'}_L(\Phi_\tau)\otimes_R R_\mathfrak{p}$ is free of rank one (Exercise 3.1.8). Using Lemma 2.1.4, there exists $w\in R$ such that $\operatorname{Cor}^{L'}_L(\Phi_\tau)\otimes_R R_w\cong{}_1(L_w)_\tau$ as $L_w\otimes_{R_w}L_w$-modules. Upon restriction of scalars to R_w, the map ψ agrees with the isomorphism ϕ defined in Step 4 of the proof of Theorem 14.1.15. Since we can do this for every prime ideal $\mathfrak{p}$, there is a faithfully flat R-algebra of the form $C=R_{w_1}\oplus\cdots\oplus R_{w_r}$ and upon restriction of scalars to C, the map ψ agrees with the isomorphism ϕ defined in Step 4 of the proof of Theorem 14.1.15. From this it follows that ψ is an isomorphism onto B^{Σ_n} and we are done. □

1.7. Exercises.

Exercise 14.1.18. In the context of Section 14.1.6, let $\phi: M\to N$ be a homomorphism of invertible left $L'\otimes_S L'$-modules. Show that $\alpha_M(a)\phi(x)=\alpha_N(a)\phi(x)$ for all $a\in L'$ and $x\in M$.

Exercise 14.1.19. Let R,S,L,L' be as in Section 14.1.5. Show that there are homomorphisms connecting the seven term exact sequence for L'/S with the sequence for L/R and that the diagram

$$\begin{array}{ccccccccc}
0 & \longrightarrow & \mathrm{H}^1(G,(L')^*) & \xrightarrow{\alpha_1} & \mathrm{Pic}(S) & \xrightarrow{\alpha_2} & \mathrm{Pic}(L')^G & \xrightarrow{\alpha_3} & \mathrm{H}^2(G,(L')^*) \\
 & & \downarrow \mathrm{Cor}_L^{L'} & & \downarrow \mathrm{Cor}_R^S & & \downarrow \mathrm{Cor}_L^{L'} & & \downarrow \mathrm{Cor}_L^{L'} \\
0 & \longrightarrow & \mathrm{H}^1(G,L^*) & \xrightarrow{\alpha_1} & \mathrm{Pic}(R) & \xrightarrow{\alpha_2} & \mathrm{Pic}(L)^G & \xrightarrow{\alpha_3} & \mathrm{H}^2(G,L^*)
\end{array}$$

$$\begin{array}{ccccccc}
\mathrm{H}^2(G,(L')^*) & \xrightarrow{\alpha_4} & \mathrm{B}(L'/S) & \xrightarrow{\alpha_5} & \mathrm{H}^1(G,\mathrm{Pic}\,L') & \xrightarrow{\alpha_6} & \mathrm{H}^3(G,(L')^*) \\
\downarrow \mathrm{Cor}_L^{L'} & & \downarrow \mathrm{Cor}_R^S & & \downarrow \mathrm{Cor}_L^{L'} & & \downarrow \mathrm{Cor}_L^{L'} \\
\mathrm{H}^2(G,L^*) & \xrightarrow{\alpha_4} & \mathrm{B}(L/R) & \xrightarrow{\alpha_5} & \mathrm{H}^1(G,\mathrm{Pic}\,L) & \xrightarrow{\alpha_6} & \mathrm{H}^3(G,L^*)
\end{array}$$

commutes. The vertical maps on cohomology groups ($\mathrm{Cor}_L^{L'}$) are the corestriction homomorphisms defined in Lemma 14.1.14. The corestriction maps on Picard groups are from Theorem 14.1.8. The corestriction map on Brauer groups is from Theorem 14.1.9. (Hints: The second square is Theorem 14.1.8. The fourth square is Theorem 14.1.15. The fifth square is Theorem 14.1.17. The rest is left to the reader.)

2. A Mayer-Vietoris Sequence for the Brauer Group

We derive sufficient conditions such that a cartesian square of commutative rings gives rise to a Mayer-Vietoris sequence for the Brauer group functor. This turns out to be a powerful computational tool, which will be applied in Section 14.3.

It will be convenient to adopt some notation for the specific categories that will be appearing in this section. Each of the following categories also is equipped with a product. Let A denote an arbitrary ring, and R a commutative ring.

(1) $\mathfrak{M}(A)$ denotes ${}_A\mathfrak{M}$, the category of left A-modules and A-module homomorphisms. The product on $\mathfrak{M}(A)$ is the direct sum $(\,)\oplus(\,)$.

(2) $\mathbf{P}(A)$ denotes the category of finitely generated projective left A-modules and A-module homomorphisms. The product on $\mathbf{P}(A)$ is the direct sum $(\,)\oplus(\,)$.

(3) $\mathbf{FP}(R)$ denotes the category of R-module progenerators and R-module homomorphisms. The product on $\mathbf{FP}(R)$ is the tensor $(\,)\otimes_R(\,)$. Recall that an R-module M is an R-progenerator if and only if M is finitely generated projective and faithful (Corollary 1.1.16). Theorem 14.2.1 below gives another equivalent condition for an R-module to be a progenerator.

(4) **CRP**(R) denotes the category of finitely generated projective R-modules of constant rank and R-module homomorphisms. The product on **CRP**(R) is the tensor $(\,) \otimes_R (\,)$. Notice **CRP**(R) is a subcategory of **FP**(R).

(5) **Pic**(R) denotes the category of invertible R-modules and R-module homomorphisms. The product on **Pic**(R) is the tensor $(\,) \otimes_R (\,)$. Notice **Pic**(R) is a subcategory of **CRP**(R).

(6) **Az**(R) denotes the category of Azumaya R-algebras and R-algebra homomorphisms. The product on **Az**(R) is the tensor $(\,)\otimes_R(\,)$. Every homomorphism in **Az**(R) is a monomorphism (Theorem 7.6.1).

In this section it will be necessary to apply the following theorem of Bass.

Theorem 14.2.1 (H. Bass). *Let R be a commutative ring and M an R-module. Then M is an R-progenerator if and only if there exists an R-module N such that $M \otimes_R N \cong R^{(n)}$ for some $n > 0$.*

Proof. See [**Bas68**, Proposition (4.6), p. 476]. □

Corollary 14.2.2. *Let R be a commutative ring and M an invertible R-module (Definition 2.6.8). Then the following are equivalent.*

(1) *M belongs to the torsion subgroup of $\operatorname{Pic} R$, or in the notation of Section 5.2, there exists $d > 0$ such that $M^{\otimes d} \cong R$.*

(2) *There exists $n > 0$ such that $M \otimes_R R^{(n)} \cong R^{(n)}$.*

Proof. (1) implies (2): The R-module $P = R \oplus M \oplus M^{\otimes 2} \oplus \cdots \oplus M^{\otimes(d-1)}$ is a progenerator and $M \otimes_R P \cong P$. By Theorem 14.2.1, there exists an R-module Q such that $P \otimes_R Q \cong R^{(n)}$, for some $n > 0$. Tensoring both sides of $M \otimes_R P \cong P$ with $(\,) \otimes_R Q$ yields $M \otimes_R R^{(n)} \cong R^{(n)}$.

(2) implies (1): If $M \otimes_R R^{(n)} \cong R^{(n)}$, then $\bigwedge^n(M^{(n)}) \cong \bigwedge^n(R^{(n)})$. By properties of exterior powers, $M^{\otimes n} \cong R$ (see, for example [**Bou89a**]). □

A generalization of Corollary 14.2.2 which applies to progenerator modules of arbitrary rank $r > 0$ is proved in [**BG86**].

2.1. Milnor's Theorem. Our objective is give Milnor's proof that for a ring A defined by a fiber product diagram (see Eq. (14.31) below), the category **P**(A) is equivalent to a fiber product category (see Definition 14.2.7 below). This result, which is proved in Theorem 14.2.8, will be applied in Section 14.2.2 to construct the six term Mayer-Vietoris exact sequence for the units and Picard groups (Theorem 14.2.10). The original source for this material is [**Bas68**, Chapter IX, § 5]. For an introduction to this subject, the interested reader is referred to [**CR87**, § 42]. We prove only a portion of

Milnor's results on the K-Theory of a fiber product category. For a deeper treatment, a book on Algebraic K-Theory is recommended.

Let A, A_1, A_2, A_3 be rings and

$$\begin{array}{ccc} A & \xrightarrow{h_2} & A_2 \\ {\scriptstyle h_1}\downarrow & & \downarrow{\scriptstyle f_2} \\ A_1 & \xrightarrow[f_1]{} & A_3 \end{array} \tag{14.31}$$

a commutative diagram of rings and homomorphisms of rings. As defined in Example 1.4.6, we say that (14.31) is a cartesian square if $A \cong \{(x_1, x_2) \in A_1 \oplus A_2 \mid f_1(x_1) = f_2(x_2)\}$ and h_1, h_2 are induced by the coordinate projections. In this case, there is an exact sequence

$$0 \to A \xrightarrow{h_1 \oplus h_2} A_1 \oplus A_2 \xrightarrow{f_1 - f_2} A_3 \tag{14.32}$$

of left A-modules. Let P be a left A-module and set $P_i = A_i \otimes_A P$. If P is flat, then tensoring (14.32) with $(\,) \otimes_A P$ yields an exact sequence

$$0 \to P \xrightarrow{(h_1 \oplus h_2) \otimes 1} P_1 \oplus P_2 \xrightarrow{(f_1 - f_2) \otimes 1} P_3$$

and a cartesian square

$$\begin{array}{ccc} P & \xrightarrow{h_2 \otimes 1} & P_2 \\ {\scriptstyle h_1 \otimes 1}\downarrow & & \downarrow{\scriptstyle f_2 \otimes 1} \\ P_1 & \xrightarrow[f_1 \otimes 1]{} & P_3 \end{array} \tag{14.33}$$

of left A-modules. Notice that each P_i is a flat A_i-module (Theorem 1.3.2). Now we show how to construct an A-module provided we are given an A_1-module, an A_2-module, and an A_3-module patching isomorphism.

Definition 14.2.3. In the context of the cartesian square of rings (14.31), suppose we are given $P_1 \in \mathfrak{M}(A_1)$, $P_2 \in \mathfrak{M}(A_2)$, and $\sigma : A_3 \otimes_{A_1} P_1 \to A_3 \otimes_{A_2} P_2$ an A_3-module isomorphism. A *triple* is a subset $(P_1, \sigma, P_2) \subseteq P_1 \oplus P_2$ defined by

$$(P_1, \sigma, P_2) = \{(x_1, x_2) \in P_1 \oplus P_2 \mid \sigma(1 \otimes x_1) = 1 \otimes x_2\}.$$

If $(a_1, a_2) \in A$ and $(x_1, x_2) \in (P_1, \sigma, P_2)$, we have

$$\sigma(1 \otimes a_1 x_1) = f_1(a_1)\sigma(1 \otimes x_1) = f_2(a_2)(1 \otimes x_2) = 1 \otimes a_2 x_2.$$

It follows that the triple (P_1, σ, P_2) is an A-submodule of $P_1 \oplus P_2$ and the four corners of the commutative diagram

$$\begin{array}{ccc} (P_1,\sigma,P_2) & \xrightarrow{h_2} & P_2 \\ {\scriptstyle h_1}\downarrow & & \downarrow{\scriptstyle \phi_2} \\ P_1 \xrightarrow{\phi_1} A_3 \otimes_{A_1} P_1 & \xrightarrow{\sigma} & A_3 \otimes_{A_2} P_2 \end{array}$$

make up a cartesian square, where h_1 and h_2 are induced by the coordinate projections, and $\phi_i(x) = 1 \otimes x$. If (Q_1, τ, Q_2) is another such triple, then a *morphism from* (P_1, σ, P_2) *to* (Q_1, τ, Q_2) is an ordered pair (ϕ_1, ϕ_2) where $\phi_i \in \operatorname{Hom}_{A_i}(P_i, Q_i)$ and the diagram of A_3-modules

$$\begin{array}{ccc} A_3 \otimes_{A_1} P_1 & \xrightarrow{\sigma} & A_3 \otimes_{A_2} P_2 \\ {\scriptstyle 1\otimes\phi_1}\downarrow & & \downarrow{\scriptstyle 1\otimes\phi_2} \\ A_3 \otimes_{A_1} Q_1 & \xrightarrow{\tau} & A_3 \otimes_{A_2} Q_2 \end{array} \tag{14.34}$$

commutes.

Lemma 14.2.4. *In the context of Definition* 14.2.3, *the following are true.*

(1) *A is equal to the triple $(A_1, 1, A_2)$.*

(2) *A flat A-module P is isomorphic to the triple (P_1, γ, P_2), where $P_i = A_i \otimes_A P$ and γ is the natural isomorphism $A_3 \otimes_{A_1} P_1 \to A_3 \otimes_{A_2} P_2$.*

(3) *If $\theta : P \to Q$ is a homomorphism of flat A-modules and $\theta_i = 1 \otimes \theta : A_i \otimes_A P \to A_i \otimes_A Q$ for $i = 1, 2$, then (θ_1, θ_2) is a morphism from the triple associated to P to the triple associated to Q.*

(4) *If (ϕ_1, ϕ_2) is a morphism from (P_1, σ, P_2) to (Q_1, τ, Q_2), then there is a unique A-module homomorphism ϕ such that the diagram*

$$\begin{array}{ccc} (P_1,\sigma,P_2) & \xrightarrow{\phi} & (Q_1,\tau,Q_2) \\ {\scriptstyle \subseteq}\downarrow & & \downarrow{\scriptstyle \subseteq} \\ P_1 \oplus P_2 & \xrightarrow{\phi_1\oplus\phi_2} & Q_1 \oplus Q_2 \end{array}$$

commutes.

(5) *In (3), if both ϕ_1 and ϕ_2 are isomorphisms, then ϕ is an isomorphism.*

(6) *There is an isomorphism $(P_1, \sigma, P_2) \oplus (Q_1, \tau, Q_2) \cong (P_1 \oplus Q_1, \sigma \oplus \tau, P_2 \oplus Q_2)$ of A-modules.*

Proof. Part (1) follows from (14.31), Part (2) follows from (14.33), and Part (3) follows from $\tau(1 \otimes \phi_1(x_1)) = (1 \otimes \phi_2)(\sigma(1 \otimes x_1)) = (1 \otimes \phi_2)(1 \otimes x_2) = 1 \otimes \phi_2(x_2)$. Part (5) is left to the reader.

(4): Let $(y_1, y_2) \in (Q_1, \tau, Q_2)$. Let (x_1, x_2) be the unique element of $P_1 \oplus P_2$ such that $\phi_1(x_1) = y_1$ and $\phi_2(x_2) = y_2$. We have $\tau(1 \otimes \phi_1 x_1) = \tau(1 \otimes y_1) = 1 \otimes y_2 = 1 \otimes \phi_2 x_2$. Since all of the maps in Diagram (14.34) are isomorphisms, this implies $\sigma(1 \otimes x_1) = 1 \otimes x_2$. Thus ϕ is an isomorphism. □

Lemma 14.2.5. *In the context of the cartesian square of rings (14.31), let $\varphi_1 \in \mathrm{GL}_n(A_1)$, $\varphi_2 \in \mathrm{GL}_n(A_2)$, and $\varphi_3 \in \mathrm{GL}_n(A_3)$. Then there is an isomorphism of A-modules $(A_1^n, \varphi_3, A_2^n) \cong (A_1^n, f_2(\varphi_2)\varphi_3 f_1(\varphi_1), A_2^n)$.*

Proof. Applying the functor $\mathrm{GL}_n(\,)$ to (14.31) yields the commutative diagram

$$\begin{array}{ccc} \mathrm{GL}_n(A) & \xrightarrow{h_2} & \mathrm{GL}_n(A_2) \\ {\scriptstyle h_1}\downarrow & & \downarrow{\scriptstyle f_2} \\ \mathrm{GL}_n(A_1) & \xrightarrow{f_1} & \mathrm{GL}_n(A_3) \end{array} \tag{14.35}$$

of groups. The reader should verify that $(\varphi_1^{-1}, \varphi_2)$ is a morphism from $(A_1^n, \varphi_3, A_2^n)$ to $(A_1^n, f_2(\varphi_2)\varphi_3 f_1(\varphi_1), A_2^n)$ which by Lemma 14.2.4 (4) induces an isomorphism on A-modules. □

Lemma 14.2.6. *In the context of the cartesian square of rings (14.31), assume f_1 is onto. Then for any $\varphi \in \mathrm{GL}_n(A_3)$, the following are true.*

(1) *There is an isomorphism*

$$(A_1^n, \varphi, A_2^n) \oplus (A_1^n, \varphi^{-1}, A_2^n) \cong A^{2n}$$

of A-modules. In particular, the A-module (A_1^n, φ, A_2^n) is finitely generated and projective.

(2) *For $i = 1, 2$, there is an isomorphism*

$$A_i \otimes_A (A_1^n, \varphi, A_2^n) \cong A_i^n$$

of A_i-modules.

Proof. (1): The reader should verify that the identity

$$\begin{bmatrix} \varphi & 0 \\ 0 & \varphi^{-1} \end{bmatrix} = \begin{bmatrix} 1 & \varphi - 1 \\ 0 & 1 \end{bmatrix} \begin{bmatrix} 1 & 0 \\ 1 & 1 \end{bmatrix} \begin{bmatrix} 1 & \varphi^{-1} - 1 \\ 0 & 1 \end{bmatrix} \begin{bmatrix} 1 & 0 \\ -\varphi & 1 \end{bmatrix}$$

holds in $M_{2n}(A_3)$ and that each factor on the right hand side lifts to an element of $\mathrm{GL}_{2n}(A_1)$. Hence there exists $\varphi_1 \in \mathrm{GL}_{2n}(A_1)$ such that $f_1(\varphi_1) = \varphi \oplus \varphi^{-1}$. From Lemmas 14.2.4 and 14.2.5 we have

$$\begin{aligned} A^{2n} &\cong (A_1^{2n}, 1, A_2^{2n}) \\ &\simeq (A_1^{2n}, f_1(\varphi_1), A_2^{2n}) \\ &\cong (A_1^{2n}, \varphi \oplus \varphi^{-1}, A_2^{2n}) \\ &\cong (A_1^n, \varphi, A_2^n) \oplus (A_1^n, \varphi^{-1}, A_2^n) \end{aligned}$$

which proves (1).

(2): Consider the A-module homomorphisms

$$A^{2n} \xrightarrow{\cong} (A_1^n, \varphi, A_2^n) \oplus (A_1^n, \varphi^{-1}, A_2^n) \xrightarrow{h_i} A_i^n \oplus A_i^n$$

where h_i is defined by projecting each summand onto coordinate i. Upon tensoring with $A_i \otimes_A (\ )$ we get isomorphisms

$$A_i^{2n} \xrightarrow{\cong} A_i \otimes_A \left((A_1^n, \varphi, A_2^n) \oplus (A_1^n, \varphi^{-1}, A_2^n)\right) \xrightarrow{h_i} A_i^n \oplus A_i^n$$

which proves (2). □

Definition 14.2.7. In the context of Definition 14.2.3, the *fiber product category*, denoted $\mathbf{P}(A_1) \times_{\mathbf{P}(A_3)} \mathbf{P}(A_2)$, is the category whose objects are triples (P_1, σ, P_2) with $P_1 \in \mathbf{P}(A_1)$, $P_2 \in \mathbf{P}(A_2)$, and $\sigma : A_3 \otimes_{A_1} P_1 \to A_3 \otimes_{A_2} P_2$ an A_3-module isomorphism. The morphisms are morphisms of triples and the product is the direct sum. Assuming the rings A_1, A_2 and A_3 are commutative, the fiber product categories $\mathbf{FP}(A_1) \times_{\mathbf{FP}(A_3)} \mathbf{FP}(A_2)$, $\mathbf{Pic}(A_1) \times_{\mathbf{Pic}(A_3)} \mathbf{Pic}(A_2)$ and $\mathbf{Az}(A_1) \times_{\mathbf{Az}(A_3)} \mathbf{Az}(A_2)$ are defined in a similar fashion.

Theorem 14.2.8 (Milnor). *In the context of the cartesian square of rings (14.31), assume f_1 is onto. Let $P_1, Q_1 \in \mathbf{P}(A_1)$, $P_2, Q_2 \in \mathbf{P}(A_2)$, and $\sigma : A_3 \otimes_{A_1} P_1 \to A_3 \otimes_{A_2} P_2$ and $\tau : A_3 \otimes_{A_1} Q_1 \to A_3 \otimes_{A_2} Q_2$ isomorphisms in $\mathbf{P}(A_3)$. If $P = (P_1, \sigma, P_2)$ and $Q = (Q_1, \tau, Q_2)$ are the corresponding triples (see Definition 14.2.3), then the following are true.*

(1) *The triple $P = (P_1, \sigma, P_2)$ is a finitely generated projective A-module. That is, $P \in \mathbf{P}(A)$.*
(2) *$A_1 \otimes_A P \cong P_1$ as A_1-modules, and $A_2 \otimes_A P \cong P_2$ as A_2-modules.*
(3) *The two triples $P = (P_1, \sigma, P_2)$ and $Q = (Q_1, \tau, Q_2)$ are isomorphic as A-modules if and only if $\tau = f_2(\phi_2)\sigma f_1(\phi_1^{-1})$ for some A_1-module isomorphism $\phi_1 : P_1 \to Q_1$ and A_2-module isomorphism $\phi_2 : P_2 \to Q_2$.*
(4) *The functor from $\mathbf{P}(A)$ to $\mathbf{P}(A_1) \times_{\mathbf{P}(A_3)} \mathbf{P}(A_2)$ defined by*

$$P \mapsto (A_1 \otimes_A P, \gamma, A_2 \otimes_A P)$$

where γ is the natural isomorphism $A_3 \otimes_{A_1} P_1 \to A_3 \otimes_{A_2} P_2$, is an equivalence of categories.

Proof. (1): For $i = 1, 2$ there exist finitely generated projective A_i-modules N_i such that $P_i \oplus N_i \cong A_i^n$ (Corollary 1.1.7). There are A_3-module isomorphisms

$$
\begin{aligned}
A_3^n \oplus (A_3 \otimes_{A_1} N_1) &\xrightarrow{\cong} (A_3 \otimes_{A_2} P_2) \oplus (A_3 \otimes_{A_1} N_1) \oplus (A_3 \otimes_{A_2} N_2) \\
&\xrightarrow{\cong} (A_3 \otimes_{A_1} P_1) \oplus (A_3 \otimes_{A_1} N_1) \oplus (A_3 \otimes_{A_2} N_2) \\
&\xrightarrow{\cong} A_3^n \oplus (A_3 \otimes_{A_2} N_2)
\end{aligned}
\tag{14.36}
$$

where the center map is $\sigma^{-1} \oplus 1 \oplus 1$. Let ψ be the isomorphism defined by the composition of the maps in (14.36). We have $P_1 \oplus A_1^n \oplus N_1 \cong A_1^{2n}$ and $P_2 \oplus A_2^n \oplus N_2 \cong A_2^{2n}$. By Lemma 14.2.4 (5) we have an isomorphism of A-modules

$$
(P_1, \sigma, P_2) \oplus (A_1^n \oplus N_1, \psi, A_2^n \oplus N_2) \cong (A_1^{2n}, \sigma \oplus \psi, A_2^{2n}). \tag{14.37}
$$

By Lemma 14.2.6, the A-modules in (14.37) are finitely generated projective, from which (1) follows.

(2): This can be proved as in Lemma 14.2.6 (2).

(3): Assume $\theta : P \to Q$ is an A-module isomorphism. Let

$$
\begin{aligned}
\alpha_i &: A_i \otimes_A P \to P_i \\
\beta_i &: A_i \otimes_A Q \to Q_i
\end{aligned}
$$

be the isomorphisms from (2). Define $\phi_i = \beta_i(1 \otimes \theta)\alpha_i^{-1}$ to be the composite map

$$
P_i \xrightarrow{\alpha_i^{-1}} A_i \otimes_A P \xrightarrow{1\otimes\theta} A_i \otimes_A Q \xrightarrow{\beta_i} Q_i
$$

which is an A_i-module isomorphism. In the diagram

$$
\begin{array}{ccccccc}
A_3 \otimes_{A_1} P_1 & \xrightarrow{1\otimes\alpha_1^{-1}} & A_3 \otimes_{A_1} A_1 \otimes_A P & \xrightarrow{1\otimes 1\otimes\theta} & A_3 \otimes_{A_1} A_1 \otimes_A Q & \xrightarrow{1\otimes\beta_1} & A_3 \otimes_{A_1} Q_1 \\
\sigma\downarrow & & \downarrow & & \downarrow & & \downarrow\tau \\
A_3 \otimes_{A_2} P_2 & \xrightarrow{1\otimes\alpha_2^{-1}} & A_3 \otimes_{A_2} A_2 \otimes_A P & \xrightarrow{1\otimes 1\otimes\theta} & A_3 \otimes_{A_2} A_2 \otimes_A Q & \xrightarrow{1\otimes\beta_2} & A_3 \otimes_{A_2} P_2
\end{array}
$$

each of the three small squares commutes, hence the diagram commutes. Consequently, $\tau = f_2(\phi_2)\sigma f_1(\phi_1^{-1})$. The converse is Lemma 14.2.4 (4).

Part (4) is left to the reader. □

Theorem 14.2.9. *Assume in addition to all of the other hypotheses of Theorem 14.2.8 that the rings in the cartesian square (14.31) are commutative, and the modules P_i, Q_i are in* $\mathbf{FP}(A_i)$*. Then the following are true:*

(1) $P = (P_1, \sigma, P_2) \in \mathbf{FP}(A)$.

(2) *There is an A-module isomorphism:* $P \otimes_A Q \cong (P_1 \otimes_{A_1} Q_1, \sigma \otimes \tau, P_2 \otimes_{A_2} Q_2)$.

(3) *The assignment of Theorem* 14.2.8 *(4) induces an equivalence of categories* $\mathbf{FP}(A) \to \mathbf{FP}(A_1) \times_{\mathbf{FP}(A_3)} \mathbf{FP}(A_2)$.

Proof. (1): From Theorem 14.2.8, P is a finitely generated projective A-module. By Corollary 1.1.16, it suffices to show P is a faithful A-module. By Proposition 1.1.15, there exists $e \in \operatorname{idemp}(A)$ such that $\operatorname{annih}_A(P)$ is equal to Ae. Then $e = (e_1, e_2)$, where $e_i = h_i(e) \in \operatorname{idemp}(A_i)$. Then $e_1 P_1 = e_1(A_1 \otimes_A P) = e_1 A_1 \otimes_A P = A_1 \otimes_A eP = (0)$, which implies $e_1 \in \operatorname{annih}_{A_1} P_1 = (0)$. By Exercise 1.4.18, $P \to P_2$ is onto. Hence $e_2 \in \operatorname{annih}_{A_2} P_2 = (0)$ which proves P is faithful.

(2): Tensoring (14.31) with $P \otimes_A Q$ results in a cartesian square. A comparison of this square with the counterpart for $(P_1 \otimes_{A_1} Q_1, \sigma \otimes \tau, P_2 \otimes_{A_2} Q_2)$ results in the isomorphism.

Part (3) is left to the reader. □

2.2. Mayer-Vietoris Sequences. First we prove the existence of Milnor's six term Mayer-Vietoris exact sequence (Theorem 14.2.10). After that, we derive sufficient conditions on the rings in the cartesian square (14.38) such that the six term sequence can be extended to a nine term exact sequence including the Brauer groups. This version of the theorem was proved independently by Knus and Ojanguren [**KO74a**] and Childs [**Chi74**]. The presentation given below is based on that of Knus and Ojanguren. Throughout this section,

$$\begin{array}{ccc} R & \xrightarrow{h_2} & R_2 \\ {\scriptstyle h_1}\downarrow & & \downarrow{\scriptstyle f_2} \\ R_1 & \xrightarrow[f_1]{} & R_3 \end{array} \tag{14.38}$$

is a cartesian square of commutative rings with f_1 onto.

Theorem 14.2.10 (Milnor). *In the context of the cartesian square of commutative rings* (14.38), *there is an exact sequence of abelian groups*

$$1 \to R^* \xrightarrow{(h_1,h_2)} R_1^* \times R_2^* \xrightarrow{(f_1,1/f_2)} R_3^*$$
$$\xrightarrow{\partial} \operatorname{Pic} R \xrightarrow{(h_1,h_2)} \operatorname{Pic} R_1 \oplus \operatorname{Pic} R_2 \xrightarrow{(f_1,-f_2)} \operatorname{Pic} R_3$$

where $\partial(\alpha) = (R_1, \alpha, R_2)$.

Proof. First we show that ∂ is a well-defined homomorphism. Let $\alpha \in R_3^*$. Set $P = (R_1, \alpha, R_2)$ and $P' = (R_1, \alpha^{-1}, R_2)$. By Lemma 14.2.6, $P \oplus P' \cong R \oplus R$. Taking the second exterior power $\bigwedge^2()$ on both sides yields $P \otimes_R P' \cong R$

(see, for example [**Bou89a**]). By Proposition 1.3.4, P and P' are both R-progenerator modules. By Lemma 2.6.7, P is an invertible R-module and $P' \cong P^* = \operatorname{Hom}_R(P, R)$. Therefore, P represents a class in $\operatorname{Pic} R$. Let $\beta \in R_3^*$ and set $Q = (R_1, \beta, R_2)$. Since

$$\begin{bmatrix} \alpha & 0 \\ 0 & \beta \end{bmatrix} \begin{bmatrix} \beta & 0 \\ 0 & \beta^{-1} \end{bmatrix} = \begin{bmatrix} \alpha\beta & 0 \\ 0 & 1 \end{bmatrix}$$

it follows from Lemmas 14.2.4 and 14.2.5 that $P \oplus Q$ is isomorphic to the direct sum $(R_1, \alpha\beta, R_2) \oplus R$. Taking the second exterior power $\bigwedge^2()$ on both sides,

$$P \otimes_R Q = (R_1, \alpha, R_2) \otimes_R (R_1, \beta, R_2) \cong (R_1, \alpha\beta, R_2)$$

(see, for example [**Bou89a**]) which shows that ∂ is a homomorphism.

Now we show exactness of the sequence. By Lemma 14.2.6, the image of $(f_1, 1/f_2)$ is equal to the kernel of ∂. Exactness at R^* and at $R_1^* \times R_2^*$ follow from the exact sequence (14.32). Since the cartesian square (14.38) is commutative, the part of the sequence involving Picard groups is a complex. Assume (P_1, P_2) is in the kernel of $\operatorname{Pic} R_1 \oplus \operatorname{Pic} R_2 \to \operatorname{Pic} R_3$. Then there are isomorphisms of R_3-modules $\sigma : R_3 \otimes_{R_1} P_1 \to R_3 \otimes_{R_2} P_2$ and $\tau : R_3 \otimes_{R_1} P_1^* \to R_3 \otimes_{R_2} P_2^*$. By Theorem 14.2.9, the triples $P = (P_1, \sigma, P_2)$ and $P' = (P_1^*, \tau, P_2^*)$ are in $\mathbf{FP}(R)$ and $P \otimes_R P' \cong (P_1 \otimes_{R_1} P_1^*, \sigma \otimes \tau, P_2 \otimes_{R_2} P_2^*)$. But $P_i \otimes_{R_i} P_i^* \cong R_i$, so $\sigma \otimes \tau$ corresponds to multiplication by a unit in R_3. Thus, $P \otimes_R P'$ is in the image of ∂. By Lemma 2.6.7, P represents a class in $\operatorname{Pic} R$ which maps onto (P_1, P_2). □

Theorem 14.2.11. *In the context of the cartesian square* (14.38), *the following are true.*

(1) *If* $A_1 \in \mathbf{Az}(R_1)$, $A_2 \in \mathbf{Az}(R_2)$, *and* $\sigma : R_3 \otimes_{R_1} A_1 \to R_3 \otimes_{R_2} A_2$ *is an isomorphism in* $\mathbf{Az}(R_3)$, *then the triple* $A = (A_1, \sigma, A_2)$ *is in* $\mathbf{Az}(R)$.

(2) *The assignment of Theorem* 14.2.8 *(4) induces an equivalence of categories* $\mathbf{Az}(R) \to \mathbf{Az}(R_1) \times_{\mathbf{Az}(R_3)} \mathbf{Az}(R_2)$.

Proof. (1): It is clear that A is an R-algebra. First we assume each A_i is R_i-separable, and prove that A is R-separable. In the cartesian square

$$\begin{array}{ccccc} A = (A_1, \sigma, A_2) & & \longrightarrow & & A_2 \\ \downarrow & & & & \downarrow \phi_2 \\ A_1 & \xrightarrow{\phi_1} & R_3 \otimes_{R_1} A_1 & \xrightarrow{\sigma} & R_3 \otimes_{R_2} A_2 \end{array}$$

$\sigma\phi_1$ is onto and in the corresponding square for A^e

$$\begin{array}{ccccc} A^e = A \otimes_R A^o & & \longrightarrow & & A_2^e \\ \downarrow & & & & \downarrow{\scriptstyle \phi_2\otimes\phi_2} \\ A_1^e & \xrightarrow{\psi_1\otimes\psi_1} & R_3 \otimes_{R_1} A_1^e & \xrightarrow{\sigma\otimes\sigma} & R_3 \otimes_{R_2} A_2^e \end{array}$$

the bottom horizontal map $(\sigma \otimes \sigma)(\phi_1 \otimes \phi_1)$ is onto. Since each A_i is R_i-separable, by Proposition 4.1.2, each A_i is in $\mathbf{P}(A_i^e)$. By Theorem 14.2.8, A is in $\mathbf{P}(A^e)$. By Proposition 4.1.2, A is R-separable. Now we prove (1). Assume each A_i is R_i-Azumaya. By Lemma 4.1.4, the center of A is isomorphic to $\operatorname{Hom}_{A^e}(A, A)$ by a natural isomorphism. If $f \in \operatorname{Hom}_{A^e}(A, A)$, then by Theorem 14.2.8, f corresponds to a morphism defined by a pair (f_1, f_2) where $f_i \in \operatorname{Hom}_{A_i^e}(A_i, A_i) \cong R_i$. Hence $f \in R$ and A is R-central. This shows A is an R-Azumaya algebra.

Part (2) is left to the reader. □

Theorem 14.2.12. *In the context of the cartesian square of commutative rings* (14.38), *assume the following two hypotheses are satisfied.*

(1) *For each $L \in \mathbf{Pic}(R_3)$, there exist $Q_i \in \mathbf{FP}(R_i)$ $(i = 1, 2)$ and an isomorphism*

$$(R_3 \otimes_{R_1} Q_1) \otimes_{R_3} L \cong R_3 \otimes_{R_2} Q_2$$

in $\mathbf{FP}(R_3)$.

(2) *For each $M \in \mathbf{CRP}(R_3)$, there exist $Q_i \in \mathbf{FP}(R_i)$ $(i = 1, 2)$ and an isomorphism*

$$(R_3 \otimes_{R_1} Q_1) \otimes_{R_3} M \cong R_3 \otimes_{R_2} Q_2$$

in $\mathbf{FP}(R_3)$.

Then there exists a homomorphism ∂_1 such that

$$\begin{aligned} 1 \to R^* \xrightarrow{(h_1,h_2)} R_1^* \times R_2^* \xrightarrow{(f_1,1/f_2)} R_3^* & \\ \xrightarrow{\partial_0} \operatorname{Pic}(R) \xrightarrow{(h_1,h_2)} \operatorname{Pic}(R_1) \oplus \operatorname{Pic}(R_2) \xrightarrow{(f_1,-f_2)} \operatorname{Pic}(R_3) & \\ \xrightarrow{\partial_1} \mathrm{B}(R) \xrightarrow{(h_1,h_2)} \mathrm{B}(R_1) \oplus \mathrm{B}(R_2) \xrightarrow{(f_1,-f_2)} \mathrm{B}(R_3) & \end{aligned}$$

is an exact sequence of abelian groups where the first six terms are from Theorem 14.2.10.

Proof. The part of the sequence that comes from Theorem 14.2.10 is exact. Since the cartesian square (14.38) is commutative, the part of the sequence involving Brauer groups is a complex. It suffices to define ∂_1 and show exactness at the three terms $\operatorname{Pic}(R_3)$, $\mathrm{B}(R)$, and $\mathrm{B}(R_1) \oplus \mathrm{B}(R_2)$. The proof is divided into a series of four steps.

Step 1: The definition of ∂_1. Let $L \in \mathbf{Pic}(R_3)$ and assume by (1) that

$$\sigma : (R_3 \otimes_{R_1} Q_1) \otimes_{R_3} L \to R_3 \otimes_{R_2} Q_2$$

is an isomorphism in $\mathbf{FP}(R_3)$, with $Q_i \in \mathbf{FP}(R_i)$ $(i = 1, 2)$. For $i = 1, 2$, set $A_i = \operatorname{Hom}_{R_i}(Q_i, Q_i)$. By Proposition 7.6.2, σ induces an isomorphism

$$\bar{\sigma} : R_3 \otimes_{R_1} A_1 \to R_3 \otimes_{R_2} A_2.$$

In fact, $\bar{\sigma}$ is conjugation with σ. That is, if $g \in \operatorname{Hom}_{R_3}(R_3 \otimes_{R_1} Q_i, R_3 \otimes_{R_1} Q_i)$, then the diagram

$$\begin{array}{ccc} R_3 \otimes_{R_1} Q_1 \otimes_{R_3} L & \xrightarrow{\sigma} & R_3 \otimes_{R_2} Q_2 \\ {\scriptstyle g\otimes 1}\downarrow & & \downarrow{\scriptstyle \bar{\sigma}(g)} \\ R_3 \otimes_{R_1} Q_1 \otimes_{R_3} L & \xrightarrow{\sigma} & R_3 \otimes_{R_2} Q_2 \end{array}$$

commutes, hence $\bar{\sigma}(g) = \sigma(g \otimes 1)\sigma^{-1}$. By Theorem 14.2.11,

$$A = (A_1, \bar{\sigma}, A_2) = (\operatorname{Hom}_{R_1}(Q_1, Q_1), \bar{\sigma}, \operatorname{Hom}_{R_2}(Q_2, Q_2))$$

is an Azumaya R-algebra. Define $\partial_1(L) = [A] \in \mathrm{B}(R)$.

Now we show that the Brauer class $[A]$ does not depend on the choices of Q_1, Q_2, or σ. Assume U_1, U_2, τ are another set of similar data such that $\tau : (R_3 \otimes_{R_1} U_1) \otimes_{R_3} L \to R_3 \otimes_{R_2} U_2$ is an isomorphism in $\mathbf{FP}(R_3)$. Set $B_i = \operatorname{Hom}_{R_i}(U_i, U_i)$ and

$$B = (B_1, \bar{\tau}, B_2) = (\operatorname{Hom}_{R_1}(U_1, U_1), \bar{\tau}, \operatorname{Hom}_{R_2}(U_2, U_2))$$

where $\bar{\tau}$ is conjugation by τ. The opposite algebra of B is given by

$$B^o = (B_1^o, \bar{\tau}, B_2^o) = (\operatorname{Hom}_{R_1}(U_1, U_1)^o, \bar{\tau}, \operatorname{Hom}_{R_2}(U_2, U_2)^o).$$

By Corollary 1.5.3 (4) we identify B_i^o with $\operatorname{Hom}_{R_i}(U_i^*, U_i^*)$, where U_i^* denotes $\operatorname{Hom}_{R_i}(U_i, R_i)$, the dual module. The isomorphisms

$$R_3 \otimes_{R_2} U_2^* \xrightarrow{\tau^*} ((R_3 \otimes_{R_1} U_1) \otimes_{R_3} L)^* \cong (R_3 \otimes_{R_1} U_1^*) \otimes_{R_3} L^*$$

are the result of taking the dual of τ and using Theorem 1.3.26. It follows that

$$B^o = (B_1^o, \bar{\tau}, B_2^o) = (\operatorname{Hom}_{R_1}(U_1^*, U_1^*), \bar{\gamma}, \operatorname{Hom}_{R_2}(U_2^*, U_2^*))$$

where $\bar{\gamma}$ is conjugation by $\gamma = (\tau^*)^{-1}$. If we identify

$$A_i \otimes_{R_i} B_i^o \cong \operatorname{Hom}_{R_i}(Q_i \otimes_{R_i} U_i^*, Q_i \otimes_{R_i} U_i^*)$$

using Theorem 1.3.26, then $A \otimes_R B^o$ can be identified with the triple

$$(\operatorname{Hom}_{R_1}(Q_1 \otimes_{R_1} U_1^*, Q_1 \otimes_{R_1} U_1^*), \bar{\sigma} \otimes \bar{\gamma}, \operatorname{Hom}_{R_2}(Q_2 \otimes_{R_2} U_2^*, Q_2 \otimes_{R_2} U_2^*)).$$

Since $L \otimes_{R_3} L^* \cong R_3$, we see that $\sigma \otimes \gamma$ simplifies to

$$\begin{aligned} R_3 \otimes_{R_1} (Q_1 \otimes_{R_1} U_1^*) &\cong ((R_3 \otimes_{R_1} Q_1) \otimes_{R_3} L) \otimes_{R_3} ((R_3 \otimes_{R_1} U_1^*) \otimes_{R_3} L^*) \\ &\xrightarrow{\sigma \otimes (\tau^*)^{-1}} (R_3 \otimes_{R_2} Q_2) \otimes_{R_3} (R_3 \otimes_{R_2} U_2^*) \\ &\cong R_3 \otimes_{R_2} (Q_2 \otimes_{R_2} U_2^*). \end{aligned}$$

Then we can write $A \otimes_R B^o = \operatorname{Hom}_R(P, P)$, where

$$P = (Q_1 \otimes_{R_1} U_1^*, \sigma \otimes (\tau^*)^{-1}, Q_2 \otimes_{R_2} U_2^*).$$

But P is in $\mathbf{FP}(R)$ by Theorem 14.2.9, and Proposition 7.3.4 implies $[A] = [B]$ in $\mathrm{B}(R)$.

We have shown that ∂ is a well defined function. To complete Step 1, the reader should verify that ∂_1 is a homomorphism of groups.

Step 2: Exactness of the sequence at $\operatorname{Pic}(R_3)$. Using Step 1, the reader should verify that the image of $(f_1, -f_2)$ is contained in the kernel of ∂_1. We prove the reverse containment. Let $L \in \operatorname{Pic} R$ and assume $\partial_1(L) = [R] \in \mathrm{B}(R)$. We continue to use the notation from Step 1. Then $\partial_1(L) = [A]$, where

$$A = (A_1, \bar{\sigma}, A_2) = (\operatorname{Hom}_{R_1}(Q_1, Q_1), \bar{\sigma}, \operatorname{Hom}_{R_2}(Q_2, Q_2))$$

and $\bar{\sigma}$ is conjugation with the isomorphism $\sigma : (R_3 \otimes_{R_1} Q_1) \otimes_{R_3} L \to R_3 \otimes_{R_2} Q_2$ in $\mathbf{FP}(R_3)$. Proposition 7.3.4 implies $A \cong \operatorname{Hom}_R(M, M)$ for some $M \in \mathbf{FP}(R)$. By Theorem 14.2.9, $M \cong (M_1, \alpha, M_2)$, where $M_i \in \mathbf{FP}(R_i)$ $(i = 1, 2)$ and $\alpha : R_3 \otimes_{R_1} M_1 \to R_3 \otimes_{R_2} M_2$ is a patching isomorphism. It follows therefore, that $(\operatorname{Hom}_{R_1}(Q_1, Q_1), \bar{\sigma}, \operatorname{Hom}_{R_2}(Q_2, Q_2))$ is isomorphic to $(\operatorname{Hom}_{R_1}(M_1, M_1), \bar{\alpha} \operatorname{Hom}_{R_2}(M_2, M_2))$, where $\bar{\alpha}$ is conjugation by α. This is an isomorphism in the fiber product category, which means that there exist isomorphisms $\phi_i : \operatorname{Hom}_{R_i}(Q_i, Q_i) \cong \operatorname{Hom}_{R_i}(M_i, M_i)$ in $\mathbf{Az}(R_i)$ such that the diagram

$$\begin{array}{ccc} R_3 \otimes_{R_1} \operatorname{Hom}_{R_1}(Q_1, Q_1) & \xrightarrow{\bar{\sigma}} & R_3 \otimes_{R_2} \operatorname{Hom}_{R_2}(Q_2, Q_2) \\ {\scriptstyle 1\otimes\phi_1}\downarrow & & \downarrow{\scriptstyle 1\otimes\phi_2} \\ R_3 \otimes_{R_1} \operatorname{Hom}_{R_1}(M_1, M_1) & \xrightarrow{\bar{\alpha}} & R_3 \otimes_{R_2} \operatorname{Hom}_{R_2}(M_2, M_2) \end{array}$$

commutes. By Proposition 7.6.2, for each i, there is $L_i \in \mathbf{Pic}\,(R_i)$ such that ϕ_i is induced by an isomorphism

$$h_i : Q_i \otimes_{R_i} L_i \cong M_i$$

in $\mathbf{FP}(R_i)$. Therefore $\bar{\sigma}$ is induced by the composition

$$\begin{aligned} R_3 \otimes_{R_1} (Q_1 \otimes_{R_1} L_1) &\xrightarrow{1\otimes h_1} R_3 \otimes_{R_1} M_1 \\ &\xrightarrow{\alpha} R_3 \otimes_{R_2} M_2 \xrightarrow{(1\otimes h_2)^{-1}} R_3 \otimes_{R_2} (Q_2 \otimes_{R_2} L_2) \end{aligned}$$

or by the isomorphism

$$(R_3 \otimes_{R_1} Q_1) \otimes_{R_3} ((R_3 \otimes_{R_1} L_1) \otimes_{R_3} (R_3 \otimes_{R_1} L_2^*)) \cong R_3 \otimes_{R_2} Q_2.$$

In Proposition 7.6.2, the isomorphism of the endomorphism rings uniquely determines the element of $\operatorname{Pic}(R_3)$. We conclude that L is isomorphic to $((R_3 \otimes_{R_1} L_1) \otimes_{R_3} (R_3 \otimes_{R_1} L_2^*))$. This proves L comes from $\operatorname{Pic} R_1 \oplus \operatorname{Pic} R_2$.

Step 3: Exactness at $\mathrm{B}(R)$. Let $A \in \mathbf{Az}(R)$. By Theorem 14.2.11, A is isomorphic to a triple (A_1, σ, A_2). Then A is split by $R_1 \oplus R_2$ if and only if A_i is split by R_i $(i = 1, 2)$. Therefore, it is clear that the image of ∂_1 is in the kernel of (h_1, h_2). We prove the reverse inclusion. Assuming A_i is split by R_i, $A_i \cong \operatorname{Hom}_{R_i}(M_i, M_i)$ for some $M_i \in \mathbf{FP}(R_i)$. Then σ induces an isomorphism of R_3-algebras

$$\begin{aligned} \operatorname{Hom}_{R_3}(R_3 \otimes_{R_1} M_1, R_3 \otimes_{R_1} M_1) \xrightarrow{\cong} R_3 \otimes_{R_1} A_1 \xrightarrow{\sigma} R_3 \otimes_{R_2} A_2 \\ \xrightarrow{\cong} \operatorname{Hom}_{R_3}(R_3 \otimes_{R_2} M_2, R_3 \otimes_{R_2} M_2) \end{aligned}$$

which is induced by an isomorphism

$$h : (R_3 \otimes_{R_1} M_1) \otimes_{R_3} L \to R_3 \otimes_{R_2} M_2$$

in $\mathbf{FP}(R_3)$, for some $L \in \mathbf{Pic}\, R_3$ (Proposition 7.6.2). This proves that $[A] = \partial_1(L)$.

Step 4: Exactness at $\mathrm{B}(R_1) \oplus \mathrm{B}(R_2)$. Let $A_i \in \mathbf{Az}(R_i)$ $(i = 1, 2)$ and assume the R_3-algebras $R_3 \otimes_{R_1} A_1$ and $R_3 \otimes_{R_2} A_2$ are Brauer equivalent. Up to Brauer equivalence, we can choose A_i so that it has constant rank over R_i (Exercise 7.4.9). There are $M, N \in \mathbf{FP}(R_3)$ and an isomorphism

$$(R_3 \otimes_{R_1} A_1) \otimes_{R_3} \operatorname{Hom}_{R_3}(M, M) \cong (R_3 \otimes_{R_2} A_2) \otimes_{R_3} \operatorname{Hom}_{R_3}(N, N)$$

of R_3-algebras. Using the fact that $R_3 \otimes_{R_1} A_1$ and $R_3 \otimes_{R_2} A_2$ have constant ranks over R_3, we can tensor with some $\operatorname{Hom}_{R_3}(M', M')$, $M' \in \mathbf{FP}(R_3)$ and assume M and N are in $\mathbf{CRP}(R_3)$. Applying Bass' Theorem (Theorem 14.2.1), we can modify M and N one more time and assume moreover that $N \cong R_3^{(n)}$ is free. By Hypothesis (2), there exist $Q_i \in \mathbf{FP}(R_i)$ $(i = 1, 2)$ and an isomorphism

$$R_3 \otimes_{R_1} Q_1 \cong M \otimes_{R_3} (R_3 \otimes_{R_2} Q_2)$$

in $\mathbf{FP}(R_3)$. This induces a sequence of isomorphisms

$$\begin{aligned}
&R_3 \otimes_{R_1} (A_1 \otimes_{R_1} \operatorname{Hom}_{R_1}(Q_1, Q_1)\\
&\cong (R_3 \otimes_{R_1} A_1) \otimes_{R_3} \operatorname{Hom}_{R_3}(R_3 \otimes_{R_1} Q_1, R_3 \otimes_{R_1} Q_1)\\
&\cong (R_3 \otimes_{R_1} A_1) \otimes_{R_3} \operatorname{Hom}_{R_3}(M \otimes_{R_3} (R_3 \otimes_{R_2} Q_2), M \otimes_{R_3} (R_3 \otimes_{R_2} Q_2))\\
&\cong ((R_3 \otimes_{R_1} A_1) \otimes_{R_3} \operatorname{Hom}_{R_3}(M, M)) \otimes_{R_3} (R_3 \otimes_{R_2} \operatorname{Hom}_{R_2}(Q_2, Q_2)\\
&\cong ((R_3 \otimes_{R_2} A_2) \otimes_{R_3} \operatorname{Hom}_{R_3}(N, N)) \otimes_{R_3} (R_3 \otimes_{R_2} \operatorname{Hom}_{R_2}(Q_2, Q_2)\\
&\cong R_3 \otimes_{R_2} \left(A_2 \otimes_{R_2} \operatorname{Hom}_{R_2}(R_2^{(n)}, R_2^{(n)}) \otimes_{R_2} \operatorname{Hom}_{R_2}(Q_2, Q_2)\right)\\
&\cong R_3 \otimes_{R_2} \left(A_2 \otimes_{R_2} \operatorname{Hom}_{R_2}(Q_2^{(n)}, Q_2^{(n)})\right).
\end{aligned}$$

Call the composite map γ. Taking γ as the patching isomorphism, the triple

$$(A_1 \otimes_{R_1} \operatorname{Hom}_{R_1}(Q_1, Q_1), \gamma, A_2 \otimes_{R_2} \operatorname{Hom}_{R_2}(Q_2^{(n)}, Q_2^{(n)}))$$

represents a class in $\mathrm{B}(R)$ that maps onto $([A_1], [A_2])$. □

Corollary 14.2.13. *In the context of Theorem* 14.2.12, *assume for every finitely generated projective R_3-module M of constant rank there exists a positive integer n such that the direct sum of n copies of M is free. Then the sequence of abelian groups*

$$\cdots \to \operatorname{Pic}(R_3) \xrightarrow{\partial_1} \mathrm{B}(R) \xrightarrow{(h_1,h_2)} \mathrm{B}(R_1) \oplus \mathrm{B}(R_2) \xrightarrow{(f_1,-f_2)} \mathrm{B}(R_3)$$

is exact.

Proof. The reader should verify that Hypotheses (1) and (2) are satisfied. □

Remark 14.2.14. In Theorem 14.2.12, Hypothesis (2) was only used in Step 4 to prove exactness at $\mathrm{B}(R_1) \oplus \mathrm{B}(R_2)$. Therefore, the rest of the sequence is exact, assuming only Hypothesis (1) is satisfied.

In Corollary 14.2.15 we use the notation $(\,)_t$ to denote the subgroup of torsion elements in an abelian group.

Corollary 14.2.15. *In the context of Theorem* 14.2.12, *the following are true.*

(1) *There is a homomorphism of abelian groups $\partial_1 : (\operatorname{Pic} R_3)_t \to \mathrm{B}(R)$.*
(2) *If $\operatorname{Pic}(R_3)$ is a torsion group, then the sequence of abelian groups*

$$\cdots \to \operatorname{Pic}(R_3) \xrightarrow{\partial_1} \mathrm{B}(R) \xrightarrow{(h_1,h_2)} \mathrm{B}(R_1) \oplus \mathrm{B}(R_2)$$

is exact.

Proof. By Corollary 14.2.2, Hypothesis (1) is satisfied for any L in $(\operatorname{Pic} R_3)_t$. Therefore, Step 1 of the proof of Theorem 14.2.12 shows that the homomorphism ∂_1 can always be defined on the subgroup of torsion elements in $\operatorname{Pic} R_3$. □

2.3. Exercises.

Exercise 14.2.16. Let R be any ring and $I_1, \dots, I_n$ two-sided ideals in R such that $(0) = I_1 \cap I_2 \cap \cdots \cap I_n$. Let $\bar{R} = R/I_1 \oplus \cdots \oplus R/I_n$. For $1 \le k \le n$, set $c_k = I_k + \bigcap_{j \neq k} I_j$. If c is the conductor ideal from $\bar{R}$ to R, show that c is equal to $c_1/I_1 \oplus \cdots \oplus c_n/I_n$.

Exercise 14.2.17. This exercise is a continuation of Exercise 4.4.21. Let k be a field and $R = k[x, y]/(xy)$.

(1) Prove that $R^* = k^*$ and $\operatorname{Pic} R = (0)$.

(2) If k is a perfect field, prove that $\mathrm{B}(R) = \mathrm{B}(k)$.

Exercise 14.2.18. This exercise is a continuation of Exercise 11.1.11. Let R be the ring $\mathbb{R}[x, y]/(x^2 + y^2)$ and S the ring $\mathbb{C} \otimes_{\mathbb{R}} R = \mathbb{C}[x, y]/(x^2 + y^2)$ obtained by extension of scalars to $\mathbb{C}$.

(1) For the Galois extension S/R, compute all of the terms in the exact sequence of Theorem 13.3.1.

(2) Show that $\mathrm{B}(R)$ is isomorphic to $\mathrm{B}(\mathbb{R})$, hence is cyclic of order two. For the complete description of the Brauer group of a real curve, the interested reader is referred to [**DeM76**] and [**DK76**].

Exercise 14.2.19. This exercise is a continuation of Exercise 2.6.18. Let k be a field and $R = k[x^2, x^3]$. Prove that $\operatorname{Pic} R$ is isomorphic to the additive group of k.

Exercise 14.2.20. Let k be a field in which 2 is invertible and set $S = k[x, w]/(x^2 - (w^2 - 1)^2)$. Let $I = (x - (w^2 - 1))$ and $J = (x + (w^2 - 1))$ be the two minimal primes of S. Set $\bar{S} = S/I \oplus S/J$. Notice that we are in the context of Exercise 9.4.7, with $D = 2$. Then the conductor ideal from $\bar{S}$ to S is equal to $I + J = (x, w^2 - 1)$. Using the Mayer-Vietoris sequence of Corollary 14.2.15 associated to the cartesian square

$$\begin{array}{ccc}
S = \frac{k[x,w]}{(x^2-(w^2-1)^2)} & \xrightarrow{h_2} & \bar{S} = \frac{k[x,w]}{(x-(w^2-1))} \oplus \frac{k[x,w]}{(x+(w^2-1))} \\
{\scriptstyle h_1}\big\downarrow & & \big\downarrow{\scriptstyle f_2} \\
\frac{S}{I+J} \cong \frac{k[w]}{(w^2-1)} & \xrightarrow[f_1]{} & \frac{\bar{S}}{I+J} \cong \frac{k[w]}{(w^2-1)} \oplus \frac{k[w]}{(w^2-1)}
\end{array}$$

prove that $S^* = k^*$ and $\operatorname{Pic}(S) = k^*$. If k is a perfect field, prove that $\operatorname{B}(S) = \operatorname{B}(k)$. The ring S is the affine coordinate ring of the union of two conics. For a continuation of this exercise, see Exercise 14.3.14.

Exercise 14.2.21. Let k be an algebraically closed field of characteristic zero, $n \geq 2$, and R the finitely generated k-algebra of Example 6.5.12 with the property that $\operatorname{Pic}(R) \cong \mathbb{Z}/n$. Then for some m, there is a presentation

$$0 \to I \to A = k[x_1, \dots, x_m] \xrightarrow{\theta} R \to 0$$

of R. Let $S = \{(a, b) \in A \oplus A \mid \theta(a) = \theta(b)\}$. Then as in Exercise 4.4.20, S corresponds to the union of two copies of $\mathbb{A}^m$ glued along a common closed subvariety corresponding to $Z(I)$ and the diagram

$$\begin{array}{ccc} S & \longrightarrow & A \\ \downarrow & & \downarrow \theta \\ A & \xrightarrow[\theta]{} & R = A/I \end{array}$$

is a cartesian square of rings. Show that $\partial_1 : \operatorname{Pic}(R) \to \operatorname{B}(S)$ is an isomorphism. Therefore, $\operatorname{B}(S) \cong \mathbb{Z}/n$.

3. Brauer Groups of Some Nonnormal Domains

Let R be an integral domain with field of fractions K, $\bar{R}$ the integral closure of R in K, and $c = R : \bar{R}$ the conductor ideal from $\bar{R}$ to R. Then c is an ideal of both R and $\bar{R}$ and by Exercise 1.4.20, the diagram

$$\begin{array}{ccc} R & \longrightarrow & R/c \\ \downarrow & & \downarrow f_2 \\ \bar{R} & \xrightarrow{f_1} & \bar{R}/c \end{array} \tag{14.39}$$

is a cartesian square in which f_1 is onto. To apply the Mayer-Vietoris sequence of Theorem 14.2.12 to compute the Brauer group $\operatorname{B}(R)$, one must first know that the hypotheses (1) and (2) are satisfied. Secondly, it is necessary to verify that $(0) \neq c \neq R$ because otherwise diagram (14.39) will be trivial. Notice that $c \neq R$ if and only if R is not integrally closed. By Proposition 6.3.3 (5), we see that $c \neq (0)$ if and only if $\bar{R}$ is an R-lattice in K. Sufficient conditions for $c \neq (0)$ to be true are given in Theorems 4.6.10 and 8.4.7. The reader should be advised that in general the integral closure of a noetherian integral domain R is not even a noetherian ring. For more on this subject, the interested reader is referred to [**Mat80**, §31, §41], [**Nag62**, §33 – §36], and [**Fos73**, §4].

3.1. The Brauer Group of an Algebraic Curve. For our first example in this section we prove that an affine algebraic curve over an algebraically closed field has a trivial Brauer group. Theorem 14.3.1 is due to Grothendieck ([**Gro68c**, Corollaire (1.2)]). The proof given below is based on the reduction process used by DeMeyer in [**DeM76**]. This example illustrates the two basic applications of the Mayer-Vietoris sequence. First, to compute the Brauer group of a normalization of a noetherian integral domain. Second, to compute the Brauer group of a noetherian ring in terms of the Brauer groups of its irreducible components.

Theorem 14.3.1. *Let k be an algebraically closed field and R a finitely generated k-algebra of Krull dimension one. That is, R is the affine coordinate ring of an algebraic curve in $\mathbb{A}^n_k$, for some n. Then* $\mathrm{B}(R) = (0)$.

Proof. Case 1: R is a regular domain. Let K be the quotient field of R. By Theorem 11.3.18, $\mathrm{B}(R) \to \mathrm{B}(K)$ is one-to-one. By Tsen's Theorem (Corollary 11.1.9, $\mathrm{B}(K) = (0)$. Thus $\mathrm{B}(R) = (0)$.

Case 2: R is an integral domain. Let K be the quotient field of R and $\bar{R}$ the integral closure of R in K. Then $\bar{R}$ is a Dedekind domain, hence is regular (Theorem 6.4.5). By Case 1, assume $R \neq \bar{R}$. Let $c = R : \bar{R}$ be the conductor ideal from $\bar{R}$ to R. Then diagram (14.39) is a cartesian square and f_1 is onto. By Theorem 8.4.7, $\bar{R}$ is an R-lattice in K. By Proposition 6.3.3 (5), we see that $c \neq (0)$. Since c is a proper ideal of both R and $\bar{R}$, it follows that R/c and $\bar{R}/c$ are both artinian. By Corollary 7.9.7, $\mathrm{B}(R/c) = \mathrm{B}(\bar{R}/c) = (0)$. By Corollary 3.2.12, a finitely generated projective $\bar{R}/c$-module of constant rank is free. In particular, this implies $\mathrm{Pic}(\bar{R}/c) = (0)$. By Case 1, $\mathrm{B}(\bar{R}) = (0)$. The exact sequence of Corollary 14.2.15 proves $\mathrm{B}(R) = (0)$.

Case 3: $\mathrm{Rad}_R(0) = (0)$. By Case 2, we assume R is not an integral domain. If $P_1, \dots, P_n$ are the distinct minimal primes of R, then $(0) = \mathrm{Rad}_R(0) = \bigcap_i P_i$ and if $\bar{R} = R/P_1 \oplus \cdots \oplus R/P_n$, then $R \to \bar{R}$ is one-to-one. For $1 \leq t \leq n$, set $c_t = P_t + \bigcap_{j \neq t} P_j$. By Exercise 14.2.16, the conductor ideal from $\bar{R}$ to R is equal to $c = c_1/P_1 \oplus \cdots \oplus c_n/P_n$. Then diagram (14.39) is a cartesian square with f_1 onto. Since P_t is a minimal prime, the reader should verify that $P_t \not\supseteq \bigcap_{j \neq t} P_j$. This means $P_t \subsetneq c_t$. Hence either c_t is the unit ideal, or $\mathrm{ht}(c_t) = 1$. Thus, R/c and $\bar{R}/c$ are either the trivial ring (0), or are artinian. In any case, $\mathrm{B}(R/c) = (0)$ and $\mathrm{Pic}(\bar{R}/c) = (0)$. The ring $\bar{R}$ is a finite direct sum of copies of k (if R has any components of dimension zero) and integral domains of dimension one. Using Case 2, $\mathrm{B}(\bar{R}) = (0)$. The exact sequence of Corollary 14.2.15 implies $\mathrm{B}(R) = (0)$.

Case 4: The last case. It follows from Case 3 that $\mathrm{B}(R/\mathrm{Rad}_R(0)) = (0)$. By Corollary 7.9.3, $\mathrm{B}(R) = \mathrm{B}(R/\mathrm{Rad}_R(0))$. □

3.2. Every Finite Abelian Group is a Brauer Group. Let $n \geq 2$ and k an algebraically closed field of characteristic zero. An example of a regular finitely generated k-algebra of Krull dimension three such that the Brauer group is cyclic of order n appeared in Example 13.6.15 above. The proof utilized Theorem 13.6.9, which is not proved in this book. In Exercise 14.2.21 there is an example of a ring S for which $\mathrm{B}(S) \cong \mathbb{Z}/n$. The ring S is not an integral domain. For these reasons, we feel justified in presenting a third example of a ring with cyclic Brauer group. The example in this section is elementary in the sense that the proof uses only theorems proved here.

We now construct an integral domain A which is a finitely generated k-algebra of Krull dimension three such that $\mathrm{B}(A) = \mathbb{Z}/n$. This example originally appeared in [**For81**]. By taking direct sums, this proves that any finite abelian group appears as the Brauer group of the affine coordinate ring of an algebraic 3-fold. By the construction below, we see that the integral closure of A has trivial Brauer group. Thus, $\mathrm{B}(A) \to \mathrm{B}(K)$ is the zero map. In other words, every Azumaya A-algebra is generically split.

Start with the ring R of Example 13.6.12. Then R is a regular integral domain, a finitely generated k-algebra, $\dim(R) = 2$, $\mathrm{B}(R) = (0)$, and $\mathrm{Pic}(R) = \mathbb{Z}/n$. Now let

$$A_1 = k[x, y]/(y^2 - x^2(x + 1))$$

be the affine coordinate ring of a nodal cubic curve (see Exercise 8.1.34). The reader should verify that A_1 is an integral domain and the integral closure is $\bar{A}_1 = k[x, w]/(w^2 - x - 1)$, where $y = xw$. By the identity $x = w^2 - 1$, we get $\bar{A}_1 = k[w]$. The conductor ideal from $\bar{A}_1$ to A_1 is $c = (y, x)$. Hence, $A_1/c = k$ and $\bar{A}_1/c = k[w]/(w^2 - 1) = k \oplus k$. The diagram

$$(14.40)\qquad \begin{array}{ccc} A_1 = \frac{k[x,y]}{(y^2 - x^2(x+1))} & \longrightarrow & \frac{A_1}{c} = k \\ \downarrow & & \downarrow \\ \bar{A}_1 = k[w] & \longrightarrow & \frac{\bar{A}_1}{c} = k[w]/(w^2 - 1) = k \oplus k \end{array}$$

is a cartesian square of commutative k-algebras. Upon tensoring everything in (14.40) with $R \otimes_k (\,)$ we get another cartesian square

$$(14.41)\qquad \begin{array}{ccc} A = \frac{R[x,y]}{(y^2 - x^2(x+1))} & \xrightarrow{h_2} & \frac{A}{c} = R \\ {\scriptstyle h_1}\downarrow & & \downarrow{\scriptstyle f_2} \\ \bar{A} = R[w] & \xrightarrow{f_1} & \frac{\bar{A}}{c} = R \oplus R \end{array}$$

where $h_1(x) = w^2 - 1$, $h_1(y) = w(w^2 - 1)$, $h_2(x) = h_2(y) = 0$, $f_2(a) = (a, a)$, and $f_1(p(w)) = (p(1), p(-1))$. Since f_1 is onto and $\mathrm{Pic}(\bar{A}/c) \cong \mathbb{Z}/n \oplus \mathbb{Z}/n$, we are in the context of Corollary 14.2.15. By Theorem 13.6.3, $\mathrm{B}(R[w]) = \mathrm{B}(R) = (0)$. Therefore, the Mayer-Vietoris sequence becomes

$$\operatorname{Pic}(A) \xrightarrow{(h_1,h_2)} \operatorname{Pic}(\bar{A}) \oplus \operatorname{Pic}(A/c) \xrightarrow{(f_1,-f_2)} \operatorname{Pic}(\bar{A}/c) \xrightarrow{\partial_1} \mathrm{B}(A) \to 0.$$

All of the rings in (14.41) are R-algebras. Moreover, the structure homomorphism $R \to A$ followed by h_2 is the identity map. Therefore, $\operatorname{Pic}(R) \to \operatorname{Pic}(A)$ embeds $\operatorname{Pic}(R)$ as a direct summand of $\operatorname{Pic}(A)$. By Theorem 6.5.3, $\operatorname{Pic}(\bar{A}) = \operatorname{Pic}(R)$. This proves that in the sequence

$$\mathbb{Z}/n \xrightarrow{(h_1,h_2)} \mathbb{Z}/n \oplus \mathbb{Z}/n \xrightarrow{(f_1,-f_2)} \mathbb{Z}/n \oplus \mathbb{Z}/n \xrightarrow{\partial_1} \mathrm{B}(A) \to 0$$

the image of (h_1, h_2) is the diagonal subgroup. It is clear that f_2 maps onto the diagonal subgroup of $\mathbb{Z}/n \oplus \mathbb{Z}/n$. Because $\operatorname{Pic}(R) \to \operatorname{Pic}(\bar{A})$ is onto we see that the image of f_1 is also the diagonal subgroup of $\operatorname{Pic}(R) \oplus \operatorname{Pic}(R)$. This proves that the image of ∂_1, which is $\mathrm{B}(A)$, is isomorphic to $\mathbb{Z}/n$.

3.3. A Family of Nonnormal Subrings of $k[x, y]$. Let k be an algebraically closed field of characteristic zero. We study a family of nonnormal rational algebraic surfaces over k such that the Brauer groups are nontrivial. The surfaces that we consider are based on an example that originally appeared in [**DF82**].

In the following examples we construct subrings $k \subseteq R \subseteq k[x, y]$ for which R is nonnormal, $k[x, y]$ is integral over R, and $\mathrm{B}(R) \neq (0)$. These examples indicate the significance of the Brauer group invariant in the study of integral domains which are not integrally closed. For instance, in [**Li95**], H. Li proves that if k is one of the fields $\mathbb{R}$ or $\mathbb{C}$, and $k \subseteq R \subseteq k[x, y]$ is a tower of subrings such that R is normal and the inclusion $R \subseteq k[x, y]$ is integral, then the natural maps $\mathrm{B}(k) \to \mathrm{B}(R) \to \mathrm{B}(k[x, y])$ are isomorphisms.

Let $p(x) \in k[x]$ be a polynomial of degree greater than or equal to three. Let X be the affine surface in $\mathbb{A}^3$ defined by $z^2 - y(y - p(x))^2 = 0$. Then X is a double cover of the xy-plane. The affine coordinate ring of X is $R = k[x, y, z]/(z^2 - y(y - p(x))^2)$. Following the procedure of Exercise 9.4.8, let $\bar{X}$ be the affine surface in $\mathbb{A}^3$ defined by $w^2 - y = 0$. The affine coordinate ring of $\bar{X}$ is $\bar{R} = k[x, y, w]/(w^2 - y)$. Notice that upon eliminating y, we have $\bar{R} = k[x, w]$. Define a k-algebra homomorphism

$$R = \frac{k[x, y, z]}{(z^2 - y(y - p(x))^2)} \xrightarrow{\phi} \bar{R} = \frac{k[x, y, w]}{(w^2 - y)} = k[x, w]$$

by $x \mapsto x$, $y \mapsto w^2$, and $z \mapsto w(y - p(x)) = w(w^2 - p(x))$. If we identify R with the image of ϕ, then as a subring of $k[x, w]$, R is equal to $k[x, w^2, w(w^2 - p(x))]$. Thus R and $\bar{R}$ have the same quotient field $k(x, w)$. The reader should verify (either directly, or as in Exercise 9.4.8) that $\bar{R}$ is the integral closure of R, and the conductor ideal from $\bar{R}$ to R is $c = (z, y - p(x))$. Notice that in $\bar{R}$

we have $c = (w^2 - p(x))$. Then $R/c \cong k[x]$ and $\bar{R}/c \cong k[x,w]/(w^2 - p(x))$. So $\bar{R}/c$ is the affine coordinate ring of the irreducible hyperelliptic curve $w^2 - p(x)$ and is a double cover of R/c (see Exercise 9.4.7). We have the following commutative diagram of rings and quotient fields

$$\begin{array}{ccc} \bar{R} = k[x,w] & \longrightarrow & K = k(x,w) \\ \uparrow & & \uparrow = \\ R = k[x, w^2, w(w^2 - p(x))] & \longrightarrow & K = k(x,w) \\ \uparrow & & \uparrow \\ k[x, w^2] & \longrightarrow & k(x, w^2) \end{array}$$

where an arrow represents inclusion. By Exercise 1.4.20, the conductor diagram

$$(14.42)\qquad \begin{array}{ccc} R = \frac{k[x,y,z]}{(z^2 - y(y-p(x))^2)} & \longrightarrow & k[x] = R/c \\ \phi \downarrow & & \downarrow f_2 \\ \bar{R} = k[x,w] & \xrightarrow{f_1} & \frac{k[x,w]}{(w^2 - p(x))} = \bar{R}/c \end{array}$$

is a cartesian square and f_1 is onto. We have the following groups.

$$\begin{aligned} (\bar{R})^* &= k[x,w]^* = k^* \\ (R/c)^* &= k[x]^* = k^* \\ \operatorname{Pic}(\bar{R}) &= \operatorname{Pic}(k[x,w]) = (0) && \text{by Theorem 6.5.3,} \\ \operatorname{Pic}(R/c) &= \operatorname{Pic}(k[x]) = (0) && \text{by Theorem 6.5.3,} \\ \mathrm{B}(R/c) &= \mathrm{B}(k[x]) = (0) && \text{by Theorem 13.6.9,} \\ \mathrm{B}(\bar{R}) &= \mathrm{B}(k[x,w]) = (0) && \text{by Theorem 13.6.9,} \\ \mathrm{B}(\bar{R}/c) &= (0) && \text{by Theorem 14.3.1.} \end{aligned}$$

The exact Mayer-Vietoris sequence of Theorem 14.2.10 associated to (14.42) becomes

$$(14.43)\qquad 1 \to R^* \to k^* \times k^* \to (\bar{R}/c)^* \to \operatorname{Pic} R \to (0).$$

Since all of the rings in (14.42) are k-algebras, it follows from (14.43) that $R^* = k^*$ and the sequence

$$(14.44)\qquad 1 \to k^* \to (\bar{R}/c)^* \to \operatorname{Pic} R \to (0)$$

is split exact. Although we do not prove it here, it follows from [**For14**, Proposition 3.7] that the group $\operatorname{Pic} R$ is trivial if the degree of $p(x)$ is odd, or if $p(x)$ is a sufficiently general polynomial of even degree in $k[x]$. Otherwise, $\operatorname{Pic} R$ is isomorphic to $\mathbb{Z}$. The proof of Theorem 14.3.2 uses the fact that any finitely generated projective module over $k[x, w]$ is free. This is a special case of Serre's Theorem which we do not prove in this book (see [**Rot79**, Theorem 4.62] or [**Lam06**]).

Theorem 14.3.2. *In the above context, there is an isomorphism*

$$\partial_1 : (\operatorname{Pic}(\bar{R}/c))_t \xrightarrow{\cong} \mathrm{B}(R)$$

of abelian groups, where by $(\ )_t$ *we denote the subgroup of torsion elements in an abelian group.*

Proof. By Corollary 14.2.15, $\partial_1 : (\operatorname{Pic}(\bar{R}/c))_t \to \mathrm{B}(R)$ is a well defined homomorphism of abelian groups. The identities $\operatorname{Pic}(\bar{R}) = 0$, $\operatorname{Pic}(R/c) = (0)$, and Step 2 of the proof of Theorem 14.2.12 show that ∂_1 is one-to-one.

To show that ∂_1 is onto, we use the identities $\mathrm{B}(\bar{R}) = \mathrm{B}(R/c) = (0)$ and proceed as in Step 3 of the proof of Theorem 14.2.12. So that our notation coincides with that of Theorem 14.2.12, write R_1 for $\bar{R}$, R_2 for R/c, and R_3 for $\bar{R}/c$. If $A = (A_1, \sigma, A_2)$ is in $\mathbf{Az}(R)$, then each A_i is split over the respective ring R_i. By Serre's Theorem mentioned above, for $i = 1, 2$, $\mathbf{FP}(R_i)$ consists of free modules. This implies $A_i = M_n(R_i)$, and $\sigma : M_n(R_3) \to M_n(R_3)$ is induced by an isomorphism

$$h : R_3^{(n)} \otimes_{R_3} L \to R_3^{(n)}$$

in $\mathbf{FP}(R_3)$, for some $L \in \mathbf{Pic}\, R_3$ (Proposition 7.6.2). By Corollary 14.2.2, L is in $(\operatorname{Pic} R_3)_t$. Therefore, $[A] = \partial_1(L)$. □

The rest of this section is devoted to computing $\mathrm{B}(R)$ for several different choices of $p(x)$. These examples illustrate various obstructions to the exactness of the sequence of Theorem 14.2.12. In each of the following cases, Hypotheses (1) and (2) fail to hold.

Example 14.3.3. Let k be an algebraically closed field of characteristic zero and $R = k[x, w^2, w(w^2 - x^3)]$. In the above notation, $p(x) = x^3$. Consider the k-algebra homomorphism $\psi : \bar{R} \to k[u]$ defined by $x \mapsto u^2$, $w \mapsto u^3$. Then the kernel of ψ is the principal ideal $c = (w^2 - x^3)$, so $\bar{R}/c$ can be identified with $k[u^2, u^3]$. As computed in Exercise 14.2.19, $\operatorname{Pic}(\bar{R}/c)$ is isomorphic to the additive group of k. Since $\operatorname{char} k = 0$, this is a torsion free group. As mentioned above, $\mathbf{FP}(\bar{R})$ and $\mathbf{FP}(R/c)$ consist of free modules. It follows from Corollary 14.2.2 that Hypothesis (1) of Theorem 14.2.12 is not satisfied, hence the map ∂_1 is not defined on the full Picard group $\operatorname{Pic}(\bar{R}/c)$. By Theorem 14.3.2, $\mathrm{B}(R) = (0)$. Using (14.44), the reader should verify that $\operatorname{Pic}(R) = (0)$.

Example 14.3.4. Let k be an algebraically closed field of characteristic zero and $R = k[x, w^2, w(w^2 - x^2(x+1))]$. In the above notation, $p(x) = x^2(x+1)$.

We have $\bar{R}/c = k[x,w]/(w^2 - x^2(x+1))$ which is isomorphic to the ring that was denoted A_1 in Section 14.3.2. The Mayer-Vietoris sequence associated to the cartesian square (14.40) becomes

$$1 \to (\bar{R}/c)^* \to k^* \times k^* \xrightarrow{(f_1, 1/f_2)} k^* \times k^* \to \operatorname{Pic}(\bar{R}/c) \to 0$$

and the image of $(f_1, 1/f_2)$ is the diagonal. Therefore, $\operatorname{Pic}(\bar{R}/c)$ is isomorphic to k^*. Again we see that Hypothesis (1) of Theorem 14.2.12 is not satisfied. For any $n > 0$, let μ_n denote the kernel of the nth power map on k^*. Since k is algebraically closed and $\operatorname{char} k = 0$, μ_n is a cyclic group of order n. The subgroup of k^* consisting of torsion elements is $\mu = \bigcup_{n \geq 1} \mu_n$. There is a (noncanonical) isomorphism $\mu \cong \mathbb{Q}/\mathbb{Z}$. Therefore, $\mathrm{B}(R) \cong \mathbb{Q}/\mathbb{Z}$, by Theorem 14.3.2. Using (14.44), the reader should verify that $\operatorname{Pic}(R) = (0)$.

Remark 14.3.5. In the above context, let $k = \mathbb{C}$ be the field of complex numbers. If X is the affine curve $Z(w^2 - p(x))$ with affine coordinate ring $\bar{R}/c$, then one can show that the torsion subgroup of $\operatorname{Pic}(\bar{R}/c)$ is either trivial, or a finite direct sums of copies of $\mathbb{Q}/\mathbb{Z}$. Assuming X is irreducible, $(\operatorname{Pic}(\bar{R}/c))_t$ is trivial if and only if X is rational and any singularity on X is geometrically unibranched. If X is normal and the degree of $p(x)$ is $d \geq 3$, then the torsion subgroup of $\operatorname{Pic}(\bar{R}/c)$ is isomorphic to a direct sum of t copies of $\mathbb{Q}/\mathbb{Z}$ where t is either $d-1$ or $d-2$ (see for example, [**For17**, §2.2]). One reduces to the case where X is normal using Mayer-Vietoris sequences. The proofs are left to the reader.

3.4. The Brauer Group of a Subring of a Global Field.

We say K is a *global field* if one of the following is true:

(1) K is a finitely generated algebraic extension field of $\mathbb{Q}$. In this case we also say K is an *algebraic number field*.

(2) K is a finitely generated extension field of a finite field k such that the transcendence degree $\operatorname{tr.deg}_k(K)$ is equal to one. In this case we also say K is the *function field of an algebraic curve* over the finite field k.

Let K be a global field and R a subring of K such that K is the quotient field of R. In this section we show that the Brauer group of R is isomorphic to the Brauer group of the integral closure of R in K. We state without proof a few classical results on the Brauer group of a global field. For proofs, the reader is referred to [**Rei75**, Chapter 8] and the references mentioned there.

The Brauer group of a global field K parametrizes the K-central division algebras. A K-central division algebra D is completely determined by its so-called Hasse invariants at the primes of K. In order to define these terms, we begin with the definition of a prime in a global field. Note that we are using the terminology of [**Rei75**]. In the literature, what we are calling a prime of K is also commonly referred to as a place of K.

First we define what it means to be a prime in a function field. Let K be a global field which is a function field of an algebraic curve over a finite field k. Let x be a fixed transcendental element such that K is separable (and by default, algebraic) over $k(x)$. Then $\{x\}$ is a separating transcendence base, which exists by Theorem 8.4.5. Let R be the integral closure of $k[x]$ in K. Then R is finitely generated as a $k[x]$-module and as a k-algebra (Theorem 4.6.10). Consequently, R is an integrally closed noetherian domain with Krull dimension one, hence is a Dedekind domain. Let $X_1(R)$ be the set of prime ideals in R of height one. For each $\mathfrak{p} \in X_1(R)$, $R_{\mathfrak{p}}$ is a discrete valuation ring and we denote its residue field by $k_{\mathfrak{p}}$. Denote by $\hat{R}_{\mathfrak{p}}$ the completion of $R_{\mathfrak{p}}$ and by $\hat{K}_{\mathfrak{p}}$ the field of fractions of $\hat{R}_{\mathfrak{p}}$. For $\mathfrak{p} \in X_1(R)$, the residue field $k_{\mathfrak{p}}$ is a finite field. In fact, $k_{\mathfrak{p}}$ is a finite algebraic extension of $k[x]/(\mathfrak{p} \cap k[x])$ (Theorem 3.3.5). By $X_1(K)$ we denote the set of all discrete valuation rings in K containing k as a subring. Points in $X_1(K)$ are also called *primes* of K. By associating a point $\mathfrak{p} \in X_1(R)$ with the associated valuation on K, we view $X_1(R)$ as a subset of $X_1(K)$. In Algebraic Geometry, primes in $X_1(K) - X_1(R)$ correspond to the "points at infinity" on a nonsingular projective embedding of $\operatorname{Spec} R$. By Lemma 8.4.9, $X_1(K) - X_1(R)$ is a finite set. That is, R has only finitely many points at infinity.

Now we define what it means to be a prime in an algebraic number field. Let K be an algebraic number field and R the integral closure of $\mathbb{Z}$ in K. By Theorem 4.6.10, R is finitely generated as a $\mathbb{Z}$-module. Then R is a noetherian integrally closed integral domain of Krull dimension one, hence is a Dedekind domain. Let $X_1(R)$ be the set of prime ideals in R of height one. For each $\mathfrak{p} \in X_1(R)$, $R_{\mathfrak{p}}$ is a discrete valuation ring with residue field which will be denoted $k_{\mathfrak{p}}$. Denote by $\hat{R}_{\mathfrak{p}}$ the completion of $R_{\mathfrak{p}}$ and by $\hat{K}_{\mathfrak{p}}$ the field of fractions of $\hat{R}_{\mathfrak{p}}$. For $\mathfrak{p} \in X_1(R)$, the residue field $k_{\mathfrak{p}}$ is a finite field. In fact, $k_{\mathfrak{p}}$ is a finite algebraic extension of $\mathbb{Z}/(\mathfrak{p} \cap \mathbb{Z})$ (Theorem 3.3.5). In addition to the discrete valuations, K will have a finite number of so-called *archimedean valuations* which define the *infinite primes* of K and correspond to embeddings $K \to \mathbb{C}$. Because the Brauer group of $\mathbb{C}$ is trivial, it is sufficient for our purposes to limit ourselves to the real primes, which correspond to the embeddings $K \to \mathbb{R}$. Let r be the number of distinct real embeddings $K \to \mathbb{R}$. By $X_1(K)$ we denote the set of all *primes* of K, which includes the infinite primes as well as the points in $X_1(R)$.

A local field is the field of fractions $\hat{K}_{\mathfrak{p}}$ of a complete local ring at a prime in K. If the prime is infinite, then the local field is either $\mathbb{R}$, or $\mathbb{C}$.

Since $\mathbb{C}$ is algebraically closed, $B(\mathbb{C})$ is the trivial group (Theorem 7.5.1). By Exercise 13.3.13, $B(\mathbb{R})$ is cyclic of order two, and is generated by the Brauer class of the ring of real quaternions, $\mathbb{H}$. Otherwise, $\hat{K}_\mathfrak{p}$ is the field of fractions of $\hat{R}_\mathfrak{p}$, and $\hat{R}_\mathfrak{p}$ is a complete DVR with finite residue field.

Now suppose K is a global field, $\mathfrak{p}$ is a finite prime, $\hat{R}_\mathfrak{p}$ is the complete discrete valuation ring, and $\hat{K}_\mathfrak{p}$ is the local field. If π is a local parameter for $\hat{R}_\mathfrak{p}$, then $\hat{K}_\mathfrak{p}$ is equal to the localization $\hat{R}_\mathfrak{p}[\pi^{-1}]$ (Theorem 6.1.15). The residue field $k_\mathfrak{p}$ is a finite field, hence $B(k_\mathfrak{p}) = (0)$ (Theorem 7.5.4). A finite Galois extension of fields $L/k_\mathfrak{p}$ is cyclic, with Galois group $\langle\varphi\rangle$ generated by the Frobenius homomorphism. By Exercise 12.5.8, L lifts to a unique cyclic Galois extension $S/\hat{R}_\mathfrak{p}$ with group $\langle\varphi\rangle$. Then $W = S \otimes_{\hat{R}_\mathfrak{p}} \hat{K}_\mathfrak{p}$ is a cyclic Galois extension of $\hat{K}_\mathfrak{p}$ with group $\langle\varphi\rangle$. By Section 13.4, the cyclic crossed product $\Delta(W/\hat{K}_\mathfrak{p}, \varphi, \pi)$ defines an element of $B(\hat{K}_\mathfrak{p})$ annihilated by n.

Theorem 14.3.6. *In the above context, let K be a global field, $\mathfrak{p}$ a finite prime with local parameter π and finite residue field $k_\mathfrak{p}$. Let $\hat{K}_\mathfrak{p}$ be the corresponding local field and $W/\hat{K}_\mathfrak{p}$ the cyclic Galois extension with group $G = \langle\varphi\rangle$ corresponding to a cyclic extension of the residue field with $[G : 1] = n$.*

(1) $H^2(\langle\varphi\rangle, W^*)$ *is a cyclic group of order n, generated by π.*

(2) *The cyclic crossed product $\Delta(W/\hat{K}_\mathfrak{p}, \varphi, \pi)$ is a division algebra of index n.*

(3) *If D is a central division algebra over $\hat{K}_\mathfrak{p}$ with index n, then D is isomorphic to a cyclic crossed product $\Delta(W/\hat{K}_\mathfrak{p}, \varphi, \pi^i)$ for a unique i such that $1 \le i < n$ and $\gcd(i, n) = 1$.*

(4) *The image of the crossed product homomorphism*

$$H^2(\langle\varphi\rangle, W^*) \to B(W/\hat{K}_\mathfrak{p})$$

is equal to the subgroup of $B(\hat{K}_\mathfrak{p})$ *annihilated by n.*

(5) *There is an isomorphism of abelian groups*

$$\mathrm{Inv} : B(\hat{K}_\mathfrak{p}) \to \mathbb{Q}/\mathbb{Z}$$

which associates a central division algebra D to its so-called Hasse invariant, $\mathrm{Inv}([D]) = [i/n]$, *where i and n are as in (3).*

Proof. For example, see [**Rei75**, Chapter 7]. □

For compatibility with the notation for finite primes, the homomorphism $\mathrm{Inv} : B(\mathbb{R}) \to \mathbb{Q}/\mathbb{Z}$ is defined by mapping the Brauer class of $\mathbb{H}$ to the unique element of order 2 in $\mathbb{Q}/\mathbb{Z}$, and $\mathrm{Inv} : B(\mathbb{C}) \to \mathbb{Q}/\mathbb{Z}$ is the zero map. Theorem 14.3.7 is usually attributed to Albert, Brauer, Hasse, and E. Noether.

Theorem 14.3.7. *In the above context, let K be a global field. Then there is an exact sequence of abelian groups*

$$0 \to \mathrm{B}(K) \to \bigoplus_{\mathfrak{p} \in X_1(K)} \mathrm{B}(K_{\mathfrak{p}}) \xrightarrow{\mathrm{Inv}} \mathbb{Q}/\mathbb{Z} \to 0$$

where the direct sum is over the set $X_1(K)$ of all primes of K and Inv *is the sum over all of the Hasse invariants.*

Proof. For example, see [**Rei75**, Chapter 8]. □

The Brauer group of a ring of integers in a number field is described by the following theorem, which is due to Grothendieck [**Gro68c**, Proposition (2.4)]. For a version of Theorem 14.3.8 when the ring R is nonnormal, see Theorem 14.3.10.

Theorem 14.3.8. *Let K be an algebraic number field and R the integral closure of $\mathbb{Z}$ in K.*

(1) *The sequence of abelian groups*

$$0 \to \mathrm{B}(R) \to \mathrm{B}(K) \to \bigoplus_{\mathfrak{p} \in X_1(R)} \mathrm{B}(\hat{K}_{\mathfrak{p}})$$

is exact.

(2) *Let r be the number of embeddings $K \to \mathbb{R}$. If $r \le 1$, then $\mathrm{B}(R) = (0)$. Otherwise, $\mathrm{B}(R)$ is an elementary 2-group of order 2^{r-1}.*

Proof. (1): Let $\mathfrak{p} \in X_1(R)$. From Proposition 11.3.21 and Theorem 14.3.7, there is an exact sequence

$$0 \to \bigcap_{\mathfrak{p} \in X_1(R)} \mathrm{B}(R_{\mathfrak{p}}) \to \mathrm{B}(K) \to \bigoplus_{\mathfrak{p} \in X_1(R)} \mathrm{B}(\hat{K}_{\mathfrak{p}})$$

which combines with Theorem 11.3.19 to prove (1).

(2): The diagram

$$\begin{array}{ccccccccc}
0 & \longrightarrow & \mathrm{B}(R) & \longrightarrow & \mathrm{B}(K) & \xrightarrow{\eta} & \mathrm{B}(K)/\mathrm{B}(R) & \longrightarrow & 0 \\
 & & \downarrow{\scriptstyle \alpha} & & \downarrow{\scriptstyle \beta} & & \downarrow{\scriptstyle \gamma} & & \\
0 & \longrightarrow & \bigoplus_{i=1}^{r} \mathrm{B}(\mathbb{R}) & \longrightarrow & \bigoplus_{\mathfrak{p} \in X_1(K)} \mathrm{B}(\hat{K}_{\mathfrak{p}}) & \longrightarrow & \bigoplus_{\mathfrak{p} \in X_1(R)} \mathrm{B}(\hat{K}_{\mathfrak{p}}) & \longrightarrow & 0 \\
 & & \downarrow{\scriptstyle \sigma} & & \downarrow{\scriptstyle \mathrm{Inv}} & & & & \\
 & & \mathbb{Q}/\mathbb{Z} & & \mathbb{Q}/\mathbb{Z} & & & &
\end{array}$$

commutes and the rows are exact. The map γ is from Part (1) and is one-to-one. The center column is from Theorem 14.3.7, and the direct sum is over all primes in K, including the r real primes. Therefore, β is one-to-one and Inv is the summation over all Hasse invariants. By Theorem 1.4.1, α is one-to-one. In the first column, σ is the summation map. If $r \leq 1$, then σ is one-to-one, and $\mathrm{B}(R) = 0$. If $r > 1$, then the image of σ has order two and the image of α is an elementary 2-group of order 2^{r-1}. □

Theorem 14.3.9 is from [**DeM76**] (see also [**DK76**]). For a version of Theorem 14.3.9 when the ring R is nonnormal, see Theorem 14.3.12.

Theorem 14.3.9. *Let K be a function field of an algebraic curve over a finite field k. Let x be a transcendental element over k such that K is separable over $k(x)$. Let R be the integral closure of $k[x]$ in K.*

(1) *The sequence of abelian groups*

$$0 \to \mathrm{B}(R) \to \mathrm{B}(K) \to \bigoplus_{\mathfrak{p} \in X_1(R)} \mathrm{B}(\hat{K}_{\mathfrak{p}})$$

is exact.

(2) *Let r be the number of primes of K at infinity. If $r \leq 1$, then $\mathrm{B}(R) = 0$. Otherwise, $\mathrm{B}(R)$ is isomorphic to a direct sum of $r - 1$ copies of $\mathbb{Q}/\mathbb{Z}$.*

Proof. (1): The proof is the same as Theorem 14.3.8 (1).

(2): Let $\mathfrak{p}_1, \ldots, \mathfrak{p}_r$ be the primes of K at infinity. The diagram

$$\begin{array}{ccccccccc}
0 & \longrightarrow & \mathrm{B}(R) & \longrightarrow & \mathrm{B}(K) & \xrightarrow{\eta} & \mathrm{B}(K)/\mathrm{B}(R) & \longrightarrow & 0 \\
 & & \downarrow \alpha & & \downarrow \beta & & \downarrow \gamma & & \\
0 & \longrightarrow & \bigoplus_{i=1}^{r} \mathrm{B}(\hat{K}_{\mathfrak{p}_i}) & \longrightarrow & \bigoplus_{\mathfrak{p} \in X_1(K)} \mathrm{B}(\hat{K}_{\mathfrak{p}}) & \longrightarrow & \bigoplus_{\mathfrak{p} \in X_1(R)} \mathrm{B}(\hat{K}_{\mathfrak{p}}) & \longrightarrow & 0
\end{array}$$

commutes and the rows are exact. The rest is as in Theorem 14.3.8 (2). □

Let k and R be as in Theorem 14.3.9. For any $n > 1$, let ${}_n\mathrm{B}(R)$ denote the subgroup of the Brauer group $\mathrm{B}(R)$ that is annihilated by n. Then ${}_n\mathrm{B}(R)$ is a direct sum of $r - 1$ copies of $\mathbb{Z}/n$. The ring R is the affine coordinate ring of a nonsingular curve over the finite field k. The analog of Dirichlet's Unit Theorem for a curve over a finite field is proved in [**RS10**, Proposition 1.2]. In our context, the theorem says that the group of units in the ring R is finitely generated and the rank of the torsion free part is equal to $r - 1$. Let $\alpha_1, \ldots, \alpha_{r-1}$ be elements of R^* such that

$$R^* = \langle \alpha_1 \rangle \times \cdots \times \langle \alpha_{r-1} \rangle \times T \tag{14.45}$$

where T is equal to the torsion subgroup of R^*. Let L/k be a Galois extension of fields with cyclic group $G = \langle \varphi \rangle$ of order n, where φ is the Frobenius automorphism. Then the cyclic crossed product algebras

$$\Lambda_{i,n} = \Delta(L/k, \varphi, \alpha_i) \tag{14.46}$$

are Azumaya R-algebras. At each prime $\mathfrak{p}$ of K, the Hasse invariant of $\Lambda_{i,n}$ at $\mathfrak{p}$ is given by $[v_{\mathfrak{p}}(\alpha_i)/n]$ (see [**Rei75**, § 31]). Therefore, if $\mathfrak{p}_1, \dots, \mathfrak{p}_r$ are the primes in $X_1(K) - X_1(R)$, then the image of

$$_n\operatorname{B}(R) \xrightarrow{a} \bigoplus_{i=1}^{r} \operatorname{B}(\hat{K}_{\mathfrak{p}})$$

is a free $\mathbb{Z}/n$-module of rank $r-1$ which is generated by the images of the Brauer classes $[\Lambda_{i,n}]$. Thus, the group $_n\operatorname{B}(R)$ is generated by $[\Lambda_{i,n}]$, for $1 \le i \le r-1$.

We end this chapter with the computation of the Brauer group of a nonnormal subring of a global field. The following theorems and corollary are due to L. Childs [**Chi74**].

Theorem 14.3.10. *Let K be an algebraic number field and $\bar{R}$ the integral closure of $\mathbb{Z}$ in K. If R is a subring of $\bar{R}$ such that K is the field of fractions of R, then $\operatorname{B}(R) \to \operatorname{B}(\bar{R})$ is an isomorphism.*

Proof. We have the commutative diagram of rings and subrings

$$\begin{array}{ccccc} R & \longrightarrow & \bar{R} & \longrightarrow & K \\ \uparrow & \nearrow & & & \uparrow \\ \mathbb{Z} & & \longrightarrow & & \mathbb{Q} \end{array}$$

where each arrow represents set inclusion. Then $\bar{R}$ is the integral closure of R in K. By Theorem 4.6.10, $\bar{R}$ is a $\mathbb{Z}$-lattice in K. Hence $\bar{R}$ is also an R-lattice in K. Let $c = R : \bar{R}$ be the conductor ideal from $\bar{R}$ to R. By Proposition 6.3.3 (5), we know $c \neq (0)$. Assuming $R \neq \bar{R}$, we can assume c is a proper ideal of both R and $\bar{R}$. By an application of Theorem 3.3.5 (see [**Mat80**, (13.C), Theorem 20]), it follows that $\dim(\mathbb{Z}) = \dim(R) = \dim(\bar{R}) = 1$. In both R and $\bar{R}$ we have $\operatorname{ht}(c) = 1$. Therefore, R/c and $\bar{R}/c$ both have Krull dimension zero. Hence R/c and $\bar{R}/c$ are both artinian. By Corollary 3.2.12, every module in $\mathbf{CRP}(\bar{R}/c)$ is free. In particular, $\operatorname{Pic}(\bar{R}/c) = (0)$. Thus we are in the setting of Corollary 14.2.13, and the sequence

$$0 \to \operatorname{B}(R) \to \operatorname{B}(\bar{R}) \oplus \operatorname{B}(R/c) \to \operatorname{B}(\bar{R}/c)$$

is exact. But R/c is integral over $\mathbb{Z}$ and $\operatorname{Spec}(R/c) \to \operatorname{Spec}\mathbb{Z}$ is not onto. By Theorem 3.3.5, $c \cap \mathbb{Z} \neq (0)$ and R/c is a finite ring. Likewise, $\bar{R}/c$ is a finite ring. By Corollary 7.9.6, $(0) = \mathrm{B}(R/c) = \mathrm{B}(\bar{R}/c)$, and the proof follows. □

The proof of Theorem 14.3.11 utilizes a result of E. Davis which is not proved here.

Theorem 14.3.11. *Let R be an integral domain such that the quotient field of R is an algebraic number field K. If $\bar{R}$ is the integral closure of R in K, then $\mathrm{B}(R) \to \mathrm{B}(\bar{R})$ is an isomorphism.*

Proof. Let R_0 be the integral closure of $\mathbb{Z}$ in K. As in Theorem 14.3.10, R_0 is a finitely generated $\mathbb{Z}$-module and we can write $R_0 = \mathbb{Z}[a_1, \dots, a_n]$ where each a_i is integral over $\mathbb{Z}$. Then $S = R[a_1, \dots, a_n]$ is a finitely generated R-module. As proved in [**Dav65**, Theorem A], $\bar{R} = W^{-1}R_0$ is a localization for some multiplicative set $W \subseteq R_0$. We have the commutative diagram of rings and subrings

$$\begin{array}{ccccccc} R & \longrightarrow & S & \longrightarrow & \bar{R} = W^{-1}R_0 & \longrightarrow & K \\ \uparrow & & \uparrow & \nearrow & & & \\ \mathbb{Z} & \longrightarrow & R_0 & & & & \end{array}$$

where each arrow represents set inclusion. Since $\bar{R}$ is integral over S, for any prime ideal $\mathfrak{p} \in \operatorname{Spec} S$, there exists $\mathfrak{q} \in \operatorname{Spec} \bar{R}$ such that $\mathfrak{p} = \mathfrak{q} \cap S$. (Theorem 3.3.5). Given $w \in W$, no prime $\mathfrak{q}$ of $\bar{R}$ contains w. Hence, no prime $\mathfrak{p}$ of S contains w. Therefore, $W \subseteq S^*$ and $W^{-1}R_0 \subseteq S$. This implies $S = \bar{R}$, and $\bar{R}$ is a finitely generated R-module. The rest of the proof is as in Theorem 14.3.10. □

Theorem 14.3.12. *Let k be a finite field and R a finitely generated k-algebra which is an integral domain of Krull dimension one. Let K be the quotient field of R and $\bar{R}$ the integral closure of R in K. Then $\mathrm{B}(R) \to \mathrm{B}(\bar{R})$ is an isomorphism and $\mathrm{B}(R) \to \mathrm{B}(K)$ is one-to-one.*

Proof. By Theorem 8.4.7, $\bar{R}$ is an R-lattice in K. The proof that $\mathrm{B}(R) = \mathrm{B}(\bar{R})$ follows by an argument similar to that used in Theorem 14.3.10 and is left to the reader. Since $\bar{R}$ is a Dedekind domain, $\mathrm{B}(\bar{R}) \to \mathrm{B}(K)$ is one-to-one (Theorem 11.3.18). □

Corollary 14.3.13. *Let R be any integral domain whose quotient field is a global field K. Let $\bar{R}$ be the integral closure of R in K. Then $\mathrm{B}(R) \to \mathrm{B}(K)$ is one-to-one.*

Proof. If K is an algebraic number field, Theorem 14.3.11 applies. We prove the case when K is the function field of a curve over a finite field k. Let A be an Azumaya R-algebra. By Proposition 7.7.2, there is a finitely generated

k-subalgebra of R, say R_0, and an Azumaya R_0-algebra, say A_0 such that $A \cong A_0 \otimes_{R_0} R$. Since K is a finitely generated algebraic extension of $k(x)$, for some transcendental element $x \in R$, there is a finite set of elements $a_1, \dots, a_n$ in R such that K is equal to the quotient field of $R_1 = R_0[a_1, \dots, a_n]$. Set $A_1 = A_0 \otimes_{R_0} R_1$, and let $\bar{R}_1$ be the integral closure of R_1 in K. Consider the commutative diagram

$$\begin{array}{ccc} \mathrm{B}(R_1) & \longrightarrow & \mathrm{B}(R) \\ {\scriptstyle\alpha}\downarrow & & \downarrow{\scriptstyle\gamma} \\ \mathrm{B}(\bar{R}_1) & \xrightarrow{\beta} & \mathrm{B}(K) \end{array}$$

of abelian groups. It follows from Theorem 14.3.12 that α is an isomorphism and by Theorem 11.3.18, β is one-to-one. Since $[A]$ is an arbitrary class in $\mathrm{B}(R)$, this proves γ is one-to-one. $\square$

3.5. Exercises.

Exercise 14.3.14. Let k and $A_2 = k[v,w]/(v^2 - (w^2-1)^2)$ be as in Exercise 14.2.20. In Exercise 12.9.21 we showed that the ring A_2 can be made into a Galois extension of $A_1 = k[x,y]/(y^2 - x^2(x+1))$, the affine coordinate ring of a nodal cubic curve (see also Exercise 8.1.34). It follows from Example 14.3.4 that $A_1^* = k^*$ and $\operatorname{Pic}(A_1) = k^*$. For the Galois extension $\beta : A_1 \to A_2$, compute all of the terms in the seven term exact sequence of Galois cohomology (Theorem 13.3.1). Show that the kernel of $\operatorname{Res}_{A_1}^{A_2} : \operatorname{Pic}(A_1) \to \operatorname{Pic}(A_2)$ is a cyclic group of order two. Show that the corestriction map $\operatorname{Cor}_{A_1}^{A_2} : \operatorname{Pic}(A_2) \to \operatorname{Pic}(A_1)$ is an isomorphism. See Exercise 14.3.16 for a generalization of this example.

Exercise 14.3.15. In the notation of Section 14.3.2, let $n \geq 2$, k an algebraically closed field of characteristic zero, R the ring of Example 13.6.12, A_1 the coordinate ring of a nodal cubic, and $A = R \otimes_k A_1$. Let $\beta : A_1 \to A_2$ be the étale covering of Exercise 12.9.21 and set $B = R \otimes_k A_2$. Hence β induces an étale covering $\beta : A \to B$. Prove the following.

(1) $\mathrm{B}(B) \cong \mathbb{Z}/n$.
(2) $\operatorname{Res}_A^B : \mathrm{B}(A) \to \mathrm{B}(B)$ is an isomorphism.
(3) The corestriction map $\operatorname{Cor}_A^B : \mathrm{B}(B) \to \mathrm{B}(A)$ is defined by $\alpha \mapsto \alpha^2$.

Exercise 14.3.16. In this exercise we construct an algebraic version of a cyclic covering space of a topological 1-sphere. Let k be a field and assume 2 is invertible in k. Let $m \geq 2$ and $G = \langle \sigma \rangle$ a cyclic group of order m. Let $T_m = E_G(k[w]) = \bigoplus_{i=1}^m k[w]$ be the trivial G-Galois extension of $k[w]$. Let $f_1 : T_m \to \bigoplus_{i=1}^m k[w]/(w^2-1)$ be the natural ring homomorphism.

Let $U_m = E_G(k) = \bigoplus_{i=1}^m k$ be the trivial G-Galois extension of k. By the Chinese Remainder Theorem, identify

$$\frac{k[w]}{(w^2-1)} = \frac{k[w]}{(w-1)} \bigoplus \frac{k[w]}{(w+1)} = k \oplus k.$$

Define $f_2 : U_m \to \bigoplus_{i=1}^m k[w]/(w^2-1) = \bigoplus_{i=1}^m (k \oplus k)$ on m-tuples by

$$f_2 \begin{bmatrix} a_1 \\ a_2 \\ a_3 \\ \vdots \\ a_m \end{bmatrix} = \begin{bmatrix} a_1 & a_m \\ a_2 & a_1 \\ a_3 & a_2 \\ \vdots & \\ a_m & a_{m-1} \end{bmatrix}.$$

Let A_m be the ring defined by the pullback diagram

$$\begin{array}{ccc} A_m & \xrightarrow{\quad h_2 \quad} & U_m = \bigoplus_{i=1}^m k \\ {\scriptstyle h_1}\downarrow & & \downarrow{\scriptstyle f_2} \\ T_m = \bigoplus_{i=1}^m k[w] & \xrightarrow[\quad f_1 \quad]{} & \bigoplus_{i=1}^m \frac{k[w]}{(w^2-1)} = \bigoplus_{i=1}^m (k \oplus k). \end{array}$$

(1) Show that A_m can be identified with the subring of T_m consisting of all m-tuples of polynomials $(P_1, \dots, P_m)$ satisfying the equations: $P_1(1) = P_2(-1)$, $P_2(1) = P_3(-1)$, $\dots$, $P_m(1) = P_1(-1)$. In Algebraic Geometry, the ring A_m is the affine coordinate ring corresponding to m copies of the affine line $\mathbb{A}_k^1$ that have been glued together to form a "loop".

(2) Show that the action by the group G on T_m and U_m induces an action by G as a group of automorphisms on A_m.

(3) Let $A_1 = k[x,y]/(y^2 - x^2(x+1))$ be the affine coordinate ring of a nodal cubic curve (see also Exercise 14.3.14). Using diagram (14.40), show that A_1 can be identified with the subring of $k[w]$ consisting of all polynomials P such that $P(1) = P(-1)$, hence there is an embedding $A_1 \to A_m^G$.

(4) Show that $A_1 = A_m^G$. (Hint: The functor $(\ )^G$ maps the cartesian square defining A_m onto the cartesian square (14.40).)

(5) Show that A_m is a Galois extension of A_1 with cyclic group G.

(6) Show that $A_m^* = k^*$ and $\operatorname{Pic}(A_m) = k^*$.

(7) Assume k is a perfect field. Show that $\mathrm{B}(A_m) = \mathrm{B}(k)$. Assume m is invertible in k, and that k contains a primitive mth root of unity, ζ. For the Galois extension A_m/A_1, compute all of the terms in the seven term exact sequence of Galois cohomology (Theorem 13.3.1).

Exercise 14.3.17. In the notation of Section 14.3.2, let $n \geq 2$, k an algebraically closed field of characteristic zero, R the ring of Example 13.6.12, A_1 the coordinate ring of a nodal cubic, and $A = R \otimes_k A_1$. Let $\beta : A_1 \to A_m$ be the cyclic Galois covering of Exercise 14.3.16, and set $B = R \otimes_k A_m$. Hence β induces a cyclic Galois covering $\beta : A \to B$ of degree m. Prove the following.

(1) $\mathrm{B}(B) \cong \mathbb{Z}/n$.

(2) $\mathrm{Res}_A^B : \mathrm{B}(A) \to \mathrm{B}(B)$ is an isomorphism.

(3) The corestriction map $\mathrm{Cor}_A^B : \mathrm{B}(B) \to \mathrm{B}(A)$ is defined by $\alpha \mapsto \alpha^m$.

Exercise 14.3.18. Determine $\mathrm{B}(R)$ for the following rings.

(1) $R = \mathbb{Z}$, (2) $R = \mathbb{Z}[x]/(x^2 - 10)$,

(3) $R = \mathbb{Z}[x]/(x^3 - 10x - 5)$, (4) $R = \mathbb{Z}[x]/(x^3 - 10x + 15)$,

(5) $R = \mathbb{Z}[x]/(x^4 - 10x - 5)$.

Exercise 14.3.19. Let k be a finite field of characteristic p, where $p \neq 2$ and $p \neq 3$.

(1) Determine $\mathrm{B}(R)$ for the following rings.

(a) $R = k[x, y](xy - 1)$, (b) $R = k[x, y](x^2y - 1)$,

(c) $R = k[x, y](y^2 - x^3 + x)$, (d) $R = k[x, z](z - x^3 + xz^2)$.

(2) For the rings in (1) which have nontrivial Brauer groups, find a basis $\alpha_1, \dots, \alpha_{r-1}$, for the group of units R^* (modulo torsion), as in Eq.(14.45). Use this to find cyclic crossed products in the form of Eq.(14.46) which generate ${}_n\mathrm{B}(R)$.

Exercise 14.3.20. Let $R = \mathbb{R}[x, y]/(y^2 - x^2(x - 1))$ and $S = \mathbb{C} \otimes_{\mathbb{R}} R = \mathbb{C}[x, y]/(y^2 - x^2(x - 1))$. Then S/R is a quadratic Galois extension with group $G = \langle \sigma \rangle$, where σ is complex conjugation.

(1) Compute R^*, $\mathrm{Pic}\, R$, $\mathrm{B}(R)$, R^*, $\mathrm{Pic}\, S$, and $\mathrm{B}(S)$.

(2) Compute all of the terms in the exact sequence of Galois cohomology for S/R (Theorem 13.3.1).

(Hints: If $\bar{R}$ is the normalization of R, then the normalization of S is $\bar{S} = \mathbb{C} \otimes_{\mathbb{R}} R$. Let c denote the conductor from $\bar{R}$ to R, as well as the conductor from $\bar{S}$ to S. For (1) use Mayer-Vietoris sequences. For (2), note that $\bar{S}/c$ is the trivial Galois extension of $\bar{R}/c$ for the group G. Use Exercise 13.3.12 and the Mayer-Vietoris sequence to show: $(\mathrm{Pic}\, S)^G \cong \mathbb{C}^*/\mathbb{R}^*$, $\mathrm{H}^1(G, (\bar{S}/c)^*) = \langle 1 \rangle$, $\mathrm{H}^1(G, \mathrm{Pic}\, S) \cong \mathrm{H}^2(G, \mathbb{C}^*) \cong \mathbb{Z}/2$.)

Acronyms

ACC Ascending Chain Condition
DCC Descending Chain Condition
PID Principal Ideal Domain
UFD Unique Factorization Domain
DVR Discrete Valuation Ring

Glossary of Notations

$(a,b)_{n,R,\zeta}$	symbol algebra, 528
$\operatorname{annih}_R M$	annihilator of M in R, 9
$\operatorname{Az}(R)$	set of isomorphism classes of Azumaya R-algebras, 252
$\mathrm{B}(R)$	Brauer group of R, 252
$\mathrm{B}(S/R)$	relative Brauer group of S over R, 255
$\check{\mathrm{H}}^n_{\mathrm{et}}(R,\mathfrak{F})$	Čech étale cohomology of R with coefficients in $\mathfrak{F}$, 396
$\mathrm{Cl}(R)$	class group of R, 207
$\operatorname{coh.dim}(R)$	global cohomological dimension of R, 202
$\operatorname{Cor}^S_R(M)$	corestriction of M from S to R, 558
$\Delta(S/R,\Phi,f)$	generalized crossed product algebra, 502
$\Delta(A,D,u)$	differential crossed product algebra, 304
$\Delta(S/R,\sigma,a)$	cyclic crossed product algebra, 526
$\Delta(S/R,G,f)$	crossed product algebra, 446
$\Delta(T/A,I,g)$	generalized cyclic crossed product algebra, 534
δ_{ij}	Kronecker delta function, 5
$\operatorname{depth}_I(M)$	I-depth of an R-module M, 201
$\operatorname{Der}_R(A,M)$	set of all R-derivations from A to M, 288
$\dim(R)$	Krull dimension of a commutative ring R, 200
$\operatorname{Div}: K^* \to \operatorname{Div}(R)$	divisor homomorphism, 207
$\operatorname{Div} R$	group of Weil divisors of R, 207
$\ell(M)$	length of a composition series of M, 78
$\operatorname{Frac}(R)$	set of all fractional ideals of R, 227

$\mathrm{Gal}(R,G)$ commutative Galois extensions of R with group G, 487

$\mathrm{GL}_n(S)$ general linear group, 197

$\mathrm{gr}_I(R)$ graded ring associated to an I-adic filtration, 110

$\hat{M} = \varprojlim M/I^iM$ I-adic completion of M, 44

$\mathrm{ht}(I)$ height of an ideal I, 200

$\mathrm{H}^n(A,M)$ Hochschild cohomology group of A with coefficients in M, 188

$\mathrm{H}^n(G,A)$ cohomology group of G with coefficients in A, 160

$\mathrm{H}^n(S/R,\mathfrak{F})$ Amitsur cohomology group of S/R with coefficients in $\mathfrak{F}$, 192

$\mathrm{Hom}_R(M,N)$ set of R-module homomorphisms, 29

$\mathrm{idemp}(R)$ idempotents of R, 62

$\mathrm{Ind}_H^G(T)$ induced Galois extension, 457

$\mathrm{Inn.Der}_R(A,M)$ set of inner derivations, 290

$\mathrm{Inv}([D])$ Hasse invariant of D, 607

$\mathrm{Invert}(R)$ set of all invertible fractional ideals of R, 232

$\mathrm{J}(R)$ Jacobson radical of R, 91

$\mathrm{Map}(M,R)$ set of functions $f : M \to R$, 14

$\mathbb{A}_k^n$ affine n-space over k, 326

$\mathbb{G}_a(R)$ additive abelian group of the ring R, 192

$\mathbb{G}_m(R)$ group of units of the ring R, 197

$\mathbb{H}$ ring of real quaternions, 2

$\mathfrak{T}_R(M)$ trace ideal of M in R, 7

$\mathrm{Max}\, R$ maximal ideal spectrum of R, 58

$\nabla(S,G)$ the S-algebra of functions $v : G \to S$, 449

$\Omega_{A/R}$ module of Kähler differentials, 308

$\mathrm{PGL}_n(S)$ projective general linear group, 398

$\mathrm{Pic}_R(A)$ Picard group of invertible left $A \otimes_R A^o$-modules, 276

$\mathrm{Pic}\,(R)$ Picard group of R, 86

$\mathrm{Prin}\, R$ group of principal Weil divisors, 207

$\mathrm{proj.dim}_R\, M$ projective dimension of M, 202

$\mathrm{Rad}(\mathfrak{a})$ nil radical of $\mathfrak{a}$, 59

$\mathrm{Rank}_R(M)$ rank of a free module over a commutative ring, 5

$\mathrm{Rank}_R(M)$ rank of a projective module over a commutative ring, 67

$\mathrm{Reflex}(R)$ set of all reflexive fractional ideals of R, 229

$\mathrm{Res}(f, g)$	resultant of f and g, 16
$\mathrm{Spec}\, R$	prime ideal spectrum of R, 58
$\mathrm{Supp}_R(M)$	support of M in $\mathrm{Spec}\, R$, 316
$\mathrm{Syl}(f, g)$	Sylvester matrix of f and g, 16
$\mathrm{Units}(R)$ or R^*	group of units in the ring R, 1
$\varinjlim A_i$	direct limit of a directed system $\{A_i, \phi^i_j\}$, 38
$\varprojlim A_i$	inverse limit of the inverse system $\{A_i, \phi^j_i\}$, 41
A^e	$A \otimes_R A^o$, the enveloping algebra of A, 116
$E_G(R)$	trivial G-Galois extension of R, 449
$I(Y)$	ideal of a subset of $\mathrm{Spec}(R)$, 59
$I : J$	module quotient, 87
$J_{A/R}$	kernel of $\mu : A^e \to A$, 116
$M \otimes_R N$	tensor product, 24
M^*	completion of a linear topological module M, 107
M^A	centralizer of A in M, 117
$M_n(R)$	ring of n-by-n matrices over R, 1
N^A_R	norm from A to R, 22
$R(G)$ or RG	group ring or group algebra, 1
R^h	henselization of R, 381
R^o	opposite ring of R, 1
R^{sh}	strict henselization of R, 386
T^A_R	trace from A to R, 22
$T_R(M)$, or $T(M)$	tensor algebra of the R-module M, 173
$T^{A,M}_R$	trace from A to R afforded by M, 149
$U(\alpha)$	basic open subset of $\mathrm{Spec}\, R$, 61
$V(L)$	closed subset of $\mathrm{Spec}(R)$, 58
$W^{-1}M$	localization of M at W, 53
$X_1(K)$	primes in a global field K, 606
$Z(A)$	center of A, 2
$\mathfrak{M}_R$	category of right R-modules, 3
${}_R\mathfrak{M}$	category of left R-modules, 2
${}_S\mathfrak{M}_R$	category of left S right R bimodules, 23

Bibliography

[ABGV11] Asher Auel, Eric Brussel, Skip Garibaldi, and Uzi Vishne, *Open problems on central simple algebras*, Transform. Groups **16** (2011), no. 1, 219–264. MR2785502

[AC16] Benjamin Antieau and Kenneth Chan, *Maximal orders in unramified central simple algebras*, J. Algebra **452** (2016), 94–105. MR3461057

[AG60a] Maurice Auslander and Oscar Goldman, *The Brauer group of a commutative ring*, Trans. Amer. Math. Soc. **97** (1960), 367–409. MR0121392 (22 #12130)

[AG60b] ______, *Maximal orders*, Trans. Amer. Math. Soc. **97** (1960), 1–24. MR0117252 (22 #8034)

[AGV72a] M. Artin, A. Grothendieck, and J.-L. Verdier, *Théorie des topos et cohomologie étale des schémas. Tome 1*, Lecture Notes in Mathematics, Vol. 269, Springer-Verlag, Berlin, 1972, Séminaire de Géométrie Algébrique du Bois-Marie 1963–1964 (SGA 4), Dirigé par M. Artin, A. Grothendieck et J. L. Verdier. Avec la collaboration de N. Bourbaki, P. Deligne et B. Saint-Donat.

[AGV72b] ______, *Théorie des topos et cohomologie étale des schémas. Tome 2*, Lecture Notes in Mathematics, Vol. 270, Springer-Verlag, Berlin, 1972, Séminaire de Géométrie Algébrique du Bois-Marie 1963–1964 (SGA 4), Dirigé par M. Artin, A. Grothendieck et J. L. Verdier. Avec la collaboration de N. Bourbaki, P. Deligne et B. Saint-Donat.

[AGV73] ______, *Théorie des topos et cohomologie étale des schémas. Tome 3*, Lecture Notes in Mathematics, Vol. 305, Springer-Verlag, Berlin, 1973, Séminaire de Géométrie Algébrique du Bois-Marie 1963–1964 (SGA 4), Dirigé par M. Artin, A. Grothendieck et J. L. Verdier. Avec la collaboration de P. Deligne et B. Saint-Donat. MR0354654 (50 #7132)

[Alb34] A. A. Albert, *Cyclic fields of degree p^n over F of characteristic p*, Bull. Amer. Math. Soc. **40** (1934), no. 8, 625–631. MR1562919

[Alb39] A. Adrian Albert, *Structure of Algebras*, American Mathematical Society Colloquium Publications, vol. 24, American Mathematical Society, New York, 1939. MR0000595 (1,99c)

[AM69] M. F. Atiyah and I. G. Macdonald, *Introduction to commutative algebra*, Addison-Wesley Publishing Co., Reading, Mass.-London-Don Mills, Ont., 1969. MR0242802 (39 #4129)

[Ami59] S. A. Amitsur, *Simple algebras and cohomology groups of arbitrary fields*, Trans. Amer. Math. Soc. **90** (1959), 73–112. MR0101265 (21 #78)

[And78] David F. Anderson, *Subrings of $k[X,Y]$ generated by monomials*, Canad. J. Math. **30** (1978), no. 1, 215–224. MR0463162

[Ara75] Jón Kr. Arason, *Cohomologische invarianten quadratischer Formen*, J. Algebra **36** (1975), no. 3, 448–491. MR0389761 (52 #10592)

[Art71] M. Artin, *On the joins of Hensel rings*, Advances in Math. **7** (1971), 282–296 (1971). MR0289501 (44 #6690)

[Aus66] Bernice Auslander, *The Brauer group of a ringed space*, J. Algebra **4** (1966), 220–273. MR0199213 (33 #7362)

[Aus71] Bernice L. Auslander, *Central separable algebras which are locally endomorphism rings of free modules*, Proc. Amer. Math. Soc. **30** (1971), 395–404. MR0285517 (44 #2735)

[AW14] Benjamin Antieau and Ben Williams, *Unramified division algebras do not always contain Azumaya maximal orders*, Invent. Math. **197** (2014), no. 1, 47–56. MR3219514

[Azu51] Gorô Azumaya, *On maximally central algebras*, Nagoya Math. J. **2** (1951), 119–150. MR0040287 (12,669g)

[Bas64] H. Bass, *K-theory and stable algebra*, Inst. Hautes Études Sci. Publ. Math. (1964), no. 22, 5–60. MR0174604

[Bas68] Hyman Bass, *Algebraic K-theory*, W. A. Benjamin, Inc., New York-Amsterdam, 1968. MR0249491 (40 #2736)

[BFH13] Djordje N. Bulj, Timothy J. Ford, and Drake M. Harmon, *Generically trivial Azumaya algebras on a rational surface with a non rational singularity*, Comm. Algebra **41** (2013), no. 11, 4333–4338. MR3169521

[BG86] Hyman Bass and Robert Guralnick, *Projective modules with free multiples and powers*, Proc. Amer. Math. Soc. **96** (1986), no. 2, 207–208. MR818444

[BK71] Michael Barr and Max-Albert Knus, *Extensions of derivations*, Proc. Amer. Math. Soc. **28** (1971), 313–314. MR0272829 (42 #7710)

[Bou89a] Nicolas Bourbaki, *Algebra. I. Chapters 1–3*, Elements of Mathematics (Berlin), Springer-Verlag, Berlin, 1989, Translated from the French, Reprint of the 1974 edition. MR979982 (90d:00002)

[Bou89b] ———, *Commutative algebra. Chapters 1–7*, Elements of Mathematics (Berlin), Springer-Verlag, Berlin, 1989, Translated from the French, Reprint of the 1972 edition. MR979760 (90a:13001)

[Bri13] Martin Bright, *Brauer groups of singular del Pezzo surfaces*, Michigan Math. J. **62** (2013), no. 3, 657–664. MR3102534

[Cae98] Stefaan Caenepeel, *Brauer groups, Hopf algebras and Galois theory*, K-Monographs in Mathematics, vol. 4, Kluwer Academic Publishers, Dordrecht, 1998. MR1610222

[CE56] Henri Cartan and Samuel Eilenberg, *Homological algebra*, Princeton University Press, Princeton, N. J., 1956. MR0077480 (17,1040e)

[CGO75] Lindsay N. Childs, Gerald Garfinkel, and Morris Orzech, *On the Brauer group and factoriality of normal domains*, J. Pure Appl. Algebra **6** (1975), no. 2, 111–123. MR0371885 (51 #8102)

[Chi69] L. N. Childs, *On projective modules and automorphismes of central separable algebras.*, Canad. J. Math. **21** (1969), 44–53. MR0237556

[Chi74] ______, *Mayer-Vietoris sequences and Brauer groups of nonnormal domains*, Trans. Amer. Math. Soc. **196** (1974), 51–67. MR0344240 (49 #8979)

[Chi76] Lindsay N. Childs, *On Brauer groups of some normal local rings*, Brauer groups (Proc. Conf., Northwestern Univ., Evanston, Ill., 1975), Springer, Berlin, 1976, pp. 1–15. Lecture Notes in Math., Vol. 549. MR0432620 (55 #5607)

[CHR65] S. U. Chase, D. K. Harrison, and Alex Rosenberg, *Galois theory and Galois cohomology of commutative rings*, Mem. Amer. Math. Soc. No. **52** (1965), 15–33. MR0195922 (33 #4118)

[Cip77] Michele Cipolla, *Remarks on the lifting of algebras over Henselian pairs*, Math. Z. **152** (1977), no. 3, 253–257. MR0429869 (55 #2879)

[CL17] Jesús Conde-Lago, *A short proof of smooth implies flat*, Comm. Algebra **45** (2017), no. 2, 774–775. MR3562538

[Cla66] Luther Claborn, *Every abelian group is a class group*, Pacific J. Math. **18** (1966), 219–222. MR0195889

[CR87] Charles W. Curtis and Irving Reiner, *Methods of representation theory. Vol. II*, Pure and Applied Mathematics (New York), John Wiley & Sons, Inc., New York, 1987, With applications to finite groups and orders, A Wiley-Interscience Publication. MR892316

[CTW12] Jean-Louis Colliot-Thélène and Olivier Wittenberg, *Groupe de Brauer et points entiers de deux familles de surfaces cubiques affines*, Amer. J. Math. **134** (2012), no. 5, 1303–1327. MR2975237

[CV15] Brendan Creutz and Bianca Viray, *On Brauer groups of double covers of ruled surfaces*, Math. Ann. **362** (2015), no. 3-4, 1169–1200. MR3368096

[Dav65] Edward D. Davis, *Rings of algebraic numbers and functions*, Math. Nachr **29** (1965), 1–7. MR0179160

[Del77] P. Deligne, *Cohomologie étale*, Lecture Notes in Mathematics, Vol. 569, Springer-Verlag, Berlin-New York, 1977, Séminaire de Géométrie Algébrique du Bois-Marie SGA $4\frac{1}{2}$, Avec la collaboration de J. F. Boutot, A. Grothendieck, L. Illusie et J. L. Verdier. MR0463174

[DeM69] F. R. DeMeyer, *Projective modules over central separable algebras*, Canad. J. Math. **21** (1969), 39–43. MR0234987

[DeM75] ______, *The Brauer group of polynomial rings*, Pacific J. Math. **59** (1975), no. 2, 391–398. MR0392971

[DeM76] ______, *The Brauer group of affine curves*, Brauer groups (Proc. Conf., Northwestern Univ., Evanston, Ill., 1975), Springer, Berlin, 1976, pp. 16–24. Lecture Notes in Math., Vol. 549. MR0437537 (55 #10461)

[DeM81] ______, *An action of the automorphism group of a commutative ring on its Brauer group*, Pacific J. Math. **97** (1981), no. 2, 327–338. MR641162 (83j:16008)

[DF82] F. R. DeMeyer and T. J. Ford, *On the Brauer group of surfaces and subrings of $k[x, y]$*, Brauer Groups in Ring Theory and Algebraic Geometry (Wilrijk, 1981), Lecture Notes in Math., vol. 917, Springer, Berlin-New York, 1982, pp. 211–221. MR657431

[DF93] Frank R. DeMeyer and Timothy J. Ford, *On the Brauer group of toric varieties*, Trans. Amer. Math. Soc. **335** (1993), no. 2, 559–577. MR93d:14077

[DF04] David S. Dummit and Richard M. Foote, *Abstract algebra*, third ed., John Wiley & Sons Inc., Hoboken, NJ, 2004. MR2286236 (2007h:00003)

[DFM93a] F. R. DeMeyer, T. J. Ford, and R. Miranda, *The cohomological Brauer group of a toric variety*, J. Algebraic Geom. **2** (1993), no. 1, 137–154. MR93i:14051

[DFM93b] Frank DeMeyer, Tim Ford, and Rick Miranda, *Rational singularities and the Brauer group*, J. Algebra **162** (1993), no. 2, 287–294. MR94m:14042

[DI71] F. R. DeMeyer and Edward Ingraham, *Separable algebras over commutative rings*, Lecture Notes in Mathematics, Vol. 181, Springer-Verlag, Berlin, 1971. MR0280479 (43 #6199)

[DK76] F. R. Demeyer and M. A. Knus, *The Brauer group of a real curve*, Proc. Amer. Math. Soc. **57** (1976), no. 2, 227–232. MR0412193 (54 #320)

[Dur79] Alan H. Durfee, *Fifteen characterizations of rational double points and simple critical points*, Enseign. Math. (2) **25** (1979), no. 1-2, 131–163. MR543555 (80m:14003)

[FH13] Timothy J. Ford and Drake M. Harmon, *The Brauer group of an affine rational surface with a non-rational singularity*, J. Algebra **388** (2013), 107–140. MR3061681

[For81] T. J. Ford, *Every finite abelian group is the Brauer group of a ring*, Proc. Amer. Math. Soc. **82** (1981), no. 3, 315–321. MR612710

[For84] Timothy J. Ford, *Hecke actions on Brauer groups*, J. Pure Appl. Algebra **33** (1984), no. 1, 11–17. MR750227 (85j:13005)

[For88] T. J. Ford, *On the Brauer group of a Laurent polynomial ring*, J. Pure Appl. Algebra **51** (1988), no. 1-2, 111–117. MR89h:13006

[For89] Timothy J. Ford, *On the Brauer group of* $k[x_1, \cdots, x_n, 1/f]$, J. Algebra **122** (1989), no. 2, 410–424. MR999083 (90d:13005)

[For95] T. J. Ford, *Examples of locally trivial Azumaya algebras*, K-theory and algebraic geometry: connections with quadratic forms and division algebras (Santa Barbara, CA, 1992), Proc. Sympos. Pure Math., vol. 58, Amer. Math. Soc., Providence, RI, 1995, pp. 197–216. MR1327298 (96c:16023)

[For14] Timothy J. Ford, *The group of units on an affine variety*, J. Algebra Appl. **13** (2014), no. 8, 1450065 (27 pages). MR3225132

[For16] ———, *The relative Brauer group and generalized cyclic crossed products for a ramified covering*, J. Algebra **450** (2016), 1–58. MR3449685

[For17] ———, *The Brauer group of an affine double plane associated to a hyperelliptic curve*, Comm. Algebra **45** (2017), no. 4, 1416–1442. MR3576667

[Fos73] Robert M. Fossum, *The divisor class group of a Krull domain*, Springer-Verlag, New York, 1973. MR0382254 (52 #3139)

[Frö73] A. Fröhlich, *The Picard group of noncommutative rings, in particular of orders*, Trans. Amer. Math. Soc. **180** (1973), 1–45. MR0318204 (47 #6751)

[Gab81] Ofer Gabber, *Some theorems on Azumaya algebras*, The Brauer group (Sem., Les Plans-sur-Bex, 1980), Lecture Notes in Math., vol. 844, Springer, Berlin, 1981, pp. 129–209. MR611868 (83d:13004)

[Gre69] Marvin J. Greenberg, *Lectures on forms in many variables*, W. A. Benjamin, Inc., New York-Amsterdam, 1969. MR0241358 (39 #2698)

[Gre92] Cornelius Greither, *Cyclic Galois extensions of commutative rings*, Lecture Notes in Mathematics, vol. 1534, Springer-Verlag, Berlin, 1992. MR1222646

[Gro61] A. Grothendieck, *Éléments de géométrie algébrique. III. Étude cohomologique des faisceaux cohérents. I*, Inst. Hautes Études Sci. Publ. Math. (1961), no. 11, 167. MR0163910 (29 #1209)

[Gro62] Alexander Grothendieck, *Fondements de la géométrie algébrique. [Extraits du Séminaire Bourbaki, 1957–1962.]*, Secrétariat mathématique, Paris, 1962. MR0146040

[Gro64] A. Grothendieck, *Éléments de géométrie algébrique. IV. Étude locale des schémas et des morphismes de schémas. I*, Inst. Hautes Études Sci. Publ. Math. (1964), no. 20, 259. MR0173675 (30 #3885)

[Gro65] ______, *Éléments de géométrie algébrique. IV. Étude locale des schémas et des morphismes de schémas. II*, Inst. Hautes Études Sci. Publ. Math. (1965), no. 24, 231. MR0199181 (33 #7330)

[Gro66] ______, *Éléments de géométrie algébrique. IV. Étude locale des schémas et des morphismes de schémas. III*, Inst. Hautes Études Sci. Publ. Math. (1966), no. 28, 255. MR0217086 (36 #178)

[Gro67] ______, *Éléments de géométrie algébrique. IV. Étude locale des schémas et des morphismes de schémas IV*, Inst. Hautes Études Sci. Publ. Math. (1967), no. 32, 361. MR0238860 (39 #220)

[Gro68a] Alexander Grothendieck, *Le groupe de Brauer. I. Algèbres d'Azumaya et interprétations diverses*, Dix Exposés sur la Cohomologie des Schémas, North-Holland, Amsterdam, 1968, pp. 46–66. MR0244269 (39 #5586a)

[Gro68b] ______, *Le groupe de Brauer. II. Théorie cohomologique*, Dix Exposés sur la Cohomologie des Schémas, North-Holland, Amsterdam, 1968, pp. 67–87. MR0244270 (39 #5586b)

[Gro68c] ______, *Le groupe de Brauer. III. Exemples et compléments*, Dix Exposés sur la Cohomologie des Schémas, North-Holland, Amsterdam, 1968, pp. 88–188. MR0244271 (39 #5586c)

[Gro71] Alexandre Grothendieck, *Revêtements étales et groupe fondamental*, Springer-Verlag, Berlin-New York, 1971, Séminaire de Géométrie Algébrique du Bois Marie 1960–1961 (SGA 1), Dirigé par Alexandre Grothendieck. Augmenté de deux exposés de M. Raynaud, Lecture Notes in Mathematics, Vol. 224. MR0354651 (50 #7129)

[Gro03] ______, *Revêtements étales et groupe fondamental (SGA 1)*, Documents Mathématiques (Paris) [Mathematical Documents (Paris)], 3, Société Mathématique de France, Paris, 2003, Séminaire de géométrie algébrique du Bois Marie 1960–61. [Algebraic Geometry Seminar of Bois Marie 1960-61], Directed by A. Grothendieck, With two papers by M. Raynaud, Updated and annotated reprint of the 1971 original [Lecture Notes in Math., 224, Springer, Berlin; MR0354651 (50 #7129)]. MR2017446 (2004g:14017)

[Har65] D. K. Harrison, *Abelian extensions of commutative rings*, Mem. Amer. Math. Soc. No. **52** (1965), 1–14. MR0195921

[Har77] Robin Hartshorne, *Algebraic geometry*, Springer-Verlag, New York, 1977. MR0463157 (57 #3116)

[Her75] I. N. Herstein, *Topics in algebra*, second ed., Xerox College Publishing, Lexington, Mass., 1975. MR0356988 (50 #9456)

[HN70] Motoshi Hongan and Takasi Nagahara, *A note on separable extensions of commutative rings*, Math. J. Okayama Univ. **14** (1969/1970), 13–15. MR0257065

[Hoc45] G. Hochschild, *On the cohomology groups of an associative algebra*, Ann. of Math. (2) **46** (1945), 58–67. MR0011076

[Hoc55] ———, *Restricted Lie algebras and simple associative algebras of characteristic p*, Trans. Amer. Math. Soc. **80** (1955), 135–147. MR0072428 (17,282d)

[Hoe63] Klaus Hoechsmann, *Simple algebras and derivations*, Trans. Amer. Math. Soc. **108** (1963), 1–12. MR0152548 (27 #2526)

[Hoo66] Raymond Taylor Hoobler, *A generalization of the Brauer group and Amitsur cohomology*, ProQuest LLC, Ann Arbor, MI, 1966, Thesis (Ph.D.)–University of California, Berkeley. MR2615918

[Hoo72] Raymond T. Hoobler, *Brauer groups of abelian schemes*, Ann. Sci. École Norm. Sup. (4) **5** (1972), 45–70. MR0311664 (47 #226)

[Hoo80] ———, *A cohomological interpretation of Brauer groups of rings*, Pacific J. Math. **86** (1980), no. 1, 89–92. MR586870 (81j:13003)

[Hoo82] ———, *When is* $\mathrm{Br}(X) = \mathrm{Br}'(X)$*?*, Brauer groups in ring theory and algebraic geometry (Wilrijk, 1981), Lecture Notes in Math., vol. 917, Springer, Berlin-New York, 1982, pp. 231–244. MR657433 (83g:14006)

[Hoo84] ———, *Functors of graded rings*, Methods in ring theory (Antwerp, 1983), NATO Adv. Sci. Inst. Ser. C Math. Phys. Sci., vol. 129, Reidel, Dordrecht, 1984, pp. 161–170. MR770588 (86f:16008)

[Hun80] Thomas W. Hungerford, *Algebra*, Graduate Texts in Mathematics, vol. 73, Springer-Verlag, New York, 1980, Reprint of the 1974 original. MR600654 (82a:00006)

[IS15] Evis Ieronymou and Alexei N. Skorobogatov, *Odd order Brauer-Manin obstruction on diagonal quartic surfaces*, Adv. Math. **270** (2015), 181–205. MR3286534

[Jac37] Nathan Jacobson, *p-algebras of exponent p*, Bull. Amer. Math. Soc. **43** (1937), no. 10, 667–670. MR1563614

[Jac79] ———, *Lie algebras*, Dover Publications, Inc., New York, 1979, Republication of the 1962 original. MR559927 (80k:17001)

[Jan66] G. J. Janusz, *Separable algebras over commutative rings*, Trans. Amer. Math. Soc. **122** (1966), 461–479. MR35#1585

[Kan68] Teruo Kanzaki, *On generalized crossed product and Brauer group*, Osaka J. Math. **5** (1968), 175–188. MR0241469 (39 #2809)

[Kel75] John L. Kelley, *General topology*, Springer-Verlag, New York-Berlin, 1975, Reprint of the 1955 edition [Van Nostrand, Toronto, Ont.], Graduate Texts in Mathematics, No. 27. MR0370454 (51 #6681)

[Knu70] Max-Albert Knus, *Sur le théorème de Skolem-Noether et sur les dérivations des algẽres d'Azumaya*, C. R. Acad. Sci. Paris Sér. A-B **270** (1970), A637–A639. MR0257149 (41 #1803)

[KO74a] M.-A. Knus and M. Ojanguren, *A Mayer-Vietoris sequence for the Brauer group*, J. Pure Appl. Algebra **5** (1974), 345–360. MR51#5572

[KO74b] Max-Albert Knus and Manuel Ojanguren, *Théorie de la descente et algèbres d'Azumaya*, Springer-Verlag, Berlin, 1974, Lecture Notes in Mathematics, Vol. 389. MR0417149 (54 #5209)

[KO75] ———, *A norm for modules and algebras*, Math. Z. **142** (1975), 33–45. MR0366893 (51 #3139)

[KO81] M.-A. Knus and M. Ojanguren, *Cohomologie étale et groupe de Brauer*, The Brauer group (Sem., Les Plans-sur-Bex, 1980), Lecture Notes in Math., vol. 844, Springer, Berlin-New York, 1981, pp. 210–228. MR611869

[KOS76] M. A. Knus, M. Ojanguren, and D. J. Saltman, *On Brauer groups in characteristic p*, Brauer groups (Proc. Conf., Northwestern Univ., Evanston, Ill., 1975), Springer, Berlin, 1976, pp. 25–49. Lecture Notes in Math., Vol. 549. MR0429859 (55 #2869)

[KPS90] M.-A. Knus, R. Parimala, and V. Srinivas, *Azumaya algebras with involutions*, J. Algebra **130** (1990), no. 1, 65–82. MR1045736

[Lam06] T. Y. Lam, *Serre's problem on projective modules*, Springer Monographs in Mathematics, Springer-Verlag, Berlin, 2006. MR2235330

[LB87] Lieven Le Bruyn, *A cohomological interpretation of the reflexive Brauer group*, J. Algebra **105** (1987), no. 1, 250–254. MR871757 (88j:16009)

[Li95] Hongnian Li, *Brauer groups over affine normal surfaces*, K-theory and algebraic geometry: connections with quadratic forms and division algebras (Santa Barbara, CA, 1992), Proc. Sympos. Pure Math., vol. 58, Amer. Math. Soc., Providence, RI, 1995, pp. 309–317. MR1327306

[Lip69] Joseph Lipman, *Rational singularities, with applications to algebraic surfaces and unique factorization*, Inst. Hautes Études Sci. Publ. Math. (1969), no. 36, 195–279. MR0276239 (43 #1986)

[LO82] Heisook Lee and Morris Orzech, *Brauer groups, class groups and maximal orders for a Krull scheme*, Canad. J. Math. **34** (1982), no. 4, 996–1010. MR672693 (84d:13004)

[Lor95] Martin Lorenz, *Class groups of multiplicative invariants*, J. Algebra **177** (1995), no. 1, 242–254. MR1356370 (96j:16036)

[Mag74a] Andy R. Magid, *Principal homogeneous spaces and Galois extensions*, Pacific J. Math. **53** (1974), 501–513. MR0382245

[Mag74b] ______, *The separable Galois theory of commutative rings*, Marcel Dekker, Inc., New York, 1974, Pure and Applied Mathematics, No. 27. MR0352075

[Mag78] ______, *Brauer groups of linear algebraic groups with characters*, Proc. Amer. Math. Soc. **71** (1978), no. 2, 164–168. MR0485816

[Man71] Y. I. Manin, *Le groupe de Brauer-Grothendieck en géométrie diophantienne*, Actes du Congrès International des Mathématiciens (Nice, 1970), Tome 1, Gauthier-Villars, Paris, 1971, pp. 401–411. MR0427322 (55 #356)

[Mat80] Hideyuki Matsumura, *Commutative algebra*, second ed., Mathematics Lecture Note Series, vol. 56, Benjamin/Cummings Publishing Co., Inc., Reading, Mass., 1980. MR575344 (82i:13003)

[McD84] Bernard R. McDonald, *Linear algebra over commutative rings*, Monographs and Textbooks in Pure and Applied Mathematics, vol. 87, Marcel Dekker, Inc., New York, 1984. MR769104 (86d:13008)

[McK98] Thomas McKenzie, *The separable closure of a local ring*, J. Algebra **207** (1998), no. 2, 657–663. MR1644231 (99g:13006)

[MCPA15] Akhil Mathew and The CRing Project Authors, *The CRing project*, `http://people.fas.harvard.edu/~amathew/cr.html`, 2015, A collaborative, open source textbook on commutative algebra.

[Mil80] James S. Milne, *Étale cohomology*, Princeton University Press, Princeton, N.J., 1980. MR559531 (81j:14002)

[Mum70] David Mumford, *Abelian varieties*, Tata Institute of Fundamental Research Studies in Mathematics, No. 5, Published for the Tata Institute of Fundamental Research, Bombay, 1970. MR0282985 (44 #219)

[Mum99] ———, *The red book of varieties and schemes*, expanded ed., Lecture Notes in Mathematics, vol. 1358, Springer-Verlag, Berlin, 1999, Includes the Michigan lectures (1974) on curves and their Jacobians, With contributions by Enrico Arbarello. MR1748380 (2001b:14001)

[Nag62] Masayoshi Nagata, *Local rings*, Interscience Tracts in Pure and Applied Mathematics, No. 13, Interscience Publishers a division of John Wiley & Sons New York-London, 1962. MR0155856 (27 #5790)

[Oja74] Manuel Ojanguren, *A nontrivial locally trivial algebra*, J. Algebra **29** (1974), 510–512. MR0352073 (50 #4561)

[Orz82] Morris Orzech, *Brauer groups and class groups for a Krull domain*, Brauer groups in ring theory and algebraic geometry (Wilrijk, 1981), Lecture Notes in Math., vol. 917, Springer, Berlin, 1982, pp. 66–90. MR657423 (83h:13005)

[OS75] Morris Orzech and Charles Small, *The Brauer group of commutative rings*, Marcel Dekker, Inc., New York, 1975, Lecture Notes in Pure and Applied Mathematics, Vol. 11. MR0457422 (56 #15627)

[Ray70] Michel Raynaud, *Anneaux locaux henséliens*, Lecture Notes in Mathematics, Vol. 169, Springer-Verlag, Berlin-New York, 1970. MR0277519 (43 #3252)

[Rei75] I. Reiner, *Maximal orders*, Academic Press [A subsidiary of Harcourt Brace Jovanovich, Publishers], London-New York, 1975, London Mathematical Society Monographs, No. 5. MR0393100 (52 #13910)

[Rim65] Dock Sang Rim, *An exact sequence in Galois cohomology*, Proc. Amer. Math. Soc. **16** (1965), 837–840. MR0179232 (31 #3480)

[Rot79] Joseph J. Rotman, *An introduction to homological algebra*, Pure and Applied Mathematics, vol. 85, Academic Press Inc. [Harcourt Brace Jovanovich Publishers], New York, 1979. MR538169 (80k:18001)

[RS82] Klaus Roggenkamp and Leonard Scott, *Hecke actions on Picard groups*, J. Pure Appl. Algebra **26** (1982), no. 1, 85–100. MR669845 (84c:12008)

[RS10] Fabian Reede and Ulrich Stuhler, *Division algebras and unit groups on surfaces*, Affine flag manifolds and principal bundles, Trends Math., Birkhäuser/Springer Basel AG, Basel, 2010, pp. 203–217. MR3013032

[RT85] Iain Raeburn and Joseph L. Taylor, *The bigger Brauer group and étale cohomology*, Pacific J. Math. **119** (1985), no. 2, 445–463. MR803128

[RZ61] Alex Rosenberg and Daniel Zelinsky, *Automorphisms of separable algebras*, Pacific J. Math. **11** (1961), 1109–1117. MR0148709

[Sal76] David J. Saltman, *Azumaya algebras over rings of characteristic p*, ProQuest LLC, Ann Arbor, MI, 1976 (English), Thesis (Ph.D.)–Yale University.

[Sal81] ———, *The Brauer group is torsion*, Proc. Amer. Math. Soc. **81** (1981), no. 3, 385–387. MR597646 (82c:16005)

[Sal82] ———, *Generic Galois extensions and problems in field theory*, Adv. in Math. **43** (1982), no. 3, 250–283. MR648801 (84a:13007)

[Sal99] ———, *Lectures on division algebras*, CBMS Regional Conference Series in Mathematics, vol. 94, Published by American Mathematical Society, Providence, RI, 1999. MR1692654 (2000f:16023)

[Sam64] P. Samuel, *Lectures on unique factorization domains*, Notes by M. Pavman Murthy. Tata Institute of Fundamental Research Lectures on Mathematics, No. 30, Tata Institute of Fundamental Research, Bombay, 1964. MR0214579

[Ser79] Jean-Pierre Serre, *Local fields*, Graduate Texts in Mathematics, vol. 67, Springer-Verlag, New York, 1979, Translated from the French by Marvin Jay Greenberg. MR554237 (82e:12016)

[Sto79] Robert R. Stoll, *Set theory and logic*, Dover Publications, Inc., New York, 1979, Corrected reprint of the 1963 edition. MR634799 (83e:04002)

[Tay82] Joseph L. Taylor, *A bigger Brauer group*, Pacific J. Math. **103** (1982), no. 1, 163–203. MR687968

[Tre78] Robert Treger, *Reflexive modules*, J. Algebra **54** (1978), no. 2, 444–466. MR514079 (81i:13005)

[vG05] Bert van Geemen, *Some remarks on Brauer groups of $K3$ surfaces*, Adv. Math. **197** (2005), no. 1, 222–247. MR2166182 (2006h:14025)

[VZ66] O. E. Villamayor and D. Zelinsky, *Galois theory for rings with finitely many idempotents*, Nagoya Math. J. **27** (1966), 721–731. MR34#5880

[Wei63] Edwin Weiss, *Algebraic number theory*, McGraw-Hill Book Co., Inc., New York-San Francisco-Toronto-London, 1963. MR0159805 (28 #3021)

[Wei77] Charles A. Weibel, *Homotopy in algebraic k-theory*, ProQuest LLC, Ann Arbor, MI, 1977 (English), Thesis (Ph.D.) – University of Chicago.

[Wei81] C. A. Weibel, *Mayer-Vietoris sequences and module structures on NK_**, Algebraic K-theory, Evanston 1980 (Proc. Conf., Northwestern Univ., Evanston, Ill., 1980), Lecture Notes in Math., vol. 854, Springer, Berlin, 1981, pp. 466–493. MR618317

[Wei91] Charles A. Weibel, *Pic is a contracted functor*, Invent. Math. **103** (1991), no. 2, 351–377. MR1085112

[Wey97] Hermann Weyl, *The classical groups*, Princeton Landmarks in Mathematics, Princeton University Press, Princeton, NJ, 1997, Their invariants and representations, Fifteenth printing, Princeton Paperbacks. MR1488158

[Yua67] Shuen Yuan, *On the theory of p-algebras and the Amitsur cohomology groups for inseparable field extensions*, J. Algebra **5** (1967), 280–304. MR0228548 (37 #4128)

[Yua71] ———, *Central separable algebras with purely inseparable splitting rings of exponent one*, Trans. Amer. Math. Soc. **153** (1971), 427–450. MR0268175 (42 #3074)

[Yua74a] ———, *Brauer groups for inseparable fields*, Amer. J. Math. **96** (1974), 430–447. MR0354637 (50 #7115)

[Yua74b] ———, *Reflexive modules and algebra class groups over noetherian integrally closed domains*, J. Algebra **32** (1974), 405–417. MR0357463 (50 #9931)

[ZS75a] Oscar Zariski and Pierre Samuel, *Commutative algebra. Vol. 1*, Springer-Verlag, New York-Heidelberg-Berlin, 1975, With the cooperation of I. S. Cohen, Corrected reprinting of the 1958 edition, Graduate Texts in Mathematics, No. 28. MR0384768 (52 #5641)

[ZS75b] ———, *Commutative algebra. Vol. II*, Springer-Verlag, New York-Heidelberg, 1975, Reprint of the 1960 edition, Graduate Texts in Mathematics, Vol. 29. MR0389876 (52 #10706)

Index

Selected Published Titles in This Series

183 **Timothy J. Ford,** Separable Algebras, 2017
179 **Henri Cohen and Fredrik Strömberg,** Modular Forms, 2017
178 **Jeanne N. Clelland,** From Frenet to Cartan: The Method of Moving Frames, 2017
177 **Jacques Sauloy,** Differential Galois Theory through Riemann-Hilbert Correspondence, 2016
176 **Adam Clay and Dale Rolfsen,** Ordered Groups and Topology, 2016
175 **Thomas A. Ivey and Joseph M. Landsberg,** Cartan for Beginners: Differential Geometry via Moving Frames and Exterior Differential Systems, Second Edition, 2016
174 **Alexander Kirillov Jr.,** Quiver Representations and Quiver Varieties, 2016
173 **Lan Wen,** Differentiable Dynamical Systems, 2016
172 **Jinho Baik, Percy Deift, and Toufic Suidan,** Combinatorics and Random Matrix Theory, 2016
171 **Qing Han,** Nonlinear Elliptic Equations of the Second Order, 2016
170 **Donald Yau,** Colored Operads, 2016
169 **András Vasy,** Partial Differential Equations, 2015
168 **Michael Aizenman and Simone Warzel,** Random Operators, 2015
167 **John C. Neu,** Singular Perturbation in the Physical Sciences, 2015
166 **Alberto Torchinsky,** Problems in Real and Functional Analysis, 2015
165 **Joseph J. Rotman,** Advanced Modern Algebra: Third Edition, Part 1, 2015
164 **Terence Tao,** Expansion in Finite Simple Groups of Lie Type, 2015
163 **Gérald Tenenbaum,** Introduction to Analytic and Probabilistic Number Theory, Third Edition, 2015
162 **Firas Rassoul-Agha and Timo Seppäläinen,** A Course on Large Deviations with an Introduction to Gibbs Measures, 2015
161 **Diane Maclagan and Bernd Sturmfels,** Introduction to Tropical Geometry, 2015
160 **Marius Overholt,** A Course in Analytic Number Theory, 2014
159 **John R. Faulkner,** The Role of Nonassociative Algebra in Projective Geometry, 2014
158 **Fritz Colonius and Wolfgang Kliemann,** Dynamical Systems and Linear Algebra, 2014
157 **Gerald Teschl,** Mathematical Methods in Quantum Mechanics: With Applications to Schrödinger Operators, Second Edition, 2014
156 **Markus Haase,** Functional Analysis, 2014
155 **Emmanuel Kowalski,** An Introduction to the Representation Theory of Groups, 2014
154 **Wilhelm Schlag,** A Course in Complex Analysis and Riemann Surfaces, 2014
153 **Terence Tao,** Hilbert's Fifth Problem and Related Topics, 2014
152 **Gábor Székelyhidi,** An Introduction to Extremal Kähler Metrics, 2014
151 **Jennifer Schultens,** Introduction to 3-Manifolds, 2014
150 **Joe Diestel and Angela Spalsbury,** The Joys of Haar Measure, 2013
149 **Daniel W. Stroock,** Mathematics of Probability, 2013
148 **Luis Barreira and Yakov Pesin,** Introduction to Smooth Ergodic Theory, 2013
147 **Xingzhi Zhan,** Matrix Theory, 2013
146 **Aaron N. Siegel,** Combinatorial Game Theory, 2013
145 **Charles A. Weibel,** The K-book, 2013
144 **Shun-Jen Cheng and Weiqiang Wang,** Dualities and Representations of Lie Superalgebras, 2012
143 **Alberto Bressan,** Lecture Notes on Functional Analysis, 2013

For a complete list of titles in this series, visit the
AMS Bookstore at **www.ams.org/bookstore/gsmseries/**.